T3-BNI-875

Energy Supply-Demand Integrations to the Year 2000

Global and National Studies

Energy Supply-Demand Integrations to the Year 2000

Global and National Studies

Third Technical Report of the Workshop on Alternative Energy Strategies (WAES)

Paul S. Basile
Editor

Carroll L. Wilson
WAES Project Director

The MIT Press
Cambridge, Massachusetts, and London, England

This material is based upon research supported in part by the National Science Foundation under Grant No. OEP74-13902 AO1. Any opinions, findings, and conclusions or recommendations expressed in this publication are those of the authors and do not necessarily reflect the views of the National Science Foundation.

This book was set in IBM Composer Century by Eastern Composition, Inc. and printed and bound by Halliday Lithograph Corp. in the United States of America.

Library of Congress Cataloging in Publication Data

Workshop on Alternative Energy Strategies.
Energy supply-demand integrations to the year 2000.

Includes bibliographical references.
1. Power resources. 2. Energy consumption.
3. Economic forecasting. I. Basile, Paul S.
II. Title
HD9502.A2W66 1977a 333.7 77-5717
ISBN 0-262-23083-6

Contents

Preface vii

Workshop Participants, Associates, and Staff ix

Acknowledgments xiii

Part I
Introduction and Overview 1

1
The Workshop Methodology 3

2
Summary of the Global and National Supply-Demand Integration Studies 13

Part II
Global Supply-Demand Integrations 43

3
Global Energy Data Base 45

4
Unconstrained Global Supply-Demand Integrations 76

5
Constrained Global Supply-Demand Integrations 127

6
Energy and Economic Growth Prospects for the Developing Countries 217

Part III
National Studies of Energy Demand to Year 2000 and Energy Supply-Demand Integrations 249

7
Canada 251

8
Denmark 274

9
Finland 317

10
France 335

11
German Federal Republic 352

12
Italy 390

13
Japan 445

14
Mexico 473

15
The Netherlands 491

16
Norway 520

17
Sweden 555

18
United Kingdom 600

19
United States 617

Appendix A
Illustrative Full GEMM Case 655

Appendix B
GEMM Model Formulation 701

Preface

In October 1974, leaders of business, industry, government, and academia from a dozen countries convened the first session of the Workshop on Alternative Energy Strategies (WAES). During its two-and-a-half year lifetime, the Workshop carried forward studies of energy demand and supply, of resource availability, and of the critical factors in alternative energy futures, both nationally and globally. These studies provided the essential ingredients for what we call "energy supply-demand integrations"—the comparing and contrasting of separate energy demands and supplies. This book reports the results of our integration studies. It is the third, and final, volume in a series of WAES published technical reports.[1]

The Workshop was an ad hoc international project, involving over seventy-five people from a total of fifteen countries. It set itself the goal of assembling a set of coherent global energy pictures of the future. To this end, it developed methods for estimating energy supply and demand through the year 2000 and for combining, comparing, and contrasting them.

This goal was set against a background which recognized that, according to the best estimates, world crude oil production will peak and begin to decline within the next twenty-five years. The transition from oil to other forms of energy is coming more rapidly than most people realize. WAES evolved out of the common concern of a number of influential people who believed that this transition needed to be widely understood and effectively managed in order to avoid major national and international dislocations.

During our studies and deliberations in the Workshop, it became clear that future energy problems are fundamentally problems of *imbalance*—of mismatches between limited supplies of certain fuels and "preferred" demands for such fuels. Certain fuels are best suited to meet certain types of energy demand and, for economic, thermodynamic, efficiency, or other reasons, not necessarily ideally suited to meet other types of energy demand. The WAES energy balances—or what we call "supply-demand integrations"—show this.

1. The first two technical reports are *Energy Demand Studies: Major Consuming Countries* (MIT Press, 1976) and *Energy Supply to the Year 2000: Global and National Studies* (MIT Press, 1977).

What will be the future fuel preferences throughout the world? Will the preferred fuels be available in quantity? If not, are there shifts that can be made in a nation's, or the world's, fuel-use patterns? Are such shifts sufficient to allow the demands for energy to be met?

This third technical report of the Workshop seeks to treat these issues, country by country and for the world outside Communist areas (WOCA), in a carefully constructed way, for the years 1985 and 2000. It contains details of energy supply and demand balances or projected imbalances for the base year, 1972, and for 1985 and 2000 for thirteen countries, for the aggregated developing countries, and for WOCA. These integrations are done under five different "scenarios" to 1985 and four to 2000, based on a combination of variables representing economic growth, energy price, national energy policy, and principal replacement fuel.

The Workshop's demand estimates for 1985 are contained in the first technical report, the supply estimates to 2000 in the second technical report. Projections of energy demand to 2000 can be found in the national chapters of this third volume.

The WAES final report—*ENERGY: Global Prospects 1985–2000* (McGraw-Hill, 1977)—summarizes the global character of the WAES findings. It draws heavily on the substance of the three-volume technical report series. It identifies the major long-term energy problems that appear as a result of our studies and draws some implications for nations acting together and nations acting alone.

The Workshop on Alternative Energy Strategies was an independent organization, with no official ties to any governments or private firms.[2] It was organized into three groups:

- The *Participants*, thirty-five senior decision makers from around the world, whose experience in dealing with problems involving the interplay of technology, economics, finance, and government policy made them well-suited to take an active part in the Workshop. Members included officials from organizations that are suppliers and users of fuels or that produce electricity, government officials with energy-related responsibilities, and others whose insights enriched the Workshop. The Partic-

2. Further information about the Workshop can be found in "A Global Energy Assessment 1985–2000," a brochure available from any of the Workshop members.

ipants took part in this study as individuals, not as official representatives of their organization or country. They provided or arranged for financial support for WAES activities in their own country and paid for the expenses involved in attending and sponsoring meetings.

• The *Associates*, chosen by the Participants from their own or cooperating organizations, included scientists, engineers, economists, and managers with expertise in many energy-related fields, who brought a wealth of broad experience to the Workshop. They conducted the technical studies that form the basis for the Workshop's findings and conclusions.

In some countries Workshop members recruited Reference Groups—people from government and industry who served as advisers, provided financial support, and did some of the analytical work.

• The *Secretariat* included the Project Director and a small staff at the Massachusetts Institute of Technology. The Director provided executive leadership, and the Secretariat provided project coordination. Secretariat expenses were supported by foundation grants.

This distinctive character allowed the Workshop to have two important effects. Through its technical analyses it provided the necessary quantitative basis for policy decisions. More directly, involvement in such an in-depth study influenced the members themselves, who will, it is hoped, assist in the process of turning the options identified by the Workshop into realities.

The Workshop objectives were to establish a quantitative basis for analyzing probable future energy developments through the year 2000 and then identifying feasible alternative global and national energy strategies to deal with these developments. The analyses were designed to provide information that would raise the level of public understanding and assist governments and others in making the choices and formulating the strategies necessary to ensure a proper balance of energy supply and demand during the period 1985 to 2000 and beyond.

Few nations, acting alone, have the resources to ensure their own energy supplies. Therefore, it is in the common interest to avoid placing insupportable demands on specific energy sources. For example, it makes no sense for one country to build power stations that will depend on coal from another country if the latter does not develop the needed capacity for coal production, transport, and export. The international nature of WAES allowed a global approach to such issues.

In looking at the period to the end of the century, it is important to try to determine what can be achieved in the time available. Economic growth and energy "savings" will be the main determinants of energy demand up to 1985, with oil remaining the principal fuel. Beyond 1985, alternative energy sources will be needed on an increasing scale. Because of the long lead times required, however, the size of their contributions will be determined by actions taken now or in the very near future.

A description of the overall methodology developed by the Workshop is given in chapter 1, together with a detailed description of the energy supply-demand integration studies methodology. This novel approach to the global energy setting draws partially on many traditional analytical techniques and tools, which have been modified and adapted by the Workshop Associates with the guidance and assistance of the Participants. The WAES methodology was endorsed and supported by all Workshop members. Each national study is, however, attributable solely to the Workshop members from that country, and although all studies were done in a common and consistent manner, each has its own distinctive aspects.

As with any study of the future, the projections in this volume should be taken with care. The Workshop claims no special ability to see into the future. It does claim, however, to have projected reasonable estimates for energy demand and supply, and their possible global balance or imbalance, in its member countries and in the world outside Communist areas. Such projections are based on explicit assumptions about several alternative states of the world.

There are differences among national approaches. Many of these variations in national style and decision making are revealed by the studies in this book. These differences add richness to the studies and illustrate some of the potentials for change in present-day patterns of energy use. We hope that such potentials will be carefully assessed by the many analysts, scientists, and research groups outside the Workshop who have the ability and time to do so.

Carroll L. Wilson
Project Director

Workshop Participants, Associates, and Staff

PROJECT DIRECTOR
Carroll L. Wilson
Massachusetts Institute of Technology

PARTICIPANTS

Canada
Mr. Marshall A. Crowe
Chairman, National Energy Board

Mr. Maurice F. Strong
Chairman, Petro-Canada

Denmark
Professor Bent Elbek
Niels Bohr Institute

Finland
Professor Jorma Routti
Helsinki University of Technology

France
M. Jean Couture
Conseiller du Président
Société Générale

M. Jean-Marie Martin
Director, Institut Economique et Juridique de l'Energie
Centre National de la Recherche Scientifique

Germany
Director Dr. Hans Detzer
Head of Central Planning
Badische Anilin & Soda Fabrik AG

Professor Heinrich Mandel
Member of the Board of Directors
Rheinisch-Westfälisches Elektrizitätswerk AG

Professor Hans K. Schneider
Director, Institute of Energy Economics
University of Cologne

Iran
Dr. Khodadad Farmanfarmaian

Italy
Professor Umberto Colombo
Director, Research & Development Division
Montedison

Professor Sergio Vaccà
Director, Istituto di Economia delle Fonti de Energia
Università L. Bocconi, Milano

Japan
Mr. Toshio Doko
President, Japan Federation of Economic Organizations (Keidanren)

Mr. Shuzo Inaba
President, Industrial Research Institute

Mr. Soichi Matsune
Chairman, Committee on Energy
Japan Federation of Economic Organizations (Keidanren)

Mr. Tatsuzo Mizukami
President, Japan Foreign Trade Council, Inc.

Dr. Saburo Okita
President, The Overseas Economic Cooperation Fund

Mr. Masao Sakisaka
President, National Institute for Research Advancement

Mr. Shigefumi Tamiya
Adviser to the Chairman
Enrichment and Reprocessing Group

Mexico
Ing. Juan Eibenschutz
Executive Secretary
Mexican National Energy Commission

Dr. Victor Urquidi
President, El Colégio de México

The Netherlands
Dr. A. A. T. van Rhijn
Deputy Director General for Energy
Ministry of Economic Affairs

Netherlands/U.K.
Mr. J. C. Davidson
Director, Shell International Petroleum Company, Ltd.
Royal Dutch/Shell Group

Norway
Mr. Christian Sommerfelt
Chairman
Elkem-Spigerverket A/S

Sweden
Mr. Erland Waldenström
Chairman
Gränges AB

United Kingdom
Mr. Robert Belgrave
Policy Adviser to the Board
British Petroleum Company, Ltd.

Professor Sir William Hawthorne
Master, Churchill College
University of Cambridge

United States
Mr. Thornton F. Bradshaw
President, Atlantic Richfield Company

Mr. Walker L. Cisler
Retired Chairman of the Board
Detroit Edison Company

Mr. John T. Connor
Chairman of the Board
Allied Chemical Corporation

Mr. Richard C. Gerstenberg
Director, General Motors Corporation

Dr. H. Guyford Stever[1]
Director, National Science Foundation

Venezuela
Dr. José A. Mayobre
Banco Central de Venezuela

Ing. Ulises Ramírez
Executive Secretary
National Energy Council

1. From October 1974 to October 1976.

ASSOCIATES

Canada
Mr. Marc LeClerc
Director General, Special Projects
National Energy Board

Dr. John Ralston Saul
Special Assistant to the Chairman
Petro-Canada

Finland
Mr. Seppo Hannus
Ministry of Trade and Industry
Energy Department

France
M. Bertrand Chateau
Institut Economique et Juridique de l'Energie
Centre National de la Recherche Scientifique

Dr. Maxime Kleinpeter
Eléctricité de France
Service Etudes Economique Générales

Germany
Dr. Georg Klotmann
Strategic Planning Department
Badische Anilin & Soda Fabrik AG

Dr. Dieter Schmitt
Head, Institute of Energy Economics
University of Cologne

Mr. Paul H. Suding
Economist, Institute of Energy Economics
University of Cologne

Italy
Dr. Oliviero Bernardini
Department of Technology Assessment
Research & Development Division
Montedison

Dr. Riccardo Galli
Director, Department of Technology Assessment
Research & Development Division
Montedison

Mr. William Mebane
Department of Technology Assessment
Research & Development Division
Montedison

Japan
Mr. Shinichiro Aoyama
Senior Staff Researcher, National Institute for Research Advancement

Mr. Kenichi Matsui
Senior Staff Economist
The Institute of Energy Economics

Mr. Yasuhiro Murota
Staff Economist
The Japan Economic Research Center

Mr. Mitsuo Takei
Director, Research Affairs
The Institute of Energy Economics

Mr. Hisashi Watanabe
Executive Director, Japan Alternative Energy Strategies Organization

Mexico
Ing. Gerardo Bazán
Assessor
Mexican National Energy Commission

Ing. Alberto Escofet
Assistant to the Executive Secretary
Mexican National Energy Commission

The Netherlands
Dr. André C. Sjoerdsma
Director, Future Shape of Technology Foundation

Netherlands/U.K.
Mr. Alan W. Clarke
Energy & Oil Economics Division
Shell International Petroleum Co., Ltd.
Royal Dutch/Shell Group

Drs. Hans DuMoulin
Head of Energy & Oil Economics Div.
Shell International Petroleum Co., Ltd.
Royal Dutch/Shell Group

Dr. Gareth Price
Group Planning, Long Term Future
Shell International Petroleum Co., Ltd.
Royal Dutch/Shell Group

Norway
Mr. Henrik Ager-Hanssen
Deputy Managing Director
Statoil, Den Norske Stats Oljeselskap A/S

Mr. Kai Killerud
Manager, Power Plant Engineering
Scandpower A/S

Sweden
Dr. Harry Albinsson
Energy Secretary
Federation of Swedish Industries

Mr. Bertil Eneroth
Special Assignments
Skandinaviska Enskilda Banken

United Kingdom
Mr. Michael Clegg
Manager, Systems Group, Corporate Planning
British Petroleum Company, Ltd.

Dr. Richard J. Eden
Head, Energy Research Group
Cavendish Laboratory
University of Cambridge

Mr. Andrew R. Flower
Assistant Policy Analyst
Policy Review Unit
British Petroleum Company, Ltd.

Dr. Edmund Crouch
Energy Research Group
Cavendish Laboratory
University of Cambridge

United States
Mr. Walter F. Allaire
Director, Energy Resources
Allied Chemical Corporation

Mr. Steven Carhart[2]
Assistant Scientist
National Center for Analysis of Energy Systems
Brookhaven National Laboratory

Dr. Paul P. Craig[3]
Director, Energy & Resources Council
University of California

Dr. Henry L. Duncombe, Jr.
Chief Economist
General Motors Corporation

2. Consultant on Demand Studies, October 1974 to December 1975; Adviser on U.S. Studies, January 1976 to February 1977.
3. Associate from October 1974 to April 1976.

Ms. Sandra Fucigna[4]
Policy Analyst
National Science Foundation

Mr. Edward D. Griffith
Senior Consultant—Policy Analysis and Forecasting
Atlantic Richfield Company

Dr. Kenneth Hoffman[5]
Associate Chairman for Energy Programs
Head, National Center for Analysis of Energy Systems
Brookhaven National Laboratory

Dr. H. Paul Root
Director of Economic Studies
General Motors Corporation

Dr. David Sternlight
Chief Economist
Atlantic Richfield Company

Venezuela
Dr. Felix Rossi-Guerrero
Minister Counselor for Petroleum Affairs
Venezuelan Embassy, Washington, D.C.

4. Associate from April 1976 to October 1976.
5. Adviser on U.S. Studies from January 1976 to February 1977.

STAFF

MIT Program Staff
Mr. Robert P. Greene
Program/Administrative Officer

Mr. Paul S. Basile
Program Officer

Mr. William F. Martin
Program Officer

MIT Research Assistants
Mr. Richard Cheston
Mr. Miles Harbur

MIT Support Staff
Ms. Elaine R. Goldberg
Ms. Susan M. Leland
Ms. Hedy Walsh

European Support Staff
Ms. Karin Berntsen
Ms. Adriana Cavagna
Ms. Dalia Jackbo
Ms. Margaret Martirosi

Acknowledgments

CONTRIBUTIONS TO THIS REPORT

The studies and analyses upon which this report is based result from substantial contributions by many in WAES. The national reports, chapters 7 through 19, represent extensive efforts by all national teams over the course of the Workshop. Each of these chapters is the responsibility of the Associates from the given country.

The chapters in Part II were prepared by various Associates, although they present work which was the product of contributions by all Associates. Edward Griffith developed the Global Energy Data Base of chapter 3, which summarizes the data supplied by national teams for global integrations. Richard Eden and Edmund Crouch prepared chapter 4, the description and results of the WAES unconstrained global integrations. David Sternlight and Edward Griffith prepared chapter 5, the description and results of the WAES constrained global integrations. Andrew Flower contributed importantly in the coordination of the two global integration techniques. Chapter 6, which describes the energy prospects of the developing countries, was prepared by William F. Martin and Frank J. P. Pinto (a consultant to WAES and the World Bank), who gratefully acknowledge the assistance of Nicholas Carter, John Foster, and William Humphrey of the World Bank and Alan Strout of the MIT Energy Laboratory.

FINANCIAL SUPPORT

Many organizations and institutions have provided direct financial support and other services for WAES. The Workshop gratefully acknowledges the following sponsors for their generous support of the Secretariat and related activities:

The Allied Chemical Foundation

The Atlantic Richfield Foundation

The Edna McConnell Clark Foundation

The General Motors Corporation

The Ford Foundation

The German Marshall Fund of the United States

The Andrew W. Mellon Foundation

The National Science Foundation

The Rockefeller Brothers Fund

The Rockefeller Foundation

The Alfred P. Sloan Foundation

The Workshop also gratefully acknowledges the many individuals and institutions in each country that are contributing financial and/or professional support for the Workshop and/or the WAES national studies, including:

Canada
National Energy Board
Petro-Canada

Denmark
Danish Research Council for Science
Niels Bohr Institute

Finland
Academy of Finland
Helsinki University of Technology
Ministry of Trade and Industry

France
Centre National de la Recherche Scientifique
Eléctricité de France
Société Générale

Germany
Badische Anilin & Soda Fabrik AG
Energiewirtschaftliches Institut, University of Cologne
Rheinisch-Westfälisches Elektrizitätswerk AG

Italy
Azienda Municipale Nettezza Urbana, Milano
Azienda Servizi Municipalizzati, Brescia
CISE
Confindustria
EGAM
ENI
FIAT

Italy (continued)

Istituto Economia Fonti di Energia (IEFE), Università L. Bocconi, Milano
Istituto di Economia e Politica Industriale, Università di Bologna
Istituto de Fisica, Università de Napoli
Istituto Internazionale per le Ricerche Geotermiche, Pisa
IRI
Montedison
SNAM
Zanussi

Japan

Electric Machinery Industry Federation
Electric Power Industry Federation
Industrial Research Institute
Industrial Research Institute, Japan
Institute of Energy Economics
Institute for Future Technology
Japan Economic Research Center
Japan EXPO Fund
Japan Foreign Trade Council, Inc.
Japan Iron and Steel Federation
Japan Techno-Economic Society
Mitsubishi Research Institute
National Institute for Research Advancement
Nomura Research Institute
Petroleum Association of Japan

Mexico

Mexican National Energy Commission

The Netherlands

AKZO Zoutchemie
Centraal Planbureau
ESTEL Hoesch-Hoogovens
Ministerie van Economische Zaken
OGEM Holding
Paktank
Stichting Toekomstbeeld der Techniek
Verenigde Machinefabrieken

Netherlands/U.K.

Shell International Petroleum Company, Ltd.
Royal Dutch/Shell Group

Norway

Royal Norwegian Council for Scientific and Industrial Research
Elkem-Spigerverket A/S
Scandpower A/S
Statoil, Den norske stats oljeselskap A/S
A reference committee with observers from the Ministries and other institutions

Sweden

Gränges AB
Skandinaviska Enskilda Banken
Federation of Swedish Industries
Kraangede AB
National Board of Industry
Secretariat for Future Studies
Royal Academy of Engineering Sciences
State Power Board

United Kingdom

British Petroleum Company, Ltd.
Department of Energy
University of Cambridge:
—Churchill College
—Cavendish Laboratory
—Engineering Department

United States

Allied Chemical Corporation
Atlantic Richfield Company
Brookhaven National Laboratory
General Motors Corporation
National Science Foundation
Overseas Advisory Associates, Inc.

Venezuela

Ministry of Energy and Mines

Part I
Introduction and Overview

1.1 THE CONCEPTUAL FRAMEWORK

1.1.1 Introduction

No widely accepted methodology exists for projecting energy supply and energy demand for periods of ten to twenty-five years into the future. Therefore, the first major tasks of the Workshop were to identify and agree on the major determinants of future energy supply and demand, to select a range of likely values for these determinants, and to develop an internally consistent and comprehensible framework for synthesizing the various national and global studies.

The period under study was divided into two time frames: (1) the period from 1972 to 1985, in which the contribution of each energy source is largely determined by existing infrastructure and projects already under development; and (2) the period from 1985 to 2000, when pressures on oil and natural gas supplies will increase, alternative energy supply systems could make major contributions, and energy conservation measures now only beginning could reduce energy demand—provided that these systems are started and developed soon enough and implemented on a sustained basis.

No single forecast can encompass all of the uncertainties of the next twenty-five years. The Workshop therefore identified a range of different but plausible futures; that is, a "scenario" approach was adopted. "Scenario" is the term used to represent a plausible future—for our purposes, with particular reference to energy. Several scenarios were defined and analyzed by the Workshop For each scenario, values of key variables were specified and the consequent energy picture evaluated. Scenarios are not forecasts; they are chosen to span a wide range of possible futures that lead to different estimates of energy demand and supply.

1.1.2 The Overall Framework[1]

The real world[2] of energy decisions is highly interactive and too complex to be simulated precisely. So WAES constructed a model—a simplified conceptual framework—of the real energy world (Figure 1.1). Its essential feature is that it separates those global factors not under control of individual, energy-consuming nations from those factors under national control. The method used in WAES involves seven basic steps, shown in Figure 1.1 from top to bottom.

Step 1 defines in broad terms the sorts of "energy worlds" that might evolve between now and 2000 by making various assumptions about world economic growth and the world price of oil or energy, the key determinants of energy supply and demand outside the control of individual national decision-making processes. The values we selected are not predictions but plausible values that adequately span the WAES assessment of the likely future states of the world—given the assumption that the past trends of world economic growth may be extrapolated into the future.

Step 2 introduces, as a third global scenario variable, a factor that nations can control—the national

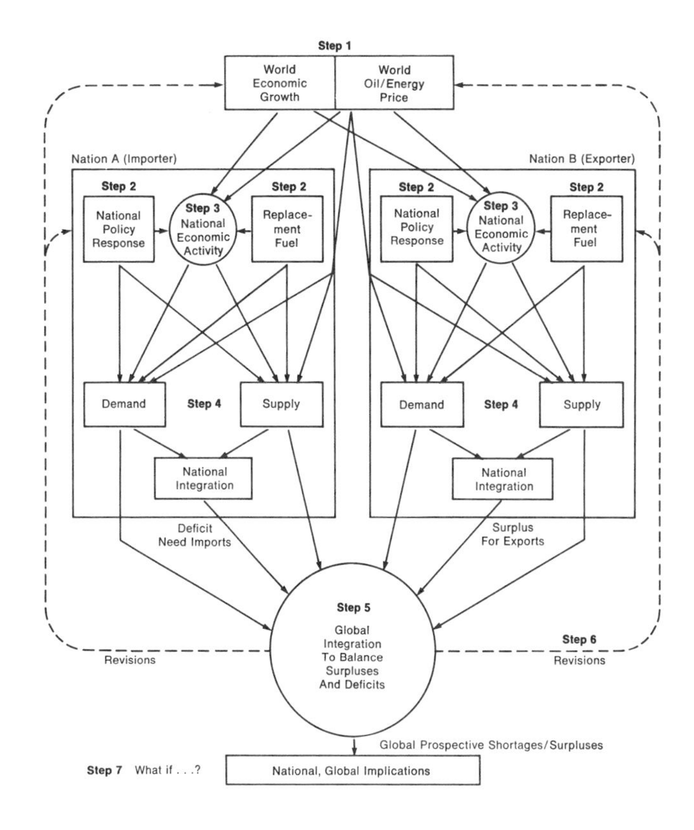

Figure 1.1
The WAES conceptual framework for world energy futures.

1. A discussion of the WAES approach in a general, global context can be found in the WAES report, *ENERGY: Global Prospects 1985–2000* (McGraw-Hill, 1977).
2. Members of the Workshop come from energy-consuming nations in the world outside Communist areas. Although the USSR and China are major world energy producers and consumers, their trade in fuels with the countries in the world outside Communist areas (WOCA) has to date been relatively small. In the WAES projections, this situation was assumed to continue through the end of the century.

public policy response to the emerging energy situation. It is assumed that until 1985 each nation has either a vigorous or a restrained national energy policy response: for example, a vigorous policy might include an active search for new sources of supply or a campaign to promote energy conservation. From 1985 to 2000 energy policies of all nations are assumed to be vigorous. Also during the 1985 to 2000 period, a major energy policy question facing nations could well be framed as: What will be the principal replacement fuel for oil as it declines in availability? This is the fourth global scenario variable for this period: whether coal or nuclear will be the principal replacement fuel. Nations may make different choices, but such choices can be added to obtain global (WOCA) totals.

Step 3 moves from these global scenarios to corresponding regional and national scenarios. Each national team individually translated and applied the global assumptions to its own national economic assumptions, taking account of such factors as trade capability and industrial infrastructure.

Step 4 analyzes in detail national energy supply and demand under world, regional, and national assumptions. Each WAES team calculated independent estimates of desired demand and potential supply. Desired demand is what each nation would like to use (assuming that supplies for each kind of fuel are available) at the given price, economic growth, and policy assumptions. Maximum potential supply is the amount that could be produced within a country given the scenario assumptions. These demand and supply estimates do not take into account the availability or unavailability of fuel imports or markets for potential exports. These demand and supply studies are done independently and are then matched in the national supply-demand integration process.

Many existing national energy forecasts stop at step 4 and simply assume that any gap between demand and domestic fuel production can be filled by imports. *Step 5* goes on to compare the fuel imports and exports of the national supply-demand integrations with global fuel availability by means of an international balancing calculation.

The global integration technique will be described later in this chapter. For now it is only necessary to state that WAES has used two kinds of integration: *unconstrained* and *constrained.*

From the global integrations, an overall energy balance (or imbalance) is obtained—the total energy supplies and demands, including shortfalls and surpluses, for each major fuel and for each scenario case under both the unconstrained and constrained alternatives. The feasibility, consistency, and likelihood of each scenario can then be evaluated for each country.

Step 6 in the WAES framework allows a major choice. Either the analysis can end here, presenting the results as a final projection indicative of prospective shortages or surpluses of fuels in the future; or, if the projection points to severe prospective energy shortages, or other major problems—in fact, to an impossible or highly undesirable future—assumptions can be altered or new scenarios can be calculated. This revision process is shown by the dashed lines in Figure 1.1. Such revisions might be motivated by, for example, balance-of-payments problems in oil-importing countries—causing increased development of domestic resources and greater fuel switching. WAES has not studied balance-of-payments questions in depth.

Step 7 lies outside the main conceptual framework. Here WAES could ask, in effect, what would happen if the main assumptions proved incorrect? What if Saudi Arabian oil production were to be held near present levels? What if there were a moratorium on nuclear power? What if solar energy rapidly became cheaper than expected? And so on. Such questions are partly explorations of the need for strategies, but they also test the impact of major uncertainties on WAES scenarios.

The next two subsections define more explicitly the scenario variables used by WAES and their combinations into the scenario cases selected for analysis.

1.1.3 The Scenario Variables

For both demand and supply projections, WAES defined the global variables given below.

World Energy Price

To 1985 energy price assumptions were based on the price of oil (in constant 1975 U.S. dollars) because oil will continue to be the dominant as well as the balancing fuel and other internationally traded fuels will therefore be priced in relation to oil. Three alternative price assumptions were made.

- Constant real price: No change from the January

1, 1976, price of $11.50 per barrel.[3] This reflects the possibility that the oil price, adjusted for inflation, will remain constant to 1985.[4]

• Rising real price: Prices rise steadily to reach $17.25 per barrel by 1985. This price was chosen arbitrarily to be 50 percent above the constant price.

• Falling real price: Prices fall steadily to reach $7.66 per barrel in 1985. This price increased by 50 percent equals the constant price.

In both the rising and falling price cases, prices were assumed to change linearly with time; the actual price changes might more realistically be a step function with increases (or decreases) every one or two years.

For 1985 to 2000 only the constant and rising price cases were used because in our study of five cases to 1985, the effect of the falling price assumption proved to be major, with the low oil price helping to drive up energy demand and hold back the development of all fuel supplies. As a result, demand exceeded available supplies. In the 1985 to 2000 period an energy price is specified, as oil may no longer be the balancing source of supply. Two energy prices for the 1985 to 2000 period were taken.

• Constant energy price: This is equivalent to $11.50 per barrel of oil (in constant 1975 U.S. dollars).

• Rising energy price: This is equivalent in 2000 to $17.25 (in constant 1975 U.S. dollars) per barrel of oil.

World Economic Growth

To 1985, world economic growth was assumed to have two values, high and low. In the high case, an average growth rate of 6 percent per year (1977 to 1985) was assumed for the world outside Communist areas. This is an upper boundary of what is likely, with the world returning to the high economic growth and flourishing international trade of the 1960s and having resolved most of its recent economic problems.

In the low case, an average economic growth rate of 3.5 percent per year (1977 to 1985) was assumed. This growth rate corresponds to the world we have seen during 1974 to 1976 and is sufficiently above population growth to allow an advance in real global GNP per capita. Growth significantly lower than this could result in periods of stagnation or recession more severe than usual in some countries and some measure of political and social instability. In the Third World, where material hopes are closely linked to world trade levels and to economic activity in the developed, industrialized nations, the economic impact could be particularly severe. WAES national studies done under these two assumptions of 6 percent and 3.5 percent global economic growth, when aggregated, actually resulted in global rates of 5.2 percent and 3.4 percent for the high and low cases, respectively, from 1977 to 1985 (see Table 1.1). WAES did not attempt to recalculate national growth rates in light of the modified aggregated global rate. If there had been time to do so, the results of such lower projections would have been taken into account. Their impact on the conclusions of the study, however, would not have been significant.

For 1985 to 2000 the growth rates were reduced slightly. In the high growth case they dropped from 6 to 5 percent because it seemed unlikely that the developed nations could maintain the high rates of the 1960s and 1970s into this period with its declining population growth rates.

Similarly, the rate in the low growth case was reduced from 3.5 to 3 percent per year for 1985 to 2000. This rate is just enough above the population growth estimates to allow some increase in the real GNP per capita in most countries. WAES national studies done under these assumptions, when aggregated, actually resulted in global economic growth rates of 4.0 percent and 2.8 percent for the high and low cases, respectively, from 1985 to 2000. Again, WAES did not attempt to recalculate national growth rates under these lower global rates, but if they had, the main results would have been only marginally affected.

As explained above, within these global boundary assumptions, each WAES national team determined corresponding national high and low growth projections. National interpretations varied widely,

3. All oil or energy prices are in constant 1975 U.S. dollars per barrel of Arabian light crude oil, FOB Persian Gulf.

4. The $11.50 per barrel price in October of 1975 represents the actual world price at that time. The $11.50 per barrel price in 1975 U.S. dollars is roughly equivalent to a $9 per barrel price in 1972 U.S. dollars, and the $7.66 per barrel price in 1975 U.S. dollars is roughly equivalent to a $6 per barrel price in 1972 U.S. dollars. These WAES price assumptions are therefore quite similar to the price assumptions used in the OECD study, *Energy Prospects to 1985*, and in several national energy studies such as the U.S. Federal Energy Administration's *1976 National Energy Outlook*.

Table 1.1
Summary of economic growth rate (EGR) assumptions by major regions

	Historical 1960-1976					1977-1985				1985-2000			
	1960 GNP	1960-1972 EGR	1972 GNP	1972-1976 EGR	1976 GNP	EGR high	1985 GNP high	EGR low	1985 GNP low	EGR high	2000 GNP high	EGR low	2000 GNP low
North America	764	4.3	1264	1.6	1345	4.3	1970	3.0	1760	3.7	3388	2.6	2576
Western Europe	533	4.7	924	2.0	1003	4.6	1500	2.7	1278	3.3	2425	2.2	1774
Japan	75	10.5	248	3.3	284	7.9	563	4.8	434	4.5	1091	2.7	646
Australia and New Zealand	26	4.5	46	3.6	53	5.0	82	3.6	73	4.3	155	2.8	111
Industrialized countries	1398	4.9	2482	2.0	2685	4.9	4115	3.1	3545	3.7	7059	2.5	5107
Lower-income developing countries	61	3.7	94	2.3	103	4.4	152	2.8	132	3.1	240	2.5	191
Middle- and upper-income developing countries	158	6.2	325	5.9	409	6.6	727	4.5	608	4.9	1490	3.9	1079
All non-OPEC developing countries	219	5.6	419	5.1	512	6.2	879	4.2	740	4.6	1730	3.7	1270
OPEC countries	34	7.2	78	12.5	125	7.2	235	5.5	202	6.5	604	4.3	380
Total WOCA	1651	5.0	2979	2.8	3322	5.2	5229	3.4	4487	4.0	9393	2.8	6757

All GNP values are given in billions (10^9) of constant 1972 U.S. dollars. All growth rates are given as percent per year increases in constant dollar value of GNP. All data for 1977-2000 for North America, Western Europe, and Japan result from WAES national team estimates; data for other regions are from special studies performed by the World Bank on the basis of WAES global economic scenario estimates. Data for 1976 are preliminary.

North America consists of Canada and the United States. Western Europe is divided into WAES participants (see contents page) and non-WAES countries (Austria, Belgium, Greece, Iceland, Ireland, Luxembourg, Portugal, Spain, and Switzerland). A listing of OPEC countries and the definitions of developing country regions may be found in Table 6.1 (p. 219).

as shown in Table 1.1, with the most significant changes occurring in Japan and the OPEC countries. The WAES approach is designed so as not to lose sight of these important national and regional differences.

National Policy Response

Two types of national energy policy response are assumed as the third scenario variable. For 1972 to 1985 either *restrained* or *vigorous* policies are assumed. Only vigorous policy is assumed from 1985 to the year 2000. A restrained policy assumes the least possible government action on conservation and supply that can be expected under the oil price and economic growth assumptions of the case. Market forces are assumed to act appropriately with the price and economic growth assumptions. Global stability is assumed. Some changes in social and individual requirements may be induced by national policy responses, but these should not extend beyond what is likely.

The definitions of vigorous and restrained policy actions are flexible, thus permitting WAES national teams to adopt individual approaches. While recognizing that price and policy effects are related, the national studies attempt to consider independently the influence of each. Explicit policies that have an effect beyond the results of price changes can be considered. Such policies have been designed in the national studies to improve the efficiency of energy use or expand supply potentials without altering the economic activities already established for any case.

"Conservation" in these studies refers primarily to improvements in the efficiency of energy use rather than to reductions of energy-using activities. The potential environmental desirability of an energy policy and the costs of meeting environmental standards have also been included by each WAES team in their definitions of national policy response.

Oil Discoveries and Production Limits

The methodology used to develop the oil production profile is described in chapter 3 of the second WAES technical report, *Energy Supply to the Year 2000: Global and National Studies* (MIT Press, 1977). Ultimately recoverable world oil resources were estimated to be 2000 billion barrels (of which about 75 to 80 percent is assumed to be in WOCA), but the possibility of higher or lower resources was also considered. However, more critical than the assessment of the total potential resources is the rate at which oil reserves are actually enlarged by new discoveries, extensions to known fields, and better recovery techniques in the period to the year 2000. Here we assumed two average rates of "gross additions" to oil reserves—a high average of 20 billion barrels per year and a low average of 10 billion barrels per year. In both cases we have somewhat arbitrarily assumed that 50 percent of these additions will be in OPEC and 50 percent in non-OPEC countries.

Theoretical oil production curves were developed, limited only by the assumption that production will increase to meet demand as long as the global reserve-to-production ratio does not fall below 15 to 1. However, since some OPEC countries may be unwilling to see oil production reach maximum theoretical levels determined by such a ratio because of a desire to extend the life of their oil reserves, we assumed that limits on total OPEC production would be 40 million barrels per day (MBD) or 45 MBD. These levels are some 10–15 MBD higher than OPEC production in 1973 (the highest annual level that OPEC production has reached to date).

Principal Replacement Fuel

The last global variable for the 1985 to 2000 period is the principal replacement fuel for oil—with coal or nuclear power as the possibilities selected. Other replacements such as oil from oil sands and shales and renewable sources such as solar, wind, and geothermal are of course also considered. Indeed, renewable energy forms represent a rapidly growing and important (if small-scale) source in many national scenarios for 2000. But their contribution was left to the analysis and judgment of national teams, and they were not included as overall scenario variables.

1.1.4 The WAES Scenarios

Figure 1.2 summarizes the WAES scenario variables and the cases selected for each period. For each scenario, separate estimates of energy demand and energy supply were made.

Five scenarios were selected to 1985, and two of these (Cases C and D in Figure 1.2) were coupled to the period 1985 to 2000. Each of these two has a coal and a nuclear variant in the latter period, thus giving four scenarios for the year 2000. The choice of extending Cases C and D through to 2000 was made after the results of the projections to 1985 became clear (see chapter 2). Case E becomes inconsistent before 1985, with demand exceeding supply. Cases A and B show that the high oil price ($17.25 by 1985) is probably higher than needed to maintain sufficient supplies to 1985. Cases C and D, with closely matched supply and demand in 1985, appeared on first analysis to be the most useful to study beyond 1985. Figure 1.2 shows this coupling and the 1985 to 2000 scenarios.

1.2 SUPPLY-DEMAND INTEGRATION METHODOLOGY

1.2.1 Introduction

Separate projections of energy demand and of energy supply were made under the WAES scenario assumptions. The supply estimates to the year 2000 are presented in *Energy Supply to the Year 2000.* The demand estimates to 1985 are presented in the first WAES technical report, *Energy Demand Studies: Major Consuming Countries* (MIT Press, 1976). The demand estimates to 2000 are described within the context of the national supply-demand integration studies (chapters 7 to 19).

Because there are great uncertainties in projecting twenty-five years into the future, the demand studies for the period 1985 to 2000 are, in general, less detailed than those for 1985. Some national teams chose to use only rather coarse measures of energy demand to 2000, based on ratios such as GNP to total energy consumption and energy use per capita. These ratios have changed with time and will change in the future, reflecting shifts in economic and industrial structures, trading capabilities, energy conservation, new technologies, and so forth. An attempt was made to take such changes into account.

Some national teams chose to do a more detailed study, so WAES developed a uniform system for condensing the original sixty-nine economic sectors (see *Energy Demand Studies*) into seventeen sectors. Other teams felt that it might be possible to perform year 2000 studies with the full sixty-nine items; the WAES framework allowed that to be done where national teams chose to do so. The seventeen sectors used by some national teams in their projections to 2000 are the following:

- Transportation: autos; air; other passenger; truck; rail; shipping
- Residential: space heat (fossil); space heat (electric); appliances
- Commercial and public: space heat (fossil); space heat (electric); other

	Factors That Influence Future Economy	Variables		
1977-1985	World economic growth rate*: high (6%) or low (3.5%)	High		Low
	Oil price: rising ($17.25) constant ($11.50) or falling ($7.66)	17.25	11.50	7.66
	National policy response: vigorous or restrained	Vig		Res
1985-2000	World economic growth rate: high (5%) or low (3%)	High		Low
	Energy price: rising ($17.25/BOE) or constant ($11.50)	17.25	11.50	
	Gross additions to oil reserves: 20 BB/YR or 10 BB/YR	20		10
	OPEC oil ceiling: 45 MBD or 40 MBD	45		40
	Principal replacement fuel: coal or nuclear	Coal		Nuc

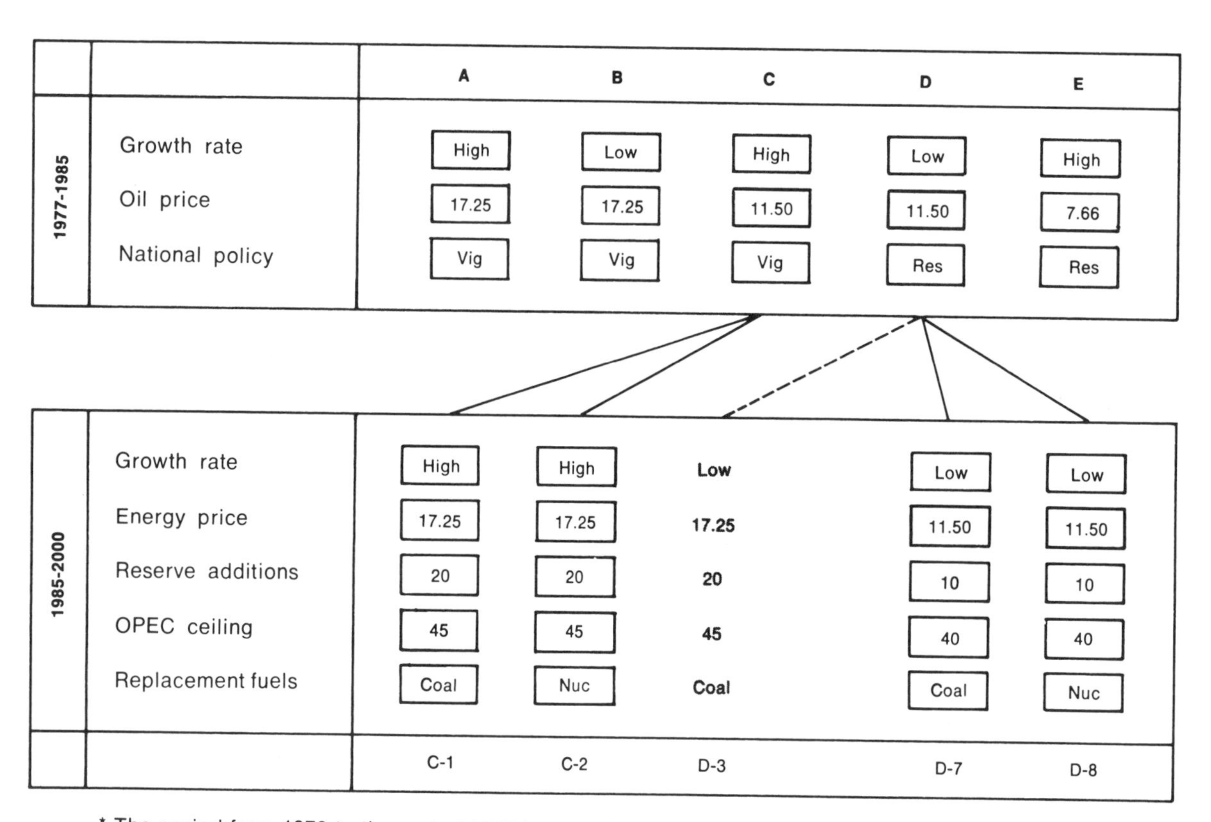

* The period from 1973 to the end of 1975 is assumed to correspond to actual world economic conditions, with recovery to 1973 levels by the end of 1976 postulated.

National economic studies done under these assumptions, when summed, actually result in global rates of 5.2 and 3.4% to 1985 and 4.0 and 2.8% from 1985 to 2000.

We assume that the scenario variables are approximately independent of each other, within a certain range of values. Any combination of values *may*, then, be possible and no combination is automatically excluded from consideration.

Figure 1.2
The WAES scenario cases, 1972–1985–2000.

- Industry: metals; chemicals; agriculture, mining, and construction; other energy-intensive industry; non-energy-intensive industry

Because WAES handles energy demand and supply separately, the two must be compared, or "integrated." This is an essential step in the WAES methodology. When supply and demand are integrated, prospective gaps (or surpluses) between the totals of import requirements and the availability of exports from other countries are discovered under particular scenario assumptions. Such gaps or surpluses cannot occur in the real world. A scenario that fails to balance is a signal that some assumptions require modification.

WAES global integrations are of two types: unconstrained and constrained. The objectives and methods of these two procedures are outlined below. Their results are presented in chapters 4 and 5.

Unconstrained integration compares desired energy demand and fuel mix with projected potential supply to see what the shortages and surpluses might be. This integration is unconstrained, or unaffected, by the possibility of insufficient imports to meet preferred fuel demands.

In reality, of course, there can never be actual shortfalls of fuels over extended periods of time. Prices will rise, inducing a higher supply level, lower demand level, and changes in fuel preferences; policy actions will reduce demands, increase supplies, or induce fuel switching. Yet, consideration of these imaginary shortfalls, or prospective shortages, as well as prospective surpluses, can be extremely instructive. It indicates where problems are likely to be and their possible size and timing.

The second type of global integration used by WAES, the constrained integration, forces substitutions of available fuels for shortfall fuels on a regional basis (when necessary and within the limits considered feasible). The aim is to reduce the projected shortages and surpluses by using all available fuels in the most efficient way possible and at the least total cost to consumers. This integration process is a reasonable approximation of the real world to the extent that energy decisions over the long term, and within physical and political constraints, will be made on a least-cost, most-efficient basis. The integration is constrained by a series of limits on supply availabilities, which may not be sufficient to meet preferred fuel demands, and by specific ranges of demand preferences for major fuels.

This integration procedure has multiple properties.

- It can suggest the cost, infrastructure, timing, and balance-of-payments consequences of efforts needed to balance supply and demand.
- It can reveal difficulties in a particular WAES case, such as shifts beyond market desires in the mix of fuels needed to reach a supply-demand balance.
- It can reveal the most economically efficient (least total cost to world consumers) adjustments, within tight physical and political constraints, for closing supply-demand gaps.
- It permits policy analysis, in that policies, costs, technical potentials, demand targets, and timings can all be varied.

The major difference between the two procedures used by WAES for global integration is that the first reveals supply-demand imbalances; the second attempts to balance supply and demand and reveals the actions needed and the consequences of these actions. More detailed descriptions of the two integration processes follow.

1.2.2 Unconstrained Integrations

Unconstrained global integrations begin with the preferred national demands for each fuel and the indigenous supplies for each fuel, as seen in Figure 1.3. These rarely balance on a national basis, and the difference gives the net imports or exports (usually imports for WAES countries). These imports and exports of fuels are calculated in national supply-demand integration studies of the flows of energy from raw resource (supply potential) to end use (desired demand) and the energy losses at each stage along the way. These national studies, contained in chapters 7 to 19, reveal national import desires and export potentials.

Integration is thus first a national summing process—adding demands by sector (transport, residential, industrial, commercial and public) and by kind of fuel (coal, oil, gas, electricity) in each country. This gives total end-use demand or energy consumed at the point of final use. To this must be added, for each country, the energy lost when raw or primary fuels are refined or converted and trans-

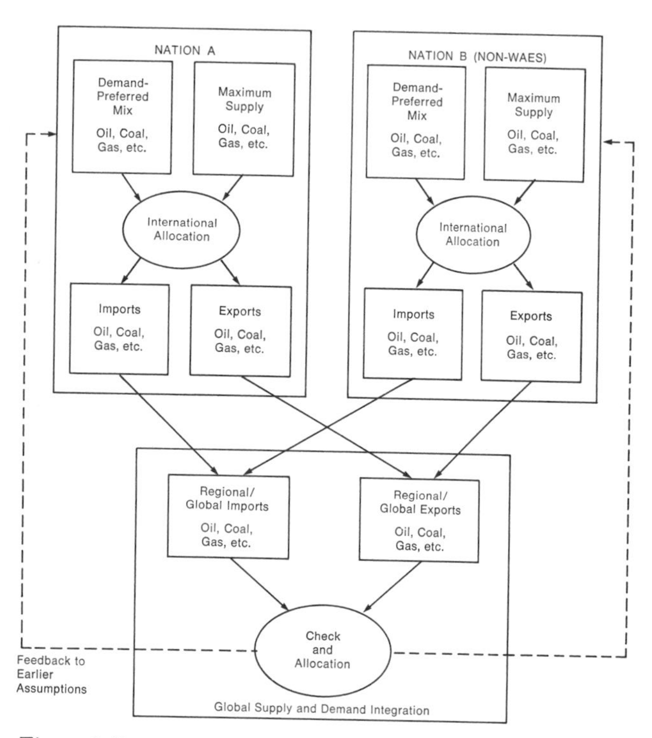

Figure 1.3
Supply-demand balance calculations: unconstrained integrations.

ported to the user. In addition to WAES national studies, import expectations and export capabilities are also assessed for non-WAES regions. (For a more complete description of the analysis of the energy prospects of the developing countries, see chapter 6.)

Aggregating these integrations on a regional and then on a global basis reveals whether demand and supply are globally in balance: Are export potentials of each fuel sufficient to satisfy import expectations? If they are, fuels may be allocated in world trade so that the potential exports that could enter world trade are based on national import desires. Traditional export-import patterns are important because of geography, transportation capabilities (e.g., docks), refinery sites, and many other historical factors.

If demand and supply do not balance—if there is a global shortfall of any fuel or of total energy—no world trade allocation is made. The size of each prospective gap (or surplus) for each fuel is noted. Such gaps should be seen by nations as signals for actions of the proper scale, timing, and duration to avoid or minimize the gap.

Unconstrained integrations are summations of national desires for fuels compared with global maximum supplies. They are done to reveal prospective gaps between national expectations and the realities of globally limited fuels, under the assumptions of each case. The interaction between supply and demand is not of central concern here. There is no attempt to balance supply and demand; the results of the unconstrained integrations are heavily influenced by the assumed fuel substitution rates. But the gaps revealed are highly instructive and result directly from national preferences and expectations. The unconstrained global integrations were done for WAES by the Energy Research Group, Cavendish Laboratory, Cambridge University, England. The results of these integrations are presented in chapter 4.

1.2.3 Constrained Integrations

In the second approach, constrained integration, shifts in the fuel mix beyond the range of consumer preferences were "forced" in order to minimize global supply-demand imbalances.

Constrained integrations are done not to reveal gaps, but to see if it is possible to close them, and also to indicate details of required expansion for the energy system and related costs. The main objective of constrained integrations is to balance total energy supply and demand within maximum and minimum constraints, with fuel preference ranges satisfied to the extent possible. The procedure requires many thousands of calculations and was carried out by WAES Associates at the Atlantic Richfield Company using a specially designed, highly constrained linear programming model—the Global Energy Mini-Model (GEMM)—with detailed inputs, constraints, and adjustments provided by WAES national teams through their national supply and demand studies.

This constrained integration approach incorporates all the advantages and disadvantages of linear programming applied to a dynamic environment. The results follow from the use of a simplified rule for decision making—least total cost to world consumers—subject to all the imposed constraints being satisfied. While such a criterion may not conform precisely to reality, particularly in the short term, it represents a reasonable approximation of the real world. Such integrations provide a wealth of data, aggregated globally, for the potentials, constraints, and costs of gap closing, fuel switching, and other measures.

The constrained process starts off with the same demand and supply inputs as the unconstrained approach. The major difference lies in the global and regional uses of fuels. "Preferences" for certain fuels may not be met when global supplies (exports) are constrained. In addition, account is taken of the expansion of needed infrastructure in each region of the world and of energy trade flows.

The procedure is shown in Figure 1.4. End-use energy demands of both WAES and non-WAES countries are summed into regions. These regional demands are compared with regional supply potentials to calculate an energy balance with its own fuel mix, costs, and infrastructure requirements.

The calculation steps start with base year infrastructure (the existing number of oil wells, tankers, coal mines, power plants, etc.) from which a pattern of supply, processing, transportation, and consumption capacity is calculated to meet specified 1985 or 2000 demands. In establishing these requirements, existing capacity is depreciated and new capacity is added within a maximum potential limit. The particular pattern of supply that results from these calculations takes account of operating costs for all capacity used and capital costs for new capacity. The pattern calculated is the one that yields least total costs to world consumers within specified maximum and minimum constraints. Such an approach also leads to a most efficient use of energy mix. It takes account of losses and of infrastructure expansion possibilities at each stage of processing, from resource extraction through delivery to the consumer.

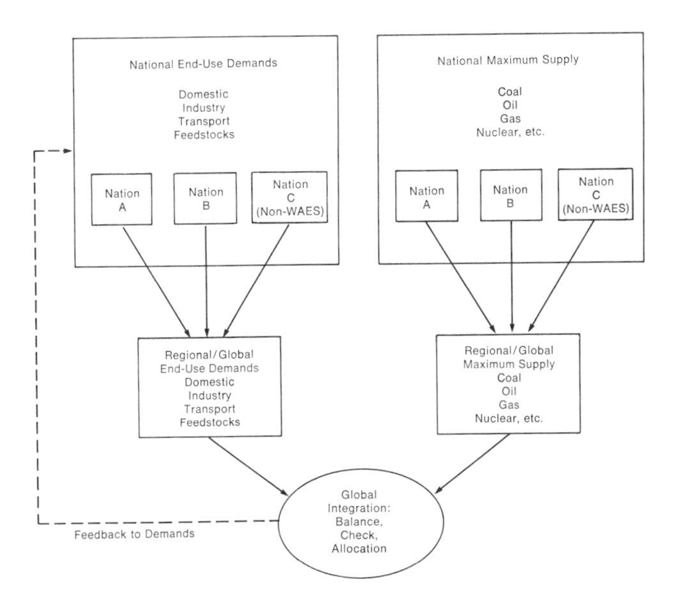

Figure 1.4
Supply-demand balance calculations: constrained integrations.

The primary rule for decision making in these calculations is to meet all end-use energy demands, that is, to make total energy supply and demand balance at the lowest total cost to world energy consumers. To achieve this balance in scenarios where some (preferred) fuels are in relative shortage while other (not preferred) fuels are in relative surplus, substantial fuel switching beyond desired mixes may be necessary. That is, the constrained integrations calculate major shifts in fuel use patterns (when there is a shortfall of particular preferred fuels and a surplus of less desired fuels) regionally and globally in order that all available supplies will be used to meet end-use demands. This fuel switching is done so that all available energy from each fuel is utilized; that is, losses in the processing and conversion of fuels are minimized. The results of the constrained global integrations are presented in chapter 5.

The real value of the two types of global integrations is what they show us. The integrations expose gaps between expected or desired imports and what is likely or potentially possible to be available in each scenario. These are starting points for revising perceptions of the future and for seeing what decisions and actions are needed, and where and when they will be necessary.

1.3 UNDERSTANDING SUPPLY-DEMAND INTEGRATION RESULTS

While every attempt has been made to bring a high degree of consistency to the various supply-demand integration results in this report, a few clarifying and qualifying notes should be kept in mind.

There are some minor inconsistencies between some of the data in Part II and other WAES reports. However, all discrepancies are the result of rounding, energy conversions, or minor adjustments relevant to the supply-demand integration process. In no case do they affect the results of the analysis or the conclusions of the Workshop.

Unconstrained integrations assume the availability of desired energy imports; constrained integrations attempt to force the fuel mix to conform to global fuel availability within maximum and minimum constraints. Nearly all national integrations presented in Part III are of the former type; national integrations that *do* take account of the limitations on import availabilities are clearly stated as such in each chapter. Thus, national integrations (which are really national "preferences" under the scenario conditions) should not be seen as predictions. There is no reason to believe that they should be globally consistent. Their purpose is to show the volume of imports required on a national basis to satisfy energy requirements.

The WAES analyses reveal prospective shortages of preferred fuels. But WAES has taken no comprehensive, detailed approach to closing such prospective gaps. Time did not permit it. Yet, several national studies report their "responses" to this global situation. WAES has developed a method for calculating and measuring such responses. Energy *action programs*—packages of energy supply, conservation, and fuel substitution measures—could be developed to reduce the consumption of oil and natural gas, improve the efficiency of energy-using devices, or increase energy supply.

Each package could be added to the year 2000 scenarios. Each action program could be a standard size, equivalent to adding supplies or reducing demand by 100,000 barrels per day of oil equivalent (0.1 MBDOE). Thus, a single deep sea oil field (provided it can be found) with this output would form one action program; the insulation of 10 million houses to save energy equivalent to 100,000 BD would be another.

One can make up combination packages of supply addition and demand reduction and add them to gauge their impact in closing supply-demand gaps. Some examples, with estimates of key characteristics such as capital and operating costs, final costs of energy produced or saved, lead times, technical efficiencies, etc., are included in the national supply-demand integration chapters in this volume. (A slightly different methodology, which shows the consequences of several factors on alternative national goals, is illustrated in the Swedish integration, chapter 17.) A great many additional action programs are needed to assess the scale, scope, duration, and impacts of actions to save or to add energy or to shift fuels used in a significant way. WAES invites, and urges, others to pursue the analysis and development of such programs.

Summary of the Global and National Supply-Demand Integration Studies

2.1 INTRODUCTION

This chapter contains a brief summary of the overall results of WAES global supply-demand integrations, which are extracted from chapters 4 and 5 of Part II and from the national reports (chapters 7 to 19) of Part III. All of these results are set in the general context of the aggregate WAES findings as set forth in *ENERGY: Global Prospects 1985–2000* (McGraw-Hill, 1977).

All data in Part II of this report are given in millions of barrels per day of oil equivalent (MBDOE) or exajoules (10^{18} J). In the individual national chapters, data are given in the appropriate units for the particular country, or in MBDOE. The Workshop uses a standard set of conversions from volumetric to energy units. A discussion of energy units, and a table with the conversions used by WAES, can be found below. Section 2.3 contains a brief summary of the global integrations. Summaries of the national supply-demand integration studies are collected in section 2.4. Section 2.5 contains summary tables for the analyses from thirteen WAES national teams: Canada, Denmark, Finland, France, the Federal Republic of Germany, Italy, Japan, Mexico, The Netherlands, Norway, Sweden, the United Kingdom, and the United States.

Part III contains the thirteen complete reports of the WAES national teams. Each report consists of an overview section, a description of the specific methodology used to derive the energy demand projections, a discussion of the results, and a collection of detailed worksheets describing energy supply-demand integrations for the base year, 1972, for 1985, and for the year 2000.

2.2 ENERGY UNITS

Like many other commodities, energy and fuels are measured by a seemingly innumerable variety of units. Attempts to achieve wide acceptance and use of a common convention—for example, the International System of Units (SI)—have met with some degree of success. But different people are accustomed to different units of measure; changes in training, habit, and usage in these matters come slowly.

Not only are different systems of units in use, but there is also little consensus on precisely how to convert from one system to another. Given the nearly infinite variety in the qualities and characteristics of energy and fuels in the world, perhaps this is to be expected. WAES has used several units of measure—each national team has done its study in the units most familiar to it—and has adopted one conversion system, shown in Table 2.1.

Two broad classes of units are in use today: energy units and physical units. Energy units are, for example, calories, joules, and British thermal units (Btu). One Btu is the energy required to raise the temperature of one pound (0.454 kg) of water from 64.5°F to 65.5°F. It is equivalent to 1055 joules; one calorie is 4.2 joules. Physical units are sizes and weights—barrels, gallons, tons, etc. A barrel of crude oil, for example, has a certain energy content; so does a ton of coal. Often energy units are expressed as the energy equivalent of a barrel of oil, or ton of oil, or ton of coal.

In this report the rate of energy use (both for fuels and electricity delivered to consumers and for supply and demand of primary energy) is generally expressed in terms of millions of barrels per day of oil equivalent (MBDOE). This measure is based upon the conventional unit, barrel of oil equivalent, with a gross calorific value of 5.8 million Btu.

In many places throughout the text we have also employed other units of measurement, for example exajoules (10^{18} joules), quadrillion Btu (10^{15} Btu or "Quads"), or units applicable to particular energy sources, such as tons of coal or cubic meters of gas.

The calorific value of the various types and sources of coal, crude oil, and natural gas varies widely. The full equivalents of the basic energy units used by WAES are given in Table 2.1.

To produce an electrical output of 620×10^9 kWh per year (1 MBDOE), one would need power stations of 100 GWe (100,000 megawatts) installed capacity, given an average load factor of 70 percent. If the average efficiency of the power stations was 35 percent, one would need a fuel input of 1/0.35 (approximately 2.8) MBDOE. The difference between the input of energy as fuel and the output of energy as electricity (in this example, 1.8 MBDOE) is the transformation loss in electricity generation,

Table 2.1
Standard energy conversions (equivalent values lie in vertical columns)

Unit									
Barrels per day of oil equivalent[a]	—	—	—	—	—	—	—	—	—
Tons of oil equivalent[b]	—	—	—	—	—	—	—	0.022	0.023
Metric tons of coal equivalent[c]	—	—	—	—	—	—	—	0.034	0.036
Barrels of oil equivalent[d]	—	—	—	—	—	0.0064	0.02	0.16	0.17
Cubic meters of "average" natural gas[e]	—	—	—	0.027	0.09	1	2.7	25	27
Kilowatthours	—	—	—	0.3	1	11	29	280	293
Cubic feet of "average" natural gas[e]	—	—	—	1	3.4	37.3	100	950	1000
Kilocalories	0.24	0.25	1	252	860	9400	25,200	0.24 million	0.25 million
British thermal units	0.95	1	4.0	1000	3400	37,300	1 therm	0.95 million	1 million
Kilojoules	1	1.06	4.2	1055	3600	39,400	105,500	1 million	1.06 million

Calorific values are measured gross. Rounded equivalents only are given.
[a] Equivalents in other units are shown on a per annum basis.
[b] of 43 million Btu (≈ 10,000 kcal/kg net cal. val.).
[c] of 12,000 Btu/lb = 7000 kcal/kg.
[d] of 5.8 million Btu
[e] of 1000 Btu/ft^3 or 9400 kcal/m^3.

Source: Energy Conversion Equivalents Table, Shell International Petroleum Co., Ltd.

								Abbreviation
—	0.003	0.013	0.02	1	2.7	18,000	470 million	BDOE
0.09	0.13	0.65	1	50	135	0.9 million	23×10^{9}	TOE
0.14	0.21	1	1.5	76	209	1.3 million	36×10^{9}	MTCE
0.68	1	4.8	7.4	365	1 TBOE	6.4 million	170×10^{9}	BOE
106	155	745	1150	57,000	0.155 million	1×10^{9}	27×10^{12}	Nm^3 NG
1160	1700	8140	12,600	0.62 million	1.7 million	11×10^{9}	290×10^{12}	kWh
4000	5800	27,800	43,000	2.1 million	5.8 million	37.3×10^{9}	1×10^{15}	ft^3 NG
1 million	1.5 million	7 million	10.8 million	530 million	1.5×10^{9}	9.4×10^{12}	250×10^{15}	kcal
4 million	5.8 million	27.8 million	43 million	2.1×10^{9}	5.8×10^{9}	37.3×10^{12}	1 Q	Btu
4.2 million	6.1 million	29.3 million	45.4 million	2.2×10^{9}	6.1×10^{9}	39.4×10^{12}	1.06×10^{18}	kJ

— = insignificant
1 therm = 100,000 Btu
1 TBOE = 1000 BOE
1 MBDOE = 10^6 BDOE
1 Q = 10^{18} Btu
1 Quad = 10^{15} Btu

which is included in the WAES data under the heading of "processing losses," along with the losses that occur in all other energy conversion processes.

When presenting data for primary energy supply and demand, WAES adopted the convention of expressing electricity from primary sources (nuclear, hydro, geothermal, etc.) in terms of the fuel input (generally expressed in MBDOE) that would be required to produce the equivalent amount of electricity output in fossil-fueled power stations.

2.3 GLOBAL SUMMARY

The overall findings from the WAES analyses center around the results of the integrations of supply and demand. Energy supply and demand, when taken together over the period 1972 to 2000, paint a disconcerting picture of growing shortages of oil. This principal WAES finding is illustrated for a high-demand, high-supply set of assumptions in Figure 2.1. Resource and production limitations begin to restrict oil supply in the period 1985 to 1990. Further increases in oil demand beyond 1990 must be satisfied from other fuels. The prospective oil shortfall must be filled or eliminated. Oil demand and supply must always balance.

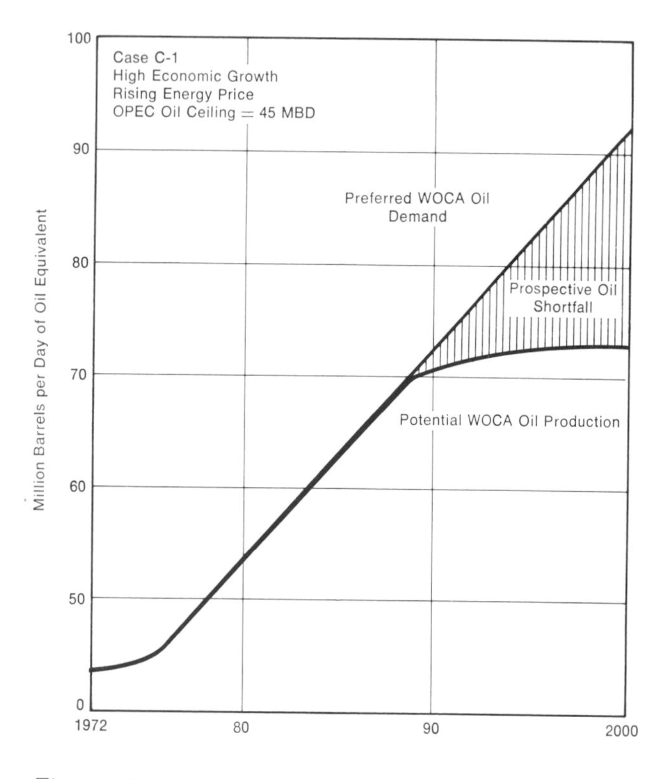

Figure 2.1
Oil supply and demand (WOCA) for Case C-1. Preferred oil demand is the sum of national preferences in each WAES case. Potential oil production is the total WOCA potential oil supply (OPEC limit is 45 MBD).

Filling the prospective oil shortfall becomes increasingly difficult as we approach the end of the century. What begins as a minor discrepancy widens rapidly to a gaping deficit before the year 2000. The picture persists under a variety of assumptions for the year 2000. Figure 2.2 illustrates the bands of estimates of both unconstrained oil demand and potential oil supply over a range of assumptions. The problem does not go away within the range of the WAES economic growth, energy price, and national policy assumptions.

2.3.1 Oil Imports

The world oil problem, in simple terms, is the difference between desired oil imports and potential oil exports. Table 2.2 is a summary of WAES oil imports and exports estimated for the year 2000. It is based on individual national "unconstrained" supply-demand integrations from WAES national teams and similar assessments for non-WAES countries and regions (including OPEC). The WAES national unconstrained supply-demand integrations take into account consumer preferences and the resistance of these preferences to change (i. e. they recognize the lead times involved in consumers' changing from one energy source to

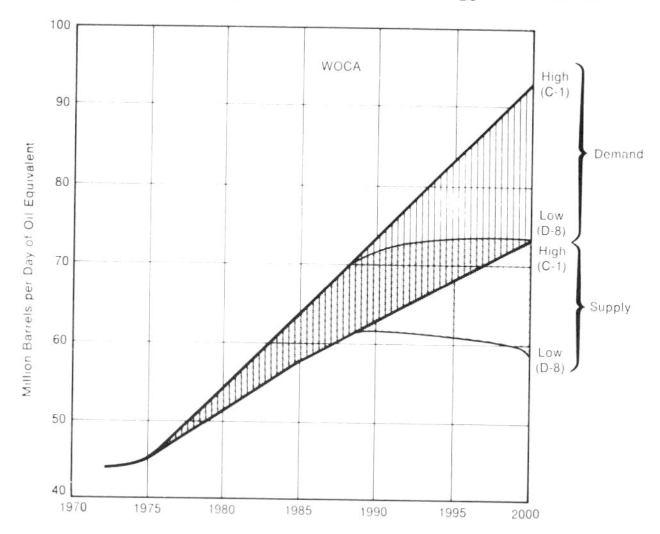

Figure 2.2
Desired demand for and potential supply of oil. The range of values for preferred oil demand comes from the WAES scenario results. The range of potential oil supply comes from the WAES maximum and minimum values. It is clear that only when the lowest of the WAES preferred oil demand estimates is matched to the highest of the WAES potential oil supply estimates is there no projected oil shortage. It is not at all certain, however, that such a coupling of estimates is based on consistent assumptions.

Table 2.2
Summary of oil balance in the year 2000 (MBD)

Economic growth Energy price (1985-2000) Principal replacement fuel WAES case	High Rising Coal C-1	High Rising Nuclear C-2	Low Constant Coal D-7	Low Constant Nuclear D-8
Major importers' desired imports[a]				
North America[b]	10.4	10.7	15.8	15.8
Western Europe	16.5	16.4	13.2	12.5
Japan	15.2	14.4	8.2	7.9
Non-OPEC rest of WAES	11.2	9.5	9.6	9.0
International bunkers[c]	5.4	5.4	4.5	4.5
Total desired imports	58.7	56.4	51.3	49.7
OPEC potential exports[d]	38.7	37.2	35.2	34.5
Prospective shortage	20.0	19.2	16.1	15.2
Prospective shortage as a percentage of total WOCA potential oil production	27%	26%	28%	26%

[a]For the definitions of areas, see the note to Table 1.1.
[b]Figures take account of domestic production of oil shale and oil sands in addition to conventional oil.
[c]International bunkers represent the oil used in international shipping.
[d]OPEC potential exports equal OPEC potential production minus OPEC internal demand.

another). Desired oil imports exceed potential oil exports; under WAES assumptions, the "prospective oil shortage" is between 15 and 20 MBD.

The shortage is equivalent to an oil flow greater than half of OPEC's maximum production to date (30.5 MBD in 1973). It is approximately equivalent to the total amount of energy used in Western Europe today.

In each of the scenario cases this oil shortfall is nearly 30 percent of total WOCA potential oil production in the year 2000. Note that in the low economic growth, constant price cases (D-7 and D-8), desired oil imports by Western Europe, Japan, and non-OPEC Rest of WOCA countries are less than their desired oil imports in the high growth, rising price cases (C-1 and C-2). This is due to the lower demand levels in low growth Cases D-7 and D-8 than in high growth Cases C-1 and C-2. For North America, desired oil imports are greater in D-7 and D-8 than in C-1 and C-2. This is due to the higher potential oil production levels of C-1 and C-2 (the rising price cases) in North America, which offset the greater demand in these cases. Among the major importers shown in Table 2.2, North America is only one that is also a significant producer, and thus the only one in which production level differences can have noticeable impacts, from case to case, on desired oil imports.

The desired oil imports for 1972, for two of the five 1985 cases, and for two of the four year 2000 cases are shown in Figure 2.3 for three major consuming regions. There is a continuing high level of dependence on oil imports by Western Europe and Japan. In North America, imports grow as a fraction of the total in all cases due to rising demand and dwindling United States and Canadian production. Yet, even though oil imports represent only a relatively small proportion of the total

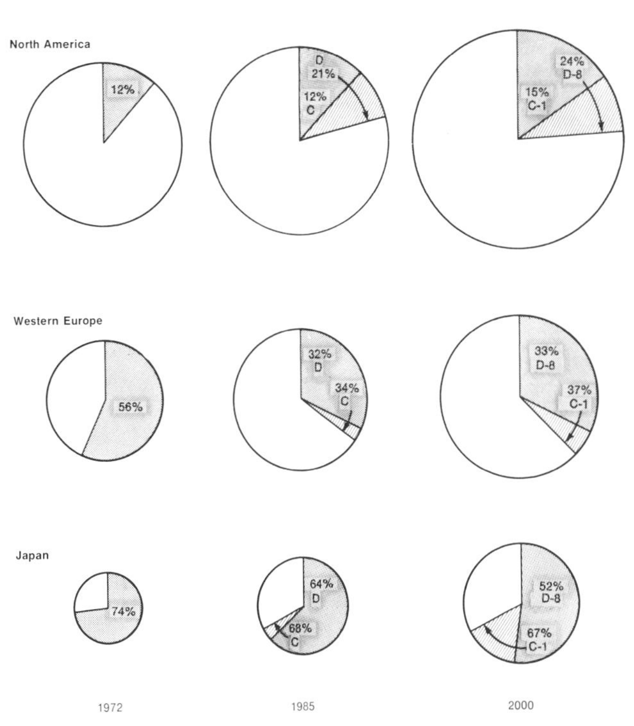

Figure 2.3
Required oil imports shown as shares of energy demand for North America. For a definition of North America and Western Europe, see the note to Table 1.1 (p. 6).

primary energy requirements in North America by the year 2000, the import volume is growing. This growing North American drain on world oil trade has important consequences for other regions.

The central results are that in spite of reduced overall demand growth, in spite of strong conservation measures, in spite of higher oil prices and vigorous actions to bring on additional supplies, demands for oil continue to grow. The sum of national expectations for oil imports is very large in the year 2000—larger than the WAES aggregate estimates of the maximum potential oil exports. The result is prospective oil shortages.

2.3.2 Gas and Coal Imports and Exports

Nations also have import and export desires for natural gas and coal. These demands, which are calculated from unconstrained national integrations, are presented in Tables 2.3 and 2.4.

Table 2.3 shows the required expansion in international gas trade resulting from the WAES studies. These required trade expansions range from 12 to 17 percent of the total desired gas demand, or about 40–50 percent of total desired gas imports. The estimated imports result from assumptions in the national demand and supply studies. The exports from the USSR and from OPEC and other Rest of WOCA countries are estimated on the basis of maximum potential pipeline and LNG (liquefied natural gas) projects as of 1985. The reserves of gas in OPEC are sufficient for the required trade expansion to occur; however, ways must be found to move the gas to markets, or else demands for gas must be reduced through conservation or acceptance of substitute fuels.

Table 2.4 shows the prospective coal surpluses of the WAES year 2000 cases. Coal is not a "preferred" fuel. The maximum potential for coal production exceeds desired coal demands by some 13–17 percent of total coal demand. However, nearly all coal demand estimates are based on known indigenous production estimates in each country. It is coal imports that are not preferred—so that potential global exports are somewhat in excess of desired imports. Table 2.4 shows, in fact, that the prospective coal surpluses are about the same size as total desired coal imports.

If demands for coal could be increased, as a substitute for oil and gas, the potentially available coal exports could be used to help close the oil and gas gaps. Strategies for the use, transport, and conversion of coal are obviously of critical importance.

The complete results of the WAES unconstrained global supply-demand integrations—showing the prospective fuel gaps and surpluses in each case—are presented in chapter 4.

2.3.4 Fuel Switching for Energy Balancing

The energy imbalances resulting from the WAES supply-demand integration analyses cannot occur in the real world. Solutions will be found. But the cure could be more harmful than the ailment. Rapidly accelerating prices, rationing, and economic stagnation all successfully "reduce demand" and

Table 2.3
Summary of natural gas balance in the year 2000 (MBDOE)

Economic growth Energy price (1985–2000) Principal replacement fuel WAES case	High Rising Coal C-1	High Rising Nuclear C-2	Low Constant Coal D-7	Low Constant Nuclear D-8
Major Importers' Desired Imports				
North America	2.8	3.0	3.5	3.4
Western Europe	4.1	3.2	3.6	2.7
Japan	1.5	1.5	1.5	1.5
Total desired imports	8.4	7.7	8.6	7.6
Operational and Planned Gas Trade[a]				
From USSR	1.0	1.0	1.0	1.0
From OPEC and rest of WOCA	3.6	3.6	3.6	3.6
Total potential trade	4.6	4.6	4.6	4.6
Further required trade to satisfy desired imports	3.8	3.1	4.0	3.0

[a]Potential trade is based on those projects operational or under discussion at the end of 1976. These projects were all planned to be operational by 1985. Further trade is required to satisfy consumers' desired imports by 2000. In both 1985 and 2000 there could be a shortfall in gas supplies if projects are not developed at the required rate. Such shortfalls would be caused by lack of infrastructure rather than lack of resources.

Table 2.4
Summary of coal balance in the year 2000 (MBDOE)

Economic growth Energy price (1985–2000) Principal replacement fuel WAES case	High Rising Coal C-1	High Rising Nuclear C-2	Low Constant Coal D-7	Low Constant Nuclear D-8
Major Importers' Desired Imports				
Western Europe	4.2	2.1	2.0	1.1
Japan	2.3	2.3	1.7	1.3
Total desired imports	6.5	4.4	3.7	2.4
Major Exporters' Potential Exports				
North America	8.6	3.7	6.9	4.6
Australia, New Zealand, S. Africa	2.8	1.0	3.4	1.7
Developing countries	2.2	1.6	1.1	1.0
Total potential exports	13.6	6.3	11.4	7.3
Prospective surplus	7.1	1.9	7.7	4.9
as % of total WOCA potential coal demand	21%	6%	30%	23%

thus balance energy supply and demand. Yet, such dislocations and disruptions are very undesirable "solutions." They may be "energy disasters" instead of strategies designed to smooth the transition from oil to other energy forms.

One broad approach to reducing energy imbalances is to ask: How might fuels be allocated so that, within reasonable constraints, the energy content of each fuel is put to maximum use? The WAES constrained global integrations try to answer this by squeezing every drop of energy out of each available fuel—minimizing the inevitable "losses" of energy content in processing, refining, and conversion—by using each fuel without conversion to other fuels, wherever possible. For example, the substantial heat loss associated with the conversion of fossil fuel into electricity can be reduced by using fossil fuels in end-use markets as fuels—not in electric power plants—while using other options such as nuclear for electricity generation. District heating from power plants is another example of a substantial efficiency improvement in this sense. Several national integrations incorporate such uses in their projections.

Account is taken in the WAES constrained integrations of fixed, nonsubstitutable uses of oil, such as transport and petrochemical feedstocks. For other uses, because of the severe constraints on energy availabilities, ranges of consumer preference in the constrained cases were often violated. Nonetheless, the results are useful to show a potential mix of energy sources and uses that can contribute to solving the problem of prospective energy shortages.

The mix of fuels is important. That total energy supply and demand in the world outside Communist areas differ only by a small percentage (between 5 and 8 percent) is only a part of the problem. Small adjustments will not correct the imbalance because the real imbalance is between the supply and demand of preferred fuels such as oil. The WAES constrained integration procedure has made an effort to push substitutions for oil to the limit, in order to make up for the deficit in its supply. The resulting fuel mix is very different from what we are accustomed to today.

In the constrained integrations, use of oil and natural gas in electricity generation in 2000 is marginal (about 10–15 percent) compared to an average of 35 percent in 1972; the balance is made up by a massive shift to nuclear and coal. Oil is essentially removed from use in the domestic sector and is replaced by electricity and by coal under various new technologies. Oil is also largely taken out of the industrial sector, except for use in feedstocks together with natural gas, and is replaced by coal to a large extent and electricity to a small extent. The use of oil is confined almost totally to transport and petrochemical feedstocks. Fuel substitution could not be pushed further under even the most optimistic technologies and vigorous assumptions about the speed of capital stock turnover—and yet supply does not quite meet demand.

The constrained integrations do not represent predictions of energy sources and uses. Rather, they show the case of using all fuels in their most efficient way, within reasonable physical constraints and given the particular WAES case assumptions. They show what might be done, with concerted fuel-switching actions. They show the broad *directions* for action, by matching supplies to demands, to obtain an energy balance at the end of the century.

These integrations show that, given all of the assumptions and conditions (already described) of the constrained integration analysis, the potential coal surplus, synthetic oil from coal, small exports from Communist areas, and processing losses saved by lowered electricity generation all combine to reduce the prospective energy gap.

In addition to these integrated cases, WAES made a special study of Case D-3 to the year 2000 (see Figure 1.2—the WAES scenario chart).[1] Case D-3 is a low economic growth (2.8 percent per year from 1985 to 2000), rising energy price (to $17.25 per barrel by 2000) case. It might be labeled a "low demand, high supply" case—at least relative to the other WAES cases.

In Case D-3, the availability of oil and gas is probably sufficient to meet preferred demands for these fuels in the year 2000. This result assumes, of course, that the WAES high OPEC production assumptions will occur along with the other assumptions of the case. The small margins of deficit or surplus in this case are probably well within the limits of uncertainty of the numbers themselves. Yet even if this case does balance—even if potential supplies are in fact sufficient to meet desired demands in 2000—this case just stretches the time scale. The same mismatch between supply and demand as in other cases might appear by 2005 or 2010. This postponement might allow more time to adjust to the eventual transition, more time to develop new technologies and renewable energy forms. But these technologies must be developed and this time must be used.

2.3.5 Conclusions

The crux of the problem revealed by the WAES integration analyses is that preferred demand for oil in the world outside Communist areas exceeds maximum potential production of oil by an amount that grows, starting in the late 1980s, at a rate of about 1.8 MBD per year and increases to some 20 MBD by 2000. Using different assumptions in other cases, the size of the prospective oil gap varies from 15 to 20 MBD, and the gap shows up as early as 1981 or as late as 2004. But the same rapidly growing shortfall results. Actions are required on a large scale to avoid shortages. The size of the gap is large and its inception is soon.

Among the WAES cases studied to 1985, only Case E with low oil prices and high economic growth has insufficient supply to meet total energy demand within expected fuel mix preferences. But, while four of the WAES cases balance in 1985, it should be remembered that the estimates assume maximum potential supplies consistent with case assumptions. Given lead times of five to ten years or more for many projects, failure to make necessary near-term commitments, to resolve a variety of current restraints on production, and to develop future supplies may foreclose some options for 1985.

The broad directions for such action are outlined in the results of the WAES national studies of demand reduction, supply addition, and fuel substitution. The elements built into the supply-demand integration procedures indicate that the energy and fuel shortfalls could be partially filled through major fuel substitutions if nearly all fossil fuels were removed from use in electric power plants to reduce processing losses, synthetic crude oil were made from available coal supplies, in order to meet essential demands for liquid fuel, and industrial and domestic sectors were to use significantly more coal and less electricity than current preferred patterns.

In the real world, any number of factors could combine to eliminate the gaps: growth could be lower, energy prices could be higher, and energy policies could be *supervigorous*—lowering demand and increasing supply. But the consequences could be most undesirable. Balancing energy supply and demand by 2000 in an acceptable manner requires great effort.

2.3.6 Implications

The analysis has two major implications:

- All energy resources and conservation measures must be developed vigorously to meet total projected demand in the year 2000. Low levels of coal or nuclear development, or restrained conservation, would result in the failure of energy supplies to meet projected demand levels.
- Incremental supplies in the 1990s depend on decisions taken in the next few years—a period likely

1. This study was based on a highly aggregated analysis, compared to the more detailed analysis for the other four WAES 2000 cases. WAES has not analyzed, in detail, the full implications of this case.

to be one without severe energy imbalances. It is critically important to look beyond the next decade.

With this global picture, individual countries will face a wide range of problems, of which the more significant are:

- Western Europe and Japan will continue to be largely dependent on imported energy.
- Within Western Europe there is a striking contrast between Norway, which shows energy self-sufficiency to the year 2000; the United Kingdom, which shows self-sufficiency by 1985 but will be importing up to 40 percent of its energy needs by 2000; and the rest of Western Europe.
- The capital investment requirements for increasingly costly energy developments may necessitate a higher proportion of investment by nations. In some of the developing countries, institutional support for these investments may become increasingly important. WAES has not studied this subject in detail.
- The criteria for assessing these capital investments during periods without severe energy stresses need to be revised to reflect the lag between market price signals and the lead times required to implement needed energy supply and conservation programs.

The national situations for the various WAES cases from which the global study was derived are summarized in the next section and reported more fully in Part III.

2.4 SUMMARIES OF NATIONAL SUPPLY-DEMAND INTEGRATIONS

2.4.1 Canada

Canadian energy requirements have been assessed in relation to domestic availability for five 1985 scenarios and for four 2000 scenarios. These reflect high and low economic growth rates, three levels of oil (and energy) prices, and the estimated extent of government policy initiatives in the energy field. The Canadian summary worksheet for energy supply-demand integration in section 2.5 summarizes the results of the Canadian supply-demand balances for the various scenarios.

Under the assumptions of scenario E, Canadian fossil fuel requirements in 1985 cannot be met from domestic production if low energy prices are maintained and the government does not undertake vigorous policy initiatives that would affect both the demands for and supplies of energy. However, there could be sufficient electrical generating capacity to permit exports under this scenario. The remaining scenarios for 1985 (A, B, C, and D) indicate that Canadian energy requirements can be met by domestic production if the mid-1975 world oil price is achieved in Canada and vigorous government policy measures are adopted.

In these scenarios, however, the production and deliverability of oil from Canada's frontier areas is assumed by 1985. It should be noted that the prospects for such production are highly speculative. Canada's frontier areas have simply not undergone sufficient development for conclusive evaluation. If all Canadian frontier oil production is excluded from the scenarios for 1985, the supply-demand balances show an oil deficit (i.e., a requirement for oil imports) in all of the cases.

With respect to natural gas, the exclusion of frontier supplies could cause a deficit in the supply-demand balance (required imports) in scenario D, make the surplus in scenario C very marginal, and reduce the surplus in scenarios A and B to just over 1 trillion cubic feet per year. Moreover, to the extent that vigorous government policy initiatives in the energy field are not fully realized, the size of the surpluses in scenarios A, B, and C would be reduced accordingly.

For the year 2000, the results of the supply-demand balances for the four scenarios indicate that Canada will become a major importer of natural gas and oil by that time if additional supplies are not forthcoming from Canada's frontier regions. The studies show that there would be sufficient domestic coal production to meet domestic requirements if coal is used as the principal replacement fuel (scenarios C-1 and D-7). When nuclear generation is assumed as the principal replacement fuel (scenarios C-2 and D-8), the studies indicate that substantial exports of coal would occur. In the electrical generation sector, the studies indicate that substantial exports would occur in the four scenarios.

The Canadian study takes into consideration several policy initiatives by the federal government that could result in a more favorable energy climate in Canada by 1985. These policy alternatives are expected to have an impact on both supply and demand.

However, for the year 2000, additional government policies may be required. These could include government initiatives that would encourage further interfuel substitution so as to reduce Canada's import dependence on oil and natural gas and promote use of domestically generated electricity. In addition, further government policy measures could be introduced to stimulate further exploration and development of Canada's frontier areas and reduce the rapid rate of build-up of electrical generation capacity. Policy initiatives of this nature would seem appropriate in light of the results of the four scenarios to the year 2000 since they indicate that Canada would have more than sufficient electrical generation capacity for domestic use.

A number of government-induced conservation measures through the year 2000 are already incorporated into the scenarios. However, additional measures might be desirable in the late 1980s and 1990s to further slow the rate of growth in Canadian primary energy demand. Under the four scenarios for the 1985 to 2000 period, Canadian energy demand is projected to grow at 4.0 percent or 3.3 percent per year.

2.4.2 Denmark

The common feature of all the scenario cases studied for Denmark is a reduced dependence on imported oil. Oil at present accounts for around 88 percent of the primary energy supply, the remainder being imported coal. The scenarios also envisage a considerable reduction in the growth rate of primary energy demand, which falls from 6 percent per year in the 1960s to 2–3 percent by 1985 and further to 1–2 percent per year in the period 1985 to 2000.

The reduced dependence on oil and the lower energy demand growth are the result of combined efforts in the demand, supply, and energy conversion sectors.

Energy demand from 1985 to 2000 was projected by detailed calculations for 19 major demand sectors. The calculations are based partly on economic indicators and partly on independent projections for physical quantities such as the number of automobiles, dwellings, and appliances. In this way certain saturation effects could be introduced in the scenarios and questionable exponential extrapolations avoided.

Energy demand is strongly affected by efficiency improvements in all major demand sectors. In the short term the greatest savings are obtained from insulation; in the long term efficiency improvements in transport and industry also have considerable impact. These improvements are the combined result of the impact of energy price, energy policy, and structural changes in the economy. A 50 percent energy price increase and vigorous government policy have about the same impact on energy efficiency. Such impacts are by and large considered additive.

The supply of domestic (North Sea) oil and gas is estimated to increase from about 1 percent of primary energy in 1975 to about 15 percent in 1985. A decline must be expected toward the end of the century unless new discoveries are made. Uranium from Greenland is estimated to supply up to 27 percent of the primary energy input in year 2000.

The supply-demand integrations all reflect, although to different degrees, some common trends in the energy system: An increased use of electricity as a partial substitute for oil, the introduction of natural gas, the use of nuclear energy for generation of electricity and heat, and an increased use of coal. In all scenarios there is a strong reliance on district heat delivered from power stations.

The resulting primary energy demand for year 2000 ranges from 145 percent (Case D-3) to 183 percent (Case C-2), when expressed relative to 1972. Oil demand is, in absolute terms, slightly less than in 1972. In the high nuclear cases, the primary energy input for year 2000 is 30 percent coal, 46 percent oil, 11 percent gas, and 8 percent nuclear.

Renewable energy sources are not expected to contribute significantly under scenario price assumptions. Solar water heating and wind generation of electricity are, however, considered as possible supply inputs. Together, they could substitute for around 5 percent of the total primary fuel. The impact on oil and gas demand would be somewhat less.

Compared to the reference cases, where all efficiencies were kept at 1972 levels, the scenarios envisage overall energy savings of 5 to 25 percent, depending on the scenario case. The energy savings in the demand sectors are considerably larger, but these savings are partially offset because of the energy cost of substituting electricity for oil.

2.4.3 Finland

The long-term problems of the Finnish energy sector

are closely related to the situation in other industrialized, energy-importing countries. There are, however, certain features peculiar to Finland that determine the range of plausible futures.

Because of its geographic location, cold climate, and population distribution, average energy consumption in Finland is high. A fairly large share of industry involves processing the most significant indigenous resource, wood, and is thus energy-intensive. Most of the available hydro energy has already been developed for electricity generation; there are no high-grade fossil fuel resources in Finland. Thus, an increasing share of the energy supply will be dependent on the international energy markets.

The Finnish WAES studies are based on economic growth rates of 3.25–4.4 percent in 1976 to 1985 and 2.0–3.5 percent in 1985 to 2000. The total energy demand starts saturating in certain consumption sectors. First, residential energy demand is expected to grow very slowly, because of low population growth and more efficient housing and heating systems, including district heating. The increase in the number of automobiles and the demand for energy for private transportation are also starting to saturate during the period analyzed. The share of industrial energy demand is projected to increase.

The most remarkable shift among energy supply sources is the introduction of nuclear energy. Peat will also be more widely used, keeping the share of indigenous fuels at about the present level. Otherwise, the indigenous share of total energy input would greatly decrease.

The total GNP is projected to grow by 2.5–3.7 percent per year from 1972 to 2000, and the total primary energy demand by 2.4–3.3 percent per year. Thus, the ratio of energy growth to GNP growth is falling below unity, and this is especially true during the latter part of the period. This is due not only to better consumption efficiencies but also to good conversion efficiencies. Both district heating and cogeneration of heat and electricity will be widespread.

2.4.4 France

According to the WAES global assumptions, French GNP growth from 1985 to 2000 is assumed to be 3.5 percent per year in the C-1 and C-2 scenarios (linked to rising energy price), and 2.5 percent per year in D-7 and D-8 (linked to constant energy price).

Many uncertainties exist about the possible cumulative effects of the price mechanism and conservation policy in the transportation sector and, to a smaller degree, in the industrial sector; thus, two levels of final energy demand have been projected for each scenario.

The supply-demand integrations are directly linked to WAES assumptions concerning the main replacement fuel for oil: nuclear energy in Cases C-2 and D-8, coal and natural gas in Cases C-1 and D-7. In addition, two different supply-demand integrations have been suggested for Cases C-1 and D-7 to take into account uncertainties concerning the possible contribution of new energy sources.

Thus, for France, six cases have been studied. C-2 and D-8, drawn directly from the French Planning Commission studies, require a total primary energy input of 390 MTOE and 330 MTOE, respectively, of which nuclear energy supplies 35 percent and new energy sources supply 10 MTOE. C-1 and D-7 are derived from the 1985 cases, but nuclear energy is assumed to contribute little, the difference being filled by coal and natural gas. Due to the different supply structure and the different processing losses, the total primary energy input required is 384 MTOE in C-1 and 315 MTOE in D-7. In C-1′ and D-7′, in which price mechanism and conservation policy effects are considered, the level of final energy demand is reduced, and low nuclear development coupled with high new energy source development (30 MTOE) is assumed. The total primary energy input required in these two scenarios is 369 MTOE in C-1′ and 300 MTOE in D-7′.

2.4.5 German Federal Republic

The scenarios of energy supply and demand to 2000 in the Federal Republic of Germany are based on a mix of several assumptions.

1. Between 1976 and 2000 economic growth will develop at an average rate of 3.4 percent or 2.2 percent per year compared to 4.6 percent between 1960 and 1972.

2. National energy price levels will either remain constant (at the 1975 level) or increase by about 50 percent.

3. National energy policy consists of actual programs (either restrained or enforced vigorously),

particularly in demand sectors where noncommercial behavior does not prevail. Vigorous policy will have little effect on supply and demand in the Federal Republic of Germany.

4. Nuclear energy either develops favorably (high nuclear) and is restricted only by electricity base load demand or must face considerable drawbacks compared to previous expectations, resulting in a nuclear capacity of 50 GW in 2000 (low nuclear).

According to scenarios written on these assumptions, primary energy demand in 2000 ranges between 580 and 690 million tons of coal equivalent (compared to 350 million TCE in 1975). Energy demand growth rates in the 2000 scenarios are, on the average, lower than in the past not only because of lower economic growth but also because of the ratio between energy supply and economic growth.

Electricity demand continues to grow at a faster rate than fossil fuel final energy demand, though these growth rates are clearly lower than in the past. Oil demand remains more or less constant after 1972; therefore, oil import requirements also remain fairly stable. Gas demand increases sharply. Only if considerable coal gasification is achieved by 2000 will import requirements be kept down to the levels projected for 1985. In the low nuclear cases, import requirements for coal will rise considerably. For a high economic growth rate scenario, 50 GW nuclear capacity would mean coal imports of 55 million tons per year.

The global integration of the national studies indicates that, given the assumed conditions in the Federal Republic of Germany, import requirements will be difficult to meet and the estimated gas and coal imports will require enormous developments of infrastructure, particularly in energy trade.

2.4.6 Italy

In the WAES scenarios C and D, taken as basic development paths in the period 1972 to 2000, economic growth in Italy is assumed to average 3.5 and 2.4 percent per year, respectively.

In this period energy demand in the end-use sector resulting from the scenario hypotheses increases at an average annual rate of between 1.9 percent in D-8 and 3.0 percent in C-1. Growth in electricity demand is projected to average 3.0 percent in D-7 and 5.4 percent in C-2.

The substantial decrease in growth rates relative to the period 1953 to 1972 (8.6 percent per year for total net energy demand and 8.2 percent for electricity demand) is due basically to the WAES scenario assumptions of lower economic growth (5.4 percent per year was the average between 1952 and 1973), saturation effects accompanying economic growth, specified increases in the price of oil, and vigorous government policies to save energy through improvements in end-use efficiency.

In 1972 oil covered about 75 percent of the primary energy needs for internal consumption. Given the declining role that this fuel is bound to play on a world scale, major strategic policy decisions must be taken in the near future that will allow the share of oil to decrease to more manageable proportions. A number of strategic replacement options have been analyzed and assessed, including nuclear energy, coal, solar and geothermal energy, and conservation.

The time available to the end of the century is too short for any *one* of these options, other than nuclear energy and coal, to have more than a limited effect in reducing oil dependence. On the other hand, national policy choices must take account of the long-term strategic importance of new technologies still in the early stages of research and development.

Two major policy options are considered. The first focuses on nuclear energy and envisages a program of rapid nuclear expansion, which in scenario C-2 leads to nuclear coverage of 29 percent of the total primary energy input in the year 2000. In the second option, nuclear energy in the period to the year 2000 is assumed to be primarily a transition fuel easing the way to an energy system driven increasingly by renewable and unlimited energy sources. Solar and geothermal energy are given particular prominence in this strategy since Italy is favorably endowed with these resources. In this option the flexibility necessary to allow a fairly rapid penetration of new sources in the period after 2000 is achieved by deemphasizing the role of nuclear energy through increased use of coal in electricity generation.

The results of the projections indicate that the share of oil in total primary energy input for internal consumption can decrease by around 50 percent by the end of the century, if strategic policy actions are taken in time. Oil dependence can be reduced more rapidly by means of a high nuclear program, but the difference is not great: 52 percent in C-1, 53 percent in D-7, 46 percent in C-2, and 49 percent in D-8. The opportunity for high nuclear development may, moreover, lose some of its thrust if the first quarter of the twenty-first century is included.

It is almost certain that the Italian economy will not be in a position to develop at the rates postulated in this study unless strategic energy policy decisions, carefully timed and planned, are taken in the immediate future. For example, if the nuclear programs specified fail to materialize, if a major energy-savings program is not enacted shortly, and if solar and geothermal energy are not given needed government backing, then additional oil, coal, and gas of up to 140 MTOE will have to be imported in the year 2000 (equal to the total primary energy input in 1976). Assuming the economic growth specified in the scenarios can be maintained and the international price of oil does not increase above the scenario levels, this situation corresponds to an increase in import costs of up to $16 billion (in 1975 U.S. dollars) per year at the turn of the century, a burden that the country would be unable to bear without great sacrifices.

2.4.7 Japan

The gross national product of Japan has grown with an elasticity of 1.9 compared to the world GNP growth rate in the last fifteen years. However, the Japanese economy will probably grow more slowly in the future, and the elasticity will drop to around 1.3 in the next decade and then still further to 0.9. Accordingly, the Japanese GNP is assumed to increase by 8.0 percent per year between 1977 and 1985, which corresponds to the WAES high growth case of 6.0 percent per year. And for the WAES low growth case of 3.5 percent, a growth rate of 4.5 percent is assumed. Between 1985 and 2000, it is assumed that Japan will grow by 4.5 percent and 2.7 percent per year, corresponding to the WAES high growth rate case of 5.0 percent and low growth case of 3.0 percent, respectively.

Given these projected growth rates, the Japanese GNP is estimated to amount to 192.9×10^{12} Yen (in 1970 prices) in the WAES high growth case for 1985; 148.2×10^{12} Yen in the low growth case; and for 2000, 373.2×10^{12} Yen in the high growth case and 220.9×10^{12} Yen in the low growth case.

According to the WAES boundary scenario approach, the estimated volume of energy demand and supply ranges from about 550×10^{13} kcal to 750×10^{13} kcal in 1985 and from 850×10^{13} kcal to 1300×10^{13} kcal in 2000 due to the scenario assumptions.

Because Japan's domestic fossil fuel and hydroelectricity supply is very limited and future supplies will be no less limited whatever the scenario assumptions are, greater demand will require increased import of fossil fuels and continued expansion of nuclear power output.

Since only one figure for nuclear power plant installed capacity (30 million kW) is assumed in the 1985 cases, increased demand will cause a greater dependence on fossil fuel imports. Thus Case B in 1985 (the smallest demand case) projects that the import share of total energy supply will be 82.5 percent, while in Case E (the largest demand case) import dependence will be 87.0 percent.

Different nuclear power plant installed capacities are projected for the various year 2000 cases, and Case D-8, which has lower demand coupled with higher nuclear capacity, shows the smallest import dependence. High demand Cases C-1 and C-2 are estimated to have fossil fuel import dependences of 85 and 82 percent, respectively, and low demand Cases D-7 and D-8 have import dependences of 77 and 72 percent, respectively.

Oil will remain the most important fossil fuel import, but the volume of gas and steam coal import is expected to increase rapidly. The desired level of oil import ranges from 350×10^{13} kcal to 530×10^{13} kcal in 1985 and from 470×10^{13} kcal to 900×10^{13} kcal in 2000, depending on the scenario chosen.

2.4.8 Mexico

Mexico's total energy demand has risen at an annual rate of 8.2 percent between 1964 and 1974. The projection of demand growth indicates that the rate will be between 6.7 and 7.7 percent between 1985 and 2000, due largely to energy conservation in the industrial and transportation sectors. By 2000 Mexico's energy needs will be between 2993 and 3875×10^{12} kcal, five to seven times more than the 570×10^{12} kcal consumed in 1975.

The investment required to satisfy energy needs during the period 1976–2000 is an estimated 1.6×10^9 pesos at 1975 prices; this figure includes the costs of exploring for and exploiting resources and installing production, transport, and distribution systems.

Because this growth rate for Mexico's energy sector is high, the strategy is to maintain economic growth while directing efforts toward more equitable distribution of wealth.

Mexico's national energy sector is highly dependent on hydrocarbons. Hydrocarbons provide 86 percent of Mexico's energy supply, while hydroelectricity accounts for only 7 percent, coal 6 percent, and other sources 1 percent. No major changes in this mix are expected between now and 2000. Our basic reserves are estimated at 54×10^9 barrels. It is projected that hydrocarbons will provide at least 70 percent of the total supply in 2000.

2.4.9 The Netherlands

Energy consumption in The Netherlands in the year 2000 has been worked out in detail for two high growth alternatives (Cases C-1 and C-2) and two low growth alternatives (Cases D-7 and D-8).

The economic projections are based on assumptions concerning the development of both labor productivity and labor potential. The year 2000 cases may be distinguished, as far as major activities are concerned, by the degree to which—during full-employment conditions—the increase in productivity is converted into either more production or more leisure. GNP growth during the period 1985-2000 has been calculated at 2.8 percent per year in the high growth case and 1.5 percent per year in the low growth case. In a number of minor activities, developmental saturation will occur by 2000. A stabilization of population growth by the end of this century is anticipated. During the period 1985-2000, the present rate of increase in the number of cars, the housing available, and allied items in the transport and residential sectors will also decline.

It is assumed that high energy prices and government energy policy carried out in anticipation of higher prices will further improve efficiency. The fuel mix, both in final sectors and in the energy sector itself—particularly for the generation of electric power—is quite obviously changing. These shifts reflect the impact of the decline in production of domestic natural gas, on the one hand, and the limited oil and gas supplies in the international energy market, on the other. Coal or nuclear energy or both will be the principal replacement fuels. The following trends are anticipated.

1. Primary energy input increases from approximately 3×10^{18} J in 1972 to approximately 7×10^{18} J in the high growth case and 5×10^{18} J in the low growth case in 2000.

2. Expansion of coal imports and nuclear energy capacity as replacements for oil raises the share of these two together to 20-25 percent of the primary energy input.

3. Despite the expansion of nuclear energy, energy supply will be less than in 1972. This is caused by the drop in Dutch natural gas production.

4. The energy shortage, arising from increased demand, will be made up by a steep rise in oil and coal imports. The net import quota in 2000 will amount to 60-70 percent as compared to 30 percent in 1972.

5. Coal import will rise from approximately 3 million tonnes in 1972 to almost 40 million tonnes in high growth Case C-1, an increase of over 10 percent per year.

6. Thus, oil dependency by 2000 will exceed the 55 percent level of 1972. Oil consumption will rise in absolute terms from 40 million tonnes in 1972 to over 90 million tonnes in the high growth case.

It should be emphasized that the figures given above for coal and nuclear energy are target quantities. The expected gap in import opportunities for oil may, together with the continuing rise in energy prices, constitute a stimulus for more drastic energy-saving steps or supply additions in the so-called supervigorous policies. However, current options seem rather limited. Some demand reductions of a more or less realistic nature might be feasible in transport and residential use. The total of the options handled amounts to over 200×10^{15} J. Extra supply by additional utilization of solar heat provides another 50×10^{15} J, while coal gasification may be applied as a substitute for oil, whatever the constraints may be. In any case, the feasibility of the suggested options should be further examined.

2.4.10 Norway

Energy balance in Norway is expected to be dominated by the production of oil and gas from the North Sea fields throughout the period to 2000. The estimate of indigenous supplies for year 2000 indicates a total supply potential of 4.1-5.1 million

TJ (1.9–2.3 MBDOE). Approximately 75 percent of this is oil and gas from the North Sea, and the rest is mainly hydro power.

A comparison of these figures with those of the estimated primary energy demand, which ranges from 2.0 to 2.2 million TJ (0.9 to 1.0 MBDOE), shows that all desired demands for oil and gas can be easily satisfied in Norway. In addition, substantial amounts of oil and gas will be available for export.

Norway has an abundant supply of hydro power, and all electricity is now produced from this energy source. In the early 1990s, hydro power will be developed to such a degree that further expansion will be unlikely within the economic and environmental limitations that can be expected to exist at that time. New demands on electricity could then be met by using oil and/or gas as fuel in conventional power plants. It has been assumed, however, that the scarce global availability of these fuels at this time would influence the choice of energy sources to be used for this purpose and lead to production of electricity mainly by nuclear power.

As late as 1972 Norway covered the largest part of its energy demand by oil imports. This situation has changed rapidly, and Norway is now self-sufficient for oil and gas. Production will continue to increase in the future, reaching a level of about 2.1 million TJ (1.0 MBDOE) in 1985 and 2.1–2.9 million TJ (1.0–1.3 MBDOE) in 2000.

As a result of the favorable supply situation, no dramatic shifts among fuels in the energy demand sector are expected in Norway during the period 1972–2000. However, some shift away from electricity toward fossil fuels is apparent in industry because of the significantly higher growth rate in the petrochemicals industry, which uses oil and gas as feedstocks, than in the traditional electrometallurgical industry.

2.4.11 Sweden

The demand projections to 2000 are based on the forecasts made by the Government Energy Forecast Commission in 1974, adjusted to conform to WAES scenario parameters.

The selected supply-demand integrations are based on spring 1976 proposals regarding the probable expansion of hydro power, nuclear power district heating, etc. For WAES high coal cases for 1985 to 2000, almost no nuclear power expansion is assumed after 1985.

In 1975 the government expressed its intention to achieve zero energy growth in the 1990s for conventional energy sources, which would imply a final energy demand of approximately 630 TWh in the year 2000. Such a goal would require more energy savings than we have assumed, especially in Cases C-1 and C-2.

The energy policy situation in Sweden has changed since the autumn 1976 parliamentary election. The Energy Commission appointed by the government in January 1977 will review all relevant energy issues and disseminate its findings in 1978. Parliament will then decide Sweden's future energy policy. It is therefore possible that nuclear power will be eliminated in the 1980s. This possibility has not, however, been taken into account in these studies.

There are many uncertainties regarding projections to 2000. For instance, uncertainties in the level of impact of WAES parameters would probably make the demand projections to 2000 valid only within a margin of 10–15 percent. Also, domestic production of peat and shale oil and the introduction of natural gas to Sweden could be possible in the period 1985–2000. These possibilities have not been taken into account, but domestic production of uranium in the period 1985–2000 to match domestic demand in the supply-demand integrations is calculated in global energy import-export estimates.

These considerations, together with the time limits of WAES, imply that the Swedish projections should be regarded as "soft." Since Sweden's import energy in a global context is quite small, a revision of projected oil import figures, etc., would not change the global picture. However, another methodology has been developed, which takes into account different possible developments of domestic energy resources within the constraint that the total costs of the energy system, defined in chapter 17, should not be more than 10 percent above the selected cases for WAES global integrations. The results of this methodology imply that a series of potential supply mixes exist in each WAES scenario case with corresponding different fuel import and balance-of-payments effects.

If Sweden should choose a policy of independence, for instance, it would be possible to reduce oil import by 15–50 percent and the corresponding impact on balance of payments by 20–60 percent without increasing the costs of the energy system by more than 10 percent above the costs for the selected WAES cases. The possibilities of reducing oil import are of course greater in the higher oil price cases, since the cost differences between oil and other alternatives are less.

The impact on balance of payments, especially in Cases C-1 and C-2, which assume high economic growth and 50 percent higher energy prices, could trigger a movement toward a national goal of independence. The main policies that would reduce fossil fuel import are increases in hydro power, wind power, solar heat, and domestic production of peat and shale oil. The degree to which this can be done depends on the availability, acceptability, and costs of these energy resources.

2.4.12 United Kingdom

The supply-demand integrations indicate that the U.K. may become a net exporter of energy in the form of oil during the 1980s, but will once again require substantial imports of energy by the year 2000, while the global results show that by this time the world's supply of all types of fuel will be strained. As a response to such a situation, the options available to the U.K. include simultaneously maximizing the production of those indigenous resources that are not limited in their reserves, namely coal and nuclear electricity, increasing the use made by the U.K. of the limited reserve of oil and gas, and adopting policies to decrease internal demand.

It is estimated that, with the adoption of suitable policies, coal production could be increased to 140–150 megatonnes per year, while nuclear capacity could reach 70 GWe by 2000 (compared with maximum figures of 140 megatonnes and 58 GWe in the WAES scenarios). In addition, by slowing down the development of oil production during the 1980s, the period of oil self-sufficiency could be extended at the expense of exports in this period, while higher current estimates of gas reserves suggest that the WAES scenario gas production rates might be increased by 10 megatonnes of oil equivalent per year.

The possibilities for reducing demand are investigated by using a model based on the price of energy to the final consumer, with plausible increases in this price leading to energy demand reductions of up to 6 percent in 2000 and cumulative reductions, realized either as extra reserves or as the extra balance-of-payments advantage from increased exports or decreased imports of up to 250 megatonnes of oil equivalent by 2000.

2.4.13 United States

Given the range of WAES assumptions, desired oil and gas imports to the United States increase over the 1972 to 2000 period. This results from continued, if moderated, growth in demand for oil and gas coupled with declining indigenous U.S. production of these fuels. At the same time, the vast potential for U.S. coal production results in possible coal exports—coal production in excess of demand for coal in the United States—in the WAES cases to 2000.

Oil imports, which were under 5 MBD in 1972, reached nearly 10 MBD in 1976. By 1985, under the WAES assumptions of rising oil price and vigorous national energy policy, desired oil imports would be at about 1972 levels. With constant or falling oil price and with restrained national energy policy, oil imports would grow—reaching a 1985 high of 18 MBD in WAES Case E. By 2000 desired oil imports range from about 9 MBD in the rising energy price cases to about 15 MBD in the constant price cases. These latter cases (D-7 and D-8), while having somewhat lower demands than Cases C-1 and C-2, have significantly lower indigenous supply projections, resulting in the large oil import figures.

Gas imports are expected to be modest and would primarily represent imports of liquefied natural gas (LNG).

U.S. potential coal exports range from 3.4 to 8.6 MBDOE (about 250 to 650 million tons) in the year 2000 cases. These major *potential* exports of steam coal would require expansion of coal-moving infrastructure in the United States and markets abroad. The implications of this potential export could be important, to the world and to the United States, in the period of transition away from oil and gas that may well occur between now and the end of the century.

2.5 SUMMARY TABLES

The following pages contain supply-demand integration summary tables for the thirteen WAES national studies included in this book.

WAES NATIONAL ENERGY SUPPLY/DEMAND INTEGRATION STUDIES, 1972-1985-2000
ENERGY SUPPLY/DEMAND INTEGRATION SUMMARY—IMPORTS AND EXPORTS OF PRIMARY ENERGY

Country: CANADA

Scenario Assumptions**		1972	1985					2000			
WAES CASE		Base Year	A	B	C	D	E	C-1	C-2	D-7	D-8
GNP 1976-85; 1985-2000		---	High	Low	High	Low	High	High	High	Low	Low
Oil/Energy Price*		---	11.50-17.25	11.50-17.25	11.50	11.50	11.50-7.66	11.50 17.25	11.50 17.25	11.50	11.50
National Policy Response		---	Vig	Vig	Vig	Res	Res	Vig	Vig	Res	Res
Principal Replacement Fuel		---	--	--	--	--	--	Coal	Nuclear	Coal	Nuclear
(Import) & Export	**Units**										
Oil	10^{15} Btu	(1.964) 2.333	2.253	2.416	1.407	.292	(1.903)	(1.304)	(1.304)	(1.160)	(1.160)
Natural Gas	10^{15} Btu	1.009	1.625	1.692	.525	.078	(.258)	(2.202)	(2.202)	(2.358)	(2.358)
Coal	10^{15} Btu	(.476) .250	.474	.502	.168	(.034)	(.251)	.323	.936	.024	1.020
Nuclear) Hydroelectric)	10^{15} Btu	(.024) .110	.328	.438	.233	.344	.438	1.036	1.860	.950	1.390
Total Net (Imports) or Exports		(2.464) 3.702	4.680	5.048	2.333	.748	(1.97)	(2.15)	(.71)	(2.54)	(1.11)

*1975 US dollars per barrel of Arabian light crude oil, fob Persian Gulf.
All Figures are primary production prior to any processing or conversion.

Footnotes:

**The oil and gas supply estimates assume reserves will be found in lightly explored frontier areas.

WAES NATIONAL ENERGY SUPPLY/DEMAND INTEGRATION STUDIES, 1972-1985-2000
ENERGY SUPPLY/DEMAND INTEGRATION SUMMARY—IMPORTS AND EXPORTS OF PRIMARY ENERGY

Country: Denmark

Scenario Assumptions	1972	1985					2000			
WAES CASE	Base Year	A	B	C	D	E	C-1	C-2	D-7	D-8
GNP 1976-85; 1985-2000		High	Low	High	Low	High	High	High	Low	Low
Oil/Energy Price*		11.50-17.25	11.50-17.25	11.50	11.50	11.50-7.66	11.50 17.25	11.50 17.25	11.50	11.50
National Policy Response		Vig	Vig	Vig	Res	Res	Vig	Vig	Vig	Vig
Principal Replacement Fuel							Coal	Nuclear	Coal	Nuclear
(Import) & Export **Units** PJ/y										
Oil	(767)	(551)	(481)	(613)	(756)	(917)	(582)	(700)	(592)	(646)
Natural Gas	0	(32)	(32)	(32)	82	82	(97)	(97)	(97)	(97)
Coal	(58)	(195)	(174)	(208)	(162)	(200)	(435)	(185)	(333)	(130)
Nuclear		167	167	167	167	167	334	50	334	50
Hydroelectric										
Total Net (Imports) or Exports	(825)	(611)	(520)	(686)	(669)	(868)	(780)	(932)	(688)	(823)

*1975 US dollars per barrel of Arabian light crude oil, fob Persian Gulf.
All Figures are primary production prior to any processing or conversion.

WAES NATIONAL ENERGY SUPPLY/DEMAND INTEGRATION STUDIES, 1972-1985-2000
ENERGY SUPPLY/DEMAND INTEGRATION SUMMARY—IMPORTS AND EXPORTS OF PRIMARY ENERGY

Country: Finland

Scenario Assumptions	1972	1985					2000			
WAES CASE	Base Year	A	B	C	D	E	C-1	C-2	D-7	D-8
GNP 1976-85; 1985-2000		High	Low	High	Low	High	High	High	Low	Low
Oil/Energy Price*		11.50-17.25	11.50-17.25	11.50	11.50	11.50-7.66	11.50 17.25	11.50 17.25	11.50	11.50
National Policy Response[1]							Vig	Vig	Vig	Vig
Principal Replacement Fuel							Coal	Nuclear	Coal	Nuclear
(Import)[2] **Units**	10^{15} Joules									
Oil	478	623	554	673	613	748	740	740	672	672
Natural Gas		47	42	45	42	47	75	75	58	58
Coal	98	138	111	154	119	149	371	261	228	228
Nuclear[3]										
Electric[4]	15	14	14	14	14	14	14	14	14	14
Total Net (Imports) or Exports	591	822	721	886	788	958	1200	1090	972	972

*1975 US dollars per barrel of Arabian light crude oil, fob Persian Gulf.
All Figures are primary production prior to any processing or conversion.

Footnotes:
[1]National policy response not analyzed.
[2]All figures are imports.
[3]Nuclear energy is considered indigenous although the fuel cycle is not.
[4]It is not known or relevant how the imported electricity has been generated.

WAES NATIONAL ENERGY SUPPLY/DEMAND INTEGRATION STUDIES, 1972-1985-2000
ENERGY SUPPLY/DEMAND INTEGRATION SUMMARY—IMPORTS AND EXPORTS OF PRIMARY ENERGY

Country: France

Scenario Assumptions	1972	1985					2000			
WAES CASE	Base Year	A	B	C	D	E	C-1	C-2	D-7	D-8
GNP 1976-85; 1985-2000		High	Low	High	Low	High	High	High	Low	Low
Oil/Energy Price*		11.50-17.25	11.50-17.25	11.50	11.50	11.50-7.66	11.50 17.25	11.50 17.25	11.50	11.50
National Policy Response		Vig	Vig	Vig	Res	Res	Vig	Vig	Vig	Vig
Principal Replacement Fuel							Coal	Nuclear	Coal	Nuclear
(Import) & Export **Units** Mtoe										
Oil[1]	(105)	(83)	(72)	(96)	(88)	(178)	(95–130)	(130)	(85–110)	(110)
Natural Gas[1]	(6)	(30)	(28)	(30)	(27)	(28)	(65)	(65)	(52)	(49)
Coal[1]	(9)	(14)	(4)	(17)	(14)	(4)	(42)	(7)	(20–32)	(7)
Nuclear										
Hydroelectric										
Total Net (Imports) or Exports	(120)	(127)	(104)	(143)	(129)	(210)	(201–236)	(202)	(157–195)	(166)

*1975 US dollars per barrel of Arabian light crude oil, fob Persian Gulf.
All Figures are primary production prior to any processing or conversion.

Footnotes:

[1]Indigenous production of coal = 13 Mtoe. These figures do not take into account an eventual discovery of oil or natural gas in France.

WAES NATIONAL ENERGY SUPPLY/DEMAND INTEGRATION STUDIES, 1972-1985-2000
ENERGY SUPPLY/DEMAND INTEGRATION SUMMARY—IMPORTS AND EXPORTS OF PRIMARY ENERGY

Country: Federal Republic of Germany

Scenario Assumptions	1972	1985					2000				
WAES CASE	Base Year	A	B	C	D	E	C-1	C-2	D-7	D-8	D-3
GNP 1976-85; 1985-2000		High	Low	High	Low	High	High	High	Low	Low	Low
Oil/Energy Price*		11.50-17.25	11.50-17.25	11.50	11.50	11.50-7.66	11.50 17.25	11.50 17.25	11.50	11.50	11.50-17.25
National Policy Response		Vig	Vig	Vig	Res	Res	Vig	Vig	Vig	Vig	Vig
Principal Replacement Fuel							Coal	Nuclear	Coal	Nuclear	Coal
(Import) & Export **Units** 10^{15} J											
Oil	(5971) 311	(6260) 300	(5745) 300	(6833) 300	(6443) 300	(7437) 300	(6575)	(5880)	(5855)	(5841)	(5436)
Natural Gas	(319) 20	(1891) 20	(1832) 20	(1868) 20	(1848) 20	(1868) 20	(3153)	(1936)	(2999)	(1615)	(2586)
Coal	(308) 685	(256) 850	(64) 700	(141) 850	(189) 700	(498) 850	(1570) 800	(369) 800	(656) 700	(518) 700	(1177)
Nuclear											
Hydroelectric balance	(118)	(75)	(75)	(75)	(75)	(75)	(75)	(75)	(75)	(75)	(75)
Total Net (Imports) or Exports	(6716) 1198	(8482) 1170	(7716) 1020	(8917) 470	(8555) 1020	(9878) 1170	(11373) 800	(8255) 800	(9585) 700	(8049) 700	(9274)

*1975 US dollars per barrel of Arabian light crude oil, fob Persian Gulf.
All Figures are primary production prior to any processing or conversion.

WAES NATIONAL ENERGY SUPPLY/DEMAND INTEGRATION STUDIES, 1972-1985-2000
ENERGY SUPPLY/DEMAND INTEGRATION SUMMARY—IMPORTS AND EXPORTS OF PRIMARY ENERGY

Country: Italy

Scenario Assumptions	1972	1985					2000			
WAES CASE	Base Year	A	B	C	D	E	C-1	C-2	D-7	D-8
GNP 1976-85; 1985-2000		High	Low	High	Low	High	High	High	Low	Low
Oil/Energy Price*		11.50-17.25	11.50-17.25	11.50	11.50	11.50-7.66	11.50 17.25	11.50 17.25	11.50	11.50
National Policy Response		Vig	Vig	Vig	Res	Res	Vig	Vig	Vig	Vig
Principal Replacement Fuel							Coal	Nuclear	Coal	Nuclear
(Import) & Export Units 10^{12} Kcal										
Oil	975	1097	817	1262	1201	1694	1632	1473	1193	1160
Natural Gas	11	153	76	178	104	209	382	364	260	312
Coal	74	102	84	112	102	136	338	166	178	114
Nuclear	8	144	144	144	144	144	435	939	290	579
Hydroelectric										
Total Net (Imports) or Exports	1068	1496	1121	1696	1551	2183	2787	2942	1921	2165

*1975 US dollars per barrel of Arabian light crude oil, fob Persian Gulf.
All Figures are primary production prior to any processing or conversion.

Footnotes:

1. Includes only imports of primary energy for internal consumption and bunkerage.
2. Coal includes wood.

WAES NATIONAL ENERGY DEMAND STUDIES, 1972-1985-2000
ENERGY DEMAND SUMMARY

Country: JAPAN

Scenario Assumptions	1972	1985					2000			
WAES CASE	Base Year	A	B	C	D	E	C-1	C-2	D-7	D-8
GNP 1976-85; 1985-2000		High	Low	High	Low	High	High	High	Low	Low
Oil/Energy Price*		11.50-17.25	11.50-17.25	11.50	11.50	11.50-7.66	11.50-17.25	11.50-17.25	11.50	11.50
National Policy Response		Vig	Vig	Vig	Res	Res	Vig	Vig	Vig	Vig
Principal Replacement Fuel							Coal	Nuclear	Coal	Nuclear
Primary Energy Demand — **Units**	10^{13}Kcal	"	"	"	"	"	"	"	"	"
Oil	261.6	467.2	351.4	492.3	372.8	527.4	924.9	890.9	502.6	490.7
Natural Gas	4.0	48.4	45.6	48.4	45.6	48.4	84.0	84.0	84.0	84.0
Coal	57.2	91.2	73.6	91.2	73.6	91.2	136.4	113.2	106.9	83.6
Nuclear	2.3	45.1	45.1	45.1	45.1	45.1	105.0	167.0	105.0	150.2
Hydroelectric	21.3	29.2	28.4	29.1	28.4	29.1	44.4	44.4	42.1	42.1
Geothermal	-	1.0	1.0	1.0	1.0	1.0	4.8	4.8	4.8	4.8
Solar Thermal	-	-	-	-	-	-	-	-	-	-
Solar Electric	-	-	-	-	-	-	-	-	-	-
TOTAL PRIMARY ENERGY DEMAND	346.4	682.0	545.1	707.1	566.5	742.2	1299.5	1304.3	845.4	855.4

*1975 US dollars per barrel of Arabian light crude oil, fob Persian Gulf.
All Figures are primary production prior to any processing or conversion.

WAES NATIONAL ENERGY SUPPLY/DEMAND INTEGRATION STUDIES, 1972-1985-2000
ENERGY SUPPLY/DEMAND INTEGRATION SUMMARY—IMPORTS AND EXPORTS OF PRIMARY ENERGY

Country: MEXICO

Scenario Assumptions	1972	1985					2000			
WAES CASE	Base Year	A	B	C	D	E	C-1	C-2	D-7	D-8
GNP 1976-85; 1985-2000		High	Low	High	Low	High	High	High	Low	Low
Oil/Energy Price*		11.50-17.25	11.50-17.25	11.50	11.50	11.50-7.66	11.50 17.25	11.50 17.25	11.50	11.50
National Policy Response		Vig	Vig	Vig	Res	Res	Vig	Vig	Vig	Vig
Principal Replacement Fuel							Coal	Nuclear	Coal	Nuclear
(Import) & Export **Units** Kcal x 10^{12}										
Oil	(29)	93	12	10	0	0	0	0	0	0
Natural Gas	2	0	0	0	0	0	0	0	0	0
Coal	(12)	(17)	0	(12)	0	0	0	0	0	0
Nuclear										
Hydroelectric										
Total Net (Imports) or Exports	(39)	76	12	(2)	0	0	0	0	0	0

*1975 US dollars per barrel of Arabian light crude oil, fob Persian Gulf.
All Figures are primary production prior to any processing or conversion.

WAES NATIONAL ENERGY SUPPLY/DEMAND INTEGRATION STUDIES, 1972-1985-2000
ENERGY SUPPLY/DEMAND INTEGRATION SUMMARY—IMPORTS AND EXPORTS OF PRIMARY ENERGY

Country: The Netherlands

Scenario Assumptions	1972	1985					2000			
WAES CASE	Base Year	A	B	C	D	E	C-1	C-2	D-7	D-8
GNP 1976-85; 1985-2000		High	Low	High	Low	High	High	High	Low	Low
Oil/Energy Price*		11.50-17.25	11.50-17.25	11.50	11.50	11.50-7.66	11.50 17.25	11.50 17.25	11.50	11.50
National Policy Response		Vig	Vig	Vig	Res	Res	Vig	Vig	Vig	Vig
Principal Replacement Fuel							Coal	Nuclear	Coal	Nuclear
(Import) & Export Units										
Oil	(1612)			(2647)	(1963)		(4002)	(4221)	(2771)	(2947)
Natural Gas	855	1677	1677	1677	1677	1677	(352)	(352)	(352)	(352)
Coal	(47)			(246)	(163)		(1153)	(508)	(722)	(316)
Nuclear										
Hydroelectric										
Oil + coal		(2613)	(1742)			(3411)				
Total Net (Imports) or Exports	804	(936)	(65)	(1216)	(449)	(1734)	(5507)	(5081)	(3791)	(3535)

*1975 US dollars per barrel of Arabian light crude oil, fob Persian Gulf.
All Figures are primary production prior to any processing or conversion.

WAES NATIONAL ENERGY SUPPLY/DEMAND INTEGRATION STUDIES, 1972-1985-2000
ENERGY SUPPLY/DEMAND INTEGRATION SUMMARY—IMPORTS AND EXPORTS OF PRIMARY ENERGY

Country: NORWAY

Scenario Assumptions	1972	1985					2000			
WAES CASE	Base Year	A	B	C	D	E	C-1	C-2	D-7	D-8
GNP 1976-85; 1985-2000		High	Low	High	Low	High	High	High	Low	Low
Oil/Energy Price*		11.50- 17.25	11.50- 17.25	11.50	11.50	11.50- 7.66	11.50 17.25	11.50 17.25	11.50	11.50
National Policy Response		Vig	Vig	Vig	Res	Res	Vig	Vig	Vig	Vig
Principal Replacement Fuel							Coal	Nuclear	Coal	Nuclear
(Import) & Export Units 10^3 TJ										
Oil	(366)			1205	1206		1759	1766	1219	1238
Natural Gas	(1.3)			818	820		1193	1194	926	926
Coal	(25)			(9)	(9)		(33)	(32)	(30)	(30)
Nuclear	–			–	–		–	–	–	–
Hydroelectric **	48			121	101		–	–	–	–
Total Net (Imports) or Exports	(344)			2135	2118		2919	2928	2115	2134

*1975 US dollars per barrel of Arabian light crude oil, fob Persian Gulf.
All Figures are primary production prior to any processing or conversion.

Footnotes:

** For all types of electric power plants: η = 35 percent.

WAES NATIONAL ENERGY SUPPLY/DEMAND INTEGRATION STUDIES, 1972-1985-2000
ENERGY SUPPLY/DEMAND INTEGRATION SUMMARY—IMPORTS AND EXPORTS OF PRIMARY ENERGY

Country: Sweden

Scenario Assumptions	1972	1985					2000			
WAES CASE	Base Year	A	B	C	D	E	C-1	C-2	D-7	D-8
GNP 1976-85; 1985-2000		High	Low	High	Low	High	High	High	Low	Low
Oil/Energy Price*		11.50-17.25	11.50-17.25	11.50	11.50	11.50-7.66	11.50 17.25	11.50 17.25	11.50	11.50
National Policy Response		Vig	Vig	Vig	Res	Res	Vig	Vig	Vig	Vig
Principal Replacement Fuel							Coal	Nuclear	Coal	Nuclear
(Import) & Export **Units** TWh										
Oil	(374)	(401)	(340)	(454)	(413)	(569)	(566)	(471)	(484)	(449)
Natural Gas [4)]										
Coal [5)]	(16)	(40)	(36)	(44)	(38)	(50)	(159)	(60)	(100)	(50)
Nuclear	(4)	2)	2)	2)	2)	2)	2)	2)	2)	2)
Hydroelectric [3)]										
Total Net [1)] (Imports) or Exports	(368)	(431)	(346)	(458)	(422)	(579)	(685)	(491)	(549)	(464)

*1975 US dollars per barrel of Arabian light crude oil, fob Persian Gulf.
All Figures are primary production prior to any processing or conversion.

Footnotes:
1) Sweden exports a small amount of refined oil products, which is why this figure is less than the sum of the above import figures.
2) Domestic production of uranium to match domestic demand is assumed. This assumption could be discussed and will not be resolved until 1978 when the Parliament will decide Sweden's future energy policy. It is not unlikely that nuclear power will be eliminated in the 1980s.
3) Small amounts of imports and exports exist among the Scandinavian countries. These have not been considered in the integrations.
4) It is not impossible that Sweden will introduce natural gas in the 1980s.
5) The figures for 1985 to 2000 may be on the high side since the activity levels in iron and steel might well be smaller than we have projected and the extent to which coal might be used for electricity and heat generation could be discussed.

WAES NATIONAL ENERGY SUPPLY/DEMAND INTEGRATION STUDIES, 1972-1985-2000
ENERGY SUPPLY/DEMAND INTEGRATION SUMMARY—IMPORTS AND EXPORTS OF PRIMARY ENERGY

Country: United Kingdom

Scenario Assumptions	1972	1985					2000			
WAES CASE	Base Year	A	B	C	D	E	C-1	C-2	D-7	D-8
GNP 1976-85; 1985-2000		High	Low	High	Low	High	High	High	Low	Low
Oil/Energy Price*		11.50-17.25	11.50-17.25	11.50	11.50	11.50-7.66	11.50 17.25	11.50 17.25	11.50	11.50
National Policy Response		Vig	Vig	Vig	Res	Res	Vig	Vig	Vig	Vig
Principal Replacement Fuel							Coal	Nuclear	Coal	Nuclear
(Import) & Export **Units**	10^{18} Joules									
Oil	(4.68)	1.95	2.52	1.69	2.38	1.55	(4.26)	(4.18)	(2.83)	(2.89)
Natural Gas [a]	(0.03)	(0.40)	(0.40)	(0.40)	(0.40)	(0.40)	(0.66)	(0.66)	(0.40)	(0.40)
Coal	(0.13)	--	--	--	--	--	(1.33)	(1.32)	(0.54)	(0.39)
Nuclear										
Hydroelectric										
Total Net (Imports) or Exports	(4.84)	1.55	2.12	1.29	1.98	1.15	(6.26)	(6.16)	(3.77)	(3.68)

*1975 US dollars per barrel of Arabian light crude oil, fob Persian Gulf.
All Figures are primary production prior to any processing or conversion.

Footnotes:
[a]see text

WAES NATIONAL ENERGY SUPPLY/DEMAND INTEGRATION STUDIES, 1972-1985-2000
ENERGY SUPPLY/DEMAND INTEGRATION SUMMARY—IMPORTS AND EXPORTS OF PRIMARY ENERGY

Country: United States

Scenario Assumptions	1972	1985					2000			
WAES CASE	Base Year	A	B	C	D	E	C-1	C-2	D-7	D-8
GNP 1976-85; 1985-2000		High	Low	High	Low	High	High	High	Low	Low
Oil/Energy Price*		11.50-17.25	11.50-17.25	11.50	11.50	11.50-7.66	11.50 17.25	11.50 17.25	11.50	11.50
National Policy Response		Vig	Vig	Vig	Res	Res	Vig	Vig	Vig	Vig
Principal Replacement Fuel							Coal	Nuclear	Coal	Nuclear
(Import) & Export **Units**		MBDOE								
Oil	(4.8)	(5.5)	(4.1)	(7.6)	(12.2)	(18.8)	(9.4)	(9.7)	(15.2)	(15.2)
Natural Gas	(0.5)	(0.8)	–	(1.3)	(1.0)	(2.8)	(1.5)	(1.7)	(2.4)	(2.4)
Coal	0.7	5.3	5.8	3.3	–	–	8.5	3.4	6.9	4.1
Nuclear	–	–	–	–	–	–	–	–	–	–
Hydroelectric	–	–	–	–	–	–	–	–	–	–
Total Net (Imports) or Exports	(4.6)	(0.9)	1.7	(5.5)	(13.2)	(21.6)	(2.4)	(8.0)	(10.7)	(13.5)

*1975 US dollars per barrel of Arabian light crude oil, fob Persian Gulf.
All Figures are primary production prior to any processing or conversion.

Part II
Global Supply-Demand Integrations

3 Global Energy Data Base

3.1 INTRODUCTION

The global energy data base is the accumulation of WAES national supply and demand data submitted by WAES national teams for each WAES scenario. It also includes the results of special WAES assessments of energy supply and demand in non-WAES countries and regions. In the case of the non-WAES rest of WOCA region, the demand data were derived from a special study of economic growth and energy use in the developing countries made available to WAES by the World Bank. Supply estimates were obtained from the WAES supply studies for oil, gas, coal, and nuclear energy. Non-WAES Europe energy demand for each sector is estimated as 26 percent and 28 percent of WAES Europe demand in 1985 and 2000, respectively, following the convention adopted by the WAES Associates. Non-WAES Europe supply estimates were provided by knowledgeable European Associates. The data are presented in the format used to aggregate national data into regional data for input into the global energy supply-demand integration process.

The data are expressed in millions of barrels per day of oil equivalent (MBDOE). All data expressed in other units in the WAES national studies were converted using the factors in Table 2.1. There are minor inconsistencies between some of the data in this data base and those of other WAES reports. However, all discrepancies are the result of rounding, energy unit conversions, or minor adjustments relevant to the supply-demand integration process. In no case do they produce material differences in the results of the analysis or have an impact on the conclusions of the Workshop.

3.2 DATA

This section contains summary tables in the following order:

WAES demand data for all cases, 1985 and 2000 (Tables 3.1–3.9)

WAES supply data for all cases, 1985 and 2000 (Tables 3.10–3.18)

WAES maximum technological development constraints for all cases, 1985 and 2000 (Tables 3.19–3.25)

Fuel preference constraints for all cases, 1985 and 2000 (Tables 3.26–3.28)

Consumer preference constraints for the individual and domestic markets for all cases, 1985 and 2000 (Tables 3.29–3.34).

Table 3.1
WAES demand data, 1985, Case A (final energy demand in MBDOE)

	Industrial heat and power	Residential heat and power	Transport on distillate	Transport on electricity	Chemical feedstocks	Nonenergy bunkers	Total
Canada	0.91	1.25	0.93	—	0.40		3.49
United States	11.06	7.06[a]	8.72	0.01	3.91	0.20	30.96
Total North America	11.97	8.31	9.65	0.01	4.31		34.45
Denmark	0.11	0.14	0.08	—	—		0.33
Finland	0.16[b]	0.13	0.07	—	—		0.36
France	0.97	1.26	0.82	0.01	0.29		3.35
Federal Republic of Germany	1.68	1.45	0.71	0.02	0.50	0.04	4.40
Italy	1.19	0.65	0.31	0.01	0.39	0.23	2.78
The Netherlands	0.50	0.57	0.17	—	0.31	0.24	1.79
Norway	0.18	0.08	0.08	—	0.03	0.03	0.40
Sweden	0.45	0.22	0.15	0.01	0.06		0.89
United Kingdom	1.26	1.16	0.71	0.01	0.44		3.58
Total WAES Western Europe	6.50	5.66	3.10	0.06	2.02		17.88
Non-WAES Western Europe[c]	1.69	1.47	0.81	0.02	0.53		4.52
Total Western Europe	8.19	7.13	3.91	0.08	2.55		22.40
Japan	5.46	1.77	1.53	0.04	0.93	0.79	10.52
Mexico	0.84	0.39	0.77		0.11		2.11
Rest of WOCA[d]	8.65	3.07	6.17	0.02	1.34		19.25
Total WOCA	34.27	20.28	21.26	0.15	9.13	1.53	86.62

[a] Excludes 0.1 MBDOE solar heating and cooling.
[b] Excludes 0.06 MBDOE other indigenous supply.
[c] Estimated as 26 percent of WAES Western Europe.
[d] Includes Mexico's totals.

Table 3.2
WAES demand data, 1985, Case B (final energy demand in MBDOE)

	Industrial heat and power	Residential heat and power	Transport on distillate	Transport on electricity	Chemical feedstocks	Nonenergy bunkers	Total
Canada	0.90	1.18	0.89	—	0.40		3.37
United States	9.40	6.94[a]	8.34	0.01	3.26	0.20	28.15
Total North America	10.30	8.12	9.23	0.01	3.66		31.52
Denmark	0.10	0.13	0.07	—	—		0.30
Finland	0.13[b]	0.12	0.06	—	—		0.31
France	0.79	1.28	0.62	0.01	0.24		2.94
Federal Republic of Germany	1.47	1.38	0.68	0.02	0.47	0.04	4.06
Italy	0.94	0.59	0.26	0.01	0.25	0.23	2.28
The Netherlands	0.27	0.45	0.34	—	0.15	0.20	1.41
Norway	0.18	0.07	0.07	—	0.03	0.03	0.38
Sweden	0.36	0.23	0.13	0.01	0.06		0.79
United Kingdom	1.23	1.10	0.67	0.01	0.38		3.39
Total WAES Western Europe	5.47	5.35	2.90	0.06	1.58		15.86
Non-WAES Western Europe[c]	1.42	1.39	0.75	0.02	0.41		3.99
Total Western Europe	6.89	6.74	3.65	0.08	1.99		19.85
Japan	4.08	1.59	1.29	0.04	0.75	0.57	8.32
Mexico	0.59	0.26	0.54		0.08		1.47
Rest of WOCA[d]	7.31	2.62	5.72	0.02	1.12		16.79
Total WOCA	28.58	19.07	19.89	0.15	7.52	1.27	76.48

[a] Excludes 0.1 MBDOE solar heating and cooling.
[b] Excludes 0.06 MBDOE other indigenous supply.
[c] Estimated as 26 percent of WAES Western Europe.
[d] Includes Mexico's totals.

Table 3.3
WAES demand data, 1985, Case C (final energy demand in MBDOE)

	Industrial heat and power	Residential heat and power	Transport on distillate	Transport on electricity	Chemical feedstocks	Nonenergy bunkers	Total
Canada	0.96	1.26	0.96	—	0.40		3.58
United States	11.45	7.80[a]	8.89	0.01	3.87	0.20	32.22
Total North America	12.41	9.06	9.85	0.01	4.27		35.80
Denmark	0.11	0.16	0.09	—	—		0.36
Finland	0.17[b]	0.14	0.07	—	—		0.38
France	1.27	1.25	0.84	0.02	0.29		3.67
Federal Republic of Germany	1.73	1.57	0.74	0.02	0.53	0.04	4.63
Italy	1.29	0.82	0.32	0.01	0.43	0.23	3.10
The Netherlands	0.53	0.60	0.18	—	0.33	0.26	1.90
Norway	0.18	0.08	0.08	—	0.03	0.03	0.40
Sweden	0.43	0.26	0.16	0.01	0.07		0.93
United Kingdom	1.28	1.18	0.72	0.01	0.45		3.64
Total WAES Western Europe	6.99	6.06	3.20	0.07	2.13		19.01
Non-WAES Western Europe[c]	1.82	1.58	0.83	0.02	0.55		4.80
Total Western Europe	8.81	7.64	4.03	0.09	2.68		23.81
Japan	5.60	1.83	1.59	0.04	0.93	0.79	10.78
Mexico	0.70	0.31	0.60		0.09		1.70
Rest of WOCA[d]	8.65	3.07	6.17	0.02	1.34		19.25
Total WOCA	35.47	21.60	21.64	0.16	9.22	1.55	89.64

[a] Excludes 0.1 MBDOE solar heating and cooling.
[b] Excludes 0.06 MBDOE other indigenous supply.
[c] Estimated as 26 percent of WAES Western Europe.
[d] Includes Mexico's totals.

Table 3.4
WAES demand data, 1985, Case D (final energy demand in MBDOE)

	Industrial heat and power	Residential heat and power	Transport on distillate	Transport on electricity	Chemical feedstocks	Nonenergy bunkers	Total
Canada	1.01	1.23	1.04	—	0.42		3.70
United States	10.21	9.23	9.34	0.01	3.35	0.20	32.34
Total North America	11.22	10.46	10.38	0.01	3.77		36.04
Denmark	0.11	0.16	0.09	—	—		0.36
Finland	0.15[a]	0.13	0.07	—	—		0.35
France	1.15	1.19	0.78	0.01	0.27		3.40
Federal Republic of Germany	1.52	1.48	0.72	0.02	0.49	0.04	4.27
Italy	1.06	0.95	0.43	0.01	0.30	0.23	2.98
The Netherlands	0.35	0.58	0.44	—	0.19	0.26	1.82
Norway	0.18	0.09	0.08	—	0.03	0.03	0.41
Sweden	0.45	0.28	0.15	0.01	0.06		0.95
United Kingdom	1.23	1.11	0.68	0.01	0.39		3.42
Total WAES Western Europe	6.20	5.97	3.44	0.06	1.73		17.96
Non-WAES Western Europe[b]	1.61	1.55	0.89	0.02	0.45		4.52
Total Western Europe	7.81	7.52	4.33	0.08	2.18		22.48
Japan	4.24	1.68	1.40	0.03	0 75	0.57	8.67
Mexico	0.65	0.26	0.59		0.08		1.58
Rest of WOCA[c]	7.31	2.62	5.72	0.02	1.12		16.79
Total WOCA	30.58	22.28	21.83	0.14	7.82	1.33	83.98

[a]Excludes 0.06 MBDOE other indigenous supply.
[b]Estimated as 26 percent of WAES Western Europe.
[c]Includes Mexico's totals.

Table 3.5
WAES demand data, 1985, Case E (final energy demand in MBDOE)

	Industrial heat and power	Residential heat and power	Transport on distillate	Transport on electricity	Chemical feedstocks	Nonenergy bunkers	Total
Canada	1.12	1.37	1.14	—	0.42		4.05
United States	12.40	9.86	10.00	0.01	3.88	0.20	36.35
Total North America	13.52	11.23	11.14	0.01	4.30		40.40
Denmark	0.14	0.20	0.11	—	—		0.45
Finland	0.19[a]	0.14	0.08	—	—		0.41
France	1.45	1.54	1.07	0.02	0.30		4.38
Federal Republic of Germany	1.81	1.60	0.76	0.02	0.57	0.04	4.80
Italy	1.42	1.14	0.58	0.01	0.49	0.23	3.87
The Netherlands[b]	0.53	0.52	0.26	—	0.34	0.26	1.91
Norway	0.19	0.09	0.08	—	0.03	0.03	0.42
Sweden	0.48	0.31	0.19	0.01	0.08		1.07
United Kingdom	1.30	1.19	0.73	0.01	0.44		3.67
Total WAES Western Europe	7.51	6.73	3.86	0.07	2.25		20.98
Non-WAES Western Europe[c]	1.95	1.75	1.00	0.02	0.59		5.31
Total Western Europe	9.46	8.48	4.86	0.09	2.84		26.29
Japan	5.87	1.95	1.74	0.04	0.93	0.82	11.35
Mexico	0.64	0.26	0.57	—	0.08		1.55
Rest of WOCA[d]	8.65	3.07	6.17	0.02	1.34		19.25
Total WOCA	37.50	24.73	23.91	0.16	9.41	1.58	97.29

[a] Excludes 0.06 MBDOE other indigenous supply.
[b] Estimated.
[c] Estimated as 26 percent of WAES Western Europe.
[d] Includes Mexico's totals.

Table 3.6
WAES demand data, 2000, Case C-1 (final energy demand in MBDOE)

	Industrial heat and power	Residential heat and power	Transport on distillate	Transport on electricity	Chemical feedstocks	Nonenergy bunkers	Total
Canada	1.91	2.24	2.34	—	0.64		7.13
United States	16.94	7.49[a]	9.63	0.01	6.29	0.44	40.80
Total North America	18.85	9.73	11.97	0.01	6.93	0.30	48.23
Denmark	0.15	0.19	0.11	—	—		0.45
Finland	0.32[b]	0.15	0.10	—	—		0.57
France	1.98	1.74	1.45	0.02	0.59		5.78
Federal Republic of Germany	2.37	1.99	0.72	0.03	0.89		6.00
Italy	1.98	0.85	0.66	0.03	0.68		4.20
The Netherlands	1.01	0.61	0.25	—	0.52	0.27	2.66
Norway	0.28	0.10	0.20	—	0.04	0.05	0.67
Sweden	0.66	0.26	0.20	0.01	0.10		1.23
United Kingdom	1.72	1.51	0.97	0.01	0.75		4.96
Total WAES Western Europe	10.47	7.40	4.66	0.10	3.57		26.52
Non-WAES Western Europe[c]	2.93	2.07	1.30	0.03	1.00		7.33
Total Western Europe	13.40	9.47	5.96	0.13	4.57		33.85
Japan	9.87	3.93	3.23	0.09	1.03	1.74	19.89
Mexico	1.96	0.91	1.85	—	0.30		5.02
Rest of WOCA[d]	16.89	6.56	13.41	0.04	3.14		40.04
Total WOCA	59.01	29.69	34.57	0.27	15.67	2.80	142.01

[a]Excludes 1.11 MBDOE solar heating and cooling.
[b]Excludes 0.07 MBDOE other indigenous supply.
[c]Estimated as 28 percent of WAES Western Europe.
[d]Includes Mexico's totals.

Table 3.7
WAES demand data, 2000, Case C-2 (final energy demand in MBDOE)

	Industrial heat and power	Residential heat and power	Transport on distillate	Transport on electricity	Chemical feedstocks	Nonenergy bunkers	Total
Canada	1.91	2.24	2.34	—	0.64		7.13
United States	16.94	7.49[a]	9.63	0.01	6.29	0.44	40.80
Total North America	18.85	9.73	11.97	0.01	6.93		47.93
Denmark	0.16	0.19	0.11	—	—		0.46
Finland	0.32[b]	0.15	0.10	—	—		0.57
France	1.80	1.68	1.42	0.03	0.59		5.52
Federal Republic of Germany	2.37	1.79	0.69	0.05	0.89		5.79
Italy	1.92	0.94	0.65	0.03	0.65		4.19
The Netherlands	1.01	0.61	0.25	—	0.52	0.27	2.66
Norway	0.28	0.10	0.20	—	0.04	0.05	0.67
Sweden	0.68	0.26	0.20	0.01	0.10		1.25
United Kingdom	1.65	1.51	0.97	0.01	0.75		4.89
Total WAES Western Europe	10.19	7.23	4.59	0.13	3.54		26.00
Non-WAES Western Europe[c]	2.85	2.02	1.29	0.04	0.99		7.19
Total Western Europe	13.04	9.25	5.88	0.17	4.53		33.19
Japan	9.87	3.93	3.23	0.09	1.03	1.74	19.89
Mexico	2.04	0.98	1.85	—	0.30		5.17
Rest of WOCA[d]	17.59	6.83	13.91	0.04	2.99		41.36
Total WOCA	59.35	29.74	34.99	0.31	15.48	2.50	142.37

[a]Excludes 1.11 MBDOE solar heating and cooling.
[b]Excludes 0.07 MBDOE other indigenous supply.
[c]Estimated as 28 percent of WAES Western Europe.
[d]Includes Mexico's totals.

Table 3.8
WAES demand data, 2000, Case D-7 (final energy demand in MBDOE)

	Industrial heat and power	Residential heat and power	Transport on distillate	Transport on electricity	Chemical feedstocks	Nonenergy bunkers	Total
Canada	1.60	1.83	1.68	—	0.57		5.68
United States	13.52	9.54[a]	9.93	0.01	4.95	0.37	38.32
Total North America	15.12	11.37	11.61	0.01	5.52		44.00
Denmark	0.14	0.18	0.10	—	—		0.42
Finland	0.22[b]	0.13	0.08	—	—		0.43
France	1.65	1.70	1.24	0.02	0.49		5.10
Federal Republic of Germany	1.98	1.77	0.71	0.03	0.74	0.05	5.28
Italy	1.48	0.87	0.42	0.02	0.36		3.15
The Netherlands	0.53	0.61	0.20	—	0.27	0.27	1.88
Norway	0.25	0.10	0.17	—	0.02	0.05	0.59
Sweden	0.52	0.31	0.19	0.01	0.28		1.31
United Kingdom	1.47	1.29	0.83	0.01	0.56		4.16
Total WAES Western Europe	8.24	6.96	3.94	0.09	2.72		22.32
Non-WAES Western Europe[c]	2.31	1.95	1.10	0.03	0.76		6.15
Total Western Europe	10.55	8.91	5.04	0.12	3.48		28.47
Japan	6.03	2.88	2.16	0.05	0.92	0.93	12.97
Mexico	1.87	0.79	1.71	—	0.23		4.60
Rest of WOCA[d]	12.59	4.67	9.54	0.03	2.12		28.95
Total WOCA	44.29	27.83	28.35	0.21	12.04	1.67	114.39

[a] Excludes 0.56 MBDOE solar heating and cooling.
[b] Excludes 0.06 MBDOE other indigenous supply.
[c] Estimated as 28 percent of WAES Western Europe.
[d] Includes Mexico's totals.

Table 3.9
WAES demand data, 2000, Case D-8 (final energy demand in MBDOE)

	Industrial heat and power	Residential heat and power	Transport on distillate	Transport on electricity	Chemical feedstocks	Nonenergy bunkers	Total
Canada	1.60	1.83	1.68	—	0.57		5.68
United States	13.52	9.54[a]	9.93	0.01	4.95	0.37	38.32
Total North America	15.12	11.37	11.61	0.01	5.52		44.00
Denmark	0.15	0.20	0.11	—			0.46
Finland	0.22[b]	0.13	0.08	—	—		0.43
France	1.49	1.49	1.21	0.03	0.49		4.71
Federal Republic of Germany	1.98	1.77	0.70	0.03	0.74	0.05	5.27
Italy	1.45	0.95	0.42	0.02	0.35		3.19
The Netherlands	0.53	0.61	0.20	—	0.27	0.27	1.88
Norway	0.25	0.10	0.18	—	0.02	0.05	0.60
Sweden	0.52	0.28	0.19	0.01	0.08		1.08
United Kingdom	1.41	1.29	0.83	0.01	0.56		4.10
Total WAES Western Europe	8.00	6.82	3.92	0.10	2.51		21.72
Non-WAES Western Europe[c]	2.24	1.91	1.10	0.03	0.70		5.98
Total Western Europe	10.24	8.73	5.02	0.13	3.21		27.70
Japan	6.03	2.88	2.16	0.05	0.92	0.93	12.97
Mexico	1.89	0.81	1.71	—	0.23		4.64
Rest of WOCA[d]	12.68	4.85	9.83	0.03	2.21		29.60
Total WOCA	44.07	27.83	28.62	0.22	11.86	1.67	114.27

[a] Excludes 0.56 MBDOE solar heating and cooling.
[b] Excludes 0.06 MBDOE other indigenous supply.
[c] Estimated as 28 percent of WAES Western Europe.
[d] Includes Mexico's totals.

Table 3.10
WAES supply data, 1985, Case A (MBDOE)

	Crude Oil					Coal		Gas	
	Onshore light	Onshore heavy	Offshore light	Offshore heavy	Tar sands/ Oil shale	Surface	Deep	Onshore	Offshore
Canada	1.21	—	1.59	—	0.40	0.54	0.12	1.58	0.21
United States	5.00	4.00	3.00	—	0.40	7.61	4.66	5.76	3.86
Total North America	6.21	4.00	4.59	—	0.80	8.15	4.78	7.34	4.07
Denmark	—	—	0.01	—	—	—	—	—	—
Finland	—	—	—	—	—	0.02[a]	—	—	—
France	—	—	—	—	—	—	0.26	—	0.18
Federal Republic of Germany	0.04	0.04	—	—	—	0.47	1.22	0.28	—
Italy	—	0.01	0.05	—	—	—	0.02	0.10	0.13
The Netherlands	—	0.04	—	—	—	—	—	1.21	0.30
Norway	—	—	0.77	—	—	—	0.02	—	0.38
Sweden	—	—	—	—	—	—	—	—	—
United Kingdom	—	—	3.15	—	—	0.10	1.40	—	0.95
Total WAES Western Europe	0.04	0.09	3.98	—	—	0.59	2.92	1.59	1.94
Non-WAES Western Europe	—	—	—	—	—	0.50	0.30	—	—
Total Western Europe	0.04	0.09	3.98	—	—	1.09	3.22	1.59	1.94
Japan	—	0.01	0.07	—	—	—	0.26	—	0.07
Mexico	—	—	1.87	—	—	—	0.13	—	0.55
Rest of WOCA	3.00	1.20	3.00[b]	—	—	—	—	1.00	2.03[b]
OPEC	21.00	12.00	6.00	—	—	—	—	3.40	2.00
OCEC	—	—	—	—	—	2.00	3.70[b]	—	—
Communist Area Exports	2.00	—	—	—	—	—	0.50	0.50	
Total WOCA	32.25	17.30	17.64	—	0.80	11.24	12.46	13.83	10.11

[a]Peat.
[b]Includes Mexico's total.

Table 3.11
WAES supply data, 1985, Case B (MBDOE)

	Crude Oil					Coal		Gas	
	Onshore light	Onshore heavy	Offshore light	Offshore heavy	Tar sands/ Oil shale	Surface	Deep	Onshore	Offshore
Canada	1.21	—	1.59	—	0.40	0.54	0.12	1.58	0.21
United States	5.00	4.00	3.00	—	0.40	6.95	4.26	5.76	3.86
Total North America	6.21	4.00	4.59	—	0.80	7.49	4.38	7.34	4.07
Denmark	—	—	0.01	—	—	—	—	—	—
Finland	—	—	—	—	—	0.02[a]	—	—	—
France	—	—	—	—	—	—	0.26	—	—
Federal Republic of Germany	0.04	0.04	—	—	—	0.47	1.22	0.28	0.13
Italy	—	0.01	0.05	—	—	—	0.02	0.10	0.13
The Netherlands	—	0.04	—	—	—	—	—	1.21	0.30
Norway	—	—	0.77	—	—	—	0.02	—	0.38
Sweden	—	—	—	—	—	—	—	—	—
United Kingdom	—	—	3.15	—	—	0.10	1.40	—	0.87
Total WAES Western Europe	0.04	0.09	3.98	—	—	0.59	2.92	1.59	1.81
Non-WAES Western Europe	—	—	—	—	—	0.50	0.30	—	—
Total Western Europe	0.04	0.09	3.98	—	—	1.09	3.22	1.59	1.81
Japan	—	0.01	0.07	—	—	—	0.26	—	0.07
Mexico	—	—	1.33	—	—	—	0.13	—	0.43
Rest of WOCA	3.00	1.20	3.00[b]	—	—	—	—	1.00	1.70[b]
OPEC	21.00	12.00	6.00	—	—			3.00	2.00
OCEC	—	—	—	—	—	2.00	2.91[b]	—	—
Communist Area Exports	2.00	—	—	—	—	—	0.50	0.50	—
Total WOCA	32.25	17.30	17.64	—	0.80	10.58	11.27	13.43	9.65

[a]Peat.
[b]Includes Mexico's total.

Table 3.12
WAES supply data, 1985, Case C (MBDOE)

	Crude Oil					Coal		Gas	
	Onshore light	Onshore heavy	Offshore light	Offshore heavy	Tar sands/ Oil shale	Surface	Deep	Onshore	Offshore
Canada	1.21	—	1.23	—	0.40	0.42	0.11	1.12	0.19
United States	4.80	3.70	2.50	—	0.20	6.95	4.26	5.49	3.65
Total North America	6.01	3.70	3.73	—	0.60	7.37	4.37	6.61	3.84
Denmark	—	—	0.01	—	—	—	—	—	—
Finland	—	—	—	—	—	0.02[a]	—	—	—
France	—	—	—	—	—	—	0.26	—	0.13
Federal Republic of Germany	0.04	0.04	—	—	—	0.47	1.22	0.28	—
Italy	—	0.01	0.05	—	—	—	0.02	0.10	0.13
The Netherlands	—	0.04	—	—	—	—	—	1.21	0.30
Norway	—	—	0.77	—	—	—	0.02	—	0.38
Sweden	—	—	—	—	—	—	—	—	—
United Kingdom	—	—	3.15	—	—	0.10	1.40	—	0.95
Total WAES Western Europe	0.04	0.09	3.98	—	—	0.59	2.92	1.59	1.89
Non-WAES Western Europe	—	—	—	—	—	0.50	0.30	—	—
Total Western Europe	0.04	0.09	3.98	—	—	1.09	3.22	1.54	1.89
Japan	—	0.01	0.07	—	—	—	0.26	—	0.07
Mexico	—	—	1.52	—	—	—	0.15	—	0.52
Rest of WOCA	3.00	1.20	3.00[b]	—	—	—	—	1.00	2.03[b]
OPEC	21.00	12.00	6.00	—	—	—	—	3.40	2.00
OCEC	—	—	—	—	—	2.00	3.70[b]	—	—
Communist Area Exports	2.00	—	—	—	—	—	0.50	0.50	—
Total WOCA	32.05	17.00	16.78	—	0.60	10.46	12.05	13.10	9.83

[a]Peat.
[b]Includes Mexico's total.

Table 3.13
WAES supply data, 1985, Case D (MBDOE)

	Crude Oil					Coal		Gas	
	Onshore light	Onshore heavy	Offshore light	Offshore heavy	Tar sands/ Oil shale	Surface	Deep	Onshore	Offshore
Canada	1.21	—	0.99	—	0.20	0.34	0.11	1.15	—
United States	4.00	3.50	2.00	—	—	5.85	3.58	4.63	3.03
Total North America	5.21	3.50	2.99	—	0.20	6.19	3.69	5.78	3.03
Denmark	—	—	0.01	—	—	—	—	—	—
Finland	—	—	—	—	—	0.02[a]	—	—	—
France	—	—	—	—	—	—	0.26	—	0.13
Federal Republic of Germany	0.04	0.04	—	—	—	0.47	1.04	0.28	—
Italy	—	0.01	0.05	—	—	—	0.02	0.10	0.13
The Netherlands	—	0.04	—	—	—	—	—	1.21	0.30
Norway	—	—	0.77	—	—	—	0.02	—	0.38
Sweden	—	—	—	—	—	—	—	—	—
United Kingdom	—	—	3.15	—	—	0.10	1.40	—	0.87
Total WAES Western Europe	0.04	0.09	3.98	—	—	0.59	2.74	1.59	1.81
Non-WAES Western Europe	—	—	—	—	—	0.50	0.30	—	—
Total Western Europe	0.04	0.09	3.98	—	—	1.09	3.04	1.59	1.81
Japan	—	0.01	0.07	—	—	—	0.26	—	0.07
Mexico	—	—	1.38	—	—	—	0.13	—	0.47
Rest of WOCA	2.00	1.00	2.60[b]	—	—	—	—	1.00	1.70[b]
OPEC	20.00	10.00	6.00	—	—	—	—	3.00	2.00
OCEC	—	—	—	—	—	2.00	2.91[b]	—	—
Communist Area Exports	2.00	—	—	—	—	—	0.50	0.50	—
Total WOCA	29.25	14.60	15.64	—	0.20	9.28	10.40	11.87	8.61

[a] Peat.
[b] Includes Mexico's total.

Table 3.14
WAES supply data, 1985, Case E (MBDOE)

	Crude Oil					Coal		Gas	
	Onshore light	Onshore heavy	Offshore light	Offshore heavy	Tar sands/ Oil shale	Surface	Deep	Onshore	Offshore
Canada	1.21	—	0.31	—	0.06	0.29	0.10	1.11	—
United States	3.50	3.00	1.50	—	—	5.85	3.58	4.20	2.54
Total North America	4.71	3.00	1.81	—	0.06	6.14	3.68	5.31	2.54
Denmark	—	—	0.01	—	—	—	—	—	—
Finland	—	—	—	—	—	0.02[a]	—	—	—
France	—	—	—	—	—	—	0.26	—	0.13
Federal Republic of Germany	0.04	0.04	—	—	—	0.47	1.04	0.28	—
Italy	—	0.01	0.05	—	—	—	0.02	0.10	0.13
The Netherlands	—	0.04	—	—	—	—	—	1.21	0.30
Norway	—	—	0.77	—	—	—	0.02	—	0.38
Sweden	—	—	—	—	—	—	—	—	—
United Kingdom	—	—	3.15	—	—	0.10	1.40	—	0.95
Total WAES Western Europe	0.04	0.09	3.98	—	—	0.59	2.74	1.59	1.89
Non-WAES Western Europe	—	—	—	—	—	0.50	0.30	—	—
Total Western Europe	0.04	0.09	3.98	—	—	1.09	3.04	1.59	1.89
Japan	—	0.01	0.07	—	—	—	0.26	—	0.07
Mexico	—	—	1.35	—	—	—	0.13	—	0.47
Rest of WOCA	2.00	1.00	2.60[b]	—	—	—	—	1.00	2.03[b]
OPEC	20.00	10.00	6.00	—	—	—	—	1.25	0.75
OCEC	—	—	—	—	—	2.00	3.70[b]	—	—
Communist Area Exports	2.00	—	—	—	—	—	0.50	0.50	—
Total WOCA	28.75	14.10	14.46	—	0.06	9.23	11.18	9.65	7.28

[a]Peat.
[b]Includes Mexico's total.

Table 3.15
WAES supply data, 2000, Case C-1 (MBDOE)

	Crude Oil					Coal		Gas	
	Onshore light	Onshore heavy	Offshore light	Offshore heavy	Tar sands/ Oil shale	Surface	Deep	Onshore	Offshore
Canada	0.63	—	1.54	—	0.89	1.30	0.13	0.58	0.28
United States	3.00	2.00	2.00	—	2.00	14.86	7.83	4.05	3.17
Total North America	3.63	2.00	3.54	—	2.89	16.16	7.96	4.63	3.45
Denmark	—	—	—	—	—	—	—	—	—
Finland	—	—	—	—	—	0.09[a]	—	—	—
France	—	—	—	—	—	—	0.26	—	0.13
Federal Republic of Germany	0.04	0.01	—	—	—	0.47	1.46	0.13	—
Italy	—	—	0.03	—	—	—	0.04	0.06	0.09
The Netherlands	—	0.04	—	—	—	—	—	0.28	0.07
Norway	—	—	1.18	—	—	—	0.02	—	0.59
Sweden	—	—	—	—	0.02	—	—	—	—
United Kingdom	—	—	1.54	—	—	0.20	1.42	—	0.50
Total WAES Western Europe	0.04	0.05	2.75	—	0.02	0.76	3.36	0.47	1.38
Non-WAES Western Europe	—	—	—	—	—	0.50	0.30	—	—
Total Western Europe	0.04	0.05	2.75	—	0.02	1.26	3.66	0.47	1.38
Japan	—	0.01	0.37	—	—	—	0.26	—	0.08
Mexico	—	—	3.53	—	—	—	0.50	—	1.11
Rest of WOCA	5.00	3.00	4.00[b]	—	—	—	—	2.00	3.30[b]
OPEC	23.00	14.00	8.00	—	—	—	—	8.62	5.00
OCEC	—	—	—	—	—	5.00	7.37[b]	—	—
Communist Area Exports	2.00	—	—	—	—	—	0.50	1.00	—
Total WOCA	33.67	19.06	18.66	—	2.91	22.42	19.75	16.72	13.21

[a]Peat.
[b]Includes Mexico's total.

Table 3.16
WAES supply data, 2000, Case C-2 (MBDOE)

	Crude Oil					Coal		Gas	
	Onshore light	Onshore heavy	Offshore light	Offshore heavy	Tar sands/ Oil shale	Surface	Deep	Onshore	Offshore
Canada	0.63	—	1.54	—	0.89	1.06	0.13	0.58	0.28
United States	3.00	2.00	2.00	—	2.00	10.53	5.66	4.05	3.17
Total North America	3.63	2.00	3.54	—	2.89	11.59	5.79	4.63	3.45
Denmark	—	—	—	—	—	—	—	—	—
Finland	—	—	—	—	—	0.08[a]	—	—	—
France	—	—	—	—	—	—	0.26	—	0.13
Federal Republic of Germany	0.04	0.01	—	—	—	0.47	1.22	0.13	—
Italy	—	—	0.03	—	—	—	0.02	0.06	0.09
The Netherlands	—	0.04	—	—	—	—	—	0.28	0.07
Norway	—	—	1.18	—	—	—	0.02	—	0.59
Sweden	—	—	—	—	0.02	—	—	—	—
United Kingdom	—	—	1.54	—	—	0.10	0.94	—	0.50
Total WAES Western Europe	0.04	0.05	2.75	—	0.02	0.65	2.46	0.47	1.38
Non-WAES Western Europe	—	—	—	—	—	0.50	0.30	—	—
Total Western Europe	0.04	0.05	2.75	—	0.02	1.15	2.76	0.47	1.38
Japan	—	0.01	0.37	—	—	—	0.26	—	0.08
Mexico	—	—	3.53	—	—	—	0.39	—	1.11
Rest of WOCA	5.00	3.00	4.00[b]	—	—	—	—	2.00	3.20[b]
OPEC	23.00	14.00	8.00	—	—	—	—	6.60	5.00
OCEC	—	—	—	—	—	4.10	6.20[b]	—	—
Communist Area Exports	2.00	—	—	—	—	—	0.50	1.00	—
Total WOCA	33.67	19.06	18.66	—	2.91	16.84	15.51	14.70	13.11

[a]Peat.
[b]Includes Mexico's total.

Table 3.17
WAES supply data, 2000, Case D-7 (MBDOE)

	Crude Oil					Coal		Gas	
	Onshore light	Onshore heavy	Offshore light	Offshore heavy	Tar sands/ Oil shale	Surface	Deep	Onshore	Offshore
Canada	0.63	—	1.05	—	0.89	1.30	0.13	0.56	0.17
United States	2.50	2.00	1.50	—	—	11.04	5.87	3.25	2.17
Total North America	3.13	2.00	2.55	—	0.89	12.34	6.00	3.81	2.24
Denmark	—	—	—	—	—	—	—	—	—
Finland	—	—	—	—	—	0.04[a]	—	—	—
France	—	—	—	—	—	—	0.26	—	0.13
Federal Republic of Germany	0.02	0.01	—	—	—	0.47	1.30	0.13	—
Italy	—	—	0.03	—	—	—	0.04	0.06	0.09
The Netherlands	—	0.04	—	—	—	—	—	0.28	0.07
Norway	—	—	0.91	—	—	—	0.02	—	0.45
Sweden	—	—	—	—	0.02	—	—	—	—
United Kingdom	—	—	1.54	—	—	0.10	1.30	—	0.80
Total WAES Western Europe	0.02	0.05	2.48	—	0.02	0.61	2.92	0.47	1.54
Non-WAES Western Europe	—	—	—	—	—	0.50	0.30	—	—
Total Western Europe	0.02	0.05	2.48	—	0.02	1.11	3.22	0.47	1.54
Japan	—	0.01	0.37	—	—	—	0.26	—	0.08
Mexico	—	—	3.29	—	—	—	0.50	—	1.09
Rest of WOCA	3.00	1.50	3.00[b]	—	—	—	—	2.00	2.25[b]
OPEC	21.00	12.00	6.00	—	—	—	—	7.30	5.00
OCEC	—	—	—	—	—	4.00	6.37[b]	—	—
Communist Area Exports	2.00	—	—	—	—	—	0.50	1.00	—
Total WOCA	29.15	15.56	14.40	—	0.91	17.45	16.35	14.58	11.11

[a]Peat.
[b]Includes Mexico's total.

Table 3.18
WAES supply data, 2000, Case D-8 (MBDOE)

	Crude Oil					Coal		Gas	
	Onshore light	Onshore heavy	Offshore light	Offshore heavy	Tar sands/ Oil shale	Surface	Deep	Onshore	Offshore
Canada	0.63	—	1.05	—	0.89	1.06	0.13	0.56	0.17
United States	2.50	2.00	1.50	—	—	8.77	4.32	3.25	2.17
Total North America	3.13	2.00	2.55	—	0.89	9.83	4.45	3.81	2.24
Denmark	—	—	—	—	—	—	—	—	—
Finland	—	—	—	—	—	0.04[a]	—	—	—
France	—	—	—	—	—	—	0.26	—	0.13
Federal Republic of Germany	0.02	0.01	—	—	—	0.47	1.04	0.13	—
Italy	—	—	0.03	—	—	—	0.02	0.06	0.09
The Netherlands	—	0.04	—	—	—	—	—	0.28	0.07
Norway	—	—	0.91	—	—	—	0.02	—	0.45
Sweden	—	—	—	—	0.02	—	—	—	—
United Kingdom	—	—	1.54	—	—	0.10	0.80	—	0.44
Total WAES Western Europe	0.02	0.05	2.48	—	0.02	0.61	2.14	0.47	1.18
Non-WAES Western Europe	—	—	—	—	—	0.50	0.30	—	—
Total Western Europe	0.02	0.05	2.48	—	0.02	1.11	2.44	0.47	1.18
Japan	—	0.01	0.37	—	—	—	0.26	—	0.08
Mexico	—	—	3.29	—	—	—	0.48	—	1.09
Rest of WOCA	3.00	1.50	3.00[b]	—	—	—	—	2.00	2.00[b]
OPEC	21.00	12.00	6.00	—	—	—	—	5.30	5.00
OCEC	—	—	—	—	—	3.14	5.20[b]	—	—
Communist Area Exports	2.00	—	—	—	—	—	0.50	1.00	—
Total WOCA	29.15	15.58	14.40	—	0.91	14.08	12.85	12.58	10.50

[a]Peat.
[b]Includes Mexico's total.

Table 3.19
WAES maximum technological development constraints, 1985, Cases A and B (available capacity in MBDOE)

	Nuclear electricity	Coal gasification	Coal liquefaction	Gas liquefaction	Regasi-fication	Hydro	Other
Canada	0.29	—	—	—	—	1.38	—
United States	4.02	0.66	—	1.00	2.00	2.18	0.24
Total North America	4.31	0.66	—	1.00	2.00	3.56	0.24
Denmark	0.02	—	—	0.01	0.01	—	—
Finland	0.07	—	—	—	—	0.06	—
France	1.09	—	—	—	—	0.28	0.06
Federal Republic of Germany	0.80	0.21	—	—	—	0.08	—
Italy	0.23	—	—	—	—	0.15	0.01
The Netherlands	0.08	—	—	—	—	—	—
Norway	—	—	—	—	—	0.46	—
Sweden	0.18	0.01	0.01	0.01	0.01	0.30	0.07
United Kingdom	0.36	—	—	—	—	0.02	—
Total WAES Western Europe	2.83	0.22	0.01	0.02	0.02	1.34	0.14
Non-WAES Western Europe	0.75	—	—	—	—	0.70	—
Total Western Europe	3.58	0.22	0.01	0.02	1.00[b]	2.05	0.14
Japan	0.97	0.02	—	—	2.00	0.54	0.02
Mexico	0.07	—	—	—	—	0.09	—
Rest of WOCA	1.24[a]	1.00	0.50	0.70	1.00	2.01[a]	0.01
OPEC	—	—	—	3.30	—	—	—
OCEC	—	—	—	—	—	—	—
Total WOCA	10.10	1.90	0.51	5.02	6.00	8.16	0.41

[a]Includes Mexico's total.
[b]Estimate for model purposes.

Table 3.20
WAES maximum technological development constraints, 1985, Case C (available capacity in MBDOE)

	Nuclear electricity	Coal gasification	Coal liquefaction	Gas liquefaction	Regasi-fication	Hydro	Other
Canada	0.29	—	—	—	—	1.38	—
United States	4.02	0.28	—	1.00	2.00	2.18	0.24
Total North America	4.31	0.28	—	1.00	2.00	3.56	0.24
Denmark	0.02	—	—	0.01	0.01	—	—
Finland	0.07	—	—	—	—	0.06	—
France	1.09	—	—	—	—	0.28	0.06
Federal Republic of Germany	0.80	0.21	—	—	—	0.08	—
Italy	0.23	—	—	—	—	0.15	0.01
The Netherlands	0.08	—	—	—	—	—	—
Norway	—	—	—	—	—	0.46	—
Sweden	0.18	0.01	0.01	0.01	0.01	0.30	0.07
United Kingdom	0.36	—	—	—	—	0.02	—
Total WAES Western Europe	2.83	0.22	0.01	0.02	0.02	1.35	0.14
Non-WAES Western Europe	0.75	—	—	—	—	0.70	—
Total Western Europe	3.58	0.22	0.01	0.02	1.00[b]	2.05	0.14
Japan	0.97	0.02	—	—	2.00	0.55	0.02
Mexico	0.07	—	—	—	—	0.09	—
Rest of WOCA	1.24[a]	1.00	0.50	0.70	1.00	2.01[a]	0.01
OPEC	—	—	—	3.30	—	—	—
OCEC	—	—	—	—	—	—	—
Total WOCA	10.10	1.52	0.51	5.02	6.00	8.17	0.41

[a]Includes Mexico's total.
[b]Estimate for model purposes.

Table 3.21
WAES maximum technological development constraints, 1985, Cases D and E (available capacity in MBDOE)

	Nuclear electricity	Coal gasification	Coal liquefaction	Gas liquefaction	Regasi-fication	Hydro	Other
Canada	0.19	—	—	—	—	1.38	—
United States	3.07	0.09	—	1.00	1.50	1.62	0.10
Total North America	3.27	0.09	—	1.00	1.50	3.00	0.10
Denmark	0.02	—	—	—	—	—	—
Finland	0.07	—	—	—	—	0.06	—
France	0.85	—	—	—	—	0.28	0.06
Federal Republic of Germany	0.51	0.20	—	—	—	0.08	—
Italy	0.13	—	—	—	—	0.11	0.01
The Netherlands	0.05	—	—	—	—	—	—
Norway	—	—	—	—	—	0.46	—
Sweden	0.09	0.01	0.01	0.01	0.01	0.30	0.06
United Kingdom	0.34	—	—	—	—	0.02	—
Total WAES Western Europe	2.06	0.21	0.01	0.01	0.01	1.31	0.13
Non-WAES Western Europe	0.56	—	—	—	—	0.70	—
Total Western Europe	2.62	0.21	0.01	0.01	1.00[b]	2.01	0.13
Japan	0.61	0.02	—	—	2.00	0.54	0.02
Mexico	0.05	—	—	—	—	0.09	—
Rest of WOCA	0.59[a]	0.50	—	0.70	1.00	2.08[a]	0.01
OPEC	—	—	—	3.30	—	—	—
OCEC	—	—	—	—	—	—	—
Total WOCA	7.09	0.82	0.01	5.01	5.50	7.63	0.26

[a]Includes Mexico's total.
[b]Estimate for model purposes.

Table 3.22
WAES maximum technological development constraints, 2000, Case C-1 (available capacity in MBDOE)

	Nuclear electricity	Coal gasification	Coal liquefaction	Gas liquefaction	Regasi-fication	Hydro	Other
Canada	1.17	—	—	—	—	1.84	—
United States	12.26	2.83	2.00	2.50	—	2.17	2.13
Total North America	13.43	2.83	2.00	2.50	∞	4.01	2.13
Denmark	0.05	—	—	—	—	—	—
Finland	0.16	—	—	—	—	0.06	—
France	2.36	—	—	—	—	0.32	—
Federal Republic of Germany	1.31	0.53	—	—	—	0.09	—
Italy	0.52	—	—	—	—	0.61	0.02
The Netherlands	0.31	—	—	—	—	—	—
Norway	0.10	—	—	—	—	0.47	—
Sweden	0.26	0.10	0.10	0.10	0.10	0.35	0.06
United Kingdom	0.87	0.29	—	—	—	0.02	—
Total WAES Western Europe	5.94	0.92	0.10	0.10	0.10	1.92	0.08
Non-WAES Western Europe	1.89	—	—	—	—	0.70	0.10
Total Western Europe	7.83	0.92	0.10	∞	∞	2.62	0.18
Japan	1.96	0.02	—	—	∞	0.83	0.09
Mexico	1.05	—	—	—	—	0.41	0.01
Rest of WOCA	4.55[a]	2.00	1.00	1.00	∞	4.59[a]	0.02[a]
OPEC	—	—	—	7.50	—	—	—
OCEC	—	—	—	—	—	—	—
Total WOCA	27.77	5.77	3.10	11.00	∞	12.05	2.42

[a]Includes Mexico's total.
∞ = no constraint assumed.

Table 3.23
WAES maximum technological development constraints, 2000, Case C-2 (available capacity in MBDOE)

	Nuclear electricity	Coal gasification	Coal liquefaction	Gas liquefaction	Regasi-fication	Hydro	Other
Canada	1.94	—	—	—	—	1.84	—
United States	20.01	2.83	2.00	2.50	—	2.17	2.13
Total North America	21.95	2.83	2.00	2.50	∞	4.01	2.13
Denmark	0.13	—	—	—	—	—	—
Finland	0.22	—	—	—	—	0.06	—
France	3.67	—	—	—	—	0.32	—
Federal Republic of Germany	3.15	0.53	—	—	—	0.09	—
Italy	1.57	—	—	—	—	0.68	0.02
The Netherlands	0.50	—	—	—	—	—	—
Norway	0.10	—	—	—	—	0.47	—
Sweden	0.52	0.05	0.05	0.05	0.05	0.35	0.06
United Kingdom	1.52	0.29	—	—	—	0.02	—
Total WAES Western Europe	11.38	0.87	0.05	0.05	0.05	1.99	0.08
Non-WAES Western Europe	2.46	—	—	—	—	0.70	0.10
Total Western Europe	13.84	0.87	0.05	∞	∞	2.69	0.18
Japan	3.15	0.02	—	—	∞	0.83	0.09
Mexico	2.10	—	—	—	—	0.41	0.01
Rest of WOCA	9.38[a]	2.00	1.00	1.00	∞	3.83[a]	0.02[a]
OPEC	—	—	—	6.60	—	—	—
OCEC	—	—	—	—	—	—	—
Total WOCA	48.32	5.72	3.05	10.10	∞	11.36	2.42

[a]Includes Mexico's total.
∞ = no constraint assumed.

Table 3.24
WAES maximum technological development constraints, 2000, Case D-7 (available capacity in MBDOE)

	Nuclear electricity	Coal gasification	Coal liquefaction	Gas liquefaction	Regasi-fication	Hydro	Other
Canada	1.17	—	—	—	—	1.49	—
United States	12.26	1.89	—	1.50	—	1.74	0.80
Total North America	13.43	1.89	—	1.50	∞	3.23	0.80
Denmark	0.05	—	—	—	—	—	—
Finland	0.11	—	—	—	—	0.06	—
France	1.84	—	—	—	—	0.32	—
Federal Republic of Germany	1.31	0.53	—	—	—	0.09	—
Italy	0.52	—	—	—	—	0.51	0.02
The Netherlands	0.20	—	—	—	—	—	—
Norway	0.06	—	—	—	—	0.47	—
Sweden	0.26	0.10	0.10	0.10	0.10	0.35	0.06
United Kingdom	0.76	0.18	—	—	—	0.02	—
Total WAES Western Europe	5.12	0.81	0.10	0.10	0.10	1.82	0.08
Non-WAES Western Europe	1.89	—	—	—	—	0.70	0.10
Total Western Europe	7.01	0.81	0.10	∞	∞	2.52	0.18
Japan	1.97	0.02	—	—	∞	0.79	0.09
Mexico	0.52	—	—	—	—	0.41	0.01
Rest of WOCA	3.13[a]	1.00	—	1.00	∞	3.24[a]	0.02[a]
OPEC	—	—	—	7.80	—	—	—
OCEC	—	—	—	—	—	—	—
Total WOCA	25.54	3.72	0.10	10.30	∞	9.78	1.09

[a]Includes Mexico's total.
∞ = no constraint assumed.

Table 3.25
WAES maximum technological development constraints, 2000, Case D-8 (available capacity in MBDOE)

	Nuclear electricity	Coal gasification	Coal liquefaction	Gas liquefaction	Regasi-fication	Hydro	Other
Canada	1.79	—	—	—	—	1.49	—
United States	20.01	—	—	1.50	—	1.74	0.80
Total North America	21.80	—	—	1.50	∞	3.23	0.80
Denmark	0.13	—	—	—	—	—	—
Finland	0.11	—	—	—	—	0.06	—
France	2.62	—	—	—	—	0.32	—
Federal Republic of Germany	2.10	0.53	—	—	—	0.09	—
Italy	1.57	—	—	—	—	0.56	0.02
The Netherlands	0.31	—	—	—	—	—	—
Norway	0.08	—	—	—	—	0.47	—
Sweden	0.52	0.05	0.05	0.05	0.05	0.35	0.06
United Kingdom	1.31	—	—	—	—	0.02	—
Total WAES Western Europe	8.75	0.58	0.05	0.05	0.05	1.87	0.08
Non-WAES Western Europe	2.46	—	—	—	—	0.70	0.10
Total Western Europe	11.21	0.58	0.05	∞	∞	2.57	0.18
Japan	2.81	0.02	—	—	∞	0.79	0.09
Mexico	0.79	—	—	—	—	—	—
Rest of WOCA	6.57[a]	1.00	—	1.00	∞	2.63	0.02
OPEC	—	—	—	6.40	—	—	—
OCEC	—	—	—	—	—	—	—
Total WOCA	42.39	1.60	0.05	8.90	∞	9.22	1.09

[a]Includes Mexico's total.
∞ = no constraint assumed.

Table 3.26
Fuel preference constraints for electrical generation, 1985, all cases (minimum and maximum percentages of electrical output generated from each source)

	Oil		Coal		Gas		Nuclear		Hydro		Other	
	Min	Max	Min	Max	Min	Max	Min	Max	Min	Max	Min	Max
Canada	2	3	23	25	1	2	13	16	56	57		
United States	10	20	20	60	5	20	10	50	10	30	0	5
Total North America	10	20	20	60	5	20	10	50	10	30	0	5
Denmark	35	60	5	30	0	20	33	33	0	0		
Finland												
France												
Federal Republic of Germany	5	10	20	30	15	20	20	40–45	4–5	5–6		
Italy	20	50	0	30	0	5	7	16	23	46		
The Netherlands	66	73	10	10	0	0	17	24	0	0	0	0
Norway	0.2	5	0	0	0	5	0	0	96	100		
Sweden	5	20	54	65	0	0	25	30	1	1		
United Kingdom												
Total Western Europe	10	30	15	40	4	10	25	35	15	22	0	5
Japan	25	60	5	15	10	25	10	30	15[a]	25	0	5
Mexico	30	40	5	10	0	20	5	50	30	40		
Rest of WOCA[b]	10	30	20	60	3	15	5	35	25	40	0	5

[a] 12 percent in Case A.
[b] Includes Mexico.

Table 3.27
Fuel preference constraints for electrical generation, 2000, Cases C-1 and D-7 (minimum and maximum percentages of electrical output generated from each source)

	Oil		Coal		Gas		Nuclear		Hydro		Other	
	Min	Max	Min	Max	Min	Max	Min	Max	Min	Max	Min	Max
Canada												
United States	10	20	20	60	5	20	10	50	10	30	0	10
Total North America	10	20	20	60	5	20	10	50	10	30	0	10
Denmark	35	60	5	30	0	20	33	33	0	0		
Finland												
France												
Federal Republic of Germany	0	10	20	35	0	15	20	50–65	4–5	5–6	3	5
Italy	20	50	0	30	0	5	7	30	23	46		
The Netherlands	29	53	10	30	0	0	37	41	0	0	0	0
Norway	0.2	5	0	0	0	5	0	0	95	100		
Sweden	5	20	54	65	0	0	25	30	1	1		
United Kingdom												
Total Western Europe	10	30	15	40	0	10	25	45	15	22	0	10
Japan	25	60	5	15	10	25	10	30	10	25	0	10
Mexico	30	40	5	10	0	20	5	50	30	40		
Rest of WOCA[a]	10	30	20	60	3	15	5	35	25	40	0	10

[a] Includes Mexico.

Table 3.28
Fuel preference constraints for electrical generation, 2000, Cases C-2 and D-8 (minimum and maximum percentages of electrical output generated from each source)

	Oil		Coal		Gas		Nuclear		Hydro		Other	
	Min	Max	Min	Max	Min	Max	Min	Max	Min	Max	Min	Max
Canada												
United States	10	20	20	60	5	20	10	70	10	30	0	10
Total North America	10	20	20	40	5	20	10	70	10	30	0	10
Denmark	35	60	5	30	0	20	33	33	0	0		
Finland												
France												
Federal Republic of Germany	0	10	15	30	0	15	65	80	4–5	5–6	3	5
Italy	20	50	0	30	0	5	50	60	23	46		
The Netherlands	3	33	10	30	0	0	57	66	0	0	0	0
Norway	0.7	5	0	0	0	5	0	0	95	100		
Sweden	5	20	54	65	0	0	25	30	1	1		
United Kingdom												
Total Western Europe	10	30	15	35	0	10	25	60	15	22	0	10
Japan	25	60	5	15	10	25	10	30	10	25	0	10
Mexico	30	40	5	10	0	20	5	50	30	40		
Rest of WOCA[a]	10	30	20	40	3	15	5	55	25	40	0	10

[a]Includes Mexico.

Table 3.29
Consumer preference constraints, industrial market, 1985, all cases (minimum and maximum percentage of final demand)

	Oil		Coal		Gas		Electricity		Distillates in liquid fuel (%)
	Min	Max	Min	Max	Min	Max	Min	Max	
Canada	17	18	19	22	26	28	35	36	
United States	10	25	25	40	25	35	12	20	33
Total North America	10	25	25	40	25	35	12	25	33
Denmark	33	87	0	20	0	54	13	19	20/50
Finland									
France									
Federal Republic of Germany	20	45	12	30	15	30	15	25	
Italy	42	70	8	14	14	23	11	19	
The Netherlands	44	61	6	6	25	41	7	7	
Norway	30	45	12	13	4	5	40	55	35
Sweden	30	60	10	20	0	10	15	35	
United Kingdom									
Total Western Europe	30	60	10	30	10	30	15	30	33
Japan	40	70	10	25	0	10	10	30	25/30
Mexico	21	24	14	16	20	50	11	18	40
Rest of WOCA[a]	40	60	20	40	5	20	10	25	33

[a]Includes Mexico.

Table 3.30
Consumer preference constraints, industrial market, 2000, Cases C-1 and D-7 (minimum and maximum percentages of final demand)

	Oil		Coal		Gas		Electricity		Distillates in liquid fuel (%)
	Min	Max	Min	Max	Min	Max	Min	Max	
Canada	17	18	19	22	26	28	35	36	
United States	10	25	25	50	15	25	18	30	33
Total North America	10	25	25	50	20	25	20	30	33
Denmark	33	87	0	20	0	54	13	19	20/50
Finland									
France									
Federal Republic of Germany	20	40	10	30	15	35	20	30	30/40
Italy	42	70	8	14	14	23	11	19	
The Netherlands	62	64	17	17	4	6	15	15	
Norway	30	45	12	13	4	5	40	55	35
Sweden	30	60	10	20	0	10	15	35	
United Kingdom									
Total Western Europe	30	60	10	40	10	30	16	30	33
Japan	40	70	10	35	0	10	12	30	25/30
Mexico	21	24	14	16	20	50	11	18	40
Rest of WOCA[a]	40	60	20	40	5	20	15	25	33

[a]Includes Mexico.

Table 3.31
Consumer preference constraints, industrial market, 2000, Cases C-2 and D-8 (minimum and maximum percentages of final demand)

	Oil		Coal		Gas		Electricity		Distillates in liquid fuel (%)
	Min	Max	Min	Max	Min	Max	Min	Max	
Canada	17	18	19	22	26	28	35	36	
United States	10	25	25	40	10	20	25	35	33
Total North America	10	25	25	40	15	25	25	35	33
Denmark	33	87	0	20	0	54	13	19	20/50
Finland									
France									
Federal Republic of Germany	20	40	10	20	15	35	20	30	30/40
Italy	42	70	8	14	14	23	11	19	
The Netherlands	71	73	8	8	4	6	15	15	
Norway	30	45	12	13	4	5	40	55	35
Sweden	30	60	10	20	0	10	15	35	
United Kingdom									
Total Western Europe	30	60	10	30	10	30	20	40	33
Japan	40	70	10	25	0	10	15	30	25/30
Mexico	21	24	14	16	20	50	11	18	40
Rest of WOCA[a]	40	60	20	40	5	20	20	35	33

[a]Includes Mexico.

Table 3.32
Consumer preference constraints, residential sector, 1985, all cases (minimum and maximum percentages of final demand)

	Oil		Coal		Gas		Electricity		Distillates in liquid fuel (%)
	Min	Max	Min	Max	Min	Max	Min	Max	
Canada	47	49	7	8	23	24	20	22	
United States	20	45	0	5	25	40	25	45	80
Total North America	25	45	0	5	25	40	25	40	80
Denmark	50	90	0	30	0	40	10	30	70/40
Finland									
France									
Federal Republic of Germany	20	50	0	10	20	40	20	35	80/90
Italy	63	83	3	8	6	17	5	14	
The Netherlands	14	15	0	0	74	75	10	10	
Norway	37	50	0	0	0	0	47	60	35
Sweden	40	65	0	10	0	10	25	40	
United Kingdom									
Total Western Europe	30	65	0	20	10	35	10	30	70
Japan	50	80	0	0.5	10	20	15	30	65/70
Mexico	30	35	—	—	50	57	10	20	70
Rest of WOCA[a]	40	70	5	20	5	20	20	35	70

[a]Includes Mexico.

Table 3.33
Consumer preference constraints, residential sector, 2000, Cases C-1 and D-7 (minimum and maximum percentages of final demand)

	Oil		Coal		Gas		Electricity		Distillates in liquid fuel (%)
	Min	Max	Min	Max	Min	Max	Min	Max	
Canada	47	49	7	8	23	24	20	22	
United States	20	45	0	5	25	40	35	45	80
Total North America	25	45	0	5	25	40	30	40	80
Denmark	50	90	0	30	0	40	10	30	70/40
Finland									
France									
Federal Republic of Germany	20	50	0	20	40	40	15	35	80/90
Italy	63	83	3	8	6	17	5	14	
The Netherlands	10	10	0	0	75	76	13	14	
Norway	37	50	0	0	0	0	47	60	35
Sweden	40	65	0	10	0	10	25	40	
United Kingdom									
Total Western Europe	30	65	0	20	10	35	15	30	70
Japan	50	80	0	0.5	10	20	16	30	65/70
Mexico	30	35	—	—	50	57	10	20	70
Rest of WOCA[a]	40	70	5	20	5	20	20	35	70

[a]Includes Mexico.

Table 3.34
Consumer preference constraints, residential sector, 2000, Cases C-2 and D-8 (minimum and maximum percentages of final demand)

	Oil		Coal		Gas		Electricity		Distillates in liquid fuel (%)
	Min	Max	Min	Max	Min	Max	Min	Max	
Canada	47	49	7	8	23	24	20	22	
United States	15	45	0	5	25	40	40	50	80
Total North America	20	45	0	5	25	40	38	50	80
Denmark	50	90	0	30	0	40	10	30	70/40
Finland									
France									
Federal Republic of Germany	30	60	0	10	10	30	10	25	
Italy	63	83	3	8	6	17	5	14	
The Netherlands	10	10	0	0	75	76	13	14	
Norway	37	50	0	0	0	0	47	60	35
Sweden	40	65	0	10	0	10	25	40	
United Kingdom									
Total Western Europe	30	65	0	20	10	35	20	40	70
Japan	50	80	0	0.5	10	20	20	30	65/70
Mexico	30	35	—	—	50	57	10	20	70
Rest of WOCA[a]	40	70	5	20	5	20	25	40	70

[a]Includes Mexico.

4 Unconstrained Global Supply-Demand Integrations

4.1 INTRODUCTION

The results of the WAES unconstrained global supply-demand integrations will be given by region, including all the data provided by the WAES national teams, together with the methods adopted to obtain these integrations. This unconstrained integration compares the available supplies of the fossil primary fuels with the preferred demand for those fuels, highlights any mismatches that may occur, and so pinpoints the strains that might arise in the supply and demand of energy under the conditions of the WAES scenarios. The data base used for this integration consists of estimates and projections from the first two technical reports of WAES, *Energy Demand Studies: Major Consuming Countries* (1972 and 1985) and *Energy Supply to the Year 2000: Global and National Studies* (including the global studies of fossil and other fuel supplies); national supply-demand integrations contained in this report; and information from other statistical sources for non-WAES countries and regions. The results are presented in the form of summary tables of regional and global energy demand and supply under the various WAES scenario assumptions and a set of supply-demand integration worksheets giving details at a regional level.

While the summary tables give a consistent and easily assimilable picture of supply and demand on a global basis, the supply-demand worksheets on which they are based are more difficult to interpret, containing as they do a large number of assumptions and conventions. After a short section on the results, an explanation of the methodology, which consists mainly of a definition of conventions, will be given.

4.2 RESULTS

The summary results of the unconstrained integrations are presented in Table 4.1 in exajoules and in Table 4.2 in MBDOE. This shows primary energy demand and supply by region and for the world outside Communist areas (WOCA), divided into the various fuels considered—coal, oil, gas, hydro, nuclear, and other. (For more exact definitions see the discussion below. The conventions used for the primary energy equivalent of hydroelectricity and nuclear power are also explained later, but since supply and demand for these are in balance, these conventions do not affect the arguments.) Ideally the base year total supply and demand should balance, as will actually happen in any practical situation. The slight imbalances here are due to statistical errors plus any errors in the data used to compile them, together with errors due to differences in conventions. (Examples of such possible sources of errors are the difficulty of taking account of stocking and destocking, the definition of bunkers, and the calorific values used for different fuels in current statistical compilations.)

Table 4.1 shows that for 1985 Cases A and B, with rising oil price and vigorous government policies, supplies of all fuels are sufficient to meet demand with ease. The WAES methodology for oil supplies assumes that supply will equal demand for as long as possible, but in these cases, as indeed with Cases C and D in 1985, there are sufficient reserves of oil for there to be unused oil supply potential. In the 1985 Cases C and D, corresponding to constant (in real terms) oil price, both coal and oil supply and demand are essentially in balance, with gas supply and demand being sufficiently close that these scenarios appear to be feasible. Finally, Case E, corresponding to a falling (in real terms) oil price and restrained policy, appears to become infeasible before 1985, with demand for every type of fossil fuel outstripping supply.

These results to 1985 formed the basis for the choice of scenarios to study to the year 2000. Since Cases A and B led to a large surplus and Case E to a large deficit in supply, these cases would already be inconsistent by 1985, with the supply surplus in Cases A and B, for example, probably causing a fall in price. Thus the requirement that only consistent scenarios be studied led to the choice of Cases C and D as those to be extended to 2000. The number of cases was extended to four by specifying coal or nuclear power as a major replacement fuel for oil; the resulting cases are labeled C-1, C-2, D-7, and D-8. The first two correspond to high world economic growth rate and rising price, the latter two to low economic growth rate and constant price, with C-1 and D-7 associated with a vigorous coal policy and C-2 and D-8 associated with a vigorous policy in favor of nuclear power. These assumptions are reflected in the figures for supply and demand in Tables 4.1 and 4.2.

The summary results for 2000 require care in their interpretation. First, it is evident that the potential supply of coal (solid fuels) substantially exceeds the preferred demand, so that a shift to solid fuels from oil and gas is possible. Second, the supply of gas is shown as being in balance with demand—actually in excess, to account for losses in international LNG transport not included elsewhere—because in 2000 the world reserves of gas are sufficient to support this level of production (see chapter 4 on natural gas in the second WAES technical report). This does not imply that the producing countries would wish to produce such quantities, but merely that it is physically possible, although it would require an international trade in gas amounting in quantity to more than half the 1976 world gas consumption. Finally, in all cases the supply of oil (liquid fuels) is far outstripped by demand. This statement requires qualification, since, using the WAES methodology (see chapter 3 on oil in the second technical report), the supply of oil is not directly related to the scenario variables that affect demand but is directly related to the discovery rate of new oil reserves and to any limits placed on production. Thus, it is possible that if a high discovery rate were achieved, producer limits were high, and economic growth were low, the higher oil supply might be available in Cases D-7 and D-8. If there were also a switch from oil to coal on the demand side and if sufficient gas to meet demand were produced, these two scenarios could be feasible.

In summary, 1985 Case E, corresponding to high world economic growth and falling oil price coupled with little government response, seems infeasible, while the 1985 cases based on constant (real) prices appear feasible, so that the cases with prices increasing by 50 percent appear unlikely. To 2000, both C-1 and C-2 appear infeasible, while D-7 and D-8 would require changes in demand preferences for types of fuels, coupled with higher discovery rates for new oil reserves than have been assumed, to remain feasible for that long.

4.3 METHODOLOGY OF REGIONAL INTEGRATIONS

In the three WAES technical reports, national teams have provided demand and supply studies, together with unconstrained national supply-demand integrations, for the various WAES scenarios covered. (But see also "Estimation Techniques," p. 86.) The individual national unconstrained supply-demand integrations provide details of the preferred use of fuel in each sector of the economy, this being the breakdown of fossil fuels expected in that country under the given WAES scenario, taking into account the price structure, energy policy, supply infrastructure, installed equipment, fuel availability, and other factors relevant to the particular country. In any such tabulation, it is necessary to apply a large number of conventions, which may differ in different countries' supply-demand integrations. Thus, the conversion to a standardized set, the conventions of which are described below, is the first step in the aggregation of such tabulations. Next, aggregation by region (North America, Europe, Japan, and Rest of WOCA) is undertaken. These regions are further defined as follows:

North America: The sum of the United States and Canada, both of which are members of WAES and both of which have supplied the data used.

Europe: The sum of the two subregions, WAES Europe and Non-WAES Europe, defined as follows:

WAES Europe: Denmark, Finland, France, Federal Republic of Germany, Italy, The Netherlands, Norway, Sweden, and the United Kingdom.

Non-WAES Europe: Austria, Greece, Ireland, Portugal, Switzerland, Belgium, Iceland, Luxembourg, and Spain.

National teams for each of the WAES Europe countries provided data used in this integration, so the supply-demand integration worksheets for WAES Europe are obtained as the sum of unconstrained supply-demand integrations for individual countries. For non-WAES Europe, a set of standardized supply-demand worksheets was prepared and aggregated for the base year, using United Nations, OECD, and other statistical sources. This allowed the preparation of a base year integration for Europe. For 1985 and 2000 integrations a procedure was adopted, described below, which assumed that non-WAES Europe changed in essentially the same way as WAES Europe.

From the 1972 base year Europe and WAES Europe integrations, a set of multipliers was computed, one for each demand sector and fuel considered, each multiplier being the ratio of the

Table 4.1
Summary results of unconstrained integrations in exajoules (10^{18} J)

	Demand				Supply and Demand			Supply			
	Coal	Oil	Gas	Total	Hydro	Nuclear	Other	Coal	Oil	Gas	Total
1972 (Base Year)											
N. America	13.99	39.19	25.92	85.23	5.40	0.73	—	15.38	29.30	25.85	76.66
Europe	10.18	27.57	5.10	47.66	3.86	0.69	0.26[a]	8.85	0.69	4.68	19.08
Japan	2.39	10.44	0.17	13.99	0.89	0.10	—	0.77	0.03	0.11	1.90
Rest of WOCA	5.74	15.31	2.52	26.10	2.48	0.05	—	6.90	68.42	9.02[b]	86.87
Bunkers	—	6.00	—	6.00	—	—	—	—	—	—	—
Total	32.30	98.51	33.71	179.0	12.63	1.57	0.26	31.90	98.44	39.66	184.5
1985A											
N. America	15.80	43.24	25.12	104.8	7.31	12.49	0.84	29.13	34.78	25.12	109.7
Europe	9.82	30.26	10.78	65.26	4.44	9.57	0.39	8.50	10.07	7.84	40.82
Japan	3.82	18.19	2.02	27.18	1.22	1.89	0.04	0.66	0.06	0.16	4.02
Rest of WOCA	10.28	31.64	8.22	57.42	4.49	2.77	0.02	12.72	85.9	18.11	124.0
Bunkers	—	7.50	—	7.50	—	—	—	—	—	—	—
Total	39.72	130.8	46.14	262.2	17.46	26.72	1.29	51.01	130.8	51.23	278.5
1985B											
N. America	12.30	40.54	22.69	96.17	7.31	12.49	0.84	26.81	34.78	25.12	107.3
Europe	8.78	25.76	10.47	57.56	4.39	7.80	0.36	8.50	10.07	7.71	38.83
Japan	3.08	13.45	1.91	21.55	1.19	1.89	0.04	0.66	0.06	0.16	3.99
Rest of WOCA	8.65	24.76	6.70	46.10	4.65	1.32	0.02	10.96	66.1	16.58	99.63
Bunkers	—	6.50	—	6.50	—	—	—	—	—	—	—
Total	32.81	111.0	41.77	227.9	17.54	23.50	1.26	46.93	111.0	49.57	249.8
1985C											
N. America	18.47	44.92	25.24	109.3	7.31	12.49	0.84	26.52	31.35	22.99	101.5
Europe	10.04	33.05	11.07	68.56	4.44	9.56	0.40	8.44	10.07	7.86	40.77
Japan	3.82	18.86	2.03	27.85	1.22	1.89	0.04	0.66	0.06	0.16	4.02
Rest of WOCA	10.85	34.01	8.55	60.69	4.49	2.77	0.02	12.72	98.12	18.43	136.6
Bunkers	—	8.80	—	8.80	—	—	—	—	—	—	—
Total	43.18	139.6	46.89	275.2	17.46	26.71	1.30	48.34	139.6	49.44	282.9
1985D											
N. America	21.45	50.37	21.54	111.4	6.19	11.55	0.36	23.32	26.57	19.38	86.37
Europe	9.36	30.17	10.67	62.69	4.39	7.73	0.37	8.03	10.07	7.58	38.16
Japan	3.08	14.34	1.91	22.45	1.19	1.89	0.04	0.66	0.06	0.16	3.99
Rest of WOCA	9.51	28.63	7.06	51.18	4.65	1.32	0.02	10.96	93.8	16.94	127.7
Bunkers	—	7.00	—	7.00	—	—	—	—	—	—	—
Total	43.40	130.5	41.18	254.8	16.42	22.49	0.79	41.97	130.5	44.06	256.2
1985E											
N. America	23.46	60.43	23.63	126.2	6.63	11.73	0.36	22.19	21.45	17.17	79.53
Europe	9.29	40.99	11.63	75.61	4.39	8.93	0.38	8.03	10.07	7.75	39.55
Japan	3.82	20.25	2.03	29.24	1.22	1.89	0.04	0.66	0.06	0.16	4.03
Rest of WOCA	10.41	40.63	8.96	66.77	4.49	2.26	0.02	12.72	113.0	18.84	151.3
Bunkers	—	10.00	—	10.00	—	—	—	—	—	—	—
Total	46.98	172.3	46.25	307.9	16.73	24.81	0.80	43.60	144.6	43.92	274.4

Table 4.1 (continued)
Summary results of unconstrained integrations in exajoules (10^{18} J)

	Demand				Supply and Demand			Supply			
	Coal	Oil	Gas	Total	Hydro	Nuclear	Other	Coal	Oil	Gas	Total
2000 C-1											
N. America	34.65	50.26	23.99	155.6	8.97	30.50	7.23	53.80	26.99	17.77	145.2
Europe	18.72	43.36	13.69	99.83	4.78	17.69	1.59	9.33	6.44	4.49	44.33
Japan	5.71	34.84	3.52	50.51	1.86	4.39	0.20	0.66	0.84	0.17	8.12
Rest of WOCA	16.59	66.05	20.63	123.7	10.25	10.16	0.05	27.62	127.3	44.09	219.5
Bunkers	—	12.10	—	12.10	—	—	—	—	—	—	—
Total	75.67	206.6	61.83	441.8	25.86	62.74	9.07	91.41	161.6	66.52	417.2
2000 C-2											
N. America	30.48	50.92	24.39	164.3	8.97	42.34	7.23	38.82	26.99	17.77	142.1
Europe	12.23	42.93	11.69	104.3	4.61	31.61	1.24	7.51	6.44	4.49	55.90
Japan	5.71	33.07	3.52	50.64	1.86	6.28	0.20	0.66	0.84	0.17	10.01
Rest of WOCA	16.97	65.43	18.29	130.2	8.55	20.95	0.05	22.9	127.3	39.75	219.5
Bunkers	—	12.10	—	12.10	—	—	—	—	—	—	—
Total	65.39	204.5	57.89	461.6	23.99	101.2	8.72	69.89	161.6	62.18	427.5
2000 D-7											
N. America	25.48	54.39	21.51	142.1	7.22	30.50	3.04	40.96	19.14	13.76	114.6
Europe	13.05	35.15	12.35	82.06	4.78	15.49	1.24	8.56	5.79	4.31	40.18
Japan	4.47	18.97	3.52	33.32	1.76	4.39	0.20	0.66	0.84	0.17	8.02
Rest of WOCA	13.58	46.34	15.25	89.43	7.24	6.99	0.05	23.5	103.8	39.18	180.76
Bunkers	—	10.00	—	10.00	—	—	—	—	—	—	—
Total	56.58	164.9	52.63	356.9	21.00	57.37	4.53	73.68	129.6	57.42	343.6
2000 D-8											
N. America	21.72	54.43	21.52	148.0	7.22	40.10	3.04	31.89	19.14	13.76	115.2
Europe	9.40	33.62	10.23	84.68	4.61	25.83	1.00	6.89	5.79	4.31	48.43
Japan	3.50	18.48	3.52	33.74	1.76	6.28	0.20	0.66	0.84	0.17	9.91
Rest of WOCA	12.68	46.72	12.77	92.85	5.96	14.67	0.05	19.57	103.8	34.06	178.1
Bunkers	—	10.00	—	10.00	—	—	—	—	—	—	—
Total	47.30	163.3	48.04	369.3	19.55	86.88	4.29	59.01	129.6	52.30	351.6

[a]Includes 0.05 EJ exported.
[b]Includes 5.64 EJ flared or vented.

Note: Oil supply for Cases A, B, and E is assumed as given in the chapter on oil in the WAES Second Technical Report for Case C, except that an OPEC limit of 40 MBDOE (89.3 EJ/yr) is assumed in Case E. In the 2000 cases, OPEC oil production is taken as 45 MBDOE (100.5 EJ/yr) for Cases C-1 and C-2 and 39 MBDOE (87.1 EJ/yr) for Cases D-7 and D-8.

Gas supply for 1985 from Rest of WOCA is assumed to meet internal demand plus maximum potential exports (3.6 MBDOE—see the chapter on gas in the WAES second report) plus losses in transport (25 percent for LNG, 10 percent for pipeline). For 2000 it is assumed to be large enough to satisfy Rest of WOCA internal demand, import requirements of the other regions, and losses (assumed at 25 percent) on those import requirements. See the chapter on gas in the second report for possible constraints.

Coal supply for Rest of WOCA is taken as the production given in the coal chapter of the second report: 433 × 10^6 tonnes coal equivalent (12.72 EJ) for Case C (also assumed for Cases A and E) and 373 × 10^6 tonnes coal equivalent (10.96 EJ) for Cases D and B. Coal supply for Rest of WOCA in 2000 is from the coal chapter of the second report.

This table assumes zero net imports to or exports from Communist areas. WAES estimates for the possible exports to WOCA from these areas are

oil: 2 MBDOE = 4.47 EJ in all cases 1985 and 2000

gas: 0.5 MBDOE = 1.12 EJ in all 1985 cases
1.0 MBDOE = 2.23 EJ in all 2000 cases

coal: 0.5 MBDOE = 1.12 EJ in all cases 1985 and 2000

Such exports would reduce the exports required from Rest of WOCA by the same amount.

Table 4.2
Summary results of unconstrained integrations in MBDOE

	Demand				Supply and Demand			Supply			
	Coal	Oil	Gas	Total	Hydro	Nuclear	Other	Coal	Oil	Gas	Total
1972 (Base Year)											
N. America	6.27	17.55	11.61	38.17	2.42	0.33	—	6.89	13.12	11.58	34.33
Europe	4.56	12.35	2.29	21.34	1.73	0.31	0.12[a]	3.96	0.31	2.10	8.54
Japan	1.07	4.68	0.08	6.27	0.40	0.04	—	0.34	0.01	0.05	0.85
Rest of WOCA	2.57	6.86	1.13	11.69	1.11	0.02	—	3.09	30.64	4.04[b]	38.90
Bunkers	—	2.69	—	2.69	—	—	—	—	—	—	—
Total	14.47	44.13	15.11	80.19	5.66	0.70	0.12	14.28	44.08	17.77	82.61
1985A											
N. America	7.07	19.37	11.25	46.93	3.27	5.59	0.38	13.05	15.58	11.25	49.11
Europe	4.40	13.55	4.83	29.23	1.99	4.29	0.17	3.81	4.51	3.51	18.28
Japan	1.71	8.15	0.91	12.17	0.55	0.85	0.02	0.29	0.03	0.07	1.80
Rest of WOCA	4.60	14.17	3.68	25.71	2.01	1.24	0.01	5.70	38.47	8.11	55.53
Bunkers	—	3.36	—	3.36	—	—	—	—	—	—	—
Total	17.78	58.60	20.67	117.4	7.82	11.97	0.58	22.85	58.59	22.94	124.7
1985B											
N. America	5.51	18.16	10.16	43.07	3.27	5.59	0.38	12.01	15.58	11.25	48.07
Europe	3.93	11.53	4.69	25.78	1.97	3.49	0.16	3.81	4.51	3.45	17.39
Japan	1.38	6.02	0.85	9.65	0.53	0.85	0.02	0.30	0.03	0.07	1.79
Rest of WOCA	3.87	11.09	3.00	20.64	2.08	0.59	0.01	4.91	29.60	7.42	44.61
Bunkers	—	2.91	—	2.91	—	—	—	—	—	—	—
Total	14.69	49.71	18.70	102.0	7.85	10.52	0.57	21.03	49.72	22.19	111.9
1985C											
N. America	8.27	20.12	11.30	48.93	3.27	5.59	0.38	11.87	14.04	10.30	45.45
Europe	4.50	14.80	4.96	30.70	1.99	4.28	0.18	3.78	4.51	3.52	18.26
Japan	1.71	8.45	0.91	12.47	0.55	0.85	0.02	0.29	0.03	0.07	1.80
Rest of WOCA	4.86	15.23	3.83	27.18	2.01	1.24	0.01	5.70	43.94	8.26	61.15
Bunkers	—	3.94	—	3.94	—	—	—	—	—	—	—
Total	19.34	62.54	21.00	123.2	7.82	11.96	0.59	21.64	62.52	22.15	126.7
1985D											
N. America	9.60	22.56	9.64	49.91	2.77	5.17	0.16	10.00	11.90	8.68	38.68
Europe	4.19	13.51	4.78	28.07	1.96	3.46	0.16	3.59	4.51	3.39	17.09
Japan	1.38	6.42	0.85	10.05	0.53	0.85	0.02	0.30	0.03	0.07	1.79
Rest of WOCA	4.26	12.82	3.16	22.92	2.08	0.59	0.01	4.91	42.01	7.59	57.18
Bunkers	—	3.13	—	3.13	—	—	—	—	—	—	—
Total	19.43	58.44	18.43	114.1	7.34	10.07	0.35	18.80	58.45	19.73	114.7

Table 4.2 (continued)
Summary results of unconstrained integrations in MBDOE

	Demand				Supply and Demand			Supply			
	Coal	Oil	Gas	Total	Hydro	Nuclear	Other	Coal	Oil	Gas	Total
1985E											
N. America	10.51	27.06	10.58	56.53	2.97	5.29	0.16	9.94	9.61	7.69	35.61
Europe	4.16	18.36	5.21	33.86	1.97	4.00	0.17	3.59	4.51	3.47	17.71
Japan	1.71	9.07	0.91	13.09	0.55	0.85	0.02	0.30	0.03	0.07	1.80
Rest of WOCA	4.66	18.20	4.01	29.90	2.01	1.01	0.01	5.70	50.60	8.44	67.77
Bunkers	—	4.48	—	4.48	—	—	—	—	—	—	—
Total	21.04	77.17	20.71	137.9	7.50	11.15	0.36	19.53	64.75	19.67	122.9
2000 C-1											
N. America	15.52	22.51	10.74	69.68	4.02	13.66	3.24	24.09	12.09	7.96	65.05
Europe	8.38	19.42	6.13	44.71	2.14	7.92	0.71	4.18	2.88	2.01	19.85
Japan	2.56	15.60	1.57	22.62	0.83	1.97	0.09	0.30	0.37	0.07	3.64
Rest of WOCA	7.43	29.58	9.24	55.41	4.59	4.55	0.02	12.37	57.0	19.74	98.28
Bunkers	—	5.42	—	5.42	—	—	—	—	—	—	—
Total	33.89	92.53	27.68	197.8	11.58	28.10	4.06	40.94	72.34	29.78	186.8
2000 C-2											
N. America	13.65	22.80	10.92	73.59	4.02	18.96	3.24	17.39	12.09	7.96	63.64
Europe	5.48	19.23	5.23	46.71	2.06	14.16	0.56	3.36	2.88	2.01	25.03
Japan	2.56	14.81	1.57	22.68	0.83	2.81	0.09	0.30	0.37	0.07	4.48
Rest of WOCA	7.60	29.30	8.19	58.32	3.83	4.38	0.02	10.26	57.00	17.80	98.29
Bunkers	—	5.42	—	5.42	—	—	—	—	—	—	—
Total	29.29	91.56	25.91	206.7	10.74	45.31	3.91	31.31	72.34	27.84	191.4
2000 D-7											
N. America	11.41	24.36	9.63	63.65	3.23	13.66	1.36	18.34	8.57	6.16	51.33
Europe	5.84	15.74	5.53	36.75	2.14	6.94	0.56	3.83	2.59	1.93	17.99
Japan	2.00	8.50	1.57	14.92	0.79	1.97	0.09	0.30	0.37	0.07	3.59
Rest of WOCA	6.08	20.75	6.83	40.05	3.24	3.13	0.02	10.55	46.50	17.55	80.99
Bunkers	—	4.48	—	4.48	—	—	—	—	—	—	—
Total	25.33	73.83	23.56	159.9	9.40	25.70	2.03	33.02	58.03	25.71	153.9
2000 D-8											
N. America	9.73	24.37	9.64	66.29	3.23	17.96	1.36	14.28	8.57	6.16	51.57
Europe	4.21	15.06	4.58	37.92	2.06	11.57	0.45	3.08	2.59	1.93	21.68
Japan	1.57	8.27	1.57	15.11	0.79	2.81	0.09	0.30	0.37	0.07	4.44
Rest of WOCA	5.68	20.92	5.72	41.58	2.67	6.57	0.02	8.76	46.50	15.25	79.76
Bunkers	—	4.48	—	4.48	—	—	—	—	—	—	—
Total	21.19	73.10	21.51	165.4	8.75	38.91	1.92	26.42	58.03	23.51	157.4

[a] Includes 0.02 MBDOE exported.
[b] Includes 2.53 MBDOE flared or vented.

See note to Table 4.1.

demand in Europe to that in WAES Europe for that sector and fuel. The multipliers obtained for the fossil fuels (all of order 1.2) were then applied to the 1985 and 2000 WAES Europe fossil fuel demands to generate a set of fossil fuel demands for Europe in the 1985 and 2000 cases. For nonenergy uses of coal, the factor was chosen as 1.1. The WAES Europe sector demands for electricity were all multiplied by a factor, depending on the particular scenario case, to obtain the Europe sector demands. The multipliers used are shown in Table 4.3. This procedure defined the sectoral demands in Europe, and hence the final energy demand.

The electricity generation sector of non-WAES Europe could then be defined by first assuming that the electricity losses in Europe were the same fraction of final demand as in WAES Europe. Adding these losses to final electricity demand gave the electricity generated in non-WAES Europe as the difference between electricity generated in Europe and in WAES Europe. The amount of electricity generated by hydro power and nuclear power in non-WAES Europe was estimated by reference to the base year for hydro power and by assuming a load factor of 65 percent for nuclear capacity estimates for non-WAES Europe (41 GWe for Cases A and C, 25 GWe for Cases B and D, 35 GWe for Case E, 76 GWe for Cases C-1 and D-7, and 122 GWe for Cases C-2 and D-8; the Case E figure is the same fractional distance between A and C and B and D as occurs in WAES Europe). Because these figures are earlier estimates, they differ from those in the chapter on nuclear energy in the second technical report. This left the quantity of electricity to be generated by fossil fuels at an assumed efficiency of 35 percent (1985) or 37.5 percent (2000), the breakdown into different fossil fuels being taken from the base year for the 1985 cases (0.359 coal, 0.48 petroleum, 0.161 gas) and equally split between coal and petroleum in the 2000 cases. The multipliers for electricity demand in Table 4.3 were chosen to give a total fossil fuel burn in power stations in non-WAES Europe within a factor of 1.5 of the 1972 base year figure. For comparison, the equivalent 1972 multiplier is also given.

The demand and conversion sectors of Europe were completed by assuming no production of synthetic gas and no demand for, or public supply of, heat in non-WAES Europe, so the figures given in the tables for these energy forms are only those for WAES Europe. Similarly, non-WAES Europe was assumed to have no demand for or supply of geothermal and other energies.

The losses in the various fuel supply industries of Europe were calculated by assuming that they would be the same fraction of primary energy input as obtained for WAES Europe.

The indigenous supply of fossil fuels for non-WAES Europe was estimated country by country from the known 1972 supplies. In every case these supplies were small, so for 1985 and 2000 they were assumed in every case to be the 1972 figures increased by about 20 percent.

This procedure completely defines the methods used to estimate a supply-demand integration for Europe from that for WAES Europe. Applying it to the WAES scenarios gives the results in Tables 4.1–4.3.

Japan: The supply-demand integrations were supplied by the WAES national team for Japan except for any standardization of conventions (see section 4.4.2).

Rest of WOCA: This consists of all countries in the rest of the world outside Communist areas that are not included in the three regions defined above. It is subdivided into Less Developed Countries (LDCs) and Australia, New Zealand, and South Africa.

Table 4.3
Demand multipliers for the estimation of total Western European demand from WAES European demand

	(a) Coal	(b) Petroleum	(d) Natural gas	(g) Electricity
(1) Transport	1.31	1.23	—	1.19 1972
(2) Industry	1.18	1.23	1.30	1.20 1985 B, D, E
(3) Agriculture, Mining, and Construction				1.22 1985 A, C
(4) Commercial	1.14	1.17	1.06	1.20 2000 C-1
(5) Public				1.25 2000 C-2, D-7
(6) Residential				1.30 2000 D-8
(7) Nonenergy Uses	1.10	1.11	1.03	

(Rows 3–6 are bracketed together for coal, petroleum and natural gas; rows 1–7 are bracketed together for electricity.)

The base year (1972) supply-demand integrations for the LDCs are based on the UN series J statistics together with World Bank statistics on India and data provided by WAES Associates. Energy demand in LDCs in 1985 Cases C and D and in the 2000 cases are based on a special study prepared for WAES by the World Bank. This special study is reported in chapter 6.

The 1972 and 1985 Case C supply-demand integrations for Australia, New Zealand, and South Africa were provided by WAES Associates. Case D was estimated from C by scaling in proportion to the GNP growth rates in the two cases, while the 2000 cases were derived by using energy to GNP ratios.

Addition of the above resulted in 1985 Cases C and D together with all the 2000 cases for the rest of WOCA. The 1985 Cases A, B, and E were derived from Cases C and D as follows: All sector demands were changed in the same ratio as occurred in Europe, using C as a base for A and E, and D as a base for B. Coal demand for transport and synthetic gas, synthetic crude, and hydroelectric production were assumed to be the same for Cases A, E, and C, and for B and D, while nuclear electric production was assumed to shift in the same way as in Europe. Electricity generated from fossil fuel was supplied from coal, oil, and gas in the same proportions as in the relevant base case (C or D), assuming a generation efficiency of 35 percent.

The supplies of fossil fuel from the rest of WOCA were obtained in a manner differing from that used for other regions (where they are the sum of WAES national teams' estimates). The base year (1972) figures were derived from published statistics—mainly UN series J and U.S. Bureau of Mines. For 1985 and 2000, the oil supply figures were derived by using the WAES oil supply methodology described in chapter 3 of the second technical report. Coal supply figures were obtained from chapter 5 of that report. Finally, for gas, following the arguments given in chapter 4 in the second technical report, the supply available in 1985 was set by expected LNG and pipeline projects, while for 2000 it was assumed that sufficient supplies would be available to meet both internal demand and the import requirements of the consuming regions. So the production for the rest of WOCA was assumed to be sufficient to supply these demands with losses in liquefying, shipping, and regasifying LNG (25 percent of the amount shipped), although such large supplies may not be forthcoming.

The supply-demand integration worksheets are presented in section 4.5. Interpretation of these sheets requires a knowledge of the conventions assumed for them, which are described below.

4.4 INTERPRETATION OF AND CONVENTIONS FOR THE SUPPLY-DEMAND WORKSHEETS

4.4.1 Interpretation

The supply-demand integration worksheets are designed to provide a summary of the flow of energy in an economy. The economy is divided into a set of separate sectors, which are listed in the left-hand column, so that each row of the worksheet corresponds to a sector of the economy and every activity in the economy is incorporated in one row. Lines 1 to 7 list the demand sectors, line 8 is the final energy demand (the sum of lines 1 to 7), while lines 9 to 13 list the energy conversion industries considered in this study. Line 13, labeled "Energy Sector Self-Consumption and Conversion Losses," is essentially a summary of the primary energy industries (this is clarified below). The columns of the table list the types of fuel considered in the analysis.

Each entry in the worksheet in lines 1 to 7, columns a to k, represents the *net* flow of the type of fuel *into* the economic sector. Thus, a positive number represents a net consumption by the sector, a negative number a net production, but the actual flows into and out of any given sector are not separately recorded. Similar definitions hold in the energy conversion sectors (lines 9 to 13), where negative numbers occur more frequently since the objective of these sectors is to produce one fuel from others.

As mentioned above, line 8 is the final energy demand, the sum of lines 1 to 7, and thus represents the net demand for energy by all the demand sectors, which has to be supplied by the conversion industries. Adding to this the net demand of the conversion industries, lines 9 to 13, then gives the total primary energy input (line 14), which is the net demand for energy of the whole economy, i.e., the net flow of energy into the country or region considered. Finally, the difference between primary energy input and the indigenous supply (line 15), the amount of energy available as raw fuel within the region considered, gives the net imports or exports (lines 16 and 17).

It should be noted that every number entered on the worksheet is the net flow of energy into a sector. The worksheet gives no information on the total inflows and outflows except their arithmetic difference, and it also gives no information on how the energy is used within the sector. The conventions used to define each sector of the economy thus may affect the values recorded in the worksheet. A good example of this is the convention for electricity generated by private concerns within industry—which could be included in the industrial sector of the economy (as in this compilation) or within the electricity generation sector. A description of the major conventions used is given below.

4.4.2 Conventions

Units

The units used throughout this chapter are exajoules (10^{18} joules) in most instances, and millions of barrels per day of oil equivalent (MBDOE) for one summary table. Some conversion factors, which may be useful for those accustomed to other units, are given below.

1 exajoule = 0.948 Quads

1 Quad = 1.055 exajoules

1 exajoule = 22.7 TOE (tonnes of oil equivalent)

1 TOE = 0.044 exajoules

1 exajoule = 0.163×10^9 BOE (barrels of oil equivalent)

10^9 BOE = 6.12 exajoules

1 exajoule/year = 0.448 MBDOE

1 MBDOE = 2.233 exajoules/year

These conversions assume 1 BOE = 5.8×10^6 Btu and 1 TOE = 44 GJ.

Economic Sectors

The sectors of the economy were chosen to conform as closely as national statistics in the WAES countries allowed to the International Standard Industrial Classification. At the level of aggregation employed in the supply-demand integrations, the classification is fairly self-evident. The worksheets in this chapter are aggregations of data from many countries, and minor discrepancies between individual country classifications should have a negligible effect on the totals. The worksheets in this chapter were compiled from a set of individual country worksheets edited to conform (as far as possible) to the following specific conventions:

- **Transport (Line 1):** Excludes international bunkers, which are included separately in the world summaries (Table 4.1), where they are estimated for 1985 and 2000 as approximately proportional to total world trade. The 1972 figures for bunkers come from published statistics (UN series J).
- **Nonenergy Uses (Line 7):** Includes all fuel products not used as fuels, including chemical feedstocks, lubricants, bitumens, wax, etc. It excludes the coal used in steel making, which is included in the industrial sector.
- **Electricity (Line 9):** The public supply industry, or the set of utilities with a major concern in producing and selling electricity (but see also hydro energy below). It excludes any electricity generated by industry for its own consumption (since this is an activity entirely within the industrial sector). This sector may also produce by-product heat. Its output is considered to be all the electricity actually generated within the industry ("at the busbar").
- **Syn. Gas (Line 10):** The public supply industry, or the set of utilities with a major concern in producing gas from other fuels and selling it. This row generally excludes the production of blast furnace and coke oven gas except insofar as these may be inputs to gas works, but it should be noted that it was not possible to apply this convention consistently. Thus the synthetic gas sectors in the worksheets may include a component of production of blast furnace and coke oven gas that should have been included in the industrial sector. This explains the apparently high efficiencies of production of synthetic gas in some cases, the inefficiencies being included in industrial demand.
- **Syn. Liquids (Line 11):** The public supply industry or the set of utilities with a major concern in producing liquid fuels from solid or gaseous fuels and selling them.
- **Heat (Line 12):** The public supply industry or the set of utilities whose major product is heat, generally in the form of steam or hot water. A by-product of this sector may be electricity.
- **Energy Sector Self-Consumption and Conversion Losses (Line 13):** Included in this sector are the net demands of the primary energy supply indus-

tries and the losses in transmission within the electricity, gas, and heat supply industries. The inputs to the primary energy supply industries (coal, petroleum, and gas) are considered to be the production at mine mouth or wellhead, or imports at the frontier or port, while the outputs are the fuels supplied to final energy demand and to the conversion sectors. Thus, line 13 records the following:

Coal (Column a): Losses in transport and handling, coal used in mines and in coal industry establishments, as well as coal used in the oil and gas industries (assuming this is not recorded in lines 10 and 11). Coal used in the electricity sector is usually included in line 9.

Petroleum (Oil) (Columns b and c): Losses in transport and handling, oil used in oil fields, refineries, and oil industry establishments, and oil used in coal and gas industries (with the same caveat as for coal). Oil used in synthetic liquids plants should be included in line 11.

Gas (Columns d and e): Gas used in gas fields and in gas industry establishments (except that used in synthetic gas plants, which is included in line 10), losses in the distribution of gas to final users, and gas used in the coal and oil industries (assuming this is not recorded in line 11).

Heat (Column f): This records the distribution losses—the difference between heat produced and that consumed by final demand.

Electricity (Column g): Losses in distribution, the amount used in the electricity industry (including use at power stations), and the amount supplied to the coal, oil, and gas industries.

Types of Fuel

The types of fuel considered were aggregated with those shown on the supply-demand worksheets. In every case the conversion to the units used here (exajoules) was done at the national level using calorific values appropriate to the fuel considered in the country concerned, so that there is no single calorific value to be quoted for any particular fuel. Other conventions are as follows:

- Coal means coal-derived solid fuels including hard coal, brown coal, coke, patent fuels, briquettes, etc.
- Petroleum is synonymous with oil in this chapter and means all liquid fuels including petroleum products, natural gas liquids, and coal-derived liquid fuels. Demand for synthetic liquids was not distinguished from nonsynthetic liquids in these summary tables, so column c is merged with column b.
- Natural gas includes all types of gaseous fuels. The synthetic gaseous fuels are not distinguished in these tables, so column e is merged with column d.
- Heat is generally steam or hot water.
- Hydro energy records the use of hydro energy for the production of electricity only. In a few cases there are privately owned hydro power schemes, but these are included in the electricity supply sector. The convention used in these worksheets is to record an equivalent primary energy for the hydroelectric energy produced, by assuming an efficiency of 30 percent in 1972, 35 percent in 1985, and 37.5 percent in 2000. Since supply and demand for hydro energy are equal and since it is the electrical output of hydro schemes that is estimated in the supply studies, this convention does not affect the demand for fossil fuels. It is used merely to give some measure of the amount of primary fossil fuel that would be required to substitute for hydro energy in electricity generation. Similarly, the efficiencies chosen are conventions and do not imply that these are the actual fossil-fuel-to-electricity conversion efficiencies in those years.
- Nuclear energy conventions are similar to those for hydro energy (although in some cases nuclear energy provides a small amount of heat). It was considered more convenient to use conventional rather than actual efficiencies, again to give a measure of the amount of fossil fuel substituted by nuclear energy. Actual thermal efficiencies of nuclear reactors may well be substantially different. Where supply studies have not specified the output of nuclear electricity, it has been assumed that the estimated amount of installed capacity (see chapter 6 on nuclear energy in the second WAES technical report) could operate at a load factor of 65 percent. Nuclear energy produced within a country or region was recorded as an indigenous supply, even though uranium and the rest of the fuel cycle might be imported.
- Geothermal energy and other includes solar energy (where explicitly included in the studies), hot water, and steam from geothermal sources and various commercially traded, indigenous fuels. In the electricity generation sector, the conventions for efficiency noted under hydro energy (above)

were used to obtain an equivalent primary energy.

Estimation Techniques

As noted in the section on methodology, WAES national teams provided the national unconstrained supply-demand integrations that formed the data base for the regional and global integrations of Table 4.1 and section 4.5. However, in some cases the complete supply-demand integration worksheet was not provided for every scenario studied. In such cases, the minimum data available were the demand, by sector, for electricity and for all fossil fuels combined. The most common scenario cases with such limited information were 1985 A, B, and E. The following technique was adopted to estimate the complete supply-demand worksheet in such cases.

For each worksheet to be estimated, the closest worksheet (with all data known) of the same country was chosen as a reference. "Closest" refers to similarity of scenario variables, so that 1985 Cases A, C, and E were grouped (usually Case C was the reference), similarly 1985 Cases B, D (D as reference), 2000 C-1, D-7 (either as reference), and 2000 C-2, D-8 (either as reference). The demand sectors were estimated by assuming the given demands for electricity, heat, and "other" fuels (in the geothermal and other column) equal to the reference case, and splitting the remaining (known) total fossil fuel demand among coal, oil, and gas in the same ratio as in the corresponding sector of the reference case. This gave a worksheet completed down to the final energy demand (line 8).

By assuming the same percentage losses in electricity as in the reference case, the total electricity generated was fixed. The hydroelectricity and nuclear electricity generated (together with any other unconventional generation) were found from the WAES supply studies, with the remainder of the specified electricity demand to be generated by fossil fuels. The fossil fuel electrical generation efficiency implied by the reference case (or explicitly stated in the supply studies) together with the relative quantities of each fuel used in electricity generation, again from the reference case, then allowed complete definition of the electricity sector. Heat, synthetic gas, and synthetic liquids sectors were assumed unchanged from the reference case, while the losses were taken as the same percentage of primary energy input.

Last, the indigenous supplies of fossil fuels could be found from the supply studies, thus completing the estimation procedure.

4.5 SUPPLY-DEMAND INTEGRATION WORKSHEETS

This section consists of forty summary worksheets for the WAES constrained global supply-demand integrations. The sheets are arranged in the following order:

Base year (1972)

1985 cases: A, B, C, D, E

2000 cases: C-1, C-2, D-7, D-8

For each of the cases considered there are four sheets: one each for North America, Europe, Japan, and the Rest of WOCA.

NATIONAL INPUT WORKSHEET FOR SUPPLY/DEMAND INTEGRATION

Country: NORTH AMERICA

Year: 1972

Case: BASE YEAR

Units: EXAJOULES(10^{18} J)

	(a) Coal	(b) Petro-leum	(c) (Syn. Liquids)	(d) Nat. Gas	(e) (Syn. Gas)	(f) (Heat)	(g) (Electri-city)††	(h) Hydro Energy	(i) Nuclear Energy	(j) Geothermal energy and other	(k) Total
(1) Transport	0.01	19.32					0.01				19.34
(2) Industry (3) Agric., Mining, Construction	4.37	3.72		8.18			2.42				18.69
(4) Commercial (5) Public	0.21	1.34		2.04			1.70				5.28
(6) Residential	0.01	4.68		6.31			2.16				13.17
(7) Non-energy Uses	0.13	3.91		0.71							4.74
(8) Final Energy Demand*	4.72	32.96		17.24			6.30				61.22
(9) Electricity**	8.56	3.50		4.44			-7.45	5.40	0.73		15.19
(10) Syn. Gas											
(11) Syn. Liquids											
(12) Heat											
(13) Energy Sector Self Consumption & conversion losses	0.71	2.72		4.24			1.16				8.83
(14) Primary Energy Input	13.99	39.19		25.92				5.40	0.73		85.23
(15) Indigenous Supply	15.38	29.30		25.85				5.40	0.73		76.66
(16) Imports***		9.89		0.07							8.57
(17) Exports	1.39										

*Includes non-energy uses.

**Blocks 9g, 10e 11c, and 12f must be negative to avoid double counting of energy from primary fuels.

***Includes imported products.

††Includes only electricity to be purchased by each sector.

NATIONAL INPUT WORKSHEET FOR SUPPLY/DEMAND INTEGRATION

Country: EUROPE

Year: 1972

Case: BASE YEAR

Units: EXAJOULES (10^{18}J)

	(a)	(b)	(c)	(d)	(e)	(f)	(g)	(h)	(i)	(j)	(k)
	Coal	Petro-leum	(Syn. Liquids)	Nat. Gas	(Syn. Gas)	(Heat)	(Electri-city)††	Hydro Energy	Nuclear Energy	Geothermal energy and other	Total
(1) Transport	0.06	6.44		0.01			0.11				6.62
(2) Industry	2.51	5.85		2.18		0.04	1.92			0.20	12.70
(3) Agric., Mining, Construction }											
(4) Commercial }	0.32	2.90		0.45		0.04	0.67			0.02	4.40
(5) Public }											
(6) Residential	1.31	4.39		1.44		0.19	1.04			0.05	8.41
(7) Non-energy Uses	0.07	2.30		0.33							2.70
(8) Final Energy Demand*	4.26	21.88		4.41		0.27	3.74			0.26	34.82
(9) Electricity**	4.67	3.50		1.12			-4.40	3.86	0.69	0.04	9.48
(10) Syn. Gas	0.70	0.12		-0.79							0.03
(11) Syn. Liquids											
(12) Heat	0.17	0.17				-0.29	-0.01				0.04
(13) Energy Sector Self Consumption & conversion losses	0.39	1.91		0.37		0.02	0.61				3.30
(14) Primary Energy Input	10.18	27.57		5.10			-0.05	3.86	0.69	0.30	47.66
(15) Indigenous Supply	8.85	0.69		4.68				3.86	0.69	0.30	19.08
(16) Imports***	1.33	26.88		0.42							28.59
(17) Exports							0.05				

*Includes non-energy uses.

**Blocks 9g, 10e 11c, and 12f must be negative to avoid double counting of energy from primary fuels.

***Includes imported products.

††Includes only electricity to be purchased by each sector.

NATIONAL INPUT WORKSHEET FOR SUPPLY/DEMAND INTEGRATION

Country: JAPAN

Year: 1972

Case: BASE YEAR

Units: EXAJOULES (10^{18} J)

	(a)	(b)	(c)	(d)	(e)	(f)	(g)	(h)	(i)	(j)	(k)
	Coal	Petro-leum	(Syn. Liquids)	Nat. Gas	(Syn. Gas)	(Heat)	(Electri-city)††	Hydro Energy	Nuclear Energy	Geothermal energy and other	Total
(1) Transport		1.76					0.04				1.80
(2) Industry	1.82	3.12		0.09			0.70				5.73
(3) Agric., Mining, Construction	0.03	0.48					0.03				0.54
(4) Commercial } (5) Public }	0.04	0.70		0.04			0.21				0.99
(6) Residential	0.03	0.63		0.16			0.16				0.98
(7) Non-energy Uses		1.14									1.14
(8) Final Energy Demand*	1.90	7.83		0.30			1.15				11.18
(9) Electricity**	0.24	2.14		0.06			-1.27	0.89	0.10		2.15
(10) Syn. Gas	0.11	0.11		-0.21							0.00
(11) Syn. Liquids											
(12) Heat											
(13) Energy Sector Self Consumption & conversion losses	0.15	0.36		0.03			0.13				0.66
(14) Primary Energy Input	2.39	10.44		0.17				0.89	0.10		13.99
(15) Indigenous Supply	0.77	0.03		0.11				0.89	0.10		1.90
(16) Imports***	1.63	10.41		0.06							12.09
(17) Exports											

*Includes non-energy uses.

**Blocks 9g, 10e 11c, and 12f must be negative to avoid double counting of energy from primary fuels.

***Includes imported products.

††Includes only electricity to be purchased by each sector.

NATIONAL INPUT WORKSHEET FOR SUPPLY/DEMAND INTEGRATION

Country: REST OF WOCA

Year: 1972

Case: BASE YEAR

Units: EXAJOULES (10^{18}J)

	(a)	(b)	(c)	(d)	(e)	(f)	(g)	(h)	(i)	(j)	(k)
	Coal	Petroleum	(Syn. Liquids)	Nat. Gas	(Syn. Gas)	(Heat)	(Electricity)††	Hydro Energy	Nuclear Energy	Geothermal energy and other	Total
(1) Transport	0.60	5.54		0.02							6.16
(2) Industry; (3) Agric., Mining, Construction	2.52	3.57		1.05			1.05				8.20
(4) Commercial; (5) Public; (6) Residential	0.45	2.12		0.31			0.60				3.48
(7) Non-energy Uses		0.98									0.98
(8) Final Energy Demand*	3.57	12.22		1.39			1.65				18.83
(9) Electricity**	1.94	1.85		0.31			-1.95	2.48	0.05		4.69
(10) Syn. Gas	0.05	0.02		-0.07							
(11) Syn. Liquids	0.05	-0.01									0.04
(12) Heat											
(13) Energy Sector Self Consumption & conversion losses	0.13	1.23		0.89			0.29				2.55
(14) Primary Energy Input	5.74	15.31		2.52				2.48	0.05		26.10
(15) Indigenous Supply	6.90	68.42		9.02[1]				2.48	0.05		86.87
(16) Imports***											
(17) Exports	1.16	53.12		0.85							55.13

*Includes non-energy uses.

**Blocks 9g, 10e 11c, and 12f must be negative to avoid double counting of energy from primary fuels.

***Includes imported products.

††Includes only electricity to be purchased by each sector.

Footnotes:

[1] of which 5.64 was flared or vented

NATIONAL INPUT WORKSHEET FOR SUPPLY/DEMAND INTEGRATION

Country: NORTH AMERICA

Year: 1985

Case: A

Units: EXAJOULES(10^{18}J)

	(a) Coal	(b) Petro-leum	(c) (Syn. Liquids)	(d) Nat. Gas	(e) (Syn. Gas)	(f) (Heat)	(g) (Electri-city)††	(h) Hydro Energy	(i) Nuclear Energy	(j) Geothermal energy and other	(k) Total
(1) Transport		21.57					0.02				21.59
(2) Industry, (3) Agric., Mining, Construction	8.66	3.63		10.85			3.59				26.72
(4) Commercial, (5) Public, (6) Residential	0.01	4.64		7.96			5.75			0.21	18.57
(7) Non-energy Uses	0.27	7.72		1.65							9.63
(8) Final Energy Demand*	8.94	37.56		20.45			9.36			0.21	76.52
(9) Electricity**	6.62	1.98		2.47			-10.89	7.31	12.49	0.63	20.61
(10) Syn. Gas											
(11) Syn. Liquids											
(12) Heat											
(13) Energy Sector Self Consumption & conversion losses	0.24	3.71		2.19			1.53				7.67
(14) Primary Energy Input	15.80	43.24		25.12				7.31	12.49	0.84	104.8
(15) Indigenous Supply	29.13	34.78		25.12				7.31	12.49	0.84	109.7
(16) Imports***		8.46									
(17) Exports	13.33										4.87

*Includes non-energy uses.

**Blocks 9g, 10e 11c, and 12f must be negative to avoid double counting of energy from primary fuels.

***Includes imported products.

††Includes only electricity to be purchased by each sector.

NATIONAL INPUT WORKSHEET FOR SUPPLY/DEMAND INTEGRATION

Country: EUROPE

Year: 1985

Case: A

Units: EXAJOULES (10^{18}J)

	(a)	(b)	(c)	(d)	(e)	(f)	(g)	(h)	(i)	(j)	(k)
	Coal	Petro-leum	(Syn. Liquids)	Nat. Gas	(Syn. Gas)	(Heat)	(Electri-city)††	Hydro Energy	Nuclear Energy	Geothermal energy and other	Total
(1) Transport		8.47					0.16				8.62
(2) Industry	2.94	5.91		4.08		−0.16	3.43			0.28	16.47
(3) Agric., Mining, Construction }											
(4) Commercial }	0.12	2.83		1.11		0.10	1.52			0.01	5.67
(5) Public }											
(6) Residential	0.63	3.21		3.30		0.46	2.26			0.01	9.86
(7) Non-energy Uses	0.09	3.93		0.83							4.85
(8) Final Energy Demand*	3.76	24.35		9.32		0.40	7.36			0.30	45.48
(9) Electricity**	5.19	3.37		1.29		−0.06	−8.52	4.44	9.57	0.05	15.35
(10) Syn. Gas	0.48	0.31		−0.76							0.03
(11) Syn. Liquids											
(12) Heat	0.16	0.43		0.07		−0.44	−0.10			0.04	0.15
(13) Energy Sector Self Consumption & conversion losses	0.23	1.79		0.86		0.10	1.26				4.25
(14) Primary Energy Input	9.82	30.26		10.78				4.44	9.57	0.39	65.26
(15) Indigenous Supply	8.50	10.07		7.84				4.44	9.57	0.39	40.82
(16) Imports***	1.32	20.18		2.94							24.44
(17) Exports											

*Includes non-energy uses.

**Blocks 9g, 10e 11c, and 12f must be negative to avoid double counting of energy from primary fuels.

***Includes imported products.

††Includes only electricity to be purchased by each sector.

NATIONAL INPUT WORKSHEET FOR SUPPLY/DEMAND INTEGRATION

Country: JAPAN

Year: 1985

Case: A

Units: EXAJOULES (10^{18}J)

	(a) Coal	(b) Petroleum	(c) (Syn. Liquids)	(d) Nat. Gas	(e) (Syn. Gas)	(f) (Heat)	(g) (Electricity)††	(h) Hydro Energy	(i) Nuclear Energy	(j) Geothermal energy and other	(k) Total
(1) Transport		3.39					0.10				3.49
(2) Industry	3.16	6.60		0.15			1.46				11.37
(3) Agric., Mining, Construction		0.71					0.04				0.75
(4) Commercial } (5) Public		1.67		0.10			0.31				2.07
(6) Residential		1.12		0.35			0.39				1.85
(7) Non-energy Uses		2.06									2.06
(8) Final Energy Demand*	3.16	15.56		0.59			2.29				21.60
(9) Electricity**	0.46	1.80		1.53			-2.54	1.22	1.89	0.04	4.39
(10) Syn. Gas	0.05	0.14		-0.15							0.04
(11) Syn. Liquids											
(12) Heat											
(13) Energy Sector Self Consumption & conversion losses	0.15	0.70		0.05			0.26				1.15
(14) Primary Energy Input	3.82	18.19		2.02				1.22	1.89	0.04	27.18
(15) Indigenous Supply	0.66	0.06		0.16				1.22	1.89	0.04	4.02
(16) Imports***	3.16	18.13		1.87							23.16
(17) Exports											

*Includes non-energy uses.

**Blocks 9g, 10e 11c, and 12f must be negative to avoid double counting of energy from primary fuels.

***Includes imported products.

††Includes only electricity to be purchased by each sector.

NATIONAL INPUT WORKSHEET FOR SUPPLY/DEMAND INTEGRATION

Country: REST OF WOCA

Year: 1985

Case: A

Units: EXAJOULES (10^{18} J)

	(a)	(b)	(c)	(d)	(e)	(f)	(g)	(h)	(i)	(j)	(k)
	Coal	Petroleum	(Syn. Liquids)	Nat. Gas	(Syn. Gas)	(Heat)	(Electricity)††	Hydro Energy	Nuclear Energy	Geothermal energy and other	Total
(1) Transport	0.29	14.15					0.14				14.58
(2) Industry (3) Agric., Mining, Construction	4.33	6.33		4.27			2.79				17.71
(4) Commercial (5) Public (6) Residential	0.35	3.42		0.91			1.73				6.41
(7) Non-energy Uses		2.21		0.57							2.78
(8) Final Energy Demand*	4.97	26.12		5.75			4.65				41.49
(9) Electricity**	4.76	2.49		1.60			-5.47	4.49	2.77	0.02	10.65
(10) Syn. Gas	0.07	0.02		-0.07							0.02
(11) Syn. Liquids	0.27	-0.09									0.18
(12) Heat											
(13) Energy Sector Self Consumption & conversion losses	0.21	3.10		0.95			0.82				5.07
(14) Primary Energy Input	10.28	31.64		8.22				4.49	2.77	0.02	57.42
(15) Indigenous Supply	12.72	85.90		18.11				4.49	2.77	0.02	124.0
(16) Imports***											
(17) Exports	2.44	54.26		9.88							66.59

*Includes non-energy uses.

**Blocks 9g, 10e 11c, and 12f must be negative to avoid double counting of energy from primary fuels.

***Includes imported products.

††Includes only electricity to be purchased by each sector.

NATIONAL INPUT WORKSHEET FOR SUPPLY/DEMAND INTEGRATION

Country: NORTH AMERICA

Year: 1985

Case: B

Units: EXAJOULES(10^{18}J)

	(a)	(b)	(c)	(d)	(e)	(f)	(g)	(h)	(i)	(j)	(k)
	Coal	Petroleum	(Syn. Liquids)	Nat. Gas	(Syn. Gas)	(Heat)	(Electricity)[††]	Hydro Energy	Nuclear Energy	Geothermal energy and other	Total
(1) Transport		20.62					0.02				20.65
(2) Industry, (3) Agric., Mining, Construction	7.71	2.90		9.22			3.18				23.01
(4) Commercial, (5) Public, (6) Residential	0.01	5.97		6.52			5.45			0.21	18.16
(7) Non-energy Uses	0.18	5.64		2.35							8.17
(8) Final Energy Demand*	7.90	35.13		18.09			8.66			0.21	69.98
(9) Electricity**	4.25	1.97		2.62			-10.07	7.31	12.49	0.63	19.20
(10) Syn. Gas											
(11) Syn. Liquids											
(12) Heat											
(13) Energy Sector Self Consumption & conversion losses	0.15	3.45		1.98			1.41				6.99
(14) Primary Energy Input	12.30	40.54		22.69				7.31	12.49	0.84	96.17
(15) Indigenous Supply	26.81	34.78		25.12				7.31	12.49	0.84	107.3
(16) Imports***		5.76									
(17) Exports	14.51			2.43							11.18

*Includes non-energy uses.

**Blocks 9g, 10e 11c, and 12f must be negative to avoid double counting of energy from primary fuels.

***Includes imported products.

††Includes only electricity to be purchased by each sector.

NATIONAL INPUT WORKSHEET FOR SUPPLY/DEMAND INTEGRATION

Country: EUROPE

Year: 1985

Case: B

Units: EXAJOULES (10^{18}J)

	(a) Coal	(b) Petro-leum	(c) (Syn. Liquids)	(d) Nat. Gas	(e) (Syn. Gas)	(f) (Heat)	(g) (Electri-city)††	(h) Hydro Energy	(i) Nuclear Energy	(j) Geothermal energy and other	(k) Total
(1) Transport		7.46					0.15				7.62
(2) Industry	2.49	4.78		3.83		−0.15	2.78			0.26	13.99
(3) Agric., Mining, Construction; (4) Commercial; (5) Public	0.09	2.77		1.13		0.12	1.34				5.45
(6) Residential	0.61	3.27		3.28		0.40	1.90			0.01	9.48
(7) Non-energy Uses	0.08	3.05		0.80							3.93
(8) Final Energy Demand*	3.27	21.33		9.03		0.37	6.17			0.28	40.45
(9) Electricity**	4.68	2.25		1.30		−0.06	−7.20	4.39	7.80	0.05	13.20
(10) Syn. Gas	0.46	0.30		−0.73							0.03
(11) Syn. Liquids											
(12) Heat	0.15	0.34		0.07		−0.41	−0.05			0.04	0.13
(13) Energy Sector Self Consumption & conversion losses	0.23	1.53		0.81		0.10	1.08				3.75
(14) Primary Energy Input	8.78	25.76		10.47				4.39	7.80	0.36	57.56
(15) Indigenous Supply	8.50	10.07		7.71				4.39	7.80	0.36	38.83
(16) Imports***	0.28	15.68		2.76							18.72
(17) Exports											

*Includes non-energy uses.

**Blocks 9g, 10e 11c, and 12f must be negative to avoid double counting of energy from primary fuels.

***Includes imported products.

††Includes only electricity to be purchased by each sector.

NATIONAL INPUT WORKSHEET FOR SUPPLY/DEMAND INTEGRATION

Country: JAPAN

Year: 1985

Case: B

Units: EXAJOULES (10^{18}J)

	(a) Coal	(b) Petro-leum	(c) (Syn. Liquids)	(d) Nat. Gas	(e) (Syn. Gas)	(f) (Heat)	(g) (Electri-city)††	(h) Hydro Energy	(i) Nuclear Energy	(j) Geothermal energy and other	(k) Total
(1) Transport		2.85					0.08				2.93
(2) Industry	2.55	4.65		0.13			1.07				8.41
(3) Agric., Mining, Construction		0.62					0.03				0.65
(4) Commercial } (5) Public		1.60		0.08			0.24				1.91
(6) Residential		0.98		0.29			0.34				1.61
(7) Non-energy Uses		1.66									1.66
(8) Final Energy Demand*	2.55	12.36		0.50			1.76				17.17
(9) Electricity**	0.37	0.36		1.53			-1.95	1.19	1.89	0.04	3.43
(10) Syn. Gas	0.05	0.16		-0.17							0.05
(11) Syn. Liquids											
(12) Heat											
(13) Energy Sector Self Consumption & conversion losses	0.11	0.56		0.04			0.20				0.91
(14) Primary Energy Input	3.08	13.45		1.91				1.19	1.89	0.04	21.55
(15) Indigenous Supply	0.66	0.06		0.16				1.19	1.89	0.04	3.99
(16) Imports***	2.42	13.39		1.75							17.56
(17) Exports											

*Includes non-energy uses.

**Blocks 9g, 10e 11c, and 12f must be negative to avoid double counting of energy from primary fuels.

***Includes imported products.

††Includes only electricity to be purchased by each sector.

NATIONAL INPUT WORKSHEET FOR SUPPLY/DEMAND INTEGRATION

Country: REST OF WOCA

Year: 1985

Case: B

Units: EXAJOULES (10^{18}J)

	(a) Coal	(b) Petroleum	(c) (Syn. Liquids)	(d) Nat. Gas	(e) (Syn. Gas)	(f) (Heat)	(g) (Electricity)††	(h) Hydro Energy	(i) Nuclear Energy	(j) Geothermal energy and other	(k) Total
(1) Transport	0.27	10.80					0.14				11.20
(2) Industry (3) Agric., Mining, Construction	3.68	5.14		3.42			2.29				14.53
(4) Commercial (5) Public (6) Residential	0.28	2.71		0.81			1.42				5.22
(7) Non-energy Uses		1.77		0.56							2.32
(8) Final Energy Demand*	4.22	20.42		4.79			3.85				33.28
(9) Electricity**	3.93	2.00		1.30			-4.52	4.65	1.32	0.02	8.69
(10) Syn. Gas	0.07	0.02		-0.07							0.02
(11) Syn. Liquids	0.25	-0.09									0.16
(12) Heat											
(13) Energy Sector Self Consumption & conversion losses	0.18	2.42		0.68			0.68				3.96
(14) Primary Energy Input	8.65	24.76		6.70				4.65	1.32	0.02	46.10
(15) Indigenous Supply	10.96	66.10		16.58				4.65	1.32	0.02	99.62
(16) Imports*** (17) Exports	2.31	41.34		9.88							53.52

*Includes non-energy uses.

**Blocks 9g, 10e 11c, and 12f must be negative to avoid double counting of energy from primary fuels.

***Includes imported products.

††Includes only electricity to be purchased by each sector.

NATIONAL INPUT WORKSHEET FOR SUPPLY/DEMAND INTEGRATION

Country: NORTH AMERICA

Year: 1985

Case: C

Units: EXAJOULES (10^{18} J)

	(a)	(b)	(c)	(d)	(e)	(f)	(g)	(h)	(i)	(j)	(k)
	Coal	Petro-leum	(Syn. Liquids)	Nat. Gas	(Syn. Gas)	(Heat)	(Electri-city)††	Hydro Energy	Nuclear Energy	Geothermal energy and other	Total
(1) Transport		21.99					0.02				22.02
(2) Industry (3) Agric., Mining, Construction	8.94	3.76		11.25			3.77				27.72
(4) Commercial (5) Public (6) Residential	0.01	5.09		8.82			6.13			0.21	20.26
(7) Non-energy Uses	0.26	7.65		1.63							9.55
(8) Final Energy Demand*	9.21	38.49		21.70			9.93			0.21	79.55
(9) Electricity**	8.97	2.58		1.33			-11.55	7.31	12.49	0.63	21.76
(10) Syn. Gas											
(11) Syn. Liquids											
(12) Heat											
(13) Energy Sector Self Consumption & conversion losses	0..29	3.85		2.20			1.63				7.96
(14) Primary Energy Input	18.47	44.92		25.24				7.31	12.49	0.84	109.3
(15) Indigenous Supply	26.52	31.35		22.99				7.31	12.49	0.84	101.5
(16) Imports***		13.57		2.24							7.77
(17) Exports	8.05										

*Includes non-energy uses.

**Blocks 9g, 10e 11c, and 12f must be negative to avoid double counting of energy from primary fuels.

***Includes imported products.

††Includes only electricity to be purchased by each sector.

NATIONAL INPUT WORKSHEET FOR SUPPLY/DEMAND INTEGRATION

Country: EUROPE

Year: 1985

Case: C

Units: EXAJOULES (10^{18} J)

	(a) Coal	(b) Petro-leum	(c) (Syn. Liquids)	(d) Nat. Gas	(e) (Syn. Gas)	(f) (Heat)	(g) (Electri-city)††	(h) Hydro Energy	(i) Nuclear Energy	(j) Geothermal energy and other	(k) Total
(1) Transport		8.74					0.17				8.91
(2) Industry	3.23	6.71		4.31		−0.17	3.52			0.29	17.88
(3) Agric., Mining, Construction; (4) Commercial; (5) Public	0.11	3.02		1.12		0.12	1.57			0.01	5.94
(6) Residential	0.59	3.96		3.32		0.43	2.19			0.01	10.49
(7) Non-energy Uses	0.09	4.32		0.82							5.23
(8) Final Energy Demand*	4.01	26.74		9.57		0.37	7.45			0.31	48.45
(9) Electricity**	5.15	3.63		1.34		−0.06	−8.64	4.44	9.56	0.05	15.48
(10) Syn. Gas	0.48	0.35		−0.80							0.04
(11) Syn. Liquids											
(12) Heat	0.15	0.39		0.07		−0.42	−0.08			0.04	0.14
(13) Energy Sector Self Consumption & conversion losses	0.25	1.93		0.89		0.10	1.27				4.45
(14) Primary Energy Input	10.04	33.05		11.07				4.44	9.56	0.40	68.56
(15) Indigenous Supply	8.44	10.07		7.86				4.44	9.56	0.40	40.77
(16) Imports***	1.61	22.97		3.21							27.79
(17) Exports											

*Includes non-energy uses.

**Blocks 9g, 10e 11c, and 12f must be negative to avoid double counting of energy from primary fuels.

***Includes imported products.

††Includes only electricity to be purchased by each sector.

NATIONAL INPUT WORKSHEET FOR SUPPLY/DEMAND INTEGRATION

Country: JAPAN

Year: 1985

Case: C

Units: EXAJOULES (10^{18}J)

	(a) Coal	(b) Petroleum	(c) (Syn. Liquids)	(d) Nat. Gas	(e) (Syn. Gas)	(f) (Heat)	(g) (Electricity)††	(h) Hydro Energy	(i) Nuclear Energy	(j) Geothermal energy and other	(k) Total
(1) Transport		3.52					0.09				3.61
(2) Industry	3.16	6.86		0.15			1.49				11.66
(3) Agric., Mining, Construction		0.72					0.04				0.77
(4) Commercial } (5) Public }		1.73		0.10			0.32				2.14
(6) Residential		1.17		0.35			0.40				1.92
(7) Non-energy Uses		2.06									2.06
(8) Final Energy Demand*	3.16	16.07		0.59			2.34				22.16
(9) Electricity**	0.46	1.95		1.53			-2.60	1.22	1.89	0.04	4.49
(10) Syn. Gas	0.05	0.14		-0.15							0.04
(11) Syn. Liquids											
(12) Heat											
(13) Energy Sector Self Consumption & conversion losses	0.15	0.70		0.05			0.26				1.15
(14) Primary Energy Input	3.82	18.86		2.03				1.22	1.89	0.04	27.85
(15) Indigenous Supply	0.66	0.06		0.16				1.22	1.89	0.04	4.02
(16) Imports***	3.16	18.80		1.87							23.82
(17) Exports											

*Includes non-energy uses.

**Blocks 9g, 10e 11c, and 12f must be negative to avoid double counting of energy from primary fuels.

***Includes imported products.

††Includes only electricity to be purchased by each sector.

NATIONAL INPUT WORKSHEET FOR SUPPLY/DEMAND INTEGRATION

Country: REST OF WOCA

Year: 1985

Case: C

Units: EXAJOULES (10^{18}J)

	(a) Coal	(b) Petro-leum	(c) (Syn. Liquids)	(d) Nat. Gas	(e) (Syn. Gas)	(f) (Heat)	(g) (Electri-city)††	(h) Hydro Energy	(i) Nuclear Energy	(j) Geothermal energy and other	(k) Total
(1) Transport	0.29	14.60					0.16				15.05
(2) Industry, (3) Agric., Mining, Construction	4.76	7.19		4.51			2.86				19.32
(4) Commercial, (5) Public, (6) Residential	0.34	3.95		0.92			1.72				6.92
(7) Non-energy Uses		2.43		0.56							2.99
(8) Final Energy Demand*	5.38	28.18		5.99			4.73				44.28
(9) Electricity**	4.91	2.57		1.65			-5.57	4.49	2.77	0.02	10.84
(10) Syn. Gas	0.07	0.02		-0.07							0.02
(11) Syn. Liquids	0.27	-0.09									0.18
(12) Heat											
(13) Energy Sector Self Consumption & conversion losses	0.22	3.33		0.98			0.84				5.37
(14) Primary Energy Input	10.85	34.01		8.55				4.49	2.77	0.02	60.69
(15) Indigenous Supply	12.72	98.12		18.43				4.49	2.77	0.02	136.6
(16) Imports***											
(17) Exports	1.87	64.11		9.88							75.86

*Includes non-energy uses.

**Blocks 9g, 10e 11c, and 12f must be negative to avoid double counting of energy from primary fuels.

***Includes imported products.

††Includes only electricity to be purchased by each sector.

NATIONAL INPUT WORKSHEET FOR SUPPLY/DEMAND INTEGRATION

Country: NORTH AMERICA

Year: 1985

Case: D

Units: EXAJOULES (10^{18}J)

	(a)	(b)	(c)	(d)	(e)	(f)	(g)	(h)	(i)	(j)	(k)
	Coal	Petro- leum	(Syn. Liquids)	Nat. Gas	(Syn. Gas)	(Heat)	(Electri- city)††	Hydro Energy	Nuclear Energy	Geothermal energy and other	Total
(1) Transport		23.18					0.02				23.21
(2) Industry (3) Agric., Mining, Construction	8.83	3.83		8.91			3.50				25.06
(4) Commercial (5) Public (6) Residential	0.01	8.27		7.94			7.04			0.11	23.36
(7) Non-energy Uses	0.22	6.74		1.47							8.43
(8) Final Energy Demand*	9.06	42.02		18.32			10.56			0.11	80.06
(9) Electricity**	12.11	3.80		1.53			-12.29	6.19	11.55	0.25	23.13
(10) Syn. Gas		0.31		-0.19							0.12
(11) Syn. Liquids											
(12) Heat											
(13) Energy Sector Self Consumption & conversion losses	0.28	4.24		1.88			1.73				8.13
(14) Primary Energy Input	21.45	50.37		21.54				6.19	11.55	0.36	111.4
(15) Indigenous Supply	22.32	26.57		19.38				6.19	11.55	0.36	86.37
(16) Imports***		23.80		2.16							25.07
(17) Exports	0.88										

*Includes non-energy uses.

**Blocks 9g, 10e 11c, and 12f must be negative to avoid double counting of energy from primary fuels.

***Includes imported products.

††Includes only electricity to be purchased by each sector.

NATIONAL INPUT WORKSHEET FOR SUPPLY/DEMAND INTEGRATION

Country: EUROPE

Year: 1985

Case: D

Units: EXAJOULES (10^{18}J)

	(a) Coal	(b) Petroleum	(c) (Syn. Liquids)	(d) Nat. Gas	(e) (Syn. Gas)	(f) (Heat)	(g) (Electricity)††	(h) Hydro Energy	(i) Nuclear Energy	(j) Geothermal energy and other	(k) Total
(1) Transport		8.58					0.15				8.73
(2) Industry	2.84	5.63		4.09		-0.11	2.93			0.26	15.64
(3) Agric., Mining, Construction; (4) Commercial; (5) Public	0.09	3.07		1.12		0.11	1.37				5.76
(6) Residential	0.59	4.33		3.13		0.37	1.99			0.01	10.42
(7) Non-energy Uses	0.07	3.36		0.80							4.23
(8) Final Energy Demand*	3.58	24.97		9.14		0.37	6.44			0.28	44.78
(9) Electricity**	4.96	2.70		1.39		-0.04	-7.51	4.39	7.73	0.05	13.67
(10) Syn. Gas	0.43	0.33		-0.73							0.03
(11) Syn. Liquids											
(12) Heat	0.14	0.35		0.07		-0.42	-0.05			0.04	0.13
(13) Energy Sector Self Consumption & conversion losses	0.24	1.82		0.80		0.09	1.12				4.07
(14) Primary Energy Input	9.36	30.17		10.67				4.39	7.73	0.37	62.69
(15) Indigenous Supply	8.03	10.07		7.58				4.39	7.73	0.37	38.16
(16) Imports***	1.33	20.10		3.10							24.53
(17) Exports											

*Includes non-energy uses.

**Blocks 9g, 10e 11c, and 12f must be negative to avoid double counting of energy from primary fuels.

***Includes imported products.

††Includes only electricity to be purchased by each sector.

NATIONAL INPUT WORKSHEET FOR SUPPLY/DEMAND INTEGRATION

Country: JAPAN

Year: 1985

Case: D

Units: EXAJOULES (10^{18}J)

	(a) Coal	(b) Petroleum	(c) (Syn. Liquids)	(d) Nat. Gas	(e) (Syn. Gas)	(f) (Heat)	(g) (Electricity)††	(h) Hydro Energy	(i) Nuclear Energy	(j) Geothermal energy and other	(k) Total
(1) Transport		3.11					0.07				3.19
(2) Industry	2.55	4.94		0.13			1.11				8.73
(3) Agric., Mining, Construction		0.64					0.04				0.67
(4) Commercial } (5) Public }		1.70		0.08			0.25				2.02
(6) Residential		1.05		0.29			0.35				1.70
(7) Non-energy Uses		1.66									1.66
(8) Final Energy Demand*	2.55	13.09		0.50			1.82				17.96
(9) Electricity**	0.37	0.53		1.53			−2.02	1.19	1.89	0.04	3.53
(10) Syn. Gas	0.05	0.16		−0.17							0.05
(11) Syn. Liquids											
(12) Heat											
(13) Energy Sector Self Consumption & conversion losses	0.11	0.56		0.04			0.20				0.91
(14) Primary Energy Input	3.08	14.34		1.91				1.19	1.89	0.04	22.45
(15) Indigenous Supply	0.66	0.06		0.16				1.19	1.89	0.04	3.99
(16) Imports***	2.42	14.28		1.75							18.46
(17) Exports											

*Includes non-energy uses.

**Blocks 9g, 10e 11c, and 12f must be negative to avoid double counting of energy from primary fuels.

***Includes imported products.

††Includes only electricity to be purchased by each sector.

NATIONAL INPUT WORKSHEET FOR SUPPLY/DEMAND INTEGRATION

Country: REST OF WOCA

Year: 1985

Case: D

Units: EXAJOULES (10^{18}J)

	(a)	(b)	(c)	(d)	(e)	(f)	(g)	(h)	(i)	(j)	(k)
	Coal	Petro-leum	(Syn. Liquids)	Nat. Gas	(Syn. Gas)	(Heat)	(Electri-city)††	Hydro Energy	Nuclear Energy	Geothermal energy and other	Total
(1) Transport	0.27	12.42					0.13				12.82
(2) Industry (3) Agric., Mining, Construction	4.20	6.05		3.66			2.41				16.32
(4) Commercial (5) Public (6) Residential	0.27	3.33		0.78			1.47				5.85
(7) Non-energy Uses		1.94		0.56							2.50
(8) Final Energy Demand*	4.73	23.74		5.00			4.02				37.49
(9) Electricity**	4.27	2.17		1.41			−4.73	4.65	1.32	0.02	9.09
(10) Syn. Gas	0.07	0.02		−0.07							0.02
(11) Syn. Liquids	0.25	−0.09									0.16
(12) Heat											
(13) Energy Sector Self Consumption & conversion losses	0.20	2.79		0.72			0.71				4.42
(14) Primary Energy Input	9.51	28.63		7.06				4.65	1.32	0.02	51.18
(15) Indigenous Supply	10.96	93.80		16.94				4.65	1.32	0.02	127.7
(16) Imports***											
(17) Exports	1.45	65.17		9.88							76.50

*Includes non-energy uses.

**Blocks 9g, 10e 11c, and 12f must be negative to avoid double counting of energy from primary fuels.

***Includes imported products.

††Includes only electricity to be purchased by each sector.

NATIONAL INPUT WORKSHEET FOR SUPPLY/DEMAND INTEGRATION

Country: NORTH AMERICA

Year: 1985

Case: E

Units: EXAJOULES (10^{18} J)

	(a)	(b)	(c)	(d)	(e)	(f)	(g)	(h)	(i)	(j)	(k)
	Coal	Petroleum	(Syn. Liquids)	Nat. Gas	(Syn. Gas)	(Heat)	(Electricity)††	Hydro Energy	Nuclear Energy	Geothermal energy and other	Total
(1) Transport		24.89					0.03				24.92
(2) Industry (3) Agric., Mining, Construction	11.36	4.11		10.59			4.14				30.20
(4) Commercial (5) Public (6) Residential	0.01	11.38		5.55			8.04			0.11	25.08
(7) Non-energy Uses	0.26	7.70		1.66							9.62
(8) Final Energy Demand*	11.63	48.09		17.79			12.20			0.11	89.81
(9) Electricity**	11.47	7.25		3.78			-14.20	6.63	11.73	0.25	26.90
(10) Syn. Gas											
(11) Syn. Liquids											
(12) Heat											
(13) Energy Sector Self Consumption & conversion losses	0.36	5.09		2.06			2.01				9.52
(14) Primary Energy Input	23.46	60.43		23.63				6.63	11.73	0.36	126.2
(15) Indigenous Supply	22.19	21.45		17.17				6.63	11.73	0.36	79.53
(16) Imports***	1.27	38.97		6.46							46.71
(17) Exports											

*Includes non-energy uses.

**Blocks 9g, 10e 11c, and 12f must be negative to avoid double counting of energy from primary fuels.

***Includes imported products.

††Includes only electricity to be purchased by each sector.

NATIONAL INPUT WORKSHEET FOR SUPPLY/DEMAND INTEGRATION

Country: EUROPE

Year: 1985

Case: E

Units: EXAJOULES (10^{18}J)

	(a)	(b)	(c)	(d)	(e)	(f)	(g)	(h)	(i)	(j)	(k)
	Coal	Petro-leum	(Syn. Liquids)	Nat. Gas	(Syn. Gas)	(Heat)	(Electri-city)††	Hydro Energy	Nuclear Energy	Geothermal energy and other	Total
(1) Transport		10.30					0.17				10.48
(2) Industry	2.55	8.47		4.42		−0.13	3.60			0.28	19.20
(3) Agric., Mining, Construction; (4) Commercial; (5) Public	0.12	3.41		1.19		0.12	1.63			0.01	6.47
(6) Residential	0.62	5.42		3.54		0.39	2.26			0.01	12.25
(7) Non-energy Uses	0.08	4.58		0.82							5.47
(8) Final Energy Demand*	3.38	32.18		9.97		0.37	7.66			0.30	53.87
(9) Electricity**	5.09	5.63		1.45		−0.04	−8.91	4.39	8.93	0.05	16.59
(10) Syn. Gas	0.46	0.36		−0.79							0.04
(11) Syn. Liquids											
(12) Heat	0.15	0.41		0.07		−0.43	−0.08			0.03	0.14
(13) Energy Sector Self Consumption & conversion losses	0.23	2.41		0.92		0.09	1.32				4.97
(14) Primary Energy Input	9.29	40.99		11.63				4.39	8.93	0.38	75.61
(15) Indigenous Supply	8.03	10.07		7.75				4.39	8.93	0.38	39.55
(16) Imports***	1.27	30.92		3.88							36.06
(17) Exports											

*Includes non-energy uses.

**Blocks 9g, 10e 11c, and 12f must be negative to avoid double counting of energy from primary fuels.

***Includes imported products.

††Includes only electricity to be purchased by each sector.

NATIONAL INPUT WORKSHEET FOR SUPPLY/DEMAND INTEGRATION

Country: JAPAN

Year: 1985

Case: E

Units: EXAJOULES (10^{18} J)

	(a) Coal	(b) Petroleum	(c) (Syn. Liquids)	(d) Nat. Gas	(e) (Syn. Gas)	(f) (Heat)	(g) (Electricity)††	(h) Hydro Energy	(i) Nuclear Energy	(j) Geothermal energy and other	(k) Total
(1) Transport		3.86					0.08				3.95
(2) Industry	3.16	7.36		0.15			1.56				12.23
(3) Agric., Mining, Construction		0.75					0.04				0.79
(4) Commercial } (5) Public		1.83		0.09			0.34				2.26
(6) Residential		1.28		0.36			0.43				2.06
(7) Non-energy Uses		2.06									2.06
(8) Final Energy Demand*	3.16	17.15		0.59			2.45				23.35
(9) Electricity**	0.46	2.26		1.53			-2.72	1.22	1.89	0.04	4.68
(10) Syn. Gas	0.05	0.15		-0.16							0.05
(11) Syn. Liquids											
(12) Heat											
(13) Energy Sector Self Consumption & conversion losses	0.15	0.70		0.05			0.27				1.17
(14) Primary Energy Input	3.82	20.25		2.03				1.22	1.89	0.04	29.24
(15) Indigenous Supply	0.66	0.06		0.16				1.22	1.89	0.04	4.03
(16) Imports***	3.16	20.19		1.87							25.21
(17) Exports											

*Includes non-energy uses.

**Blocks 9g, 10e 11c, and 12f must be negative to avoid double counting of energy from primary fuels.

***Includes imported products.

††Includes only electricity to be purchased by each sector.

NATIONAL INPUT WORKSHEET FOR SUPPLY/DEMAND INTEGRATION

Country: REST OF WOCA

Year: 1985

Case: E

Units: EXAJOULES (10^{18}J)

	(a)	(b)	(c)	(d)	(e)	(f)	(g)	(h)	(i)	(j)	(k)
	Coal	Petroleum	(Syn. Liquids)	Nat. Gas	(Syn. Gas)	(Heat)	(Electricity)††	Hydro Energy	Nuclear Energy	Geothermal energy and other	Total
(1) Transport	0.29	17.22					0.16				17.66
(2) Industry (3) Agric., Mining, Construction	3.76	9.08		4.63			2.93				20.39
(4) Commercial (5) Public (6) Residential	0.36	5.01		0.98			1.78				8.12
(7) Non-energy Uses		2.58		0.56							3.14
(8) Final Energy Demand*	4.41	33.88		6.16			4.86				49.31
(9) Electricity**	5.45	2.85		1.83			-5.72	4.49	2.26	0.02	11.19
(10) Syn. Gas	0.07	0.02		-0.07							0.02
(11) Syn. Liquids	0.27	-0.09									0.18
(12) Heat											
(13) Energy Sector Self Consumption & conversion losses	0.21	3.97		1.03			0.86				6.07
(14) Primary Energy Input	10.41	40.63		8.96				4.49	2.26	0.02	66.77
(15) Indigenous Supply	12.72	113.0		18.84				4.49	2.26	0.02	151.3
(16) Imports***											
(17) Exports	2.32	72.37		9.88							84.56

*Includes non-energy uses.

**Blocks 9g, 10e 11c, and 12f must be negative to avoid double counting of energy from primary fuels.

***Includes imported products.

††Includes only electricity to be purchased by each sector.

NATIONAL INPUT WORKSHEET FOR SUPPLY/DEMAND INTEGRATION

Country: NORTH AMERICA

Year: 2000

Case: C1

Units: EXAJOULES (10^{18}J)

	(a)	(b)	(c)	(d)	(e)	(f)	(g)	(h)	(i)	(j)	(k)
	Coal	Petro-leum	(Syn. Liquids)	Nat. Gas	(Syn. Gas)	(Heat)	(Electri-city)††	Hydro Energy	Nuclear Energy	Geothermal energy and other	Total
(1) Transport		27.76					0.03				27.79
(2) Industry (3) Agric., Mining, Construction	19.43	3.27		11.28			8.06				42.03
(4) Commercial (5) Public (6) Residential	0.42	4.01		6.79			8.66			2.48	22.37
(7) Non-energy Uses		12.50		2.55							15.05
(8) Final Energy Demand*	19.85	47.53		20.63			16.76			2.48	107.2
(9) Electricity**	7.52	3.35		1.19			-19.41	8.97	30.50	4.75	36.87
(10) Syn. Gas											
(11) Syn. Liquids	6.62	-4.43									2.18
(12) Heat											
(13) Energy Sector Self Consumption & conversion losses	0.66	3.81		2.17			2.66				9.30
(14) Primary Energy Input	34.65	50.26		23.99				8.97	30.50	7.23	155.6
(15) Indigenous Supply	53.80	26.99		17.77				8.97	30.50	7.23	145. 2
(16) Imports***		23.27		6.22							10.34
(17) Exports	19.15										

*Includes non-energy uses.

**Blocks 9g, 10e 11c, and 12f must be negative to avoid double counting of energy from primary fuels.

***Includes imported products.

††Includes only electricity to be purchased by each sector.

NATIONAL INPUT WORKSHEET FOR SUPPLY/DEMAND INTEGRATION

Country: EUROPE

Year: 2000

Case: C1

Units: EXAJOULES $(10^{18} J)$

	(a)	(b)	(c)	(d)	(e)	(f)	(g)	(h)	(i)	(j)	(k)
	Coal	Petro-leum	(Syn. Liquids)	Nat. Gas	(Syn. Gas)	(Heat)	(Electri-city)††	Hydro Energy	Nuclear Energy	Geothermal energy and other	Total
(1) Transport		12.83					0.27				13.10
(2) Industry	5.31	9.93		6.41		0.89	6.19			0.49	29.21
(3) Agric., Mining, Construction; (4) Commercial; (5) Public	0.10	3.08		1.41		0.44	2.54			0.08	7.64
(6) Residential	0.40	3.07		3.78		0.92	3.66			0.45	12.28
(7) Non-energy Uses	0.40	7.09		1.00							8.49
(8) Final Energy Demand*	6.22	36.00		12.59		2.24	12.65			1.02	70.72
(9) Electricity**	9.21	3.69		0.83		-1.04	-13.83	4.78	17.69	0.47	21.80
(10) Syn. Gas	0.40	0.35		-0.72							0.03
(11) Syn. Liquids											
(12) Heat	2.63	0.30		0.02		-1.50	-0.77		0.07	0.10	0.85
(13) Energy Sector Self Consumption & conversion losses	0.27	3.01		0.97		0.30	1.95				6.50
(14) Primary Energy Input	18.72	43.36		13.69				4.78	17.69	1.59	99.83
(15) Indigenous Supply	9.33	6.44		4.49				4.78	17.69	1.59	44.33
(16) Imports***	9.39	36.92		9.19							55.51
(17) Exports											

*Includes non-energy uses.

**Blocks 9g, 10e 11c, and 12f must be negative to avoid double counting of energy from primary fuels.

***Includes imported products.

††Includes only electricity to be purchased by each sector.

NATIONAL INPUT WORKSHEET FOR SUPPLY/DEMAND INTEGRATION

Country: JAPAN

Year: 2000

Case: C1

Units: EXAJOULES (10^{18} J)

	(a)	(b)	(c)	(d)	(e)	(f)	(g)	(h)	(i)	(j)	(k)
	Coal	Petro-leum	(Syn. Liquids)	Nat. Gas	(Syn. Gas)	(Heat)	(Electri-city)[††]	Hydro Energy	Nuclear Energy	Geothermal energy and other	Total
(1) Transport		7.16					0.20				7.36
(2) Industry	3.56	14.60		0.14			2.46				20.77
(3) Agric., Mining, Construction		1.05					0.06				1.11
(4) Commercial, (5) Public		3.54		0.18			0.94				4.67
(6) Residential		2.59		0.82			0.64				4.05
(7) Non-energy Uses		2.28									2.28
(8) Final Energy Demand*	3.56	31.23		1.15			4.31				40.24
(9) Electricity**	1.95	2.13		2.52			-4.79	1.86	4.39	0.20	8.26
(10) Syn. Gas	0.05	0.27		-0.26							0.06
(11) Syn. Liquids											
(12) Heat											
(13) Energy Sector Self Consumption & conversion losses	0.15	1.21		0.11			0.48				1.95
(14) Primary Energy Input	5.71	34.84		3.52				1.86	4.39	0.20	50.51
(15) Indigenous Supply	0.66	0.84		0.17				1.86	4.39	0.20	8.12
(16) Imports***	5.05	34.00		3.35							42.39
(17) Exports											

*Includes non-energy uses.

**Blocks 9g, 10e 11c, and 12f must be negative to avoid double counting of energy from primary fuels.

***Includes imported products.

††Includes only electricity to be purchased by each sector.

NATIONAL INPUT WORKSHEET FOR SUPPLY/DEMAND INTEGRATION

Country: REST OF WOCA

Year: 2000

Case: C1

Units: EXAJOULES(10^{18}J)

	(a)	(b)	(c)	(d)	(e)	(f)	(g)	(h)	(i)	(j)	(k)
	Coal	Petro-leum	(Syn. Liquids)	Nat. Gas	(Syn. Gas)	(Heat)	(Electri-city)††	Hydro Energy	Nuclear Energy	Geothermal energy and other	Total
(1) Transport	0.58	28.49					0.87				29.95
(2) Industry; (3) Agric., Mining, Construction	8.15	11.99		11.05			6.52				37.71
(4) Commercial; (5) Public; (6) Residential	0.92	7.73		2.23			3.77				14.65
(7) Non-energy Uses		4.73		2.28							7.01
(8) Final Energy Demand*	9.65	52.94		15.56			11.17				89.32
(9) Electricity**	6.63	6.45		3.19			-13.14	10.25	10.16	0.05	23.60
(10) Syn. Gas											
(11) Syn. Liquids											
(12) Heat											
(13) Energy Sector Self Consumption & conversion losses	0.31	6.65		1.88			1.97				10.81
(14) Primary Energy Input	16.59	66.05		20.63				10.25	10.16	0.05	123.7
(15) Indigenous Supply	27.62	127.3		44.09				10.25	10.16	0.05	219.5
(16) Imports***											
(17) Exports	11.03	61.25		23.46							95.73

*Includes non-energy uses.

**Blocks 9g, 10e 11c, and 12f must be negative to avoid double counting of energy from primary fuels.

***Includes imported products.

††Includes only electricity to be purchased by each sector.

NATIONAL INPUT WORKSHEET FOR SUPPLY/DEMAND INTEGRATION

Country: NORTH AMERICA

Year: 2000

Case: C2

Units: EXAJOULES (10^{18} J)

	(a) Coal	(b) Petroleum	(c) (Syn. Liquids)	(d) Nat. Gas	(e) (Syn. Gas)	(f) (Heat)	(g) (Electricity)††	(h) Hydro Energy	(i) Nuclear Energy	(j) Geothermal energy and other	(k) Total
(1) Transport		27.76					0.03				27.79
(2) Industry (3) Agric., Mining, Construction	18.59	3.30		11.26			8.78				41.93
(4) Commercial (5) Public (6) Residential		3.58		6.79			10.67			2.48	23.52
(7) Non-energy Uses	0.42	12.50		2.55							15.48
(8) Final Energy Demand*	19.02	47.13		20.60			19.48			2.48	108.7
(9) Electricity**	4.11	4.45		1.61			-22.61	8.97	42.34	4.75	43.61
(10) Syn. Gas											
(11) Syn. Liquids	6.62	-4.43									2.18
(12) Heat											
(13) Energy Sector Self Consumption & conversion losses	0.74	3.77		2.17			3.14				9.82
(14) Primary Energy Input	30.48	50.92		24.39				8.97	42.34	7.23	164.3
(15) Indigenous Supply	38.82	26.99		17.77				8.97	42.34	7.23	142.1
(16) Imports***		23.93		6.62							22.21
(17) Exports	8.34										

*Includes non-energy uses.

**Blocks 9g, 10e 11c, and 12f must be negative to avoid double counting of energy from primary fuels.

***Includes imported products.

††Includes only electricity to be purchased by each sector.

NATIONAL INPUT WORKSHEET FOR SUPPLY/DEMAND INTEGRATION

Country: EUROPE

Year: 2000

Case: C2

Units: EXAJOULES (10^{18}J)

	(a)	(b)	(c)	(d)	(e)	(f)	(g)	(h)	(i)	(j)	(k)
	Coal	Petro-leum	(Syn. Liquids)	Nat. Gas	(Syn. Gas)	(Heat)	(Electri-city)††	Hydro Energy	Nuclear Energy	Geothermal energy and other	Total
(1) Transport		12.69					0.36				13.05
(2) Industry	3.91	10.16		5.86		0.90	7.33			0.47	28.63
(3) Agric., Mining, Construction; (4) Commercial; (5) Public	0.10	3.06		1.43		0.42	2.83			0.04	7.88
(6) Residential	0.35	2.88		3.65		0.93	4.27			0.28	12.34
(7) Non-energy Uses	0.29	7.44		0.79							8.52
(8) Final Energy Demand*	4.65	36.23		11.72		2.25	14.79			0.78	70.42
(9) Electricity**	5.58	3.13		0.40		-1.07	-16.31	4.61	29.24	0.40	25.97
(10) Syn. Gas	0.95	0.31		-1.36			-0.02		0.52		0.39
(11) Syn. Liquids											
(12) Heat	0.87	0.27		0.03		-1.48	-0.75		1.86	0.06	0.85
(13) Energy Sector Self Consumption & conversion losses	0.20	3.01		0.89		0.30	2.29				6.69
(14) Primary Energy Input	12.23	42.93		11.69				4.61	31.61	1.24	104.3
(15) Indigenous Supply	7.51	6.44		4.49				4.61	31.61	1.24	55.90
(16) Imports***	4.72	36.50		7.19							48.41
(17) Exports											

*Includes non-energy uses.

**Blocks 9g, 10e 11c, and 12f must be negative to avoid double counting of energy from primary fuels.

***Includes imported products.

††Includes only electricity to be purchased by each sector.

NATIONAL INPUT WORKSHEET FOR SUPPLY/DEMAND INTEGRATION

Country: JAPAN

Year: 2000

Case: C2

Units: EXAJOULES(10^{18}J)

	(a) Coal	(b) Petroleum	(c) (Syn. Liquids)	(d) Nat. Gas	(e) (Syn. Gas)	(f) (Heat)	(g) (Electricity)††	(h) Hydro Energy	(i) Nuclear Energy	(j) Geothermal energy and other	(k) Total
(1) Transport		7.16					0.20				7.36
(2) Industry	3.56	14.16		0.58			2.46				20.77
(3) Agric., Mining, Construction		1.05					0.06				1.11
(4) Commercial } (5) Public }		3.54		0.18			0.94				4.67
(6) Residential		2.51		0.90			0.64				4.05
(7) Non-energy Uses		2.28									2.28
(8) Final Energy Demand*	3.56	30.71		1.66			4.31				40.24
(9) Electricity**	1.95	0.89		2.00			−4.79	1.86	6.28	0.20	8.40
(10) Syn. Gas	0.05	0.27		−0.26							0.06
(11) Syn. Liquids											
(12) Heat											
(13) Energy Sector Self Consumption & conversion losses	0.15	1.20		0.11			0.48				1.94
(14) Primary Energy Input	5.71	33.07		3.52				1.86	6.28	0.20	50.64
(15) Indigenous Supply	0.66	0.84		0.17				1.86	6.28	0.20	10.01
(16) Imports***	5.05	32.24		3.35							40.63
(17) Exports											

*Includes non-energy uses.

**Blocks 9g, 10e 11c, and 12f must be negative to avoid double counting of energy from primary fuels.

***Includes imported products.

††Includes only electricity to be purchased by each sector.

NATIONAL INPUT WORKSHEET FOR SUPPLY/DEMAND INTEGRATION

Country: REST OF WOCA

Year: 2000

Case: C2

Units: EXAJOULES (10^{18}J)

	(a)	(b)	(c)	(d)	(e)	(f)	(g)	(h)	(i)	(j)	(k)
	Coal	Petroleum	(Syn. Liquids)	Nat. Gas	(Syn. Gas)	(Heat)	(Electricity)††	Hydro Energy	Nuclear Energy	Geothermal energy and other	Total
(1) Transport	0.60	29.68					0.78				31.06
(2) Industry (3) Agric., Mining, Construction	8.28	13.18		10.38			7.44				39.28
(4) Commercial (5) Public (6) Residential	0.67	7.97		2.30			4.31				15.25
(7) Non-energy Uses		4.27		2.41							6.68
(8) Final Energy Demand*	9.56	55.09		15.10			12.53				92.27
(9) Electricity**	7.10	3.80		1.38			-14.74	8.55	20.95	0.05	27.09
(10) Syn. Gas											
(11) Syn. Liquids											
(12) Heat											
(13) Energy Sector Self Consumption & conversion losses	0.31	6.54		1.81			2.21				10.88
(14) Primary Energy Input	16.97	65.43		18.29				8.55	20.95	0.05	130.2
(15) Indigenous Supply	20.74	127.3		39.75				8.55	20.95	0.05	217.3
(16) Imports***											
(17) Exports	3.77	61.87		21.46							87.1

*Includes non-energy uses.

**Blocks 9g, 10e 11c, and 12f must be negative to avoid double counting of energy from primary fuels.

***Includes imported products.

††Includes only electricity to be purchased by each sector.

NATIONAL INPUT WORKSHEET FOR SUPPLY/DEMAND INTEGRATION

Country: NORTH AMERICA

Year: 2000

Case: D7

Units: EXAJOULES $(10^{18}J)$

	(a)	(b)	(c)	(d)	(e)	(f)	(g)	(h)	(i)	(j)	(k)
	Coal	Petro-leum	(Syn. Liquids)	Nat. Gas	(Syn. Gas)	(Heat)	(Electri-city)††	Hydro Energy	Nuclear Energy	Geothermal energy and other	Total
(1) Transport		26.80					0.03				26.83
(2) Industry (3) Agric., Mining, Construction	15.59	4.78		6.78			6.86				34.00
(4) Commercial (5) Public (6) Residential		5.41		9.57			8.67			1.25	24.89
(7) Non-energy Uses	0.32	9.95		2.04							12.31
(8) Final Energy Demand*	15.91	46.93		18.39			15.56			1.25	98.03
(9) Electricity**	9.15	3.28		1.19			-18.10	7.22	30.50	1.79	35.03
(10) Syn. Gas											
(11) Syn. Liquids											
(12) Heat											
(13) Energy Sector Self Consumption & conversion losses	0.43	4.18		1.93			2.54				9.07
(14) Primary Energy Input	25.48	54.39		21.51				7.22	30.50	3.04	142.1
(15) Indigenous Supply	40.96	19.14		13.76				7.22	30.50	3.04	114.6
(16) Imports***		35.25		7.75							27.52
(17) Exports	15.48										

*Includes non-energy uses.

**Blocks 9g, 10e 11c, and 12f must be negative to avoid double counting of energy from primary fuels.

***Includes imported products.

††Includes only electricity to be purchased by each sector.

NATIONAL INPUT WORKSHEET FOR SUPPLY/DEMAND INTEGRATION

Country: EUROPE

Year: 2000

Case: D7

Units: EXAJOULES (10^{18} J)

	(a)	(b)	(c)	(d)	(e)	(f)	(g)	(h)	(i)	(j)	(k)
	Coal	Petro-leum	(Syn. Liquids)	Nat. Gas	(Syn. Gas)	(Heat)	(Electri-city)††	Hydro Energy	Nuclear Energy	Geothermal energy and other	Total
(1) Transport		10.86					0.24				11.09
(2) Industry	3.83	7.18		5.64		0.66	4.53			0.37	22.20
(3) Agric., Mining, Construction; (4) Commercial; (5) Public	0.07	3.10		1.37		0.31	2.03			0.06	6.95
(6) Residential	0.40	3.78		3.70		0.78	3.10			0.34	12.10
(7) Non-energy Uses	0.25	5.34		0.65							6.24
(8) Final Energy Demand*	4.55	30.26		11.35		1.75	9.90			0.78	58.58
(9) Electricity**	6.22	1.72		0.80		-0.93	-10.92	4.78	15.42	0.40	17.51
(10) Syn. Gas	0.40	0.36		-0.72							0.04
(11) Syn. Liquids											
(12) Heat	1.67	0.31		0.02		-1.07	-0.51		0.07	0.07	0.55
(13) Energy Sector Self Consumption & conversion losses	0.21	2.51		0.89		0.24	1.54				5.39
(14) Primary Energy Input	13.05	35.15		12.35				4.78	15.49	1.24	82.06
(15) Indigenous Supply	8.56	5.79		4.31				4.78	15.49	1.24	40.18
(16) Imports***	4.49	29.37		8.03							41.89
(17) Exports											

*Includes non-energy uses.

**Blocks 9g, 10e 11c, and 12f must be negative to avoid double counting of energy from primary fuels.

***Includes imported products.

††Includes only electricity to be purchased by each sector.

NATIONAL INPUT WORKSHEET FOR SUPPLY/DEMAND INTEGRATION

Country: JAPAN

Year: 2000

Case: D7

Units: EXAJOULES (10^{18} J)

	(a) Coal	(b) Petroleum	(c) (Syn. Liquids)	(d) Nat. Gas	(e) (Syn. Gas)	(f) (Heat)	(g) (Electricity)††	(h) Hydro Energy	(i) Nuclear Energy	(j) Geothermal energy and other	(k) Total
(1) Transport		4.78					0.11				4.90
(2) Industry	3.38	5.70		1.96			1.52				12.56
(3) Agric., Mining, Construction		0.77					0.04				0.81
(4) Commercial } (5) Public }		2.94		0.11			0.39				3.44
(6) Residential		1.84		0.50			0.60				2.94
(7) Non-energy Uses		2.04									2.04
(8) Final Energy Demand*	3.38	18.06		2.57			2.67				26.68
(9) Electricity**	0.98			0.91			-2.97	1.76	4.39	0.20	5.28
(10) Syn. Gas	0.05	0.01		-0.03							0.03
(11) Syn. Liquids											
(12) Heat											
(13) Energy Sector Self Consumption & conversion losses	0.07	0.90		0.07			0.30				1.34
(14) Primary Energy Input	4.47	18.97		3.52				1.76	4.39	0.20	33.32
(15) Indigenous Supply	0.66	0.84		0.17				1.76	4.39	0.20	8.02
(16) Imports***	3.81	18.14		3.35							25.30
(17) Exports											

*Includes non-energy uses.

**Blocks 9g, 10e 11c, and 12f must be negative to avoid double counting of energy from primary fuels.

***Includes imported products.

††Includes only electricity to be purchased by each sector.

NATIONAL INPUT WORKSHEET FOR SUPPLY/DEMAND INTEGRATION

Country: REST OF WOCA

Year: 2000

Case: D7

Units: EXAJOULES (10^{18}J)

	(a) Coal	(b) Petroleum	(c) (Syn. Liquids)	(d) Nat. Gas	(e) (Syn. Gas)	(f) (Heat)	(g) (Electricity)††	(h) Hydro Energy	(i) Nuclear Energy	(j) Geothermal energy and other	(k) Total
(1) Transport	0.40	20.30					0.63				21.33
(2) Industry (3) Agric., Mining, Construction	6.12	8.11		9.27			4.65				28.14
(4) Commercial (5) Public (6) Residential	0.76	5.45		1.54			2.75				10.50
(7) Non-energy Uses		3.24		1.50							4.73
(8) Final Energy Demand*	7.28	37.09		12.30			8.02				64.69
(9) Electricity**	6.03	4.56		1.54			-9.43	7.24	6.99	0.05	16.96
(10) Syn. Gas											
(11) Syn. Liquids											
(12) Heat											
(13) Energy Sector Self Consumption & conversion losses	0.27	4.69		1.41			1.42				7.78
(14) Primary Energy Input	13.58	46.34		15.25				7.24	6.99	0.05	89.43
(15) Indigenous Supply	23.56	103.8		39.18				7.24	6.99	0.05	180.82
(16) Imports***											
(17) Exports	9.98	57.47		23.93							91.38

*Includes non-energy uses.

**Blocks 9g, 10e 11c, and 12f must be negative to avoid double counting of energy from primary fuels.

***Includes imported products.

††Includes only electricity to be purchased by each sector.

NATIONAL INPUT WORKSHEET FOR SUPPLY/DEMAND INTEGRATION

Country: NORTH AMERICA

Year: 2000

Case: D8

Units: EXAJOULES (10^{18}J)

	(a) Coal	(b) Petro- leum	(c) (Syn. Liquids)	(d) Nat. Gas	(e) (Syn. Gas)	(f) (Heat)	(g) (Electri- city)††	(h) Hydro Energy	(i) Nuclear Energy	(j) Geothermal energy and other	(k) Total
(1) Transport		26.80					0.03				26.83
(2) Industry (3) Agric., Mining, Construction	15.59	4.97		6.16			7.49				34.20
(4) Commercial (5) Public (6) Residential		4.28		9.82			10.14			1.25	25.48
(7) Non-energy Uses	0.32	9.95		2.04							12.31
(8) Final Energy Demand*	15.91	45.99		18.02			17.66			1.25	98.82
(9) Electricity**	5.40	4.36		1.61			-20.56	7.22	40.10	1.79	39.93
(10) Syn. Gas											
(11) Syn. Liquids											
(12) Heat											
(13) Energy Sector Self Consumption & conversion losses	0.41	4.07		1.89			2.90				9.27
(14) Primary Energy Input	21.72	54.43		21.52				7.22	40.10	3.04	148.0
(15) Indigenous Supply	31.89	19.14		13.76				7.22	40.10	3.04	115.1
(16) Imports***		35.29		7.76							32.88
(17) Exports	10.17										

*Includes non-energy uses.

**Blocks 9g, 10e 11c, and 12f must be negative to avoid double counting of energy from primary fuels.

***Includes imported products.

††Includes only electricity to be purchased by each sector.

NATIONAL INPUT WORKSHEET FOR SUPPLY/DEMAND INTEGRATION

Country: EUROPE

Year: 2000

Case: D8

Units: EXAJOULES (10^{18} J)

	(a)	(b)	(c)	(d)	(e)	(f)	(g)	(h)	(i)	(j)	(k)
	Coal	Petro-leum	(Syn. Liquids)	Nat. Gas	(Syn. Gas)	(Heat)	(Electri-city)††	Hydro Energy	Nuclear Energy	Geothermal energy and other	Total
(1) Transport		10.77					0.29				11.07
(2) Industry	3.38	7.09		5.03		0.66	5.20			0.36	21.72
(3) Agric., Mining, Construction (4) Commercial (5) Public	0.07	2.81		1.39		0.31	2.37			0.03	6.98
(6) Residential	0.40	3.07		3.54		0.78	3.66			0.19	11.65
(7) Non-energy Uses	0.20	5.35		0.69							6.24
(8) Final Energy Demand*	4.05	29.10		10.64		1.75	11.53			0.59	57.65
(9) Electricity**	3.84	1.44		0.13		-0.93	-12.79	4.61	23.98	0.36	20.65
(10) Syn. Gas	0.95	0.36		-1.40			-0.02		0.52		0.39
(11) Syn. Liquids											
(12) Heat	0.40	0.33		0.02		-1.07	-0.51		1.33	0.05	0.55
(13) Energy Sector Self Consumption & conversion losses	0.16	2.40		0.85		0.24	1.79				5.45
(14) Primary Energy Input	9.40	33.62		10.23				4.61	25.83	1.00	84.68
(15) Indigenous Supply	6.89	5.79		4.31				4.61	25.83	1.00	48.42
(16) Imports***	2.51	27.83		5.92							36.26
(17) Exports											

*Includes non-energy uses.

**Blocks 9g, 10e 11c, and 12f must be negative to avoid double counting of energy from primary fuels.

***Includes imported products.

††Includes only electricity to be purchased by each sector.

NATIONAL INPUT WORKSHEET FOR SUPPLY/DEMAND INTEGRATION

Country: JAPAN

Year: 2000

Case: D8

Units: EXAJOULES (10^{18} J)

	(a) Coal	(b) Petroleum	(c) (Syn. Liquids)	(d) Nat. Gas	(e) (Syn. Gas)	(f) (Heat)	(g) (Electricity)††	(h) Hydro Energy	(i) Nuclear Energy	(j) Geothermal energy and other	(k) Total
(1) Transport		4.78					0.11				4.90
(2) Industry	3.38	6.04		1.62			1.52				12.56
(3) Agric., Mining, Construction		0.77					0.04				0.81
(4) Commercial } (5) Public }		2.94		0.11			0.39				3.44
(6) Residential		1.00		1.34			0.60				2.94
(7) Non-energy Uses		2.04									2.04
(8) Final Energy Demand*	3.38	17.56		3.07			2.67				26.68
(9) Electricity**				0.19			-2.97	1.76	6.28	0.20	5.47
(10) Syn. Gas	0.05	0.01		0.10							0.16
(11) Syn. Liquids											
(12) Heat											
(13) Energy Sector Self Consumption & conversion losses	0.07	0.90		0.16			0.30				1.43
(14) Primary Energy Input	3.50	18.48		3.52				1.76	6.28	0.20	33.74
(15) Indigenous Supply	0.66	0.84		0.17				1.76	6.28	0.20	9.91
(16) Imports***	2.84	17.64		3.35							23.82
(17) Exports											

*Includes non-energy uses.

**Blocks 9g, 10e 11c, and 12f must be negative to avoid double counting of energy from primary fuels.

***Includes imported products.

††Includes only electricity to be purchased by each sector.

NATIONAL INPUT WORKSHEET FOR SUPPLY/DEMAND INTEGRATION

Country: REST OF WOCA

Year: 2000

Case: D8

Units: EXAJOULES (10^{18}J)

	(a) Coal	(b) Petroleum	(c) (Syn. Liquids)	(d) Nat. Gas	(e) (Syn. Gas)	(f) (Heat)	(g) (Electricity)††	(h) Hydro Energy	(i) Nuclear Energy	(j) Geothermal energy and other	(k) Total
(1) Transport	0.42	20.88					0.65				21.95
(2) Industry, (3) Agric., Mining, Construction	6.34	9.38		7.28			5.29				28.29
(4) Commercial, (5) Public, (6) Residential	0.47	5.69		1.61			3.06				10.83
(7) Non-energy Uses		3.37		1.56							4.94
(8) Final Energy Demand*	7.24	39.32		10.45			9.00				66.01
(9) Electricity**	5.20	2.66		1.05			−10.59	5.96	14.67	0.05	19.00
(10) Syn. Gas											
(11) Syn. Liquids											
(12) Heat											
(13) Energy Sector Self Consumption & conversion losses	0.25	4.73		1.27			1.59				7.84
(14) Primary Energy Input	12.68	46.72		12.77				5.96	14.67	0.05	92.85
(15) Indigenous Supply	19.57	103.8		34.06				5.96	14.67	0.05	178.1
(16) Imports***											
(17) Exports	6.88	57.09		21.29							85.26

*Includes non-energy uses.

**Blocks 9g, 10e 11c, and 12f must be negative to avoid double counting of energy from primary fuels.

***Includes imported products.

††Includes only electricity to be purchased by each sector.

5 Constrained Global Supply-Demand Integrations

5.1 INTRODUCTION

The WAES methodology leads to estimates of a country's or region's desired demand for energy and its potential supply, given each scenario's assumptions about price, real economic growth, and national policy. WAES has developed two techniques for comparing demand and supply: the *unconstrained* and the *constrained* approaches. The unconstrained approach matches supply to demand using national preferences for each type of fuel. Although supply availability may implicitly affect these preferences, this approach does not explicitly account for global supply limitations; it assumes that needed imports will be available.

The constrained approach, on the other hand, incorporates global supply limitations, as well as numerous additional features. This approach allocates the available supply of each fuel according to a range of preferences for that fuel in each region and energy market and does so in a way that minimizes the total cost of energy to world consumers.

To allocate supply in this way, WAES uses the Global Energy Mini-Model (GEMM), a linear programming model developed by WAES Associates at Atlantic Richfield Company. For each scenario, GEMM incorporates maximum and minimum fuel preferences and the maximum potential supply of fossil fuels. For nuclear, solar, geothermal, shale oil, and oil sands, as well as "other" fuels, it is constrained by their maximum technical potential. Technological constraints on refining are included in the form of regional refinery optimizations. Within the remaining and limited choices, GEMM calculates fuel market allocations taking into account the operating costs of mining, transporting, converting, distributing, and using fuels, as well as the capital costs of expanding the infrastructure whenever necessary. Finally, GEMM uses export costs or prices and international shipping costs to allocate international energy flows among regions. Such choices are made so that the total cost to the consumer is minimized. As a result of incorporating these constraints, GEMM provides a global energy pattern that reflects the most efficient use of energy within the cost and constraint structure.

There were two stages in the WAES approach to supply-demand integrations. First an analysis of desired demand and potential supply at the assumed values of the scenario variables was prepared and national and global unconstrained and constrained supply-demand integrations were made. The implications of such integrations for national policy response, as well as for the direction of needed shifts in the assumed scenario variables, were then considered.

This approach eliminated the need for a fully dynamic model that could vary prices internally to produce a supply-demand balance at some price. Fully dynamic models that make internal price changes but ignore explicit and visible treatment of policy options conceal policy issue detail that is essential to the WAES approach. Since one WAES objective is to make alternative energy strategies visible, we felt it appropriate to treat them explicitly and overtly. As the state of the art advances, many other approaches may become suitable.

5.2 THE ECONOMICS OF THE OPEC CARTEL

The constrained approach to supply-demand integration is especially appropriate and useful in accounting for the economic effects of the OPEC (Organization of Petroleum-Exporting Countries) cartel because GEMM makes it possible to account for OPEC's price-setting ability under a variety of demand-supply combinations. Because OPEC can maintain set prices under broad supply and demand conditions, a partieular OPEC price assumption can be one basis for a variety of different national and global case analyses. Such OPEC-set prices will come under pressure in any of three cases: (1) if OPEC cannot adjust output to meet demand profitably (taking account of revenue needs and investment choices); (2) if consumers are unwilling or unable to adjust demand to a level that OPEC finds profitable (in the above sense); or (3) if the cartel members fail to agree on prices, leading to important price competition among members. These possibilities may be examined in the light of particular case outcomes rather than as initial constraints on an analysis.

This section explains the broad ability of OPEC to set oil prices over a range of supply and demand—an ability that, in part, makes the WAES constrained integration approach possible. It also relates this ability to national policies to reduce the severity of prospective physical or cartel-imposed shortages of energy.

To begin, consider the curves representing a consuming country's internal oil supply and demand shown in Figure 5.1. At the *self-sufficiency price*, demand is low enough and indigenous supply high enough for the two to balance. This occurs at the self-sufficiency quantity, i.e., the quantity for which imports are not needed. The demand curve may be rising very steeply near this point. It may be shifted to the left or right by income or policy changes (which are assumed constant for purposes of explanation).

A country can also import fuels at various prices, as illustrated in Figure 5.2. Indigenous supply is the quantity measured between the price axis and the indigenous supply curve at a given price. Imports (OPEC exports) are represented by the region between the indigenous and the total supply curves. Total supply available at a given price is then the sum of indigenous supplies and imported supplies at that price. For the same demand curve, supply and demand balance at this price, which is below the self-sufficiency price. If fuels were priced above the balance price, imported supply would be phased out as demand fell and indigenous supply rose, assuming that policy and/or price advantages were available for indigenous resources.

Figure 5.3 shows in greater detail what happens when the actual price is lower than the self-sufficiency price. At a high self-sufficiency price (and implied low quantity), no OPEC supplies flow and OPEC revenue from exports is zero. Such a situation is unstable without import barriers and an assumed acceptable level of economic activity at such a price and quantity. If prices are somewhat lower than the self-sufficiency price (shown as "inadequate OPEC revenue"), more oil is available

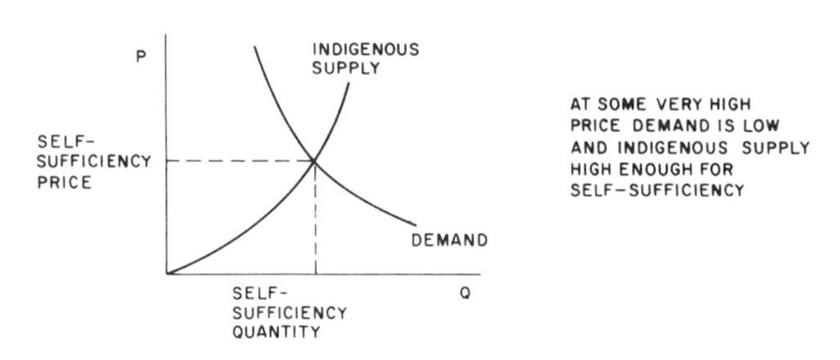

Figure 5.1
National supply-demand balance without imports.

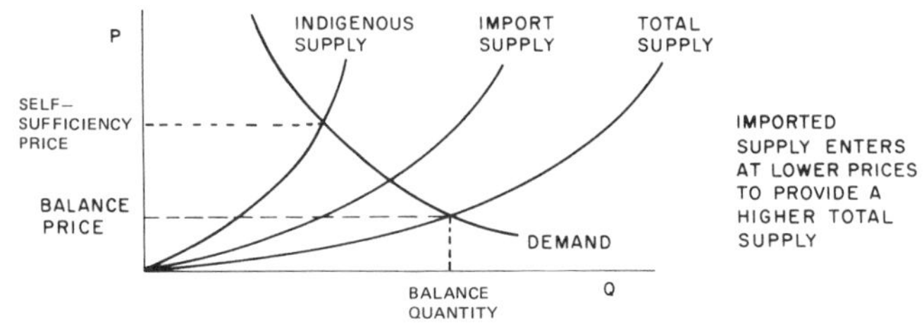

Figure 5.2
National supply-demand balance with imports.

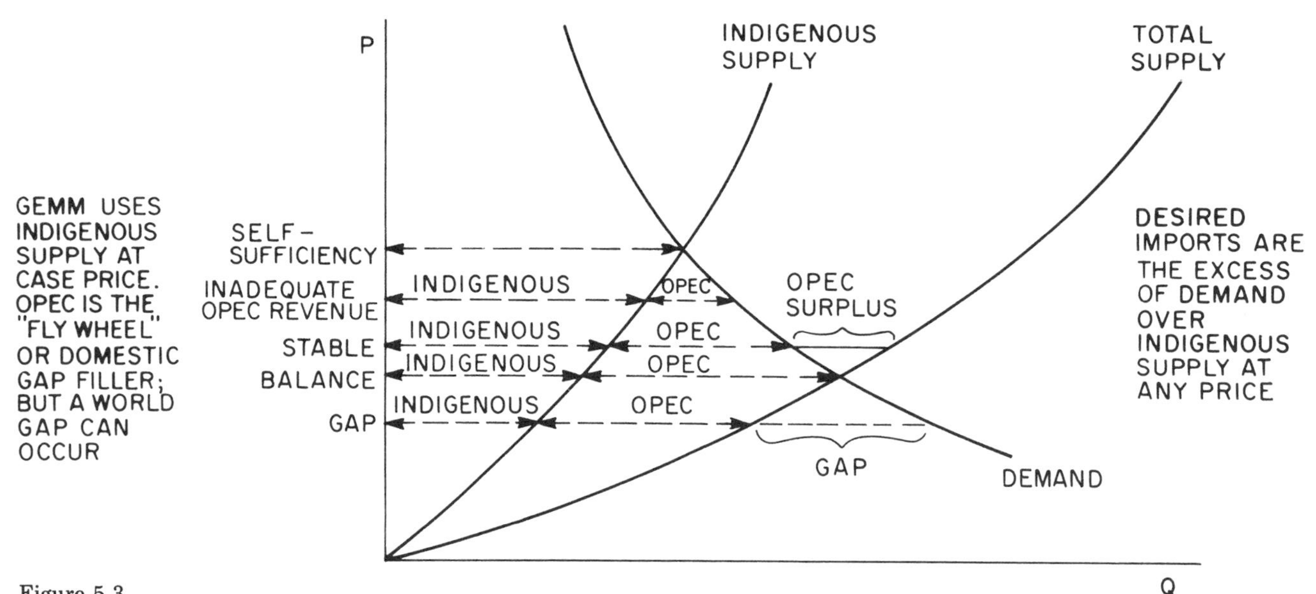

Figure 5.3
Supply and demand effects at various prices: prospective shortfalls and surpluses.

through imports, which expand to take up the slack caused by lower indigenous production at the lower price. (Note that as prices rise the OPEC supply curve may bend backward.) The amount of imports is represented by the difference between the indigenous supply curve and the demand curve at a given price. The increase in imports is more rapid than the decrease in indigenous production since the lower price also induces a greater quantity demanded. The reverse is also true. As prices rise, imports decrease by the quantity of additional supply induced plus the quantity of demand reduced by the price rise. Under such conditions it may be disadvantageous for exporting countries to raise prices beyond some level. Since they are the marginal supplier, their revenue loss will equal the sum of the value of increased production and reduced demand in consuming countries. Prices are likely to fall below the inadequate revenue price if competition arises in an attempt by individual producers to increase their revenues.

At a price lower than the "inadequate OPEC revenue" price, the so-called stable price, OPEC exports are yet higher, represented by the distance between the indigenous supply curve and the demand curve. Indigenous supply at this point is also higher than at the balance price. OPEC surplus production potential must be suppressed, but not as much as in the inadequate revenue case. Finally, at a still lower "gap" price, suppliers are unwilling or unable to supply the full quantity of oil demanded. Excess demand (gap) exists between the total supply curve and the demand curve. These conditions are maintained only through regulations and controls since unfettered market pressures would raise the price to the stable point. OPEC producers can raise their prices to this point and increase their volume at the same time.

How can energy policy improve this situation? Figure 5.4 shows the effect of vigorous supply policies within the energy-importing countries. New indigenous supply can be added at unchanged prices by easing regulatory and related limitations. Transfer payments, tax relief, or focused subsidies can have a similar effect, as can technological improvement. *Supply action programs*—additional increments of higher-cost supply not included in the old indigenous supply curve—can also add to supply but will probably also increase prices. The indigenous supply curve shifts to the right, producing a new indigenous supply curve and hence a new total supply curve. The result is a lower self-sufficiency price, a lower stable price, and a smaller gap at the old gap price. Reduced import dependence results, and the gap (at the assumed old gap price) is almost filled. (Of course the same gap would exist at a still lower gap price than that assumed in Figure 5.3.)

Finally, Figure 5.5 shows the effect of *conservation action programs*—reduced energy demand stimulated by policy, regulatory, and other actions in addition to those induced by market forces already included in the old demand curve. The old demand curve is shifted downward and to the left. At the arbitrarily chosen old gap price of Figure 5.3, demand reduction and supply augmentation have combined to match desired demand with available total supply; the gap is closed at this new (and lower) stable price. Thus energy policy can have a significant effect in remedying an

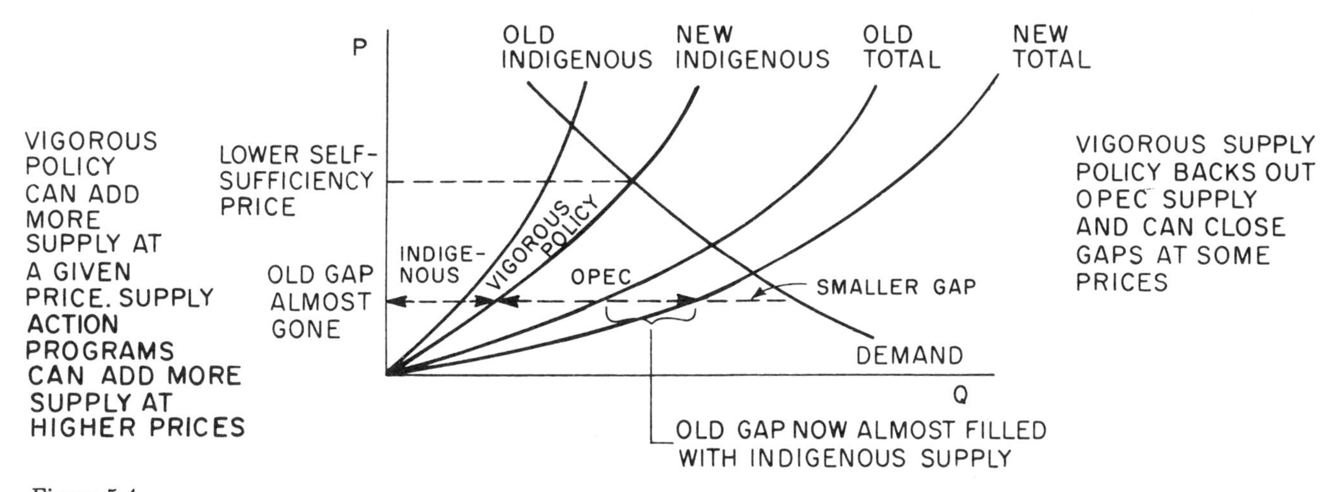

Figure 5.4
Effects of vigorous policies on supply.

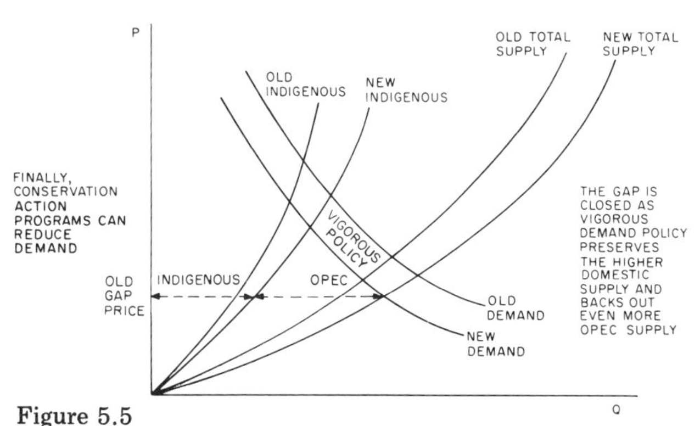

Figure 5.5
Effects of vigorous conservation policies on demand.

energy supply shortfall, or *prospective gap*, independent of whether this gap arises from physical or political production limits in OPEC. Policy can help considerably in avoiding the full price rise implications of such a gap.

The remainder of this chapter describes the constrained global supply-demand integration analysis and presents information on the inputs and results of the process, including some selected case output. Appendix A of this volume presents the full results of a selected case and describes the various tables from a GEMM printout. Appendix B describes the formulation and flow of logic in the model.

Inputs to the constrained global integrations come from the WAES global data base (chapter 3), an evolutionary collection of supply, demand, and infrastructure information that was prepared and refined as the project proceeded on the basis of the various cases analyzed.

5.3 DEMAND DATA

For each scenario, demand by country and by market sector (i.e., industrial, residential, commercial, transport on distillate, transport on electricity, and chemical feedstocks) was obtained from the national studies. Nonenergy bunkers (fuel for shipping goods other than energy materials) are also estimated. A special study for developing countries produced similar data (see chapter 6). The demand data are aggregated into regions (North America, Europe, Japan, and Rest of WOCA) to be used as input. Note that particular fuels are unspecified at this point, except for the transport sector.

Fuel preference constraints by case and market sector are provided by country and aggregated by region. These maximum and minimum percentages are provided for electricity generation by oil, coal, gas, nuclear, hydro, and "other" to summarize fuel preferences for base, intermediate, and peak loads. Preferences for industrial and domestic use of oil, coal, gas, and electricity are also provided, as is the percentage of distillates in liquid fuels. GEMM uses these preferences to perform tradeoffs between fuels and markets in each region and to take account of availability and costs. Existing fuel-using equipment is aggregated for each market, and the cost of any new equipment required for fuel switching is taken into account.

5.4 SUPPLY DATA

Supply data by primary fuel are derived from national team studies and special studies for OPEC and other non-WAES countries. Net export data are estimated for Communist areas. Regional aggregations are made for use in GEMM. Maximum potential production (or politically or technically constrained levels when appropriate) are specified for five crude oil types: onshore light, onshore heavy, offshore light, offshore heavy, and oil sands and oil shale. For coal, surface and deep mining are treated; for gas, onshore and offshore potentials are estimated.

Maximum technical development constraints are provided in each case for nuclear electricity, coal gasification and liquefaction, LNG regasification, hydroelectricity, and "other" (including solar and geothermal). Data from national studies are once again aggregated into regions. In the supply studies, two additional supply-only regions are used: OPEC and OCEC (other Coal-Exporting countries—those not in North America, Europe, or Japan). Demand for fuels in OPEC and OCEC is treated as part of Rest of WOCA demand.

5.5 INFRASTRUCTURE DATA AND COSTS

In addition to supply and demand data, GEMM uses a description of existing infrastructure in each region for the base year (1975), obtained from published sources and Associates' estimates for each stage of energy flow and each region. Average operating costs for each element and capital costs for infrastructure expansion are also used. These costs were estimated from WAES and other available sources; regional cost estimates were not always available from WAES national teams. These data were reviewed by all Associates. Although cost data sometimes vary with the level of operation of a physical process, estimates were made in the range of activity level expected for each process in each region. In the case results, all processes operated in the ranges expected. If they had not, cost estimates would have required review and revision and cases would have had to be rerun.

5.6 CASE RUN PERIODS—DYNAMICS

Since GEMM makes calculations from a base year to a case year, we found it essential to substitute the results of each 1975 to 1985 case into the model before running the corresponding 1985 to 2000 cases. It would have been equally possible to run 1975 to 2000 cases, and we did so in some initial analyses. But the step approach allowed different choices to be used in each period. This was especially useful since many fuel constraints do not take effect until after 1985. For example, in some WAES cases high demand meant that by the year 2000 coal could not meet all the incremental demand without some coal being switched from electricity generation to end use, because of the physical limits on the rate of coal infrastructure development between 1985 and 2000. The step approach also allows different price, policy, and economic growth assumptions for each part of the 1975 to 2000 interval. The size and number of steps reflect a trade-off between within-step choices adequate for the short term and the need to take account of subsequent developments.

5.7 THE CONSTRAINED INTEGRATION CALCULATION

GEMM is derived in part from earlier models. The Shell International Petroleum Company World Energy Mini-Model, the U.S. Brookhaven Energy System Optimization Model (BESOM), the U.S. Atlantic Richfield Company refinery models, and many others formed essential bases for the work. The Shell model, in particular, provided useful formats for model input and output.

The constrained energy supply-demand integration calculation for each WAES case takes desired demand in each region as a "target." Given supply potential, it depreciates base year infrastructure to the case year and then adds infrastructure as needed to find a pattern within processing, transportation, and consumption constraints that meets the target demand at the least total cost to world consumers, including capital charges for infrastructure expansion. If any notional gaps result they appear in the market sectors and regions having highest marginal costs.

This distribution is accomplished by allowing each market sector in each region to have access to a *penalty fuel* at a cost so high that none will be used until all other possibilities have been exhausted. (It is assumed that conservation possibilities costing less than incremental supply have already been included in case input data or will be considered when the implications of any supply-demand gaps are analyzed.) In such cases, reported marginal costs for many activities become more relevant to the (arbitrarily set) cost of penalty fuel than to other costs; and thus the calculation results are more illustrative of the notional gap itself than of any possible real world allocation. Industrial nations may, particularly in the short term, compete for scarcer fuels in such cases in ways that, for a time, are unrelated to true cost differences. In the longer run, prices will change, supply agreements will run out or be rewritten, and a nation's policy will become based on other factors—including the ability to raise funds from nonenergy international trade activities to pay for higher-cost scarce energy.

5.8 CONSTRAINED INTEGRATION RESULTS

Constrained integration results report the outcome of the process; in addition, by displaying associated constraint and cost information, these results clarify many of the reasons (constraints, costs) for particular allocations. Before reviewing summaries of case results, the reader may wish to refer to the illustrative GEMM output in Appendix A and the accompanying description of results. The case run used as an illustration is WAES 1985 Case D, with assumptions of low (3.5 percent per year) economic growth, an $11.50/bbl (1975 U.S.$) oil price, and a restrained national policy response.

Integration of the various WAES scenarios with energy supplies constrained to the maximum levels attainable under the given assumptions of each case led to the energy mixes shown in Table 5.1 for the year 1985 and Table 5.2 for 2000. These tables show the magnitudes of global energy production by fuel for 1985 and 2000. The major differences between cases are in oil production, reflecting the rate of new additions to reserves and

Table 5.1
WOCA primary energy supply, 1985 (MBDOE)

Economic growth Oil price National policy response WAES case	High $17.25 Vig. A	Low $17.25 Vig. B	High $11.50 Vig. C	Low $11.50 Rest. D	High $7.66 Rest. E
Fuel					
Oil	63.6	54.9	66.2	59.6	57.3
Coal	22.2	19.2	21.0	19.9	19.8
Gas	19.2	16.3	18.4	19.9	19.0
Nuclear	9.4	7.9	9.0	7.0	7.0
Hydro and other	8.0	7.0	8.0	7.7	7.7
Total primary energy	122.3	105.2	122.5	114.1	110.8
Potential energy surplus or (deficit)	10.8	26.7	6.9	—	(10.1)

Table 5.2
WOCA primary energy supply, 2000 (MBDOE)

Economic growth Energy price Principal replacement fuel WAES case	High $17.25 Coal C-1	High $17.25 Nuclear C-2	Low $11.50 Coal D-7	Low $11.50 Nuclear D-8
Fuel				
Oil	74.3	74.3	60.0	60.0
Coal	42.2	30.4	35.9	27.2
Gas	24.8	23.7	20.5	19.5
Nuclear	27.7	48.2	25.6	42.5
Hydro and other	14.4	13.7	10.6	10.3
Total primary energy	183.4	109.4	152.9	159.5
Potential energy deficit	7.2	13.6	3.2	6.5

assumed OPEC production limits, and the swing between relative quantities of coal and nuclear, reflecting scenario assumptions about principal replacement fuels.

In the 1985 WAES cases integrated with constrained fuel supplies, world energy demands in Cases A and B can be met within preferred fuel mix ranges without using total potential supplies of any fuel. We can thus conclude that given Case A and B assumptions of rising real energy prices and vigorous national policy response, the world will have surplus energy through the year 1985.

In Case C, energy demands are again met within desired fuel mix preferences, but remaining surpluses of potential fuel supplies are small, indicating that energy supply and demand are close to equilibrium. In Case D as in Case C, constant real energy prices are assumed, but national policy is restrained and economic growth is lower, and although energy demands are again met by production of all available supplies, some minor violations of fuel mix preferences occur, especially in electricity generation. Energy supply and demand are again close to equilibrium, with restrained energy policies and lower economic growth working to reduce both supply and demand by amounts similar to those in Case C.

In Case E, which assumes high economic growth, falling real energy prices, and restrained policy, all potential energy supplies are produced but fail to meet expected demand, even following extensive fuel substitutions in violation of assumed preferences in consumer markets. Thus the assumptions of this case lead to a global shortfall of energy supply prior to the year 1985.

In year 2000 Cases C-1, C-2, D-7, and D-8, the quantity of fossil fuels produced and consumed is the maximum available, and no additional supplies of oil, gas, or coal are obtainable consistent with the case assumptions. The levels of nuclear, hydroelectric, and other primary energy are also at a maximum for each case. Thus, in each of the cases, by the year 2000 all available energy consistent with the scenario assumptions is being produced. Yet these supplies are not sufficient to meet total demand. As discussed in chapter 2, there are also prospective shortages and surpluses of particular fuels in relation to desired demand—most notably shortages of oil and surpluses of coal. Even though all potential fuel supplies are made available, the mix of these fuels does not meet the preferred fuel mix of the world's energy consumers. Total available energy, even when used in its most efficient manner and in violation of consumers' preferred fuel mixes, fails to meet the total specified final energy demand in each case, by quantities ranging from 3.2 to 13.6 MBDOE. While these differences are relatively small, they emphasize the degree of stress in the four WAES scenarios to the year 2000. Even if all available energy is produced and massive fuel substitutions are assumed in markets to compensate for the large notional gaps in oil supply, there is still not enough energy to meet specified demand given the scenario assumptions about energy prices and economic growth.

In addition to the four cases to 2000 described above, a special analysis using the rising energy price, low economic growth assumptions, Case D-3, was undertaken. This analysis showed that the appearance of the notional oil gap could be deferred until after the year 2000. In this case all oil and most gas and coal available under the case assumptions are utilized. The surpluses available in nuclear, hydroelectric, and other primary energy are somewhat larger due to the higher marginal cost of new electric capacity, but the magnitude of the total primary energy surplus is relatively small and suggests that the postponement of the oil shortage under the case assumptions will only be for a few years beyond 2000.

Key tables giving the results of the constrained integrations are included in the following pages for those who may wish to make more detailed comparisons between cases. There are twelve tables for each of the WAES cases, A through E for 1985 and C-1, C-2, D-7, and D-8 for 2000. These include production tables by fuel and region, market input summaries, demand-supply matrices for WOCA and for regions, and summaries of capital charges, costs, and revenues for consumers and producers. The contents of these and the other detailed tables produced by GEMM are described in Appendix A.

Note: Output page numbers are indicated by large bold numbers in the upper right-hand corners of the output sheets. There are twelve sheets for each case, starting on the following pages:

1985 Case A p. 134
Case B p. 143
Case C p. 151
Case D p. 160
Case E p. 171

2000 Case C-1 p. 182
Case C-2 p. 192
Case D-7 p. 201
Case D-8 p. 209

VOLUMETRIC SUMMARY

PROBLEM: WAES85A BASE YR: 1975.
CASE: 31JAN77 CASE YR: 1985.

1

PRODUCTION

	OIL	COAL	GAS	NUCLEAR	HYDRO	TOTAL
NORTH AMERICA	15.6	12.9	11.3	4.3	3.6	47.7
EUROPE	4.1	2.8	3.6	2.9	2.0	15.3
JAPAN	0.1	0.3	0.1	1.0	0.6	1.9
OPEC	35.7	0.0	1.5	0.0	0.0	37.1
OCEC	0.0	5.7	0.0	0.0	0.0	5.7
REST WOCA	6.2	0.0	2.3	1.2	1.9	11.5
SUBTOTAL	61.6	21.7	18.7	9.4	8.0	119.3
WOCA EXPORTS	2.0	0.5	0.5	0.0	0.0	3.0
TOTAL	63.6	22.2	19.2	9.4	8.0	122.3

TOTAL WORLD ENERGY PRODUCTION (EXCLUDING COMMUNIST AREA INTERNAL): 122.3

CONVERSION(INPUT)

	OIL	COAL	GAS	NUCLEAR	HYDRO
OIL REFINING	63.6				
COAL GASIFICATION		0.0			
COAL LIQUEFACTION		0.0			
GAS LIQUEFACTION (NOT INCL. COMMUNIST AREA)				0.0	
ELECTRICITY GENERATION	4.5	7.5	2.2	9.4	8.0

MARKET INPUT

	OIL	COAL	GAS	ELECTRICITY
INDUSTRIAL				
HEAT, LIGHT AND POWER	11.3	11.7	6.9	4.8
NON ENERGY	7.9	0.2	1.0	
DOMESTIC				
HEAT, LIGHT AND POWER	9.4	1.2	4.8	4.4
TRANSPURT AND BUNKERS	25.0			0.1

ALL QUANTITIES ARE GIVEN IN MILLION BARREL PER DAY OIL EQUIVALENT

2

PRODUCTION AND TRANSPORTS CRUDE OIL

PROBLEM: WAES85A BASE YR: 1975.
CASE: 31JAN77 CASE YR: 1985.

FROM \ TO	NORTH AMERICA	EUROPE	JAPAN	REST WOCA	OPEC	PROD'N
NORTH AMERICA	15.60	.	.	.	.	15.60
EUROPE	.	4.07	.	.	.	4.07
JAPAN	.	.	0.08	.	.	0.08
REST WOCA	.	.	.	6.20	.	6.20
OPEC	2.69	11.28	9.64	9.37	2.69	35.66
COMMUNIST AREAS	.	2.00	.	.	.	2.00
TOTAL SUPPLY	18.29	17.35	9.72	15.57	2.69	63.61

PRODUCTION AND TRANSPORTS OIL PRODUCTS

PROBLEM: WAES85A BASE YR: 1975.
CASE: 31JAN77 CASE YR: 1985.

F=FUEL OIL, D=DISTILLATES

FROM \ TO		NORTH AMERICA	EUROPE	JAPAN	REST WOCA	OPEC	PROD'N
NORTH AMERICA	F	3.32	.	.	.	.	3.32
	D	14.32	.	.	.	.	14.32
EUROPE	F	.	6.03	.	.	.	6.03
	D	.	10.27	.	.	.	10.27
JAPAN	F	.	.	4.56	.	.	4.56
	D	.	.	4.51	.	.	4.51
REST WOCA	F	.	.	.	4.89	.	4.89
	D	.	.	.	10.33	.	10.33
OPEC	F	.	.	.	0.70	.	0.70
	D	1.80	.	.	.	0.00	1.80
TOTAL SUPPLY	F	3.32	6.03	4.56	5.58	.	19.48
	D	16.12	10.27	4.51	10.33	0.00	41.24

3

PRODUCTION AND TRANSPORTS COAL — PROBLEM: WAES85A — CASE: 31JAN77 — BASE YR: 1975. — CASE YR: 1985.

FROM \ TO	NORTH AMERICA	EUROPE	JAPAN	REST WOCA	LOSS	PROD'N
NORTH AMERICA	9.20	2.61	0.86	.	0.25	12.92
EUROPE	.	2.73	.	.	0.05	2.79
JAPAN	.	.	0.25	.	0.01	0.26
OCEC	.	.	0.90	4.69	0.11	5.70
COMMUNIST AREAS	.	0.50	.	.	.	0.50
TOTAL SUPPLY	9.20	5.84	2.02	4.69	0.42	22.17

PRODUCTION AND TRANSPORTS GAS — PROBLEM: WAES85A — CASE: 31JAN77 — BASE YR: 1975. — CASE YR: 1985.

P=BY PIPE L=AS LNG

FROM \ TO		NORTH AMERICA	EUROPE	JAPAN	REST WOCA	OPEC	LOSS	PROD'N
NORTH AMERICA	P	10.17	.	.	.	.	1.13	11.30
EUROPE	P	.	3.21	.	.	.	0.36	3.57
JAPAN	P	.	.	0.06	.	.	0.01	0.07
REST WOCA	P	.	.	.	2.07	.	0.23	2.30
OPEC	P	.	0.18	.	1.08	0.04	0.15	1.45
	L	.	.	0.04	.	.	0.01	.
COMMUNIST AREAS	P	.	.	.	.	.	.	.
	L	.	.	0.50	.	.	.	0.50
TOTAL SUPPLY	P	10.17	3.39	0.06	3.15	0.04	1.87	18.69
	L	.	.	0.54	.	.	0.01	0.50

4

MARKET INPUT SUMMARY

PROBLEM: WAES85A BASE YR: 1975.
CASE: 31JAN77 CASE YR: 1985.

INDUSTRIAL HEAT, LIGHT AND POWER		OIL	COAL	GAS	ELECTRICITY
NORTH AMERICA	:	1.6	4.8	4.2	1.4
EUROPE	:	2.9	2.6	1.8	1.4
JAPAN	:	3.4	1.4	0.0	0.7
REST WOCA	:	3.5	3.0	0.9	1.3
		11.3	11.7	6.9	4.8

INDUSTRIAL NON ENERGY		OIL	COAL	GAS
NORTH AMERICA	:	3.5	0.1	0.7
EUROPE	:	2.2	0.1	0.3
JAPAN	:	0.9	.	.
REST WOCA	:	1.3	.	.
		7.9	0.2	1.0

DOMESTIC HEAT,LIGHT AND POWER		OIL	COAL	GAS	ELECTRICITY
NORTH AMERICA	:	2.1	0.4	3.3	2.5
EUROPE	:	4.4	0.6	0.7	1.0
JAPAN	:	1.3	0.0	0.2	0.3
REST WOCA	:	1.7	0.2	0.6	0.6
		9.4	1.2	4.8	4.4

TRANSPORT		OIL	ELECTRICITY
NORTH AMERICA	:	10.0	0.0
EUROPE	:	5.2	0.1
JAPAN	:	1.9	0.0
REST WOCA	:	7.3	0.0
		24.5	0.1

DEMAND			IND H,L&P	IND NONEN	DOM H,L&P	TRANSPORT
NORTH AMERICA						
	DEMAND	:	12.0	4.3	8.3	10.0
	SUPPLY	:	12.0	4.3	8.3	10.0
	DEFICIT	:	0.0	0.0	0.0	0.0
EUROPE						
	DEMAND	:	8.6	2.6	6.7	5.3
	SUPPLY	:	8.6	2.6	6.7	5.3
	DEFICIT	:	0.0	0.0	0.0	0.0
JAPAN						
	DEMAND	:	5.5	0.9	1.8	2.0
	SUPPLY	:	5.5	0.9	1.8	2.0
	DEFICIT	:	0.0	0.0	0.0	0.0
REST WOCA						
	DEMAND	:	8.6	1.3	3.1	7.4
	SUPPLY	:	8.6	1.3	3.1	7.4
	DEFICIT	:	0.0	0.0	0.0	0.0

W O C A

DEMAND/SUPPLY MATRIX

PROBLEM: WAES85A
CASE: 31 JAN77

BASE YR: 1975.
CASE YR: 1985.

5

	OIL	COAL	GAS	ELECTRICITY	NUCLEAR	HYDRO-ETC	TOTAL
TRANSPORT	24.53			0.15			24.68
INDUSTRY	11.33	11.71	6.91	4.77			34.72
DOMESTIC	9.39	1.22	4.85	4.44			19.89
TOTAL INTO MARKETS	45.25	12.92	11.76	9.36			79.29
TRANSM./DISTR.LOSSES*	3.04	0.87	2.96	1.65			8.51
CONV. THERMAL ELECTR.	1.58	2.61	0.75	-4.94			
USES+LOSSES	2.93	4.85	1.40				9.18
TOTAL INPUT	4.51	7.47	2.15				
GAS MANUFACTURE		0.0	0.0				
USES+LOSSES		0.0					0.0
TOTAL INPUT		0.0					
PRIMARY ENERGY DEMAND							
OUTPUT	52.79	21.26	16.87		3.28	2.79	96.98
INPUT					9.36	7.97	108.25
CHEMICAL FEED	7.93	0.21	1.02				9.16
LUBES & ASPHALTS	0.0						0.0
TOTAL REFINED PRODUCT	60.72						
REFINERY USE & PROCESS LOSS*	2.89	0.71	1.30				4.90
SYNCRUDE MANUFACTURE	0.0	0.0					
USES+LOSSES		0.0					0.0
TOTAL INPUT		0.0					
PRIMARY ENERGY SUPPLY	63.61	22.18	19.19		9.36	7.97	122.31

*INCLUDES OPEC & OCEC

ALL QUANTITIES ARE GIVEN IN MILLION BARREL PER DAY OIL EQUIVALENT

6

DEMAND/SUPPLY MATRIX NORTH AMERICA — PROBLEM: WAES85A — BASE YR: 1975.
CASE: 31JAN77 — CASE YR: 1985.

	OIL	COAL	GAS	ELECTRICITY		TOTAL
TRANSPORT	10.02			0.01		10.03
INDUSTRY	1.56	4.79	4.19	1.44		11.97
DOMESTIC	2.08	0.42	3.32	2.49		8.31
TOTAL INTO MARKETS	13.65	5.20	7.51	3.94		30.31
TRANSM./DISTR.LOSSES*	0.97	0.44	1.74	0.70		3.85
CONV. THERMAL ELECTR.	0.46	1.18	0.23	-1.88		
USES+LOSSES	0.86	2.20	0.43			3.49
TOTAL INPUT	1.32	3.39	0.66			
GAS MANUFACTURE		0.0	0.0			
USES+LOSSES		0.0				0.0
TOTAL INPUT		0.0				
PRIMARY ENERGY DEMAND				NUCLEAR	HYDRO-ETC	
OUTPUT	15.95	9.03	9.92	1.51	1.25	37.65
INPUT				4.31	3.56	42.77
CHEMICAL FEED	3.49	0.13	0.69			4.31
LUBES & ASPHALTS	0.0					0.0
TOTAL REFINED PRODUCT	19.44					
REFINERY USE & PROCESS LOSS*	0.65	0.30	0.69			1.64
SYNCRUDE MANUFACTURE	0.0	0.0				
USES+LOSSES		0.0				0.0
TOTAL INPUT		0.0				
PRIMARY ENERGY SUPPLY	20.09	9.46	11.30	4.31	3.56	48.72

ALL QUANTITIES ARE GIVEN IN MILLION BARREL PER DAY OIL EQUIVALENT

7

DEMAND/SUPPLY MATRIX EUROPE — PROBLEM: WAES85A — BASE YR: 1975.
CASE: 31JAN77 — CASE YR: 1985.

	OIL	COAL	GAS	ELECTRICITY		TOTAL
TRANSPORT	5.23			0.08		5.31
INDUSTRY	2.87	2.59	1.79	1.38		8.64
DOMESTIC	4.38	0.62	0.73	1.01		6.74
TOTAL INTO MARKETS	12.48	3.21	2.52	2.47		20.69
TRANSM./DISTR.LOSSES*	0.81	0.17	0.56	0.44		1.98
CONV. THERMAL ELECTR.	0.29	0.79	0.12	-1.20		
USES+LOSSES	0.54	1.48	0.22			2.23
TOTAL INPUT	0.83	2.27	0.33			
GAS MANUFACTURE		0.0	0.0			
USES+LOSSES		0.0				0.0
TOTAL INPUT		0.0				
PRIMARY ENERGY DEMAND				NUCLEAR	HYDRO-ETC	
OUTPUT	14.13	5.66	3.41	1.02	0.69	24.91
INPUT				2.91	1.97	28.08
CHEMICAL FEED	2.17	0.08	0.34			2.58
LUBES & ASPHALTS	0.0					0.0
TOTAL REFINED PRODUCT	16.30					
REFINERY USE & PROCESS LOSS*	1.05	0.17	0.0			1.22
SYNCRUDE MANUFACTURE	0.0	0.0				
USES+LOSSES		0.0				0.0
TOTAL INPUT		0.0				
PRIMARY ENERGY SUPPLY	17.35	5.90	3.75	2.91	1.97	31.88

ALL QUANTITIES ARE GIVEN IN MILLION BARREL PER DAY OIL EQUIVALENT

DEMAND/SUPPLY MATRIX JAPAN | PROBLEM: WAES85A | BASE YR: 1975. | 8

CASE: 31JAN77 | CASE YR: 1985.

	OIL	COAL	GAS	ELECTRICITY	NUCLEAR	HYDRO-ETC	TOTAL
TRANSPORT	1.95			0.04			1.99
INDUSTRY	3.44	1.36	0.0	0.66			5.46
DOMESTIC	1.27	0.0	0.18	0.32			1.77
TOTAL INTO MARKETS	6.66	1.36	0.18	1.01			9.22
TRANSM./DISTR.LOSSES*	0.45	0.05	0.08	0.18			0.76
CONV. THERMAL ELECTR.	0.36	0.18	0.12	-0.66			
USES+LOSSES	0.67	0.33	0.22				1.22
TOTAL INPUT	1.03	0.51	0.34				
GAS MANUFACTURE		0.0	0.0				
USES+LOSSES		0.0					0.0
TOTAL INPUT		0.0					
PRIMARY ENERGY DEMAND					NUCLEAR	HYDRO-ETC	
OUTPUT	8.14	1.92	0.61		0.34	0.20	11.20
INPUT					0.97	0.56	12.19
CHEMICAL FEED	0.93	0.0	0.0				0.93
LUBES & ASPHALTS	0.0						0.0
TOTAL REFINED PRODUCT	9.07						
REFINERY USE & PROCESS LOSS*	0.65	0.10	0.0				0.75
SYNCRUDE MANUFACTURE	0.0	0.0					
USES+LOSSES		0.0					0.0
TOTAL INPUT		0.0					
PRIMARY ENERGY SUPPLY	9.72	2.02	0.61		0.97	0.56	13.87

ALL QUANTITIES ARE GIVEN IN MILLION BARREL PER DAY OIL EQUIVALENT

DEMAND/SUPPLY MATRIX REST WOCA | PROBLEM: WAES85A | BASE YR: 1975. | 9

CASE: 31JAN77 | CASE YR: 1985.

	OIL	COAL	GAS	ELECTRICITY	NUCLEAR	HYDRO-ETC	TOTAL
TRANSPORT	7.33			0.02			7.35
INDUSTRY	3.46	2.96	0.93	1.30			8.65
DOMESTIC	1.66	0.18	0.61	0.61			3.07
TOTAL INTO MARKETS	12.45	3.14	1.55	1.93			19.07
TRANSM./DISTR.LOSSES*	0.80	0.09	0.42	0.34			1.65
CONV. THERMAL ELECTR.	0.46	0.45	0.29	-1.20			
USES+LOSSES	0.86	0.84	0.53				2.24
TOTAL INPUT	1.33	1.30	0.82				
GAS MANUFACTURE		0.0	0.0				
USES+LOSSES		0.0					0.0
TOTAL INPUT		0.0					
PRIMARY ENERGY DEMAND					NUCLEAR	HYDRO-ETC	
OUTPUT	14.57	4.54	2.78		0.41	0.66	22.96
INPUT					1.17	1.88	24.94
CHEMICAL FEED	1.34	0.0	0.0				1.34
LUBES & ASPHALTS	0.0						0.0
TOTAL REFINED PRODUCT	15.91						
REFINERY USE & PROCESS LOSS*	0.35	0.15	0.60				1.10
SYNCRUDE MANUFACTURE	0.0	0.0					
USES+LOSSES		0.0					0.0
TOTAL INPUT		0.0					
PRIMARY ENERGY SUPPLY	16.26	4.69	3.38		1.17	1.88	27.38

ALL QUANTITIES ARE GIVEN IN MILLION BARREL PER DAY OIL EQUIVALENT

SUMMARY
(ANNUAL CAPITAL CHARGES - CASE YEAR)

PROBLEM: WAES85A BASE YR: 1975.
CASE: 31JAN77 CASE YR: 1985.

10

PRODUCERS			OPEC	OCE(
OIL				
PRODUCTION	:	B$	0.82	0.0
REFINING	:	B$	0.34	0.0
COAL				
PRODUCTION	:	B$	0.0	2.01
CONVERSION	:	B$	0.0	0.0
GAS				
PRODUCTION	:	B$	0.0	0.0
LIQUEFACTION	:	B$	0.0	0.0
TOTAL	:	B$	1.16	2.01

CONSUMERS			NORTH AMERICA	EUROPE	JAPAN	REST WOCA
OIL						
PRODUCTION	:	B$	12.62	4.95	0.10	5.27
REFINING	:	B$	3.17	2.73	2.44	3.25
DISTRIBUTION	:	B$	0.63	0.56	0.32	0.71
TOTAL	:	B$	16.42	8.23	2.86	9.23
COAL						
PRODUCTION	:	B$	2.16	0.09	0.08	0.0
CONVERSION	:	B$	0.0	0.0	0.0	0.0
DISTRIBUTION	:	B$	0.18	0.10	0.07	0.16
TOTAL	:	B$	2.35	0.19	0.15	0.16
GAS						
PRODUCTION	:	B$	12.96	3.89	0.12	4.38
REGASIFICATION	:	B$	0.0	0.0	0.09	0.0
DISTRIBUTION	:	B$	1.32	0.50	0.07	1.67
TOTAL	:	B$	14.28	4.39	0.27	6.05
ELECTRICITY						
FUEL OIL	:	B$	0.69	0.0	0.55	0.0
COAL	:	B$	0.0	0.0	0.68	1.35
GAS	:	B$	0.0	0.05	0.40	0.78
NUCLEAR	:	B$	8.12	5.64	1.95	2.41
HYDRO ELEC 1	:	B$	1.98	0.28	0.31	1.26
OTHER	:	B$	0.0	0.65	0.09	0.55
DISTRIBUTION	:	B$	0.10	0.67	0.10	0.0
TOTAL	:	B$	10.90	7.30	4.09	6.36
IND HEATING EQP	:	B$	3.15	2.59	1.90	3.26
DOM HEATING EQP	:	B$	8.78	4.78	1.60	2.27
TOTAL CAPEX	:	B$	55.88	27.47	10.87	27.32

11

SUMMARY

(PRODUCING REGIONS REVENUES)

PROBLEM: WAES85A CASE: 31JAN77

BASE YR: 1975. CASE YR: 1985.

PRODUCERS			OPEC	OCEC
GOVERNMENT TAXES *				
CRUDE OIL			11.28	0.0
COAL			0.0	2.00
GAS			8.32	0.0
TARIFFS ***				
CRUDE OIL			18.25	18.25
DISTILLATES			20.73	20.73
FUEL OIL			13.29	13.29
COAL			11.68	11.68
GAS			20.73	20.73
GOVT. TAKE(FROM TAXES)				
OIL	:	B$	146.03	0.0
COAL	:	B$	0.0	4.08
GAS	:	B$	3.94	0.0
TOTAL GOVT. TAKE	:	B$	149.98	4.08
REVENUES FROM TARIFFS				
CRUDE OIL	:	B$	219.63	0.0
DISTILLATES	:	B$	13.61	0.0
FUEL OIL	:	B$	3.38	0.0
COAL	:	B$	0.0	23.81
GAS	:	B$	9.83	0.0
TOTAL REVENUES	:	B$	246.46	23.81

* DOLLAR/BOE

*** DOLLAR/BOE. AVERAGE WORLD PRICE LAID-IN AT PORT OF ENTRY

12

SUMMARY

(CONSUMING REGIONS COSTS AND REVENUES)

PROBLEM: WAES85A CASE: 31JAN77

BASE YR: 1975. CASE YR: 1985.

CONSUMERS (TOTAL COSTS)			NORTH AMERICA	EUROPE	JAPAN	REST WOCA
IMPORT BILL						
CRUDE OIL	:	B$	17.91	88.45	64.20	62.39
DISTILLATES	:	B$	13.61	0.0	0.0	0.0
FUEL OIL	:	B$	0.0	0.0	0.0	3.38
COAL	:	B$	0.0	13.27	7.50	19.98
GAS	:	B$	0.0	1.51	4.05	9.11
TOTAL IMPORT BILL	:	B$	31.53	103.23	75.76	94.86
INTERNAL COSTS **						
OIL	:	B$	33.28	28.52	8.18	18.98
COAL	:	B$	23.84	10.36	1.58	2.08
GAS	:	B$	19.13	5.90	0.49	7.44
ELECTRICITY	:	B$	15.56	10.30	5.42	8.69
TOTAL	:	B$	91.80	55.07	15.67	37.18
ENERGY CONS EQUIP	:	B$	11.93	7.37	3.50	5.53
TOTAL INTERNAL	:	B$	103.73	62.44	19.17	42.71
REVENUES FROM EXPORTS						
CRUDE OIL	:	B$	0.0	0.0	0.0	0.0
DISTILLATES	:	B$	0.0	0.0	0.0	0.0
FUEL OIL	:	B$	0.0	0.0	0.0	0.0
COAL	:	B$	14.81	0.0	0.0	0.0
GAS	:	B$	0.0	0.0	0.0	0.0
TOTAL REVENUES	:	B$	14.81	0.0	0.0	0.0
NET CONSUM REG. COST	:	B$	120.45	165.67	94.93	137.58
IMPORT/EXPORT BALANCE						
EQUILIBRIUM	:	B$	16.72	103.23	75.76	94.86
LOANS AT 15.0% PA	:	B$	0.0	0.0	0.0	0.0
SHORTFALL	:	B$	0.0	0.0	0.0	0.0

** CAPEX CHARGED AS A CONSTANT COST CHARGED OVER 100.0% OF THE USEFUL PHYSICAL LIFETIME OF A PLANT AT 15.0% ANNUAL COST OF CAPITAL.

1

VOLUMETRIC SUMMARY
====================

PROBLEM: WAES85B BASE YR: 1975.
CASE: 31JAN77 CASE YR: 1985.

PRODUCTION

	OIL	COAL	GAS	NUCLEAR	HYDRO	TOTAL
NORTH AMERICA	15.6	11.9	9.9	3.9	3.3	44.5
EUROPE	4.1	1.1	3.4	2.2	1.4	12.2
JAPAN	0.1	0.2	0.1	0.7	0.6	1.6
OPEC	26.9	0.0	0.2	0.0	0.0	27.1
OCEC	0.0	5.5	0.0	0.0	0.0	5.5
REST WOCA	6.2	0.0	2.2	1.2	1.7	11.3
SUBTOTAL	52.9	18.7	15.8	7.9	7.0	102.2
WOCA EXPORTS	2.0	0.5	0.5	0.0	0.0	3.0
TOTAL	54.9	19.2	16.3	7.9	7.0	105.2

TOTAL WORLD ENERGY PRODUCTION (EXCLUDING COMMUNIST AREA INTERNAL): 105.2

CONVERSION(INPUT)

	OIL	COAL	GAS	NUCLEAR	HYDRO
OIL REFINING	54.9				
COAL GASIFICATION		0.0			
COAL LIQUEFACTION		0.0			
GAS LIQUEFACTION (NOT INCL. COMMUNIST AREA)				0.0	
ELECTRICITY GENERATION	2.7	5.1	1.2	7.9	7.0

MARKET INPUT

	OIL	COAL	GAS	ELECTRICITY
INDUSTRIAL				
HEAT, LIGHT AND POWER	9.1	10.3	6.2	3.5
NON ENERGY	6.6	0.2	0.9	
DOMESTIC				
HEAT, LIGHT AND POWER	8.5	2.2	4.7	3.4
TRANSPORT AND BUNKERS	23.0			0.1

ALL QUANTITIES ARE GIVEN IN MILLION BARREL PER DAY OIL EQUIVALENT

2

PRODUCTION AND TRANSPORTS CRUDE OIL — PROBLEM: WAES85B, CASE: 31JAN77 — BASE YR: 1975. CASE YR: 1985.

FROM \ TO	NORTH AMERICA	EUROPE	JAPAN	REST WOCA	OPEC	PROD"N
NORTH AMERICA	15.60	.	.	.	.	15.60
EUROPE	.	4.07	.	.	.	4.07
JAPAN	.	.	0.08	.	.	0.08
REST WOCA	.	.	.	6.20	.	6.20
OPEC	2.09	8.51	6.58	6.33	3.39	26.91
COMMUNIST AREAS	.	2.00	.	.	.	2.00
TOTAL SUPPLY	17.69	14.58	6.66	12.53	3.39	54.86

PRODUCTION AND TRANSPORTS OIL PRODUCTS — PROBLEM: WAES85B, CASE: 31JAN77 — BASE YR: 1975. CASE YR: 1985.

F=FUEL OIL, D=DISTILLATES

FROM \ TO		NORTH AMERICA	EUROPE	JAPAN	REST WOCA	OPEC	PROD"N
NORTH AMERICA	F	2.97	.	.	.	.	2.97
	D	13.80	.	.	.	.	13.80
EUROPE	F	.	4.96	.	.	.	4.96
	D	.	8.74	.	.	.	8.74
JAPAN	F	.	.	2.51	.	.	2.51
	D	.	.	3.75	.	.	3.75
REST WOCA	F	.	.	.	3.95	.	3.95
	D	.	.	.	8.30	.	8.30
OPEC	F	.	.	0.88	.	.	0.88
	D	1.55	.	.	0.72	0.00	2.27
TOTAL SUPPLY	F	2.97	4.96	3.39	3.95	.	15.27
	D	15.35	8.74	3.75	9.02	0.00	36.86

3

PRODUCTION AND TRANSPORTS COAL — PROBLEM: WAES85B, CASE: 31JAN77 — BASE YR: 1975. CASE YR: 1985.

FROM \ TO	NORTH AMERICA	EUROPE	JAPAN	REST WOCA	LOSS	PROD"N
NORTH AMERICA	7.27	4.11	0.25	.	0.23	11.87
EUROPE	.	1.07	.	.	0.02	1.09
JAPAN	.	.	0.18	.	0.00	0.18
OCEC	.	.	0.94	4.47	0.11	5.52
COMMUNIST AREAS	.	0.50	.	.	.	0.50
TOTAL SUPPLY	7.27	5.68	1.37	4.47	0.37	19.15

PRODUCTION AND TRANSPORTS GAS — PROBLEM: WAES85B, CASE: 31JAN77 — BASE YR: 1975. CASE YR: 1985.

P=BY PIPE L=AS LNG

FROM \ TO		NORTH AMERICA	EUROPE	JAPAN	REST WOCA	OPEC	LOSS	PROD"N
NORTH AMERICA	P	8.91	.	.	.	.	0.99	9.90
EUROPE	P	.	3.10	.	.	.	0.34	3.44
JAPAN	P	.	.	0.06	.	.	0.01	0.07
REST WOCA	P	.	.	.	1.99	.	0.22	2.21
OPEC	P	.	0.17	.	.	.	0.02	0.19
	L	.	.	.	.	.	.	.
COMMUNIST AREAS	P	.	0.06	.	.	.	0.01	0.06
	L	.	.	0.44	.	.	.	0.44
TOTAL SUPPLY	P	8.91	3.33	0.06	1.99	.	1.59	15.87
	L	.	.	0.44	.	.	.	0.44

4

MARKET INPUT SUMMARY

PROBLEM: WAES858 BASE YR: 1975.
CASE: 31JAN77 CASE YR: 1985.

INDUSTRIAL HEAT, LIGHT AND POWER	OIL	COAL	GAS	ELECTRICITY
NORTH AMERICA :	1.3	4.1	3.6	1.2
EUROPE :	2.3	2.3	1.9	1.1
JAPAN :	2.6	1.0	0.0	0.4
REST WOCA :	2.9	2.9	0.7	0.7
	9.1	10.3	6.2	3.5

INDUSTRIAL NON ENERGY	OIL	COAL	GAS
NORTH AMERICA :	3.0	0.1	0.6
EUROPE :	1.7	0.1	0.3
JAPAN :	0.8	.	.
REST WOCA :	1.1	.	.
	6.6	0.2	0.9

DOMESTIC HEAT,LIGHT AND POWER	OIL	COAL	GAS	ELECTRICITY
NORTH AMERICA :	2.4	0.4	3.2	2.0
EUROPE :	3.8	1.3	0.7	0.6
JAPAN :	1.2	0.0	0.2	0.2
REST WOCA :	1.1	0.5	0.5	0.5
	8.5	2.2	4.7	3.4

TRANSPORT	OIL	ELECTRICITY
NORTH AMERICA :	9.6	0.0
EUROPE :	4.6	0.1
JAPAN :	1.7	0.0
REST WOCA :	6.8	0.0
	22.6	0.1

DEMAND

		IND H,L&P	IND NONEN	DOM H,L&P	TRANSPORT
NORTH AMERICA					
	DEMAND :	10.3	3.7	8.1	9.6
	SUPPLY :	10.3	3.7	8.1	9.6
	DEFICIT :	0.0	0.0	0.0	0.0
EUROPE					
	DEMAND :	7.5	2.0	6.5	4.7
	SUPPLY :	7.5	2.0	6.5	4.7
	DEFICIT :	0.0	0.0	0.0	0.0
JAPAN					
	DEMAND :	4.1	0.8	1.6	1.7
	SUPPLY :	4.1	0.8	1.6	1.7
	DEFICIT :	0.0	0.0	0.0	0.0
REST WOCA					
	DEMAND :	7.3	1.1	2.6	6.8
	SUPPLY :	7.3	1.1	2.6	6.8
	DEFICIT :	0.0	0.0	0.0	0.0

5

W O C A — DEMAND/SUPPLY MATRIX — PROBLEM: WAES85B — CASE: 31JAN77 — BASE YR: 1975. — CASE YR: 1985.

	OIL	COAL	GAS	NUCLEAR	ELECTRICITY	HYDRO-ETC	TOTAL
TRANSPORT	22.60				0.15		22.75
INDUSTRY	9.15	10.32	6.25		3.50		29.22
DOMESTIC	8.50	2.19	4.68		3.44		18.81
TOTAL INTO MARKETS	40.24	12.51	10.93		7.09		70.78
TRANSM./DISTR.LOSSES*	2.61	0.75	2.50		1.25		7.11
CONV. THERMAL ELECTR.	0.96	1.79	0.41		-3.15		
USES+LOSSES	1.78	3.33	0.75				5.86
TOTAL INPUT	2.73	5.12	1.16				
GAS MANUFACTURE		0.0	0.0				
USES+LOSSES		0.0					0.0
TOTAL INPUT		0.0					
PRIMARY ENERGY DEMAND							
OUTPUT	45.58	18.38	14.59	2.76		2.43	83.74
INPUT				7.89		6.95	93.38
CHEMICAL FEED	6.56	0.17	0.85				7.58
LUBES & ASPHALTS	0.0						0.0
TOTAL REFINED PRODUCT	52.14						
REFINERY USE & PROCESS LOSS*	2.72	0.61	0.87				4.20
SYNCRUDE MANUFACTURE	0.0	0.0					
USES+LOSSES		0.0					0.0
TOTAL INPUT		0.0					
PRIMARY ENERGY SUPPLY	54.86	19.16	16.31	7.89		6.95	105.17

*INCLUDES OPEC & OCEC

ALL QUANTITIES ARE GIVEN IN MILLION BARREL PER DAY OIL EQUIVALENT

6

DEMAND/SUPPLY MATRIX NORTH AMERICA — PROBLEM: WAES85B — CASE: 31JAN77 — BASE YR: 1975. — CASE YR: 1985.

	OIL	COAL	GAS	NUCLEAR	ELECTRICITY	HYDRO-ETC	TOTAL
TRANSPORT	9.57				0.01		9.58
INDUSTRY	1.34	4.12	3.60		1.24		10.30
DOMESTIC	2.44	0.41	3.25		2.03		8.12
TOTAL INTO MARKETS	13.35	4.53	6.85		3.28		28.00
TRANSM./DISTR.LOSSES*	0.92	0.38	1.52		0.58		3.40
CONV. THERMAL ELECTR.	0.39	0.77	0.19		-1.35		
USES+LOSSES	0.72	1.43	0.36				2.51
TOTAL INPUT	1.10	2.20	0.55				
GAS MANUFACTURE		0.0	0.0				
USES+LOSSES		0.0					0.0
TOTAL INPUT		0.0					
PRIMARY ENERGY DEMAND							
OUTPUT	15.36	7.11	8.93	1.35		1.16	33.91
INPUT				3.85		3.30	38.56
CHEMICAL FEED	2.96	0.11	0.59				3.66
LUBES & ASPHALTS	0.0						0.0
TOTAL REFINED PRODUCT	18.33						
REFINERY USE & PROCESS LOSS*	0.91	0.28	0.38				1.58
SYNCRUDE MANUFACTURE	0.0	0.0					
USES+LOSSES		0.0					0.0
TOTAL INPUT		0.0					
PRIMARY ENERGY SUPPLY	19.24	7.51	9.90	3.85		3.30	43.80

ALL QUANTITIES ARE GIVEN IN MILLION BARREL PER DAY OIL EQUIVALENT

7

DEMAND/SUPPLY MATRIX EUROPE — PROBLEM: WAES85B — CASE: 31JAN77 — BASE YR: 1975. — CASE YR: 1985.

	OIL	COAL	GAS	ELECTRICITY		TOTAL
TRANSPORT	4.60			0.08		4.68
INDUSTRY	2.26	2.26	1.88	1.13		7.53
DOMESTIC	3.81	1.30	0.73	0.65		6.48
TOTAL INTO MARKETS	10.66	3.55	2.61	1.86		18.69
TRANSM./DISTR.LOSSES*	0.68	0.14	0.54	0.33		1.69
CONV. THERMAL ELECTR.	0.22	0.63	0.09	-0.94		
USES+LOSSES	0.41	1.18	0.16			1.75
TOTAL INPUT	0.62	1.81	0.25			
GAS MANUFACTURE		0.0	0.0			
USES+LOSSES		0.0				0.0
TOTAL INPUT		0.0				
PRIMARY ENERGY DEMAND				NUCLEAR	HYDRO-ETC	
OUTPUT	11.97	5.50	3.40	0.76	0.48	22.12
INPUT				2.19	1.37	24.44
CHEMICAL FEED	1.72	0.06	0.27			2.05
LUBES & ASPHALTS	0.0					0.0
TOTAL REFINED PRODUCT	13.70					
REFINERY USE & PROCESS LOSS*	0.88	0.14	0.0			1.02
SYNCRUDE MANUFACTURE	0.0	0.0				
USES+LOSSES		0.0				0.0
TOTAL INPUT		0.0				
PRIMARY ENERGY SUPPLY	14.58	5.70	3.67	2.19	1.37	27.51

ALL QUANTITIES ARE GIVEN IN MILLION BARREL PER DAY OIL EQUIVALENT

8

DEMAND/SUPPLY MATRIX JAPAN — PROBLEM: WAES85B — CASE: 31JAN77 — BASE YR: 1975. — CASE YR: 1985.

	OIL	COAL	GAS	ELECTRICITY		TOTAL
TRANSPORT	1.67			0.04		1.71
INDUSTRY	2.63	1.02	0.03	0.41		4.08
DOMESTIC	1.16	0.01	0.18	0.24		1.59
TOTAL INTO MARKETS	5.46	1.03	0.21	0.69		7.38
TRANSM./DISTR.LOSSES*	0.36	0.03	0.07	0.12		0.58
CONV. THERMAL ELECTR.	0.20	0.09	0.08	-0.37		
USES+LOSSES	0.37	0.16	0.15			0.69
TOTAL INPUT	0.58	0.25	0.23			
GAS MANUFACTURE		0.0	0.0			
USES+LOSSES		0.0				0.0
TOTAL INPUT		0.0				
PRIMARY ENERGY DEMAND				NUCLEAR	HYDRO-ETC	
OUTPUT	6.39	1.31	0.51	0.24	0.20	8.65
INPUT				0.69	0.56	9.46
CHEMICAL FEED	0.75	0.0	0.0			0.75
LUBES & ASPHALTS	0.0					0.0
TOTAL REFINED PRODUCT	7.14					
REFINERY USE & PROCESS LOSS*	0.40	0.06	0.0			0.47
SYNCRUDE MANUFACTURE	0.0	0.0				
USES+LOSSES		0.0				0.0
TOTAL INPUT		0.0				
PRIMARY ENERGY SUPPLY	7.55	1.37	0.51	0.69	0.56	10.68

ALL QUANTITIES ARE GIVEN IN MILLION BARREL PER DAY OIL EQUIVALENT

9

DEMAND/SUPPLY MATRIX REST WOCA — PROBLEM: WAES85B — BASE YR: 1975.

CASE: 31JAN77 — CASE YR: 1985.

	OIL	COAL	GAS	NUCLEAR	ELECTRICITY	HYDRO-ETC	TOTAL
TRANSPORT	6.76				0.02		6.78
INDUSTRY	2.92	2.92	0.73		0.73		7.31
DOMESTIC	1.09	0.48	0.52		0.52		2.62
TOTAL INTO MARKETS	10.77	3.40	1.26		1.27		16.71
TRANSM./DISTR.LOSSES*	0.65	0.09	0.34		0.22		1.30
CONV. THERMAL ELECTR.	0.15	0.30	0.04		-0.49		
USES+LOSSES	0.28	0.56	0.08				0.92
TOTAL INPUT	0.43	0.86	0.13				
GAS MANUFACTURE		0.0	0.0				
USES+LOSSES		0.0					0.0
TOTAL INPUT		0.0					
PRIMARY ENERGY DEMAND							
OUTPUT	11.85	4.35	1.72	0.40		0.60	18.93
INPUT				1.16		1.71	20.80
CHEMICAL FEED	1.12	0.0	0.0				1.12
LUBES & ASPHALTS	0.0						0.0
TOTAL REFINED PRODUCT	12.97						
REFINERY USE & PROCESS LOSS*	0.28	0.12	0.49				0.89
SYNCRUDE MANUFACTURE	0.0	0.0					
USES+LOSSES		0.0					0.0
TOTAL INPUT		0.0					
PRIMARY ENERGY SUPPLY	13.25	4.47	2.21	1.16		1.71	22.81

ALL QUANTITIES ARE GIVEN IN MILLION BARREL PER DAY OIL EQUIVALENT

SUMMARY (ANNUAL CAPITAL CHARGES - CASE YEAR)

PROBLEM: WAES85B BASE YR: 1975.
CASE: 31JAN77 CASE YR: 1985.

10

PRODUCERS			OPEC	OCEC		
OIL						
PRODUCTION	:	B$	0.44	0.0		
REFINING	:	B$	0.54	0.0		
COAL						
PRODUCTION	:	B$	0.0	1.89		
CONVERSION	:	B$	0.0	0.0		
GAS						
PRODUCTION	:	B$	0.0	0.0		
LIQUEFACTION	:	B$	0.0	0.0		
TOTAL	:	B$	0.98	1.89		
CONSUMERS			NORTH AMERICA	EUROPE	JAPAN	REST WOCA
OIL						
PRODUCTION	:	B$	12.62	4.95	0.10	5.27
REFINING	:	B$	2.97	1.94	1.33	2.39
DISTRIBUTION	:	B$	0.55	0.38	0.19	0.51
TOTAL	:	B$	16.14	7.27	1.62	8.16
COAL						
PRODUCTION	:	B$	1.85	0.06	0.0	0.0
CONVERSION	:	B$	0.0	0.0	0.0	0.0
DISTRIBUTION	:	B$	0.08	0.09	0.03	0.15
TOTAL	:	B$	1.93	0.15	0.03	0.15
GAS						
PRODUCTION	:	B$	9.05	3.59	0.12	4.13
REGASIFICATION	:	B$	0.0	0.0	0.07	0.0
DISTRIBUTION	:	B$	0.37	0.45	0.0	0.79
TOTAL	:	B$	9.42	4.05	0.19	4.92
ELECTRICITY						
FUEL OIL	:	B$	0.35	0.0	0.0	0.0
COAL	:	B$	0.0	0.0	0.24	0.60
GAS	:	B$	0.0	0.0	0.26	0.0
NUCLEAR	:	B$	7.17	4.12	1.37	2.38
HYDRO ELEC 1	:	B$	1.61	0.0	0.31	1.20
OTHER	:	B$	0.0	0.0	0.09	0.0
DISTRIBUTION	:	B$	0.0	0.11	0.0	0.0
TOTAL	:	B$	9.13	4.23	2.27	4.18
IND HEATING EQP	:	B$	2.40	1.99	1.18	2.62
DOM HEATING EQP	:	B$	8.39	5.64	1.30	2.01
TOTAL CAPEX	:	B$	47.42	23.32	6.60	22.04

11

S U M M A R Y
===================
(PRODUCING REGIONS REVENUES)

PROBLEM: WAES85B CASE: 31JAN77 BASE YR: 1975. CASE YR: 1985.

PRODUCERS			OPEC	OCEC
GOVERNMENT TAXES *				
CRUDE OIL			11.28	0.0
COAL			0.0	2.00
GAS			8.32	0.0
TARIFFS ***				
CRUDE OIL			18.25	18.25
DISTILLATES			20.73	20.73
FUEL OIL			13.29	13.29
COAL			11.68	11.68
GAS			20.73	20.73
GOVT. TAKE(FROM TAXES)				
OIL	:	B$	109.80	0.0
COAL	:	B$	0.0	3.95
GAS	:	B$	0.53	0.0
TOTAL GOVT. TAKE	:	B$	110.33	3.95
REVENUES FROM TARIFFS				
CRUDE OIL	:	B$	156.62	0.0
DISTILLATES	:	B$	17.21	0.0
FUEL OIL	:	B$	4.28	0.0
COAL	:	B$	0.0	23.06
GAS	:	B$	1.32	0.0
TOTAL REVENUES	:	B$	179.43	23.06

* DOLLAR/BOE

*** DOLLAR/BOE, AVERAGE WORLD PRICE LAID-IN AT PORT OF ENTRY

12

S U M M A R Y
===================
(CONSUMING REGIONS COSTS AND REVENUES)

PROBLEM: WAES85B CASE: 31JAN77 BASE YR: 1975. CASE YR: 1985.

CONSUMERS (TOTAL COSTS)			NORTH AMERICA	EUROPE	JAPAN	REST WOCA
IMPORT BILL						
CRUDE OIL	:	B$	13.91	70.00	43.86	42.17
DISTILLATES	:	B$	11.73	0.0	0.0	5.48
FUEL OIL	:	B$	0.0	0.0	4.28	0.0
COAL	:	B$	0.0	19.67	5.07	19.07
GAS	:	B$	0.0	1.93	3.32	0.0
TOTAL IMPORT BILL	:	B$	25.64	91.60	56.52	66.71
INTERNAL COSTS **						
OIL	:	B$	32.41	26.06	5.66	16.23
COAL	:	B$	21.18	4.24	1.02	1.98
GAS	:	B$	13.67	5.51	0.38	5.88
ELECTRICITY	:	B$	12.87	6.51	3.11	5.54
TOTAL	:	B$	80.13	42.32	10.17	29.62
ENERGY CONS EQUIP	:	B$	10.79	7.63	2.48	4.63
TOTAL INTERNAL	:	B$	90.92	49.95	12.65	34.25
REVENUES FROM EXPORTS						
CRUDE OIL	:	B$	0.0	0.0	0.0	0.0
DISTILLATES	:	B$	0.0	0.0	0.0	0.0
FUEL OIL	:	B$	0.0	0.0	0.0	0.0
COAL	:	B$	18.61	0.0	0.0	0.0
GAS	:	B$	0.0	0.0	0.0	0.0
TOTAL REVENUES	:	B$	18.61	0.0	0.0	0.0
NET CONSUM REG. COST	:	B$	97.96	141.55	69.17	100.96
IMPORT/EXPORT BALANCE						
EQUILIBRIUM	:	B$	7.04	91.60	56.52	66.71
LOANS AT 15.0% PA	:	B$	0.0	0.0	0.0	0.0
SHORTFALL	:	B$	0.0	0.0	0.0	0.0

** CAPEX CHARGED AS A CONSTANT COST CHARGED OVER 100.0% OF THE USEFUL PHYSICAL LIFETIME OF A PLANT AT 15.0% ANNUAL COST OF CAPITAL.

1

VOLUMETRIC SUMMARY

PROBLEM: WAES85C BASE YR: 1975.
CASE: 31JAN77 CASE YR: 1985.

PRODUCTION

	OIL	COAL	GAS	NUCLEAR	HYDRO	TOTAL
NORTH AMERICA	14.0	11.7	10.4	4.3	3.8	44.3
EUROPE	4.1	2.8	3.5	2.6	1.8	14.7
JAPAN	0.1	0.3	0.1	0.9	0.6	1.9
OPEC	39.8	0.0	1.5	0.0	0.0	41.3
OCEC	0.0	5.7	0.0	0.0	0.0	5.7
REST WOCA	6.2	0.0	2.3	1.2	1.9	11.5
SUBTOTAL	64.2	20.5	17.9	9.0	8.0	119.5
WOCA EXPORTS	2.0	0.5	0.5	0.0	0.0	3.0
TOTAL	66.2	21.0	18.4	9.0	8.0	122.5

TOTAL WORLD ENERGY PRODUCTION (EXCLUDING COMMUNIST AREA INTERNAL): 122.5

CONVERSION(INPUT)

	OIL	COAL	GAS	NUCLEAR	HYDRO
OIL REFINING	66.2				
COAL GASIFICATION		0.0			
COAL LIQUEFACTION		0.0			
GAS LIQUEFACTION (NOT INCL. COMMUNIST AREA)				0.0	
ELECTRICITY GENERATION	3.6	6.1	1.4	9.0	8.0

MARKET INPUT

	OIL	COAL	GAS	ELECTRICITY
INDUSTRIAL				
HEAT, LIGHT AND POWER	12.2	11.7	7.7	4.3
NON ENERGY	8.0	0.2	1.0	
DOMESTIC				
HEAT, LIGHT AND POWER	11.1	1.4	4.9	3.9
TRANSPORT AND BUNKERS	25.0			0.2

ALL QUANTITIES ARE GIVEN IN MILLION BARREL PER DAY OIL EQUIVALENT

2

PRODUCTION AND TRANSPORTS CRUDE OIL — PROBLEM: WAES85C, CASE: 31JAN77 — BASE YR: 1975., CASE YR: 1985.

FROM \ TO	NORTH AMERICA	EUROPE	JAPAN	REST WOCA	OPEC	PROD"N
NORTH AMERICA	14.04	.	.	.	.	14.04
EUROPE	.	4.07	.	.	.	4.07
JAPAN	.	.	0.08	.	.	0.08
REST WOCA	.	.	.	6.20	.	6.20
OPEC	6.13	12.36	9.31	9.37	2.63	39.79
COMMUNIST AREAS	.	2.00	.	.	.	2.00
TOTAL SUPPLY	20.17	18.43	9.39	15.57	2.63	66.18

PRODUCTION AND TRANSPORTS OIL PRODUCTS — PROBLEM: WAES85C, CASE: 31JAN77 — BASE YR: 1975., CASE YR: 1985.

F=FUEL OIL, D=DISTILLATES

FROM \ TO		NORTH AMERICA	EUROPE	JAPAN	REST WOCA	PROD"N
NORTH AMERICA	F	3.68	.	.	.	3.68
	D	15.07	.	.	.	15.07
EUROPE	F	.	6.41	.	.	6.41
	D	.	10.88	.	.	10.88
JAPAN	F	.	.	4.04	.	4.04
	D	.	.	4.71	.	4.71
REST WOCA	F	.	.	.	4.83	4.83
	D	.	.	.	10.33	10.33
OPEC	F	.	.	0.43	0.26	0.68
	D	1.76	.	.	.	1.76
TOTAL SUPPLY	F	3.68	6.41	4.47	5.08	19.65
	D	16.84	10.88	4.71	10.33	42.75

3

PRODUCTION AND TRANSPORTS COAL — PROBLEM: WAES85C, CASE: 31JAN77 — BASE YR: 1975., CASE YR: 1985.

FROM \ TO	NORTH AMERICA	EUROPE	JAPAN	REST WOCA	LOSS	PROD"N
NORTH AMERICA	8.32	3.03	0.15	.	0.23	11.74
EUROPE	.	2.73	.	.	0.05	2.79
JAPAN	.	.	0.25	.	0.01	0.26
OCEC	.	.	1.57	4.02	0.11	5.70
COMMUNIST AREAS	.	0.50	.	.	.	0.50
TOTAL SUPPLY	8.32	6.26	1.98	4.02	0.40	20.98

PRODUCTION AND TRANSPORTS GAS — PROBLEM: WAES85C, CASE: 31JAN77 — BASE YR: 1975., CASE YR: 1985.

P=BY PIPE L=AS LNG

FROM \ TO		NORTH AMERICA	EUROPE	JAPAN	REST WOCA	OPEC	LOSS	PROD"N
NORTH AMERICA	P	9.40	.	.	.	.	1.04	10.45
EUROPE	P	.	3.17	.	.	.	0.35	3.52
JAPAN	P	.	.	0.06	.	.	0.01	0.07
REST WOCA	P	.	.	.	2.07	.	0.23	2.30
OPEC	P	.	0.18	.	1.19	.	0.15	1.52
	L	.	.	.	.	.	.	.
COMMUNIST AREAS	P	.	0.02	.	.	.	0.00	0.02
	L	.	.	0.48	.	.	.	0.48
TOTAL SUPPLY	P	9.40	3.36	0.06	3.26	.	1.79	17.88
	L	.	.	0.48	.	.	.	0.48

4

MARKET INPUT SUMMARY
===================

PROBLEM: WAES85C BASE YR: 1975.
CASE: 31JAN77 CASE YR: 1985.

INDUSTRIAL HEAT, LIGHT AND POWER		OIL	COAL	GAS	ELECTRICITY
NORTH AMERICA	:	1.8	5.0	4.2	1.5
EUROPE	:	3.3	2.8	1.8	1.4
JAPAN	:	3.6	1.4	0.0	0.6
REST WOCA	:	3.5	2.6	1.7	0.9
		12.2	11.7	7.7	4.3

INDUSTRIAL NON ENERGY		OIL	COAL	GAS
NORTH AMERICA	:	3.5	0.1	0.7
EUROPE	:	2.3	0.1	0.4
JAPAN	:	0.9	.	.
REST WOCA	:	1.3	.	.
		8.0	0.2	1.0

DOMESTIC HEAT, LIGHT AND POWER

		OIL	COAL	GAS	ELECTRICITY
NORTH AMERICA	:	3.3	0.1	3.4	2.3
EUROPE	:	4.7	1.1	0.7	0.7
JAPAN	:	1.4	0.0	0.2	0.3
REST WOCA	:	1.7	0.2	0.6	0.6
		11.1	1.4	4.9	3.9

TRANSPORT

		OIL	ELECTRICITY
NORTH AMERICA	:	9.7	0.0
EUROPE	:	5.4	0.1
JAPAN	:	2.0	0.0
REST WOCA	:	7.4	0.0
		24.5	0.2

DEMAND

			IND H,L&P	IND NONEN	DOM H,L&P	TRANSPORT
NORTH AMERICA						
	DEMAND	:	12.4	4.3	9.1	9.7
	SUPPLY	:	12.4	4.3	9.1	9.7
	DEFICIT	:	0.0	0.0	0.0	0.0
EUROPE						
	DEMAND	:	9.3	2.7	7.3	5.5
	SUPPLY	:	9.3	2.7	7.2	5.5
	DEFICIT	:	0.0	0.0	0.0	0.0
JAPAN						
	DEMAND	:	5.6	0.9	1.8	2.1
	SUPPLY	:	5.6	0.9	1.8	2.1
	DEFICIT	:	0.0	0.0	0.0	0.0
REST WOCA						
	DEMAND	:	8.6	1.3	3.1	7.4
	SUPPLY	:	8.6	1.3	3.1	7.4
	DEFICIT	:	0.0	0.0	0.0	0.0

W O C A ====== DEMAND/SUPPLY MATRIX ==================== PROBLEM: WAESB5C CASE: 31JAN77 BASE YR: 1975. CASE YR: 1985. **5**

	OIL	COAL	GAS		ELECTRICITY		TOTAL
TRANSPORT	24.47				0.16		24.63
INDUSTRY	12.20	11.74	7.69		4.30		35.93
DOMESTIC	11.05	1.40	4.88		3.88		21.21
TOTAL INTO MARKETS	47.72	13.14	12.57		8.34		81.77
TRANSM./DISTR.LOSSES*	3.12	0.82	2.82		1.47		8.23
CONV. THERMAL ELECTR.	1.25	2.14	0.49		-3.87		
USES+LOSSES	2.31	3.97	0.91				7.19
TOTAL INPUT	3.56	6.10	1.40				
GAS MANUFACTURE		0.0	0.0				
USES+LOSSES		0.0					0.0
TOTAL INPUT		0.0					
PRIMARY ENERGY DEMAND				NUCLEAR		HYDRO-ETC	
OUTPUT	54.40	20.06	16.79	3.13		2.81	97.19
INPUT				8.96		8.02	108.23
CHEMICAL FEED	8.00	0.21	1.03				9.24
LUBES & ASPHALTS	0.0						0.0
TOTAL REFINED PRODUCT	62.40						
REFINERY USE & PROCESS LOSS*	3.79	0.71	0.54				5.04
SYNCRUDE MANUFACTURE	0.0	0.0					
USES+LOSSES		0.0					0.0
TOTAL INPUT		0.0					
PRIMARY ENERGY SUPPLY	66.18	20.99	18.36	8.96		8.02	122.51

*INCLUDES OPEC & OCEC

ALL QUANTITIES ARE GIVEN IN MILLION BARREL PER DAY OIL EQUIVALENT

DEMAND/SUPPLY MATRIX NORTH AMERICA ==================================== PROBLEM: WAES85C CASE: 31JAN77 BASE YR: 1975. CASE YR: 1985. **6**

	OIL	COAL	GAS		ELECTRICITY		TOTAL
TRANSPORT	9.67				0.01		9.68
INDUSTRY	1.79	4.96	4.17		1.49		12.41
DOMESTIC	3.31	0.13	3.36		2.26		9.06
TOTAL INTO MARKETS	14.77	5.10	7.52		3.76		31.15
TRANSM./DISTR.LOSSES*	1.03	0.40	1.61		0.66		3.70
CONV. THERMAL ELECTR.	0.44	0.93	0.22		-1.59		
USES+LOSSES	0.82	1.72	0.41				2.95
TOTAL INPUT	1.27	2.64	0.63				
GAS MANUFACTURE		0.0	0.0				
USES+LOSSES		0.0					0.0
TOTAL INPUT		0.0					
PRIMARY ENERGY DEMAND				NUCLEAR		HYDRO-ETC	
OUTPUT	17.06	8.14	9.77	1.51		1.33	37.81
INPUT				4.31		3.80	43.08
CHEMICAL FEED	3.46	0.13	0.68				4.27
LUBES & ASPHALTS	0.0						0.0
TOTAL REFINED PRODUCT	20.52						
REFINERY USE & PROCESS LOSS*	1.41	0.29	0.0				1.70
SYNCRUDE MANUFACTURE	0.0	0.0					
USES+LOSSES		0.0					0.0
TOTAL INPUT		0.0					
PRIMARY ENERGY SUPPLY	21.93	8.56	10.45	4.31		3.80	49.05

ALL QUANTITIES ARE GIVEN IN MILLION BARREL PER DAY OIL EQUIVALENT

7

DEMAND/SUPPLY MATRIX EUROPE — PROBLEM: WAES85C — BASE YR: 1975.
CASE: 31JAN77 — CASE YR: 1985.

	OIL	COAL	GAS	NUCLEAR	ELECTRICITY	HYDRO-ETC	TOTAL
TRANSPORT	5.39				0.09		5.48
INDUSTRY	3.31	2.78	1.79		1.39		9.27
DOMESTIC	4.71	1.08	0.73		0.72		7.25
TOTAL INTO MARKETS	13.42	3.87	2.51		2.21		22.00
TRANSM./DISTR.LOSSES*	0.86	0.18	0.55		0.39		1.99
CONV. THERMAL ELECTR.	0.26	0.70	0.10		-1.07		
USES+LOSSES	0.48	1.31	0.19				1.98
TOTAL INPUT	0.74	2.01	0.30				
GAS MANUFACTURE		0.0	0.0				
USES+LOSSES		0.0					0.0
TOTAL INPUT		0.0					
PRIMARY ENERGY DEMAND				NUCLEAR		HYDRO-ETC	
OUTPUT	15.02	6.06	3.37	0.91		0.62	25.97
INPUT				2.59		1.77	28.81
CHEMICAL FEED	2.27	0.08	0.35				2.70
LUBES & ASPHALTS	0.0						0.0
TOTAL REFINED PRODUCT	17.29						
REFINERY USE & PROCESS LOSS*	1.14	0.18	0.0				1.32
SYNCRUDE MANUFACTURE	0.0	0.0					
USES+LOSSES		0.0					0.0
TOTAL INPUT		0.0					
PRIMARY ENERGY SUPPLY	18.43	6.32	3.72	2.59		1.77	32.83

ALL QUANTITIES ARE GIVEN IN MILLION BARREL PER DAY OIL EQUIVALENT

8

DEMAND/SUPPLY MATRIX JAPAN — PROBLEM: WAES85C — BASE YR: 1975.
CASE: 31JAN77 — CASE YR: 1985.

	OIL	COAL	GAS	NUCLEAR	ELECTRICITY	HYDRO-ETC	TOTAL
TRANSPORT	2.02				0.04		2.06
INDUSTRY	3.64	1.40	0.0		0.56		5.60
DOMESTIC	1.37	0.0	0.18		0.27		1.83
TOTAL INTO MARKETS	7.03	1.40	0.18		0.87		9.49
TRANSM./DISTR.LOSSES*	0.46	0.04	0.08		0.15		0.73
CONV. THERMAL ELECTR.	0.26	0.15	0.10		-0.52		
USES+LOSSES	0.49	0.29	0.19				0.97
TOTAL INPUT	0.75	0.44	0.29				
GAS MANUFACTURE		0.0	0.0				
USES+LOSSES		0.0					0.0
TOTAL INPUT		0.0					
PRIMARY ENERGY DEMAND				NUCLEAR		HYDRO-ETC	
OUTPUT	8.25	1.89	0.55	0.31		0.20	11.19
INPUT				0.88		0.57	12.13
CHEMICAL FEED	0.93	0.0	0.0				0.93
LUBES & ASPHALTS	0.0						0.0
TOTAL REFINED PRODUCT	9.18						
REFINERY USE & PROCESS LOSS*	0.64	0.10	0.0				0.74
SYNCRUDE MANUFACTURE	0.0	0.0					
USES+LOSSES		0.0					0.0
TOTAL INPUT		0.0					
PRIMARY ENERGY SUPPLY	9.82	1.98	0.55	0.88		0.57	13.81

ALL QUANTITIES ARE GIVEN IN MILLION BARREL PER DAY OIL EQUIVALENT

DEMAND/SUPPLY MATRIX REST WOCA PROBLEM: WAES85C BASE YR: 1975.
CASE: 31JAN77 CASE YR: 1985.

	OIL	COAL	GAS	NUCLEAR	ELECTRICITY	HYDRO-ETC	TOTAL
TRANSPORT	7.38				0.02		7.40
INDUSTRY	3.46	2.60	1.73		0.86		8.65
DOMESTIC	1.66	0.18	0.61		0.61		3.07
TOTAL INTO MARKETS	12.50	2.78	2.34		1.50		19.12
TRANSM./DISTR.LOSSES*	0.77	0.08	0.43		0.26		1.54
CONV. THERMAL ELECTR.	0.28	0.35	0.06		-0.70		
USES+LOSSES	0.52	0.66	0.12				1.29
TOTAL INPUT	0.80	1.01	0.18				
GAS MANUFACTURE		0.0	0.0				
USES+LOSSES		0.0					0.0
TOTAL INPUT		0.0					
PRIMARY ENERGY DEMAND				NUCLEAR		HYDRO-ETC	
OUTPUT	14.07	3.87	2.95	0.41		0.66	21.96
INPUT				1.17		1.88	23.94
CHEMICAL FEED	1.34	0.0	0.0				1.34
LUBES & ASPHALTS	0.0						0.0
TOTAL REFINED PRODUCT	15.41						
REFINERY USE & PROCESS LOSS*	0.41	0.15	0.54				1.10
SYNCRUDE MANUFACTURE	0.0	0.0					
USES+LOSSES		0.0					0.0
TOTAL INPUT		0.0					
PRIMARY ENERGY SUPPLY	15.82	4.02	3.49	1.17		1.88	26.38

ALL QUANTITIES ARE GIVEN IN MILLION BARREL PER DAY OIL EQUIVALENT

SUMMARY
(ANNUAL CAPITAL CHARGES - CASE YEAR)

PROBLEM: WAES85C BASE YR: 1975.
CASE: 31JAN77 CASE YR: 1985.

10

PRODUCERS			OPEC	OCEC		
OIL						
PRODUCTION	:	B$	1.01	0.0		
REFINING	:	B$	0.33	0.0		
COAL						
PRODUCTION	:	B$	0.0	2.01		
CONVERSION	:	B$	0.0	0.0		
GAS						
PRODUCTION	:	B$	0.0	0.0		
LIQUEFACTION	:	B$	0.0	0.0		
TOTAL	:	B$	1.33	2.01		
CONSUMERS			NORTH AMERICA	EUROPE	JAPAN	REST WOCA
OIL						
PRODUCTION	:	B$	10.30	4.95	0.10	5.27
REFINING	:	B$	3.55	3.11	2.40	3.25
DISTRIBUTION	:	B$	0.70	0.63	0.33	0.68
TOTAL	:	B$	14.55	8.68	2.82	9.19
COAL						
PRODUCTION	:	B$	1.82	0.09	0.08	0.0
CONVERSION	:	B$	0.0	0.0	0.0	0.0
DISTRIBUTION	:	B$	0.14	0.12	0.06	0.13
TOTAL	:	B$	1.96	0.21	0.14	0.13
GAS						
PRODUCTION	:	B$	10.65	3.78	0.12	4.38
REGASIFICATION	:	B$	0.0	0.0	0.08	0.0
DISTRIBUTION	:	B$	0.75	0.48	0.03	1.75
TOTAL	:	B$	11.40	4.26	0.23	6.13
ELECTRICITY						
FUEL OIL	:	B$	0.60	0.0	0.15	0.0
COAL	:	B$	0.0	0.0	0.57	0.56
GAS	:	B$	0.0	0.01	0.34	0.0
NUCLEAR	:	B$	8.12	4.98	1.76	2.41
HYDRO ELEC 1	:	B$	1.98	0.0	0.33	1.26
OTHER	:	B$	1.11	0.65	0.09	0.55
DISTRIBUTION	:	B$	0.0	0.43	0.0	0.0
TOTAL	:	B$	11.82	6.06	3.24	5.09
IND HEATING EQP	:	B$	3.28	2.92	1.97	3.12
DOM HEATING EQP	:	B$	9.56	6.54	1.70	2.27
TOTAL CAPEX	:	B$	52.56	28.67	10.11	25.93

SUMMARY

PROBLEM: WAES85C — CASE: 31JAN77 — BASE YR: 1975. — CASE YR: 1985.

(PRODUCING REGIONS REVENUES)

PRODUCERS		OPEC	OCEC
GOVERNMENT TAXES *			
CRUDE OIL		11.28	0.0
COAL		0.0	2.00
GAS		8.32	0.0
TARIFFS ***			
CRUDE OIL		12.50	12.50
DISTILLATES		14.20	14.20
FUEL OIL		9.10	9.10
COAL		8.00	8.00
GAS		14.20	14.20
GOVT. TAKE(FROM TAXES)			
OIL	: B$	163.08	0.0
COAL	: B$	0.0	4.08
GAS	: B$	4.17	0.0
TOTAL GOVT. TAKE	: B$	167.25	4.08
REVENUES FROM TARIFFS			
CRUDE OIL	: B$	169.56	0.0
DISTILLATES	: B$	9.14	0.0
FUEL OIL	: B$	2.27	0.0
COAL	: B$	0.0	16.31
GAS	: B$	7.11	0.0
TOTAL REVENUES	: B$	188.08	16.31

* DOLLAR/BOE

*** DOLLAR/BOE, AVERAGE WORLD PRICE LAID-IN AT PORT OF ENTRY

12

S U M M A R Y ==================== PROBLEM: WAES85C BASE YR: 1975.
CASE: 31JAN77 CASE YR: 1985.

(CONSUMING REGIONS COSTS AND REVENUES)

CONSUMERS (TOTAL COSTS)			NORTH AMERICA	EUROPE	JAPAN	REST WOCA
IMPORT BILL						
CRUDE OIL	:	B$	27.95	65.51	42.50	42.73
DISTILLATES	:	B$	9.14	0.0	0.0	0.0
FUEL OIL	:	B$	0.0	0.0	1.41	0.86
COAL	:	B$	0.0	10.30	5.03	11.73
GAS	:	B$	0.0	1.13	2.50	6.86
TOTAL IMPORT BILL	:	B$	37.08	76.94	51.44	62.18
INTERNAL COSTS **						
OIL	:	B$	31.06	29.57	8.18	18.70
COAL	:	B$	21.48	10.55	1.56	1.77
GAS	:	B$	15.88	5.74	0.43	7.56
ELECTRICITY	:	B$	16.14	8.74	4.35	6.75
TOTAL	:	B$	84.55	54.60	14.53	34.78
ENERGY CONS EQUIP	:	B$	12.84	9.46	3.67	5.39
TOTAL INTERNAL	:	B$	97.39	64.06	18.20	40.17
REVENUES FROM EXPORTS						
CRUDE OIL	:	B$	0.0	0.0	0.0	0.0
DISTILLATES	:	B$	0.0	0.0	0.0	0.0
FUEL OIL	:	B$	0.0	0.0	0.0	0.0
COAL	:	B$	9.29	0.0	0.0	0.0
GAS	:	B$	0.0	0.0	0.0	0.0
TOTAL REVENUES	:	B$	9.29	0.0	0.0	0.0
NET CONSUM REG. COST	:	B$	125.19	141.00	69.64	102.35
IMPORT/EXPORT BALANCE						
EQUILIBRIUM	:	B$	27.80	76.94	51.44	62.18
LOANS AT 15.0% PA	:	B$	0.0	0.0	0.0	0.0
SHORTFALL	:	B$	0.0	0.0	0.0	0.0

** CAPEX CHARGED AS A CONSTANT COST CHARGED OVER 100.0% OF THE USEFUL PHYSICAL LIFETIME OF A PLANT AT 15.0% ANNUAL COST OF CAPITAL.

1

VOLUMETRIC SUMMARY
=====================

PROBLEM: WAES85D BASE YR: 1975.
CASE: 31JAN77 CASE YR: 1985.

PRODUCTION

	OIL	COAL	GAS	NUCLEAR	HYDRO	TOTAL
NORTH AMERICA	11.9	9.9	8.8	3.3	3.1	37.0
EUROPE	4.1	4.3	3.4	2.6	2.1	16.5
JAPAN	0.1	0.3	0.1	0.6	0.6	1.6
OPEC	36.0	0.0	4.8	0.0	0.0	40.8
OCEC	0.0	4.9	0.0	0.0	0.0	4.9
REST WOCA	5.6	0.0	2.3	0.5	1.9	10.3
SUBTOTAL	57.6	19.4	19.4	7.0	7.7	111.1
WOCA EXPORTS	2.0	0.5	0.5	0.0	0.0	3.0
TOTAL	59.6	19.9	19.9	7.0	7.7	114.1

TOTAL WORLD ENERGY PRODUCTION (EXCLUDING COMMUNIST AREA INTERNAL): 114.1

CONVERSION(INPUT)

	OIL	COAL	GAS	NUCLEAR	HYDRO
OIL REFINING	59.6				
COAL GASIFICATION		0.0			
COAL LIQUEFACTION		0.0			
GAS LIQUEFACTION (NOT INCL. COMMUNIST AREA)				2.5	
ELECTRICITY GENERATION	1.9	5.8	1.1	7.0	7.7

MARKET INPUT

	OIL	COAL	GAS	ELECTRICITY
INDUSTRIAL				
HEAT, LIGHT AND POWER	11.0	10.2	7.2	2.7
NON ENERGY	6.8	0.2	0.9	
DOMESTIC				
HEAT, LIGHT AND POWER	9.0	2.5	6.3	4.1
TRANSPORT AND BUNKERS	25.5			0.1

ALL QUANTITIES ARE GIVEN IN MILLION BARREL PER DAY OIL EQUIVALENT

2

PRODUCTION AND TRANSPORTS CRUDE OIL

PROBLEM: WAES85D BASE YR: 1975.
CASE: 31JAN77 CASE YR: 1985.

FROM \ TO	NORTH AMERICA	EUROPE	JAPAN	REST WOCA	OPEC	PROD"N
NORTH AMERICA	11.90	.	.	.	.	11.90
EUROPE	.	4.07	.	.	.	4.07
JAPAN	.	.	0.08	.	.	0.08
REST WOCA	.	.	.	5.60	.	5.60
OPEC	7.31	8.84	6.24	7.70	5.91	36.00
COMMUNIST AREAS	.	2.00	.	.	.	2.00
TOTAL SUPPLY	19.21	14.91	6.32	13.30	5.91	59.65

PRODUCTION AND TRANSPORTS OIL PRODUCTS

PROBLEM: WAES85D BASE YR: 1975.
CASE: 31JAN77 CASE YR: 1985.

F=FUEL OIL, D=DISTILLATES

FROM \ TO		NORTH AMERICA	EUROPE	JAPAN	REST WOCA	OPEC	PROD"N
NORTH AMERICA	F	4.10	.	.	.	.	4.10
	D	13.81	.	.	.	.	13.81
EUROPE	F	.	4.57	.	.	.	4.57
	D	.	9.43	.	.	.	9.43
JAPAN	F	.	.	1.95	.	.	1.95
	D	.	.	3.96	.	.	3.96
REST WOCA	F	.	.	.	4.06	.	4.06
	D	.	.	.	8.99	.	8.99
OPEC	F	.	.	1.54	.	.	1.54
	D	3.96	.	.	.	0.00	3.96
TOTAL SUPPLY	F	4.10	4.57	3.49	4.06	.	16.22
	D	17.77	9.43	3.96	8.99	0.00	40.15

3

PRODUCTION AND TRANSPORTS COAL
================================

PROBLEM: WAES85D CASE: 31JAN77

BASE YR: 1975. CASE YR: 1985.

TO / FROM	NORTH AMERICA	EUROPE	JAPAN	REST WOCA	LOSS	PROD'N
NORTH AMERICA	8.36	1.32	.	.	0.19	9.88
EUROPE	.	4.24	.	.	0.08	4.33
JAPAN	.	.	0.25	.	0.01	0.26
OCEC	.	0.11	1.05	3.65	0.10	4.91
COMMUNIST AREAS	.	0.50	.	.	.	0.50
TOTAL SUPPLY	8.36	6.17	1.31	3.65	0.38	19.87

PRODUCTION AND TRANSPORTS GAS
==============================

PROBLEM: WAES85D CASE: 31JAN77

BASE YR: 1975. CASE YR: 1985.

P=BY PIPE L=AS LNG

TO / FROM		NORTH AMERICA	EUROPE	JAPAN	REST WOCA	OPEC	LOSS	PROD'N
NORTH AMERICA	P	7.93	.	.	.	.	0.88	8.81
EUROPE	P	.	3.10	.	.	.	0.34	3.44
JAPAN	P	.	.	0.06	.	.	0.01	0.07
REST WOCA	P	.	.	.	2.07	.	0.23	2.30
OPEC	P	.	0.18	.	1.62	2.50	0.48	4.78
	L	0.96	1.04	.	.	.	0.50	.
COMMUNIST AREAS	P	.	.	.	.	.	.	.
	L	.	.	0.50	.	.	.	0.50
TOTAL SUPPLY	P	7.93	3.28	0.06	3.69	2.50	1.94	19.40
	L	0.96	1.04	0.50	.	.	0.50	0.50

4

MARKET INPUT SUMMARY
===================
PROBLEM: WAES85D BASE YR: 1975.
CASE: 31JAN77 CASE YR: 1985.

INDUSTRIAL HEAT. LIGHT AND POWER

	OIL	COAL	GAS	ELECTRICITY
NORTH AMERICA :	2.8	4.5	3.5	0.4
EUROPE :	2.5	2.5	2.1	1.2
JAPAN :	2.8	1.1	0.1	0.3
REST WOCA :	2.9	2.2	1.5	0.7
	11.0	10.2	7.2	2.7

INDUSTRIAL NON ENERGY

	OIL	COAL	GAS
NORTH AMERICA :	3.1	0.1	0.6
EUROPE :	1.8	0.1	0.3
JAPAN :	0.8	.	.
REST WOCA :	1.1	.	.
	6.8	0.2	0.9

DOMESTIC HEAT.LIGHT AND POWER

	OIL	COAL	GAS	ELECTRICITY
NORTH AMERICA :	3.2	0.5	4.2	2.6
EUROPE :	3.6	1.4	1.4	0.7
JAPAN :	1.2	0.0	0.2	0.3
REST WOCA :	1.0	0.5	0.5	0.5
	9.0	2.5	6.3	4.1

TRANSPORT

	OIL	ELECTRICITY
NORTH AMERICA :	10.7	0.0
EUROPE :	5.4	0.1
JAPAN :	1.8	0.0
REST WOCA :	6.9	0.0
	24.9	0.1

DEMAND

		IND H.L&P	IND NONEN	DOM H.L&P	TRANSPORT
NORTH AMERICA					
	DEMAND :	11.2	3.8	10.5	10.8
	SUPPLY :	11.2	3.8	10.5	10.8
	DEFICIT :	0.0	0.0	0.0	0.0
EUROPE					
	DEMAND :	8.3	2.2	7.1	5.4
	SUPPLY :	8.3	2.2	7.1	5.4
	DEFICIT :	0.0	0.0	0.0	0.0
JAPAN					
	DEMAND :	4.2	0.8	1.7	1.9
	SUPPLY :	4.2	0.8	1.7	1.9
	DEFICIT :	0.0	0.0	0.0	0.0
REST WOCA					
	DEMAND :	7.3	1.1	2.6	7.0
	SUPPLY :	7.3	1.1	2.6	7.0
	DEFICIT :	0.0	0.0	0.0	0.0

5

W O C A DEMAND/SUPPLY MATRIX PROBLEM: WAES85D BASE YR: 1975.
CASE: 31JAN77 CASE YR: 1985.

	OIL	COAL	GAS	NUCLEAR	ELECTRICITY	HYDRO-ETC	TOTAL
TRANSPORT	24.88				0.14		25.02
INDUSTRY	10.98	10.24	7.16		2.72		31.09
DOMESTIC	9.05	2.48	6.26		4.10		21.89
TOTAL INTO MARKETS	44.90	12.72	13.42		6.96		78.00
TRANSM./DISTR.LOSSES*	2.82	0.78	3.67		1.23		8.49
CONV. THERMAL ELECTR.	0.66	2.01	0.38		-3.05		
USES+LOSSES	1.23	3.74	0.70				5.67
TOTAL INPUT	1.89	5.75	1.08				
GAS MANUFACTURE		0.0	0.0				
USES+LOSSES		0.0					0.0
TOTAL INPUT		0.0					
PRIMARY ENERGY DEMAND							
OUTPUT	49.61	19.24	18.17	2.46		2.68	92.16
INPUT				7.02		7.66	101.70
CHEMICAL FEED	6.76	0.18	0.89				7.83
LUBES & ASPHALTS	0.0						0.0
TOTAL REFINED PRODUCT	56.37						
REFINERY USE & PROCESS LOSS*	3.28	0.45	0.84				4.57
SYNCRUDE MANUFACTURE	0.0	0.0					
USES+LOSSES		0.0					0.0
TOTAL INPUT		0.0					
PRIMARY ENERGY SUPPLY	59.65	19.88	19.90	7.02		7.66	114.11

*INCLUDES OPEC & OCEC

ALL QUANTITIES ARE GIVEN IN MILLION BARREL PER DAY OIL EQUIVALENT

DEMAND/SUPPLY MATRIX NORTH AMERICA PROBLEM: WAES85D BASE YR: 1975.
CASE: 31JAN77 CASE YR: 1985.

	OIL	COAL	GAS	ELECTRICITY			TOTAL
TRANSPORT	10.75			0.01			10.76
INDUSTRY	2.80	4.49	3.51	0.41			11.22
DOMESTIC	3.15	0.52	4.17	2.61			10.46
TOTAL INTO MARKETS	16.71	5.01	7.68	3.04			32.44
TRANSM./DISTR.LOSSES*	1.09	0.36	1.48	0.54			3.48
CONV. THERMAL ELECTR.	0.36	0.99	0.0	-1.35			
USES+LOSSES	0.66	1.84	0.0				2.50
TOTAL INPUT	1.02	2.82	0.0				
GAS MANUFACTURE		0.0	0.0				
USES+LOSSES		0.0					0.0
TOTAL INPUT		0.0					
PRIMARY ENERGY DEMAND					NUCLEAR	HYDRO-ETC	
OUTPUT	18.82	8.20	9.17		1.14	1.08	38.42
INPUT					3.27	3.10	42.56
CHEMICAL FEED	3.05	0.11	0.60				3.77
LUBES & ASPHALTS	0.0						0.0
TOTAL REFINED PRODUCT	21.87						
REFINERY USE & PROCESS LOSS*	1.30	0.25	0.0				1.54
SYNCRUDE MANUFACTURE	0.0	0.0					
USES+LOSSES		0.0					0.0
TOTAL INPUT		0.0					
PRIMARY ENERGY SUPPLY	23.17	8.56	9.77		3.27	3.10	47.87

ALL QUANTITIES ARE GIVEN IN MILLION BARREL PER DAY OIL EQUIVALENT

7

DEMAND/SUPPLY MATRIX EUROPE — PROBLEM: WAES85D — CASE: 31JAN77 — BASE YR: 1975. — CASE YR: 1985.

	OIL	COAL	GAS	ELECTRICITY			TOTAL
TRANSPORT	5.36			0.08			5.44
INDUSTRY	2.50	2.50	2.08	1.25			8.32
DOMESTIC	3.61	1.43	1.38	0.71			7.13
TOTAL INTO MARKETS	11.46	3.92	3.46	2.04			20.89
TRANSM./DISTR.LOSSES*	0.70	0.21	0.68	0.36			1.95
CONV. THERMAL ELECTR.	0.0	0.67	0.08	-0.75			
USES+LOSSES	0.0	1.25	0.15				1.40
TOTAL INPUT	0.0	1.92	0.23				
GAS MANUFACTURE		0.0	0.0				
USES+LOSSES		0.0					0.0
TOTAL INPUT		0.0					
PRIMARY ENERGY DEMAND					NUCLEAR	HYDRO-ETC	
OUTPUT	12.16	6.05	4.38		0.91	0.73	24.24
INPUT					2.61	2.10	27.30
CHEMICAL FEED	1.84	0.07	0.28				2.19
LUBES & ASPHALTS	0.0						0.0
TOTAL REFINED PRODUCT	14.00						
REFINERY USE & PROCESS LOSS*	0.90	0.14	0.0				1.05
SYNCRUDE MANUFACTURE	0.0	0.0					
USES+LOSSES		0.0					0.0
TOTAL INPUT		0.0					
PRIMARY ENERGY SUPPLY	14.91	6.26	4.66		2.61	2.10	30.54

ALL QUANTITIES ARE GIVEN IN MILLION BARREL PER DAY OIL EQUIVALENT

8

DEMAND/SUPPLY MATRIX JAPAN — PROBLEM: WAES85D — CASE: 31JAN77 — BASE YR: 1975. — CASE YR: 1985.

	OIL	COAL	GAS	ELECTRICITY			TOTAL
TRANSPORT	1.83			0.03			1.86
INDUSTRY	2.75	1.06	0.11	0.32			4.24
DOMESTIC	1.24	0.01	0.18	0.25			1.68
TOTAL INTO MARKETS	5.82	1.07	0.29	0.60			7.78
TRANSM./DISTR.LOSSES*	0.37	0.03	0.08	0.11			0.59
CONV. THERMAL ELECTR.	0.18	0.05	0.07	-0.30			
USES+LOSSES	0.33	0.10	0.13				0.56
TOTAL INPUT	0.51	0.15	0.20				
GAS MANUFACTURE		0.0	0.0				
USES+LOSSES		0.0					0.0
TOTAL INPUT		0.0					
PRIMARY ENERGY DEMAND					NUCLEAR	HYDRO-ETC	
OUTPUT	6.70	1.25	0.57		0.21	0.20	8.93
INPUT					0.61	0.56	9.69
CHEMICAL FEED	0.75	0.0	0.0				0.75
LUBES & ASPHALTS	0.0						0.0
TOTAL REFINED PRODUCT	7.45						
REFINERY USE & PROCESS LOSS*	0.41	0.06	0.0				0.47
SYNCRUDE MANUFACTURE	0.0	0.0					
USES+LOSSES		0.0					0.0
TOTAL INPUT		0.0					
PRIMARY ENERGY SUPPLY	7.86	1.31	0.57		0.61	0.56	10.91

ALL QUANTITIES ARE GIVEN IN MILLION BARREL PER DAY OIL EQUIVALENT

9

DEMAND/SUPPLY MATRIX REST WOCA — PROBLEM: WAES85D — BASE YR: 1975.
CASE: 31JAN77 — CASE YR: 1985.

	OIL	COAL	GAS	ELECTRICITY	NUCLEAR	HYDRO-ETC	TOTAL
TRANSPORT	6.95			0.02			6.97
INDUSTRY	2.92	2.19	1.46	0.73			7.31
DOMESTIC	1.05	0.52	0.52	0.52			2.62
TOTAL INTO MARKETS	10.92	2.72	1.99	1.27			16.90
TRANSM./DISTR.LOSSES*	0.65	0.07	0.45	0.22			1.40
CONV. THERMAL ELECTR.	0.12	0.30	0.22	-0.65			
USES+LOSSES	0.23	0.56	0.42				1.21
TOTAL INPUT	0.36	0.86	0.64				
GAS MANUFACTURE		0.0	0.0				
USES+LOSSES		0.0					0.0
TOTAL INPUT		0.0					
PRIMARY ENERGY DEMAND					NUCLEAR	HYDRO-ETC	
OUTPUT	11.93	3.65	3.08		0.19	0.66	19.50
INPUT					0.53	1.90	21.08
CHEMICAL FEED	1.12	0.0	0.0				1.12
LUBES & ASPHALTS	0.0						0.0
TOTAL REFINED PRODUCT	13.05						
REFINERY USE & PROCESS LOSS*	0.26	0.0	0.84				1.10
SYNCRUDE MANUFACTURE	0.0	0.0					
USES+LOSSES		0.0					0.0
TOTAL INPUT		0.0					
PRIMARY ENERGY SUPPLY	13.30	3.65	3.92		0.53	1.90	23.30

ALL QUANTITIES ARE GIVEN IN MILLION BARREL PER DAY OIL EQUIVALENT

S U M M A R Y
(ANNUAL CAPITAL CHARGES - CASE YEAR)

PROBLEM: WAES85D CASE: 31JAN77 BASE YR: 1975. CASE YR: 1985.

10

PRODUCERS			OPEC	OCEC		
OIL						
PRODUCTION	:	B$	0.84	0.0		
REFINING	:	B$	1.25	0.0		
COAL						
PRODUCTION	:	B$	0.0	1.51		
CONVERSION	:	B$	0.0	0.0		
GAS						
PRODUCTION	:	B$	2.86	0.0		
LIQUEFACTION	:	B$	1.25	0.0		
TOTAL	:	B$	6.20	1.51		
CONSUMERS			NORTH AMERICA	EUROPE	JAPAN	REST WOCA
OIL						
PRODUCTION	:	B$	6.98	4.95	0.10	4.52
REFINING	:	B$	3.11	2.03	1.33	2.98
DISTRIBUTION	:	B$	0.79	0.40	0.21	0.52
TOTAL	:	B$	10.88	7.38	1.64	8.01
COAL						
PRODUCTION	:	B$	1.28	1.78	0.08	0.0
CONVERSION	:	B$	0.0	0.0	0.0	0.0
DISTRIBUTION	:	B$	0.14	0.11	0.03	0.11
TOTAL	:	B$	1.42	1.89	0.11	0.11
GAS						
PRODUCTION	:	B$	6.31	3.59	0.12	4.38
REGASIFICATION	:	B$	0.15	0.17	0.08	0.0
DISTRIBUTION	:	B$	0.30	1.14	0.04	2.07
TOTAL	:	B$	6.76	4.90	0.24	6.45
ELECTRICITY						
FUEL OIL	:	B$	0.24	0.0	0.0	0.0
COAL	:	B$	0.0	0.0	0.07	0.60
GAS	:	B$	0.0	0.0	0.23	0.57
NUCLEAR	:	B$	5.94	5.01	1.19	1.07
HYDRO ELEC 1	:	B$	1.16	0.49	0.31	1.40
OTHER	:	B$	0.46	0.60	0.09	0.23
DISTRIBUTION	:	B$	0.0	0.28	0.0	0.0
TOTAL	:	B$	7.80	6.38	1.90	3.86
IND HEATING EQP	:	B$	3.01	2.39	1.25	2.44
DOM HEATING EQP	:	B$	13.11	7.29	1.46	2.17
TOTAL CAPEX	:	B$	42.98	30.24	6.62	23.05

S U M M A R Y ==================== PROBLEM: WAES85D BASE YR: 1975.
CASE: 31JAN77 CASE YR: 1985.

(PRODUCING REGIONS REVENUES)

PRODUCERS			OPEC	OCEC
GOVERNMENT TAXES *				
CRUDE OIL			11.28	0.0
COAL			0.0	2.00
GAS			8.32	0.0
TARIFFS ***				
CRUDE OIL			12.50	12.50
DISTILLATES			14.20	14.20
FUEL OIL			9.10	9.10
COAL			8.00	8.00
GAS			14.20	14.20
GOVT. TAKE(FROM TAXES)				
OIL	:	B$	146.51	0.0
COAL	:	B$	0.0	3.51
GAS	:	B$	11.54	0.0
TOTAL GOVT. TAKE	:	B$	158.05	3.51
REVENUES FROM TARIFFS				
CRUDE OIL	:	B$	137.27	0.0
DISTILLATES	:	B$	20.53	0.0
FUEL OIL	:	B$	5.10	0.0
COAL	:	B$	0.0	14.05
GAS	:	B$	19.70	0.0
TOTAL REVENUES	:	B$	182.60	14.05

* DOLLAR/BOE

*** DOLLAR/BOE, AVERAGE WORLD PRICE LAID-IN AT PORT OF ENTRY

12

S U M M A R Y
===================
(CONSUMING REGIONS COSTS AND REVENUES)

PROBLEM: WAES85D
CASE: 31JAN77

BASE YR: 1975.
CASE YR: 1985.

CONSUMERS (TOTAL COSTS)			NORTH AMERICA	EUROPE	JAPAN	REST WOCA
IMPORT BILL						
CRUDE OIL	:	B$	33.34	49.44	28.47	35.14
DISTILLATES	:	B$	20.53	0.0	0.0	0.0
FUEL OIL	:	B$	0.0	0.0	5.10	0.0
COAL	:	B$	0.0	5.63	3.08	10.65
GAS	:	B$	4.97	6.44	2.59	9.33
TOTAL IMPORT BILL	:	B$	58.84	61.51	39.24	55.12
INTERNAL COSTS **						
OIL	:	B$	26.74	26.35	5.83	16.21
COAL	:	B$	18.41	17.66	1.26	1.60
GAS	:	B$	10.87	6.72	0.45	8.04
ELECTRICITY	:	B$	11.32	8.68	2.62	5.21
TOTAL	:	B$	67.33	59.40	10.16	31.05
ENERGY CONS EQUIP	:	B$	16.12	9.68	2.71	4.61
TOTAL INTERNAL	:	B$	83.45	69.09	12.87	35.66
REVENUES FROM EXPORTS						
CRUDE OIL	:	B$	0.0	0.0	0.0	0.0
DISTILLATES	:	B$	0.0	0.0	0.0	0.0
FUEL OIL	:	B$	0.0	0.0	0.0	0.0
COAL	:	B$	3.85	0.0	0.0	0.0
GAS	:	B$	0.0	0.0	0.0	0.0
TOTAL REVENUES	:	B$	3.85	0.0	0.0	0.0
NET CONSUM REG. COST	:	B$	138.45	130.59	52.12	90.79
IMPORT/EXPORT BALANCE						
EQUILIBRIUM	:	B$	54.99	61.51	39.24	55.12
LOANS AT 15.0% PA	:	B$	0.0	0.0	0.0	0.0
SHORTFALL	:	B$	0.0	0.0	0.0	0.0

** CAPEX CHARGED AS A CONSTANT COST CHARGED OVER 100.0% OF
THE USEFUL PHYSICAL LIFETIME OF A PLANT AT 15.0% ANNUAL COST OF CAPITAL.

1

VOLUMETRIC SUMMARY

PROBLEM: WAES85E BASE YR: 1975.
CASE: 31JAN77 CASE YR: 1985.

PRODUCTION

	OIL	COAL	GAS	NUCLEAR	HYDRO	TOTAL
NORTH AMERICA	9.6	9.8	7.8	3.3	3.1	33.6
EUROPE	4.1	4.3	3.5	2.6	2.1	16.6
JAPAN	0.1	0.3	0.1	0.6	0.6	1.6
OPEC	36.0	0.0	4.8	0.0	0.0	40.8
OCEC	0.0	4.9	0.0	0.0	0.0	4.9
REST WOCA	5.6	0.0	2.3	0.5	1.9	10.3
SUBTOTAL	55.3	19.3	18.5	7.0	7.7	107.8
WOCA EXPORTS	2.0	0.5	0.5	0.0	0.0	3.0
TOTAL	57.3	19.8	19.0	7.0	7.7	110.8

TOTAL WORLD ENERGY PRODUCTION (EXCLUDING COMMUNIST AREA INTERNAL): 110.8

CONVERSION(INPUT)

	OIL	COAL	GAS	NUCLEAR	HYDRO
OIL REFINING	57.3				
COAL GASIFICATION		0.0			
COAL LIQUEFACTION		0.0			
GAS LIQUEFACTION (NOT INCL. COMMUNIST AREA)				2.5	
ELECTRICITY GENERATION	0.0	0.0	0.0	7.0	7.7

MARKET INPUT

	OIL	COAL	GAS	ELECTRICITY
INDUSTRIAL				
HEAT, LIGHT AND POWER	14.3	14.5	8.3	0.7
NON ENERGY	8.1	0.2	1.1	
DOMESTIC				
HEAT, LIGHT AND POWER	11.6	3.9	5.5	3.5
TRANSPORT AND BUNKERS	18.0			0.2

ALL QUANTITIES ARE GIVEN IN MILLION BARREL PER DAY OIL EQUIVALENT

2

PRODUCTION AND TRANSPORTS CRUDE OIL

PROBLEM: WAES85E CASE: 31JAN77

BASE YR: 1975. CASE YR: 1985.

FROM \ TO	NORTH AMERICA	EUROPE	JAPAN	REST WOCA	OPEC	PROD"N
NORTH AMERICA	9.58	.	.	.	.	9.58
EUROPE	.	4.07	.	.	.	4.07
JAPAN	.	.	0.08	.	.	0.08
REST WOCA	.	.	.	5.60	.	5.60
OPEC	3.05	13.20	8.85	7.75	3.15	36.00
COMMUNIST AREAS	.	2.00	.	.	.	2.00
TOTAL SUPPLY	12.63	19.27	8.93	13.35	3.15	57.33

PRODUCTION AND TRANSPORTS OIL PRODUCTS

PROBLEM: WAES85E CASE: 31JAN77

BASE YR: 1975. CASE YR: 1985.

F=FUEL OIL, D=DISTILLATES

FROM \ TO		NORTH AMERICA	EUROPE	JAPAN	REST WOCA	PROD"N
NORTH AMERICA	F	3.63	.	.	.	3.63
	D	8.22	.	.	.	8.22
EUROPE	F	.	5.98	.	.	5.98
	D	.	12.11	.	.	12.11
JAPAN	F	.	.	3.22	.	3.22
	D	.	.	5.10	.	5.10
REST WOCA	F	.	.	.	4.19	4.19
	D	.	.	.	8.86	8.86
OPEC	F	.	.	0.82	.	0.82
	D	0.96	.	.	1.15	2.11
TOTAL SUPPLY	F	3.63	5.98	4.03	4.19	17.83
	D	9.18	12.11	5.10	10.01	36.40

3

PRODUCTION AND TRANSPORTS COAL PROBLEM: WAES85E BASE YR: 1975.
CASE: 31JAN77 CASE YR: 1985.

FROM \ TO	NORTH AMERICA	EUROPE	JAPAN	REST WOCA	LOSS	PROD''N
NORTH AMERICA	9.29	0.22	.	0.11	0.19	9.82
EUROPE	.	4.24	.	.	0.08	4.33
JAPAN	.	.	0.25	.	0.01	0.26
OCEC	.	.	1.35	3.47	0.10	4.91
COMMUNIST AREAS	.	0.50	.	.	.	0.50
TOTAL SUPPLY	9.29	4.97	1.60	3.58	0.38	19.81

PRODUCTION AND TRANSPORTS GAS PROBLEM: WAES85E BASE YR: 1975.
CASE: 31JAN77 CASE YR: 1985.

P=BY PIPE L=AS LNG

FROM \ TO		NORTH AMERICA	EUROPE	JAPAN	REST WOCA	OPEC	LOSS	PROD''N
NORTH AMERICA	P	7.06	.	.	.	.	0.78	7.85
EUROPE	P	.	3.17	.	.	.	0.35	3.52
JAPAN	P	.	.	0.06	.	.	0.01	0.07
REST WOCA	P	.	.	.	2.07	.	0.23	2.30
OPEC	P	.	0.18	.	1.62	2.50	0.48	4.78
	L	0.96	1.04	.	.	.	0.50	.
COMMUNIST AREAS	P	.	.	.	.	.	.	.
	L	.	.	0.50	.	.	.	0.50
TOTAL SUPPLY	P	7.06	3.35	0.06	3.69	2.50	1.85	18.52
	L	0.96	1.04	0.50	.	.	0.50	0.50

MARKET INPUT SUMMARY PROBLEM: WAES85E BASE YR: 1975. 4

CASE: 31JAN77 CASE YR: 1985.

INDUSTRIAL HEAT, LIGHT AND POWER

	OIL	COAL	GAS	ELECTRICITY
NORTH AMERICA :	3.4	6.8	3.4	0.0
EUROPE :	3.5	2.9	2.9	0.5
JAPAN :	4.0	1.5	0.3	0.1
REST WOCA :	3.5	3.3	1.8	0.1
	14.3	14.5	8.3	0.7

INDUSTRIAL NON ENERGY

	OIL	COAL	GAS
NORTH AMERICA :	3.5	0.1	0.7
EUROPE :	2.4	0.1	0.4
JAPAN :	0.9	.	.
REST WOCA :	1.3	.	.
	8.1	0.2	1.1

DOMESTIC HEAT,LIGHT AND POWER

	OIL	COAL	GAS	ELECTRICITY
NORTH AMERICA :	3.9	2.1	3.4	1.9
EUROPE :	5.0	1.7	0.8	0.8
JAPAN :	1.6	0.0	0.2	0.2
REST WOCA :	1.2	0.2	1.0	0.6
	11.6	3.9	5.5	3.5

TRANSPORT

	OIL	ELECTRICITY
NORTH AMERICA :	1.4	0.0
EUROPE :	6.3	0.1
JAPAN :	2.2	0.0
REST WOCA :	7.5	0.0
	17.4	0.2

DEMAND

		IND H,L&P	IND NONEN	DOM H,L&P	TRANSPORT
NORTH AMERICA					
	DEMAND :	13.5	4.3	11.2	11.6
	SUPPLY :	13.5	4.3	11.2	1.5
	DEFICIT :	0.0	0.0	0.0	10.1
EUROPE					
	DEMAND :	9.8	2.8	8.3	6.4
	SUPPLY :	9.8	2.8	8.3	6.4
	DEFICIT :	0.0	0.0	0.0	0.0
JAPAN					
	DEMAND :	5.9	0.9	1.9	2.2
	SUPPLY :	5.9	0.9	1.9	2.2
	DEFICIT :	0.0	0.0	0.0	0.0
REST WOCA					
	DEMAND :	8.6	1.3	3.1	7.5
	SUPPLY :	8.6	1.3	3.1	7.5
	DEFICIT :	0.0	0.0	0.0	0.0

5

W O C A — DEMAND/SUPPLY MATRIX — PROBLEM: WAES85E — BASE YR: 1975.
CASE: 31JAN77 — CASE YR: 1985.

	OIL	COAL	GAS	ELECTRICITY		TOTAL
TRANSPORT	17.43			0.16		17.59
INDUSTRY	14.33	14.49	8.31	0.71		37.84
DOMESTIC	11.61	3.94	5.47	3.50		24.52
TOTAL INTO MARKETS	43.38	18.43	13.78	4.37		79.95
TRANSM./DISTR.LOSSES*	2.71	0.77	3.54	0.77		7.79
CONV. THERMAL ELECTR.	0.0	0.0	0.0	0.0		
USES+LOSSES	0.0	0.0	0.0			0.0
TOTAL INPUT	0.0	0.0	0.0			
GAS MANUFACTURE		0.0	0.0			
USES+LOSSES		0.0				0.0
TOTAL INPUT		0.0				
PRIMARY ENERGY DEMAND				NUCLEAR	HYDRO-ETC	
OUTPUT	46.09	19.20	17.31	2.46	2.68	87.75
INPUT				7.02	7.66	97.29
CHEMICAL FEED	8.14	0.21	1.06			9.41
LUBES & ASPHALTS	0.0					0.0
TOTAL REFINED PRODUCT	54.23					
REFINERY USE & PROCESS LOSS*	3.10	0.40	0.65			4.15
SYNCRUDE MANUFACTURE	0.0	0.0				
USES+LOSSES		0.0				0.0
TOTAL INPUT		0.0				
PRIMARY ENERGY SUPPLY	57.33	19.82	19.02	7.02	7.66	110.85

*INCLUDES OPEC & OCEC

ALL QUANTITIES ARE GIVEN IN MILLION BARREL PER DAY OIL EQUIVALENT

6

DEMAND/SUPPLY MATRIX NORTH AMERICA

PROBLEM: WAES85E BASE YR: 1975.
CASE: 31JAN77 CASE YR: 1985.

	OIL	COAL	GAS	ELECTRICITY		TOTAL
TRANSPORT	1.45			0.01		1.46
INDUSTRY	3.38	6.76	3.38	0.0		13.52
DOMESTIC	3.85	2.09	3.40	1.89		11.23
TOTAL INTO MARKETS	8.68	8.85	6.78	1.90		26.21
TRANSM./DISTR.LOSSES*	0.64	0.38	1.34	0.33		2.69
CONV. THERMAL ELECTR.	0.0	0.0	0.0	0.0		
USES+LOSSES	0.0	0.0	0.0			0.0
TOTAL INPUT	0.0	0.0	0.0			
GAS MANUFACTURE		0.0	0.0			
USES+LOSSES		0.0				0.0
TOTAL INPUT		0.0				
PRIMARY ENERGY DEMAND				NUCLEAR	HYDRO-ETC	
OUTPUT	9.32	9.23	8.12	1.14	1.08	28.90
INPUT				3.27	3.10	33.04
CHEMICAL FEED	3.48	0.13	0.69			4.30
LUBES & ASPHALTS	0.0					0.0
TOTAL REFINED PRODUCT	12.80					
REFINERY USE & PROCESS LOSS*	0.79	0.12	0.0			0.91
SYNCRUDE MANUFACTURE	0.0	0.0				
USES+LOSSES		0.0				0.0
TOTAL INPUT		0.0				
PRIMARY ENERGY SUPPLY	13.59	9.49	8.81	3.27	3.10	38.26

ALL QUANTITIES ARE GIVEN IN MILLION BARREL PER DAY OIL EQUIVALENT

7

DEMAND/SUPPLY MATRIX EUROPE

PROBLEM: WAES85E BASE YR: 1975.
CASE: 31JAN77 CASE YR: 1985.

	OIL	COAL	GAS	ELECTRICITY		TOTAL
TRANSPORT	6.32			0.09		6.41
INDUSTRY	3.51	2.94	2.85	0.50		9.80
DOMESTIC	4.97	1.65	0.83	0.81		8.27
TOTAL INTO MARKETS	14.80	4.59	3.68	1.40		24.48
TRANSM./DISTR.LOSSES*	0.90	0.19	0.69	0.25		2.03
CONV. THERMAL ELECTR.	0.0	0.0	0.0	0.0		
USES+LOSSES	0.0	0.0	0.0			0.0
TOTAL INPUT	0.0	0.0	0.0			
GAS MANUFACTURE		0.0	0.0			
USES+LOSSES		0.0				0.0
TOTAL INPUT		0.0				
PRIMARY ENERGY DEMAND				NUCLEAR	HYDRO-ETC	
OUTPUT	15.70	4.78	4.37	0.91	0.73	26.51
INPUT				2.61	2.10	29.57
CHEMICAL FEED	2.39	0.09	0.37			2.84
LUBES & ASPHALTS	0.0					0.0
TOTAL REFINED PRODUCT	18.09					
REFINERY USE & PROCESS LOSS*	1.18	0.19	0.0			1.36
SYNCRUDE MANUFACTURE	0.0	0.0				
USES+LOSSES		0.0				0.0
TOTAL INPUT		0.0				
PRIMARY ENERGY SUPPLY	19.27	5.05	4.74	2.61	2.10	33.77

ALL QUANTITIES ARE GIVEN IN MILLION BARREL PER DAY OIL EQUIVALENT

DEMAND/SUPPLY MATRIX JAPAN PROBLEM: WAES85E BASE YR: 1975.
CASE: 31JAN77 CASE YR: 1985.

	OIL	COAL	GAS		ELECTRICITY		TOTAL
TRANSPORT	2.21				0.04		2.25
INDUSTRY	3.98	1.47	0.30		0.12		5.87
DOMESTIC	1.56	0.01	0.19		0.19		1.95
TOTAL INTO MARKETS	7.75	1.48	0.49		0.35		10.07
TRANSM./DISTR.LOSSES*	0.46	0.04	0.08		0.06		0.63
CONV. THERMAL ELECTR.	0.0	0.0	0.0		0.0		
USES+LOSSES	0.0	0.0	0.0				0.0
TOTAL INPUT	0.0	0.0	0.0				
GAS MANUFACTURE		0.0	0.0				
USES+LOSSES		0.0					0.0
TOTAL INPUT		0.0					
PRIMARY ENERGY DEMAND				NUCLEAR		HYDRO-ETC	
OUTPUT	8.21	1.51	0.57	0.21		0.20	10.70
INPUT				0.61		0.56	11.46
CHEMICAL FEED	0.93	0.0	0.0				0.93
LUBES & ASPHALTS	0.0						0.0
TOTAL REFINED PRODUCT	9.14						
REFINERY USE & PROCESS LOSS*	0.61	0.09	0.0				0.70
SYNCRUDE MANUFACTURE	0.0	0.0					
USES+LOSSES		0.0					0.0
TOTAL INPUT		0.0					
PRIMARY ENERGY SUPPLY	9.75	1.61	0.57	0.61		0.56	13.09

ALL QUANTITIES ARE GIVEN IN MILLION BARREL PER DAY OIL EQUIVALENT

9

DEMAND/SUPPLY MATRIX REST WOCA PROBLEM: WAES85E BASE YR: 1975.
CASE: 31JAN77 CASE YR: 1985.

	OIL	COAL	GAS	ELECTRICITY	NUCLEAR	HYDRO-ETC	TOTAL
TRANSPORT	7.46			0.02			7.48
INDUSTRY	3.46	3.33	1.78	0.09			8.65
DOMESTIC	1.23	0.18	1.05	0.61			3.07
TOTAL INTO MARKETS	12.15	3.51	2.82	0.72			19.20
TRANSM./DISTR.LOSSES*	0.71	0.07	0.45	0.13			1.36
CONV. THERMAL ELECTR.	0.0	0.0	0.0	0.0			
USES+LOSSES	0.0	0.0	0.0				0.0
TOTAL INPUT	0.0	0.0	0.0				
GAS MANUFACTURE		0.0	0.0				
USES+LOSSES		0.0					0.0
TOTAL INPUT		0.0					
PRIMARY ENERGY DEMAND							
OUTPUT	12.86	3.58	3.27		0.19	0.66	20.56
INPUT					0.53	1.90	22.14
CHEMICAL FEED	1.34	0.0	0.0				1.34
LUBES & ASPHALTS	0.0						0.0
TOTAL REFINED PRODUCT	14.20						
REFINERY USE & PROCESS LOSS*	0.30	0.0	0.65				0.95
SYNCRUDE MANUFACTURE	0.0	0.0					
USES+LOSSES		0.0					0.0
TOTAL INPUT		0.0					
PRIMARY ENERGY SUPPLY	14.50	3.58	3.92		0.53	1.90	24.43

ALL QUANTITIES ARE GIVEN IN MILLION BARREL PER DAY OIL EQUIVALENT

S U M M A R Y
(ANNUAL CAPITAL CHARGES - CASE YEAR)

PROBLEM: WAES85E BASE YR: 1975.
CASE: 31JAN77 CASE YR: 1985.

10

PRODUCERS			OPEC	OCEC
OIL				
PRODUCTION	:	B$	0.84	0.0
REFINING	:	B$	0.47	0.0
COAL				
PRODUCTION	:	B$	0.0	1.51
CONVERSION	:	B$	0.0	0.0
GAS				
PRODUCTION	:	B$	2.86	0.0
LIQUEFACTION	:	B$	1.25	0.0
TOTAL	:	B$	5.42	1.51

CONSUMERS			NORTH AMERICA	EUROPE	JAPAN	REST WOCA
OIL						
PRODUCTION	:	B$	3.90	4.95	0.10	4.52
REFINING	:	B$	1.16	3.31	2.25	2.61
DISTRIBUTION	:	B$	0.18	0.68	0.33	0.59
TOTAL	:	B$	5.24	8.93	2.67	7.72
COAL						
PRODUCTION	:	B$	1.26	1.78	0.08	0.0
CONVERSION	:	B$	0.0	0.0	0.0	0.0
DISTRIBUTION	:	B$	0.19	0.05	0.04	0.10
TOTAL	:	B$	1.45	1.83	0.12	0.10
GAS						
PRODUCTION	:	B$	4.34	3.78	0.12	4.38
REGASIFICATION	:	B$	0.15	0.17	0.08	0.0
DISTRIBUTION	:	B$	0.0	1.19	0.04	2.07
TOTAL	:	B$	4.49	5.14	0.24	6.45
ELECTRICITY						
FUEL OIL	:	B$	0.0	0.0	0.0	0.0
COAL	:	B$	0.0	0.0	0.0	0.0
GAS	:	B$	0.0	0.0	0.0	0.0
NUCLEAR	:	B$	5.94	5.01	1.19	1.07
HYDRO ELEC 1	:	B$	1.16	0.49	0.31	1.40
OTHER	:	B$	0.46	0.60	0.09	0.23
DISTRIBUTION	:	B$	0.0	0.0	0.0	0.0
TOTAL	:	B$	7.57	6.10	1.60	2.70
IND HEATING EQP	:	B$	4.66	3.07	2.13	3.31
DOM HEATING EQP	:	B$	17.06	9.44	1.94	2.48
TOTAL CAPEX	:	B$	40.45	34.52	8.71	22.76

S U M M A R Y PROBLEM: WAES85E BASE YR: 1975.
CASE: 31JAN77 CASE YR: 1985.

(PRODUCING REGIONS REVENUES)

PRODUCERS			OPEC	OCEC
GOVERNMENT TAXES *				
CRUDE OIL			11.28	0.0
COAL			0.0	2.00
GAS			8.32	0.0
TARIFFS ***				
CRUDE OIL			8.66	8.66
DISTILLATES			9.84	9.84
FUEL OIL			6.30	6.30
COAL			5.54	5.54
GAS			9.84	9.84
GOVT. TAKE(FROM TAXES)				
OIL	:	B$	147.31	0.0
COAL	:	B$	0.0	3.51
GAS	:	B$	11.54	0.0
TOTAL GOVT. TAKE	:	B$	158.85	3.51
REVENUES FROM TARIFFS				
CRUDE OIL	:	B$	103.83	0.0
DISTILLATES	:	B$	7.58	0.0
FUEL OIL	:	B$	1.88	0.0
COAL	:	B$	0.0	9.73
GAS	:	B$	13.65	0.0
TOTAL REVENUES	:	B$	126.95	9.73

* DOLLAR/BOE

*** DOLLAR/BOE, AVERAGE WORLD PRICE LAID-IN AT PORT OF ENTRY

S U M M A R Y
===================
(CONSUMING REGIONS COSTS AND REVENUES)

PROBLEM: WAES85E
CASE: 31JAN77

BASE YR: 1975.
CASE YR: 1985.

CONSUMERS (TOTAL COSTS)			NORTH AMERICA	EUROPE	JAPAN	REST WOCA
IMPORT BILL						
CRUDE OIL	:	B$	9.65	48.04	27.97	24.49
DISTILLATES	:	B$	3.45	0.0	0.0	4.14
FUEL OIL	:	B$	0.0	0.0	1.88	0.0
COAL	:	B$	0.0	1.46	2.72	7.24
GAS	:	B$	3.44	4.46	1.80	6.46
TOTAL IMPORT BILL	:	B$	16.54	53.95	34.37	42.33
INTERNAL COSTS **						
OIL	:	B$	14.84	30.26	7.98	16.37
COAL	:	B$	18.74	17.10	1.39	1.57
GAS	:	B$	8.18	6.99	0.45	8.04
ELECTRICITY	:	B$	9.24	7.38	1.91	3.18
TOTAL	:	B$	51.01	61.74	11.73	29.15
ENERGY CONS EQUIP	:	B$	21.71	12.51	4.07	5.79
TOTAL INTERNAL	:	B$	72.73	74.25	15.80	34.94
REVENUES FROM EXPORTS						
CRUDE OIL	:	B$	0.0	0.0	0.0	0.0
DISTILLATES	:	B$	0.0	0.0	0.0	0.0
FUEL OIL	:	B$	0.0	0.0	0.0	0.0
COAL	:	B$	0.68	0.0	0.0	0.0
GAS	:	B$	0.0	0.0	0.0	0.0
TOTAL REVENUES	:	B$	0.68	0.0	0.0	0.0
NET CONSUM REG. COST	:	B$	88.59	128.21	50.18	77.26
IMPORT/EXPORT BALANCE						
EQUILIBRIUM	:	B$	15.87	53.95	34.37	42.33
LOANS AT 15.0% PA	:	B$	0.0	0.0	0.0	0.0
SHORTFALL	:	B$	0.0	0.0	0.0	0.0

** CAPEX CHARGED AS A CONSTANT COST CHARGED OVER 100.0% OF
THE USEFUL PHYSICAL LIFETIME OF A PLANT AT 15.0% ANNUAL COST OF CAPITAL.

VOLUMETRIC SUMMARY

PROBLEM: WAESC1 BASE YR: 1985.
CASE: 4FEB77 CASE YR: 2000.

1

PRODUCTION

	OIL	COAL	GAS	NUCLEAR	HYDRO	TOTAL
NORTH AMERICA	12.1	24.1	8.1	13.4	6.1	63.8
EUROPE	2.8	4.9	2.1	7.8	2.8	20.4
JAPAN	0.4	0.3	0.1	2.0	0.9	3.6
OPEC	45.0	0.0	9.5	0.0	0.0	54.5
OCEC	0.0	12.4	0.0	0.0	0.0	12.4
REST WOCA	12.0	0.0	4.0	4.5	4.6	25.2
SUBTOTAL	72.3	41.7	23.8	27.7	14.4	179.9
WOCA EXPORTS	2.0	0.5	1.0	0.0	0.0	3.5
TOTAL	74.3	42.2	24.8	27.7	14.4	183.4

TOTAL WORLD ENERGY PRODUCTION (EXCLUDING COMMUNIST AREA INTERNAL): 183.4

CONVERSION(INPUT)

	OIL	COAL	GAS	NUCLEAR	HYDRO
OIL REFINING	74.3				
COAL GASIFICATION		0.0			
COAL LIQUEFACTION		0.0			
GAS LIQUEFACTION (NOT INCL. COMMUNIST AREA)				3.2	
ELECTRICITY GENERATION	0.0	0.0	0.0	27.7	14.4

MARKET INPUT

	OIL	COAL	GAS	ELECTRICITY
INDUSTRIAL				
HEAT, LIGHT AND POWER	15.5	29.9	8.6	6.0
NON ENERGY	13.7	0.3	1.7	
DOMESTIC				
HEAT, LIGHT AND POWER	5.8	9.8	7.3	6.3
TRANSPORT AND BUNKERS	33.5			0.3

ALL QUANTITIES ARE GIVEN IN MILLION BARREL PER DAY OIL EQUIVALENT

2

PRODUCTION AND TRANSPORTS CRUDE OIL

PROBLEM: WAESC1 CASE: 4FEB77 BASE YR: 1985. CASE YR: 2000.

FROM \ TO	NORTH AMERICA	EUROPE	JAPAN	REST WOCA	OPEC	PROD'N
NORTH AMERICA	12.06	.	.	.	.	12.06
EUROPE	.	2.82	.	.	.	2.82
JAPAN	.	.	0.38	.	.	0.38
REST WOCA	.	.	.	12.00	.	12.00
OPEC	.	15.85	10.76	14.43	3.96	45.00
COMMUNIST AREAS	.	2.00	.	.	.	2.00
TOTAL SUPPLY	12.06	20.67	11.14	26.43	3.96	74.26

PRODUCTION AND TRANSPORTS OIL PRODUCTS

PROBLEM: WAESC1 CASE: 4FEB77 BASE YR: 1985. CASE YR: 2000.

F=FUEL OIL, D=DISTILLATES

FROM \ TO		NORTH AMERICA	EUROPE	JAPAN	REST WOCA	OPEC	PROD'N
NORTH AMERICA	F	1.99	.	.	.	.	1.99
	D	9.21	.	.	.	.	9.21
EUROPE	F	.	6.92	.	.	.	6.92
	D	.	13.10	.	.	.	13.10
JAPAN	F	.	.	3.37	.	.	3.37
	D	.	.	7.05	.	.	7.05
REST WOCA	F	.	.	.	6.20	.	6.20
	D	.	.	.	19.90	.	19.90
OPEC	F	.	.	1.03	.	0.00	1.03
	D	2.41	.	.	0.24	0.00	2.65
TOTAL SUPPLY	F	1.99	6.92	4.40	6.20	0.00	19.52
	D	11.62	13.10	7.05	20.14	0.00	51.92

PRODUCTION AND TRANSPORTS COAL PROBLEM: WAESC1 BASE YR: 1985.
CASE: 4FEB77 CASE YR: 2000.

FROM \ TO	NORTH AMERICA	EUROPE	JAPAN	REST WOCA	LOSS	PROD"N
NORTH AMERICA	15.38	5.82	2.43	.	0.47	24.10
EUROPE	.	4.84	.	.	0.10	4.94
JAPAN	.	.	0.25	.	0.01	0.26
OCEC	.	.	3.89	8.23	0.24	12.37
COMMUNIST AREAS	.	0.50	.	.	.	0.50
TOTAL SUPPLY	15.38	11.16	6.58	8.23	0.82	42.16

PRODUCTION AND TRANSPORTS GAS PROBLEM: WAESC1 BASE YR: 1985.
CASE: 4FEB77 CASE YR: 2000.

P=BY PIPE L=AS LNG

FROM \ TO		NORTH AMERICA	EUROPE	JAPAN	REST WOCA	OPEC	LOSS	PROD"N
NORTH AMERICA	P	7.27	.	.	.	.	0.81	8.08
EUROPE	P	.	1.91	.	.	.	0.21	2.12
JAPAN	P	.	.	0.07	.	.	0.01	0.08
REST WOCA	P	.	.	.	3.60	.	0.40	4.00
OPEC	P	.	0.45	.	4.95	3.15	0.95	9.50
	L	0.38	2.14	.	.	.	0.63	.
COMMUNIST AREAS	P	.	.	.	.	.	.	.
	L	0.17	.	0.83	.	.	0.00	1.00
TOTAL SUPPLY	P	7.27	2.36	0.07	8.55	3.15	2.38	23.78
	L	0.55	2.14	0.83	.	.	0.63	1.00

4

MARKET INPUT SUMMARY ===================
PROBLEM: WAESC1 CASE: 4FEB77
BASE YR: 1985. CASE YR: 2000.

INDUSTRIAL HEAT, LIGHT AND POWER	OIL	COAL	GAS	ELECTRICITY
NORTH AMERICA :	1.9	10.3	3.8	2.9
EUROPE :	4.3	6.9	1.4	1.6
JAPAN :	3.9	5.9	0.0	0.1
REST WOCA :	5.4	6.8	3.4	1.4
	15.5	29.9	8.6	6.0

INDUSTRIAL NON ENERGY	OIL	COAL	GAS
NORTH AMERICA :	5.6	0.2	1.1
EUROPE :	4.0	0.1	0.6
JAPAN :	1.0	.	.
REST WOCA :	3.1	.	.
	13.7	0.3	1.7

DOMESTIC HEAT,LIGHT AND POWER

	OIL	COAL	GAS	ELECTRICITY
NORTH AMERICA :	0.0	4.4	2.4	2.9
EUROPE :	2.7	3.6	1.3	1.4
JAPAN :	2.0	0.5	0.8	0.7
REST WOCA :	1.2	1.3	2.8	1.3
	5.8	9.8	7.3	6.3

TRANSPORT

	OIL	ELECTRICITY
NORTH AMERICA :	5.4	0.0
EUROPE :	8.0	0.1
JAPAN :	3.9	0.1
REST WOCA :	15.3	0.0
	32.8	0.3

DEMAND

	IND H,L&P	IND NONEN	DOM H,L&P	TRANSPORT
NORTH AMERICA				
DEMAND :	18.8	6.9	9.7	12.6
SUPPLY :	18.8	6.9	9.7	5.4
DEFICIT :	0.0	0.0	0.0	7.2
EUROPE				
DEMAND :	14.3	4.7	9.1	8.2
SUPPLY :	14.3	4.7	9.1	8.2
DEFICIT :	0.0	0.0	0.0	0.0
JAPAN				
DEMAND :	9.9	1.0	3.9	4.0
SUPPLY :	9.9	1.0	3.9	4.0
DEFICIT :	0.0	0.0	0.0	0.0
REST WOCA				
DEMAND :	16.9	3.1	6.6	15.4
SUPPLY :	16.9	3.1	6.6	15.4
DEFICIT :	0.0	0.0	0.0	0.0

W O C A
=======

DEMAND/SUPPLY MATRIX
=====================

PROBLEM: WAESC1 CASE: 4FEB77 BASE YR: 1985. CASE YR: 2000.

5

	OIL	COAL	GAS	ELECTRICITY		TOTAL
TRANSPORT	32.76			0.27		33.03
INDUSTRY	15.51	29.86	8.58	5.97		59.92
DOMESTIC	5.85	9.81	7.34	6.30		29.30
TOTAL INTO MARKETS	54.12	39.67	15.92	12.54		122.25
TRANSM./DISTR.LOSSES*	3.57	1.66	4.57	2.21		12.01
CONV. THERMAL ELECTR.	0.0	0.0	0.0	0.0		
USES+LOSSES	0.0	0.0	0.0			0.0
TOTAL INPUT	0.0	0.0	0.0			
GAS MANUFACTURE		0.0	0.0			
USES+LOSSES		0.0				0.0
TOTAL INPUT		0.0				
PRIMARY ENERGY DEMAND				NUCLEAR	HYDRO-ETC	
OUTPUT	57.69	41.33	20.50	9.70	5.05	134.27
INPUT				27.72	14.43	161.67
CHEMICAL FEED	13.75	0.35	1.72			15.82
LUBES & ASPHALTS	0.0					0.0
TOTAL REFINED PRODUCT	71.44					
REFINERY USE & PROCESS LOSS*	2.82	0.50	2.56			5.88
SYNCRUDE MANUFACTURE	0.0	0.0				
USES+LOSSES		0.0				0.0
TOTAL INPUT		0.0				
PRIMARY ENERGY SUPPLY	74.26	42.18	24.78	27.72	14.43	183.37

*INCLUDES OPEC & OCEC

ALL QUANTITIES ARE GIVEN IN MILLION BARREL PER DAY OIL EQUIVALENT

DEMAND/SUPPLY MATRIX NORTH AMERICA
==================================

PROBLEM: WAESC1 CASE: 4FEB77 BASE YR: 1985. CASE YR: 2000.

6

	OIL	COAL	GAS	ELECTRICITY		TOTAL
TRANSPORT	5.43			0.01		5.44
INDUSTRY	1.88	10.30	3.77	2.89		18.85
DOMESTIC	0.0	4.38	2.43	2.92		9.73
TOTAL INTO MARKETS	7.32	14.68	6.20	5.82		34.02
TRANSM./DISTR.LOSSES*	0.68	0.79	1.32	1.03		3.82
CONV. THERMAL ELECTR.	0.0	0.0	0.0	0.0		
USES+LOSSES	0.0	0.0	0.0			0.0
TOTAL INPUT	0.0	0.0	0.0			
GAS MANUFACTURE		0.0	0.0			
USES+LOSSES		0.0				0.0
TOTAL INPUT		0.0				
PRIMARY ENERGY DEMAND				NUCLEAR	HYDRO-ETC	
OUTPUT	8.00	15.47	7.52	4.70	2.15	37.84
INPUT				13.43	6.14	50.56
CHEMICAL FEED	5.61	0.21	1.11			6.93
LUBES & ASPHALTS	0.0					0.0
TOTAL REFINED PRODUCT	13.61					
REFINERY USE & PROCESS LOSS*	0.86	0.18	0.0			1.04
SYNCRUDE MANUFACTURE	0.0	0.0				
USES+LOSSES		0.0				0.0
TOTAL INPUT		0.0				
PRIMARY ENERGY SUPPLY	14.47	15.86	8.63	13.43	6.14	58.53

ALL QUANTITIES ARE GIVEN IN MILLION BARREL PER DAY OIL EQUIVALENT

7

DEMAND/SUPPLY MATRIX EUROPE — PROBLEM: WAESC1 — BASE YR: 1985.
CASE: 4FEB77 — CASE YR: 2000.

	OIL	COAL	GAS	NUCLEAR	ELECTRICITY	HYDRO-ETC	TOTAL
TRANSPORT	8.04				0.13		8.17
INDUSTRY	4.29	6.94	1.43		1.64		14.31
DOMESTIC	2.72	3.65	1.35		1.36		9.08
TOTAL INTO MARKETS	15.06	10.59	2.78		3.14		31.56
TRANSM./DISTR.LOSSES*	1.00	0.32	0.64		0.55		2.52
CONV. THERMAL ELECTR.	0.0	0.0	0.0		0.0		
USES+LOSSES	0.0	0.0	0.0				0.0
TOTAL INPUT	0.0	0.0	0.0				
GAS MANUFACTURE		0.0	0.0				
USES+LOSSES		0.0					0.0
TOTAL INPUT		0.0					
PRIMARY ENERGY DEMAND							
OUTPUT	16.06	10.91	3.42	2.72		0.97	34.08
INPUT				7.78		2.76	40.93
CHEMICAL FEED	3.96	0.14	0.61				4.72
LUBES & ASPHALTS	0.0						0.0
TOTAL REFINED PRODUCT	20.03						
REFINERY USE & PROCESS LOSS*	0.64	0.21	0.68				1.53
SYNCRUDE MANUFACTURE	0.0	0.0					
USES+LOSSES		0.0					0.0
TOTAL INPUT		0.0					
PRIMARY ENERGY SUPPLY	20.67	11.26	4.71	7.78		2.76	47.18

ALL QUANTITIES ARE GIVEN IN MILLION BARREL PER DAY OIL EQUIVALENT

DEMAND/SUPPLY MATRIX REST WOCA

PROBLEM: WAESC1 CASE: 4FEB77 BASE YR: 1985. CASE YR: 2000.

	OIL	COAL	GAS	ELECTRICITY		TOTAL
TRANSPORT	15.35			0.04		15.39
INDUSTRY	5.38	6.76	3.38	1.37		16.89
DOMESTIC	1.16	1.31	2.78	1.31		6.56
TOTAL INTO MARKETS	21.89	8.07	6.16	2.73		38.84
TRANSM./DISTR.LOSSES*	1.32	0.16	0.91	0.48		2.88
CONV. THERMAL ELECTR.	0.0	0.0	0.0	0.0		
USES+LOSSES	0.0	0.0	0.0			0.0
TOTAL INPUT	0.0	0.0	0.0			
GAS MANUFACTURE		0.0	0.0			
USES+LOSSES		0.0				0.0
TOTAL INPUT		0.0				
PRIMARY ENERGY DEMAND				NUCLEAR	HYDRO-ETC	
OUTPUT	23.21	8.23	7.07	1.59	1.61	41.71
INPUT				4.55	4.61	47.67
CHEMICAL FEED	3.14	0.0	0.0			3.14
LUBES & ASPHALTS	0.0					0.0
TOTAL REFINED PRODUCT	26.35					
REFINERY USE & PROCESS LOSS*	0.33	0.0	1.88			2.21
SYNCRUDE MANUFACTURE	0.0	0.0				
USES+LOSSES		0.0				0.0
TOTAL INPUT		0.0				
PRIMARY ENERGY SUPPLY	26.67	8.23	8.95	4.55	4.61	53.02

ALL QUANTITIES ARE GIVEN IN MILLION BARREL PER DAY OIL EQUIVALENT

DEMAND/SUPPLY MATRIX JAPAN PROBLEM: WAESC1 BASE YR: 1985.
CASE: 4FEB77 CASE YR: 2000.

	OIL	COAL	GAS	ELECTRICITY		TOTAL
TRANSPORT	3.94			0.09		4.03
INDUSTRY	3.95	5.86	0.0	0.06		9.87
DOMESTIC	1.96	0.47	0.79	0.71		3.93
TOTAL INTO MARKETS	9.85	6.33	0.79	0.86		17.83
TRANSM./DISTR.LOSSES*	0.57	0.14	0.12	0.15		0.98
CONV. THERMAL ELECTR.	0.0	0.0	0.0	0.0		
USES+LOSSES	0.0	0.0	0.0			0.0
TOTAL INPUT	0.0	0.0	0.0			
GAS MANUFACTURE		0.0	0.0			
USES+LOSSES		0.0				0.0
TOTAL INPUT		0.0				
				NUCLEAR	HYDRO-ETC	
PRIMARY ENERGY DEMAND						
OUTPUT	10.42	6.47	0.91	0.69	0.32	18.81
INPUT				1.96	0.92	20.68
CHEMICAL FEED	1.03	0.0	0.0			1.03
LUBES & ASPHALTS	0.0					0.0
TOTAL REFINED PRODUCT	11.45					
REFINERY USE & PROCESS LOSS*	0.72	0.11	0.0			0.83
SYNCRUDE MANUFACTURE	0.0	0.0				
USES+LOSSES		0.0				0.0
TOTAL INPUT		0.0				
PRIMARY ENERGY SUPPLY	12.17	6.58	0.91	1.96	0.92	22.54

ALL QUANTITIES ARE GIVEN IN MILLION BARREL PER DAY OIL EQUIVALENT

S U M M A R Y
(ANNUAL CAPITAL CHARGES - CASE YEAR)

PROBLEM: WAESC1 BASE YR: 1985.
CASE: 4FEB77 CASE YR: 2000.

10

PRODUCERS			OPEC	OCEC		
OIL						
PRODUCTION	:	B$	1.56	0.0		
REFINING	:	B$	0.77	0.0		
COAL						
PRODUCTION	:	B$	0.0	5.45		
CONVERSION	:	B$	0.0	0.0		
GAS						
PRODUCTION	:	B$	11.05	0.0		
LIQUEFACTION	:	B$	1.71	0.0		
TOTAL	:	B$	15.09	5.45		
CONSUMERS			NORTH AMERICA	EUROPE	JAPAN	REST WOCA
OIL						
PRODUCTION	:	B$	13.75	1.78	0.45	11.03
REFINING	:	B$	0.83	3.66	1.92	6.10
DISTRIBUTION	:	B$	0.27	0.81	0.48	1.29
TOTAL	:	B$	14.85	6.24	2.85	18.42
COAL						
PRODUCTION	:	B$	5.11	3.63	0.14	0.0
CONVERSION	:	B$	0.0	0.0	0.0	0.0
DISTRIBUTION	:	B$	0.59	0.42	0.29	0.32
TOTAL	:	B$	5.70	4.05	0.43	0.32
GAS						
PRODUCTION	:	B$	9.61	1.46	0.10	8.01
REGASIFICATION	:	B$	0.09	0.36	0.10	0.0
DISTRIBUTION	:	B$	0.99	1.52	0.35	4.74
TOTAL	:	B$	10.69	3.34	0.56	12.76
ELECTRICITY						
FUEL OIL	:	B$	0.0	0.0	0.0	0.0
COAL	:	B$	0.0	0.0	0.0	0.0
GAS	:	B$	0.0	0.0	0.0	0.0
NUCLEAR	:	B$	23.20	13.33	3.10	8.19
HYDRO ELEC 1	:	B$	2.02	2.02	0.62	4.83
OTHER	:	B$	9.02	0.35	0.35	0.0
DISTRIBUTION	:	B$	2.80	1.38	0.21	1.46
TOTAL	:	B$	37.05	17.08	4.27	14.49
IND HEATING EQP	:	B$	8.01	6.15	4.65	7.29
DOM HEATING EQP	:	B$	20.76	16.25	6.56	12.38
TOTAL CAPEX	:	B$	97.04	53.11	19.32	65.66

11

S U M M A R Y
===================
(PRODUCING REGIONS REVENUES)

PROBLEM: WAESC1 CASE: 4FEB77
BASE YR: 1985. CASE YR: 2000.

PRODUCERS			OPEC	OCEC
GOVERNMENT TAXES *				
CRUDE OIL			11.28	0.0
COAL			0.0	2.00
GAS			8.32	0.0
TARIFFS ***				
CRUDE OIL			18.25	18.25
DISTILLATES			20.73	20.73
FUEL OIL			13.29	13.29
COAL			11.68	11.68
GAS			20.73	20.73
GOVT. TAKE(FROM TAXES)				
OIL	:	B$	184.13	0.0
COAL	:	B$	0.0	8.85
GAS	:	B$	24.05	0.0
TOTAL GOVT. TAKE	:	B$	208.18	8.85
REVENUES FROM TARIFFS				
CRUDE OIL	:	B$	273.39	0.0
DISTILLATES	:	B$	20.06	0.0
FUEL OIL	:	B$	4.99	0.0
COAL	:	B$	0.0	51.68
GAS	:	B$	59.93	0.0
TOTAL REVENUES	:	B$	358.37	51.68

* DOLLAR/BOE

*** DOLLAR/BOE, AVERAGE WORLD PRICE LAID-IN AT PORT OF ENTRY

12

S U M M A R Y
===================
(CONSUMING REGIONS COSTS AND REVENUES)

PROBLEM: WAESC1 CASE: 4FEB77
BASE YR: 1985. CASE YR: 2000.

CONSUMERS (TOTAL COSTS)			NORTH AMERICA	EUROPE	JAPAN	REST WOCA
IMPORT BILL						
CRUDE OIL	:	B$	0.0	118.92	71.68	96.12
DISTILLATES	:	B$	18.22	0.0	0.0	1.84
FUEL OIL	:	B$	0.0	0.0	4.99	0.0
COAL	:	B$	0.0	26.94	26.95	35.10
GAS	:	B$	4.15	19.99	6.27	41.62
TOTAL IMPORT BILL	:	B$	22.37	165.84	109.89	174.68
INTERNAL COSTS **						
OIL	:	B$	29.27	25.35	9.50	34.82
COAL	:	B$	44.46	24.28	3.73	3.69
GAS	:	B$	14.36	5.13	0.88	16.29
ELECTRICITY	:	B$	43.03	20.37	5.14	16.86
TOTAL	:	B$	131.11	75.13	19.26	71.65
ENERGY CONS EQUIP	:	B$	28.76	22.40	11.21	19.68
TOTAL INTERNAL	:	B$	159.88	97.53	30.47	91.33
REVENUES FROM EXPORTS						
CRUDE OIL	:	B$	0.0	0.0	0.0	0.0
DISTILLATES	:	B$	0.0	0.0	0.0	0.0
FUEL OIL	:	B$	0.0	0.0	0.0	0.0
COAL	:	B$	35.18	0.0	0.0	0.0
GAS	:	B$	0.0	0.0	0.0	0.0
TOTAL REVENUES	:	B$	35.18	0.0	0.0	0.0
NET CONSUM REG. COST	:	B$	147.08	263.37	140.36	266.00
IMPORT/EXPORT BALANCE						
EQUILIBRIUM	:	B$	-12.80	165.84	109.89	174.68
LOANS AT 15.0% PA	:	B$	0.0	0.0	0.0	0.0
SHORTFALL	:	B$	0.0	0.0	0.0	0.0

** CAPEX CHARGED AS A CONSTANT COST CHARGED OVER 100.0% OF THE USEFUL PHYSICAL LIFETIME OF A PLANT AT 15.0% ANNUAL COST OF CAPITAL.

VOLUMETRIC SUMMARY

PROBLEM: WAESC2 BASE YR: 1985. CASE: 04FEB77 CASE YR: 2000.

1

PRODUCTION

	OIL	COAL	GAS	NUCLEAR	HYDRO	TOTAL
NORTH AMERICA	12.1	17.4	8.1	21.9	6.1	65.6
EUROPE	2.8	3.9	2.1	13.8	2.8	25.5
JAPAN	0.4	0.3	0.1	3.1	0.9	4.8
OPEC	45.0	0.0	8.5	0.0	0.0	53.5
OCEC	0.0	8.3	0.0	0.0	0.0	8.3
REST WOCA	12.0	0.0	4.0	9.4	3.8	29.2
SUBTOTAL	72.3	29.9	22.7	48.2	13.7	186.9
WOCA EXPORTS	2.0	0.5	1.0	0.0	0.0	3.5
TOTAL	74.3	30.4	23.7	48.2	13.7	190.4

TOTAL WORLD ENERGY PRODUCTION (EXCLUDING COMMUNIST AREA INTERNAL): 190.4

CONVERSION(INPUT)

	OIL	COAL	GAS	NUCLEAR	HYDRO
OIL REFINING	74.3				
COAL GASIFICATION		0.0			
COAL LIQUEFACTION		0.0			
GAS LIQUEFACTION (NOT INCL. COMMUNIST AREA)				3.1	
ELECTRICITY GENERATION	0.0	0.0	0.0	48.2	13.7

MARKET INPUT

	OIL	COAL	GAS	ELECTRICITY
INDUSTRIAL				
HEAT, LIGHT AND POWER	18.8	23.6	7.9	10.0
NON ENERGY	13.6	0.3	1.7	
DOMESTIC				
HEAT, LIGHT AND POWER	7.8	4.8	8.6	8.1
TRANSPORT AND BUNKERS	27.4			0.3

ALL QUANTITIES ARE GIVEN IN MILLION BARREL PER DAY OIL EQUIVALENT

2

PRODUCTION AND TRANSPORTS CRUDE OIL

PROBLEM: WAESC2 CASE: 04FEB77 BASE YR: 1985. CASE YR: 2000.

FROM \ TO	NORTH AMERICA	EUROPE	JAPAN	REST WOCA	OPEC	PROD"N
NORTH AMERICA	12.06	.	.	.	.	12.06
EUROPE	.	2.82	.	.	.	2.82
JAPAN	.	.	0.38	.	.	0.38
REST WOCA	.	.	.	12.00	.	12.00
OPEC	.	14.00	10.32	11.07	9.62	45.00
COMMUNIST AREAS	.	2.00	.	.	.	2.00
TOTAL SUPPLY	12.06	18.82	10.70	23.07	9.62	74.26

PRODUCTION AND TRANSPORTS OIL PRODUCTS

PROBLEM: WAESC2 CASE: 04FEB77 BASE YR: 1985. CASE YR: 2000.

F=FUEL OIL, D=DISTILLATES

FROM \ TO		NORTH AMERICA	EUROPE	JAPAN	REST WOCA	OPEC	PROD"N
NORTH AMERICA	F	3.37	.	.	.	.	3.37
	D	7.94	.	.	.	.	7.94
EUROPE	F	.	6.11	.	.	.	6.11
	D	.	11.45	.	.	.	11.45
JAPAN	F	.	.	2.88	.	.	2.88
	D	.	.	7.17	.	.	7.17
REST WOCA	F	.	.	.	7.34	.	7.34
	D	.	.	.	15.21	.	15.21
OPEC	F	.	0.70	1.80	.	0.00	2.50
	D	0.74	.	.	5.70	.	6.44
TOTAL SUPPLY	F	3.37	6.82	4.68	7.34	0.00	22.20
	D	8.68	11.45	7.17	20.91	.	48.21

3

PRODUCTION AND TRANSPORTS COAL

PROBLEM: WAESC2 CASE: 04FEB77

BASE YR: 1985. CASE YR: 2000.

FROM \ TO	NORTH AMERICA	EUROPE	JAPAN	REST WOCA	LOSS	PROD"N
NORTH AMERICA	9.82	3.63	3.58	.	0.34	17.37
EUROPE	.	3.85	.	.	0.08	3.93
JAPAN	.	.	0.25	.	0.01	0.26
OCEC	.	.	1.84	6.33	0.16	8.34
COMMUNIST AREAS	.	0.50	.	.	.	0.50
TOTAL SUPPLY	9.82	7.98	5.67	6.33	0.59	30.40

PRODUCTION AND TRANSPORTS GAS

PROBLEM: WAESC2 CASE: 04FEB77

BASE YR: 1985. CASE YR: 2000.

P=BY PIPE L=AS LNG

FROM \ TO		NORTH AMERICA	EUROPE	JAPAN	REST WOCA	OPEC	LOSS	PROD"N
NORTH AMERICA	P	7.27	.	.	.	.	0.81	8.08
EUROPE	P	.	1.91	.	.	.	0.21	2.12
JAPAN	P	.	.	0.07	.	.	0.01	0.08
REST WOCA	P	.	.	.	3.60	.	0.40	4.00
OPEC	P	.	0.45	.	4.05	3.12	0.85	8.47
	L	.	2.50	.	.	.	0.62	.
COMMUNIST AREAS	P	.	.	.	.	.	.	.
	L	.	.	1.00	.	.	.	1.00
TOTAL SUPPLY	P	7.27	2.36	0.07	7.65	3.12	2.27	22.75
	L	.	2.50	1.00	.	.	0.62	1.00

4

MARKET INPUT SUMMARY PROBLEM: WAESC2 BASE YR: 1985. CASE: 04FEB77 CASE YR: 2000.

INDUSTRIAL HEAT, LIGHT AND POWER	OIL	COAL	GAS	ELECTRICITY
NORTH AMERICA :	3.3	8.1	2.8	4.6
EUROPE :	4.2	5.6	1.4	2.8
JAPAN :	4.3	5.1	0.1	0.3
REST WOCA :	7.0	4.8	3.5	2.2
	18.8	23.6	7.9	10.0

INDUSTRIAL NON ENERGY	OIL	COAL	GAS
NORTH AMERICA :	5.6	0.2	1.1
EUROPE :	3.9	0.1	0.6
JAPAN :	1.0	.	.
REST WOCA :	3.0	.	.
	13.6	0.3	1.7

DOMESTIC HEAT, LIGHT AND POWER	OIL	COAL	GAS	ELECTRICITY
NORTH AMERICA :	1.9	1.2	2.9	3.7
EUROPE :	2.7	1.9	2.4	2.0
JAPAN :	2.0	0.4	0.8	0.8
REST WOCA :	1.2	1.4	2.6	1.7
	7.8	4.8	8.6	8.1

TRANSPORT	OIL	ELECTRICITY
NORTH AMERICA :	0.6	0.0
EUROPE :	6.6	0.2
JAPAN :	3.9	0.1
REST WOCA :	15.6	0.0
	26.7	0.3

DEMAND

		IND H.L&P	IND NONEN	DOM H.L&P	TRANSPORT
NORTH AMERICA					
	DEMAND :	18.8	6.9	9.7	12.6
	SUPPLY :	18.8	6.9	9.7	0.6
	DEFICIT :	0.0	0.0	0.0	12.0
EUROPE					
	DEMAND :	14.0	4.7	8.9	8.1
	SUPPLY :	14.0	4.7	8.9	6.7
	DEFICIT :	0.0	0.0	0.0	1.4
JAPAN					
	DEMAND :	9.9	1.0	3.9	4.0
	SUPPLY :	9.9	1.0	3.9	4.0
	DEFICIT :	0.0	0.0	0.0	0.0
REST WOCA					
	DEMAND :	17.6	3.0	6.8	15.9
	SUPPLY :	17.6	3.0	6.8	15.7
	DEFICIT :	0.0	0.0	0.0	0.2

W O Č A — DEMAND/SUPPLY MATRIX — PROBLEM: WAESC2 — CASE: 04FEB77 — BASE YR: 1985. — CASE YR: 2000.

	OIL	COAL	GAS	ELECTRICITY	NUCLEAR	HYDRO-ETC	TOTAL
TRANSPORT	26.73			0.31			27.04
INDUSTRY	18.82	23.63	7.89	9.98			60.33
DOMESTIC	7.77	4.82	8.61	8.15			29.35
TOTAL INTO MARKETS	53.33	28.45	16.50	18.44			116.72
TRANSM./DISTR.LOSSES*	3.52	1.19	4.41	3.25			12.37
CONV. THERMAL ELECTR.	0.0	0.0	0.0	0.0			
USES+LOSSES	0.0	0.0	0.0				0.0
TOTAL INPUT	0.0	0.0	0.0				
GAS MANUFACTURE		0.0	0.0				
USES+LOSSES		0.0					0.0
TOTAL INPUT		0.0					
PRIMARY ENERGY DEMAND					NUCLEAR	HYDRO-ETC	
OUTPUT	56.85	29.65	20.91		16.88	4.81	129.10
INPUT					48.24	13.74	169.38
CHEMICAL FEED	13.56	0.35	1.72				15.63
LUBES & ASPHALTS	0.0						0.0
TOTAL REFINED PRODUCT	70.41						
REFINERY USE & PROCESS LOSS*	3.85	0.41	1.12				5.38
SYNCRUDE MANUFACTURE	0.0	0.0					
USES+LOSSES		0.0					0.0
TOTAL INPUT		0.0					
PRIMARY ENERGY SUPPLY	74.26	30.41	23.75		48.24	13.74	190.40

*INCLUDES OPEC & OCEC

ALL QUANTITIES ARE GIVEN IN MILLION BARREL PER DAY OIL EQUIVALENT

6

DEMAND/SUPPLY MATRIX NORTH AMERICA — PROBLEM: WAESC2 — CASE: 04FEB77 — BASE YR: 1985. — CASE YR: 2000.

	OIL	COAL	GAS	ELECTRICITY		TOTAL
TRANSPORT	0.63			0.01		0.64
INDUSTRY	3.26	8.11	2.83	4.65		18.85
DOMESTIC	1.95	1.19	2.90	3.70		9.73
TOTAL INTO MARKETS	5.83	9.30	5.73	8.36		29.22
TRANSM./DISTR.LOSSES*	0.60	0.54	1.24	1.47		3.87
CONV. THERMAL ELECTR.	0.0	0.0	0.0	0.0		
USES+LOSSES	0.0	0.0	0.0			0.0
TOTAL INPUT	0.0	0.0	0.0			
GAS MANUFACTURE		0.0	0.0			
USES+LOSSES		0.0				0.0
TOTAL INPUT		0.0				
				NUCLEAR	HYDRO-ETC	
PRIMARY ENERGY DEMAND						
OUTPUT	6.44	9.85	6.97	7.68	2.15	33.09
INPUT				21.95	6.14	51.34
CHEMICAL FEED	5.61	0.21	1.11			6.93
LUBES & ASPHALTS	0.0					0.0
TOTAL REFINED PRODUCT	12.05					
REFINERY USE & PROCESS LOSS*	0.75	0.12	0.0			0.87
SYNCRUDE MANUFACTURE	0.0	0.0				
USES+LOSSES		0.0				0.0
TOTAL INPUT		0.0				
PRIMARY ENERGY SUPPLY	12.80	10.17	8.08	21.95	6.14	59.14

ALL QUANTITIES ARE GIVEN IN MILLION BARREL PER DAY OIL EQUIVALENT

7

DEMAND/SUPPLY MATRIX EUROPE — PROBLEM: WAESC2 — CASE: 04FEB77 — BASE YR: 1985. — CASE YR: 2000.

	OIL	COAL	GAS	ELECTRICITY		TOTAL
TRANSPORT	6.55			0.17		6.72
INDUSTRY	4.21	5.62	1.40	2.81		14.04
DOMESTIC	2.66	1.88	2.37	1.96		8.86
TOTAL INTO MARKETS	13.42	7.49	3.77	4.94		29.62
TRANSM./DISTR.LOSSES*	0.91	0.24	0.69	0.87		2.71
CONV. THERMAL ELECTR.	0.0	0.0	0.0	0.0		
USES+LOSSES	0.0	0.0	0.0			0.0
TOTAL INPUT	0.0	0.0	0.0			
GAS MANUFACTURE		0.0	0.0			
USES+LOSSES		0.0				0.0
TOTAL INPUT		0.0				
				NUCLEAR	HYDRO-ETC	
PRIMARY ENERGY DEMAND						
OUTPUT	14.34	7.73	4.46	4.82	0.99	32.33
INPUT				13.76	2.83	43.12
CHEMICAL FEED	3.93	0.14	0.61			4.68
LUBES & ASPHALTS	0.0					0.0
TOTAL REFINED PRODUCT	18.27					
REFINERY USE & PROCESS LOSS*	1.25	0.19	0.0			1.44
SYNCRUDE MANUFACTURE	0.0	0.0				
USES+LOSSES		0.0				0.0
TOTAL INPUT		0.0				
PRIMARY ENERGY SUPPLY	19.52	8.06	5.07	13.76	2.83	49.24

ALL QUANTITIES ARE GIVEN IN MILLION BARREL PER DAY OIL EQUIVALENT

8

DEMAND/SUPPLY MATRIX JAPAN

PROBLEM: WAESC2 BASE YR: 1985.
CASE: 04FEB77 CASE YR: 2000.

	OIL	COAL	GAS	ELECTRICITY		TOTAL
TRANSPORT	3.93			0.09		4.02
INDUSTRY	4.32	5.06	0.15	0.33		9.87
DOMESTIC	1.96	0.39	0.79	0.79		3.93
TOTAL INTO MARKETS	10.22	5.46	0.93	1.21		17.82
TRANSM./DISTR.LOSSES*	0.59	0.12	0.15	0.21		1.07
CONV. THERMAL ELECTR.	0.0	0.0	0.0	0.0		
USES+LOSSES	0.0	0.0	0.0			0.0
TOTAL INPUT	0.0	0.0	0.0			
GAS MANUFACTURE		0.0	0.0			
USES+LOSSES		0.0				0.0
TOTAL INPUT		0.0				
PRIMARY ENERGY DEMAND				NUCLEAR	HYDRO-ETC	
OUTPUT	10.81	5.57	1.08	1.10	0.32	18.89
INPUT				3.15	0.92	21.54
CHEMICAL FEED	1.03	0.0	0.0			1.03
LUBES & ASPHALTS	0.0					0.0
TOTAL REFINED PRODUCT	11.84					
REFINERY USE & PROCESS LOSS*	0.65	0.10	0.0			0.75
SYNCRUDE MANUFACTURE	0.0	0.0				
USES+LOSSES		0.0				0.0
TOTAL INPUT		0.0				
PRIMARY ENERGY SUPPLY	12.49	5.68	1.08	3.15	0.92	23.32

ALL QUANTITIES ARE GIVEN IN MILLION BARREL PER DAY OIL EQUIVALENT

9

DEMAND/SUPPLY MATRIX REST WOCA

PROBLEM: WAESC2 BASE YR: 1985.
CASE: 04FEB77 CASE YR: 2000.

	OIL	COAL	GAS	ELECTRICITY		TOTAL
TRANSPORT	15.62			0.04		15.66
INDUSTRY	7.03	4.84	3.51	2.19		17.57
DOMESTIC	1.20	1.37	2.56	1.71		6.83
TOTAL INTO MARKETS	23.85	6.21	6.07	3.94		40.06
TRANSM./DISTR.LOSSES*	1.41	0.13	0.86	0.69		3.09
CONV. THERMAL ELECTR.	0.0	0.0	0.0	0.0		
USES+LOSSES	0.0	0.0	0.0			0.0
TOTAL INPUT	0.0	0.0	0.0			
GAS MANUFACTURE		0.0	0.0			
USES+LOSSES		0.0				0.0
TOTAL INPUT		0.0				
PRIMARY ENERGY DEMAND				NUCLEAR	HYDRO-ETC	
OUTPUT	25.26	6.33	6.93	3.28	1.35	43.15
INPUT				9.38	3.85	51.75
CHEMICAL FEED	2.99	0.0	0.0			2.99
LUBES & ASPHALTS	0.0					0.0
TOTAL REFINED PRODUCT	28.25					
REFINERY USE & PROCESS LOSS*	0.52	0.0	1.12			1.65
SYNCRUDE MANUFACTURE	0.0	0.0				
USES+LOSSES		0.0				0.0
TOTAL INPUT		0.0				
PRIMARY ENERGY SUPPLY	28.77	6.33	8.05	9.38	3.85	56.38

ALL QUANTITIES ARE GIVEN IN MILLION BARREL PER DAY OIL EQUIVALENT

SUMMARY (ANNUAL CAPITAL CHARGES - CASE YEAR)

PROBLEM: WAESC2 CASE: 04FEB77 BASE YR: 1985. CASE YR: 2000.

10

PRODUCERS		OPEC	OCEC		
OIL					
PRODUCTION	: B$	1.56	0.0		
REFINING	: B$	2.37	0.0		
COAL					
PRODUCTION	: B$	0.0	3.14		
CONVERSION	: B$	0.0	0.0		
GAS					
PRODUCTION	: B$	9.78	0.0		
LIQUEFACTION	: B$	1.70	0.0		
TOTAL	: B$	15.41	3.14		
CONSUMERS		NORTH AMERICA	EUROPE	JAPAN	REST WOCA
OIL					
PRODUCTION	: B$	13.75	1.78	0.45	11.03
REFINING	: B$	1.27	3.27	1.64	4.54
DISTRIBUTION	: B$	0.16	0.69	0.51	1.42
TOTAL	: B$	15.17	5.74	2.60	16.98
COAL					
PRODUCTION	: B$	3.20	2.59	0.14	0.0
CONVERSION	: B$	0.0	0.0	0.0	0.0
DISTRIBUTION	: B$	0.30	0.26	0.24	0.23
TOTAL	: B$	3.51	2.85	0.39	0.23
GAS					
PRODUCTION	: B$	9.61	1.46	0.10	8.01
REGASIFICATION	: B$	0.0	0.42	0.13	0.0
DISTRIBUTION	: B$	0.61	1.77	0.47	4.07
TOTAL	: B$	10.22	3.65	0.70	12.08
ELECTRICITY					
FUEL OIL	: B$	0.0	0.0	0.0	0.0
COAL	: B$	0.0	0.0	0.0	0.0
GAS	: B$	0.0	0.0	0.0	0.0
NUCLEAR	: B$	41.07	25.87	5.59	18.32
HYDRO ELEC 1	: B$	2.02	2.12	0.62	3.72
OTHER	: B$	9.02	0.35	0.35	0.0
DISTRIBUTION	: B$	5.13	3.03	0.53	2.57
TOTAL	: B$	57.24	31.37	7.09	24.61
IND HEATING EQP	: B$	7.81	5.86	4.46	7.34
DOM HEATING EQP	: B$	12.22	12.99	6.41	12.85
TOTAL CAPEX	: B$	106.17	62.46	21.64	74.09

SUMMARY | PROBLEM: WAESC2 | BASE YR: 1985. | **11**
CASE: 04FEB77 | CASE YR: 2000.

(PRODUCING REGIONS REVENUES)

PRODUCERS			OPEC	OCEC
GOVERNMENT TAXES *				
CRUDE OIL			11.28	0.0
COAL			0.0	2.00
GAS			8.32	0.0
TARIFFS ***				
CRUDE OIL			18.25	18.25
DISTILLATES			20.73	20.73
FUEL OIL			13.29	13.29
COAL			11.68	11.68
GAS			20.73	20.73
GOVT. TAKE(FROM TAXES)				
OIL	:	B$	182.49	0.0
COAL	:	B$	0.0	5.97
GAS	:	B$	21.25	0.0
TOTAL GOVT. TAKE	:	B$	203.75	5.97
REVENUES FROM TARIFFS				
CRUDE OIL	:	B$	235.70	0.0
DISTILLATES	:	B$	48.75	0.0
FUEL OIL	:	B$	12.12	0.0
COAL	:	B$	0.0	34.84
GAS	:	B$	52.95	0.0
TOTAL REVENUES	:	B$	349.52	34.84

* DOLLAR/BOE

*** DOLLAR/BOE. AVERAGE WORLD PRICE LAID-IN AT PORT OF ENTRY

SUMMARY | PROBLEM: WAESC2 | BASE YR: 1985. | **12**
CASE: 04FEB77 | CASE YR: 2000.

(CONSUMING REGIONS COSTS AND REVENUES)

CONSUMERS (TOTAL COSTS)			NORTH AMERICA	EUROPE	JAPAN	REST WOCA
IMPORT BILL						
CRUDE OIL	:	B$	0.0	106.58	68.72	73.73
DISTILLATES	:	B$	5.59	0.0	0.0	43.15
FUEL OIL	:	B$	0.0	3.41	8.71	0.0
COAL	:	B$	0.0	17.61	23.09	27.00
GAS	:	B$	0.0	22.69	7.57	34.05
TOTAL IMPORT BILL	:	B$	5.59	150.29	108.09	177.93
INTERNAL COSTS **						
OIL	:	B$	28.80	23.89	9.34	33.87
COAL	:	B$	30.97	18.42	3.32	2.82
GAS	:	B$	13.69	5.56	1.09	15.28
ELECTRICITY	:	B$	66.36	36.88	8.40	28.65
TOTAL	:	B$	139.83	84.75	22.15	80.61
ENERGY CONS EQUIP	:	B$	20.03	18.85	10.86	20.19
TOTAL INTERNAL	:	B$	159.86	103.60	33.01	100.80
REVENUES FROM EXPORTS						
CRUDE OIL	:	B$	0.0	0.0	0.0	0.0
DISTILLATES	:	B$	0.0	0.0	0.0	0.0
FUEL OIL	:	B$	0.0	0.0	0.0	0.0
COAL	:	B$	30.73	0.0	0.0	0.0
GAS	:	B$	0.0	0.0	0.0	0.0
TOTAL REVENUES	:	B$	30.73	0.0	0.0	0.0
NET CONSUM REG. COST	:	B$	134.72	253.89	141.10	278.73
IMPORT/EXPORT BALANCE						
EQUILIBRIUM	:	B$	-25.13	150.29	108.09	177.93
LOANS AT 15.0% PA	:	B$	0.0	0.0	0.0	0.0
SHORTFALL	:	B$	0.0	0.0	0.0	0.0

** CAPEX CHARGED AS A CONSTANT COST CHARGED OVER 100.0% OF
THE USEFUL PHYSICAL LIFETIME OF A PLANT AT 15.0% ANNUAL COST OF CAPITAL.

1

VOLUMETRIC SUMMARY

PROBLEM: WAESD7 BASE YR: 1985.
CASE: 04FEB77 CASE YR: 2000.

PRODUCTION

	OIL	COAL	GAS	NUCLEAR	HYDRO	TOTAL
NORTH AMERICA	8.6	18.3	6.2	13.4	4.0	50.6
EUROPE	2.5	4.5	2.3	7.1	2.7	19.0
JAPAN	0.4	0.3	0.1	2.0	0.9	3.6
OPEC	39.0	0.0	7.9	0.0	0.0	46.9
OCEC	0.0	12.4	0.0	0.0	0.0	12.4
REST WOCA	7.5	0.0	3.0	3.1	3.3	16.9
SUBTOTAL	58.0	35.4	19.5	25.6	10.8	149.4
WOCA EXPORTS	2.0	0.5	1.0	0.0	0.0	3.5
TOTAL	60.0	35.9	20.5	25.6	10.8	152.9

TOTAL WORLD ENERGY PRODUCTION (EXCLUDING COMMUNIST AREA INTERNAL): 152.9

CONVERSION(INPUT)

	OIL	COAL	GAS	NUCLEAR	HYDRO
OIL REFINING	60.0				
COAL GASIFICATION		0.0			
COAL LIQUEFACTION		0.0			
GAS LIQUEFACTION (NOT INCL. COMMUNIST AREA)				3.2	
ELECTRICITY GENERATION	0.0	0.0	0.0	25.6	10.8

MARKET INPUT

	OIL	COAL	GAS	ELECTRICITY
INDUSTRIAL				
HEAT, LIGHT AND POWER	11.4	22.3	6.7	4.6
NON ENERGY	10.4	0.3	1.3	
DOMESTIC				
HEAT, LIGHT AND POWER	3.1	11.2	7.0	6.1
TRANSPORT AND BUNKERS	30.4			0.2

ALL QUANTITIES ARE GIVEN IN MILLION BARREL PER DAY OIL EQUIVALENT

2

PRODUCTION AND TRANSPORTS CRUDE OIL

PROBLEM: WAESD7 CASE: 04FEB77 BASE YR: 1985. CASE YR: 2000.

FROM \ TO	NORTH AMERICA	EUROPE	JAPAN	REST WOCA	OPEC	PROD"N
NORTH AMERICA	8.57	.	.	.	.	8.57
EUROPE	.	2.53	.	.	.	2.53
JAPAN	.	.	0.38	.	.	0.38
REST WOCA	.	.	.	7.50	.	7.50
OPEC	7.87	11.43	7.34	11.63	0.74	39.00
COMMUNIST AREAS	.	2.00	.	.	.	2.00
TOTAL SUPPLY	16.44	15.96	7.72	19.13	0.74	59.98

PRODUCTION AND TRANSPORTS OIL PRODUCTS

PROBLEM: WAESD7 CASE: 04FEB77 BASE YR: 1985. CASE YR: 2000.

F=FUEL OIL, D=DISTILLATES

FROM \ TO		NORTH AMERICA	EUROPE	JAPAN	REST WOCA	OPEC	PROD"N
NORTH AMERICA	F	1.62	.	.	.	.	1.62
	D	13.58	.	.	.	.	13.58
EUROPE	F	.	5.33	.	.	.	5.33
	D	.	10.25	.	.	.	10.25
JAPAN	F	.	.	2.58	.	.	2.58
	D	.	.	4.62	.	.	4.62
REST WOCA	F	.	.	.	4.79	.	4.79
	D	.	.	.	14.06	.	14.06
OPEC	F	.	.	0.19	.	0.00	0.19
	D	0.50	.	.	.	0.00	0.50
TOTAL SUPPLY	F	1.62	5.33	2.77	4.79	0.00	14.50
	D	14.07	10.25	4.62	14.06	0.00	43.01

3

PRODUCTION AND TRANSPORTS COAL

PROBLEM: WAESD7 CASE: 04FEB77 BASE YR: 1985. CASE YR: 2000.

FROM \ TO	NORTH AMERICA	EUROPE	JAPAN	REST WOCA	LOSS	PROD"N
NORTH AMERICA	14.68	3.29	.	.	0.36	18.33
EUROPE	.	4.40	.	.	0.09	4.49
JAPAN	.	.	0.25	.	0.01	0.26
OCEC	.	1.38	4.40	6.35	0.24	12.37
COMMUNIST AREAS	.	0.50	.	.	.	0.50
TOTAL SUPPLY	14.68	9.57	4.65	6.35	0.70	35.95

PRODUCTION AND TRANSPORTS GAS

PROBLEM: WAESD7 CASE: 04FEB77 BASE YR: 1985. CASE YR: 2000.

P=BY PIPE L=AS LNG

FROM \ TO		NORTH AMERICA	EUROPE	JAPAN	REST WOCA	OPEC	LOSS	PROD"N
NORTH AMERICA	P	5.62	.	.	.	.	0.62	6.25
EUROPE	P	.	2.05	.	.	.	0.23	2.28
JAPAN	P	.	.	0.07	.	.	0.01	0.08
REST WOCA	P	.	.	.	2.70	.	0.30	3.00
OPEC	P	.	0.45	.	3.51	3.17	0.79	7.92
	L	0.94	1.59	.	.	.	0.63	.
COMMUNIST AREAS	P	.	.	.	.	.	.	.
	L	0.75	.	0.25	.	.	.	1.00
TOTAL SUPPLY	P	5.62	2.50	0.07	6.21	3.17	1.95	19.53
	L	1.69	1.59	0.25	.	.	0.63	1.00

MARKET INPUT SUMMARY
===================

PROBLEM: WAESD7 CASE: 04FEB77 BASE YR: 1985. CASE YR: 2000.

INDUSTRIAL HEAT, LIGHT AND POWER		OIL	COAL	GAS	ELECTRICITY
NORTH AMERICA	:	1.5	8.8	3.0	1.8
EUROPE	:	3.3	5.2	1.1	1.5
JAPAN	:	2.4	3.3	0.0	0.3
REST WOCA	:	4.1	5.0	2.5	0.9
		11.4	22.3	6.7	4.6

INDUSTRIAL NON ENERGY		OIL	COAL	GAS
NORTH AMERICA	:	4.5	0.2	0.9
EUROPE	:	2.9	0.1	0.5
JAPAN	:	0.9	.	.
REST WOCA	:	2.1	.	.
		10.4	0.3	1.3

DOMESTIC HEAT, LIGHT AND POWER		OIL	COAL	GAS	ELECTRICITY
NORTH AMERICA	:	0.0	5.1	2.8	3.4
EUROPE	:	1.7	3.9	1.5	1.3
JAPAN	:	0.9	1.2	0.3	0.5
REST WOCA	:	0.5	0.9	2.3	0.9
		3.1	11.2	7.0	6.1

TRANSPORT		OIL	ELECTRICITY
NORTH AMERICA	:	8.9	0.0
EUROPE	:	6.8	0.1
JAPAN	:	2.8	0.0
REST WOCA	:	11.2	0.0
		29.7	0.2

DEMAND			IND H.L&P	IND NONEN	DOM H.L&P	TRANSPORT
NORTH AMERICA						
	DEMAND	:	15.1	5.5	11.4	12.1
	SUPPLY	:	15.1	5.5	11.4	8.9
	DEFICIT	:	0.0	0.0	0.0	3.2
EUROPE						
	DEMAND	:	11.1	3.5	8.5	6.9
	SUPPLY	:	11.1	3.5	8.5	6.9
	DEFICIT	:	0.0	0.0	0.0	0.0
JAPAN						
	DEMAND	:	6.0	0.9	2.9	2.8
	SUPPLY	:	6.0	0.9	2.9	2.8
	DEFICIT	:	0.0	0.0	0.0	0.0
REST WOCA						
	DEMAND	:	12.6	2.1	4.7	11.2
	SUPPLY	:	12.6	2.1	4.7	11.2
	DEFICIT	:	0.0	0.0	0.0	0.0

W O C A DEMAND/SUPPLY MATRIX PROBLEM: WAESD7 BASE YR: 1985. **5**

CASE: 04FEB77 CASE YR: 2000.

	OIL	COAL	GAS	ELECTRICITY		TOTAL
TRANSPORT	29.70			0.21		29.91
INDUSTRY	11.36	22.31	6.66	4.55		44.88
DOMESTIC	3.14	11.19	6.97	6.07		27.38
TOTAL INTO MARKETS	44.21	33.50	13.63	10.84		102.17
TRANSM./DISTR.LOSSES*	2.88	1.41	3.92	1.91		10.12
CONV. THERMAL ELECTR.	0.0	0.0	0.0	0.0		
USES+LOSSES	0.0	0.0	0.0			0.0
TOTAL INPUT	0.0	0.0	0.0			
GAS MANUFACTURE		0.0	0.0			
USES+LOSSES		0.0				0.0
TOTAL INPUT		0.0				
PRIMARY ENERGY DEMAND				NUCLEAR	HYDRO-ETC	
OUTPUT	47.08	34.91	17.55	8.96	3.79	112.29
INPUT				25.60	10.83	135.97
CHEMICAL FEED	10.43	0.27	1.33			12.03
LUBES & ASPHALTS	0.0					0.0
TOTAL REFINED PRODUCT	57.51					
REFINERY USE & PROCESS LOSS*	2.47	0.78	1.64			4.89
SYNCRUDE MANUFACTURE	0.0	0.0				
USES+LOSSES		0.0				0.0
TOTAL INPUT		0.0				
PRIMARY ENERGY SUPPLY	59.98	35.96	20.53	25.60	10.83	152.90

*INCLUDES OPEC & OCEC

ALL QUANTITIES ARE GIVEN IN MILLION BARREL PER DAY OIL EQUIVALENT

DEMAND/SUPPLY MATRIX NORTH AMERICA PROBLEM: WAESD7 BASE YR: 1985. **6**

CASE: 04FEB77 CASE YR: 2000.

	OIL	COAL	GAS	ELECTRICITY		TOTAL
TRANSPORT	8.92			0.01		8.93
INDUSTRY	1.51	8.81	3.02	1.77		15.12
DOMESTIC	0.0	5.12	2.84	3.41		11.37
TOTAL INTO MARKETS	10.43	13.93	5.87	5.19		35.42
TRANSM./DISTR.LOSSES*	0.78	0.66	1.19	0.92		3.55
CONV. THERMAL ELECTR.	0.0	0.0	0.0	0.0		
USES+LOSSES	0.0	0.0	0.0			0.0
TOTAL INPUT	0.0	0.0	0.0			
GAS MANUFACTURE		0.0	0.0			
USES+LOSSES		0.0				0.0
TOTAL INPUT		0.0				
PRIMARY ENERGY DEMAND				NUCLEAR	HYDRO-ETC	
OUTPUT	11.22	14.59	7.05	4.70	1.41	38.97
INPUT				13.43	4.03	50.32
CHEMICAL FEED	4.47	0.17	0.88			5.52
LUBES & ASPHALTS	0.0					0.0
TOTAL REFINED PRODUCT	15.69					
REFINERY USE & PROCESS LOSS*	1.24	0.29	0.0			1.54
SYNCRUDE MANUFACTURE	0.0	0.0				
USES+LOSSES		0.0				0.0
TOTAL INPUT		0.0				
PRIMARY ENERGY SUPPLY	16.93	15.05	7.94	13.43	4.03	57.38

ALL QUANTITIES ARE GIVEN IN MILLION BARREL PER DAY OIL EQUIVALENT

7

DEMAND/SUPPLY MATRIX EUROPE

PROBLEM: WAESD7 CASE: 04FEB77 BASE YR: 1985. CASE YR: 2000.

	OIL	COAL	GAS		ELECTRICITY		TOTAL
TRANSPORT	6.83				0.12		6.95
INDUSTRY	3.34	5.18	1.11		1.51		11.14
DOMESTIC	1.71	3.94	1.54		1.27		8.46
TOTAL INTO MARKETS	11.89	9.12	2.65		2.89		26.55
TRANSM./DISTR.LOSSES*	0.78	0.28	0.59		0.51		2.16
CONV. THERMAL ELECTR.	0.0	0.0	0.0		0.0		
USES+LOSSES	0.0	0.0	0.0				0.0
TOTAL INPUT	0.0	0.0	0.0				
GAS MANUFACTURE		0.0	0.0				
USES+LOSSES		0.0					0.0
TOTAL INPUT		0.0					
PRIMARY ENERGY DEMAND				NUCLEAR		HYDRO-ETC	
OUTPUT	12.66	9.40	3.24	2.47		0.93	28.71
INPUT				7.07		2.66	35.04
CHEMICAL FEED	2.91	0.10	0.45				3.47
LUBES & ASPHALTS	0.0						0.0
TOTAL REFINED PRODUCT	15.58						
REFINERY USE & PROCESS LOSS*	0.38	0.16	0.63				1.17
SYNCRUDE MANUFACTURE	0.0	0.0					
USES+LOSSES		0.0					0.0
TOTAL INPUT		0.0					
PRIMARY ENERGY SUPPLY	15.96	9.66	4.32	7.07		2.66	39.67

ALL QUANTITIES ARE GIVEN IN MILLION BARREL PER DAY OIL EQUIVALENT

8

DEMAND/SUPPLY MATRIX JAPAN

PROBLEM: WAESD7 CASE: 04FEB77 BASE YR: 1985. CASE YR: 2000.

	OIL	COAL	GAS		ELECTRICITY		TOTAL
TRANSPORT	2.76				0.05		2.81
INDUSTRY	2.41	3.28	0.0		0.34		6.03
DOMESTIC	0.93	1.20	0.29		0.46		2.88
TOTAL INTO MARKETS	6.10	4.48	0.29		0.85		11.72
TRANSM./DISTR.LOSSES*	0.37	0.10	0.05		0.15		0.66
CONV. THERMAL ELECTR.	0.0	0.0	0.0		0.0		
USES+LOSSES	0.0	0.0	0.0				0.0
TOTAL INPUT	0.0	0.0	0.0				
GAS MANUFACTURE		0.0	0.0				
USES+LOSSES		0.0					0.0
TOTAL INPUT		0.0					
PRIMARY ENERGY DEMAND				NUCLEAR		HYDRO-ETC	
OUTPUT	6.47	4.58	0.33	0.69		0.31	12.38
INPUT				1.97		0.88	14.23
CHEMICAL FEED	0.92	0.0	0.0				0.92
LUBES & ASPHALTS	0.0						0.0
TOTAL REFINED PRODUCT	7.39						
REFINERY USE & PROCESS LOSS*	0.52	0.08	0.0				0.60
SYNCRUDE MANUFACTURE	0.0	0.0					
USES+LOSSES		0.0					0.0
TOTAL INPUT		0.0					
PRIMARY ENERGY SUPPLY	7.91	4.66	0.33	1.97		0.88	15.75

ALL QUANTITIES ARE GIVEN IN MILLION BARREL PER DAY OIL EQUIVALENT

9

DEMAND/SUPPLY MATRIX REST WOCA
==

PROBLEM: WAESD7 CASE: 04FEB77

BASE YR: 1985. CASE YR: 2000.

	OIL	COAL	GAS	NUCLEAR	ELECTRICITY	HYDRO-ETC	TOTAL
TRANSPORT	11.19				0.03		11.22
INDUSTRY	4.10	5.04	2.52		0.94		12.59
DOMESTIC	0.50	0.93	2.31		0.93		4.67
TOTAL INTO MARKETS	15.79	5.97	4.82		1.90		28.48
TRANSM./DISTR.LOSSES*	0.94	0.13	0.67		0.34		2.08
CONV. THERMAL ELECTR.	0.0	0.0	0.0		0.0		
USES+LOSSES	0.0	0.0	0.0				0.0
TOTAL INPUT	0.0	0.0	0.0				
GAS MANUFACTURE		0.0	0.0				
USES+LOSSES		0.0					0.0
TOTAL INPUT		0.0					
PRIMARY ENERGY DEMAND							
OUTPUT	16.73	6.10	5.50	1.10		1.14	30.56
INPUT				3.13		3.26	34.71
CHEMICAL FEED	2.12	0.0	0.0				2.12
LUBES & ASPHALTS	0.0						0.0
TOTAL REFINED PRODUCT	18.85						
REFINERY USE & PROCESS LOSS*	0.28	0.25	1.01				1.54
SYNCRUDE MANUFACTURE	0.0	0.0					
USES+LOSSES		0.0					0.0
TOTAL INPUT		0.0					
PRIMARY ENERGY SUPPLY	19.13	6.35	6.51	3.13		3.26	38.37

ALL QUANTITIES ARE GIVEN IN MILLION BARREL PER DAY OIL EQUIVALENT

S U M M A R Y (ANNUAL CAPITAL CHARGES - CASE YEAR)

PROBLEM: WAESD7 BASE YR: 1985. CASE: 04FEB77 CASE YR: 2000.

10

PRODUCERS			OPEC	OCEC
OIL				
PRODUCTION	:	B$	1.29	0.0
REFINING	:	B$	0.0	0.0
COAL				
PRODUCTION	:	B$	0.0	5.70
CONVERSION	:	B$	0.0	0.0
GAS				
PRODUCTION	:	B$	7.74	0.0
LIQUEFACTION	:	B$	1.72	0.0
TOTAL	:	B$	10.76	5.70

CONSUMERS			NORTH AMERICA	EUROPE	JAPAN	REST WOCC
OIL						
PRODUCTION	:	B$	6.74	1.42	0.45	6.36
REFINING	:	B$	3.07	2.71	1.50	3.86
DISTRIBUTION	:	B$	0.39	0.61	0.26	0.86
TOTAL	:	B$	10.19	4.74	2.21	11.08
COAL						
PRODUCTION	:	B$	3.73	2.31	0.14	0.0
CONVERSION	:	B$	0.0	0.0	0.0	0.0
DISTRIBUTION	:	B$	0.55	0.34	0.21	0.24
TOTAL	:	B$	4.28	2.65	0.35	0.24
GAS						
PRODUCTION	:	B$	6.67	1.78	0.10	5.22
REGASIFICATION	:	B$	0.21	0.18	0.00	0.0
DISTRIBUTION	:	B$	0.85	0.79	0.0	2.76
TOTAL	:	B$	7.72	2.76	0.10	7.98
ELECTRICITY						
FUEL OIL	:	B$	0.0	0.0	0.0	0.0
COAL	:	B$	0.0	0.0	0.0	0.0
GAS	:	B$	0.0	0.0	0.0	0.0
NUCLEAR	:	B$	24.40	11.82	3.43	5.95
HYDRO ELEC 1	:	B$	1.48	1.50	0.57	2.75
OTHER	:	B$	3.32	0.39	0.35	0.0
DISTRIBUTION	:	B$	2.69	1.27	0.37	0.88
TOTAL	:	B$	31.89	14.98	4.72	9.58
IND HEATING EQP	:	B$	6.34	4.57	2.63	5.28
DOM HEATING EQP	:	B$	23.39	15.43	6.00	8.25
TOTAL CAPEX	:	B$	83.81	45.13	16.03	42.41

11

SUMMARY | PROBLEM: WAESD7 | BASE YR: 1985.
CASE: 04FEB77 | CASE YR: 2000.

(PRODUCING REGIONS REVENUES)

PRODUCERS			OPEC	OCEC
GOVERNMENT TAXES *				
CRUDE OIL			11.28	0.0
COAL			0.0	2.00
GAS			8.32	0.0
TARIFFS ***				
CRUDE OIL			12.50	12.50
DISTILLATES			14.20	14.20
FUEL OIL			9.10	9.10
COAL			8.00	8.00
GAS			14.20	14.20
GOVT. TAKE(FROM TAXES)				
OIL	:	B$	160.36	0.0
COAL	:	B$	0.0	8.85
GAS	:	B$	19.72	0.0
TOTAL GOVT. TAKE	:	B$	180.08	8.85
REVENUES FROM TARIFFS				
CRUDE OIL	:	B$	174.56	0.0
DISTILLATES	:	B$	2.57	0.0
FUEL OIL	:	B$	0.64	0.0
COAL	:	B$	0.0	35.40
GAS	:	B$	33.66	0.0
TOTAL REVENUES	:	B$	211.43	35.40

* DOLLAR/BOE

*** DOLLAR/BOE, AVERAGE WORLD PRICE LAID-IN AT PORT OF ENTRY

12

SUMMARY | PROBLEM: WAESD7 | BASE YR: 1985.
CASE: 04FEB77 | CASE YR: 2000.

(CONSUMING REGIONS COSTS AND REVENUES)

CONSUMERS (TOTAL COSTS)			NORTH AMERICA	EUROPE	JAPAN	REST WOCA
IMPORT BILL						
CRUDE OIL	:	B$	35.89	61.27	33.47	53.05
DISTILLATES	:	B$	2.57	0.0	0.0	0.0
FUEL OIL	:	B$	0.0	0.0	0.64	0.0
COAL	:	B$	0.0	15.10	12.85	18.53
GAS	:	B$	8.75	10.84	1.32	20.21
TOTAL IMPORT BILL	:	B$	47.21	87.22	48.28	91.79
INTERNAL COSTS **						
OIL	:	B$	22.97	20.51	6.63	22.74
COAL	:	B$	34.96	20.66	2.87	2.83
GAS	:	B$	11.00	4.42	0.23	10.57
ELECTRICITY	:	B$	37.49	17.99	5.59	11.22
TOTAL	:	B$	106.42	63.59	15.31	47.36
ENERGY CONS EQUIP	:	B$	29.72	20.00	8.64	13.53
TOTAL INTERNAL	:	B$	136.14	83.59	23.94	60.89
REVENUES FROM EXPORTS						
CRUDE OIL	:	B$	0.0	0.0	0.0	0.0
DISTILLATES	:	B$	0.0	0.0	0.0	0.0
FUEL OIL	:	B$	0.0	0.0	0.0	0.0
COAL	:	B$	9.62	0.0	0.0	0.0
GAS	:	B$	0.0	0.0	0.0	0.0
TOTAL REVENUES	:	B$	9.62	0.0	0.0	0.0
NET CONSUM REG. COST	:	B$	173.73	170.80	72.22	152.68
IMPORT/EXPORT BALANCE						
EQUILIBRIUM	:	B$	37.59	87.22	48.28	91.79
LOANS AT 15.0% PA	:	B$	0.0	0.0	0.0	0.0
SHORTFALL	:	B$	0.0	0.0	0.0	0.0

** CAPEX CHARGED AS A CONSTANT COST CHARGED OVER 100.0% OF
THE USEFUL PHYSICAL LIFETIME OF A PLANT AT 15.0% ANNUAL COST OF CAPITAL.

1

VOLUMETRIC SUMMARY

PROBLEM: WAESD8 BASE YR: 1985.
CASE: 04FEB77 CASE YR: 2000.

PRODUCTION

	OIL	COAL	GAS	NUCLEAR	HYDRO	TOTAL
NORTH AMERICA	8.6	14.3	6.2	21.8	4.0	54.9
EUROPE	2.5	3.9	1.9	11.3	2.7	22.4
JAPAN	0.4	0.3	0.1	2.8	0.9	4.4
OPEC	39.0	0.0	7.3	0.0	0.0	46.3
OCEC	0.0	8.3	0.0	0.0	0.0	8.3
REST WOCA	7.5	0.0	3.0	6.6	2.6	19.7
SUBTOTAL	58.0	26.7	18.5	42.5	10.3	156.0
WOCA EXPORTS	2.0	0.5	1.0	0.0	0.0	3.5
TOTAL	60.0	27.2	19.5	42.5	10.3	159.5

TOTAL WORLD ENERGY PRODUCTION (EXCLUDING COMMUNIST AREA INTERNAL): 159.5

CONVERSION(INPUT)

	OIL	COAL	GAS	NUCLEAR	HYDRO
OIL REFINING	60.0				
COAL GASIFICATION		0.0			
COAL LIQUEFACTION		0.0			
GAS LIQUEFACTION (NOT INCL. COMMUNIST AREA)				3.1	
ELECTRICITY GENERATION	0.0	0.0	0.0	42.5	10.3

MARKET INPUT

	OIL	COAL	GAS	ELECTRICITY
INDUSTRIAL				
HEAT, LIGHT AND POWER	10.8	19.8	6.7	7.5
NON ENERGY	4.9	0.1	0.4	
DOMESTIC				
HEAT, LIGHT AND POWER	6.3	5.8	7.3	8.0
TRANSPORT AND BUNKERS	32.7			0.2

ALL QUANTITIES ARE GIVEN IN MILLION BARREL PER DAY OIL EQUIVALENT

2

PRODUCTION AND TRANSPORTS CRUDE OIL

PROBLEM: WAESD8 CASE: 04FEB77 BASE YR: 1985. CASE YR: 2000.

FROM \ TO	NORTH AMERICA	EUROPE	JAPAN	REST WOCA	OPEC	PROD'N
NORTH AMERICA	8.57	.	.	.	.	8.57
EUROPE	.	2.53	.	.	.	2.53
JAPAN	.	.	0.38	.	.	0.38
REST WOCA	.	.	.	7.50	.	7.50
OPEC	7.03	11.65	8.05	12.10	0.18	39.00
COMMUNIST AREAS	.	2.00	.	.	.	2.00
TOTAL SUPPLY	15.60	16.18	8.43	19.60	0.18	59.98

PRODUCTION AND TRANSPORTS OIL PRODUCTS

PROBLEM: WAESD8 CASE: 04FEB77 BASE YR: 1985. CASE YR: 2000.

F=FUEL OIL, D=DISTILLATES

FROM \ TO		NORTH AMERICA	EUROPE	JAPAN	REST WOCA	OPEC	PROD'N
NORTH AMERICA	F	1.86	.	.	.	.	1.86
	D	12.58	.	.	.	.	12.58
EUROPE	F	.	4.58	.	.	.	4.58
	D	.	10.60	.	.	.	10.60
JAPAN	F	.	.	2.87	.	.	2.87
	D	.	.	4.98	.	.	4.98
REST WOCA	F	.	.	.	4.60	.	4.60
	D	.	.	.	14.76	.	14.76
OPEC	F	.	.	0.06	.	.	0.06
	D	0.11	.	.	.	0.00	0.11
TOTAL SUPPLY	F	1.86	4.58	2.93	4.60	.	13.97
	D	12.69	10.60	4.98	14.76	0.00	43.02

3

PRODUCTION AND TRANSPORTS COAL

PROBLEM: WAESD8 CASE: 04FEB77 BASE YR: 1985. CASE YR: 2000.

FROM \ TO	NORTH AMERICA	EUROPE	JAPAN	REST WOCA	LOSS	PROD'N
NORTH AMERICA	11.36	1.67	0.96	.	0.28	14.27
EUROPE	.	3.80	.	.	0.08	3.88
JAPAN	.	.	0.25	.	0.01	0.26
OCEC	.	.	2.01	6.17	0.16	8.34
COMMUNIST AREAS	.	0.50	.	.	.	0.50
TOTAL SUPPLY	11.36	5.97	3.23	6.17	0.52	27.25

PRODUCTION AND TRANSPORTS GAS

PROBLEM: WAESD8 CASE: 04FEB77 BASE YR: 1985. CASE YR: 2000.

P=BY PIPE L=AS LNG

FROM \ TO		NORTH AMERICA	EUROPE	JAPAN	REST WOCA	OPEC	LOSS	PROD'N
NORTH AMERICA	P	5.62	.	.	.	.	0.62	6.25
EUROPE	P	.	1.73	.	.	.	0.19	1.92
JAPAN	P	.	.	0.07	.	.	0.01	0.08
REST WOCA	P	.	.	.	2.70	.	0.30	3.00
OPEC	P	.	0.45	.	2.97	3.11	0.73	7.26
	L	.	2.49	.	.	.	0.62	.
COMMUNIST AREAS	P	.	.	.	.	.	.	.
	L	.	.	1.00	.	.	.	1.00
TOTAL SUPPLY	P	5.62	2.18	0.07	5.67	3.11	1.85	18.51
	L	.	2.49	1.00	.	.	0.62	1.00

4

MARKET INPUT SUMMARY — PROBLEM: WAESD8 — CASE: 04FEB77 — BASE YR: 1985. — CASE YR: 2000.

INDUSTRIAL HEAT, LIGHT AND POWER	OIL	COAL	GAS	ELECTRICITY
NORTH AMERICA :	1.5	8.0	2.3	3.4
EUROPE :	3.3	3.9	1.5	2.2
JAPAN :	2.4	2.8	0.4	0.5
REST WOCA :	3.6	5.1	2.5	1.5
	10.8	19.8	6.7	7.5

INDUSTRIAL NON ENERGY	OIL	COAL	GAS
NORTH AMERICA :	.	.	.
EUROPE :	1.9	0.1	0.4
JAPAN :	0.9	.	.
REST WOCA :	2.1	.	.
	4.9	0.1	0.4

DOMESTIC HEAT, LIGHT AND POWER

	OIL	COAL	GAS	ELECTRICITY
NORTH AMERICA :	1.1	2.9	3.0	4.3
EUROPE :	2.5	1.7	2.3	1.9
JAPAN :	1.4	0.3	0.6	0.6
REST WOCA :	1.3	1.0	1.4	1.2
	6.3	5.8	7.3	8.0

TRANSPORT

	OIL	ELECTRICITY
NORTH AMERICA :	11.2	0.0
EUROPE :	6.8	0.1
JAPAN :	2.7	0.0
REST WOCA :	11.4	0.0
	32.1	0.2

DEMAND

		IND H, L&P	IND NONEN	DOM H, L&P	TRANSPORT
NORTH AMERICA					
	DEMAND :	15.1	5.5	11.4	12.1
	SUPPLY :	15.1	0.0	11.4	11.2
	DEFICIT :	0.0	5.5	0.0	1.0
EUROPE					
	DEMAND :	10.9	3.2	8.3	6.9
	SUPPLY :	10.9	3.2	8.3	6.9
	DEFICIT :	0.0	0.0	0.0	0.0
JAPAN					
	DEMAND :	6.0	0.9	2.9	2.8
	SUPPLY :	6.0	0.9	2.9	2.8
	DEFICIT :	0.0	0.0	0.0	0.0
REST WOCA					
	DEMAND :	12.7	2.1	4.8	11.5
	SUPPLY :	12.7	2.1	4.8	11.5
	DEFICIT :	0.0	0.0	0.0	0.0

5

W O C A — DEMAND/SUPPLY MATRIX — PROBLEM: WAESD8 — BASE YR: 1985.
CASE: 04FEB77 — CASE YR: 2000.

	OIL	COAL	GAS	NUCLEAR	ELECTRICITY	HYDRO-ETC	TOTAL
TRANSPORT	32.10				0.22		32.32
INDUSTRY	10.77	19.79	6.67		7.51		44.74
DOMESTIC	6.34	5.80	7.27		7.97		27.38
TOTAL INTO MARKETS	49.21	25.59	13.94		15.70		104.44
TRANSM./DISTR.LOSSES*	2.85	1.07	3.75		2.77		10.44
CONV. THERMAL ELECTR.	0.0	0.0	0.0		0.0		
USES+LOSSES	0.0	0.0	0.0				0.0
TOTAL INPUT	0.0	0.0	0.0				
GAS MANUFACTURE		0.0	0.0				
USES+LOSSES		0.0					0.0
TOTAL INPUT		0.0					
PRIMARY ENERGY DEMAND				NUCLEAR		HYDRO-ETC	
OUTPUT	52.06	26.65	17.70	14.87		3.59	114.88
INPUT				42.50		10.27	149.18
CHEMICAL FEED	4.93	0.10	0.42				5.44
LUBES & ASPHALTS	0.0						0.0
TOTAL REFINED PRODUCT	56.99						
REFINERY USE & PROCESS LOSS*	2.99	0.51	1.40				4.90
SYNCRUDE MANUFACTURE	0.0	0.0					
USES+LOSSES		0.0					0.0
TOTAL INPUT		0.0					
PRIMARY ENERGY SUPPLY	59.98	27.26	19.51	42.50		10.27	159.52

*INCLUDES OPEC & OCEC

ALL QUANTITIES ARE GIVEN IN MILLION BARREL PER DAY OIL EQUIVALENT

6

DEMAND/SUPPLY MATRIX NORTH AMERICA — PROBLEM: WAESD8 — BASE YR: 1985.
CASE: 04FEB77 — CASE YR: 2000.

	OIL	COAL	GAS	NUCLEAR	ELECTRICITY	HYDRO-ETC	TOTAL
TRANSPORT	11.16				0.01		11.17
INDUSTRY	1.51	7.99	2.27		3.35		15.12
DOMESTIC	1.15	2.88	3.02		4.32		11.37
TOTAL INTO MARKETS	13.82	10.87	5.29		7.68		37.66
TRANSM./DISTR.LOSSES*	0.73	0.51	0.96		1.36		3.56
CONV. THERMAL ELECTR.	0.0	0.0	0.0		0.0		
USES+LOSSES	0.0	0.0	0.0				0.0
TOTAL INPUT	0.0	0.0	0.0				
GAS MANUFACTURE		0.0	0.0				
USES+LOSSES		0.0					0.0
TOTAL INPUT		0.0					
PRIMARY ENERGY DEMAND				NUCLEAR		HYDRO-ETC	
OUTPUT	14.55	11.38	6.25	7.63		1.41	41.22
INPUT				21.80		4.03	58.01
CHEMICAL FEED	0.0	0.0	0.0				0.0
LUBES & ASPHALTS	0.0						0.0
TOTAL REFINED PRODUCT	14.55						
REFINERY USE & PROCESS LOSS*	1.15	0.26	0.0				1.42
SYNCRUDE MANUFACTURE	0.0	0.0					
USES+LOSSES		0.0					0.0
TOTAL INPUT		0.0					
PRIMARY ENERGY SUPPLY	15.70	11.64	6.25	21.80		4.03	59.43

ALL QUANTITIES ARE GIVEN IN MILLION BARREL PER DAY OIL EQUIVALENT

7

DEMAND/SUPPLY MATRIX EUROPE — PROBLEM: WAESD8 — CASE: 04FEB77 — BASE YR: 1985. — CASE YR: 2000.

	OIL	COAL	GAS	ELECTRICITY	TOTAL
TRANSPORT	6.77			0.13	6.90
INDUSTRY	3.27	3.94	1.51	2.18	10.91
DOMESTIC	2.48	1.66	2.28	1.86	8.28
TOTAL INTO MARKETS	12.53	5.60	3.79	4.17	26.09
TRANSM./DISTR.LOSSES*	0.76	0.20	0.66	0.74	2.35
CONV. THERMAL ELECTR.	0.0	0.0	0.0	0.0	
USES+LOSSES	0.0	0.0	0.0		0.0
TOTAL INPUT	0.0	0.0	0.0		
GAS MANUFACTURE		0.0	0.0		
USES+LOSSES		0.0			0.0
TOTAL INPUT		0.0			

	OIL	COAL	GAS	NUCLEAR	HYDRO-ETC	TOTAL
PRIMARY ENERGY DEMAND						
OUTPUT	13.29	5.80	4.45	3.96	0.95	28.44
INPUT				11.32	2.71	37.56
CHEMICAL FEED	1.89	0.10	0.42			2.40
LUBES & ASPHALTS	0.0					0.0
TOTAL REFINED PRODUCT	15.17					
REFINERY USE & PROCESS LOSS*	1.00	0.16	0.0			1.16
SYNCRUDE MANUFACTURE	0.0	0.0				
USES+LOSSES		0.0				0.0
TOTAL INPUT		0.0				
PRIMARY ENERGY SUPPLY	16.18	6.05	4.86	11.32	2.71	41.12

ALL QUANTITIES ARE GIVEN IN MILLION BARREL PER DAY OIL EQUIVALENT

8

DEMAND/SUPPLY MATRIX JAPAN — PROBLEM: WAESD8 — CASE: 04FEB77 — BASE YR: 1985. — CASE YR: 2000.

	OIL	COAL	GAS	ELECTRICITY	TOTAL
TRANSPORT	2.75			0.05	2.80
INDUSTRY	2.41	2.79	0.36	0.47	6.03
DOMESTIC	1.44	0.29	0.58	0.58	2.88
TOTAL INTO MARKETS	6.60	3.08	0.93	1.10	11.71
TRANSM./DISTR.LOSSES*	0.40	0.07	0.15	0.19	0.81
CONV. THERMAL ELECTR.	0.0	0.0	0.0	0.0	
USES+LOSSES	0.0	0.0	0.0		0.0
TOTAL INPUT	0.0	0.0	0.0		
GAS MANUFACTURE		0.0	0.0		
USES+LOSSES		0.0			0.0
TOTAL INPUT		0.0			

	OIL	COAL	GAS	NUCLEAR	HYDRO-ETC	TOTAL
PRIMARY ENERGY DEMAND						
OUTPUT	6.99	3.15	1.08	0.98	0.31	12.51
INPUT				2.81	0.88	14.91
CHEMICAL FEED	0.92	0.0	0.0			0.92
LUBES & ASPHALTS	0.0					0.0
TOTAL REFINED PRODUCT	7.91					
REFINERY USE & PROCESS LOSS*	0.58	0.09	0.0			0.66
SYNCRUDE MANUFACTURE	0.0	0.0				
USES+LOSSES		0.0				0.0
TOTAL INPUT		0.0				
PRIMARY ENERGY SUPPLY	8.49	3.23	1.08	2.81	0.88	16.49

ALL QUANTITIES ARE GIVEN IN MILLION BARREL PER DAY OIL EQUIVALENT

9

DEMAND/SUPPLY MATRIX REST WOCA — PROBLEM: WAESD8 — BASE YR: 1985.
CASE: 04FEB77 — CASE YR: 2000.

	OIL	COAL	GAS	ELECTRICITY		TOTAL
TRANSPORT	11.43			0.03		11.46
INDUSTRY	3.57	5.07	2.54	1.50		12.68
DOMESTIC	1.27	0.97	1.40	1.21		4.85
TOTAL INTO MARKETS	16.27	6.04	3.93	2.74		28.99
TRANSM./DISTR.LOSSES*	0.97	0.12	0.64	0.48		2.22
CONV. THERMAL ELECTR.	0.0	0.0	0.0	0.0		
USES+LOSSES	0.0	0.0	0.0			0.0
TOTAL INPUT	0.0	0.0	0.0			
GAS MANUFACTURE		0.0	0.0			
USES+LOSSES		0.0				0.0
TOTAL INPUT		0.0				
PRIMARY ENERGY DEMAND				NUCLEAR	HYDRO-ETC	
OUTPUT	17.24	6.17	4.57	2.30	0.93	31.20
INPUT				6.57	2.65	37.19
CHEMICAL FEED	2.12	0.0	0.0			2.12
LUBES & ASPHALTS	0.0					0.0
TOTAL REFINED PRODUCT	19.36					
REFINERY USE & PROCESS LOSS*	0.24	0.0	1.40			1.64
SYNCRUDE MANUFACTURE	0.0	0.0				
USES+LOSSES		0.0				0.0
TOTAL INPUT		0.0				
PRIMARY ENERGY SUPPLY	19.60	6.17	5.97	6.57	2.65	40.95

ALL QUANTITIES ARE GIVEN IN MILLION BARREL PER DAY OIL EQUIVALENT

10

S U M M A R Y
(ANNUAL CAPITAL CHARGES - CASE YEAR)

PROBLEM: WAESD8 BASE YR: 1985.
CASE: 04FEB77 CASE YR: 2000.

PRODUCERS		OPEC	OCEC		
OIL					
PRODUCTION	: B$	1.29	0.0		
REFINING	: B$	0.0	0.0		
COAL					
PRODUCTION	: B$	0.0	3.39		
CONVERSION	: B$	0.0	0.0		
GAS					
PRODUCTION	: B$	6.85	0.0		
LIQUEFACTION	: B$	1.69	0.0		
TOTAL	: B$	9.83	3.39		
CONSUMERS		NORTH AMERICA	EUROPE	JAPAN	REST WOCA
OIL					
PRODUCTION	: B$	6.74	1.42	0.45	6.36
REFINING	: B$	2.55	2.69	1.75	4.11
DISTRIBUTION	: B$	0.29	0.58	0.30	0.90
TOTAL	: B$	9.59	4.68	2.49	11.36
COAL					
PRODUCTION	: B$	2.54	1.53	0.14	0.0
CONVERSION	: B$	0.0	0.0	0.0	0.0
DISTRIBUTION	: B$	0.38	0.16	0.13	0.23
TOTAL	: B$	2.92	1.69	0.28	0.23
GAS					
PRODUCTION	: B$	6.67	1.17	0.10	5.22
REGASIFICATION	: B$	0.0	0.33	0.13	0.0
DISTRIBUTION	: B$	0.0	1.17	0.48	2.36
TOTAL	: B$	6.67	2.68	0.71	7.58
ELECTRICITY					
FUEL OIL	: B$	0.0	0.0	0.0	0.0
COAL	: B$	0.0	0.0	0.0	0.0
GAS	: B$	0.0	0.0	0.0	0.0
NUCLEAR	: B$	41.95	20.73	5.19	13.17
HYDRO ELEC 1	: B$	1.48	1.57	0.57	1.85
OTHER	: B$	3.36	0.39	0.35	0.0
DISTRIBUTION	: B$	4.97	2.44	0.60	1.65
TOTAL	: B$	51.76	25.14	6.71	16.67
IND HEATING EQP	: B$	6.33	4.26	2.52	5.35
DOM HEATING EQP	: B$	17.24	11.16	4.41	8.20
TOTAL CAPEX	: B$	94.50	49.61	17.12	49.38

11

SUMMARY

PROBLEM: WAESD8 BASE YR: 1985.
CASE: 04FEB77 CASE YR: 2000.

(PRODUCING REGIONS REVENUES)

PRODUCERS		OPEC	OCEC
GOVERNMENT TAXES *			
CRUDE OIL		11.28	0.0
COAL		0.0	2.00
GAS		8.32	0.0
TARIFFS ***			
CRUDE OIL		12.50	12.50
DISTILLATES		14.20	14.20
FUEL OIL		9.10	9.10
COAL		8.00	8.00
GAS		14.20	14.20
GOVT. TAKE(FROM TAXES)			
OIL	: B$	160.51	0.0
COAL	: B$	0.0	5.97
GAS	: B$	17.95	0.0
TOTAL GOVT. TAKE	: B$	178.46	5.97
REVENUES FROM TARIFFS			
CRUDE OIL	: B$	177.12	0.0
DISTILLATES	: B$	0.54	0.0
FUEL OIL	: B$	0.20	0.0
COAL	: B$	0.0	23.87
GAS	: B$	30.64	0.0
TOTAL REVENUES	: B$	208.50	23.87

* DOLLAR/BOE

*** DOLLAR/BOE, AVERAGE WORLD PRICE LAID-IN AT PORT OF ENTRY

12

SUMMARY

PROBLEM: WAESD8 BASE YR: 1985.
CASE: 04FEB77 CASE YR: 2000.

(CONSUMING REGIONS COSTS AND REVENUES)

CONSUMERS (TOTAL COSTS)		NORTH AMERICA	EUROPE	JAPAN	REST WOCA
IMPORT BILL					
CRUDE OIL	: B$	32.05	62.27	36.73	55.19
DISTILLATES	: B$	0.54	0.0	0.0	0.0
FUEL OIL	: B$	0.0	0.0	0.20	0.0
COAL	: B$	0.0	6.34	8.68	18.00
GAS	: B$	0.0	15.50	5.18	17.10
TOTAL IMPORT BILL	: B$	32.60	84.11	50.79	90.30
INTERNAL COSTS **					
OIL	: B$	21.72	20.26	7.24	23.33
COAL	: B$	26.64	15.74	2.21	2.75
GAS	: B$	9.35	4.50	1.09	9.96
ELECTRICITY	: B$	60.44	29.73	7.89	19.49
TOTAL	: B$	118.15	70.24	18.43	55.53
ENERGY CONS EQUIP	: B$	23.57	15.42	6.93	13.54
TOTAL INTERNAL	: B$	141.72	85.66	25.36	69.07
REVENUES FROM EXPORTS					
CRUDE OIL	: B$	0.0	0.0	0.0	0.0
DISTILLATES	: B$	0.0	0.0	0.0	0.0
FUEL OIL	: B$	0.0	0.0	0.0	0.0
COAL	: B$	7.69	0.0	0.0	0.0
GAS	: B$	0.0	0.0	0.0	0.0
TOTAL REVENUES	: B$	7.69	0.0	0.0	0.0
NET CONSUM REG. COST	: B$	166.63	169.77	76.15	159.37
IMPORT/EXPORT BALANCE					
EQUILIBRIUM	: B$	24.90	84.11	50.79	90.30
LOANS AT 15.0% PA	: B$	0.0	0.0	0.0	0.0
SHORTFALL	: B$	0.0	0.0	0.0	0.0

** CAPEX CHARGED AS A CONSTANT COST CHARGED OVER 100.0% OF THE USEFUL PHYSICAL LIFETIME OF A PLANT AT 15.0% ANNUAL COST OF CAPITAL.

6 Energy and Economic Growth Prospects for the Developing Countries

6.1 PREFACE

The Workshop on Alternative Energy Strategies has focused on projected energy supply and demand in the industrialized world. Members of WAES come from fifteen countries, most of them in the industrialized world outside of Communist areas (WOCA). We believe that the actions of these countries—or their failure to act—to alleviate possible future shortages of world energy will significantly affect the energy prospects of the developing nations.

In the WAES analyses, it is necessary to make assumptions about the energy supply and demand patterns in what is termed "Rest of WOCA"—WOCA countries outside Western Europe, North America, and Japan. These countries consist primarily of the developing countries (both OPEC and non-OPEC) but also include Australia, New Zealand, and the Republic of South Africa.

We have relied extensively on others with knowledge of these countries to help analyze their future energy supply and demand prospects. In particular, individuals from the International Bank for Reconstruction and Development (World Bank) have been most helpful in estimating developing country economic growth rates to 1990, deriving relevant income elasticities of energy demand, and providing energy supply estimates to 1980.

The energy supply estimates by fuel type for 1980 to 1985 come from World Bank and WAES sources, and the 1985 to 2000 figures are taken from the WAES global supply estimates for oil, gas, and coal.

The energy supply-demand estimates in this paper are very tentative. Historical data on energy consumption in developing countries are generally incomplete and are clouded by the fact that a significant proportion of total energy consumption comes from noncommercial sources such as firewood, cow dung, and vegetable waste. The survey also attempts to cover over ninety countries and so is exceedingly general in nature.

This report was prepared jointly by William F. Martin (WAES Staff) and Frank J.P. Pinto (Consultant to WAES and the World Bank). The authors gratefully acknowledge the written papers and verbal communications received from Nicholas Carter, John Foster, and William Humphrey of the World Bank and Alan Strout of the MIT Energy Laboratory.

This report attempts to answer the following questions, which are essential to the WAES global supply-demand analysis: (1) Given certain assumptions regarding economic growth, what is the probable range of commercial energy consumption in the developing countries during 1985 and 2000? (2) What potential domestic energy supplies are available to help these countries meet their anticipated demands? (3) What would be the probable range of desired imports (or exports) of energy by these countries? Possible answers to these questions represent the boundaries of the study. We recognize that the developing countries will need appropriate mechanisms to help them achieve desired levels of economic growth and that new arrangements may be needed to assist developing countries in meeting the rising costs of energy. Such arrangements must be part of a broad and complex system of existing economic relationships and institutions, an analysis of which is outside the scope of the WAES study.

6.2 INTRODUCTION

This paper focuses on the energy and economic growth prospects of the OPEC and non-OPEC developing economies in the world outside Communist areas (WOCA).[1]

6.2.1 Projected Shares of World Energy Consumption

Primary energy consumption in the developing countries during 1972 constituted approximately 15 percent of total WOCA energy consumption. As these countries industrialize, their share will rise relatively faster than that of the industrialized world. The WAES global supply-demand integrations suggest that the developing countries could consume as much as 25 percent of total world energy by the year 2000, as Figure 6.1 illustrates.

1. Estimates of energy supply and demand for Australia, New Zealand, and the Republic of South Africa are also included in the worksheets in section 6.6. Because these countries did not participate in WAES, it was necessary to make provisional estimates for them on the basis of published source material and information from individuals within the Workshop who were familiar with the energy prospects in these countries.

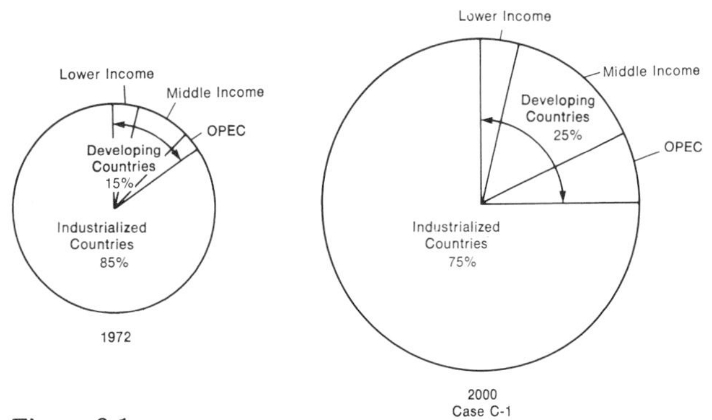

Figure 6.1
Shares of WOCA energy consumption.

6.2.2 Developing Country Coverage

The 93 developing nations in WOCA considered in this analysis are divided into two major groups: (1) the 13 OPEC countries and (2) the non-OPEC developing nations of Asia, Africa, and Latin America. The latter group is further divided into two classifications according to per capita income (the number of countries in each category is in parentheses):

- *Lower-income countries* comprising those developing economies with per capita annual income below $200. The countries in this group fall into two regions: South Asia (7) and lower-income sub-Sahara Africa (20).
- *Middle-income countries* comprising those developing economies with per capita annual income above $200. These countries can be grouped into three regions: East Asia (9); Caribbean, Central and South America (21); middle-income sub-Sahara Africa and West Asia (23). Per capita incomes range from $200 to around $1000 with the mean being about $550.[2]

These classifications are identical to those used by World Bank. A detailed list of the countries in each major group is given in Table 6.1.

6.2.3 Population Trends

An important factor in the analysis of energy consumption in the developing world is the present level and future growth rate of population. WOCA had a population of around 2.7 billion in 1975, and it is estimated that this figure will increase to 4.5 billion by the year 2000. The industrialized nations of North America, Europe, Japan, and Oceania constituted approximately 28 percent of the total population in 1975 and will probably constitute around 20 percent by the year 2000.

Table 6.2 presents estimates of the absolute levels and growth rates of population for the developed and developing nations for the years 1970, 1975, 1985, and 2000. Most of these figures have been derived from United Nations population projections. The energy problems facing the developing economies are more severe because of their 2.4–2.7 percent average population growth between now and 2000, as compared to 0.7–0.9 percent for the industrialized world.

6.2.4 WAES Scenario Assumptions

WAES uses the scenario approach in its studies to 1985 and 2000. The scenario variables include economic growth (high or low), real energy price (rising, constant, or falling), government policy (vigorous or restrained), and, for 1985 to 2000, principal replacement fuel (coal or nuclear). For the developing economies, only a constant real energy price to 1985 was examined. Furthermore, government policy response to 1985 is not included as a variable due to the uncertainty in modeling an aggregate policy response for over ninety countries. The WAES scenario cases considered in the analysis of developing country prospects are more fully specified in Table 6.3.

6.2.5 Methodology

The first step is the determination of economic growth rates for the developing nations, primarily from the World Bank's SIMLINK Model.[3] High and low economic growth rate assumptions for the developed economies, as well as the WAES oil price assumption of $11.50 (per barrel of light Arabian crude, FOB Persian Gulf, in constant 1975 U.S.$) are special inputs to the model, whose simulations then provide economic growth projections for the major developing regions of WOCA to 1985. Figure

2. These distinctions are based on data in the 1974 and 1975 *World Bank Atlas*. They relate to the year 1972.

3. "The SIMLINK Model of Trade and Growth for the Developing World," World Bank Staff Working Paper No. 220, October 1975. Also in *European Economic Review*, Vol. 7 (1976), pp. 239–255. SIMLINK is primarily a medium-term forecasting system in which exports of the non-OPEC developing countries are related to the level of economic activity in the OECD countries through a series of individual commodity models. Growth in the developing countries is linked to investment levels and imports; imports in turn depend on export earnings and inflows of foreign capital. Given assumptions as to OECD growth, the availability of foreign capital to the developing countries, and the international price of petroleum, the model may be run to determine either the import-constrained GDP growth rates to be anticipated in developing countries or the real resource transfer they would need to support a specified GDP growth target.

Table 6.1
Classification of developing countries

OPEC Countries

Algeria	Iran	Qatar
Ecuador	Iraq	Saudi Arabia
Gabon	Kuwait	United Arab Emirates
Indonesia	Libya	Venezuela
	Nigeria	

Non-OPEC Developing Countries

Lower-Income Countries [annual per capita income under $200 (1972 U.S. $)]

South Asia	Lower-Income Sub-Sahara Africa	
Afghanistan	Burundi	Niger
Bangladesh	Central African Republic	Rwanda
Burma	Chad	Sierra Leone
India	Dahomey	Somalia
Nepal	Ethiopia	Sudan
Pakistan	Guinea	Tanzania
Sri Lanka	Kenya	Togo
	Madagascar	Uganda
	Malawi	Upper Volta
	Mali	Zaire

Middle-Income Countries [annual per capita income over $200 (1972 U.S. $)]

East Asia	Middle-Income Sub-Sahara Africa and West Asia	Caribbean, Central and South America
Fiji	Angola	Argentina
Hong Kong	Bahrein	Barbados
Korea (South)	Cameroon	Bolivia
Malaysia	Congo P.R.	Brazil
Papua New Guinea	Cyprus	Chile
Philippines	Egypt	Colombia
Singapore	Ghana	Costa Rica
Taiwan	Israel	Dominican Republic
Thailand	Ivory Coast	El Salvador
	Jordan	Guatemala
	Lebanon	Guyana
	Liberia	Haiti
	Mauritania	Honduras
	Morocco	Jamaica
	Mozambique	Mexico
	Oman	Nicaragua
	Rhodesia	Panama
	Senegal	Paraguay
	Syria	Peru
	Tunisia	Trinidad and Tobago
	Turkey	Uruguay
	Yemen AR, DM	
	Zambia	

Table 6.2
Estimated developed and developing country population levels and growth rates, 1970-2000

	Population (in millions, rounded)				Population growth rate (%/yr)		
	1970	1975	1985	2000	1970-75	1975-85	1985-2000
Total population in WAES analysis	2399	2661	3310	4475	2.1	2.2	2.05
Developed economies	702	732	792	872	0.9	0.8	0.7
Developing economies	1697	1929	2518	3603	2.6	2.7	2.4
OPEC	255	292	388	566	2.8	2.9	2.6
Non-OPEC developing countries	1442	1637	2130	3037	2.6	2.7	2.4
Lower-income countries	889	1005	1301	1835	2.5	2.65	2.3
South Asia	740	835	1076	1487	2.4	2.6	2.2
Lower-income Africa	149	170	225	348	2.7	2.9	2.9
Middle-income countries	553	632	829	1202	2.7	2.75	2.5
East Asia	138	158	207	290	2.8	2.7	2.3
Middle-income Africa and West Africa	162	184	240	353	2.6	2.7	2.6
Caribbean, Central and South America	253	290	382	559	2.8	2.8	2.6

Table 6.3
WAES scenario assumptions

	Economic growth	Energy price[a]	Principal Replacement fuel
1976-1985			
C	High	$11.50	
D	Low	$11.50	
1985-2000			
C-1	High	$11.50-$17.25	Coal
C-2	High	$11.50-$17.25	Nuclear
D-7	Low	$11.50	Coal
D-8	Low	$11.50	Nuclear

[a]Price per barrel light Arabian crude FOB Persian Gulf in 1975 U.S. dollars.

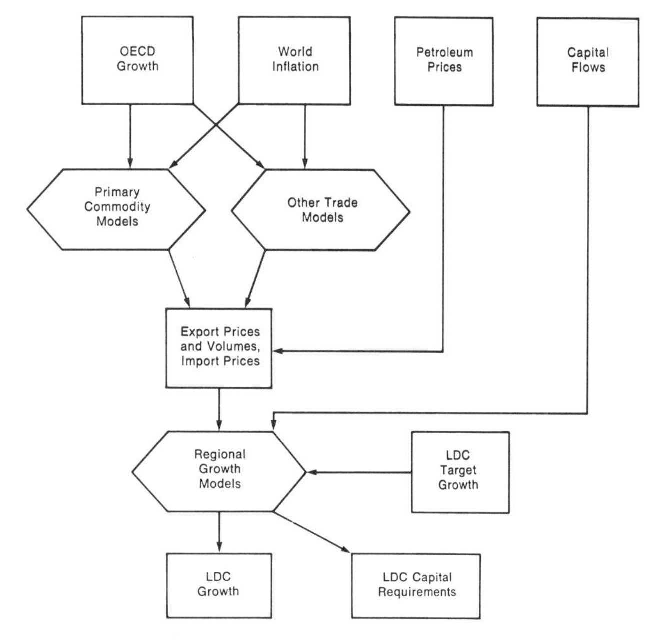

Figure 6.2
Flow diagram for SIMLINK III.

6.2 illustrates the SIMLINK flow from the exogenous input to the output.

The historical (1960 to 1972) relationship between regional economic growth and energy consumption is then examined by considering the income elasticity of demand for energy use. This is defined as the growth rate of energy consumption divided by the growth rate of real income. Since real energy prices rose substantially between 1972 and 1976 and are assumed to either remain constant (WAES Cases C and D in 1985, Cases D-7 and D-8 in 2000) or increase by 50 percent by year 2000 (WAES Cases C-1 and C-2), the historical income elasticities were revised downward for the period 1976 to 2000 to reflect the more energy-efficient use of resources in the future.

The primary energy demand growth rates are obtained from the real economic growth rates and the income elasticities of energy demand. The total primary energy consumption of developing countries in 1985 and 2000 can then be determined.

The energy supply estimates by fuel type for 1975 to 1985 come from World Bank[4] and WAES sources. The 1985 to 2000 estimates come primarily from the WAES global supply studies described in chapters 3 through 7 of the second WAES technical report.

Supply-demand integrations then balance available energy supplies with expected energy demands. The resulting figures show the range of energy exports and imports of both OPEC and non-OPEC developing countries.

4. A. Lambertini, "Energy and Petroleum in Non-OPEC Developing Countries, 1974-80," World Bank Staff Working Paper No. 229, February 1976.

6.3 DEMAND FOR ENERGY

6.3.1 Economic Growth Rate Assumptions, 1960–2000

The historical and projected economic growth rates for the developing economies for the period 1960 to 2000 are presented in Table 6.4. The 1976 to 1985 growth estimates have been derived primarily from simulation of the World Bank's SIMLINK Model using the WAES high (Case C) and low (Case D) economic growth assumptions for the OECD countries as well as the WAES constant oil price assumption of $11.50 per barrel (1975 U.S.$). Because SIMLINK is primarily a medium-term forecasting system, extrapolations were made to determine the 1985 to 2000 developing countries' economic growth rates.

Several of the middle-income countries profited by the commodity boom of 1972–1973 and should be able to achieve comparatively rapid growth over the next decade. Latin American and East Asian countries are expected to show the highest economic growth patterns among the non-OPEC developing countries due to higher capital productivity, an increase in the rate of investment, and export promotion. The lower-income countries will probably grow much slower than the middle-income countries, and they will continue to suffer from the effects of higher oil prices and agricultural shortfalls. The projected growth rates assume, however, that improvements will be achieved in domestic policies with special emphasis on increasing exports.

As a group, the non-OPEC developing countries are expected to maintain a growth rate higher than that of the OECD area. In the long run, middle-income countries are expected to grow 1–2 percent faster and low-income countries about 0.5 percent slower than the developed nations.

OPEC countries are expected to achieve high economic growth rates from 1976 to 1985, possibly as high as 7.2 percent per year in the high growth case and 5.5 percent per year in the low growth case. Several OPEC countries—mainly those with larger populations such as Algeria, Ecuador, Indonesia, Iran, Iraq, Nigeria, and Venezuela—are expected to have current-account deficits in their balance of payments by 1985. From 1985 to 2000 OPEC growth will likely decrease since many OPEC countries have undertaken large development projects that may cause foreign exchange shortages and curtailed imports after 1985.

6.3.2 Historical Growth in Energy Consumption, 1960–1972

From 1960 to 1972 the nations of the developing world more than doubled their consumption of commercial energy and increased their demand for electric power by more than 250 percent. In 1972 the developing countries accounted for around 15 percent of total world energy consumption in WOCA.

Sixteen countries accounted for about three-quarters of all developing world energy demand.[5] They are Argentina, Brazil, Chile, Colombia, Egypt, India, Indonesia, Iran, Korea, Mexico, Pakistan, Philippines, Taiwan, Thailand, Turkey, and Venezuela.

5. *Modern Power Prospects in Developing Countries*, Richard Barber Associates (1976).

Table 6.4
Real GNP growth rate assumptions, 1972–2000 (average annual percent growth)

Period	1960–1972	1972–1976	1976–1985		1985–2000	
Economic growth WAES case			High C	Low D	High C-1, C-2	Low D-7, D-8
Non-OPEC developing countries	5.6	5.1	6.1	4.1	4.6	3.6
Lower-income countries	3.7	2.3	4.4	2.8	3.1	2.5
Middle-income countries	6.2	5.9	6.6	4.5	4.9	3.9
OPEC	7.2	12.5[b]	7.2	5.5	6.5	4.3
Developed countries[a] (OECD)	4.9	2.0	4.9	3.1	3.7	2.5

[a]As derived by WAES analyses of individual countries.

[b]Preliminary estimate.

Total commercial energy demand within the developing world (including OPEC) during 1972 was approximately 9.5 MBDOE (excluding international oil bunkers), distributed among the developing regions as shown in Table 6.5.

Table 6.6 summarizes the actual quantities of commercial energy consumed by the OPEC and non-OPEC developing countries for the period 1960 to 1972 in both tonnes of coal equivalent (TCE) and millions of barrels per day of oil equivalent (MBDOE). The energy and real income growth rates for the period are also included. (For more detailed information, see the table of GNP and commercial energy consumption included in section 6.6.)

Table 6.5
Distribution of total commercial energy demand in the developing world, 1972 (%)

Lower-income countries	
South Asia	23.0
Lower-income Africa	2.0
Middle-income countries	
East Asia	13.5
Middle-income Africa and West Asia	11.0
Caribbean, Central and South America	33.0
OPEC	17.5
All developing countries	100.0

We have estimated only commercial fuel consumption. In several developing countries noncommercial energy in the form of firewood, cow dung, and vegetable waste constitutes a significant share of total energy consumption. In India, noncommercial energy was estimated at 59 percent of total energy consumption in 1960 and 48 percent in 1970; it is expected to retain a significant though decreasing percentage of total energy consumption.[6] Due to the great difficulty in obtaining accurate data, the noncommercial energy sources have been excluded from this study. A detailed study on the developing world's energy prospects would have to consider such fuels explicitly.

6.3.3 Economic Growth and Energy Growth

Table 6.7 shows our assumptions regarding the historical (1960–1972) and projected relationship between energy consumption and real economic growth. This relationship is termed the *income elasticity of demand for energy use.* The price of oil quadrupled between 1972 and 1976. A simple extrapolation of the historical (1960–1972) income elasticity would result in an overestimation of energy demand. The much higher current and

6. *Report of the Fuel Policy Committee,* Government of India, 1974.

Table 6.6
Developing country commercial energy consumption and income elasticity of energy demand, 1960–1972

	Energy consumption				Energy growth rate, 1960–1972	GNP growth rate, 1960–1972	Income elasticity of energy demand, 1960–1972
	(10^6 metric tons of coal equivalent)		(MBDOE)				
	1960	1972	1960	1972	(%/yr)	(%/yr)	
Non-OPEC developing countries	260.40	594.50	3.43	7.81	7.10	5.50	1.29
Lower-income countries	88.50	177.40	1.17	2.33	6.00	3.70	1.62
South Asia	83.40	165.10	1.10	2.17	5.80	3.60	1.61
Lower-income Asia	5.10	12.30	0.07	0.16	7.60	4.00	1.90
Middle-income countries	171.90	417.10	2.26	5.48	7.70	6.20	1.24
East Asia	27.80	97.60	0.37	1.28	11.00	7.70	1.43
Middle-income Africa and West Asia	32.90	79.30	0.43	1.04	7.60	5.80	1.31
Caribbean, Central and South America	111.20	240.20	1.46	3.16	6.60	5.90	1.12
OPEC	48.30	127.90	0.64	1.68	8.40	7.20	1.17
Total developing countries	308.70	722.40	4.07	9.49	7.40	5.80	1.28

Sources: For energy consumption: (1) U.N., *World Energy Supplies, 1950–74,* Series J, No. 19. (2) *Report of the Fuel Policy Committee,* Government of India, 1974, and also J. D. Henderson, *India: The Energy Sector,* World Bank. (3) A. Lambertini, "Energy and Petroleum in Non-OPEC Developing Countries," World Bank Staff Working Paper No. 229, February 1976. (4) William Humphrey, "Estimation of Energy Growth for Non-OPEC LDCs by Region and OPEC Countries for 1972–85 and 1985–2000," Cambridge, WAES Associates Report, January 15, 1976.
For GNP data: *World Bank Atlas,* 1975 (in constant U.S. 1973 dollars).

Table 6.7
Projected developing country income elasticity of energy demand, 1960–2000

Period WAES case Oil price	1960–1972 $2.00	1976–1985 C, D $11.50	1985–2000 C-1, C-2 $11.50–17.25	1985–2000 D-7, D-8 $11.50
Non-OPEC developing countries[a]	1.29	1.19	1.04	1.10
Lower-income countries	1.62	1.50	1.20	1.30
Middle-income countries	1.24	1.10	1.00	1.05
OPEC	1.17	1.10	1.05	1.10

[a]The ratios for the non-OPEC developing countries have been weighted using the following year-end weights:

Year WAES case	1972	1985 C	1985 D	2000 C-1, C-2	2000 D-7, D-8
Lower-income countries	0.246	0.215	0.219	0.176	0.186
Middle-income countries	0.754	0.785	0.781	0.824	0.814
Total	1.000	1.000	1.000	1.000	1.000

expected energy prices assumed in the WAES cases to 1985 would tend to reduce the energy intensiveness in economic activity. Specifically, the income elasticity of energy demand would decrease if current energy prices were maintained. For the period 1985 to 2000, the C cases assume a 50 percent increase in real energy prices, and this would serve to lower the income elasticity of energy demand even further.

From the World Bank study,[7] one can derive interesting relationships between developing-country income growth and energy consumption for the period 1970 to 1980. In the lower-income countries, per capita income elasticity of energy demand is about 40 percent higher than the corresponding total income elasticity, where

per capita income elasticity = (growth rate of energy consumption per capita)/(growth rate of real income per capita)

and

total income elasticity = (growth rate of energy consumption per capita + population growth rate)/(growth rate of real income per capita + population growth rate).

For the middle-income countries, however, per capita and total income elasticities are found to be noticeably closer. In other words, the process of industrialization that raises per capita income also results in a gradual reduction in the growth rate of energy consumption with respect to growth in real income.[8]

Table 6.8 specifies the percentage decrease assumed in the income elasticity of energy demand for the period 1976 to 1985 as compared to 1960 to 1972 and for 1985 to 2000 as compared to 1976 to 1985. The lower-income countries are dominated by India, whose energy consumption in 1972 was around 75 percent of the total, while Argentina, Brazil, and Mexico together account for around 42 percent of the energy consumption of the middle-income countries. The long-run income elasticities for the lower-income countries are consistent with those implied in the 1974 report of the Indian Fuel Policy Committee, while the income elasticities for the middle-income countries are consistent with estimates provided by the WAES Mexican team.

While the income elasticities of energy demand for non-OPEC developing countries are assumed to decline, the decrease will be more pronounced in the lower-income than in the middle-income countries. Their elasticities, however, will still be higher than the developed country elasticities, which, in the WAES analysis, have been shown to drop below unity.

7. Lambertini, "Energy and Petroleum."

8. Variances in total energy elasticity with respect to GNP at different per capita income levels have been found to disappear when price-adjusted GNPs are used as independent explanatory variables in place of official exchange-rate GNPs. While attempts are being made to determine the price-adjusted GNPs for several developing nations (see the pioneering study by Kravis, Kenessy, Heston, and Summers, *A System of International Comparisons of Gross Product and Purchasing Power,* 1975), our study uses readily available official exchange-rate GNPs.

Table 6.8
Percent change assumed in the income elasticity of energy demand over the preceding period[a]

Period WAES case Oil Price	1976-1985 C, D $11.50	1985-2000 C-1, C-2 $11.50-17.25	1985-2000 D-7, D-8 $11.50
Non-OPEC developing countries	-8	-13	-8
Lower-income countries	-7	-20	-13
Middle-income countries	-11	-9	-5
OPEC	-6	-5	nil

[a]Rounded to nearest percent. Elasticity for 1976-1985 period is compared to that of 1960-1972.

Obviously, using such a simple relationship between economic growth and energy growth is inadequate in some respects, in that it fails to include factors that may significantly affect energy consumption, such as changes in the industrial structure or increased mechanization in agriculture. Alan Strout, for instance, considers the production of a small group of key energy-intensive materials (such as iron and steel, cement, aluminum), when combined using energy weights, as an indication of the "energy-intensiveness" of a country's industry.[9] For developing countries achieving rapid industrialization, the production of energy-intensive goods would have to grow even faster than normal, and income elasticities would accordingly be higher.

This raises the important issue of whether or not developing countries will choose to develop energy-intensive industries. Some feel that a significant transfer of industry to the developing countries endowed with energy resources will occur. For example, a significant proportion of the industrialized world's aluminum processing industry might be relocated in Brazil, where abundant hydroelectric reserves would allow for large-scale, relatively inexpensive production. Likewise, many chemical plants might be "exported" to the OPEC countries where abundant gas reserves permit manufacturing of chemicals such as ammonia at lower cost. It is an assumption of this chapter that while such transfers may occur, there will be no major transfer of energy-intensive industries from the developed to the developing world.

Furthermore, the immense complications of such a transfer—for both the developed and the developing countries—suggest that most developing countries will continue to be net importers of these products. Strout has also shown that most developing countries did not develop significant energy-intensive industries when oil prices were low, and the higher prices assumed by the WAES cases would, if anything, discourage them from major investment in energy-intensive industry. This important issue needs further research and is certainly beyond the scope of this analysis.

A somewhat different approach to the problem is that of Charpentier and Beaujean, who replace the GNP indicator with a long list of nonenergy variables describing the level of development of the economy.[10] Their study uses the technique of factor analysis to sum up and reduce a given set of data by extracting factors that are linear combinations of nonenergy variables. Multiple correlation analysis then gives a series of factors. The first factor, termed *degree of development*, determines the greatest amount of variance between countries. This first factor is highly correlated with per capita steel and electricity consumption, total energy consumption, and the number of newspapers and cars. Their choice of variables was, however, biased in favor of the industrialized nations (e.g., number of newspapers and cars) and hence is not really applicable here. Selecting key variables in the development process of poorer nations would make this approach useful and relevant.

6.3.4 Estimated Developing Country Energy Consumption, 1972-2000

Table 6.9 shows the estimates of developing country consumption of energy for the principal WAES cases based on the analysis in the preceding sections. Non-OPEC developing countries are expected to increase their energy consumption from four to as high as five times their 1972 level by the year 2000. This is based on an average economic growth rate of between 4 and 5 percent per year during the period. OPEC countries are expected to

9. Alan Strout, "The Future of Nuclear Power in the Developing Countries," unpublished working paper, MIT Energy Laboratory, 1976.

10. J.P. Charpentier and J.M. Beaujean, "Toward a Better Understanding of Energy Consumption. II. Factor Analysis —A New Approach to Energy Demand?" draft paper, 1976.

Table 6.9
Primary energy consumption targets for the developing world (MBDOE)[a]

Year WAES case	1972	1985 C	1985 D	2000 C-1, C-2	2000 D-7, D-8
Non-OPEC developing countries	7.81	18.20	14.90	35.60	26.50
Lower-income countries	2.33	4.90	4.00	8.40	6.50
Middle-income countries	5.48	13.30	10.90	27.20	20.00
OPEC	1.68	4.90	4.20	13.10	8.40
All developing countries	9.49	23.10	19.10	48.70	34.90

[a]Excludes bunkers.

achieve a five- to eightfold increase in their energy consumption by the year 2000 based on an economic growth rate of between 6 and 7 percent per year from 1972 to 2000.

6.4 SUPPLY OF ENERGY

6.4.1 Estimated Developing Country Energy Supplies, 1972-2000

The OPEC and non-OPEC developing countries have the potential to significantly increase their domestic energy supplies by the year 2000. There are a number of reasons for this. First, the higher energy prices will encourage exploration for and production of fuels in many developing countries where such activities have been seen as uneconomic. This is particularly true for oil, natural gas, and coal. Developing countries with adequate reserves of coal and natural gas may choose to develop these resources rather than risk further deterioration in their balance of payments because of higher oil import bills.

It is difficult to generalize for over ninety countries. Therefore in discussing energy supply, we shall distinguish between OPEC and non-OPEC developing economies. Even within these two groups differences are substantial. OPEC countries differ widely in their economic growth potentials, development plans, population, and revenue needs. The non-OPEC developing countries also vary greatly in their energy supply potentials and revenue needs. Some, such as Mexico and Brazil, will most likely attain a high degree of energy self-sufficiency, while others, particularly the low-income countries of Africa and Asia, will continue to depend on energy imports.

Table 6.10 shows the resource base for middle- and lower-income non-OPEC developing countries. The middle-income countries are relatively well-endowed with energy resources, especially oil and natural gas. The lower-income countries do not have abundant supplies of oil and natural gas but do have plentiful coal reserves. These resource statistics indicate that, as a group, the developing countries have sufficient reserves of the major fossil fuels. Table 6.11 is a summary table of the energy production potential of the developing countries by fuel type.

6.4.2 Oil

OPEC Oil Production

OPEC countries currently account for about 80 percent of WOCA oil reserves—about 450 billion barrels. OPEC production in 1975 was about 27 million barrels per day (MBD), most of which was exported to Western Europe, Japan, and North America. OPEC oil production will continue to be of critical importance to the world oil supply, and

Table 6.10
Non-OPEC developing countries' medium-term energy resources (10^6 metric tons of oil equivalent)

	Oil[a]	Natural Gas[a]	Coal[b]	Hydro power[c]	Nuclear power[b]	Total
Middle-income	3480	1835	5320	190	350	11,175
Lower-income	180	480	9250	185	1050	11,145
Total	3660	2315	14570	375	1400	22,320

[a]Economically recoverable at current prices and costs.
[b]The measured or reasonably assured fraction of resources that could be economically exploited in the coming five years.
[c]All estimated reserves.

Source: "Energy and Petroleum in Non-OPEC Developing Countries, 1974-1980," World Bank Staff Working Paper No. 229, February 1976.

Table 6.11
Projected production of primary energy in developing countries, 1985-2000 (MBDOE)

Year	1985		2000			
WAES case	Case C	Case D	Case C-1	Case C-2	Case D-7	Case D-8
Economic growth	High	Low	High	High	Low	Low
Oil/energy price[a]	11.50	11.50	11.50-17.25	11.50-17.25	11.50	11.50
Principal replacement fuel			Coal	Nuclear	Coal	Nuclear
Oil						
Non-OPEC developing countries	6.70	5.10	11.50	11.50	7.00	7.00
OPEC	40.00	36.00	45.00	45.00	39.00	39.00
Total	46.70	41.10	56.50	56.50	46.00	46.00
Solid Fuels						
Non-OPEC developing countries	2.67	2.24	5.26	5.00	4.10	3.60
OPEC	0.42	0.30	1.46	1.30	0.80	0.80
Total	3.09	2.54	6.72	6.30	4.90	4.40
Natural Gas						
Non-OPEC developing countries	2.40	2.10	4.56	4.50	3.88	3.40
OPEC	5.47	5.08	13.60	11.60	12.30	10.20
Total	7.87	7.18	18.16	16.10	16.18	13.60
Hydroelectricity						
Non-OPEC developing countries	1.70	1.80	3.80	3.10	2.70	2.10
OPEC	0.10	0.10	0.40	0.40	0.30	0.30
Total	1.80	1.90	4.20	3.50	3.00	2.40
Nuclear Electricity						
Non-OPEC developing countries	1.10	0.40	4.00	8.30	2.80	5.90
OPEC	0.10	0.10	0.46	1.00	0.30	0.60
Total	1.20	0.50	4.46	9.30	3.10	6.50
Total Primary Energy						
Non-OPEC developing countries	14.57	11.64	29.12	32.40	20.50	22.00
OPEC	46.09	41.58	60.92	59.30	52.70	50.90
Total	60.66	53.22	90.04	91.70	73.20	72.90

[a]In 1975 U.S. dollars per barrel of Arabian light crude oil FOB Persian Gulf or equivalent.

therefore an assessment of the potential production over the next 25 years is an important part of the WAES oil supply assessment (chapter 3 in the second technical report).

If OPEC countries produce at a rate sufficient to meet the import requirements of the major consuming countries for as long as possible, limited only by a theoretical reserve-to-production (R/P) ratio of 15 to 1, they could be producing 57 MBD by 1995, assuming that additions to reserves over the next 25 years average 10 billion barrels a year.

However, it is likely that the OPEC countries will decide for economic, conservation, or other reasons to limit their production to a level below the theoretical maximum. WAES therefore chose three alternative assumptions to represent possible ceilings to total OPEC oil production:

1. 33 million barrels a day
2. 40 million barrels a day
3. 45 million barrels a day

Assumption 1 appears to be most likely if future additions to reserves are low (5 billion barrels) and oil prices remain constant, so it is coupled with the year 2000 Case D scenarios. Assumption 3 is more likely if additions to reserves are high (10 billion barrels) and oil prices rise, so it is coupled with the Case C scenarios.

These levels define the maximum potential production of oil by OPEC. Once OPEC oil production reaches such a limit, it is assumed to remain at that level until the R/P ratio falls to 15 to 1. From then on, maintaining this R/P ratio will result in declining OPEC production. The OPEC production rates given by these assumptions are shown in Table 6.12. Table 6.13 presents an overview of OPEC oil reserves and production in 1975.

Non-OPEC Oil Production

Proven oil reserves in non-OPEC developing countries are equal to about 40 billion barrels—7 percent of total WOCA reserves. However, the region is vast and is to a large extent relatively unexplored for oil. Thus, extensive exploration and development encouraged by a desire to reduce dependence on imported oil could lead to a significant increase in production. A high future discovery rate was set at 4 billion barrels a year. This would permit production to more than triple today's levels—reaching 11.5 MBD by year 2000. A less optimistic

Table 6.12
OPEC oil production rates, 1975-2000 (MBD)

	Cases C, C-1			Cases D, D-8		
	Limit 33 MBD	Limit 45 MBD	No government limit	Limit 35 MBD	Limit 40 MBD	No government limit
1975 (actual)	27	27	27	27	27	27
1980	32	32	32	29	29	29
1985	33	39	39	33	36	36
1990	33	45	47	33	40	42
1995	33	45	57	33	40	44
2000	33	45	56	33	39	34

Table 6.13
OPEC oil reserves and oil production, 1975

	Proven reserves (10^6 bbl)	Production (10^3 BD)
Algeria	7,370	965
Ecuador	2,450	160
Gabon	2,200	200
Indonesia	14,000	1,315
Iran	64,500	5,395
Iraq	34,300	2,230
Kuwait	68,000	1,840
Libya	26,100	1,490
Nigeria	20,200	1,785
Qatar	5,850	435
Saudi Arabia	148,600	6,970
United Arab Emirates	36,750	1,995
Venezuela	17,700	2,410
Total	448,020	27,190

discovery rate (around 2 billion barrels a year) would probably allow oil production to be twice as high as current levels by the year 2000.

Much of this production will probably be limited to a few countries. Walter Levy notes that three countries—Mexico, Brazil, and Egypt—could account for as much as 40 percent of total non-OPEC developing country oil production by 1985.[11] Production in Mexico alone could be 25 percent of the non-OPEC developing country production by year 2000. Thus, although production for the non-OPEC developing countries as a whole is likely to increase, there is still the problem of the non-oil-producing developing countries who will continue to face large oil import bills and corresponding balance-of-payments deficits. Table 6.14 shows crude oil reserves and oil production in 1974 for selected non-OPEC developing countries.

11. "OPEC in the Medium-Term," W. J. Levy Consultants, New York, September 1976.

Table 6.14
Selected non-OPEC developing countries' crude oil reserves and production, 1974

	Reserves (10^6 bbl)	Production (10^3 BD)
High-Income		
Argentina	2459	413
Brazil	779	178
Brunei	2000	191
Chile	210	27
Colombia	627	167
Malaysia	2700	82
Mexico	3087	575
Peru	830	77
Trinidad and Tobago	651	186
Tunisia	533	85
Middle-Income		
Angola	1409	172
Bolivia	216	46
Congo	950	46
Egypt	2761	148
Syria	2776	123
Low-Income		
Burma	117	22
India	926	183
Pakistan	30	8

Source: "Energy and Petroleum in Non-OPEC Developing Countries, 1974-1980," World Bank Staff Working Paper No. 229.

Note: These are selected countries, so the sum of these countries does not add up to total non-OPEC developing country production.

6.4.3 Gas

OPEC Gas Production

Proven reserves of natural gas in OPEC are estimated to be 140 billion BOE—about 60 percent of

WOCA reserves. There is at present only a limited demand within OPEC for natural gas, though future consumption is expected to increase substantially. Future natural gas production will depend primarily on two factors:

1. Domestic requirements for natural gas (including gas for reinjection into oil wells to improve recovery rates.

2. Expected level of exports to major consuming countries.

It is likely that OPEC countries will increasingly use this plentiful, inexpensive, and convenient fuel and that 3 to 5 MBDOE could be consumed domestically by year 2000. Iran has already announced that it plans to increase its natural gas component from 18 percent to as high as 35 percent of total primary energy consumption by 1987. If other OPEC countries follow this lead, we could expect as much as 40 percent of total OPEC primary energy consumption by year 2000 to be in the form of natural gas.

The economies of Western Europe, North America, and Japan may increasingly look to OPEC gas reserves as a supplement to their energy needs. Maximum planned and potential natural gas imports from OPEC in 1985 are estimated at about 3.5 MBDOE. If all planned projects are realized, about 2.5 MBDOE could be delivered as liquefied natural gas (LNG). This would require OPEC production of 3.3 MBDOE for export, allowing for 25 percent losses in the processing of LNG. Another 0.6 MBDOE could be delivered by pipeline from North Africa and the Middle East, resulting in a total OPEC capacity of 3.9 MBDOE of gas exports. If all import demands of the major consuming countries are to be met in year 2000 by OPEC, then these countries would need to produce as much as 9 MBDOE for export in addition to their internal demand requirements of up to 5.0 MBDOE.

Proven reserves of natural gas are sufficient to support such a large expansion in international trade. Because of its desirable qualities as a fuel compared with oil and coal, gas may command a premium price. Uncertainties about the growth of world gas trade lie in (1) the attitudes of the governments of OPEC countries toward export of gas versus domestic use, including use as a chemical feedstock, (2) availability of capital for investment in LNG systems, and (3) possible repercussions of an LNG tanker incident.

Non-OPEC Gas Production

Despite natural gas reserves of approximately 20 billion BOE, current natural gas consumption in non-OPEC developing countries is small. The World Bank reports that only about 65 percent of the gas produced in 1973 was marketed, with the remainder either flared, vented, or reinjected. Reserves have remained largely undeveloped due to lack of markets. With rare exceptions, natural gas has been produced only to meet export demand or for use in maintaining pressure in oil fields. Afghanistan, Bolivia, and Brunei have been small exporters of natural gas.

There are indications that gas consumption is rising. Between 1960 and 1974, its share in total energy consumption rose from 4 to 8 percent. It seems plausible from a resource perspective that natural gas production and consumption will continue to increase. By the year 2000 the resource base would enable production in non-OPEC developing countries to reach 4.5 MBDOE. This level would satisfy domestic consumption needs and also allow for some export.

6.4.4 Solid Fuels

Coal deposits can be found in several countries, but the largest known reserves are concentrated in comparatively few areas, notably the United States, the USSR, and China. Most of the coal discovered so far lies in the northern temperate zone where most of the exploration has been concentrated. Few of the developing countries have ever had to search for coal since historically their development processes got underway after oil was generally available. However, there could be significant reserves in many parts of the developing world. Rising oil prices may encourage non-OPEC developing countries to look toward coal to help meet domestic energy needs and, in certain cases, possibly contribute to exports.

OPEC Coal Production

Only Indonesia and Venezuela, among OPEC nations, have significant coal reserves. Abundant oil and gas reserves in other OPEC countries have discouraged exploration for coal. Therefore, future coal production is expected to be principally confined to Indonesia and Venezuela. As shown in Table 6.15, it is estimated that OPEC coal production could grow to 31 million tons by 1985. Production in the year 2000 could be as high as 110 million tons.

Non-OPEC Coal Production

Coal is not widely used in the non-OPEC developing countries. Only two countries, India and Korea, mine significant amounts. In Korea, coal accounts for about half, and in India for about 70 percent, of total commercial energy consumption.

Total coal production in the non-OPEC developing countries was about 125 million tonnes in 1972. There is increasing incentive for developing countries to expand their coal production, and the resource base could support much higher levels of coal production in many of these countries. By 1985 coal production could reach 230 million tonnes, rising to as much as 510 million metric tonnes by year 2000, with about half of this production coming from India.

These levels would require developing new coal mines and handling facilities as networks for delivery to consumers. The capital costs for such development are high, but probably not prohibitive in light of increasing oil import bills. Clearly, developing countries, especially lower-income countries with large reserves of coal, have the potential and the motivation to develop this energy source.

Table 6.16, from the World Energy Conference (1974), shows the reserves and total resources of solid fossil fuels in Africa, East and South Asia, and Latin America.

6.4.5 Electricity Generation

Electrical capacity nearly tripled in the developing world during the period 1960 to 1973, and this rapid growth is expected to continue into the future. Over the next five years, the World Bank estimates that electrical generation will increase another 50 percent in the non-OPEC developing countries.[12]

Table 6.17 summarizes electricity growth by fuel type for the WAES scenario cases. Primary electricity use is expected to increase from 2.2 MBDOE in 1972 to 5.8 MBDOE in 1985 in the high growth case (Case C). By the year 2000, total installed electrical capacity in a high nuclear growth future could be as high as 16.5 MBDOE (730 GWe installed).[13] The International Atomic Energy Agency estimates that about 11 percent of total generating capacity in developing countries would be in OPEC countries and 89 percent in non-OPEC countries.[14]

Most of this increased demand for electricity will be met by expanded hydroelectric and nuclear capacity. Reserves of hydroelectric power are abundant in developing countries, and only 4 percent of current potential is being utilized.[15] In 1972, hydroelectricity constituted about 44 percent of the non-OPEC developing countries' electrical generation. The World Bank estimates that this percentage share could increase over the next five years, and the WAES assumptions extend this growth to 1985. Hydroelectric capacity is projected to increase over fourfold by year 2000 in one of the WAES cases.

Similarly, nuclear power is expected to grow rapidly. The maximum potential nuclear installed capacity for the developing countries (OPEC and

12. Lambertini, "Energy and Petroleum."
13. 100 GW of installed capacity = 2.4 MBDOE primary energy, assuming 35 percent efficiency and a 60 percent load factor.
14. International Atomic Energy Agency, *Market Survey for Nuclear Power in Developing Countries*, 1974.
15. Lambertini, "Energy and Petroleum."

Table 6.15
Projected coal supply and demand in the developing countries (10^6 metric tons of coal equivalent)

	Cases C, C-1			Cases D, D-8		
	Production	Consumption	Export	Production	Consumption	Export
1985						
OPEC	31	31	—	23	23	—
Non-OPEC	202	182	20	170	160	10
Total developing countries	233	213	20	193	183	10
2000						
OPEC	110	35	75	61	15	46
Non-OPEC	400	306	94	274	243	31
Total developing countries	510	341	169	335	258	77

Table 6.16
Reserves and resources of solid fossil fuels in Africa, East and South Asia, and Latin America (10^6 metric tons)

	Reserves		Total
	Recoverable	Total	resources
Africa			
South Africa	10,584	24,224	44,339
Rhodesia	~1,390	1,760	6,613
Swaziland	1,820	2,022	5,022
Zaire	720	720	720
Botswana	506	506	506
Mozambique	80	100	400
Tanzania	180	309	370
Nigeria	225	449	449
Zambia	51	74	154
Morocco	15	15	96
Malagasy	39	78	92
Malawi			38
Egypt	~13	25	25
Algeria	~5	9	20
Total	15,628	30,291	58,844
Asia			
India	11,580	23,160	82,977
Japan	1,026	8,628	8,628
Turkey	2,025	2,893	7,282
Indonesia	1,060	2,123	2,533
Pakistan	172	804	1,941
Bangladesh	519	780	1,491
Korea, Rep. of	544	890	1,450
China, Rep. of	261	479	660
Iran	193	385	385
Burma	7	13	286
Thailand	~118	235	235
Philippines	~38	75	88
Afghanistan			85
Vietnam, Rep. of	~6	12	12
Total	17,549	40,477	108,053
Latin America and Greenland			
Mexico	629	5,316	12,000
Peru	105	211	6,964
Colombia	109	150	5,330
Chile	58	97	3,945
Brazil	1,790	3,256	3,256
Venezuela	11	14	871
Argentina	100	155	555
Honduras			5
Greenland	1	2	2
Total	2,803	9,201	32,928

Source: World Energy Conference, 1974.

non-OPEC) in the year 2000 could be as high as 416 GWe. This high estimate is from the IAEA/OECD survey of December 1975, but there are now indications that installed capacity may be considerably less. Our analysis assumes that in the high economic growth, high nuclear scenario case (C-1) almost all the nuclear potential is utilized. In this case, nuclear power would meet about 56 percent of the demand for electricity by the year 2000. All other cases (having scenario assumptions of low nuclear growth or low economic growth or both) are scaled downward.

6.5 ENERGY SUPPLY-DEMAND INTEGRATION

The interactions between energy demand and supply are complex. WAES has developed methods of comparing desired demands and potential supplies, termed *supply-demand integrations.* The techniques used in the developing country analysis are simpler than those used in the other national and global integrations. We have assumed certain shares of energy demand for various sectors—transport, industry, domestic and commercial, and nonenergy uses. And we have further made the simplifying assumption that energy use in all sectors will grow at about the same rate as total energy demand. These assumptions are shown in Table 6.18. Fuel mix assumptions to the year 2000 for the various sectors are shown in Table 6.19. Oil is expected to constitute almost all of transportation's energy requirements in all cases. The industrial sector's fuel needs are met by substantial amounts of oil and gas, some coal, and in the latter years, some electricity. The relatively large share of gas in the industrial sector is due largely to the probable large-scale use of gas by OPEC countries. The domestic and commercial sectors use substantial amounts of oil and electricity. Nonenergy demands for feedstocks and asphalt are made up largely of oil and some gas. Table 6.20 shows the total amount of each fuel consumed for the WAES cases when these fuel mix assumptions are incorporated.

6.5.1 Imports and Exports

By comparing the tables on energy production (Table 6.11) and energy consumption (Table 6.20), desired imports and potential exports by fuel type can be determined for OPEC and non-OPEC developing countries. Table 6.21 shows these balancing calculations.

Table 6.17
Primary electricity generation in developing countries, 1972-2000

Year	1972	1985		2000			
WAES case		C	D	C-1	C-2	D-7	D-8
Primary Electrical Generation (MBDOE)							
Oil	0.81	1.13	0.95	2.85	1.66	2.00	1.16
Gas	0.13	0.63	0.53	1.28	0.47	0.57	0.35
Coal	0.29	1.13	0.95	1.45	1.66	1.53	1.16
Hydro	0.97	1.76	1.85	4.24	3.50	3.00	2.40
Nuclear	0.01	1.17	0.53	4.46	9.24	3.10	6.50
Total	2.21	5.82	4.81	14.28	16.53	10.20	11.57
Share of Total Primary Electrical Generation (%)							
Oil	37	20	20	20	10	20	10
Gas	6	11	11	9	3	6	3
Coal	13	20	20	10	10	15	10
Hydro	44	31	38	30	21	29	21
Nuclear	—	18	11	31	56	30	56
Total	100	100	100	100	100	100	100

OPEC Exports

OPEC countries will continue to be large fuel exporters. The potential for export does, however, vary among OPEC countries. Saudi Arabia and other Arabian Peninsula countries have the potential to maintain their high export levels. Other OPEC countries, such as Venezuela and Indonesia, have growing domestic needs that will limit their exports. Total OPEC oil exports could range between 33 and 37 MBD in 1985 and between 35 and 38 MBD in 2000. These year 2000 estimates are based on OPEC production ceilings described earlier.

OPEC countries with plentiful gas reserves could export large quantities of gas, but this depends on whether they want to increase exports significantly and on whether the importing countries will construct the expensive and complicated LNG systems needed. As noted earlier in the section on natural gas, consuming countries of Western Europe, North America, and Japan could have import requirements as high as 8.5 MBDOE by year 2000. About 1 MBDOE could be imported from the USSR. The remainder would have to come largely from OPEC countries.

Non-OPEC Imports and Exports

Non-OPEC developing countries have traditionally been large importers of energy. During the period 1960 to 1972, they imported about 30 percent of their overall energy needs. Our projections show that in 1985 they will have to import 18-22 percent and by year 2000, 15-25 percent of their energy needs, depending on the scenario.

Table 6.18
Sector share of final energy demand (excluding processing losses), 1972-2000 (%)

	1972	1985	2000
Transport	34	35	34
Industrial	42	42	41
Domestic and commercial	18	16	17
Nonenergy	6	7	8

Table 6.19
Developing country fuel mix assumptions by demand sector (%)

	Oil	Coal	Gas	Electricity
1972				
Transport	91	9	—	—
Industry	48	26	15	11
Domestic and commercial	66	9	8	17
1985 C, D				
Transport	97	2	—	1
Industry	42	20	24	14
Domestic and commercial	61	3	13	23
2000 C-1				
Transport	95	2	—	3
Industry	35	17	31	17
Domestic and commercial	55	5	15	25
2000 C-2				
Transport	95	2	—	3
Industry	37	16	28	19
Domestic and commercial	55	3	15	27
2000 D-7				
Transport	95	2	—	3
Industry	31	17	36	16
Domestic and commercial	55	5	15	25
2000 D-8				
Transport	95	2	—	3
Industry	37	18	27	18
Domestic and commercial	55	3	15	27

Table 6.20
Projected developing country consumption of primary energy, 1985-2000 (MBDOE)

Year	1985		2000			
WAES case	C	D	C-1	C-2	D-7	D-8
Oil						
Non-OPEC developing countries	10.70	8.94	20.87	19.15	15.20	14.60
OPEC	3.30	2.76	6.33	7.80	3.85	4.50
Total	14.00	11.70	27.20	26.95	19.05	19.10
Solid Fuels						
Non-OPEC developing countries	2.40	2.11	4.03	4.19	3.63	3.20
OPEC	0.42	0.30	0.46	0.46	0.20	0.20
Total	2.82	2.41	4.49	4.65	3.83	3.40
Natural Gas						
Non-OPEC developing countries	1.92	1.63	3.56	3.50	2.88	2.40
OPEC	1.54	1.15	5.00	3.97	3.42	2.76
Total	3.46	2.78	8.56	7.47	6.30	5.16
Hydroelectricity						
Non-OPEC developing countries	1.70	1.80	3.80	3.10	2.70	2.10
OPEC	0.10	0.10	0.40	0.40	0.30	0.30
Total	1.80	1.90	4.20	3.50	3.00	2.40
Nuclear Electricity						
Non-OPEC developing countries	1.10	0.40	4.00	8.30	2.80	5.90
OPEC	0.10	0.10	0.46	1.00	0.30	0.60
Total	1.20	0.50	4.46	9.30	3.10	6.50
Total Primary Energy[a]						
Non-OPEC developing countries	17.82	14.88	36.26	38.24	27.21	28.20
OPEC	5.46	4.41	12.65	13.63	8.07	8.36
Total	23.28	19.29	48.91	51.87	35.28	36.56

[a]Totals are from supply-demand integration worksheets and may not exactly equal the targets in Table 6.9.

Table 6.21
Projected (imports) and exports of primary energy for all developing countries, 1985-2000 (MBDOE)

Year	1985		2000			
WAES case	C	D	C-1	C-2	D-7	D-8
Oil						
Non-OPEC developing countries	(4.00)	(3.84)	(9.37)	(7.65)	(8.20)	(7.60)
OPEC	36.70	33.24	38.67	37.20	35.15	34.50
Total	32.70	29.40	29.30	29.55	26.95	26.90
Solid Fuels						
Non-OPEC developing countries	0.27	0.13	1.23	0.81	0.47	0.40
OPEC	0	0	1.00	0.84	0.60	0.60
Total	0.27	0.13	2.23	1.65	1.07	1.00
Natural Gas						
Non-OPEC developing countries	0.48	0.47	1.00	1.00	1.00	1.00
OPEC	3.93	3.93	8.60	7.63	8.88	7.44
Total	4.41	4.40	9.60	8.63	9.88	8.44
Total Primary Energy						
Non-OPEC developing countries	(3.25)	(3.24)	(7.14)	(5.84)	(6.73)	(6.20)
OPEC	40.63	37.17	48.27	45.67	44.63	42.54
Total	37.38	33.93	41.13	39.83	37.90	36.34

The non-OPEC developing countries differ widely in their dependence on imports. Some developing countries such as Mexico and Brazil may be able to achieve energy self-sufficiency, but most will continue to depend on imports—especially of oil. These countries may require oil imports as high as 4 MBD in 1985 and between 7.6 and 9.4 MBD in the year 2000.

As described earlier, certain countries could produce substantial amounts of coal and gas. Therefore, the non-OPEC developing countries as a group will require no imports of coal or gas, and in fact may be modest exporters. However, this statistic masks the fact that, as in the case of oil, production of coal and gas will most probably be concentrated in a few countries.

6.6 WORKSHEETS AND ASSESSMENT OF REAL GNP AND COMMERCIAL ENERGY CONSUMPTION

This section consists of the following worksheets and subsidiary tables:

A three-page table showing real GNP and commercial energy consumption for the developing countries plus Australia, New Zealand, and South Africa, 1960-1972.

National input worksheets for supply/demand integration—Developing countries (7 sheets)

National input worksheets for supply/demand integration—Australia, New Zealand, South Africa (5 sheets)

Real GNP and commercial energy consumption by country and region, developing countries plus Australia, New Zealand, and South Africa, 1960–1972

	Real GNP (10^9 1973 U.S. dollars)			Commercial energy consumption (10^6 TCE)					
	1960	1972	Growth rate, 1960–1972	1960	1972	Growth rate, 1960–1972	1960 oil-adjusted	1972 oil-adjusted	Oil-adjusted growth rate, 1960–1972
South Asia									
Afghanistan	1.01	1.33		0.2	0.8				
Bangladesh	4.34	5.32			2.1				
Burma	1.63	2.34		1.2	1.7				
India	46.07	70.32		70.8	140.0				
Nepal	0.82	1.08			0.2				
Pakistan	3.63	7.35		6.1	11.6				
Sri Lanka	0.94	1.51		1.1	1.9				
Total	58.44	89.25	3.6%	79.4	158.3	5.9%	83.4	165.1	5.9%
Low-Income Sub-Sahara Africa									
Burundi	0.20	0.27			0.1				
Central African Rep.	0.21	0.27			0.1				
Chad	0.34	0.32			0.1				
Dahomey	0.21	0.32		0.1	0.2				
Ethiopia	1.29	2.22		0.2	0.9				
Guinea	0.37	0.54		0.2	0.4				
Kenya	1.07	2.06		1.2	1.8				
Madagascar	0.88	1.17		0.2	0.5				
Malawi	0.24	0.49			0.2				
Mali	0.26	0.39		0.1	0.1				
Niger	0.42	0.50			0.1				
Rwanda	0.20	0.28		0.1	0.1				
Sierra Leone	0.24	0.47		0.1	0.2				
Somalia	0.19	0.25		0.1	0.1				
Sudan	1.82	2.42		0.6	2.0				
Tanzania	0.93	1.75		0.4	1.0				
Togo	0.17	0.37			0.2				
Uganda	0.95	1.64		0.2	0.7				
Upper Volta	0.35	0.45			0.1				
Zaire	1.66	3.01		1.2	1.9				
Total	12.00	19.19	4.0%	4.7	10.8	7.2%	5.1	12.3	7.6%
East Asia									
Fiji	0.18	0.33		0.1	0.3				
Hong Kong	1.81	5.28		1.4	4.2				
Korea (South)	4.25	11.37		6.4	26.9				
Malaysia	2.73	5.89		2.0	5.8				
Papua New Guinea	0.41	0.88		0.1	0.5				
Philippines	5.46	10.16		4.0	12.9				
Singapore	1.21	3.64		0.6	4.4				
Taiwan	2.95	9.16		8.5	20.5				
Thailand	4.02	9.61		1.7	11.9				
Total	23.02	56.32	7.7%	24.8	87.4	11.1%	27.8	97.6	11.0%

Real GNP and commercial energy consumption by country and region, developing countries plus Australia, New Zealand, and South Africa, 1960–1972 (continued)

	Real GNP (10^9 1973 U.S. dollars)			Commercial energy consumption (10^6 TCE)					
	1960	1972	Growth rate, 1960–1972	1960	1972	Growth-rate, 1960–1972	1960 oil-adjusted	1972 oil-adjusted	Oil-adjusted growth rate, 1960–1972
Middle-Income Sub-Sahara Africa and West Asia									
Angola	1.34	2.38		0.4	1.2				
Bahrein	0.06	0.21		0.1	1.7				
Cameroon	0.82	1.46		0.3	0.6				
Congo PR	0.23	0.37		0.1	0.3				
Cyprus	0.40	0.86		0.5	1.2				
Egypt	5.16	8.61		7.7	11.2				
Ghana	1.93	2.69		0.7	1.4				
Israel	3.07	8.98		2.7	8.6				
Ivory Coast	0.92	2.11		0.2	1.5				
Jordan	0.47	0.88		0.3	0.8				
Lebanon	1.33	2.67		1.0	2.4				
Liberia	0.23	0.44		0.1	0.6				
Mauritania	0.11	0.24			0.2				
Morocco	3.10	5.00		1.7	3.7				
Mozambique	1.61	2.97		0.7	1.3				
Oman	0.10	0.57			0.2				
Rhodesia	1.33	2.38		2.0	3.9				
Senegal	1.18	0.27		0.4	0.6				
Syria	1.05	2.76		1.5	3.4				
Tunisia	1.16	2.48		0.7	1.7				
Turkey	10.19	21.43		6.7	20.6				
Yemen, AR, DM	0.50	0.70		0.3	0.7				
Zambia	1.20	1.98		1.1	2.3				
Total	37.49	73.44	5.8%	29.2	71.3	7.7%	32.9	79.3	7.6%
Caribbean, Central and South America									
Argentina	23.38	37.94		23.0	44.1				
Barbados	0.12	0.23		0.1	0.2				
Bolivia	0.61	1.12		0.5	1.4				
Brazil	31.10	68.74		23.3	53.7				
Chile	7.05	11.81		6.5	14.4				
Colombia	4.96	9.17		7.6	14.0				
Costa Rica	0.64	1.25		0.3	0.9				
Dominican Rep.	1.05	2.10		0.5	1.8				
El Salvador	0.65	1.32		0.3	0.8				
Guatemala	1.24	2.59		0.7	1.3				
Guyana	0.21	0.32		0.3	0.8				
Haiti	0.47	0.57		0.1	0.1				
Honduras	0.52	0.89		0.3	0.7				
Jamaica	0.94	1.93		0.7	2.8				
Mexico	21.46	49.83		27.8	61.4				
Nicaragua	0.49	1.06		0.3	0.8				

Real GNP and commercial energy consumption by country and region, developing countries plus Australia, New Zealand, and South Africa, 1960-1972 (continued)

	Real GNP (10^9 1973 U.S. dollars)			Commercial energy consumption (10^6 TCE)					
	1960	1972	Growth rate, 1960-1972	1960	1972	Growth rate, 1960-1972	1960 oil-adjusted	1972 oil-adjusted	Oil-adjusted growth rate, 1960-1972
Panama	0.55	1.45		0.5	1.3				
Paraguay	0.55	1.00		0.2	0.3				
Peru	4.55	8.63		4.5	8.7				
Trinidad and Tobago	0.85	1.36		1.5	4.6				
Uruguay	2.55	2.81		2.1	2.9				
Total	103.44	206.12	5.9%	101.1	217.0	6.6%	111.2	240.2	6.6%
OPEC									
Algeria	5.64	8.31		2.7	5.4				
Ecuador	1.22	2.21		0.9	2.0				
Gabon	0.31	0.62		0.1	0.5				
Indonesia	8.82	14.79		12.1	19.4				
Iran	8.61	24.93		5.8	30.9				
Iraq	3.34	6.41		3.4	7.6				
Kuwait	4.30	9.37		2.9	8.8				
Libya	0.03	7.27		0.3	1.4				
Nigeria	6.48	13.79		1.5	4.2				
Qatar	0.22	0.92		0.1	1.6				
Saudi Arabia	2.95	9.53		1.6	7.6				
U.A.E.	0.13	2.92			2.1				
Venezuela	10.10	19.17		13.1	26.6				
Total	52.15	120.24	7.2%	44.5	118.1	8.5%	48.3	127.9	8.4%
Developed									
Australia	30.81	54.28		39.8	73.9				
New Zealand	6.61	10.52		5.4	9.2				
South Africa	12.50	24.32		42.1	71.7				
Total	49.72	89.12	5.0%	87.3	154.8	4.9%	91.4	162.9	4.9%

Source: Real GNP data are from the 1974 and 1975 editions of the *World Bank Atlas.* Commercial energy consumption data are mainly from the U.N. *World Energy Supplies 1950-1974,* Series J, Number 19. Figures for India are from Indian government sources. To ensure consistency with other WAES studies, adjustments have been made in the U.N. oil consumption figures to reflect WAES oil-to-coal conversion factors.

NATIONAL INPUT WORKSHEET FOR SUPPLY/DEMAND INTEGRATION

Country: Developing Countries

Year: 1972

Case: Base Year

Units: MBDOE

	(a) Coal	(b) Petro-leum	(c) (Syn. Liquids)	(d) Nat. Gas	(e) (Syn. Gas)	(f) (Heat)	(g) (Electri-city)††	(h) Hydro Energy	(i) Nuclear Energy	(j) Geothermal energy and other	(k) Total
(1) Transport	.21	2.1									2.31
(2) Industry (3) Agric., Mining, Construction	.77	1.43		.45			.35				3.00
(4) Commercial (5) Public (6) Residential	.12	.87		.12			.21				1.32
(7) Non-energy Uses		.36									.36
(8) Final Energy Demand*	1.10	4.76		.57			.56				6.99
(9) Electricity**	.29	.81		.13			-.66	.97	.01		1.55
(10) Syn. Gas											
(11) Syn. Liquids											
(12) Heat											
(13) Energy Sector Self Consumption & conversion losses	.02	.46		.38			.10				.96
(14) Primary Energy Input	1.41	6.03		1.08				.97	.01		9.50
(15) Indigenous Supply	1.67	30.3		3.98				.97	.01		36.93
(16 Imports*** (17) Exports	.26	24.27		2.90							27.43

*Includes non-energy uses.

**Blocks 9g, 10e 11c, and 12f must be negative to avoid double counting of energy from primary fuels.

***Includes imported products.

††Includes only electricity to be purchased by each sector.

NATIONAL INPUT WORKSHEET FOR SUPPLY/DEMAND INTEGRATION

Country: Developing Countries

Year: 1985

Case: C

Units: MBDOE

	(a) Coal	(b) Petro-leum	(c) (Syn. Liquids)	(d) Nat. Gas	(e) (Syn. Gas)	(f) (Heat)	(g) (Electri-city)††	(h) Hydro Energy	(i) Nuclear Energy	(j) Geothermal energy and other	(k) Total
(1) Transport	.12	5.79					.06				5.97
(2) Industry } (3) Agric., Mining, Construction	1.42	3.02		1.79			1.01				7.24
(4) Commercial } (5) Public (6) Residential	.09	1.67		.36			.63				2.75
(7) Non-energy Uses		.96		.25							1.21
(8) Final Energy Demand*	1.63	11.44		2.40			1.70				17.17
(9) Electricity**	1.13	1.13		.63			−2.0	1.76	1.17		3.82
(10) Syn. Gas											
(11) Syn. Liquids											
(12) Heat											
(13) Energy Sector Self Consumption & conversion losses	.06	1.39		.38			.3				2.13
(14) Primary Energy Input	2.82	13.96		3.41				1.76	1.17		23.12
(15) Indigenous Supply	3.09	46.70		7.87				1.76	1.17		60.59
(16 Imports***											
(17) Exports	.27	32.74		4.46							37.47

*Includes non-energy uses.

**Blocks 9g, 10e 11c, and 12f must be negative to avoid double counting of energy from primary fuels.

***Includes imported products.

††Includes only electricity to be purchased by each sector.

NATIONAL INPUT WORKSHEET FOR SUPPLY/DEMAND INTEGRATION

Country: Developing Countries

Year: 1985

Case: D

Units: MBDOE

	(a)	(b)	(c)	(d)	(e)	(f)	(g)	(h)	(i)	(j)	(k)
	Coal	Petro-leum	(Syn. Liquids)	Nat. Gas	(Syn. Gas)	(Heat)	(Electri-city)††	Hydro Energy	Nuclear Energy	Geothermal energy and other	Total
(1) Transport	.11	4.88					.05				5.04
(2) Industry											
(3) Agric., Mining, Construction	1.24	2.53		1.43			.84				6.04
(4) Commercial											
(5) Public	.07	1.40		.30			.53				2.30
(6) Residential											
(7) Non-energy Uses		.75		.25							1.00
(8) Final Energy Demand*	1.42	9.56		1.98			1.42				14.38
(9) Electricity**	.95	.95		.53			-1.67	1.85	.53		3.14
(10) Syn. Gas											
(11) Syn. Liquids											
(12) Heat											
(13) Energy Sector Self Consumption & conversion losses	.05	1.16		.27			.25				1.73
(14) Primary Energy Input	2.42	11.67		2.78				1.85	.53		19.25
(15) Indigenous Supply	2.54	41.10		7.18				1.85	.53		53.20
(16 Imports***											
(17) Exports	.12	29.43		4.40							33.95

(2)–(3) and (4)–(6) are bracketed together in the original.

*Includes non-energy uses.

**Blocks 9g, 10e 11c, and 12f must be negative to avoid double counting of energy from primary fuels.

***Includes imported products.

††Includes only electricity to be purchased by each sector.

NATIONAL INPUT WORKSHEET FOR SUPPLY/DEMAND INTEGRATION

Country: Developing Countries

Year: 2000

Case: C-1

Units: MBDOE

	(a)	(b)	(c)	(d)	(e)	(f)	(g)	(h)	(i)	(j)	(k)
	Coal	Petro-leum	(Syn. Liquids)	Nat. Gas	(Syn. Gas)	(Heat)	(Electri-city)††	Hydro Energy	Nuclear Energy	Geothermal energy and other	Total
(1) Transport	.25	11.44					.36				12.05
(2) Industry (3) Agric., Mining, Construction	2.41	5.02		4.53			2.45				14.41
(4) Commercial (5) Public (6) Residential	.30	3.28		.91			1.45				5.94
(7) Non-energy Uses		1.89		1.02							2.91
(8) Final Energy Demand*	2.96	21.63		6.46			4.26				35.31
(9) Electricity**	1.45	2.85		1.28			-5.01	4.24	4.46		9.27
(10) Syn. Gas											
(11) Syn. Liquids											
(12) Heat											
(13) Energy Sector Self Consumption & conversion losses	.09	2.73		.81			.75				4.38
(14) Primary Energy Input	4.50	27.21		8.55				4.24	4.46		48.96
(15) Indigenous Supply	6.74	56.5		18.16				4.24	4.46		90.10
(16 Imports***											
(17) Exports	2.24	29.29		9.61							41.14

*Includes non-energy uses.

**Blocks 9g, 10e 11c, and 12f must be negative to avoid double counting of energy from primary fuels.

***Includes imported products.

††Includes only electricity to be purchased by each sector.

NATIONAL INPUT WORKSHEET FOR SUPPLY/DEMAND INTEGRATION

Country: Developing Countries

Year: 2000

Case: C-2

Units: MBDOE

	(a)	(b)	(c)	(d)	(e)	(f)	(g)	(h)	(i)	(j)	(k)
	Coal	Petro- leum	(Syn. Liquids)	Nat. Gas	(Syn. Gas)	(Heat)	(Electri- city)††	Hydro Energy	Nuclear Energy	Geothermal energy and other	Total
(1) Transport	.26	11.97					.32				12.55
(2) Industry (3) Agric., Mining, Construction	2.47	5.55		4.23			2.86				15.11
(4) Commercial (5) Public (6) Residential	.19	3.39		.95			1.69				6.22
(7) Non-energy Uses		1.68		1.08							2.76
(8) Final Energy Demand*	2.92	22.59		6.26			4.87				36.64
(9) Electricity**	1.66	1.66		.47			-5.73	3.48	9.29		10.83
(10) Syn. Gas											
(11) Syn. Liquids											
(12) Heat											
(13) Energy Sector Self Consumption & conversion losses	.07	2.70		.74			.86				4.37
(14) Primary Energy Input	4.65	26.95		7.47				3.48	9.29		51.84
(15) Indigenous Supply	6.30	56.50		16.10				3.48	9.29		91.67
(16 Imports*** (17) Exports	1.65	29.55		8.63							39.83

*Includes non-energy uses.

**Blocks 9g, 10e 11c, and 12f must be negative to avoid double counting of energy from primary fuels.

***Includes imported products.

††Includes only electricity to be purchased by each sector.

NATIONAL INPUT WORKSHEET FOR SUPPLY/DEMAND INTEGRATION

Country: Developing Countries

Year: 2000

Case: D-7

Units: MBDOE

	(a) Coal	(b) Petroleum	(c) (Syn. Liquids)	(d) Nat. Gas	(e) (Syn. Gas)	(f) (Heat)	(g) (Electricity)††	(h) Hydro Energy	(i) Nuclear Energy	(j) Geothermal energy and other	(k) Total
(1) Transport	.17	8.08					.26				8.51
(2) Industry (3) Agric., Mining, Construction	1.80	3.37		3.83			1.72				10.72
(4) Commercial (5) Public (6) Residential	.26	2.30		.63			1.04				4.23
(7) Non-energy Uses		1.27		.67							1.94
(8) Final Energy Demand*	2.23	15.02		5.13			3.02				25.40
(9) Electricity**	1.53	2.01		.57			-3.55	2.97	3.06		6.59
(10) Syn. Gas											
(11) Syn. Liquids											
(12) Heat											
(13) Energy Sector Self Consumption & conversion losses	.07	1.92		.56			.53				3.08
(14) Primary Energy Input	3.83	18.95		6.26				2.97	3.06		35.07
(15) Indigenous Supply	4.90	46.00		16.16				2.97	3.06		73.09
(16 Imports*** (17) Exports	1.07	27.05		9.90							38.02

*Includes non-energy uses.

**Blocks 9g, 10e 11c, and 12f must be negative to avoid double counting of energy from primary fuels.

***Includes imported products.

††Includes only electricity to be purchased by each sector.

NATIONAL INPUT WORKSHEET FOR SUPPLY/DEMAND INTEGRATION

Country: Developing Countries

Year: 2000

Case: D-8

Units: MBDOE

	(a) Coal	(b) Petro- leum	(c) (Syn. Liquids)	(d) Nat. Gas	(e) (Syn. Gas)	(f) (Heat)	(g) (Electri- city)††	(h) Hydro Energy	(i) Nuclear Energy	(j) Geothermal energy and other	(k) Total
(1) Transport	.18	8.34					.27				8.79
(2) Industry (3) Agric., Mining, Construction	1.90	3.94		2.94			2.01				10.79
(4) Commercial (5) Public (6) Residential	.13	2.41		.66			1.18				4.38
(7) Non-energy Uses		1.33		.70							2.03
(8) Final Energy Demand*	2.21	16.02		4.30			3.46				25.99
(9) Electricity**	1.16	1.16		.35			-4.07	2.40	6.50		7.50
(10) Syn. Gas											
(11) Syn. Liquids											
(12) Heat											
(13) Energy Sector Self Consumption & conversion losses	.05	1.94		.51			.61				3.11
(14) Primary Energy Input	3.42	19.12		5.16				2.40	6.50		36.60
(15) Indigenous Supply	4.40	46.00		13.60				2.40	6.50		72.90
(16 Imports*** (17) Exports	.98	26.88		8.44							36.30

*Includes non-energy uses.

**Blocks 9g, 10e 11c, and 12f must be negative to avoid double counting of energy from primary fuels.

***Includes imported products.

††Includes only electricity to be purchased by each sector.

NATIONAL INPUT WORKSHEET FOR SUPPLY/DEMAND INTEGRATION

Country: Australia, New Zealand, South Africa

Year: 1972

Case: Base Year

Units: MBDOE

	(a) Coal	(b) Petro-leum	(c) (Syn. Liquids)	(d) Nat. Gas	(e) (Syn. Gas)	(f) (Heat)	(g) (Electri-city)[††]	(h) Hydro Energy	(i) Nuclear Energy	(j) Geothermal energy and other	(k) Total
(1) Transport	.06	.38		.01							.45
(2) Industry; (3) Agric., Mining, Construction	.36	.17		.02			.12				.67
(4) Commercial; (5) Public; (6) Residential	.08	.08		.02			.06				.24
(7) Non-energy Uses		.08									.08
(8) Final Energy Demand*	.50	.71		.05			.18				1.44
(9) Electricity**	.58	.02		.01			−.21	.14	.01		.55
(10) Syn. Gas	.01	.01		(.01)							.01
(11) Syn. Liquids	.02	(.01)									.01
(12) Heat											
(13) Energy Sector Self Consumption & conversion losses	.04	.09		.02			.03				.18
(14) Primary Energy Input	1.15	.82		.07				.14	.01		2.19
(15) Indigenous Supply	1.42	.33		.07				.14	.01		1.97
(16 Imports***											.49
(17) Exports	.27	.49									.27

*Includes non-energy uses.

**Blocks 9g, 10e 11c, and 12f must be negative to avoid double counting of energy from primary fuels.

***Includes imported products.

††Includes only electricity to be purchased by each sector.

NATIONAL INPUT WORKSHEET FOR SUPPLY/DEMAND INTEGRATION

Country: ~~Australia, New Zealand,~~ South Africa (Provisional Estimate)

Year: 1985

Case: C

Units: MBDOE

	(a)	(b)	(c)	(d)	(e)	(f)	(g)	(h)	(i)	(j)	(k)
	Coal	Petro-leum	(Syn. Liquids)	Nat. Gas	(Syn. Gas)	(Heat)	(Electri-city)[††]	Hydro Energy	Nuclear Energy	Geothermal energy and other	Total
(1) Transport	.01	.75					.01				.77
(2) Industry }											
(3) Agric., Mining, Construction }	.71	.20		.23			.27				1.41
(4) Commercial }											
(5) Public }	.06	.10		.05			.14				.35
(6) Residential }											
(7) Non-energy Uses											
(8) Final Energy Demand*	.78	1.05		.28			.42				2.53
(9) Electricity**	1.07	.02		.11			−.49	.25	.07	.01	1.04
(10) Syn. Gas	.03	.01		(.03)							.01
(11) Syn. Liquids	.12	(.04)									.08
(12) Heat											
(13) Energy Sector Self Consumption & conversion losses	.04	.10		.06			.07				.27
(14) Primary Energy Input	2.04	1.14		.42				.25	.07	.01	3.93
(15) Indigenous Supply	2.63	.50		.66				.25	.07	.01	4.12
(16 Imports***		.64									.64
(17) Exports	.59			.24							.83

*Includes non-energy uses.

**Blocks 9g, 10e 11c, and 12f must be negative to avoid double counting of energy from primary fuels.

***Includes imported products.

††Includes only electricity to be purchased by each sector.

NATIONAL INPUT WORKSHEET FOR SUPPLY/DEMAND INTEGRATION

Country: Australia, New Zealand, South Africa (Provisional Estimate)

Year: 1985

Case: D

Units: MBDOE

	(a) Coal	(b) Petro-leum	(c) (Syn. Liquids)	(d) Nat. Gas	(e) (Syn. Gas)	(f) (Heat)	(g) (Electri-city)[††]	(h) Hydro Energy	(i) Nuclear Energy	(j) Geothermal energy and other	(k) Total
(1) Transport	.01	.68					.01				.70
(2) Industry, (3) Agric., Mining, Construction	.64	.18		.21			.24				1.27
(4) Commercial, (5) Public, (6) Residential	.05	.09		.05			.13				.32
(7) Non-energy Uses		.12									.12
(8) Final Energy Demand*	.70	1.07		.26			.38				2.41
(9) Electricity**	.96	.02		.10			-.44	.23	.06	.01	.94
(10) Syn. Gas	.03	.01		(.03)							.01
(11) Syn. Liquids	.11	(.04)									.07
(12) Heat											
(13) Energy Sector Self Consumption & conversion losses	.04	.09		.05			.06				.24
(14) Primary Energy Input	1.84	1.15		.38				.23	.06	.01	3.67
(15) Indigenous Supply	2.37	.51		.59				.23	.06	.01	3.77
(16 Imports***		.64									.64
(17) Exports	.53			.21							.74

*Includes non-energy uses.

**Blocks 9g, 10e 11c, and 12f must be negative to avoid double counting of energy from primary fuels.

***Includes imported products.

[††]Includes only electricity to be purchased by each sector.

NATIONAL INPUT WORKSHEET FOR SUPPLY/DEMAND INTEGRATION

Country: Australia, New Zealand, South Africa (Provisional Estimate)

Year: 2000

Case: C-1**

Units: MBDOE

	(a) Coal	(b) Petroleum	(c) (Syn. Liquids)	(d) Nat. Gas	(e) (Syn. Gas)	(f) (Heat)	(g) (Electricity)††	(h) Hydro Energy	(i) Nuclear Energy	(j) Geothermal energy and other	(k) Total
(1) Transport	.01	1.32					.03				1.36
(2) Industry											
(3) Agric., Mining, Construction	1.24	.35		.42			.47				2.48
(4) Commercial											
(5) Public	.11	.18		.08			.24				.61
(6) Residential											
(7) Non-energy Uses		.23									.23
(8) Final Energy Demand*	1.36	2.08		.50			.74				4.68
(9) Electricity**	1.52	.04		.15			−.87	.35	.09	.02	1.30
(10) Syn. Gas											
(11) Syn. Liquids											
(12) Heat											
(13) Energy Sector Self Consumption & conversion losses	.06	.23		.07			.13				.49
(14) Primary Energy Input	2.94	2.35		.72				.35	.09	.02	6.47
(15) Indigenous Supply	5.70	.50		.93				.35	.09	.02	7.59
(16 Imports***		1.85									1.85
(17) Exports	2.76			.21							2.97

*Includes non-energy uses.

**Blocks 9g, 10e 11c, and 12f must be negative to avoid double counting of energy from primary fuels.

***Includes imported products.

††Includes only electricity to be purchased by each sector.

Footnotes:

*It is likely that there will be some synthetic fuels manufactured, but we have not had the information to make an estimate.

**A worksheet for C-2 is not provided as it is identical to C-1, except potential coal production is 3.95 MBDOE, thereby resulting in a lower export figure.

NATIONAL INPUT WORKSHEET FOR SUPPLY/DEMAND INTEGRATION

Country: Australia, New Zealand, South Africa (Provisional Estimate)

Year: 2000

Case: D-8*

Units: MBDOE

	(a)	(b)	(c)	(d)	(e)	(f)	(g)	(h)	(i)	(j)	(k)
	Coal	Petro-leum	(Syn. Liquids)	Nat. Gas	(Syn. Gas)	(Heat)	(Electri-city)††	Hydro Energy	Nuclear Energy	Geothermal energy and other	Total
(1) Transport	.01	1.01					.02				1.04
(2) Industry }											
(3) Agric., Mining, Construction }	.94	.26		.32			.36				1.88
(4) Commercial }											
(5) Public }	.08	.14		.06			.19				.47
(6) Residential }											
(7) Non-energy Uses		.18									.18
(8) Final Energy Demand*	1.03	1.59		.38			.57				3.57
(9) Electricity**	1.17	.03		.12			-.67	.27	.07	.02	1.01
(10) Syn. Gas											
(11) Syn. Liquids											
(12) Heat											
(13) Energy Sector Self Consumption & conversion losses	.05	.18		.06			.10				.39
(14) Primary Energy Input	2.25	1.80		.56				.27	.07	.02	4.97
(15) Indigenous Supply	3.95	.40		.77				.27	.07	.02	5.48
(16 Imports***		1.40									1.40
(17) Exports	1.70			.21							1.91

*Includes non-energy uses.

**Blocks 9g, 10e 11c, and 12f must be negative to avoid double counting of energy from primary fuels.

***Includes imported products.

††Includes only electricity to be purchased by each sector.

Footnotes:

*A worksheet for D-7 is not provided as it is identical to D-8 except for potential coal production which is assumed to be 5.66 MBDOE, thereby resulting in higher exports.

Part III
National Studies of Energy Demand to Year 2000 and Energy Supply-Demand Integrations

7.1 OVERVIEW AND SUMMARY

7.1.1 Conclusions: Supply-Demand Balances in 1985 and 2000

Canadian energy requirements have been assessed in relation to domestic availability for the five WAES 1985 scenarios and the four WAES 2000 scenarios.

It must be stressed that these scenarios were derived solely to meet WAES requirements and are in no sense official forecasts of the National Energy Board or the Government of Canada.

For 1985, Figure 7.1 summarizes the results of the Canadian supply-demand balances for the five scenarios. The results for scenario E indicate that Canadian fossil fuel requirements in 1985 cannot be met from domestic production if low energy prices are maintained and the government does not undertake vigorous policy initiatives affecting both the demand for and the supply of energy. However, there could be sufficient electrical generating capacity to permit exports under this scenario.

Under the remaining scenarios for 1985 (A, B, C, and D), Canadian energy requirements could be met by domestic production if the mid-1975 world oil price and vigorous government policy measures were adopted in Canada. However, in these four scenarios the production and deliverability of oil from Canada's frontier areas is assumed by 1985. It should be noted that the prospects for such production are highly speculative. Canada's frontier areas have simply not undergone sufficient development for conclusive evaluation. If all Canadian frontier oil production were excluded from the supply scenarios for 1985, the supply-demand balances would show an oil deficit (i.e., a requirement for oil imports) in all five scenarios.

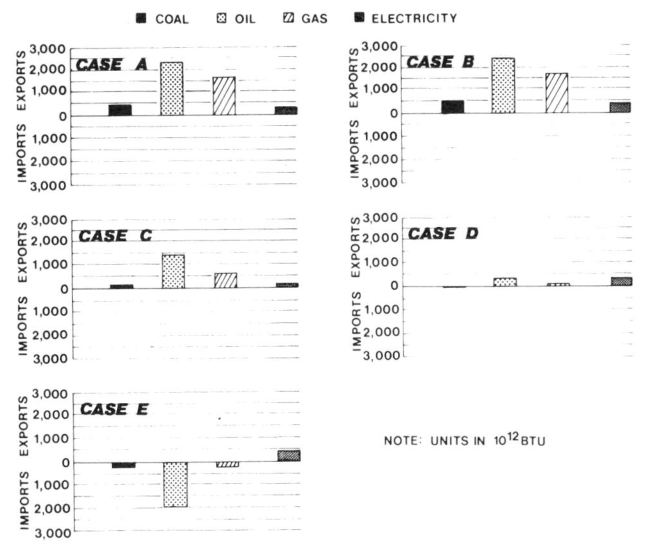

Figure 7.1
WAES Canadian supply-demand balances, 1985.

With respect to natural gas, the exclusion of frontier supplies would cause a deficit in the supply-demand balance in scenario D, make the surplus in scenario C very marginal, and reduce the surplus in scenarios A and B to just over 1 TCF per year. It must be emphasized that scenarios A, B, and C assume vigorous government energy conservation measures.

For the year 2000, the results of the supply-demand balances for the four scenarios are summarized in Figure 7.2. The studies show that Canada will be a major importer of natural gas and oil by that time if additional supplies are not forthcoming from the frontier regions. They also show that domestic coal production would be sufficient to meet domestic requirements if coal were chosen as the principal replacement fuel. If nuclear generation were chosen as the principal replacement fuel, substantial exports of coal would occur. In the electrical generation sector, substantial exports are indicated in all four scenarios.

The Canadian studies take into account several policy initiatives on the part of the federal government that could result in a more favorable energy climate in Canada by 1985. These policy initiatives are expected to have an impact on both the supply of and the demand for energy.

The studies indicate that by the year 2000 additional government measures may be required.

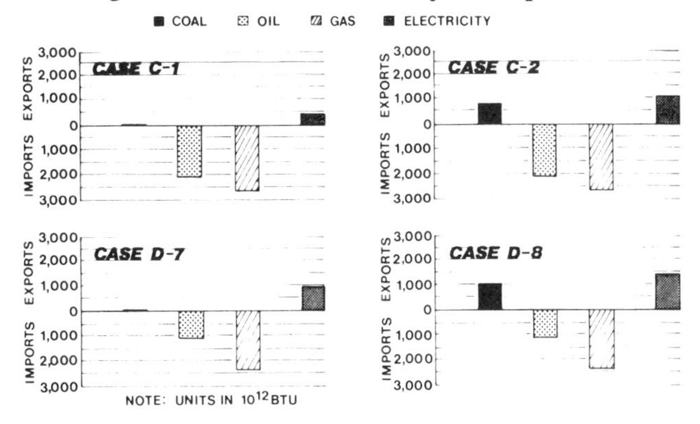

Figure 7.2
WAES Canadian supply-demand balances, 2000.

These could include initiatives to encourage further interfuel substitution, with the aim of lessening Canada's dependence on imported oil and natural gas by increasing the use of domestic electrical generation. Alternatively, measures could be introduced to direct capital expenditure toward further exploration and development of Canada's frontier areas and away from the rapid expansion of the country's electrical generation capacity. The latter policy initiatives would seem appropriate in light of the results of the four scenarios to the year 2000, which all indicate that Canada will have more than sufficient generation capacity for domestic use.

The studies have also taken into consideration some government-induced measures through the year 2000. However, additional measures might well become available in the late 1980s and 1990s that could further slow the rate of growth of Canadian primary energy demand. For the 1985 to 2000 period, Canadian energy demand is projected to grow at 4.0 or 3.3 percent per year.

7.1.2 Possible Policy Initiatives

While it must be emphasized that the scenarios presented should not be interpreted as forecasts, they do represent possible developments on the basis of specific alternative sets of assumptions. As such, the scenarios indicate that several policy initiatives on the part of the federal government could result in a more favorable energy climate in Canada by 1985. On the demand side, these initiatives include (1) an increase in domestic oil prices to at least the mid-1975 international oil price level with appropriate price increases in other sources of energy and (2) a national energy conservation program.

On the supply side, government initiatives might include (1) measures to increase exploration and development activity in the frontier regions of Canada, (2) measures to obtain more information on Canadian resources so that government planning can be done more effectively, (3) increased research and development of new energy systems, and (4) the encouragement of interfuel substitution.

It has been estimated that, at current Canadian prices, further oil sands developments and the production of oil and natural gas from Canada's frontier areas would not be economically practicable. Higher prices in Canada could reduce the average increase in energy demand from an estimated 4.5 percent to approximately 4 percent per year and also stimulate the production of additional resources from Canada's frontier areas. Probable dates when frontier resources could be made available must be tentative since they will depend on the results of drilling programs, on decisions not yet taken with regard to delivery systems, and most critically, on the assessment of the social, environmental, and economic costs of frontier resources.

The provision of additional domestic energy supplies will be expensive, requiring in the range of $180 billion worth of manpower and materials, purchased at 1975 prices, over the next fifteen years. The magnitude and timing of such investments suggest that strains on labor and capital equipment markets may result. Thus, it may be necessary for government and industry to coordinate the planning of large investment projects if they are to be appropriately phased in and completed.

The scenarios suggest that Canada's supply-demand situation for energy poses serious potential problems that could adversely affect the domestic economy and lead to an adjustment in Canadians' living standards. The successful resolution of these problems would seem to require government initiatives encouraging energy conservation and interfuel substitution. Because of the current uncertainties with regard to the magnitude and location of potential frontier resources and the anticipated environmental, social, and marketing costs, however, it is difficult to identify specific appropriate policy initiatives, particularly with regard to interfuel substitution.

For a discussion of the current direction of government policies in Canada and of recent conservation initiatives, see the Canadian submission to the first WAES technical report.

7.2 METHODOLOGY

7.2.1 Introduction

This chapter provides estimates of Canadian energy demand and supplies for the year 2000 under the four WAES scenarios. In the preparation of this study, advice and background information were sought from various representatives of Energy, Mines and Resources Canada, Statistics Canada, and the National Energy Board of Canada.

The basic tool used to prepare the Canadian submission to WAES was an integrated macroeconomic energy demand forecasting model developed by Energy, Mines and Resources Canada. Extensive use was also made of their published report, *An Energy Strategy for Canada: Policies for Self-Reliance* of April 1976. Use was also made of reports dealing with natural gas and oil issued by the National Energy Board: *Canadian Natural Gas—Supply and Requirements* of April 1975, and *Canadian Oil—Supply and Requirements* of September 1975. In addition, a background study, *Energy Conservation*, prepared for the Science Council of Canada and published in July 1975, was used as a basis for quantifying the effects on energy consumption of government-induced conservation measures. The 1972 base year estimates were derived from a Statistics Canada publication, *Detailed Energy Supply and Demand in Canada.*

7.2.2 Methodology Used in Projecting Energy Demand to 2000

Demand Forecasting Model

An integrated macroeconomic energy demand forecasting model developed by Energy, Mines and Resources Canada was used initially for providing the various scenarios requested. This analytical framework projects energy demands on the basis of historical relationships between energy use and demographic and economic activity. However, it does allow substantial scope for the adjustment of these historical relationships on the basis of anticipated reactions to changing energy prices and technology. With regard to the effect of increases in future Canadian energy price levels, it is recognized that only limited historical information is available, so that the quantification of these price effects is a difficult task. Future pricing relationships among competing energy sources are difficult to assess, as is the potential for interfuel substitutions. These critical areas where uncertainties are most pronounced are the subjects of ongoing investigation.

The projections of future Canadian energy requirements presented in this report are internally consistent, in the sense that once total energy requirements are estimated for each end-use sector of the economy, the demands for all energy sources are determined on a "market share" basis, which ensures that the sum of the energy sources equals total energy demands.

The relationships used in projecting energy demands are more detailed than simple energy/GNP or energy/population ratios. In the residential sector, for example, energy demand depends upon such factors as personal disposable income, the number of households, the split between single-family and multifamily housing, and the relative price of energy. Demand in the commercial sector also depends on the nature of the housing stock since energy used by large multifamily dwellings is priced at commercial rates and included as commercial demand. Commercial sector energy requirements also depend on the level of retail trade and the energy price. Demand for energy in the industrial sector depends not only on price and industrial production, but also on the capital/labor mix employed in the production process. In general, the relationships used reflect the structural characteristics of the sector examined to the greatest extent possible.

It should be noted that the forecasting model does not take into account explicit energy conservation measures, although the projections do reflect a more efficient use of energy that might reasonably be expected as a response to higher energy prices.

Use of the Demand Forecasting Model in the Year 2000 Studies

The integrated macroeconomic energy demand forecasting model developed by Energy, Mines and Resources Canada was used initially for estimating the impact of the various scenarios requested by WAES for the 1985 to 2000 period. Since estimates for the exogenous variables were not available for scenarios D-7 and D-8, it was necessary to extrapolate energy requirements by sector over the last decade to obtain the projections to the year 2000.

Economic Growth Rate

As with the 1985 scenarios, two global economic growth rates were established by the Workshop for the 2000 scenarios. These parameters were intended not as forecasts but as order-of-magnitude indicators to guide national teams in making estimates of national growth rates.

WAES assumed a high global economic growth rate of 5 percent per year for the 1985 to 2000 period. The corresponding Canadian high economic growth variable was assumed to be 5.2 percent per year. This is equivalent to the high growth parameter used for the 1981 to 1985 period and reflects the average growth rate experienced in the 1960s.

Scenarios C-1 and C-2 are based on Canada's sustaining a high rate of economic growth (5.2 percent) over the next 25 years. Based on historical experience, this rate of growth will certainly not continue for such an extensive period of time. In addition, these scenarios reflect a bias toward significant growth in electricity demand at the expense of fossil fuels. Scenarios C-1 and C-2 should be viewed as representing an upper range for Canadian energy requirements in the year 2000.

For the low economic growth scenarios (D-7 and D-8), Canadian annual economic growth rates of 3.6 percent for the 1980s and approximately 3 percent for the 1990s were assumed.

Energy Prices

To cushion the Canadian economy against major disruptions resulting from the very rapid increases in world oil prices, a two-price system has been adopted. Government policy is to raise the Canadian price of oil gradually, so that by 1985 the internal Canadian price will have reached the level of international oil prices.

The Workshop methodology calls for assumptions on generalized energy prices rather than oil prices in the 1985 to 2000 period, since oil will probably play a declining role in total energy supply during this period and thus will no longer be the major determinant of energy price. The constant price scenarios in the 1985 to 2000 period correspond to the mid-1975 price of the 1985 scenarios (i.e., $11.50 per barrel in constant 1975 U.S. dollars), and the higher price for the 1985 to 2000 period corresponds to the higher price in the 1985 scenarios (i.e., $17.25 per barrel).

For a detailed discussion of oil price assumptions used in the Canadian WAES scenarios to 1985, see section 3.3.3 of chapter 3 in the second WAES technical report.

Principal Replacement Fuels

WAES agreed that the 1985 to 2000 period may be characterized by a decline of oil and gas as the world's major sources of energy. It was noted that the only fuels available in sufficient quantities to fill the "prospective gap" resulting from the relative global decline of oil and gas supplies will be coal and nuclear energy.

The Canadian response to the proposed use of nuclear power as a principal replacement fuel is demonstrated in scenarios C-2 and D-8, in which it was assumed that nuclear-generated electricity would fill an increasingly greater share of Canadian electricity requirements. In scenarios C-1 and D-7, on the contrary, it was assumed that the substantial increase in demand for nuclear-generated electricity would not occur but would be replaced by a demand for electricity generated by coal. This substitution was implemented to the extent allowed by domestic coal production.

It should be noted that the above assumptions, particularly with regard to coal, were made in an effort to meet the criteria established by WAES and should not be interpreted as a forecast of the likely course of events in the Canadian energy scene.

7.2.3 Canadian Energy Supplies

Canadian supply estimates for 1985 and 2000 were based on the criteria outlined by WAES.

Projected Oil and Natural Gas Supplies

In evaluating the data supplied for WAES, it is important to recognize the highly speculative nature of the projections of exploitable hydrocarbons in the frontier areas. While it is a reasonable assumption that large volumes of oil and natural gas may exist, the fortuitous combination of geological factors required for the existence of major reserves is by no means guaranteed. Moreover, major questions with regard to environmental considerations, native land claims, provincial development policies, federal-provincial jurisdictional considerations, and transportation systems need to to be answered before the delivery of hydrocarbons from the frontier areas can begin. These considerations will play a major role in the development and timing of frontier resources. Thus, the supply projections, insofar as they relate to these areas, must be regarded as highly speculative.

It must also be recognized that the methodology adopted by WAES does not explicitly take into account the possibility of the national government's imposing export constraints on indigenous supplies that are surplus to national requirements. Therefore, the balance sheets for Canada assume that any surpluses resulting from the supply-demand balance will be available for export.

Moreover, adjustments have not been made to indigenous supply estimates to allow for contractual export commitments of natural gas, estimated at 900 BCF in 1985. Under the WAES methodology, it was assumed that after domestic requirements were met, any available surplus would be exported.

The Canadian Coal Industry

Since the early 1970s the coal industry has changed dramatically in Canada. After a severe production decline in the 1950s and 1960s as a result of conversion of the railways to diesel fuel and the expanded use of natural gas in the residential, commercial, and industrial sectors, the coal industry in Canada is again enjoying a period of expansion attributable to expanded domestic use of thermal coal supplies and long-term export contracts of metallurgical coal.

In Canada, each province has jurisdiction over its own resource developments, so that supplies of coal will obviously depend on the favorable resolution of a number of issues, and should be qualified accordingly. For example, recent concerns about the depletion of oil and gas reserves have caused western producing provinces to adopt a policy of restraint in coal development. The provinces of British Columbia and Alberta, which have the bulk of Canada's coal resources, have reassessed the environmental impact of coal operations in order to define clearly the areas where mining may be permitted and under what conditions. A report has been issued by the Environmental Conservation Authority of Alberta on land use and resource development in the eastern slopes of the Rocky Mountains. The government of Alberta is currently considering this report. In the meantime, no final approvals for coal development have been granted. Increases in royalties in the producing provinces have either been implemented or are being contemplated.

Although extensive exploratory and development drilling programs have been carried out in many areas, this period of reassessment has effectively resulted in a cessation of new coal developments since 1974. In mid-1976 the government of Alberta adopted a new coal development policy and royalty schedules. New coal policies are expected in the near future from the government of British Columbia.

Generation of Electricity

The expansion of electrical systems is primarily determined by considerations of meeting anticipated regional market demands, with acceptable reserve margins, in the most efficient manner possible. An electricity supply projection must therefore be based principally on a projection of future demand growth. The demand for electricity, in turn, depends on such factors as income, population, the availability and price of substitute fuels, capital costs of energy-using equipment, technology, convenience factors, and of course, the price of electricity. This price, in turn, will be affected by capital requirements for capacity expansion, financial characteristics of the utility concerned, and decisions of regulatory agencies.

For the 1985 scenarios, two projections of installed nuclear generation capacity were made, based on forecasts provided by Energy, Mines and Resources Canada in its publication *1975 Assessment of Canada's Uranium Supply and Demand* of June 1976. In order to obtain the high estimate for the 1985 scenario, it was assumed that the total in-service capacity forecast for 1986 would apply in 1985. For the low estimate, it was assumed that the total projected in-service capacity for 1984 would apply in 1985.

Two projections of installed hydroelectric capacity were made for the 1985 scenarios, based on estimates from *An Energy Strategy for Canada: Policies for Self-Reliance* of April 1976.

Other Energy Sources

The economics of large-scale commercial coal gasification and liquefaction plants have not yet been clearly defined. The potential to introduce large-scale (250 MCF/day) synthetic natural gas plants in Canada appears to be limited by the small number of economically mineable coal deposits large enough to support operations of this scale.

The potential for renewable energy sources, such as solar, wind, geothermal, biomass, and tidal power, has yet to be evaluated properly in Canada. Thus, while renewable energy technologies may become cost-competitive with conventional energy sources by the year 2000, the extent to which they will contribute to meeting primary energy demand is expected to be marginal. These, therefore, have been considered to be outside of the WAES time frame and will probably contribute 3 percent or less of total energy requirements.

7.3 SUPPLY-DEMAND INTEGRATIONS

7.3.1 Introduction

Detailed presentations of Canadian energy demand to 1985 and of supply to 2000 corresponding to the various WAES scenarios are included in the first two WAES technical reports. In this section the demand and supply scenarios are reviewed in very general terms; then the supply-demand integrations are presented.

7.3.2 Energy Demand Scenarios to 1985

General Observations

The 1985 energy demand projections were determined in order to indicate a range of reasonable possibilities for Canadian energy consumption in 1985. The results are presented graphically in Figure 7.3.

The range between high and low projected energy use for the five 1985 cases is about 1450×10^{12} Btu or about 17 percent of the total requirements. This range represents the overall effects of detailed assumptions of high and low economic growth rates, three levels of oil prices, and the estimated extent of government policy initiatives in the energy field. Thus, the potentials for energy savings through variations of these parameters are included. It must be borne in mind, of course, that none of the scenarios presented in this report will

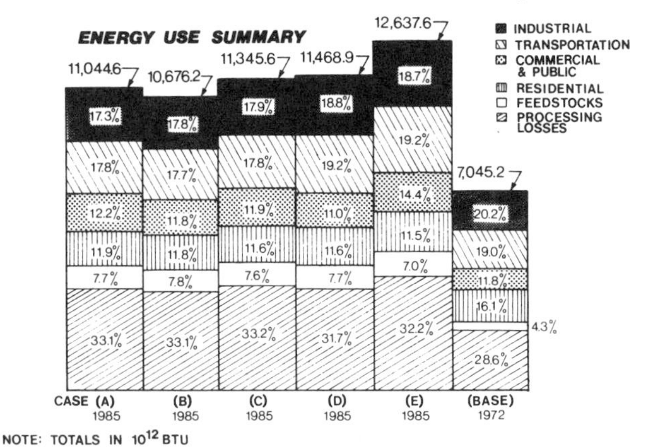

Figure 7.3
WAES Canadian energy use summary, 1972 and 1985.

necessarily occur. The WAES scenario approach attempts merely to indicate the direction in which energy demand might develop under the assumptions outlined.

The projections indicate a substantial growth in fossil fuel requirements in the feedstocks sector (i.e., nonenergy use) to 1985. A large growth of electricity use in the residential and commercial sectors from 1972 to 1985 is also indicated, while direct use of fossil fuels in these sectors actually declines in one scenario.

The distribution by end-use sector is relatively stable across all scenarios and somewhat stable over time. The share of the residential and industrial sectors is projected to decline over time as the relative importance of the transportation and feedstocks sectors increases. Part of the decline in the residential sector is attributable to the relative increase in multifamily dwelling units, many of which purchase energy at commercial rates and are accordingly classified in the commercial sector. In addition, a significant decrease in the residential sector's share of fossil fuels reflects the large growth of electricity use in this sector.

By using the national policy response variable, a preliminary indication of the potential for conservation in the Canadian economy can be obtained. The studies show that this is clearly a major determining factor of energy demand.

In general, the scenarios indicate that the conservation possibilities through price and nonprice effects could be significant in the next decade. Based on the results of the various scenarios, it would appear that while the price effects on energy demand are considerable, a government-induced nonprice conservation program could further reduce the growth in energy demand in the next decade.

As might be expected, the impact of the price and nonprice effects varies by sector. The scenarios indicate, for example, that the price effects in the commercial sector could account for virtually all of the energy savings. In the residential sector, energy savings resulting from price and nonprice effects would be approximately equal. In the transportation sector, energy savings from government-induced nonprice conservation efforts could have more than twice the impact of price effects.

In 1985 scenarios A, B, and C, the assumed savings by sector resulting from national conservation efforts combined with price effects are as shown in Table 7.1.

Review of Scenarios

The sectoral totals for final energy requirements show that scenario B—low economic growth coupled with high oil price and vigorous national policy response—results in the lowest final energy demand, while scenario E—high economic growth coupled with low oil price and restrained national policy response—results in the largest final energy demand. Scenarios C and D, the constant oil price cases, are approximately midway between scenario B and scenario E for energy use.

The projection results indicate that the improvement in energy efficiencies through a vigorous national policy response would have more of an impact on reducing energy demand than would lower economic growth. This probably results from the assumption that all differences between the two economic profiles occur after 1981, whereas the effect of vigorous national policy response begins almost immediately.

The "processing losses" (i.e., those losses incurred in the conversion of primary energy into energy delivered to the consumer) indicated in the energy use summary worksheet in section 7.4 result from national supply-demand balances and are further modified following the global supply-demand integration process. However, it is thought that these figures represent a close-to-final estimate for this category.

The energy growth rates for 1972 to 1985 indicate some general trends. The most striking is the large increase in fossil fuel requirements in the feedstocks sector (i.e., nonenergy use).

Another noticeable trend is the large growth of electricity use in the residential and commercial sectors from 1972 to 1985, while direct use of fossil fuels in these sectors actually declines in one case. For scenario C, electricity use is projected to be 26 percent of the total energy demand in the residential sector and nearly 32 percent of the total energy demand in the commercial sector in 1985. In 1972, the electricity share was 15 percent of the total energy demand in the residential sector and 23 percent of the total energy demand in the commercial sector.

Table 7.1
Energy conservation estimates in 1985 (TBOE)[a]

	Case A	Case B	Case C
Transport	126	119	129
Industry	55	54	58
Commercial and public	4	6	4
Residential	34	33	34
Nonenergy uses	21	21	21
Final energy demand	240	228	246
Percent reduction	6.4%	6.3%	6.4%

[a]Savings in response to a vigorous national conservation policy. Conversion factor: 5.8×10^6 Btu per barrel of oil.

7.3.3 Energy Demand Scenarios to 2000

Canadian energy demand estimates for the year 2000 are projected under the four WAES scenarios. The range of projected demand is 1692×10^{12} Btu or approximately 8 percent of total demand in the year 2000. The range results from varying assumptions on economic growth rates, energy prices, and the extent of government policy initiatives in the energy field.

The projections indicate an increase in secondary energy demand for electricity relative to fossil fuel demand. For example, electricity demand by end-use sector is estimated at 20.4 percent of total secondary requirements in 2000 as compared with 17.9 percent for the comparable 1985 scenario.

In the high economic growth scenarios for the 1985 to 2000 period, Canadian primary energy demand is projected to grow at 4 percent per year. In the low economic growth scenarios, primary energy requirements are projected to grow at 3.3 percent per year. While it is anticipated that additional government policy initiatives could further reduce these growth rates of energy demand, the degree to which this is possible is not demonstrated in this study.

7.3.4 Energy Supplies to 2000

A summary of detailed projections of Canadian energy supply prospects by fuel type for five scenarios in 1985 and four scenarios in the year 2000 is given in section 7.4.

The assessment of oil, natural gas, and coal supplies is based on a variety of assumptions selected to reflect the criteria established by WAES. The supply estimates and production schedules were calculated in light of world economic growth, energy price, and national policy response. The supply estimates for nonfossil fuels and their conversion processes to other energy forms were derived from available published data and information supplied by knowledgeable professionals.

It is important to recognize the highly speculative nature of the forecasts of ultimately recoverable fossil fuels. While it is reasonable to assume that large volumes of oil, natural gas, and coal do in fact exist, the fortuitous combination of geological factors required for the existence of major reserves is by no means guaranteed. Canada's frontier areas in particular have simply not undergone sufficient development to allow conclusive evaluation of the hydrocarbon potential. Thus, the results, so far as they relate to these fuels, must be regarded as highly speculative.

General Review of Supply Assessments

Canada is relatively well endowed with all five main energy resources: coal, oil, gas, hydro power, and uranium. The direction of future resource development to meet growing requirements remains unclear. Whether the estimated energy supply under the WAES scenarios to 1985 and 2000 will in fact materialize and render Canada self-sufficient in these resources will be contingent on costs of development and availability of capital, materials, technology, and labor. It is apparent that increasing Canadian self-reliance will require the development of additional supplies of indigenous oil and of other energy sources that can be substituted for oil imports. This chapter does not deal with energy development in response to national requirements. With the exception of electrical generation, it only analyzes and projects the maximum potential availability of each energy resource under the various scenario assumptions.

The Btu equivalents of the expected annual production rates for the various energy sources are summarized in section 7.4. The expected annual production rates of oil and natural gas in the year 1985 vary from a low in scenario E to a high in scenarios A and B. This variation is essentially the effect of the high price structure in scenarios A and B. In the year 2000 the production rates vary from a low in scenario D-8 to a high in scenario C-1, again reflecting sensitivity to high energy prices. Similarly, coal production varies from a low in scenario E to a high in scenarios A and B in the year 1985 in response to high energy prices. In the year 2000 coal production varies from a low in scenarios C-2 and D-8 to a high in scenarios C-1 and D-7, reflecting increased use of coal as a principal replacement fuel in the latter cases.

This study shows that the key factor affecting expected annual production rates in 1985 is the price of energy. A lesser but still important factor is national policy response.

National policy response will be governed by domestic energy demands and will require serious consideration and cooperation by various levels of government and energy resource industries. The implementation of vigorous national policy initiatives to obtain relatively high levels of production must allow for proper lead times in the development of energy resources.

In the scenarios to the year 2000, energy prices are maintained at constant or high levels after 1985. National policy response varies from restrained to vigorous in individual scenarios. An additional policy variable is the choice of coal or nuclear energy as the principal replacement fuel. In the current main producing areas, the supplies of conventionally produced crude oil and natural gas decline in all scenarios. It is only in the frontier regions that these supplies show increases, attributable to the anticipated development in these areas by 1985. Coal, uranium, and nuclear supplies show substantial increases by the year 2000. Other sources such as solar and wind energy are not expected to contribute significantly to Canada's energy supplies, accounting for 3 percent or less of total energy requirements.

Electrical Generating Capacity

Electrical systems are usually expanded to meet projected increases in regional market demands. To this is generally added a reserve margin to ensure that peak load demand is met. Any increases in generating capacity must of necessity be at prices and efficiencies comparable to those of other energy forms that could be substituted for it. It is therefore important that as close a balance as possible be struck between the electrical energy supply availability and the projected demand at a given price.

The main factors that dictate an increase in electrical demand, and therefore an increase in the generating capacity required to meet that demand, are usually the prices and availability of substitute fuels in the market area, income, population, available technology, convenience, the capital requirements of the proposed expansion, and finally the price at which this electricity will be made available to the consumer. This latter will be affected mainly by the financial characteristics of the utility systems and decisions of regulatory agencies.

Based on forecasts provided by Energy, Mines and Resources (EMR) Canada in its *1975 Assessment of Canada's Uranium Supply and Demand*, released in June 1976, two nuclear generating stations are assumed for the 1985 scenarios. Scenario E, with its low price assumption and resulting high electricity demand, is compared to the high growth EMR scenarios, yielding a generating capacity requirement of 14.7×10^3 MW. The high and medium price scenarios, A, B, C, D, and their mid-1975 projected electricity demand are compared to the EMR scenarios showing a low growth of nuclear generating capacity, estimated to be 12.5×10^3 MW. The 1986 forecasts provided in the EMR publication are assumed to apply for 1985 in order to obtain the high demand estimates for the 1985 scenarios. For estimates of 1985 low demand scenarios, the 1984 EMR figures were assumed to be valid.

The two projections of hydroelectric capacity requirements for the 1985 secnarios were based on *An Energy Strategy for Canada: Policies for Self-Reliance* of April 1976. It is estimated that 61.9×10^3 MW of hydroelectric capacity will be required to meet demand in the high growth cases and 55.5×10^3 MW to meet that in the low growth cases.

It is generally accepted by most members of WAES that oil and gas will gradually lose their importance as the world's major source of energy as we get closer to the year 2000. If the "prospective gap" between desired fuels and their availability is to be filled, it is reasonable to assume that only fuels that can be made available in sufficient quantity and are of a quality acceptable to the user can be used. Although there are some challenges to be met, at present only coal and nuclear energy appear to meet these criteria.

In scenarios C-2 and D-8 it is assumed that nuclear generated electricity will fill an increasing share of Canadian electricity requirements. In scenarios C-1 and D-7 coal was chosen as the main replacement fuel, thereby maximizing its potential for electricity generation and minimizing the electricity generated from nuclear fuel. This substitution was allowed to take place only to the extent that domestic coal production could meet this demand.

In the year 2000 scenarios, two projections of nuclear and hydro electrical capacities have been considered. The high and low nuclear generating capacities are estimated at 74.5×10^3 MW and 45.0×10^3 MW, respectively. Hydroelectric generating capacities for the high and low cases are estimated at 74.0×10^3 MW and 60.0×10^3 MW, respectively.

A quick review of the actual plans of electric utilities for increasing electrical generating capabilities reveals several interesting trends. First, increased emphasis will be placed on generating electricity from nuclear fuel, particularly toward the end of the twentieth century. Second, in the Prairie regions, because of the proximity to major coal deposits, electrical generation will use coal almost exclusively as its primary fuel. Third, the use of natural gas for power generation will be virtually eliminated, and its uses will be limited to meeting domestic demand and certain speciality purposes.

7.3.5 Results of Supply-Demand Balances

In considering the results of the supply-demand balancing procedure, it is important to bear in mind that the scenarios for 1985 and 2000 are based on specific criteria, some of which involve a high degree of uncertainty.

The results of the supply-demand balances for 1985 and 2000 are presented graphically in Figures 7.1 and 7.2. The Canadian worksheet in section 2.6 of chapter 2 outlines the numbers used in preparing those figures. The worksheet also shows the corresponding data for 1972, the base year for this study.

From this worksheet, it may be seen that the supply-demand balance for scenario E (assuming high economic growth, low oil price, and a restrained government policy) indicates that Canada could require major imports of coal and oil and minor imports of natural gas in 1985 in order to meet domestic requirements. Exports of electricity could occur in that year.

The supply-demand balance for scenario D (assuming low economic growth, mid-1975 oil price, and a restrained government policy) indicates that Canada's energy supplies and requirements could be roughly in balance in 1985.

The supply-demand balance for scenario C (assuming high economic growth, mid-1975 oil price, and vigorous government policy initiatives on both demand and supply) indicates that Canada could be an exporter of fossil fuels and electricity in 1985.

Under the assumptions of scenarios A and B (high or low economic growth, high oil price, and vigorous government policy initiatives on both demand and supply), Canada could become a major exporter of fossil fuels by 1985.

As noted in section 7.1, in the study of scenarios A, B, C, and D it was assumed that the production of oil from Canada's frontier areas would begin by 1985. This assumption must await confirmation by ongoing exploration and development and is therefore, at this time, highly speculative. If all Canadian frontier oil production were excluded from the supply scenarios for 1985, the supply-demand balances would show an oil deficit (i.e., a requirement for oil imports) in all scenarios.

With respect to natural gas, the exclusion of frontier supplies could cause a deficit in the supply-demand balance in scenario D, make the surplus in scenario C very marginal, and reduce the surplus in scenarios A and B to just over 1 TCF per year. It must be emphasized that scenarios A, B, and C assume vigorous government energy conservation measures.

For the year 2000 scenarios, the balances presented in the energy supply-demand integration worksheet in chapter 2 indicate that Canada could become a major importer of oil and natural gas in the year 2000. This results in spite of the bias exhibited on the demand side whereby electricity requirements increase at the expense of fossil fuels.

7.4 SUPPLY-DEMAND INTEGRATION WORKSHEETS

This section contains the WAES supply-demand integration worksheets for the cases presented in this chapter. The energy use summary (1972–1985) and the supply summary (1972–2000) worksheets are also included for comparison. The worksheets appear in the following sequence:

Energy use summary, 1972–1985

Energy supply summary, 1972–2000

Energy demand summary, 1972–1985–2000

National input worksheets for supply/demand integration (10 sheets, 1 each for the base year, 1972, the five WAES 1985 scenarios, and the four WAES 2000 scenarios)

WAES NATIONAL ENERGY DEMAND STUDIES, 1972-1985
ENERGY USE SUMMARY

Country CANADA

Scenario Assumptions

WAES Case	A		B		C		D		E			
GNP 1976-1985	High		Low		High		Low		High			
Oil Price*	$11.50 → $17.25		$11.50 → $17.25		$11.50		$11.50		$11.50 → $7.66			
Natl. Policy Response	Vigorous		Vigorous		Vigorous		Restrained		Restrained			
Natl. GNP Growth Rate 1976-1985 (%/yr.) **	5.4 / 5.2		5.4 / 3.6		5.4 / 5.2		5.4 / 3.6		5.4 / 5.2		1972 Base Year	
Fuel Type	Fossil Fuel	Elec.	Fossil Fuel	Elec.	Fossil Fuel	Elec.	Fossil Fuel	Elec.	Fossil Fuel	Elec.	Fossil Fuel	Elec.
Units	10^{12} Btu		10^{12} Btu		10^{12} Btu		10^{12} Btu		10^{12} Btu		10^{12} Btu	
Energy-Using Sectors												
Industrial (excluding feedstocks)	1357.8	558.3	1348.4	556.0	1442.0	593.7	1518.4	626.8	1666.4	699.6	1070.9	355.1
Transportation	1972.7	0.0	1886.9	0.0	2025.4	0.0	2196.9	0.0	2415.9	0.0	1339.3	0.0
Commercial and Public	921.9	426.4	854.3	395.9	923.7	426.4	857.0	396.3	989.8	454.1	636.3	192.4
Residential	964.4	333.1	932.8	323.1	976.8	337.4	996.5	345.1	1083.7	374.3	960.3	171.3
Fishing, Agriculture, Mining, Construction												
Feedstocks	850.1	0.0	842.2	0.0	850.1	0.0	886.5	0.0	894.8	0.0	301.6	0.0
Final Energy Demand (excluding processing losses)	6066.9	1317.8	5864.6	1275.0	6218.0	1357.5	6455.3	1368.2	7050.6	1528.0	4308.4	718.8
Processing Losses ***	3659.9		3536.6		3770.1		3645.4		4059.0		2018.0	
Primary Energy Input	11044.6		10676.2		11345.6		11468.9		12637.6		7045.2	

*Oil price 1975 US$ per barrel Arabian light crude oil FOB Persian Gulf.

** Upper figure applies to 1976-80; lower figure applies to 1981-85

*** Disaggregation provided in the supply/demand balance worksheets

WAES NATIONAL ENERGY SUPPLY STUDIES, 1972-2000
ENERGY SUPPLY SUMMARY

Country: CANADA

Scenario Assumptions **		1972	1985					2000			
WAES CASE		Base Year	A	B	C	D	E	C-1	C-2	D-7	D-8
GNP 1976-85; 1985-2000		---	High	Low	High	Low	High	High	High	Low	Low
Oil/Energy Price*		---	11.50-17.25	11.50-17.25	11.50	11.50	11.50-7.66	11.50-17.25	11.50-17.25	11.50	11.50
National Policy Response		---	Vig	Vig	Vig	Res	Res	Vig	Vig	Res	Res
Principal Replacement Fuel		---	--	--	--	--	--	Coal	Nuclear	Coal	Nuclear
Energy Supply	**Units**										
Crude Oil	10^{15}Btu	3.797	6.769	6.769	6.015	5.087	3.335	6.479	6.479	5.440	5.440
Natural Gas	10^{15}Btu	2.324	3.809	3.809	2.794	2.370	2.277	1.842	1.842	1.542	1.542
Coal	10^{15}Btu	.437	1.411	1.411	1.134	.958	.832	3.024	2.520	3.024	2.520
Nuclear)	10^{15}Btu	1.867	.820	.820	.820	.820	.970	2.950	4.890	2.950	4.890
Hydro-electric)			2.915	2.915	2.915	2.915	3.250	3.900	3.900	3.150	3.150
Geothermal											
Solar Thermal											
Solar Electric											
Total Energy Supply		8.425	15.724	15.724	13.679	12.150	10.664	18.195	19.631	16.106	17.542

*1975 US dollars per barrel of Arabian light crude oil, fob Persian Gulf.
All Figures are primary production prior to any processing or conversion.

Footnotes:

**Oil and gas supply estimates assume reserves will be found in lightly explored frontier areas.

WAES NATIONAL ENERGY DEMAND STUDIES, 1972-1985-2000
ENERGY DEMAND SUMMARY

Country: CANADA

Scenario Assumptions		1972	1985					2000			
WAES CASE		Base Year	A	B	C	D	E	C-1	C-2	D-7	D-8
GNP 1976-85; 1985-2000		---	High	Low	High	Low	High	High	High	Low	Low
Oil/Energy Price*		---	11.50-17.25	11.50-17.25	11.50	11.50	11.50-7.66	11.50-17.25	11.50-17.25	11.50	11.50
National Policy Response		---	Vig	Vig	Vig	Res	Res	Vig	Vig	Res	Res
Principal Replacement Fuel		---	--	--	--	--	--	Coal	Nuclear	Coal	Nuclear
Primary Energy Demand	**Units**										
Oil	10^{15}Btu	3.359	4.516	4.353	4.608	4.795	5.238	7.783	7.783	6.600	6.600
Natural Gas	10^{15}Btu	1.315	2.184	2.118	2.270	2.292	2.535	4.044	4.044	3.900	3.900
Coal	10^{15}Btu	.591	.937	.909	.966	.992	1.083	2.701	1.585	3.000	1.500
Nuclear)	10^{15}Btu	1.780	3.407	3.297	3.502	3.391	3.782	5.814	6.930	5.150	6.650
Hydroelectric)											
Geothermal											
Solar Thermal											
Solar Electric											
TOTAL PRIMARY ENERGY DEMAND		7.045	11.044	10.677	11.346	11.470	12.638	20.342	20.342	18.650	18.650

*1975 US dollars per barrel of Arabian light crude oil, fob Persian Gulf.
All Figures are primary production prior to any processing or conversion.

NATIONAL INPUT WORKSHEET FOR SUPPLY/DEMAND INTEGRATION

Country: CANADA

Year:

Case: 1972 - Base Year

Units: 10^{12} Btu

	(a) Coal	(b) Petroleum	(c) (Syn. Liquids)	(d) Nat. Gas	(e) (Syn. Gas)	(f) (Heat)	(g) (Electricity)[††]	(h) Hydro Energy	(i) Nuclear Energy	(j) Geothermal energy and other	(k) Total
(1) Transport	4.4	1334.9		0.0			0.0				1339.3
(2) Industry	216.0	440.9		414			355.1				1426.0
(3) Agric., Mining, Construction											
(4) Commercial	5.0	365.1		266.2			192.4				828.7
(5) Public											
(6) Residential	11.7	667.4		281.2			171.3				1131.6
(7) Non-energy Uses	0.0	226.6		75.0							301.6
(8) Final Energy Demand*	237.1	3034.9		1036.4			718.8				5027.2
(9) Electricity**	353.1	107.3		140.			(-790)	1780.3			1590.7
(10) Syn. Gas											
(11) Syn. Liquids											
(12) Heat											
(13) Energy Sector Self Consumption & conversion losses	0.5	216.7		138.9			71.2				427.3
(14) Primary Energy Input	590.7	3358.9		1315.3			-	1780.3			7045.2
(15) Indigenous Supply	436.9	3797		2324.3				1866.8			8425
(16 Imports***	475.7	1964						23.8			2463.5
(17) Exports	249.6	2333		1009				110.4			3702

*Includes non-energy uses.

**Blocks 9g, 10e 11c, and 12f must be negative to avoid double counting of energy from primary fuels.

***Includes imported products.

††Includes only electricity to be purchased by each sector.

NATIONAL INPUT WORKSHEET FOR SUPPLY/DEMAND INTEGRATION

Country: CANADA

Year:

Case: 1985 - Case A

Units: 10^{12} Btu

	(a) Coal	(b) Petroleum	(c) (Syn. Liquids)	(d) Nat. Gas	(e) (Syn. Gas)	(f) (Heat)	(g) (Electricity)††	(h) Hydro Energy	(i) Nuclear Energy	(j) Geothermal energy and other	(k) Total
(1) Transport	1.7	1971.0		0.0			0.0				1972.7
(2) Industry	295.7	378.9		683.2			558.3				1916.1
(3) Agric., Mining, Construction											
(4) Commercial) (5) Public)	3.7	459.2		459.0			426.4				1348.3
(6) Residential	4.6	504.4		455.4			333.1				1297.5
(7) Non-energy Uses		613.5		236.6							850.1
(8) Final Energy Demand*	305.7	3927.0		1834.2			1317.8				7384.7
(9) Electricity**	630.6	167.5		117.4			(-1434.5)		3407.1		2888.1
(10) Syn. Gas											
(11) Syn. Liquids											
(12) Heat											
(13) Energy Sector Self Consumption & conversion losses	1.0	421.6		232.5			116.7				771.8
(14) Primary Energy Input	937.3	4516.1		2184.1			-		3407.1		11044.6
(15) Indigenous Supply****	1411	6769		3809				2915	820		15724
(16 Imports***											
(17) Exports	473.7	2252.9		1624.9					327.9		4679.4

*Includes non-energy uses.

**Blocks 9g, 10e 11c, and 12f must be negative to avoid double counting of energy from primary fuels.

***Includes imported products.

††Includes only electricity to be purchased by each sector.

Footnotes:

****Oil and gas supply estimates assume deliverability from lightly explored frontier areas.

NATIONAL INPUT WORKSHEET FOR SUPPLY/DEMAND INTEGRATION

Country: CANADA

Year:

Case: 1985 - Case B

Units: 10^{12} Btu

	(a)	(b)	(c)	(d)	(e)	(f)	(g)	(h)	(i)	(j)	(k)
	Coal	Petroleum	(Syn. Liquids)	Nat. Gas	(Syn. Gas)	(Heat)	(Electricity)[††]	Hydro Energy	Nuclear Energy	Geothermal energy and other	Total
(1) Transport	1.6	1885.3		0.0			0.0				1886.9
(2) Industry	290.6	377.3		680.5			556.0				1904.4
(3) Agric., Mining, Construction											
(4) Commercial)	3.4	428.3		422.6			395.9				1250.2
(5) Public)											
(6) Residential	4.4	488.7		439.7			323.1				1255.9
(7) Non-energy Uses		605.6		236.6							842.2
(8) Final Energy Demand*	300.0	3785.2		1779.4			1275.0				7139.6
(9) Electricity**	607.9	161.8		112.8			(-1387.8)		3296.9		2791.6
(10) Syn. Gas											
(11) Syn. Liquids											
(12) Heat											
(13) Energy Sector Self Consumption & conversion losses	1.0	405.9		225.3			112.8				745
(14) Primary Energy Input	908.9	4352.9		2117.5			-		3296.9		10676.2
(15) Indigenous Supply****	1411	6769		3809				2915	820		15724
(16 Imports***											
(17) Exports	502.1	2416.1		1691.5					438.1		5047.8

*Includes non-energy uses.

**Blocks 9g, 10e 11c, and 12f must be negative to avoid double counting of energy from primary fuels.

***Includes imported products.

††Includes only electricity to be purchased by each sector.

Footnotes:

****Oil and gas supply estimates assume deliverability from lightly explored frontier areas.

NATIONAL INPUT WORKSHEET FOR SUPPLY/DEMAND INTEGRATION

Country: CANADA

Year:

Case: 1985 - Case C

Units: 10^{12} Btu

	(a) Coal	(b) Petroleum	(c) (Syn. Liquids)	(d) Nat. Gas	(e) (Syn. Gas)	(f) (Heat)	(g) (Electricity)††	(h) Hydro Energy	(i) Nuclear Energy	(j) Geothermal energy and other	(k) Total
(1) Transport	1.8	2023.6		0.0			0.0				2025.4
(2) Industry	297.1	397.6		747.3			593.7				2035.7
(3) Agric., Mining, Construction											
(4) Commercial)	3.7	458.2		461.8			426.4				1350.1
(5) Public)											
(6) Residential	4.7	512.6		459.5			337.4				1314.2
(7) Non-energy Uses		613.5		236.6							850.1
(8) Final Energy Demand*	307.3	4005.5		1905.2			1357.5				7575.5
(9) Electricity**	657.3	172.0		123.1			(-1477.7)	3502.1			2976.8
(10) Syn. Gas											
(11) Syn. Liquids											
(12) Heat											
(13) Energy Sector Self Consumption & conversion losses	1.0	430.3		241.8			120.2				793.3
(14) Primary Energy Input	965.6	4607.8		2270.1			-	3502.1			11345.6
(15) Indigenous Supply****	1134	6015		2795				2915	820		13679
(16 Imports***											
(17) Exports	168.4	1407.2		524.9				232.9			2333.4

*Includes non-energy uses.

**Blocks 9g, 10e 11c, and 12f must be negative to avoid double counting of energy from primary fuels.

***Includes imported products.

††Includes only electricity to be purchased by each sector.

Footnotes:

****Oil and gas supply estimates assume deliverability from lightly explored frontier areas.

NATIONAL INPUT WORKSHEET FOR SUPPLY/DEMAND INTEGRATION

Country: CANADA

Year:

Case: 1985 - Case D

Units: 10^{12} Btu

	(a)	(b)	(c)	(d)	(e)	(f)	(g)	(h)	(i)	(j)	(k)
	Coal	Petro- leum	(Syn. Liquids)	Nat. Gas	(Syn. Gas)	(Heat)	(Electri- city)††	Hydro Energy	Nuclear Energy	Geothermal energy and other	Total
(1) Transport	1.9	2195.0		0.0			0.0				2196.9
(2) Industry	309.6	419.8		789.0			626.8				2145.2
(3) Agric., Mining, Construction											
(4) Commercial)	3.4	427.7		425.9			396.3				1253.3
(5) Public)											
(6) Residential	4.7	523.8		468.0			345.1				1341.6
(7) Non-energy Uses		637.5		249.0							886.5
(8) Final Energy Demand*	319.6	4203.8		1931.9			1368.2				7823.5
(9) Electricity**	671.1	176.5		125.5			(-1484.5)		3390.6		2879.2
(10) Syn. Gas											
(11) Syn. Liquids											
(12) Heat											
(13) Energy Sector Self Consumption & conversion losses	1.0	414.4		234.5			116.3				766.2
(14) Primary Energy Input	991.7	4794.7		2291.9			-		3390.6		11468.9
(15) Indigenous Supply****	958	5087		2370				2915	820		12150
(16 Imports***	33.7										33.7
(17) Exports		292.3		78.1					344.4		714.8

*Includes non-energy uses.

**Blocks 9g, 10e 11c, and 12f must be negative to avoid double counting of energy from primary fuels.

***Includes imported products.

††Includes only electricity to be purchased by each sector.

Footnotes:
****Oil and gas supply estimates assume deliverability from lightly explored frontier areas.

NATIONAL INPUT WORKSHEET FOR SUPPLY/DEMAND INTEGRATION

Country: CANADA

Year:

Case: 1985 - Case E

Units: 10^{12} Btu

	(a) Coal	(b) Petro-leum	(c) (Syn. Liquids)	(d) Nat. Gas	(e) (Syn. Gas)	(f) (Heat)	(g) (Electri-city)††	(h) Hydro Energy	(i) Nuclear Energy	(j) Geothermal energy and other	(k) Total
(1) Transport	2.1	2413.8		0.0			0.0				2415.9
(2) Industry	317.2	468.5		880.7			699.6				2366.0
(3) Agric., Mining, Construction											
(4) Commercial)	3.9	490.7		495.2			454.1				1443.9
(5) Public)											
(6) Residential	5.2	568.7		509.8			374.3				1458.0
(7) Non-energy Uses	-	645.8		249.0			-				894.8
(8) Final Energy Demand*	328.4	4587.5		2134.7			1528.0				8578.6
(9) Electricity**	753.0	197.7		141.3			(-1657.9)		3781.8		3215.9
(10) Syn. Gas											
(11) Syn. Liquids											
(12) Heat											
(13) Energy Sector Self Consumption & conversion losses	1.1	452.7		259.4			129.9				843.9
(14) Primary Energy Input	1082.5	5237.9		2535.4			-		3781.8		12637.6
(15) Indigenous Supply****	832	3335		2277				3250	970		10664
(16 Imports***	250.5	1902.9		258.4							2411.8
(17) Exports									438.2		438.2

*Includes non-energy uses.

**Blocks 9g, 10e 11c, and 12f must be negative to avoid double counting of energy from primary fuels.

***Includes imported products.

††Includes only electricity to be purchased by each sector.

Footnotes:
****Oil and gas supply estimates assume deliverability from lightly explored frontier areas.

NATIONAL INPUT WORKSHEET FOR SUPPLY/DEMAND INTEGRATION

Country: CANADA

Year:

Case: 2000 C-1

Units: 10^{12} Btu

	(a)	(b)	(c)	(d)	(e)	(f)	(g)	(h)	(i)	(j)	(k)
	Coal	Petroleum	(Syn. Liquids)	Nat. Gas	(Syn. Gas)	(Heat)	(Electricity)††	Hydro Energy	Nuclear Energy	Geothermal energy and other	Total
(1) Transport	0	4460		0			0				4460
(2) Industry	383	554		1530			1170				3637
(3) Agric., Mining, Construction											
(4) Commercial)	0	594		1035			990				2619
(5) Public)											
(6) Residential	1.4	369		660			612				1642.4
(7) Non-energy Uses		954		261							1215
(8) Final Energy Demand*	384.4	6931		3486			2772				13573.4
(9) Electricity**	2314	123		117			(-3006)	5814			5362
(10) Syn. Gas											
(11) Syn. Liquids											
(12) Heat											
(13) Energy Sector Self Consumption & conversion losses	2.3	729		441			234				1406.3
(14) Primary Energy Input	2700.7	7783		4044			--	5814			20341.7
(15) Indigenous Supply****	3024	6479		1842				3900	2950		18195
(16 Imports***		1304		2202							3506
(17) Exports	323.3							1036			1359.3

*Includes non-energy uses.

**Blocks 9g, 10e 11c, and 12f must be negative to avoid double counting of energy from primary fuels.

***Includes imported products.

††Includes only electricity to be purchased by each sector.

Footnotes:

****Oil and gas supply estimates assume deliverability from lightly explored frontier areas.

NATIONAL INPUT WORKSHEET FOR SUPPLY/DEMAND INTEGRATION

Country: CANADA

Year:

Case: 2000 C-2

Units: 10^{12} Btu

	(a)	(b)	(c)	(d)	(e)	(f)	(g)	(h)	(i)	(j)	(k)
	Coal	Petro-leum	(Syn. Liquids)	Nat. Gas	(Syn. Gas)	(Heat)	(Electri-city)††	Hydro Energy	Nuclear Energy	Geothermal energy and other	Total
(1) Transport	0	4460		0			0				4460
(2) Industry	383	554		1530			1170				3637
(3) Agric., Mining, Construction											
(4) Commercial)	0	594		1035			990				2619
(5) Public)											
(6) Residential	1.4	369		660			612				1642.4
(7) Non-energy Uses		954		261							1215
(8) Final Energy Demand*	384.4	6931		3486			2772				13573.4
(9) Electricity**	1198.5	123		117			(-3006)	6930			5362.5
(10) Syn. Gas											
(11) Syn. Liquids											
(12) Heat											
(13) Energy Sector Self Consumption & conversion losses	1.6	729		441			234				1405.6
(14) Primary Energy Input	1584.5	7783		4044			--	6930			20341.5
(15) Indigenous Supply****	2520	6479		1842				3900	4890		19631
(16 Imports***		1304		2202							3506
(17) Exports	935.5							1860			2795.5

*Includes non-energy uses.

**Blocks 9g, 10e 11c, and 12f must be negative to avoid double counting of energy from primary fuels.

***Includes imported products.

††Includes only electricity to be purchased by each sector.

Footnotes:
****Oil and gas supply estimates assume deliverability from lightly explored frontier areas.

NATIONAL INPUT WORKSHEET FOR SUPPLY/DEMAND INTEGRATION

Country: CANADA

Year:

Case: 2000 D-7

Units: 10^{12} Btu

	(a)	(b)	(c)	(d)	(e)	(f)	(g)	(h)	(i)	(j)	(k)
	Coal	Petro-leum	(Syn. Liquids)	Nat. Gas	(Syn. Gas)	(Heat)	(Electri-city)††	Hydro Energy	Nuclear Energy	Geothermal energy and other	Total
(1) Transport											3550
(2) Industry											3380
(3) Agric., Mining, Construction											
(4) Commercial) (5) Public)											2020
(6) Residential											1850
(7) Non-energy Uses											1200
(8) Final Energy Demand*											12000
(9) Electricity**											5425
(10) Syn. Gas											
(11) Syn. Liquids											
(12) Heat											
(13) Energy Sector Self Consumption & conversion losses											1225
(14) Primary Energy Input	3000	6600		3900			--	5150			18650
(15) Indigenous Supply****	3024	5440		1542				3150	2950		16106
(16 Imports***		1160		2358							3518
(17) Exports	24							950			974

*Includes non-energy uses.

**Blocks 9g, 10e 11c, and 12f must be negative to avoid double counting of energy from primary fuels.

***Includes imported products.

††Includes only electricity to be purchased by each sector.

Footnotes:
****Oil and gas supply estimates assume deliverability from lightly explored frontier areas.

NATIONAL INPUT WORKSHEET FOR SUPPLY/DEMAND INTEGRATION

Country: CANADA

Year:

Case: 2000 D-8

Units: 10^{12} Btu

	(a)	(b)	(c)	(d)	(e)	(f)	(g)	(h)	(i)	(j)	(k)
	Coal	Petro-leum	(Syn. Liquids)	Nat. Gas	(Syn. Gas)	(Heat)	(Electri-city)††	Hydro Energy	Nuclear Energy	Geothermal energy and other	Total
(1) Transport											3550
(2) Industry											3380
(3) Agric., Mining, Construction											
(4) Commercial)											2020
(5) Public)											
(6) Residential											1850
(7) Non-energy Uses											1200
(8) Final Energy Demand*											12000
(9) Electricity**											5430
(10) Syn. Gas											
(11) Syn. Liquids											
(12) Heat											
(13) Energy Sector Self Consumption & conversion losses											1220
(14) Primary Energy Input	1500	6600		3900			--		6650		18650
(15) Indigenous Supply****	2520	5440		1542				3150	4890		17542
(16 Imports***		1160		2358							3518
(17) Exports	1020								1390		2410

*Includes non-energy uses.

**Blocks 9g, 10e 11c, and 12f must be negative to avoid double counting of energy from primary fuels.

***Includes imported products.

††Includes only electricity to be purchased by each sector.

Footnotes:
****Oil and gas supply estimates assume deliverability from lightly explored frontier areas.

Denmark

8.1 SUMMARY

The Danish energy situation always has been and probably always will be dominated by the almost complete lack of internal fuel supplies. Therefore, the conditions on the world energy market have a major impact on Danish economic development. The low oil prices during the 1950s and 1960s changed the Danish fuel mix from 75 percent solid fuel and 25 percent oil in 1950 to 10 percent solid fuel and 90 percent oil in 1970. During the same period, energy consumption increased dramatically from 275 PJ (6.6 million TOE) in 1950 to 790 PJ (18.8 million TOE) in 1970. The average energy growth rate in this period was 5.4 percent per year but came close to 8 percent per year in the 1960s.

The steep increase in energy prices and the general uncertainty about future energy supplies have strongly underlined the perils of the Danish energy situation. Any strategy for the energy supply future must attempt to lower the energy demand growth rate and provide diversification of supplies on several fuels and fuel producers.

Since the oil crisis, a number of plans for the future energy system have been put forward (1–4), including an official plan by the government. None of these plans has gained general acceptance, however, and the government has postponed the legislation necessary for the execution of its own plan for an indefinite period because of disagreements over the introduction of nuclear energy.

The scenarios presented for Denmark are determined by the WAES global parameters. A translation of these parameters into national parameters is, of course, uncertain, but less so than an isolated estimate of the likely development in Denmark. Table 8.1 summarizes the scenarios considered for Denmark.

The WAES national demand studies for the period 1972 to 1985 have already been published together with a description of the data base and the methodology(5). In this chapter, the demand studies have been extended to year 2000 for selected cases. This wider time period of course introduces larger uncertainties. Furthermore, questions that could be left open in the 1985 studies become crucial for the year 2000 forecasts.

The primary driving force behind energy demand is the global economic growth rate, which is specified in the WAES scenarios and from which the national economic growth rates are derived. For the ten to fifteen year period considered in the 1985 scenarios, the relationships between the economic growth rates by sector and the corresponding growth rates for energy demand could be established fairly well. The main assumptions concerning the effects of energy price and government policy were based upon empirical data from past trends and upon fairly detailed technical estimates of energy-saving potentials. Furthermore, the structure of the economy, although influenced by differences in economic growth rates among sectors, would not change much relative to the base year 1972.

Table 8.1
Scenario parameters for Denmark

WAES scenario		Economic growth rate (% per year)	Government policy or principal replacement fuel[a]	Energy price ($/bbl)
1969–1970		4.2		
1985	A	4.1	vigorous	11.50 → 17.25
	B	3.1	vigorous	11.50 → 17.25
	C	4.1	vigorous	11.50 constant
	D	3.1	restrained	11.50 constant
	E	4.1	restrained	11.50 → 7.67
2000	C-1	3.0	coal	high
	C-2	3.0	nuclear	high
	D-7	2.0	coal	medium
	D-8	2.0	nuclear	medium
	D-3	2.0	coal	high

[a]Government policy after 1985 is assumed to be vigorous.

A strict application of the same projection principles to 2000, or beyond, is questionable. The structure of the economy is likely to change considerably and the relationships between energy and economy could be quite different from today. It is well known that fixed exponential growth rates over extended period of time lead to absurd results. For the year 1985 exponential projections might still be within the bounds of credibility, but another fifteen years' projection beyond this could, in many cases, imply developments that are neither credible nor desirable. This argument becomes even more compelling if just a few years are added to the projection period.

For the Danish year 2000 projection, these considerations have led to modifications of the projection principles in a number of ways. The 1985 projections were intimately connected to the economic projections. For the 2000 projections, attention has been focused on descriptors more directly connected to physical conditions, such as living space, need for transportation, public services, and so on.

The WAES global economic growth rates for 1985 to 2000 are lower than those for the period 1975 to 1985. This in itself implies lower energy growth. In addition, constant to high energy prices have been assumed toward the end of the century, further slowing the energy growth. The year 2000 scenarios, in addition, are characterized by the choice of nuclear energy or coal as the main replacement fuel for oil. The choice among the replacement fuels has little influence on total energy demand but is of course reflected in the fuel mix calculations.

This paper considers five scenarios to year 2000 which are extensions of "medium" 1985 scenarios (C and D). The energy growth rate, which was 6–8 percent per year in the 1960s, is projected to decline to 2–3 percent per year in the period 1975 to 1985 and further to 1–2 percent per year in the period 1985 to 2000. The final gross energy demand around year 2000 would then be 1200–1400 PJ/year corresponding to 28–33 million TOE per year as compared to 800 PJ/year or 19 million TOE per year in 1972.

The primary fuel mix found for the high coal cases (C-1 and D-7) is approximately 30 percent coal, 46 percent oil, 11 percent natural gas, and 8 percent nuclear energy. For the high nuclear cases (C-2 and D-8), the fuel mix changes to 17 percent coal, 44 percent oil, 11 percent natural gas, and 27 percent nuclear. In all cases oil consumption is less than today.

Renewable energy sources have not been introduced in the supply studies because they are beyond the scenario energy price levels. They are, however, considered as possible national responses to the severe shortages for hydrocarbon fuels that are likely to develop toward the end of the century.

8.2 METHODOLOGY

In accordance with the general WAES methodology and the earlier 1985 projections for Denmark, energy demand E is calculated as the product of an activity level A and an energy intensity I:

$$E = AI.$$

This applies to each sector of the economy and to the economy as a whole. The energy demand is expressed in energy units per unit time, the proper physical unit being watt, but for practical purposes more often joules (J) per year or tons of oil equivalent (TOE) per year. The activity level can be expressed in monetary units per unit time (e.g., value added in 1972 U.S. dollars per year) or in physical units (m^2 floor area, number of automobile kilometers per year, etc.). The energy intensity is then measured in energy or power units per activity level unit.

The main driving force behind the growth in energy demand is growth in the economy. Economic growth determines the activity level in each sector. For the period 1972 to 1985, the activity levels were determined by reference to the economic plan PPII, which was used to estimate the relative growth rates for each of the major sectors within the total growth rate of the economy. This again was fixed by reference to the WAES global growth rates defining the scenario cases.

The other WAES scenario parameters, energy price and government policy, were interpreted as affecting energy intensities but not activity levels. This is a reasonable approximation because the total growth rate of the economy is exogenously defined and thus not influenced by the other scenario parameters.

Two different levels of aggregation were used for the 1985 projections. One was based on a breakdown of the economy into as many as 69 sectors for 1972. The activity levels in these sectors were given partially in physical terms and partially in economic terms. The energy intensities were estimated by reference to historical trends, empirical price elasticities, and technical estimates of energy efficiency improvements. The other level of aggregation had only 8 sectors corresponding rather closely to major economic sectors. The activity levels were described exclusively in monetary terms.

Experience with the 1985 projections showed that a full 69-sector projection to year 2000 would be cumbersome and necessitate a large number of quite arbitrary estimates for the individual sectors. On the other hand, the 8-sector projections, although probably reflecting the scenario conditions as truthfully as the 69-sector projections, are not well suited for detailed studies of, say, the impact of conservation strategies.

It was therefore decided to use an intermediate aggregation involving 19 major energy-consuming sectors (see Table 8.2). The same sectors have been used by a number of WAES countries and can be obtained by a simple aggregation of the 69 sectors used by others. The advantage of this method is that important activity levels, such as the number of automobiles, number of dwellings, and production in major industrial sectors, can be explicitly stated in physical rather than economic terms. This permits a better picture of the lifestyle of the projected future and helps to avoid some of the pitfalls of exponential growth. Furthermore, the energy intensity or energy efficiency can be directly given, in MJ/passenger-km or MJ/dwelling/year, for example. This type of efficiency measure is a precondition for a meaningful analysis of the impact of energy strategies whether induced by energy price or by government policy. The efficiencies also reflect a number of inherent trends in economic and technical development as described previously in connection with the 1985 projections.

Table 8.2
Major consuming sectors

Transportation	**Commercial, Public**
1. Automobiles	10. Space heat, fossil fuel
2. Airplanes	11. Space heat, electric
3. Other passenger	12. Other energy
4. Truck freight	**Industry**
5. Rail	13. Metals
6. Shipping	14. Chemicals
Agriculture, Mining, Construction	15. Energy-intensive industries
7. Agriculture	16. Non-energy-intensive industries
8. Mining	**Residential**
9. Construction	17. Space heat, fossil fuel
	18. Space heat, electric
	19. Appliances

In order to give a consistent picture of the whole 1972 to 2000 period, the five earlier projections to 1985 were recast in the 19-sector format. This new set of 1985 projections are by and large in agreement with the earlier 8-sector projections because they build on exactly the same assumptions. Some minor deviations do, however, show up when a sector is divided into several which are then projected independently. For example, in the transport sector the growth rates for automobiles and airplane travel are quite different. A disaggregated projection therefore leads to a total which deviates from that obtained from a projection on the basis of an average growth rate and average initial energy intensity. The new format for base year 1972 and for the 1985 projections is here presented together with the 2000 projections in order to facilitate comparisons. It should be stressed that their totals for all practical purposes are identical to those presented in the earlier 1985 demand study.

This paper also contains the supply-demand integrations for the base year 1972, for the five 1985 cases, and for the five 2000 cases. Supply-demand integration is the process by which supplies are matched to projected demand. This process raises a number of methodological questions. One possible method would be an optimization to find the cheapest solution at the scenario energy prices with additional assumptions about costs of capital equipment, etc. This method is essentially the one used in the "mechanized" global integrations described in Part II of this volume. An optimization for a single country like Denmark that has few options is difficult unless subject to a number of constraints. Such constraints could, for example, restrict the fuel mix in electric generation, allow a maximum or minimum amount of natural gas, or require maximum introduction of district heating even if uneconomical. These constraints reflect important goals or restrictions for the national energy policy but are more or less of a subjective nature.

The supply-demand integrations here were performed manually by trial and error and with a number of specific goals enumerated in section 8.4. The solution is not necessarily economically optimal, nor is it independent of the general expectations of world energy development. Although the integrations have not been subject to revisions because of the expected inability of the world energy supplies to meet demand, this situation has been preconceived in the formulation of the goals. Thus, there is little room left for national responses in the case of a severe fuel shortage at the end of the century.

Section 8.5 of this paper explicitly addresses this highly likely situation. This section also briefly considers the possible role of nontraditional energy sources which have not been included in the integrations presented here.

8.3 ENERGY DEMAND TO YEAR 2000

The demand projections are summarized in Tables 8.3 and 8.4. The full 19-sector projections are contained in the worksheets collected at the end of the paper. These worksheets have in turn served as input to the demand-supply integration worksheets discussed in section 8.4.

Table 8.3 gives rise to the following comments:

1. The scenarios point to a considerable lowering of the energy growth rates. This is a combined effect caused by many factors. Among the most important are increased efficiency in the consuming sectors and growth concentrated in the less energy-intensive sectors of the economy. In a number of areas, especially the residential sector but also in transport, saturation effects are becoming important.

2. The demand growth rate declines toward the end of the century. This is mostly an effect of the lower economic growth rates imposed by the lower global growth rates selected by WAES. This lowering of the economic growth rates is not entirely arbitrary. A number of factors such as lower growth of labor force, lower birth rates, and saturation in many consumer areas point in this direction. The lower growth rate can also be seen as a preconception of the impact of low energy availability and connected imbalances in world trade. These factors, however, have not explicitly entered into the

Table 8.3
Energy demand summary

WAES scenario		*A* Fossil fuel (PJ/yr)	*B* Electricity (PJ/yr)	*C* Delivered energy (PJ/yr)	*D* Primary energy (PJ/yr)	*E* Gross efficiency (%)	*F* Energy growth (%/yr)	*G* Energy coefficient
1972		600	54.6	655	825	79	6.5	1.51
1985	A	679	90.8	770	991	78	1.42	0.50
	B	617	82.8	700	900	78	0.67	0.31
	C	731	96.6	828	1066	78	1.99	0.71
	D	718	93.6	812	1049	77	1.86	0.87
	E	853	109.6	963	1248	77	3.24	1.15
2000	C-1	874	156.0	1030	1366	75	1.67	0.56
	C-2	942	181.2	1123	1518	74	2.38	0.79
	D-7	821	141.5	963	1274	76	1.30	0.65
	D-8	890	163.3	1053	1409	75	1.99	1.00
	D-3	770	136.7	906	1202	75	0.91	0.46

Note: $C = A + B$; D is from the supply-demand integration; $E = C/D$; F is the rate of growth of primary energy D; G is the ratio of primary energy growth to economic growth in the periods 1960 to 1970, 1972 to 1985, and 1985 to 2000, respectively.

Table 8.4
Sectoral percentages of primary energy demand[a]

WAES scenarios	Transport	Industry	Agriculture, mining, construction	Public, commercial	Residential	Total
1972	19	22	10	14	35	100
1985[b]	22	22	11	14	31	100
2000[b]	23	24	11	14	28	100

[a] Losses allocated to final demand sector.
[b] Average of scenarios. The distribution of primary energy by sectors is almost independent of the specific scenario case.

determination of the global growth rates, but the declining energy growth rates can be seen as a general adaptation to the global energy situation and as an approach toward societies in energy equilibrium. The time horizon of the WAES studies is unfortunately not quite wide enough to encompass this decisive transition.

3. Primary energy consumption grows faster than overall energy demand. This is illustrated by the "gross efficiency" listing in Table 8.3, which gives the ratio of energy delivered to the consumer to primary energy. This ratio is declining in the scenario period mainly because of increased use of electricity. The efforts toward utilization of waste heat from power generation, which is an important element in the supply strategies described in section 8.4, are not sufficient to offset the losses connected with increased electricity production.

4. The energy coefficient, which is the ratio of the rate of energy growth to the rate of economic growth, is for most scenarios well below unity. This indicates that the economy is becoming less energy-intensive through the whole period. This is mostly an effect of specific energy conservation measures, but structural changes in the economy are also important. One should note a considerable lowering of the growth rate of housing stock. The yearly additions fall from 40,000 in the 1960s to 29,000 in the period 1972 to 1985 and further to 21,000 in the period 1985 to 2000. This lowering seems necessary if the scenarios are not going to project the Danes into a nation of hermits. The average size of households is down to 2.1 persons in 2000.

5. The distribution of energy among the major economic sectors (see Table 8.4) does not change drastically in the periods considered. However, the general tendency is clearly a decrease in the importance of the residential sector and an increase in industry and transportation. The dominant role of comfort heat (residential, public, and commercial sectors) is reduced from 34 percent in 1972 to 27 percent in 1985 and 24 percent in 2000. This, together with an increase in the industrial energy consumption, brings the Danish consumption pattern much closer to that of other European countries.

6. In the Danish debate, there has been considerable confusion over the question of energy conservation. It has been stated by some that energy consumption easily could be reduced by one-third without harm to the economy. Others have indicated that energy use is already quite efficient and that, at present energy prices, there are no great gains to be made through conservation.

It should be stressed that all scenarios (except the 1985 falling price scenario E) assume a considerable improvement in the efficiency of energy utilization by consumers. There is, however, no unique way of stating how much energy is saved in any particular scenario because we lack knowledge of how development would have proceeded under another set of conditions. One method, often used, is to refer to a "historic growth" scenario in which energy is growing at some rate determined from, for example, the period 1960 to 1970. Measured relative to historic growth, all scenarios considered here give impressive savings on the order of 20–50 percent. However, the period 1960 to 1970 was probably unique in terms of energy because of the welfare transition which took place in these years.

A somewhat more realistic picture of the conservation impact is given in Table 8.5.

The reference for the comparisons in Table 8.5 is the energy demand as it would have been if all energy intensities were kept at the 1972 levels but modified for inherent trends that would be expected to be present independent of energy price and energy policy. The reference cases obtained this way also reflect the changing economic growth rates in the scenarios and normal structural changes in the economy that could affect energy demand.

As shown in Table 8.5, energy demand in 2000 can in this way be judged to be 5–27 percent lower

Table 8.5
Energy savings by scenario

WAES scenario		Primary energy in scenario (PJ/yr)	Primary energy at 1972 efficiency (PJ/yr)	Savings (%)[a]
1972		823	823	0
1985	A	991	1187	17
	B	900	1078	17
	C	1066	1187	10
	D	1049	1078	3
	E	1248	1187	-5
2000	C-1	1366	1876	27
	C-2	1518	1876	19
	D-7	1274	1476	14
	D-8	1409	1476	5
	D-3	1202	1476	19

[a]Savings in % relative to projection with unchanged energy efficiencies.

than normal. These numbers are probably the closest one can get to a fair estimate of the overall conservation assumptions built into the scenarios.

7. The estimates in Table 8.5 do not directly give any impression of the technical efficiency changes assumed. Some of these, especially within the heating sectors, were discussed in some detail in connection with the 1985 scenarios. Table 8.6 presents in condensed form some of the specific efficiency assumptions in scenarios C and C-1, which can be considered intermediate scenarios.

Note in Table 8.6 that not all energy intensities go down. For example, the increase in the energy intensity of freight is caused by a strong growth in the amount of air freight. The peculiar behavior of energy intensity in agriculture is the combined effect of a strong trend toward further mechanization that is counteracted in the long run by a general efficiency improvement.

Some of the most important subsectors within the transport, industry, and residential sectors are, however, assumed to undergo considerable efficiency improvements. The technical bases for some of these were discussed in detail for the 1985 projections. Some of the improvements, for example, in the heat insulation of buildings, will continue after 1985 because of slow penetration. In general, however, the efficiency improvements for 1985 to 2000 are of a more speculative nature, although by no means beyond the capability of present technology.

8.4 SUPPLY-DEMAND INTEGRATION

Once the energy demand has been calculated, the process of allocating fuels and other energy supplies to the demand sectors can proceed. This process is performed with the aid of a separate worksheet listing seven major demand sectors and ten types of primary or secondary energy. In addition, the worksheets account for losses in energy conversion and distribution. (The worksheets for the base year 1972 and the scenarios to 1985 and 2000 are collected in section 8.7.)

For the Danish cases, the possible types of energy were, prior to the integration, reduced to six: the fuels coal, oil, gas, and uranium and the secondary energies electricity and heat. The other energy sources, including renewable sources, were judged to be of minor importance in Denmark for the period studied under the assumed energy prices.

The allocation of fuels reflects some important goals for future energy policy. As stated earlier, the diversification of fuel and fuel suppliers has highest priority. The main strategies followed here are the same in kind as suggested in the government energy plan (3) but are somewhat different in degree. The main consideration is that the fuels constitute a hierarchy with oil, the most versatile and scarce fuel, at the top, followed by natural gas and coal. Uranium, which at present is useful only for the generation of electricity, is at the bottom. Any change in the energy supply structure must respect the hierarchical order in the sense that a lower-ranking fuel replaces a higher-ranking fuel.

Compared to the present situation, the scenarios contain two major changes: the introduction of nuclear energy as a principal source for electricity generation and the introduction of natural gas, mostly for residential and industrial use. Both measures, however, have a long lead time, and the fuel shifts therefore have little impact before 1990.

The demand projections separately determine the quantity of electricity that must be generated. It should be realized that this already involves a fuel allocation, especially since electric heat is an impor-

Table 8.6
Energy intensity developments 1972-2000

	1972	1985 Case C	2000 Case C-1	Change (%/year, 1972-2000)
Automobiles (MJ/km)	3.13	3.04	2.50	-0.8
Air travel (MJ/passenger-km)	3.5	3.4	2.8	-0.8
Freight (MJ/ton-km)	2.5	2.5	3.1	+0.8
Agriculture, primary energy (MJ/$ output)	32.8	44.7	38.9	+0.6
Public and commercial, heat loss ($MJ/m^2/yr$)	720	490	400	-2.1
Public and commercial, electricity ($MJ/m^2/yr$)	320	350	325	+0.0
Industry, primary energy (MJ/$ output)	41.0	35.2	31.1	-1.0
Residential, heat loss (GJ/dwelling/yr)	64	50	39	-1.8
Residential, electricity (GJ/dwelling/yr)	10	13.7	16.5	+1.8

tant substitute for oil. In the scenarios 15-20 percent of the buildings, mostly in remote areas, are electrically heated. This contributes significantly to the electricity growth rate, which, depending on the scenario, is 4-7 percent per year in the period 1975 to 1985 and 3-4 percent per year in the period 1985 to 2000. In year 2000 about 10-15 percent of the electricity is used for room heating. The electricity demand determines the primary energy input for electricity generation, which is entirely thermal in Denmark. The generation capacity increases from 3 GWe today to around 9 GWe in 2000. The high nuclear cases assume 5-6 GWe installed nuclear capacity, which is in accordance with the government's projection. In the high coal cases, nuclear generation is reduced to 2 GWe. In both cases, approximately 20 percent of the electricity generation is from oil or gas in order to maintain peak load capacity.

Electricity generation requires approximately 40 percent of the primary energy in the 2000 scenarios. In order to recoup some of the losses, all scenarios envisage a strong expansion of district heating from power stations. At present about 10 percent of the Danish heating demand is satisfied by power stations with combined cycles. This fraction is increased to around 40 percent which, by year 2000, considering the geography of the country, is estimated to be the attainable maximum (6). This contribution to the heat supply does require some extra fuel, estimated to be approximately 50 percent of the heat supplied. In other words, in the combined cycle, heat is generated with an efficiency of 200 percent whereas electricity, for ease of calculation and in accordance with a general WAES recommendation, is taken to be generated at 35 percent efficiency. The combined cycle generation conserves fuel amounting to about 4 percent of the total Danish primary energy demand.

The amount of natural gas introduced in the scenarios is in line with the government's estimate (4×10^9 m^3/yr). It is not known to what extent gas can be obtained from the Danish North Sea fields, but most likely the main supply must come from the Norwegian fields. The delivery possibilities are presently being investigated. Introduction of natural gas probably requires government investment in the transmission and distribution systems. In view of the projected scarcity of gas in Europe, the gas option may be economically unattractive, and a larger fraction of oil and coal may after all be preferable. Another possibility would be to limit gas supply to the amount available from the Danish fields.

Coal consumption is in all cases considerably larger than in 1972. In this respect, the scenarios differ considerably from the government plan, which envisages coal consumption only slightly above the present level. This difference is partially due to the fact that some of the scenarios are defined as high coal scenarios, but it is also a reflection of the perceived hydrocarbon shortage and the use of coal in several cases as a substitute for heavy fuel oil, which at present is the fuel preferred by industry and some other users.

With most of the other fuel supplies determined by the considerations above, oil is left as a balancing fuel in all sectors except transport and agriculture, which are almost completely dependent on oil. Only small amounts of oil are needed for heating purposes in 2000 because of the substitution by district heat, gas, and electricity. Some oil is still required in industry. It is perhaps questionable whether the projected massive transition to coal in industry can be accomplished. If this is the case, industry will require more oil.

Table 8.3 summarizes the gross efficiency of the energy system in the scenarios. This efficiency is steadily declining in spite of the district heating efforts. Generally, this reflects the high energy cost of most substitutes for oil. Substitution with electricity, which is one of the major strategies for the introduction of the lower-ranking fuels in the system, is especially costly.

The reshaping of the energy system needed to reach the goals outlined above will require substantial investments. The Danish WAES project has not made any independent analysis of the investments needed, but some of the major components are indicated in Table 8.7, which is based mainly on information contained in the government plan. A complete economic comparison of the scenarios has not been attempted because of the large uncertainty of fuel prices and capital costs.

8.5 NATIONAL RESPONSES

At the time the WAES global parameters were selected it was not known whether the resulting scenarios were viable in the sense that global supplies of energy would be sufficient to meet global demand. As work progressed it became clear that

Table 8.7
Investments required to the year 2000

Component	Investment ($/GJ/yr)	Energy[a] (PJ/yr)	Investment (10^9 $ (1975))				
			C-1	C-2	D-7	D-8	D-3
Gas distribution	10	164	1.6	1.6	1.6	1.6	1.6
Nuclear power	11	400	1.3	4.4	1.3	4.4	1.9
Coal power	6	127	2.3	0.8	1.7	0.5	1.5
District heating	23	105	2.3	2.4	2.0	2.3	1.7
Insulation	20	150	3.0	3.0	2.7	2.7	2.7
North Sea	5	122	0.6	0.6	0.6	0.6	0.6
Total			11.1	12.8	9.9	12.1	10.0

[a]Fuel substitutions or savings above 1972 assumed in scenario C-2.
Note: 1 U.S. $ (1975) = 6.00 D.kr.

everything was not well. In fact, none of the scenarios was viable, since they all led to energy deficits on the order of 10–20 percent of total demand at the turn of the century. This happened in spite of fairly optimistic assumptions about efficiency improvements in energy utilization, successes in the development of new supplies, and free movements of energy across borders.

A first response to this situation would be that the world must do better than assumed in the scenarios. Use less, produce more, or both. Indeed, it has been possible to invent a scenario that makes supply and demand balance with a comfortable margin. It assumes low economic growth, high energy prices, and strong efforts toward efficient energy utilization and development of supplies. The consequences of this scenario (D-3) are described in section 8.3. Compared to the other Danish scenarios, the differences are small. For the world they might be large, because the basic reason for the apparent success of this scenario is low global economic growth coupled with the conservation and production incentives that result from high energy prices. One can have severe doubts about the internal consistency of these assumptions. Under all circumstances they point toward a world economy which is energy-constrained and in which any tendency to economic upsurge will be checked by lack of energy. This situation would be especially severe for the developing countries.

What could a national response to the energy-constrained economy be in Denmark? First of all one must realize that the situations described in the scenarios are not necessarily realistic. The situation might indeed be a lot worse. The two main components of the Danish strategies behind the scenarios are rapid efficiency improvements in all sectors and rapid introduction of new supplies. The same components enter into the government's energy plan to an even higher degree.

That some of the targets in the government plan will not be met is already known. The efficiency improvements will be slower, and energy consumption will tend to grow even with a sluggish economy. We also know that nuclear expansion will be delayed considerably and that coal use will expand only for electric utility use. Finally, there are grave doubts about the prospects for natural gas in the Danish supply system.

A first national response to the perceived situation of severe energy shortages must therefore be to ensure that the goals for energy savings and supply additions and substitutions are met. This will take considerable effort because much time has already been lost. Even if the plans were realized and the scenario levels for energy consumption and energy supply were reached, Denmark could not be expected to contribute much more globally. The Danish energy growth rates would be among the lowest in the world and the per capita consumption lower than in most European countries.

Nevertheless, it might be prudent to consider other options, such as possible contributions from renewable energy sources. These have not been considered in the scenarios, partially because they would imply energy prices considerably above the scenario levels and partially because of technical constraints and uncertainties.

Most energy sources of this type are capital-intensive, and the energy price is determined by the capital costs. How these capital costs should be calculated remains a question since the amounts of investment required are unknown. The considerations in the supply study concerning renewable energy sources indicated possible increases in energy prices two to four times above their present levels (see the Danish contribution to the second WAES technical report). Normal economic considerations would therefore exclude major contributions from these sources.

The oil and gas shortage situation outlined above might very well lead to energy prices at a level where some renewable sources would become attractive. Solar water heaters for the summer are one of the more promising possibilities. Although their potential is limited to a small percentage of demand, they would, under present conditions, save high-grade fuel oil. If the district heating schemes outlined in section 8.4 are realized, solar water heaters would still be attractive in the approximately 60 percent of the dwellings heated by electricity and fossil fuels. More elaborate solar heating schemes become economically attractive only if electricity becomes very scarce and expensive (e.g., because of a rejection of nuclear energy).

Wind energy has received much attention in public discussions. A wind system that could cover 10 percent of present-day electricity consumption would require around 1000 large wind generators. Such a system would save about 2 percent of the present primary fuel consumption. The savings, however, would mostly be in coal, and the scheme becomes attractive only with the rejection of nuclear as well as coal-generated electricity.

The possible supply of low-temperature geothermal heat is presently being investigated in a number of abandoned exploratory oil wells. The economics of this heat for domestic use seems favorable, but the total potential is at present unknown.

It is difficult to estimate the maximum potential of renewable sources within the time frame considered here. The government plan envisages a 3 percent contribution at the turn of the century, mostly from combustion of waste. A recent plan that excludes the nuclear option (4) operates with a contribution from renewable sources corresponding to about 10 percent of the projected demand in the WAES scenarios.

In conclusion, it can be safely stated that there will be no easy solution of Denmark's energy problem. The WAES scenarios presented here call for considerable efforts and large investments, though hopefully not beyond credible levels. The response by government and the populace must be strong and immediate. Time will turn out to be as precious a commodity as oil.

8.6 REFERENCES

1. *Energy in Denmark 1990–2005. A Case Study*, IFIAS, The Niels Bohr Institute (1976). ISBN 87-87585-04-9.

2. Bent Elbek and Uffe Korsbech, *Dansk Energipolitik*, Teknisk Forlag (1975).

3. *Dansk Energipolitik*, Handelsministeriet, Copenhagen (1976). ISBN 87-503-1969-8.

4. *Skitse til Alternativ Energiplan*, OOA (1976). ISBN 87-87625-00-8.

5. *Energy Demand Studies: Major Consuming Countries*, MIT Press (1976).

6. IFIAS report no. 3, The Niels Bohr Institute (1976). ISBN 87-87585-01-3.

8.7 WORKSHEETS

This section contains summary tables and worksheets in the following order:

Energy demand summary 1972–1985–2000

Aggregated demand worksheets
- 1972
- 1985 (5 sheets)
- 2000 (5 sheets)

National input worksheets for supply/demand integration
- 1972
- 1985 (5 sheets)
- 2000 (5 sheets)

WAES NATIONAL ENERGY DEMAND STUDIES, 1972-1985-2000
ENERGY DEMAND SUMMARY

Country: Denmark

Scenario Assumptions	1972	1985					2000			
WAES CASE	Base Year	A	B	C	D	E	C-1	C-2	D-7	D-8
GNP 1976-85; 1985-2000		High	Low	High	Low	High	High	High	Low	Low
Oil/Energy Price*		11.50-17.25	11.50-17.25	11.50	11.50	11.50-7.66	11.50-17.25	11.50-17.25	11.50	11.50
National Policy Response		Vig	Vig	Vig	Res	Res	Vig	Vig	Vig	Vig
Principal Replacement Fuel							Coal	Nuclear	Coal	Nuclear
Primary Energy Demand **Units**	PJ/y									
Oil	771	614	544	676	819	980	641	759	651	705
Natural Gas	0	114	114	114	0	0	164	164	164	164
Coal	58	195	174	208	162	200	435	185	333	130
Nuclear	0	58	58	58	58	58	116	400	116	400
Hydroelectric										
Geothermal										
Solar Thermal										
Solar Electric										
Refuse	2	10	10	10	10	10	10	10	10	10
TOTAL PRIMARY ENERGY DEMAND	817	991	900	1066	1049	1248	1366	1518	1274	1409

*1975 US dollars per barrel of Arabian light crude oil, fob Persian Gulf.
All Figures are primary production prior to any processing or conversion.

WAES AGGREGATED DEMAND WORKSHEET

Country: Denmark

Year: 1972

WAES Case: Base

GNP: $22.7 \cdot 10^9$ $

Population: $5.0 \cdot 10^6$

Sector Transportation	Activity Level	Energy Intensity	Other PJ/y	Electric PJ/y	Foss. Fuel PJ/y	Feed-stocks PJ/y
1. Automobiles	$1.19 \cdot 10^6$ autos x 17000 km/auto/y	3.13 MJ/km			63	
2. Airtravel	$4.2 \cdot 10^9$ pass-km/y	3.5 MJ/pass-km			15	
3. Other passengers	$11.7 \cdot 10^9$ pass-km/y	1.46 MJ/pass-km		0.3	17	
4. Truck & Air freight	$13.5 \cdot 10^9$ ton-km/y	2.49 MJ/ton-km			34	
5. Rail	$1.9 \cdot 10^9$ ton-km/y	0.54 MJ/ton-km			1	
6. Shipping	$24.4 \cdot 10^9$ ton-km/y	0.63 MJ/ton-km			15	
1-6 Total				0.3	145	
Agricult., Mining, Construct.						
7. Agriculture	$1533 \cdot 10^6$ $/y output	27 MJ/$		3.1	41	
8. Mining	$26.2 \cdot 10^6$ $/y output	103 MJ/$		0.2	3	
9. Construction	$1895 \cdot 10^6$ $/y output	9.5 MJ/$		1.1	18	
7-9 Total				4.4	62	

Commercial, Public						
10 Space Heat (Fossil Fuel)		$47 \cdot 10^6$ m²	720 MJ/m²/y	0.75 Effic		45
11. – – (Electric)		m²	MJ/m²/y	Effic		
12. Other		$47 \cdot 10^6$ m²	320 MJ/m²/y	425 MJ/m²/y	15.0	20
10-12 Total					15.0	65
Industry						
13. Metals	$82.7 \cdot 10^6$ \$/y	(ton/y)	12.1 MJ/\$ (elec)	121 MJ/\$ (fos)	1.0	10
14. Chemicals	$431 \cdot 10^6$ \$/y	(ton/y)	6.3 MJ/\$ (–)	25.8 MJ/\$ (–)	2.7	11
15. Other Energy Industries	$316 \cdot 10^6$ \$/y		7.1 MJ/\$ (–)	141 MJ/\$ (–)	2.2	45
16. Non-Energy Industries	$3.2 \cdot 10^9$ \$/y		2.76 MJ/\$ (–)	17.1 MJ/\$ (–)	8.9	55
13-16 Total					14.8	121
Residential						
17. Space Heat (Fossil Fuel)		$1.844 \cdot 10^6$ Dwellings (fossil)	64 000 MJ/dw/y	0.7 Effic		169
18. – – (Electric)		$18 \cdot 10^3$ Dwellings (elect)	80 000 MJ/dw/y	1 Effic	1.4	
			21 000 MJ fossil/dwelling			
19. Appliances		$1.862 \cdot 10^6$ Dwellings	10 000 MJ elect/dwelling		18.7	39
17-19 Total					20.1	208
1-19 Total					54.6	601

WAES AGGREGATED DEMAND WORKSHEET

Country: Denmark

Year: 1985

WAES Case: A, DK 4.1%

GNP: $34 \cdot 10^9$ $

Population: $5.3 \cdot 10^6$

Sector Transportation	Activity Level	Energy Intensity	Other PJ/y	Electric PJ/y	Foss. Fuel PJ/y	Feed-stocks PJ/y
1. Automobiles	$1.76 \cdot 10^6$ autos x 17000 km/auto/y	2.75 MJ/km			82	
2. Airtravel	$9.1 \cdot 10^9$ pass-km/y	3.07 MJ/pass-km			28	
3. Other passengers	$17.3 \cdot 10^9$ pass-km/y	1.28 MJ/pass-km		0.54	22	
4. Truck & Air freight	$18.8 \cdot 10^9$ ton-km/y	2.27 MJ/ton-km			43	
5. Rail	$2.5 \cdot 10^9$ ton-km/y	0.54 MJ/ton-km			1	
6. Shipping	$34.1 \cdot 10^9$ ton-km/y	0.55 MJ/ton-km			19	
1-6 Total				0.54	195	
Agricult., Mining, Construct.						
7. Agriculture	$2060 \cdot 10^6$ $/y output	2.18MJ/$(elec) 34 MJ/$ (fos)		4.5	70	
8. Mining	$32 \cdot 10^6$ $/y output	6.0 MJ/$(-) 88.6 MJ/$ (-)		0.2	3	
9. Construction	$2310 \cdot 10^6$ $/y output	0.8 MJ/$(-) 6.25 MJ/$ (-)		1.8	14	
7-9 Total				6.5	87	

Commercial, Public						
10 Space Heat (Fossil Fuel)		$73 \cdot 10^6$ m²	450 MJ/m²/y	0.75 Effic		44
11. – – (Electric)		____ m²	____ MJ/m²/y	____ Effic		
12. Other		$73 \cdot 10^6$ m²	330 MJ/m²/y	275 MJ/m²/y	24.0	20
10-12 Total					24.0	64
Industry						
13. Metals	$158 \cdot 10^6$ $/y	____ (ton/y)	13 MJ/$ (elec)	87 MJ/$ (fos)	2.1	14
14. Chemicals	$682 \cdot 10^6$ $/y	____ (ton/y)	6.8 MJ/$ (–)	19 MJ/$ (–)	4.6	13
15. Other Energy Industries	$495 \cdot 10^6$ $/y		7.6 MJ/$ (–)	102 MJ/$ (–)	3.8	51
16. Non-Energy Industries	$5176 \cdot 10^6$ $/y		3.0 MJ/$ (–)	12.4 MJ/$ (–)	15.3	64
13-16 Total					25.8	142
Residential						
17. Space Heat (Fossil Fuel)		$2.195 \cdot 10^6$ Dwellings (fossil)	47 400 MJ/dw/y	0.8 Effic		130
18. – – (Electric)		$79 \cdot 10^3$ Dwellings (elect)	63 300 MJ/dw/y	1 Effic	5.0	
			26 800 MJ fossil/dwelling			61
19. Appliances		$2.27 \cdot 10^6$ Dwellings	12 800 MJ elect/dwelling		29.0	
17-19 Total					34.0	191
1-19 Total					90.8	679

WAES AGGREGATED DEMAND WORKSHEET

Country: Denmark

Year: 1985

WAES Case: B, DK 3.1

GNP: $31 \cdot 10^9$ \$

Population: $5.3 \cdot 10^6$

Sector	Activity Level	Energy Intensity	Other PJ/y	Electric PJ/y	Foss. Fuel PJ/y	Feed-stocks PJ/y
Transportation						
1. Automobiles	$1.6 \cdot 10^6$ autos x 17000 km/auto/y	2.75 MJ/km			75	
2. Airtravel	$8.3 \cdot 10^9$ pass-km/y	3.07 MJ/pass-km			25	
3. Other passengers	$15.7 \cdot 10^9$ pass-km/y	1.28 MJ/pass-km		0.5	20	
4. Truck & Air freight	$17.1 \cdot 10^9$ ton-km/y	2.27 MJ/ton-km			39	
5. Rail	$2.5 \cdot 10^9$ ton-km/y	0.54 MJ/ton-km			1	
6. Shipping	$30.9 \cdot 10^9$ ton-km/y	0.55 MJ/ton-km			17	
1-6 Total				0.5	177	
Agricult., Mining, Construct.						
7. Agriculture	1924 \$/y output	3.18MJ/\$(elec) 34 MJ/\$(fos)		4.2	66	
8. Mining	30 \$/y output	6.0 MJ/\$(elec) 88.6 MJ/\$(-)		0.2	3	
9. Construction	2200 \$/y output	0.80MJ/\$(-) 6.25 MJ/\$(-)		1.8	14	
7-9 Total				6.2	83	

Commercial, Public						
10 Space Heat (Fossil Fuel)		$66.3 \cdot 10^6$ m²	450 MJ/m²/y	0.75 Effic		40
11. – – (Electric)		____ m²	____ MJ/m²/y	____ Effic		
12. Other		$66.3 \cdot 10^6$ m²	330 MJ/m²/y	270 MJ/m²/y	21.9	18
10-12 Total					21.9	58
Industry						
13. Metals	$135 \cdot 10^6$ $/y	____ (ton/y)	13 MJ/$ (elec)	87 MJ/$ (fos)	1.8	12
14. Chemicals	$608 \cdot 10^6$ $/y	____ (ton/y)	6.8 MJ/$ (–)	19 MJ/$ (–)	4.1	11
15. Other Energy Industries	$441 \cdot 10^6$ $/y		7.6 MJ/$ (–)	102 MJ/$ (–)	3.4	45
16. Non-Energy Industries	$4613 \cdot 10^6$ $/y		3.0 MJ/$ (–)	12.4 MJ/$ (–)	13.7	57
13-16 Total					23.0	125
Residential						
17. Space Heat (Fossil Fuel)		$2.175 \cdot 10^6$ Dwellings (fossil)	46350 MJ/dw/y	0.8 Effic		126
18. – – (Electric)		$77 \cdot 10^3$ Dwellings (elect)	59750 MJ/dw/y	1 Effic	4.6	
			21500 MJ fossil/dwelling			48
19. Appliances		$2.25 \cdot 10^6$ Dwellings	11800 MJ elect/dwelling		26.6	
17-19 Total					31.2	174
1-19 Total					82.8	617

WAES AGGREGATED DEMAND WORKSHEET

Country: Denmark

Year: 1985

WAES Case: C, DK 4.1%

GNP: $34 \cdot 10^9$ \$

Population: $5.3 \cdot 10^6$

Sector Transportation	Activity Level	Energy Intensity	Other PJ/y	Electric PJ/y	Foss. Fuel PJ/y	Feed-stocks PJ/y
1. Automobiles	$1.76 \cdot 10^6$ autos x 17000 km/auto/y	3.04 MJ/km			91	
2. Airtravel	$9.1 \cdot 10^9$ pass-km/y	3.40 MJ/pass-km			31	
3. Other passengers	$17.3 \cdot 10^9$ pass-km/y	1.42 MJ/pass-km		0.6	24	
4. Truck & Air freight	$18.8 \cdot 10^9$ ton-km/y	2.52 MJ/ton-km			47	
5. Rail	$2.5 \cdot 10^9$ ton-km/y	0.54 MJ/ton-km			1	
6. Shipping	$34.1 \cdot 10^9$ ton-km/y	0.61 MJ/ton-km			21	
1-6 Total				0.6	215	
Agricult., Mining, Construct.						
7. Agriculture	$2060 \cdot 10^6$ \$/y output	2.43MJ/\$(elec) 37.4 MJ/\$ (fos)		5.0	77	
8. Mining	$32 \cdot 10^6$ \$/y output	6.63MJ/\$(-) 98 MJ/\$ (-)		0.2	3	
9. Construction	$2310 \cdot 10^6$ \$/y output	0.82MJ/\$(-) 6.25 MJ/\$ (-)		1.9	14	
7-9 Total				7.1	94	

Commercial, Public						
10 Space Heat (Fossil Fuel)		$73\cdot10^6$ m²	490 MJ/m²/y	0.75 Effic		48
11. – – (Electric)		____ m²	____ MJ/m²/y	____ Effic		
12. Other		$73\cdot10^6$ m²	350 MJ/m²/y	290 MJ/m²/y	25.6	21
10-12 Total					25.6	69
Industry						
13. Metals	$158\cdot10^6$ \$/y	____ (ton/y)	13.6 MJ/\$ (elec)	92 MJ/\$ (fos)	2.1	15
14. Chemicals	$682\cdot10^6$ \$/y	____ (ton/y)	7.1 MJ/\$ (–)	20 MJ/\$ (–)	4.8	13
15. Other Energy Industries	$495\cdot10^6$ \$/y		8.0 MJ/\$ (–)	107 MJ/\$ (–)	4.0	53
16. Non-Energy Industries	$5176\cdot10^6$ \$/y		3.1 MJ/\$ (–)	13.0 MJ/\$ (–)	16.1	67
13-16 Total					27.0	148
Residential						
17. Space Heat (Fossil Fuel)		$2.195\cdot10^6$ Dwellings (fossil)	49550 MJ/dw/y	0.75 Effic		145
18. – – (Electric)		$79\cdot10^3$ Dwellings (elect)	67900 MJ/dw/y	____ Effic	5.3	
			26600 MJ fossil/dwelling			60
19. Appliances		$2.27\cdot10^6$ Dwellings	13650 MJ elect/dwelling		31.0	
17-19 Total					36.3	205
1-19 Total					96.6	731

WAES AGGREGATED DEMAND WORKSHEET

Country: Denmark

Year: 1985

WAES Case: D, DK 3.1%

GNP: $31 \cdot 10^9$ \$

Population: $5.3 \cdot 10^6$

Sector Transportation	Activity Level	Energy Intensity	Other PJ/y	Electric PJ/y	Foss. Fuel PJ/y	Feed-stocks PJ/y
1. Automobiles	$1.6 \cdot 10^6$ autos x 17000 km/auto/y	3.13 MJ/km			85	
2. Airtravel	$8.3 \cdot 10^9$ pass-km/y	3.50 MJ/pass-km			29	
3. Other passengers	$15.7 \cdot 10^9$ pass-km/y	1.46 MJ/pass-km		0.56	22	
4. Truck & Air freight	$17.1 \cdot 10^9$ ton-km/y	2.59 MJ/ton-km			44	
5. Rail	$2.5 \cdot 10^9$ ton-km/y	0.54 MJ/ton-km			1	
6. Shipping	$30.9 \cdot 10^9$ ton-km/y	0.63 MJ/ton-km			20	
1-6 Total				0.56	201	

Agricult., Mining, Construct.	Activity Level	Energy Intensity	Other PJ/y	Electric PJ/y	Foss. Fuel PJ/y	Feed-stocks PJ/y
7. Agriculture	$1924 \cdot 10^6$ \$/y output	2.55 MJ/\$(elec) 39.5 MJ/\$(fos)		4.9	76	
8. Mining	$30 \cdot 10^6$ \$/y output	6.96 MJ/\$(-) 103 MJ/\$(-)		0.2	3	
9. Construction	$2200 \cdot 10^6$ \$/y output	0.84 MJ/\$(-) 6.6 MJ/\$(-)		1.9	15	
7-9 Total				7.0	94	

Commercial, Public						
10 Space Heat (Fossil Fuel)		$66.3 \cdot 10^6$ m²	540 MJ/m²/y	0.75 Effic		48
11. – – (Electric)		____ m²	____ MJ/m²/y	____ Effic		
12. Other		$66.3 \cdot 10^6$ m²	380 MJ/m²/y	320 MJ/m²/y	25.2	21
10-12 Total					25.2	69
Industry						
13. Metals	$135 \cdot 10^6$ $/y	____ (ton/y)	14.3 MJ/$ (elec)	97 MJ/$ (fos)	1.9	13
14. Chemicals	$608 \cdot 10^6$ $/y	____ (ton/y)	7.5 MJ/$ (–)	21 MJ/$ (–)	4.6	13
15. Other Energy Industries	$441 \cdot 10^6$ $/y		8.4 MJ/$ (–)	113 MJ/$ (–)	3.7	50
16. Non-Energy Industries	$4613 \cdot 10^6$ $/y		3.3 MJ/$ (–)	13.7 MJ/$ (–)	15.1	63
13-16 Total					25.3	139
Residential						
17. Space Heat (Fossil Fuel)		$2.175 \cdot 10^6$ Dwellings (fossil)	54100 MJ/dw/y	0.75 Effic		157
18. – – (Electric)		$77 \cdot 10^3$ Dwellings (elect)	72700 MJ/dw/y	1 Effic	5.6	
			25850 MJ fossil/dwelling			58
19. Appliances		$2.25 \cdot 10^6$ Dwellings	13300 MJ elect/dwelling		29.9	
17-19 Total					35.5	215
1-19 Total					93.6	718

WAES AGGREGATED DEMAND WORKSHEET

Country: Denmark

Year: 1985

WAES Case: E, DK 4.1%

GNP: $34 \cdot 10^9$$

Population: $5.3 \cdot 10^6$

Sector Transportation	Activity Level	Energy Intensity	Other PJ/y	Electric PJ/y	Foss. Fuel PJ/y	Feed-stocks PJ/y
1. Automobiles	$1.76 \cdot 10^6$ autos x 17000 km/auto/y	3.46 MJ/km			104	
2. Airtravel	$9.1 \cdot 10^9$ pass-km/y	3.86 MJ/pass-km			35	
3. Other passengers	$17.3 \cdot 10^9$ pass-km/y	1.61 MJ/pass-km		0.7	27	
4. Truck & Air freight	$18.8 \cdot 10^9$ ton-km/y	2.86 MJ/ton-km			54	
5. Rail	$2.5 \cdot 10^9$ ton-km/y	0.54 MJ/ton-km			1	
6. Shipping	$34.1 \cdot 10^9$ ton-km/y	0.70 MJ/ton-km			24	
1-6 Total				0.7	245	
Agricult., Mining, Construct.						
7. Agriculture	$2060 \cdot 10^6$ $/y output	2.82 MJ/$(elec) 43.2 MJ/$(fos)		5.8	89	
8. Mining	$32 \cdot 10^6$ $/y output	7.68 MJ/$(-) 114 MJ/$(-)		0.2	4	
9. Construction	$2310 \cdot 10^6$ $/y output	0.84 MJ/$(-) 6.6 MJ/$(-)		1.9	15	
7-9 Total				7.9	108	

Commercial, Public						
10 Space Heat (Fossil Fuel)		$73 \cdot 10^6$ m²	590 MJ/m²/y	0.75 Effic		57
11. – – (Electric)		______ m²	______ MJ/m²/y	______ Effic		
12. Other		$73 \cdot 10^6$ m²	410 MJ/m²/y	360 MJ/m²/y	29.9	26
10-12 Total					29.9	83
Industry						
13. Metals	$158 \cdot 10^6$ $/y	______ (ton/y)	15 MJ/$ (elec)	102 MJ/$ (fos)	2.4	16
14. Chemicals	$682 \cdot 10^6$ $/y	______ (ton/y)	7.8 MJ/$ (–)	22 MJ/$ (–)	5.3	15
15. Other Energy Industries	$495 \cdot 10^6$ $/y		8.8 MJ/$ (–)	119 MJ/$ (–)	4.4	59
16. Non-Energy Industries	$5176 \cdot 10^6$ $/y		3.4 MJ/$ (–)	14.4 MJ/$ (–)	17.8	75
13-16 Total					29.9	165
Residential						
17. Space Heat (Fossil Fuel)		$2.195 \cdot 10^6$ Dwellings (fossil)	64600 MJ/dw/y	0.75 Effic		189
18. – – (Electric)		$79 \cdot 10^3$ Dwellings (elect)	81000 MJ/dw/y	1 Effic	6.4	
			27700 MJ fossil/dwelling			63
19. Appliances		$2.27 \cdot 10^6$ Dwellings	15300 MJ elect/dwelling		34.8	
17-19 Total					41.2	252
1-19 Total					109.6	853

WAES AGGREGATED DEMAND WORKSHEET

Country: Denmark

Year: 2000

WAES Case: C1, DK 3.0%

GNP: $ $53 \cdot 10^9$

Population: $5.4 \cdot 10^6$

Sector Transportation	Activity Level	Energy Intensity	Other PJ/y	Electric PJ/y	Foss. Fuel PJ/y	Feed-stocks PJ/y
1. Automobiles	$2.55 \cdot 10^6$ autos x 18000 km/auto/y	2.50 MJ/km			115	
2. Airtravel	$14 \cdot 10^9$ pass-km/y	2.80 MJ/pass-km			39	
3. Other passengers	$27 \cdot 10^9$ pass-km/y	1.20 MJ/pass-km		1.6	31	
4. Truck & Air freight	$24 \cdot 10^9$ ton-km/y	3.1 MJ/ton-km			75	
5. Rail	$4.6 \cdot 10^9$ ton-km/y	0.22 MJ/ton-km		1.0	-	
6. Shipping	$52 \cdot 10^9$ ton-km/y	0.50 MJ/ton-km			26	
1-6 Total				2.6	286	
Agricult., Mining, Construct.						
7. Agriculture	$2800 \cdot 10^6$ $/y output	32.2 MJ/$		6.3	90	
8. Mining	$40 \cdot 10^6$ $/y output	78 MJ/$		0.3	3	
9. Construction	$2930 \cdot 10^6$ $/y output	5.0 MJ/$		2.6	15	
7-9 Total				9.2	108	

Commercial, Public						
10 Space Heat (Fossil Fuel)		$119\cdot10^6$ m²	400 MJ/m²/y	0.8 Effic		60
11. – – (Electric)		___ m²	___ MJ/m²/y	___ Effic		
12. Other		$119\cdot10^6$ m²	325 MJ/m²/y	210 MJ/m²y	39.0	25
10-12 Total					39.0	85
Industry						
13. Metals	$269\cdot10^6$ \$/y	___ (ton/y)	12.6 MJ/\$ (elec)	79 MJ/\$ (fos)	3.4	21
14. Chemicals	$1159\cdot10^6$ \$/y	___ (ton/y)	6.6 MJ/\$ (–)	17 MJ/\$ (–)	7.6	20
15. Other Energy Industries	$841\cdot10^6$ \$/y		7.4 MJ/\$ (–)	92 MJ/\$ (–)	6.2	77
16. Non-Energy Industries	$8.8\cdot10^9$ \$/y		2.88 MJ/\$ (–)	11.2 MJ/\$ (–)	25.3	99
13-16 Total					42.5	217
Residential						
17. Space Heat (Fossil Fuel)		$2.219\cdot10^6$ Dwellings (fossil)	39300 MJ/dw/y	0.8 Effic		109
18. – – (Electric)		$385\cdot10^3$ Dwellings (elect)	51400 MJ/dw/y	1 Effic	19.8	
			26500 MJ fossil/dwelling			69
19. Appliances		$2.604\cdot10^6$ Dwellings	16500 MJ elect/dwelling		42.9	
17-19 Total					62.7	178
1-19 Total					156.0	874

WAES AGGREGATED DEMAND WORKSHEET

Country: Denmark

Year: 2000

WAES Case: C2, DK 3.0%

GNP: $53 \cdot 10^9$ \$

Population: $5.4 \cdot 10^6$

Sector	Activity Level	Energy Intensity	Other PJ/y	Electric PJ/y	Foss. Fuel PJ/y	Feed-stocks PJ/y
Transportation						
1. Automobiles	$2.55 \cdot 10^6$ autos x 18000 km/auto/y	2.69 MJ/km			123	
2. Airtravel	$14 \cdot 10^9$ pass-km/y	3.00 MJ/pass-km			42	
3. Other passengers	$27 \cdot 10^9$ pass-km/y	1.26 MJ/pass-km		1.9	32	
4. Truck & Air freight	$24 \cdot 10^9$ ton-km/y	3.34 MJ/ton-km			80	
5. Rail	$4.6 \cdot 10^9$ ton-km/y	0.24 MJ/ton-km		1.1	-	
6. Shipping	$52 \cdot 10^9$ ton-km/y	0.54 MJ/ton-km			28	
1-6 Total				3.0	305	
Agricult., Mining, Construct.						
7. Agriculture	$2800 \cdot 10^6$ \$/y output	34.7 MJ/\$		7.3	97	
8. Mining	$40 \cdot 10^6$ \$/y output	91 MJ/\$		0.3	4	
9. Construction	$2930 \cdot 10^6$ \$/y output	5.8 MJ/\$		3.0	17	
7-9 Total				10.6	118	

Commercial, Public						
10 Space Heat (Fossil Fuel)		$119\cdot10^6$ m²	410 MJ/m²/y	0.8 Effic		61
11. – – (Electric)		____ m²	____ MJ/m²/y	____ Effic		
12. Other		$119\cdot10^6$ m²	380 MJ/m²y	240 MJ/m²y	45.2	29
10-12 Total					45.2	90
Industry						
13. Metals	$269\cdot10^6$ \$/y	____ (ton/y)	14.7 MJ/\$ (elec)	85 MJ/\$ (fos)	3.9	23
14. Chemicals	$1159\cdot10^6$ \$/y	____ (ton/y)	7.7 MJ/\$ (–)	18 MJ/\$ (–)	8.9	21
15. Other Energy Industries	$841\cdot10^6$ \$/y		8.6 MJ/\$ (–)	99 MJ/\$ (–)	7.2	83
16. Non-Energy Industries	$8.8\cdot10^9$ \$/y		3.34 MJ/\$ (–)	12.1 MJ/\$ (–)	29.4	106
13-16 Total					49.4	233
Residential						
17. Space Heat (Fossil Fuel)		$2.19\cdot10^6$ Dwellings (fossil)	42300 MJ/dw/y	0.8 Effic		116
18. – – (Electric)		$415\cdot10^3$ Dwellings (elect)	55400 MJ/dw/y	1.0 Effic	23.0	
			30800 MJ fossil/dwelling			80
19. Appliances		$2.60\cdot10^6$ Dwellings	19200 MJ elect/dwelling		50.0	
17-19 Total					73.0	196
1-19 Total					181.2	942

WAES AGGREGATED DEMAND WORKSHEET

Country: Denmark

Year: 2000

WAES Case: D7, DK 2% p.a.

GNP: $42 \cdot 10^9$ $

Population: $5.4 \cdot 10^6$

Sector	Activity Level	Energy Intensity	Other PJ/y	Electric PJ/y	Foss. Fuel PJ/y	Feed-stocks PJ/y
Transportation						
1. Automobiles	$2.15 \cdot 10^6$ autos x 17000 km/auto/y	2.77 MJ/km			101	
2. Airtravel	$11.2 \cdot 10^9$ pass-km/y	3.10 MJ/pass-km			35	
3. Other passengers	$21 \cdot 10^9$ pass-km/y	1.29 MJ/pass-km		1.6	25	
4. Truck & Air freight	$23 \cdot 10^9$ ton-km/y	3.40 MJ/ton-km			78	
5. Rail	$3.4 \cdot 10^9$ ton-km/y	0.24 MJ/ton-km		0.8	-	
6. Shipping	$42 \cdot 10^9$ ton-km/y	0.56 MJ/ton-km			24	
1-6 Total				2.4	263	
Agricult., Mining, Construct.						
7. Agriculture	$2370 \cdot 10^6$ $/y output	36.6 MJ/$		6.0	87	
8. Mining	$35 \cdot 10^6$ $/y output	96 MJ/$		0.3	3	
9. Construction	$2550 \cdot 10^6$ $/y output	6.1 MJ/$		2.5	16	
7-9 Total				8.8	106	

Commercial, Public

10 Space Heat (Fossil Fuel)	$92\cdot10^6$ m²	450 MJ/m²/y	0.8 Effic		52
11. – – (Electric)	___ m²	___ MJ/m²/y	___ Effic		
12. Other	$92\cdot10^6$ m²	380 MJ/m²y	265 MJ/m²y	35.0	24
10-12 Total				35.0	76

Industry

13. Metals	$193\cdot10^6$ \$/y	___ (ton/y)	14.3 MJ/\$ (elec)	90 MJ/\$ (fos)	2.8	17
14. Chemicals	$868\cdot10^6$ \$/y	___ (ton/y)	7.5 MJ/\$ (–)	19.5 MJ/\$ (–)	6.5	17
15. Other Energy Industries	629.10^6 \$/y		8.4 MJ/\$ (–)	105 MJ/\$ (–)	5.3	66
16. Non-Energy Industries	$6.6\cdot10^9$ \$/y		3.3 MJ/\$ (–)	12.7 MJ/\$ (–)	21.7	84
13-16 Total					36.3	184

Residential

17. Space Heat (Fossil Fuel)	$2.16\cdot10^6$ Dwellings (fossil)	45600 MJ/dw/y	0.8 Effic		123
18. – – (Electric)	$361\cdot10^3$ Dwellings (elect)	55400 MJ/dw/y	1 Effic	20.0	
		27500 MJ fossil/dwelling			69
19. Appliances	$2.522\cdot10^6$ Dwellings	15450 MJ elect/dwelling		39.0	
17-19 Total				59.0	192
1-19 Total				141.5	821

WAES AGGREGATED DEMAND WORKSHEET

Country: Denmark

Year: 2000

WAES Case: D8, DK 2% p.a.

GNP: $42 \cdot 10^9$ $

Population: $5.4 \cdot 10^6$

Sector Transportation	Activity Level	Energy Intensity	Other PJ/y	Electric PJ/y	Foss. Fuel PJ/y	Feed-stocks PJ/y
1. Automobiles	$2.15 \cdot 10^6$ autos x 17000 km/auto/y	2.99 MJ/km			109	
2. Airtravel	$11.2 \cdot 10^9$ pass-km/y	3.35 MJ/pass-km			38	
3. Other passengers	$21 \cdot 10^9$ pass-km/y	1.40 MJ/pass-km		1.8	28	
4. Truck & Air freight	$23 \cdot 10^9$ ton-km/y	3.60 MJ/ton-km			83	
5. Rail	$3.4 \cdot 10^9$ ton-km/y	0.26 MJ/ton-km		0.9	-	
6. Shipping	$42 \cdot 10^9$ ton-km/y	0.60 MJ/ton-km			25	
1-6 Total				2.7	283	
Agricult., Mining, Construct.						
7. Agriculture	$2370 \cdot 10^6$ $/y output	40 MJ/$		7.1	95	
8. Mining	$35 \cdot 10^6$ $/y output	103 MJ/$		0.3	4	
9. Construction	$2550 \cdot 10^6$ $/y output	6.6 MJ/$		2.9	17	
7-9 Total				10.3	116	

Commercial, Public					
10 Space Heat (Fossil Fuel)	$92 \cdot 10^6$ m²	480 MJ/m²/y	0.8 Effic		55
11. – – (Electric)	___ m²	___ MJ/m²/y	___ Effic		
12. Other	$92 \cdot 10^6$ m²	440 MJ/m²/y	285 MJ/m²/y	40,5	26
10-12 Total				40,5	81
Industry					
13. Metals $193 \cdot 10^6$ \$/y	___ (ton/y)	16.6 MJ/\$ (elec)	97 MJ/\$ (fos)	3.2	19
14. Chemicals $868 \cdot 10^6$ \$/y	___ (ton/y)	8.7 MJ/\$ (–)	21 MJ/\$ (–)	7.6	18
15. Other Energy Industries $629 \cdot 10^6$ \$/y		9.8 MJ/\$ (–)	113 MJ/\$ (–)	6.2	71
16. Non-Energy Industries $6.6 \cdot 10^9$ \$/y		3.8 MJ/\$ (–)	13.7 MJ/\$ (–)	25.0	90
13-16 Total				42.0	198
Residential					
17. Space Heat (Fossil Fuel)	$2.14 \cdot 10^6$ Dwellings (fossil)	49 140 MJ/dw/y	0.8 Effic		131
18. – – (Electric)	$382 \cdot 10^3$ Dwellings (elect)	59 700 MJ/dw/y	1 Effic	22.8	
		32 100 MJ fossil/dwelling			81
19. Appliances	$2.52 \cdot 10^6$ Dwellings	17 850 MJ elect/dwelling		45.0	
17-19 Total				67.8	212
1-19 Total				163.3	890

WAES AGGREGATED DEMAND WORKSHEET

Country: Denmark

Year: 2000

WAES Case: D3, DK 2.0%

GNP: $42 \cdot 10^9$ \$

Population: $5.4 \cdot 10^6$

Sector Transportation	Activity Level	Energy Intensity	Other PJ/y	Electric PJ/y	Foss. Fuel PJ/y	Feed-stocks PJ/y
1. Automobiles	$2.15 \cdot 10^6$ autos x 17000 km/auto/y	2.57 MJ/km			94	
2. Airtravel	$11.2 \cdot 10^9$ pass-km/y	2.88 MJ/pass-km			32	
3. Other passengers	$21 \cdot 10^9$ pass-km/y	1.23 MJ/pass-km		1.2	26	
4. Truck & Air freight	$23 \cdot 10^9$ ton-km/y	3.2 MJ/ton-km		-	74	
5. Rail	$3.4 \cdot 10^9$ ton-km/y	0.22 MJ/ton-km		0.8	-	
6. Shipping	$42 \cdot 10^9$ ton-km/y	0.52 MJ/ton-km		-	22	
1-6 Total				2.0	248	
Agricult., Mining, Construct.						
7. Agriculture	$2370 \cdot 10^6$ \$/y output	2.36 MJ/\$(elec) 34 MJ/\$ (fos)		5.6	81	
8. Mining	$35 \cdot 10^6$ \$/y output	2.51 MJ/\$(elec) 82 MJ/\$ (-)		0.3	3	
9. Construction	$2550 \cdot 10^6$ \$/y output	0.91 MJ/\$(-) 5.3 MJ/\$ (-)		2.3	13	
7-9 Total				8.2	97	

Commercial, Public

10 Space Heat (Fossil Fuel)		$92 \cdot 10^6$ m²	440 MJ/m²/y	0.8 Effic		51
11. – – (Electric)		____ m²	____ MJ/m²/y	____ Effic		
12. Other		$92 \cdot 10^6$ m²	350 MJ/m²/y	230 MJ/m²/y	32.2	21
10-12 Total					32.2	72
Industry						
13. Metals	$193 \cdot 10^6$ $/y	____ (ton/y)	13.2 MJ/$ (elec)	83 MJ/$ (fos)	2.5	16
14. Chemicals	$868 \cdot 10^6$ $/y	____ (ton/y)	7.0 MJ/$ (–)	18 MJ/$ (–)	6.1	16
15. Other Energy Industries	$629 \cdot 10^6$ $/y		7.8 MJ/$ (–)	97 MJ/$ (–)	4.9	61
16. Non-Energy Industries	$6{,}6 \cdot 10^9$ $/y		3.07 MJ/$ (–)	11.8 MJ/$ (–)	20.3	78
13-16 Total					33.8	171
Residential						
17. Space Heat (Fossil Fuel)		$2.16 \cdot 10^6$ Dwellings (fossil)	42900 MJ/dw/y	0.8 Effic		116
18. – – (Electric)		$361 \cdot 10^3$ Dwellings (elect)	55000 MJ/dw/y	1 Effic	19.9	-
			25800 MJ fossil/dwelling			65
19. Appliances		$2.52 \cdot 10^6$ Dwellings	16100 MJ elect/dwelling		40.6	-
17-19 Total					60.5	181
1-19 Total					136.7	769

NATIONAL INPUT WORKSHEET FOR SUPPLY/DEMAND INTEGRATION

Country: Denmark

Year: 1972

Case: Base year

Units: PJ/y

	(a)	(b)	(c)	(d)	(e)	(f)	(g)	(h)	(i)	(j)	(k)
	Coal	Petroleum	(Syn. Liquids)	Nat. Gas	(Syn. Gas)	(Heat)	(Electricity)[††]	Hydro Energy	Nuclear Energy	Geothermal energy and other	Total
(1) Transport	–	144			1	–	0.3				145
(2) Industry	7	109			3	2	14.8				136
(3) Agric., Mining, Construction	–	61			1	–	4.4				67
(4) Commercial	1	27			1	6	9.3				44
(5) Public	1	22			1	6	5.7				36
(6) Residential	2	158			7	41	20.1				228
(7) Non-energy Uses	–	–			–	–	–				
(8) Final Energy Demand*	11	521			14	55	54.6				656
(9) Electricity**	45	144			–	–23	–68.3				98
(10) Syn. Gas	2	14			–15						1
(11) Syn. Liquids											
(12) Heat		56				–46				2	12
(13) Energy Sector Self Consumption & conversion losses		36			1	14	5.4				56
(14) Primary Energy Input	58	771			0	0	–8.2			2	823
(15) Indigenous Supply		4								2	6
(16 Imports***	58	767									825
(17) Exports							8.2				8.2

*Includes non-energy uses.

**Blocks 9g, 10e 11c, and 12f must be negative to avoid double counting of energy from primary fuels.

***Includes imported products.

††Includes only electricity to be purchased by each sector.

NATIONAL INPUT WORKSHEET FOR SUPPLY/DEMAND INTEGRATION

Country: Denmark

Year: 1985

Case: A, DK 4.1%

Units: PJ/y

	(a)	(b)	(c)	(d)	(e)	(f)	(g)	(h)	(i)	(j)	(k)
	Coal	Petro-leum	(Syn. Liquids)	Nat. Gas	(Syn. Gas)	(Heat)	(Electri-city)††	Hydro Energy	Nuclear Energy	Geothermal energy and other	Total
(1) Transport	–	195		–		–	0.5				196
(2) Industry	34	96		12			25.8				168
(3) Agric., Mining, Construction	–	87		–		–	6.5				93
(4) Commercial	–	15		4		13	14.0				46
(5) Public	–	13		5		14	10.0				42
(6) Residential	–	49		64		78	34.0				225
(7) Non-energy Uses	–	–		–		–	–				
(8) Final Energy Demand*	34	455		85		105	90.8				770
(9) Electricity**	161	74		20		-55	-99.9		58		158
(10) Syn. Gas		–				–					
(11) Syn. Liquids		–				–					
(12) Heat	–	45				-50				10	5
(13) Energy Sector Self Consumption						-21					-21
& conversion losses	–	40		9		21	9.1				79
(14) Primary Energy Input	195	614		114		0	0		58	10	991
(15) Indigenous Supply	0	63		82					225	10	380
(16 Imports***	195	551		32					0	0	778
(17) Exports									167		167

*Includes non-energy uses.

**Blocks 9g, 10e 11c, and 12f must be negative to avoid double counting of energy from primary fuels.

***Includes imported products.

††Includes only electricity to be purchased by each sector.

NATIONAL INPUT WORKSHEET FOR SUPPLY/DEMAND INTEGRATION

Country: Denmark

Year: 1985

Case: B, DK 3.1%

Units: PJ/y

	(a)	(b)	(c)	(d)	(e)	(f)	(g)	(h)	(i)	(j)	(k)
	Coal	Petro- leum	(Syn. Liquids)	Nat. Gas	(Syn. Gas)	(Heat)	(Electri- city)[††]	Hydro Energy	Nuclear Energy	Geothermal energy and other	Total
(1) Transport	–	177		–		–	0.5				178
(2) Industry	30	83		12		–	23.0				148
(3) Agric., Mining, Construction		83		–		–	6.2				89
(4) Commercial		12		4		13	13.0				42
(5) Public		10		5		14	8.9				38
(6) Residential		32		64		78	31.2				205
(7) Non-energy Uses	–	–		–		–	–				–
(8) Final Energy Demand*	30	397		85		105	82.8				700
(9) Electricity**	144	66		20		-55	-91.1		58		142
(10) Syn. Gas						–					
(11) Syn. Liquids						–					
(12) Heat	–	45				-50				10	5
(13) Energy Sector Self Consumption						-21					-21
& conversion losses	–	36		9		21	8.3				74
(14) Primary Energy Input	174	544		114		0	0		58	10	900
(15) Indigenous Supply	0	63		82					225	10	380
(16 Imports***	174	481		32					0	0	687
(17) Exports									167		167

*Includes non-energy uses.

**Blocks 9g, 10e 11c, and 12f must be negative to avoid double counting of energy from primary fuels.

***Includes imported products.

[††]Includes only electricity to be purchased by each sector.

NATIONAL INPUT WORKSHEET FOR SUPPLY/DEMAND INTEGRATION

Country: Denmark

Year: 1985

Case: C, DK 4.1%

Units: PJ/y

	(a)	(b)	(c)	(d)	(e)	(f)	(g)	(h)	(i)	(j)	(k)
	Coal	Petroleum	(Syn. Liquids)	Nat. Gas	(Syn. Gas)	(Heat)	(Electricity)[††]	Hydro Energy	Nuclear Energy	Geothermal energy and other	Total
(1) Transport	-	215		-		-	0.6				216
(2) Industry	34	102		12		-	27.0				175
(3) Agric., Mining, Construction	-	94		-		-	7.1				101
(4) Commercial	-	20		4		12	15.4				51
(5) Public	-	16		5		13	10.2				44
(6) Residential	-	61		64		80	36.3				241
(7) Non-energy Uses	-	-		-		-	-				-
(8) Final Energy Demand*	34	508		85		105	96.6				828
(9) Electricity**	174	79		20		-55	-106.3		58		170
(10) Syn. Gas											
(11) Syn. Liquids											
(12) Heat	-	45				-50				10	5
(13) Energy Sector Self Consumption						-21					-21
& conversion losses		44		9		21	9.7				84
(14) Primary Energy Input	208	676		114		0	0		58	10	1066
(15) Indigenous Supply	0	63		82					225	10	380
(16 Imports***	208	613		32					0	0	853
(17) Exports									167		167

*Includes non-energy uses.

**Blocks 9g, 10e 11c, and 12f must be negative to avoid double counting of energy from primary fuels.

***Includes imported products.

††Includes only electricity to be purchased by each sector.

NATIONAL INPUT WORKSHEET FOR SUPPLY/DEMAND INTEGRATION

Country: Denmark

Year: 1985

Case: D, DK 3.1%

Units: PJ/y

	(a) Coal	(b) Petro-leum	(c) (Syn. Liquids)	(d) Nat. Gas	(e) (Syn. Gas)	(f) (Heat)	(g) (Electri-city)††	(h) Hydro Energy	(i) Nuclear Energy	(j) Geothermal energy and other	(k) Total
(1) Transport	–	201				–	0.6				202
(2) Industry	34	105				–	25.3				164
(3) Agric., Mining, Construction	–	94				–	7.0				101
(4) Commercial	–	27				10	15.4				52
(5) Public	–	21				11	9.8				42
(6) Residential	–	148				67	35.5				251
(7) Non-energy Uses	–	–				–	–				–
(8) Final Energy Demand*	34	596				88	93.6				812
(9) Electricity**	128	124				−38	−103		58		169
(10) Syn. Gas	–	–				–	–				–
(11) Syn. Liquids	–	–				–					–
(12) Heat	–	45				−50				10	5
(13) Energy Sector Self Consumption						−18					−18
& conversion losses	–	54				18	9.4				81
(14) Primary Energy Input	162	819				0	0		58	10	1049
(15) Indigenous Supply	0	63		82					225	10	380
(16 Imports***	162	756		0					0	0	918
(17) Exports				82					167		249

*Includes non-energy uses.

**Blocks 9g, 10e 11c, and 12f must be negative to avoid double counting of energy from primary fuels.

***Includes imported products.

††Includes only electricity to be purchased by each sector.

NATIONAL INPUT WORKSHEET FOR SUPPLY/DEMAND INTEGRATION

Country: Denmark

Year: 1985

Case: E, DK 4.1%

Units: PJ/y

	(a)	(b)	(c)	(d)	(e)	(f)	(g)	(h)	(i)	(j)	(k)
	Coal	Petroleum	(Syn. Liquids)	Nat. Gas	(Syn. Gas)	(Heat)	(Electricity)††	Hydro Energy	Nuclear Energy	Geothermal energy and other	Total
(1) Transport	–	245					0.7				246
(2) Industry	39	126					29.9				195
(3) Agric., Mining, Construction		108					7.9				116
(4) Commercial		34				10	18.5				63
(5) Public		28				11	11.4				50
(6) Residential		185				67	41.2				293
(7) Non-energy Uses						–					
(8) Final Energy Demand*	39	726				88	109.6				963
(9) Electricity**	161	145				-38	-120.6		58		205
(10) Syn. Gas											
(11) Syn. Liquids											
(12) Heat	–	45				-50				10	5
(13) Energy Sector Self Consumption						-18					-18
& conversion losses		64				18	11				93
(14) Primary Energy Input	200	980				0	0		58	10	1248
(15) Indigenous Supply	0	63		82					225	10	380
(16 Imports***	200	917		0					0	0	1117
(17) Exports				82					167		249

*Includes non-energy uses.

**Blocks 9g, 10e 11c, and 12f must be negative to avoid double counting of energy from primary fuels.

***Includes imported products.

††Includes only electricity to be purchased by each sector.

NATIONAL INPUT WORKSHEET FOR SUPPLY/DEMAND INTEGRATION

Country: Denmark

Year: 2000

Case: C1, DK 3%

Units: PJ/y

	(a)	(b)	(c)	(d)	(e)	(f)	(g)	(h)	(i)	(j)	(k)
	Coal	Petro- leum	(Syn. Liquids)	Nat. Gas	(Syn. Gas)	(Heat)	(Electri- city)††	Hydro Energy	Nuclear Energy	Geothermal energy and other	Total
(1) Transport		286		-		-	2.6				289
(2) Industry	100	63		42		12	42.5				259
(3) Agric., Mining, Construction		86		10		12	9.2				117
(4) Commercial		11		7		24	22.4				64
(5) Public		11		8		24	16.6				60
(6) Residential		37		57		84	62.7				241
(7) Non-energy Uses											
(8) Final Energy Demand*	100	494		124		156	156				1030
(9) Electricity**	324	95		15		-117	-172		116		261
(10) Syn. Gas											
(11) Syn. Liquids											
(12) Heat	11	10		12		- 39				10	4
(13) Energy Sector Self Consumption						- 31					-31
& conversion losses		42		13		31	16				102
(14) Primary Energy Input	435	641		164		0	0		116	10	1366
(15) Indigenous Supply	0	59		67					450	10	586
(16 Imports***	435	582		97					0		1114
(17) Exports									334		334

*Includes non-energy uses.

**Blocks 9g, 10e 11c, and 12f must be negative to avoid double counting of energy from primary fuels.

***Includes imported products.

††Includes only electricity to be purchased by each sector.

NATIONAL INPUT WORKSHEET FOR SUPPLY/DEMAND INTEGRATION

Country: Denmark

Year: 2000

Case: C2, DK 3%

Units: PJ/y

	(a) Coal	(b) Petro-leum	(c) (Syn. Liquids)	(d) Nat. Gas	(e) (Syn. Gas)	(f) (Heat)	(g) (Electri-city)††	(h) Hydro Energy	(i) Nuclear Energy	(j) Geothermal energy and other	(k) Total
(1) Transport		305					3.0				308
(2) Industry	70	107		42		14	49.4				282
(3) Agric., Mining, Construction		94		10		14	10.6				129
(4) Commercial		14		6		26	26.2				72
(5) Public		11		7		26	19.0				63
(6) Residential		61		55		80	73.0				269
(7) Non-energy Uses		-		-		-	-				-
(8) Final Energy Demand*	70	592		120		160	181.2				1123
(9) Electricity**	103	103		23		-120	-199		400		310
(10) Syn. Gas											
(11) Syn. Liquids											
(12) Heat	12	14		8		- 40				10	4
(13) Energy Sector Self Consumption						- 32					-32
& conversion losses		50		13		32	18				113
(14) Primary Energy Input	185	759		164		0	0		400	10	1518
(15) Indigenous Supply	0	59		67					450	10	586
(16 Imports***	185	700		97					0	0	982
(17) Exports									50		50

*Includes non-energy uses.

**Blocks 9g, 10e 11c, and 12f must be negative to avoid double counting of energy from primary fuels.

***Includes imported products.

††Includes only electricity to be purchased by each sector.

NATIONAL INPUT WORKSHEET FOR SUPPLY/DEMAND INTEGRATION

Country: Denmark

Year: 2000

Case: D7, DK 2%

Units: PJ/y

	(a)	(b)	(c)	(d)	(e)	(f)	(g)	(h)	(i)	(j)	(k)
	Coal	Petroleum	(Syn. Liquids)	Nat. Gas	(Syn. Gas)	(Heat)	(Electricity)[††]	Hydro Energy	Nuclear Energy	Geothermal energy and other	Total
(1) Transport		263					2.4				265
(2) Industry	90	42		42		10	36.3				220
(3) Agric., Mining, Construction		86		10		10	8.8				115
(4) Commercial		13		7		20	21.0				61
(5) Public		10		8		18	14.0				50
(6) Residential		45		65		82	59.0				251
(7) Non-energy Uses											
(8) Final Energy Demand*	90	459		132		140	141.5				963
(9) Electricity**	233	137		12		-105	-156		116		237
(10) Syn. Gas											
(11) Syn. Liquids											
(12) Heat	10	12		7		- 35				10	4
(13) Energy Sector Self Consumption						- 28					-28
& conversion losses		43		13		28	14.2				98
(14) Primary Energy Input	333	651		164		0	0		116	10	1274
(15) Indigenous Supply	0	59		67					450	10	586
(16 Imports***	333	592		97					0		1022
(17) Exports									334		334

*Includes non-energy uses.

**Blocks 9g, 10e 11c, and 12f must be negative to avoid double counting of energy from primary fuels.

***Includes imported products.

††Includes only electricity to be purchased by each sector.

NATIONAL INPUT WORKSHEET FOR SUPPLY/DEMAND INTEGRATION

Country: Denmark

Year: 2000

Case: D8, DK 2%

Units: PJ/y

	(a)	(b)	(c)	(d)	(e)	(f)	(g)	(h)	(i)	(j)	(k)
	Coal	Petroleum	(Syn. Liquids)	Nat. Gas	(Syn. Gas)	(Heat)	(Electricity)††	Hydro Energy	Nuclear Energy	Geothermal energy and other	Total
(1) Transport		283					2.7				286
(2) Industry	60	84		42		12	42.0				240
(3) Agric., Mining, Construction		94		10		12	10.3				126
(4) Commercial		14		7		22	24.2				67
(5) Public		11		8		19	16.3				54
(6) Residential		56		65		91	67.8				280
(7) Non-energy Uses		–		–		–	–				–
(8) Final Energy Demand*	60	542		132		156	163.3				1053
(9) Electricity**	58	103		12		-117	-180		400		276
(10) Syn. Gas											
(11) Syn. Liquids											
(12) Heat	12	14		7		- 39				10	4
(13) Energy Sector Self Consumption						- 31					-31
& conversion losses		46		13		31	16.2				106
(14) Primary Energy Input	130	705		164		0	0		400	10	1409
(15) Indigenous Supply	0	59		67					450	10	586
(16 Imports***	130	646		97					0	0	873
(17) Exports									50		50

*Includes non-energy uses.

**Blocks 9g, 10e 11c, and 12f must be negative to avoid double counting of energy from primary fuels.

***Includes imported products.

††Includes only electricity to be purchased by each sector.

NATIONAL INPUT WORKSHEET FOR SUPPLY/DEMAND INTEGRATION

Country: Denmark

Year: 2000

Case: D-3, DK 2.0%

Units: PJ/y

	(a)	(b)	(c)	(d)	(e)	(f)	(g)	(h)	(i)	(j)	(k)
	Coal	Petro- leum	(Syn. Liquids)	Nat. Gas	(Syn. Gas)	(Heat)	(Electri- city)††	Hydro Energy	Nuclear Energy	Geothermal energy and other	Total
(1) Transport		248		–		–	2.0				250
(2) Industry	90	36		35		10	33.8				205
(3) Agric., Mining, Construction		78		10		9	8.2				105
(4) Commercial		12		8		16	18.9				55
(5) Public		11		7		18	13.3				49
(6) Residential		44		59		78	60.5				242
(7) Non-energy Uses		–		–		–	–				
(8) Final Energy Demand*	90	249		119		131	136.7				906
(9) Electricity**	210	72		24		-100	-150		174		230
(10) Syn. Gas											
(11) Syn. Liquids											
(12) Heat	8	9		8		- 31				10	4
(13) Energy Sector Self Consumption						- 26					- 26
& conversion losses	–	36		13		26	13				88
(14) Primary Energy Input	308	546		164		0	0		174	10	1202
(15) Indigenous Supply	0	59		67					450	10	586
(16 Imports***	308	487		97					0		892
(17) Exports									276		276

*Includes non-energy uses.

**Blocks 9g, 10e 11c, and 12f must be negative to avoid double counting of energy from primary fuels.

***Includes imported products.

††Includes only electricity to be purchased by each sector.

Finland

9.1 SUMMARY

The Finnish WAES scenarios are based on the following economic assumptions: high growth rates of 4.4 percent per year for 1976 to 1985 and 3.5 percent per year for 1985 to 2000; low growth rates of 3.25 percent per year for 1976 to 1985 and 2.0 percent per year for 1985 to 2000. These growth rates are comparatively low compared to the 1960 to 1972 rate of 4.9 percent per year, reflecting a gradual slowing down and saturation of economic growth that is expected in Finland as in other industrialized countries. Because of higher energy prices and structural development of the economy, the rate of energy demand per unit output is expected to decrease from about 1.3 in the 1960s to close to 0.8 in the 1990s. Both these factors cause a significant decrease in the energy demand growth compared to the pre-oil crisis era.

At the same time that total energy demand is saturating, the economic structure is changing. Energy demand in the residential sector is expected to grow very slowly for many reasons. The growth of the residential building volume is already saturating, and this trend will become quite clear in the 1990s. At the same time houses will be better insulated, and more efficient heating methods such as district heating are spreading. On the other hand, industrial production is still expected to grow faster than the total economy. Although energy efficiencies are improving, the share of industrial energy demand is increasing. The ranges of sectoral energy demand shares in different scenarios are summarized in Table 9.1.

Table 9.1
Share of energy demand by sector and source (%)

	1972	1985	2000
Transport	13	14-15	13-14
Industry	45	46-48	51-56
Agriculture, mining, and construction	5	4-5	4
Public and commercial	10	12-13	12
Residential	27	20-23	15-19
Nonenergy	1	1	1
Fossil fuels	74	65-70	55-63
Nuclear	—	11-14	16-25
Other indigenous sources	24	18-20	19-21

The most remarkable shift among energy supply sources in Finland from 1972 to 2000 is the introduction of nuclear energy. At the same time the share of imported fossil fuels is slowly decreasing. The share of other indigenous energy sources is decreasing but will stabilize after the introduction of large-scale peat utilization. The shares of different energy sources vary within the limits shown in Table 9.1. These energy supply calculations are based on the calorific values of fuels, and thus hydroelectric energy has a lower value than its substitutes. The residual is comprised of imported electricity.

9.2 METHODOLOGY

The method used in the Finnish supply-demand integration study is in principle the same as in the first two WAES reports. First, secondary energy demand and indigenous primary energy supply are determined. Then the necessary imports are calculated and primary energy allocated to various conversion processes.

The Finnish supply-demand integration has been carried out manually and step by step. There are different criteria for determining the energy conversion processes. First, the scale of the Finnish energy sector is small compared to the size of energy supply units. Thus each major supply unit (e.g., a nuclear power plant) demands a separate analysis and cannot be analyzed on a continuous scale. Second, there are supply sources such as industrial wastes and process fuels that depend on the production volume and thus have to be analyzed together with the demand.

The indigenous energy supply is very limited and suited only to specific purposes. Thus the supply-demand integration is very much determined by the demand, if no import constraints are assumed. There are not many degrees of freedom; questions arise mainly with regard to the fuels to be used for district heating and electricity generation.

There are some additional assumptions included in the scenarios. Electricity import is maintained at the level of existing trade agreements. Crude oil refinement is assumed to take place domestically. No synthetic gas or oil industries are included in the Finnish national scenarios.

9.3 ENERGY DEMAND IN THE YEAR 2000

The WAES-Finland national energy demand scenarios have been worked out on the basis of a preliminary and unofficial study made at the Energy Department of the Ministry of Trade and Industry. In that study Finnish energy demand was disaggregated into detailed levels and analyzed up to 1990. The technical energy demand coefficients have been aggregated from this study and determined for the following sectoral division:

1. Transportation: automobiles; other transportation
2. Industry: paper and pulp; primary metal; chemicals; other industry
3. Residential
4. Public and commercial
5. Other consumption

The study made by the Finnish Energy Department was based on constant energy prices and steady 4 percent per year economic growth from 1976 to 1990. Since the WAES scenarios differ from these assumptions, the energy demand coefficients had to be modified. This modification was carried out with the aid of the WAES econometric function, i.e., using the estimated or assumed price and income elasticities.

The demand was estimated separately for the various fuels (oil, coal, gas, heat, electricity, and indigenous fuels). The fuel mix includes some implicit assumptions. The demand for oil is kept low and the share of district heat is rather high. Also, the demand for indigenous fuels such as peat is estimated to be high.

Table 9.2
Gross national product (1975 Fmk)

	1976	1985	2000
Case C	92,273,000	135,950,000	227,764,000
Case D	92,273,000	123,051,000	165,610,000

Note: 1 Fmk = 0.26 U.S. $.

Energy demand was estimated for WAES scenarios C-1, C-2, D-7, and D-8. As a simplifying assumption, however, the demand was considered to be independent of the choice between nuclear and coal. Thus, there are actually only two demand scenarios for 2000: one for high growth and high price and the other for low growth and medium price. The choice between nuclear and coal can be made in electricity and district heat generation. It has not been estimated how this choice will affect consumer prices; thus the demand projections are the same for both choices.

The demand estimates are based on the economic growth assumptions given in section 9.1. The energy price is constant in scenario D and increases by 50 percent in 2000 scenario C. Based on these assumptions, the gross national product (GNP) is as shown in Table 9.2.

9.3.1 Transportation

The estimated number of automobiles per 1000 inhabitants saturates at a comparatively low level—350—in Finland. The assumptions concerning automobiles are summarized in Table 9.3.

The electricity demand in the transportation sector is totally due to electrification of the railroad network. No electrical automobiles are included in the scenarios. The pace of electrification of railroads is mostly dependent on factors exogenous to the energy sector. Estimates of electricity used for transport are 59 GWhe for 1976; 582 GWhe for 1985; 1600 GWhe for 2000 Case C; and 1180 GWhe for 2000 Case D.

Other energy demand in the transportation sector is based on the detailed studies mentioned above. The data have been aggregated. Income elasticity is estimated to be 1.15 until 1985 and 1.35 after that. The price elasticity is estimated to be −0.2. These assumptions lead to estimates for fuel demand in transport other than automobiles of 1.33 MTOE for 1976; 2.06 MTOE for 1985

Table 9.3
Summary of automobile assumptions

		1985		2000	
	1972	C	D	C	D
Automobiles/1000 inhabitants	0.180	0.307	0.287	0.348	0.331
Number of automobiles (10^6)	0.82	1.47	1.38	1.71	1.62
Fuel consumption (liters/100 km)	9.5	9.5	9.5	7.0	7.5
Average distance/auto/year (1000 km)	17	17	17	15	16
Total fuel consumption (MTOE)	1.06	1.90	1.78	1.43	1.56

Case C; 1.85 MTOE for 1985 Case D; 3.8 MTOE for 2000 Case C; and 2.8 MTOE for 2000 Case D.

9.3.2 Industry

The industrial sector has been disaggregated into four subsectors. The volume growth rates for each scenario are given in Table 9.4. The respective values added are given in Table 9.5. The technical efficiencies are estimated for each of these sectors separately for fuels and electricity. The fuel demand estimates in MTOE/10^9 Fmk value added are given in Table 9.6. The electricity demand estimates are given in Table 9.7. The fuel mix estimated for the total industrial sector is given in Table 9.8.

The estimates determine the industrial demand for the WAES Finnish scenarios. It should be noted that industrial backpressure electricity generation is not accounted for. Only electricity purchased by each sector is included in the demand. This is contrary to the conventional method used for estimating electricity demand in Finland. That is why the demand coefficients also differ from the conventional numbers. Fuel used for backpressure electricity generation is thus included in the fuel demand of each sector. The method used is compatible with the WAES methodology.

Indigenous fuels include industrial waste materials, heat, wood, process fuels, and peat.

Table 9.4
Volume growth rates by industrial sector (%/yr)

	1976-1985		1985-2000	
	C	D	C	D
Paper and pulp	6.1	4.5	4.8	2.8
Chemicals	7.7	5.7	6.1	3.5
Primary metal	7.3	5.4	5.8	3.3
Other industries	6.1	4.5	4.8	2.8
Industry, total	6.3	4.6	5.0	2.9

Table 9.5
Value added by sector

		1985		2000	
	1976	C	D	C	D
Paper and pulp	4.11	6.97	6.08	14.09	9.13
Chemicals	2.74	5.33	4.49	13.02	7.53
Primary metal	1.31	2.46	2.09	5.72	3.41
Other industries	20.14	34.29	29.91	69.78	45.06
Industry, total	28.29	49.03	42.58	102.605	65.13

Note: Value added is given in billion (10^9) 1975 Fmk.

Table 9.6
Fuel demand estimates (MTOE/10^9 Fmk value added)

			2000	
	1976	1985	C	D
Paper and pulp	0.776	0.780	0.600	0.700
Chemicals	0.223	0.290	0.250	0.290
Primary metal	0.312	0.326	0.270	0.300
Other industries	0.081	0.060	0.037	0.045
Industry, total	0.206	0.200	0.154	0.178

9.3.3 Residential

The volume unit used in the residential sector is the physical volume of residential buildings in Mm^3 (million cubic meters); this is fast saturating. Estimations for different scenarios are 445 Mm^3 for 1976; 548 Mm^3 for 1985; 655 Mm^3 for 2000 Case C; and 617 Mm^3 for 2000 Case D.

The amount of space heating energy needed for each volume unit is determined next. This is decreasing because of better insulation and also because of better utilization of solar radiation in the future. Estimates of yearly heat loss are 72.9 kWh/m^3 for 1976; 67.8 kWh/m^3 for 1985; 55 kWh/m^3 for 2000 Case C; and 60 kWh/m^3 for 2000 Case D.

The demand for space heat is allocated to different types of energy. The most remarkable shifts are, first, the large-scale introduction of district heating in urban areas and, second, utilization of electrical heating in rural areas and small houses. The use of coal and wood for space heating practically disappears by 2000. The estimated fuel mix

Table 9.7
Electricity demand estimates (TWh/10^9 Fmk value added)

			2000	
	1976	1985	C	D
Paper and pulp	1.303	1.353	1.550	1.600
Chemicals	0.734	0.577	0.470	0.530
Primary metal	1.370	1.109	0.850	1.000
Other industries	0.147	0.149	0.150	0.155
Industry, total	0.428	0.414	0.421	0.445

Table 9.8
Fuel mix shares (%)

			2000	
	1976	1985	C	D
Oil	48.6	45.8	35.0	40.0
Coal	8.4	6.6	8.0	7.0
Gas	7.9	9.7	10.0	10.0
Indigenous fuels	35.1	37.9	47.0	43.0

for space heating is given in Table 9.9. It seems that the demand for oil for space heating is drastically decreasing. This is, however, not totally true because oil will be used for both district heat and electricity production.

The demand for electricity (other than for space heating) is also determined on the basis of residential building volumes. These estimates, like all other demand estimates, are based on a detailed study up to the year 1990 and thereafter on trend analyses and income and price elasticities. The electricity demand coefficients used (in kWh/m^3) are 9.85 for 1976; 15.00 for 1985; 20.00 for 2000 Case C; and 19.0 for 2000 Case D. These estimates together determine the residential sector energy demand for the WAES Finnish scenarios.

9.3.4 Public and Commercial

The method used to estimate the energy demand in the public and commercial sectors is very similar to that used in the residential sector. The physical volume of buildings is not, however, saturating as strongly. Building volume estimates in Mm^3 are 161 for 1976; 210 for 1985 Case C; 190 for 1985 Case D; 296 for 2000 Case C; and 238 for 2000 Case D. The yearly heat loss estimates in kWh/m^3 are 178.6 for 1976; 74.8 for 1985; 60 for 2000 Case C; and 65 for 2000 Case D.

The use of different fuel types is assumed to be the same as in the residential sector. Actually the fuel mix has been estimated for the total building stock including residential, public, and commercial buildings.

The electricity demand coefficients in kWh/m^3 are 25.9 for 1976; 41.7 for 1985; 62 for 2000 Case C; and 56 for 2000 Case D. These estimates determine the energy demand for the public and commercial sector.

9.3.5 Other Consumption

The other demand sectors have been analyzed in the same way as described earlier. In practice the demand includes only electricity and oil. The oil estimates used (in MTOE) are 0.91 for 1976; 0.94 for 1985; 1.01 for 2000 Case C; and 0.96 for 2000 Case D. The electricity demand estimates (in TWh) are 1.05 in 1976; 1.69 in 1985; 3.43 in 2000 Case C; and 2.35 in 2000 Case D. In addition, the nonenergy use of oil is included in the scenarios as

Table 9.9
Estimated fuel mix shares for space heating (%)

			2000	
	1976	1985	C	D
Oil	57.1	40.9	20	26
Wood	16.6	3.9	0	0
Coal	0.5	0.5	0	0
District heat	20.0	44.0	65	60
Electricity	5.8	10.7	15	14

0.05 MTOE for 1972; 0.1 MTOE for 1985; and 0.2 MTOE for 2000.

9.4 SUPPLY-DEMAND INTEGRATIONS

The methodology used in the Finnish supply-demand integrations was described in section 9.2. The situation in 1985 is very different from that in 2000. Most of the energy supply and conversion capacities for 1985 have already been fixed, and the problem is primarily to scale imports to match demand. For the year 2000, however, there are more degrees of freedom because new energy supply units and processes must be obtained.

As pointed out earlier, when both demand and indigenous supply are fixed, the analysis boils down to the determination of energy conversion processes, which in Finland includes fuel for district heat and electricity generation, and the fuel mix.

9.4.1 1972 Integration

The base year provides a good starting point to describe the energy flows in the Finnish energy sector. There are several characteristics worth mentioning. First, in Finland, as in many other industrialized countries, imported oil was the major energy source in every demand sector in 1972. There was some coal but no gas consumption at that time. The indigenous fuels had a notable share, too. These include wood and industrial waste materials such as bark, black liquor, process fuels, and a very small amount of peat.

In electricity and district heating, the situation was somewhat different. The main fuel was coal, although it was not very far above oil. Practically no indigenous fuel was used, but hydro energy, of course, provided a large share of the electricity. Larger district-heating plants cogenerated heat and electricity and thus made a significant contribution to the electricity supply. Also, a major share of electricity was imported.

9.4.2 1985 Integrations

Since most decisions concerning energy supply in the year 1985 have been made, there are no structural differences between the five scenarios. In many cases the difference is a matter of scaling the supply according to the level of demand. It must be repeated that national policy response has been analyzed neither in the demand nor in the supply study and is thus considered constant.

There are some structural differences compared to the base year. First, a large amount of space heat is provided by district heating. Although oil is the major fuel for district heating, considerable amounts of coal and peat and some natural gas are also used. The choice of fuel in district heating is dependent on the size and location of the plant. Smaller plants that do not include electricity generation burn oil. Larger plants meant to supply heat for 30,000 or more inhabitants include electricity generation. If the plant is located close to the coast, coal is an economical fuel, while inland closer to the peat fields peat may be used.

A major change in the electricity-generating sector is the introduction of nuclear power. The capacity of power plants under construction is 2.2 GW or about the same as the hydro power capacity. According to present estimates, the next nuclear power plant will be installed in 1985. The energy produced by this plant has not been included in the 1985 figures. Coal and oil will be used for electricity generation in plants either existing or under construction. The importation of electricity is 4 TWh per year, according to existing trade agreements. About half of the supplied peat is used for district heat and the rest for industrial heat.

9.4.3 2000 Integrations

As discussed earlier, there are only two demand scenarios for the year 2000, namely, C and D, and only three supply scenarios, C-1, C-2, and D. Even Cases C-1 and C-2 are very similar to each other. This makes the scenario analysis rather straightforward.

The trends from 1972 to 1985 will mostly continue up to 2000. The share of district heating increases by almost two-thirds of the net space heat demand. Because of cogeneration of heat and electricity, an increasing share of electricity is supplied by district-heating plants. This share goes above 20 percent by the year 2000. A new nuclear plant technology is included in the scenarios, that is, a nuclear plant producing both district heat and electricity. Both electrical and thermal capacities are planned to be about 1000 MW. In Case C-2 there are two such units, in the other scenarios only one. Concerning the fuel mix in district-heating plants, the same principles hold as for the year 1985.

9.5 ENERGY CONSERVATION

Many energy-saving measures are being practiced in Finland, and others are being investigated. Some of them will be reviewed in this section.

9.5.1 Industry

A comprehensive study concerning energy-saving potentials in wood processing and other heavy-process industries has recently been carried out. The results show that there are several means to reduce the consumption of imported energy. They can be classified as follows:

1. Conservation of heat and electricity
2. Improved efficiency of indigenous energy production
3. Increased use of cogeneration of heat and electricity
4. Increased use of peat
5. Increased use of indigenous process wastes as fuels

The potential of each of these has been assessed, including the financial requirements. The potential conservation projects have been divided into three classes according to the present stage of the project:

1. Decided conservation projects. The financial requirements are 535 million Fmk and the estimated savings are 0.58 MTOE per year.
2. Planned projects. The financial requirements are 300 million Fmk and the estimated savings are 0.27 MTOE per year.
3. Projects under investigation. The financial requirements are 850 million Fmk and the estimated savings are 0.40 MTOE per year.

It is evident that the marginal utility of conservation projects is decreasing. The cumulative ratios of the savings of these projects to the total industrial energy consumption in 1974 are 7 percent, 9 percent, and 14 percent, respectively.

9.5.2 Space Heating

There is also a large research project going on in Finland concerning energy savings in buildings. The results are not yet available. Although all buildings already have double windows and are rather well insulated, some potential for energy savings remains. This is especially true for new buildings. There are several means for decreasing the heat loss, such as:

- lower room temperature
- better insulation
- smaller windows
- better temperature regulation
- better regulation and control of heating devices
- heat exchangers

At this point in time, not much comprehensive information is available about the amount of potential savings and about the financial requirements.

9.5.3 Transportation

The number of vehicles produced in Finland is very small compared to the total fleet used for transportation. There is, however, a progressive tax on automobiles that depends on the size (cost) of the car. This taxing policy does tend to promote the use of smaller cars. There are also speed limits of 80 km per hour to 120 km per hour on every road, depending on the quality of the road and the density of traffic. The share of public transportation is already rather high, especially in the larger urban centers. The conclusion is that major energy savings in the transportation sector will be difficult to achieve without decreasing the activity level.

9.6 WORKSHEETS

This section contains summary tables and worksheets in the following order:

Energy demand summary 1972-1985-2000

Energy supply summary 1972-2000

National input worksheets for supply/demand integration (10 sheets)

WAES NATIONAL ENERGY DEMAND STUDIES, 1972-1985-2000
ENERGY DEMAND SUMMARY

Country: Finland

Scenario Assumptions	1972	1985					2000			
WAES CASE	Base Year	A	B	C	D	E	C-1	C-2	D-7	D-8
GNP 1976-85; 1985-2000		High	Low	High	Low	High	High	High	Low	Low
Oil/Energy Price*		11.50-17.25	11.50-17.25	11.50	11.50	11.50-7.66	11.50-17.25	11.50-17.25	11.50	11.50
National Policy Response		Vig	Vig	Vig	Res	Res	Vig	Vig	Vig	Vig
Principal Replacement Fuel							Coal	Nuclear	Coal	Nuclear
Primary Energy Demand Units 10^{15}J										
Oil	478	623	554	673	613	748	740	740	672	672
Natural Gas		47	42	45	42	47	75	75	58	58
Coal	98	138	111	154	119	149	371	261	228	228
Nuclear		149	149	149	149	149	355	490	248	248
Hydroelectric	56	61	61	61	61	61	61	61	61	61
Geothermal										
Solar Thermal										
Solar Electric										
Other	151	192	171	192	171	192	361	341	242	242
TOTAL PRIMARY ENERGY DEMAND	783	1210	1088	1274	1155	1346	1963	1968	1509	1509

*1975 US dollars per barrel of Arabian light crude oil, fob Persian Gulf.
All Figures are primary production prior to any processing or conversion.

WAES NATIONAL ENERGY SUPPLY STUDIES, 1972-2000
ENERGY SUPPLY SUMMARY

Country: Finland

Scenario Assumptions	1972	1985					2000			
WAES CASE	Base Year	A	B	C	D	E	C-1	C-2	D-7	D-8
GNP 1976-85; 1985-2000		High	Low	High	Low	High	High	High	Low	Low
Oil/Energy Price*		11.50-17.25	11.50-17.25	11.50	11.50	11.50-7.66	11.50-17.25	11.50-17.25	11.50	11.50
National Policy Response		Vig	Vig	Vig	Res	Res	Vig	Vig	Vig	Vig
Principal Replacement Fuel							Coal	Nuclear	Coal	Nuclear
Energy Supply **Units**	10^{15}J									
Crude Oil										
Natural Gas										
Coal										
Nuclear		149	149	149	149	149	355	490	248	248
Hydro-electric	41	47	47	47	47	47	47	47	47	47
Peat	1	55	45	55	45	55	195	180	100	100
Other Indigenous	150	137	126	137	126	137	166	161	142	142
Solar Electric										
Total Energy Supply	192	388	367	388	367	388	763	878	537	537

*1975 US dollars per barrel of Arabian light crude oil, fob Persian Gulf.
All Figures are primary production prior to any processing or conversion.

NATIONAL INPUT WORKSHEET FOR SUPPLY/DEMAND INTEGRATION

Country: Finland

Year: 1972

Date: 15.9.1976

Units: 10^{15} J

	(a) Coal	(b) Petro-leum	(c) (Syn. Liquids)	(d) Nat. Gas	(e) (Syn. Gas)	(f) (Heat)	(g) (Electri-city)[††]	(h) Hydro Energy	(i) Nuclear Energy	(j) Geothermal energy and other	(k) Total
(1) Transport	–	98									98
(2) Industry	40	129				2	39			90	300
(3) Agric., Mining, Construction	–	31					3				34
(4) Commercial, (5) Public	–	39				5	10			15	69
(6) Residential	–	114				14	15			45	188
(7) Non-energy Uses		2									2
(8) Final Energy Demand*	40	413				21	67			150	691
(9) Electricity**	38	26					–54	41			51
(10) Syn. Gas											
(11) Syn. Liquids											
(12) Heat	20	11				–22	–5			1	5
(13) Energy Sector Self Consumption & conversion losses		28				1	7				36
(14) Primary Energy Input	98	478					15	41		151	783
(15) Indigenous Supply								41		151	192
(16 Imports***	98	478					15				591
(17) Exports											

*Includes non-energy uses.

**Blocks 9g, 10e 11c, and 12f must be negative to avoid double counting of energy from primary fuels.

***Includes imported products.

††Includes only electricity to be purchased by each sector.

NATIONAL INPUT WORKSHEET FOR SUPPLY/DEMAND INTEGRATION

Country: Finland

Year: 1985

Case: A

Units: 10^{15} J

Date: 22.11.1976

	(a) Coal	(b) Petroleum	(c) (Syn. Liquids)	(d) Nat. Gas	(e) (Syn. Gas)	(f) (Heat)	(g) (Electricity)††	(h) Hydro Energy	(i) Nuclear Energy	(j) Geothermal energy and other	(k) Total
(1) Transport		148					2				150
(2) Industry	26	160		39		11	70			151	457
(3) Agric., Mining, Construction		35					7				42
(4) Commercial (5) Public	1	34				26	39			5	105
(6) Residential	1	70				59	44			12	186
(7) Non-energy Uses		4									4
(8) Final Energy Demand*	28	451		39		96	162			168	944
(9) Electricity**	69	31		3			-131	47	149		168
(10) Syn. Gas											
(11) Syn. Liquids											
(12) Heat	41	84		5		-101	-24			24	29
(13) Energy Sector Self Consumption & conversion losses		57				5	7				69
(14) Primary Energy Input	138	623		47			14	47	149	192	1210
(15) Indigenous Supply								47	149	192	388
(16 Imports***	138	623		47			14				822
(17) Exports											

*Includes non-energy uses.

**Blocks 9g, 10e 11c, and 12f must be negative to avoid double counting of energy from primary fuels.

***Includes imported products.

††Includes only electricity to be purchased by each sector.

NATIONAL INPUT WORKSHEET FOR SUPPLY/DEMAND INTEGRATION

Country: Finland

Year: 1985

Case: B

Units: 10^{15} J

Date: 22.11.1976

	(a) Coal	(b) Petroleum	(c) (Syn. Liquids)	(d) Nat. Gas	(e) (Syn. Gas)	(f) (Heat)	(g) (Electricity)††	(h) Hydro Energy	(i) Nuclear Energy	(j) Geothermal energy and other	(k) Total
(1) Transport		135					2				137
(2) Industry	23	136		34		11	61			131	396
(3) Agric., Mining, Construction		35					6				41
(4) Commercial, (5) Public	1	29				23	34			4	91
(6) Residential	1	70				59	44			12	186
(7) Non-energy Uses		4									4
(8) Final Energy Demand*	25	409		34		93	147			147	855
(9) Electricity**	47	13		3			-117	47	149		142
(10) Syn. Gas											
(11) Syn. Liquids											
(12) Heat	39	82		5		-98	-23			24	29
(13) Energy Sector Self Consumption & conversion losses		50				5	7				62
(14) Primary Energy Input	111	554		42			14	47	149	171	1088
(15) Indigenous Supply								47	149	171	367
(16 Imports***	111	554		42			14				721
(17) Exports											

*Includes non-energy uses.

**Blocks 9g, 10e 11c, and 12f must be negative to avoid double counting of energy from primary fuels.

***Includes imported products.

††Includes only electricity to be purchased by each sector.

NATIONAL INPUT WORKSHEET FOR SUPPLY/DEMAND INTEGRATION

Country: Finland

Year: 1985

Case: C

Units: 10^{15} J

Date: 22.11.1976

	(a) Coal	(b) Petroleum	(c) (Syn. Liquids)	(d) Nat. Gas	(e) (Syn. Gas)	(f) (Heat)	(g) (Electricity)††	(h) Hydro Energy	(i) Nuclear Energy	(j) Geothermal energy and other	(k) Total
(1) Transport		161					2				163
(2) Industry	26	182		37		11	73			151	480
(3) Agric., Mining, Construction		38					7				45
(4) Commercial, (5) Public	1	37				26	39			5	108
(6) Residential	1	83				59	44			12	199
(7) Non-energy Uses		4									4
(8) Final Energy Demand*	28	505		37		96	165			168	999
(9) Electricity**	85	23		3			-134	47	149		173
(10) Syn. Gas											
(11) Syn. Liquids											
(12) Heat	41	84		5		-101	-24			24	29
(13) Energy Sector Self Consumption & conversion losses		61				5	7				73
(14) Primary Energy Input	154	673		45			14	47	149	192	1274
(15) Indigenous Supply								47	149	192	388
(16 Imports***	154	673		45			14				886
(17) Exports											

*Includes non-energy uses.

**Blocks 9g, 10e 11c, and 12f must be negative to avoid double counting of energy from primary fuels.

***Includes imported products.

††Includes only electricity to be purchased by each sector.

NATIONAL INPUT WORKSHEET FOR SUPPLY/DEMAND INTEGRATION

Country: Finland

Year: 1985

Case: D

Units: 10^{15} J

Date: 22.11.1976

	(a) Coal	(b) Petroleum	(c) (Syn. Liquids)	(d) Nat. Gas	(e) (Syn. Gas)	(f) (Heat)	(g) (Electricity)††	(h) Hydro Energy	(i) Nuclear Energy	(j) Geothermal energy and other	(k) Total
(1) Transport		147					2				149
(2) Industry	23	158		34		11	64			131	421
(3) Agric., Mining, Construction		38					6				44
(4) Commercial, (5) Public	1	32				23	34			4	94
(6) Residential	1	83				59	44			12	199
(7) Non-energy Uses		4									4
(8) Final Energy Demand*	25	462		34		93	150			147	911
(9) Electricity**	55	13		3			-120	47	149		147
(10) Syn. Gas											
(11) Syn. Liquids											
(12) Heat	39	82		5		-98	-23			24	29
(13) Energy Sector Self Consumption & conversion losses		56				5	7				68
(14) Primary Energy Input	119	613		42			14	47	149	171	1155
(15) Indigenous Supply								47	149	171	367
(16 Imports***	119	613		42			14				788
(17) Exports											

*Includes non-energy uses.

**Blocks 9g, 10e 11c, and 12f must be negative to avoid double counting of energy from primary fuels.

***Includes imported products.

††Includes only electricity to be purchased by each sector.

NATIONAL INPUT WORKSHEET FOR SUPPLY/DEMAND INTEGRATION

Country: Finland

Year: 1985

Case: E

Units: 10^{15} J

Date: 22.11.1976

	(a) Coal	(b) Petroleum	(c) (Syn. Liquids)	(d) Nat. Gas	(e) (Syn. Gas)	(f) (Heat)	(g) (Electricity)††	(h) Hydro Energy	(i) Nuclear Energy	(j) Geothermal energy and other	(k) Total
(1) Transport		175					2				177
(2) Industry	26	199		39		11	79			151	505
(3) Agric., Mining, Construction		42					7				49
(4) Commercial, (5) Public	1	40				26	39			5	111
(6) Residential	1	90				59	44			12	206
(7) Non-energy Uses		4									4
(8) Final Energy Demand*	28	550		39		96	171			168	1052
(9) Electricity**	80	46		3			−140	47	149		185
(10) Syn. Gas											
(11) Syn. Liquids											
(12) Heat	41	84		5		−101	−24			24	29
(13) Energy Sector Self Consumption & conversion losses		68				5	7				80
(14) Primary Energy Input	149	748		47			14	47	149	192	1346
(15) Indigenous Supply								47	149	192	388
(16 Imports***	149	748		47			14				958
(17) Exports											

*Includes non-energy uses.

**Blocks 9g, 10e 11c, and 12f must be negative to avoid double counting of energy from primary fuels.

***Includes imported products.

††Includes only electricity to be purchased by each sector.

NATIONAL INPUT WORKSHEET FOR SUPPLY/DEMAND INTEGRATION

Country: Finland

Year: 2000

Case: C 1

Units: 10^{15} J

Date: 22.11.1976

	(a) Coal	(b) Petroleum	(c) (Syn. Liquids)	(d) Nat. Gas	(e) (Syn. Gas)	(f) (Heat)	(g) (Electricity)[††]	(h) Hydro Energy	(i) Nuclear Energy	(j) Geothermal energy and other	(k) Total
(1) Transport		212					6				218
(2) Industry	51	224		64		20	156			301	816
(3) Agric., Mining, Construction		41					12				53
(4) Commercial (5) Public		19				42	78				139
(6) Residential		39				84	67				190
(7) Non-energy Uses		8									8
(8) Final Energy Demand*	51	543		64		146	319			301	1424
(9) Electricity**	240	45		6			-246	47	290		382
(10) Syn. Gas											
(11) Syn. Liquids											
(12) Heat	80	85		5		-153	-74		65	60	68
(13) Energy Sector Self Consumption & conversion losses		67				7	15				89
(14) Primary Energy Input	371	740		75			14	47	355	361	1963
(15) Indigenous Supply								47	355	361	763
(16 Imports***	371	740		75			14				1200
(17) Exports											

*Includes non-energy uses.

**Blocks 9g, 10e 11c, and 12f must be negative to avoid double counting of energy from primary fuels.

***Includes imported products.

††Includes only electricity to be purchased by each sector.

NATIONAL INPUT WORKSHEET FOR SUPPLY/DEMAND INTEGRATION

Country: Finland

Year: 2000

Case: C 2

Units: 10^{15} J

Date: 22.11.1976

	(a) Coal	(b) Petroleum	(c) (Syn. Liquids)	(d) Nat. Gas	(e) (Syn. Gas)	(f) (Heat)	(g) (Electricity)††	(h) Hydro Energy	(i) Nuclear Energy	(j) Geothermal energy and other	(k) Total
(1) Transport		212					6				218
(2) Industry	51	224		64		20	156			301	816
(3) Agric., Mining, Construction		41					12				53
(4) Commercial, (5) Public		19				42	78				139
(6) Residential		39				84	67				190
(7) Non-energy Uses		8									8
(8) Final Energy Demand*	51	543		64		146	319			301	1424
(9) Electricity**	170	45		6			-246	47	360		382
(10) Syn. Gas											
(11) Syn. Liquids											
(12) Heat	40	85		5		-153	-74		130	40	73
(13) Energy Sector Self Consumption & conversion losses		67				7	15				89
(14) Primary Energy Input	261	740		75			14	47	490	341	1968
(15) Indigenous Supply								47	490	341	878
(16 Imports***	261	740		75			14				1090
(17) Exports											

*Includes non-energy uses.

**Blocks 9g, 10e 11c, and 12f must be negative to avoid double counting of energy from primary fuels.

***Includes imported products.

††Includes only electricity to be purchased by each sector.

NATIONAL INPUT WORKSHEET FOR SUPPLY/DEMAND INTEGRATION

Country: Finland

Year: 2000

Case: D 7

Units: 10^{15} J

Date: 22.11.1976

	(a) Coal	(b) Petro-leum	(c) (Syn. Liquids)	(d) Nat. Gas	(e) (Syn. Gas)	(f) (Heat)	(g) (Electri-city)††	(h) Hydro Energy	(i) Nuclear Energy	(j) Geothermal energy and other	(k) Total
(1) Transport		177					4				181
(2) Industry	33	188		47		15	104			202	589
(3) Agric., Mining, Construction		39					8				47
(4) Commercial, (5) Public		22				33	56				111
(6) Residential		52				80	61				193
(7) Non-energy Uses		8									8
(8) Final Energy Demand*	33	486		47		128	233			202	1129
(9) Electricity**	155	45		6			-179	47	183		257
(10) Syn. Gas											
(11) Syn. Liquids											
(12) Heat	40	80		5		-134	-50		65	40	46
(13) Energy Sector Self Consumption & conversion losses		61				6	10				77
(14) Primary Energy Input	228	672		58			14	47	248	242	1509
(15) Indigenous Supply								47	248	242	537
(16 Imports***	228	672		58			14				972
(17) Exports											

*Includes non-energy uses.

**Blocks 9g, 10e 11c, and 12f must be negative to avoid double counting of energy from primary fuels.

***Includes imported products.

††Includes only electricity to be purchased by each sector.

NATIONAL INPUT WORKSHEET FOR SUPPLY/DEMAND INTEGRATION

Country: Finland

Year: 2000

Case: D 8

Units: 10^{15} J

Date: 22.11.1976

	(a) Coal	(b) Petroleum	(c) (Syn. Liquids)	(d) Nat. Gas	(e) (Syn. Gas)	(f) (Heat)	(g) (Electricity)††	(h) Hydro Energy	(i) Nuclear Energy	(j) Geothermal energy and other	(k) Total
(1) Transport		177					4				181
(2) Industry	33	188		47		15	104			202	589
(3) Agric., Mining, Construction		39					8				47
(4) Commercial (5) Public		22				33	56				111
(6) Residential		52				80	61				193
(7) Non-energy Uses		8									8
(8) Final Energy Demand*	33	486		47		128	233			202	1129
(9) Electricity**	155	45		6			−179	47	183		257
(10) Syn. Gas											
(11) Syn. Liquids											
(12) Heat	40	80		5		−134	−50		65	40	46
(13) Energy Sector Self Consumption & conversion losses		61				6	10				77
(14) Primary Energy Input	228	672		58			14	47	248	242	1509
(15) Indigenous Supply								47	248	242	537
(16) Imports***	228	672		58			14				972
(17) Exports											

*Includes non-energy uses.

**Blocks 9g, 10e 11c, and 12f must be negative to avoid double counting of energy from primary fuels.

***Includes imported products.

††Includes only electricity to be purchased by each sector.

10.1 INTRODUCTION

The Workshop on Alternative Energy Strategies considered four scenarios for the period 1985 to 2000. The two high growth rate scenarios, C-1 and C-2, assume a yearly increase of the gross world (WOCA) product of 5 percent, and the low growth scenarios, D-7 and D-8, assume an average annual growth rate of 3 percent. For this chapter, two further scenarios, C-1′ and D-7′, have been considered; these are directly linked to C-1 and D-7 but with a higher rate of new energies.

For the situation in France, as for other industrialized countries, some lower growth figures have to be considered. The average annual growth for France for scenarios C-1 and C-2 amounts to 3.5 percent, and in scenarios D-7 and D-8 a value of 2.5 percent has been projected. Coupled with scenarios C-1 and C-2 are high energy prices ($17.25 per BOE at 1975 prices), while for scenarios D-7 and D-8 constant prices prevail ($11.50 per BOE).

Up to 1985 the energy prices are directly linked to the price of a barrel of oil. From 1985 to 2000, when oil will probably play a reduced role in energy supply, the specific price has to be established as a combination of different energies but is nevertheless expressed as the price of a barrel of oil equivalent.

In establishing the projection for 1985 to 2000, the French WAES team relied on some preliminary studies of the National Planning Office. Furthermore, a critical approach was undertaken by a research group of the Centre National de la Recherche Scientifique and the University of Grenoble.

It must be underlined that in a forecast covering a period of 25 years some figures are relatively uncertain; therefore, further explanation will be given below.

10.1.1 Demographic Evolution of France up to 2000

The French population will increase from about 52 million today to 57.7 million in 2000 (about 10 percent in the next quarter of a century). The number of households in 2000 will be around 21.8 million, corresponding to 27 million residential dwellings (vacant houses and secondary residences included).

10.2 GLOBAL PICTURE OF THE ENERGY SECTOR FROM 1985 TO 2000

The major difference between scenarios C-1 and D-7 and scenarios C-2 and D-8 is the alternative energy source (coal or nuclear) chosen to replace oil in the energy demand.

The alternative replacement energies were defined by WAES according to their global impact (high coal, high nuclear, low coal, low nuclear), each country being free to define the particular meaning of "high" and "low." It must be emphasized that for France a differentiation between replacement fossil fuels is approximate; therefore, the expression "high coal" has to be interpreted as a case in which there is a higher amount of fossil fuel (coal, gas) demand with low nuclear power development.

It is expected that coal will maintain its position and may even increase to supply electricity generation, but it will not play an important role as a replacement fuel, at least in the given scenarios (coal will be limited to 20 percent of overall energy demand in the high coal case).

The development of new energies between 1985 and 2000 will depend on the average price of energy, on the amount of capital available in the economy, on political incentives, and on the lead times required to develop these technologies.

In scenarios D-7 and D-8, in which the price of energy remains constant, the development of new energies could be slower than in scenarios C-1 and C-2, in which the price is rising.

In scenarios C-2 and D-8, in which the financial requirements of nuclear development are quite high, the development of these new energies could also be slower in France than in scenarios C-1 and D-7.

In fact, the level of development of new energies in each scenario is the result of political decisions under technical constraints; this point is further discussed in section 10.6.

10.3 ENERGY DEMAND BY SECTOR

The overall and sectoral forecasts for primary

energy demand for the four scenarios are relatively uncertain, since the effects of the relation between price and energy demand over a long period and the impact of a conservation policy are difficult to identify and measure separately. Therefore, we shall present for each scenario the upper limit of energy demand, which corresponds roughly to the assumption of inelasticity of energy demand to prices, and a lower limit in which price mechanism and conservation measures are taken into account.

10.3.1 Residential and Commercial Sectors

Energy demand in the residential sector has been investigated according to its historical evolution and the previous projections to 1985. The increase in energy demand for households will slow down to an average of 3.2 percent per year from 1973 to 2000, compared with 7.5 percent per year between 1964 and 1973. This slowdown can be explained by the saturation in household appliances that will be achieved in the early 1990s.

Supply-demand integration and primary energy demand projections for this sector depend upon the following points:

- In C-2 and D-8 (nuclear), it is assumed that in year 2000 about 6 million households will be heated by electricity; in C-1 and D-7 (coal), fewer than 2 million households use electric heat.
- In France, the heat pump will not play an important role in the residential sector before the end of the century and will not appreciably modify the energy demand.
- Great uncertainties remain concerning the possible introduction of new energies (solar, geothermal, heat recovery, etc.).
- The commercial sector will be characterized by an active development of air conditioning, heat recovery equipment, and an increasing demand for hot water. Unlike the residential sector, the heat pump will probably play a very important role in commercial energy supply, including heat recovery systems.

The efficiency improvements in thermal insulation for residential and commercial construction are taken into account. In the commercial sector, technological progress achieved in electrical appliances and the heat pumps will be included in the specific energy demand estimates below.

Thus, energy demand in the residential and commercial sectors amounts to about 135–139 million tonnes of oil equivalent (MTOE) in scenarios C-1 and C-2 and about 120 MTOE in scenarios D-7 and D-8. The main uncertainty lies less in the overall demand projection than in the sharing of this demand among electric, fossil fuel, and new energies, since the sharing is connected to an intensive energy policy and follows from the evolution of the relative prices.

10.3.2 Transportation Sector

The transportation sector is certainly the sector in which long-term forecasting of energy demand is the most uncertain, although in the near future its inertia is relatively large due to the existing infrastructure, which is difficult to remove, and to established habits long in changing.

For each scenario, two projections are given: a high one corresponding to a situation in which no price and no conservation effects occur, and a low one for which different production practices are taken into account. These practices are as follows:

- Technical improvements in cars, which could reduce the average specific consumption from 10 liters per 100 km to 8 liters per 100 km.
- Reduction of annual distance traveled per car from 15,000 km to 13,000 km through more rational consumer behavior and through a relative shift from individual cars to public transportation.
- Shift from trucks to railways for freight transportation.
- Technical progress and standardization measures in the commercial and transportation sectors.

In scenarios C-1 and C-2, it has been assumed that a relative saturation will be achieved with an average of 100 GWe per 100 households (these saturation rates take into account the fact that 15–20 percent of households own two or more cars). The highest projection of final energy demand for these scenarios reaches 77 MTOE, and the lowest 58 MTOE.

In scenarios D-7 and D-8, the saturation is assumed to be achieved with an average of 90 cars per 100 households. The final energy demand for these scenarios is projected with a range of 66 MTOE in the high case to 50 MTOE in the low case.

10.3.4 Industrial Sector

A maximum elasticity factor of 0.9 between energy demand and GNP growth has been assumed to project the high levels of energy demand of the industrial sector in all scenarios. Thus the final energy demand of this sector is expected to be, in the upper limit cases, 138 MTOE in C-1 and C-2 and 115 MTOE in D-7 and D-8. These figures can be reduced to 127 and 108 MTOE, respectively, under the pressure of three main possible evolutions in the industrial development:

1. Short-term improvement of energy efficiency for existing processes.
2. Medium-term improvements by technological changes.
3. Long-term industrial restructuring leading to a less energy-intensive economy.

10.3.5 Total Primary Energy Demand

Given the uncertainties that have been discussed for the various sectoral energy demand projections, two figures for total primary energy demand should be presented for each scenario. However, in order to simplify the presentation and to be consistent in economic and financial terms, we associated two figures only with the low nuclear scenarios C-1 and D-7. Primary energy demand projections for scenarios C-2 and D-8, derived directly from the French Planning Commission estimates, are 390 MTOE and 330 MTOE, respectively.

Total final energy demands are the same in C-1 (upper limit) and C-2 and in D-7 (upper limit) and D-8, but the differences in the supply structure and in the total amount of processing losses make the total amount of primary energy required different: 384 MTOE in C-1 and 325 MTOE in D-7. The low projections for C-1 and D-7, combined with all the reduction effects discussed above, reach 349 MTOE in C-1′ and 300 MTOE in D-7′.

10.3.6 Is There a Limit to Specific Consumption?

It is probable that the increase in the energy consumption per capita will have a strong environmental impact in the dense urban areas of Western Europe. The possibility of a limitation of energy consumption per capita due to ecological factors has already been raised by some experts. It is difficult, however, to indicate relevant figures, and no such limit has been assumed in our analysis.

10.3.7 Financial Requirements

The scenarios have been worked out under the assumption that the financial effort required for needed energy systems over the next 25 years will not overstress the economy. Therefore, the introduction of alternative energies depends on the ability of the capital markets to finance energy system needs. The same applies to the fossil fuel policy (for example, whether or not to maintain national coal production and at what level), the oil policy, the nuclear policy, and energy conservation measures.

10.4 ENERGY SUPPLY, 2000

Fossil fuels will represent, according to the various scenarios, between one-half and two-thirds of the total demand in 2000. A reduction of the percentages is possible with a vigorous conservation effort and an accelerated promotion of new energies with the needed financial effort. At least 90 percent of this fossil fuel will have to be imported. Between 85 and 130 MTOE of oil imports will be needed, the rest of the fossil fuel import being shared by coal and natural gas.

The indigenous supply consists of hydroelectricity, nuclear energy, and new energies. France relies heavily on its uranium resources and has developed advanced systems such as the breeder reactor and capacities for nuclear fuel enrichment and reprocessing, assuring a degree of independence not possible with fossil fuels.

10.5 ENERGY FORECAST WITH AN OIL SUPPLY GAP

If, after 1990, an oil gap due to rising prices should appear, French energy policy will probably examine the possibility of restricting oil to those uses for which there are no substitutes (recognizing the requisite lead times), that is, to transportation and some industrial uses. Feedstocks (for nonenergy use) of 25 to 30 MTOE are not included in the forecasts of the four scenarios.

A possible solution would be acceleration of the introduction of new energies and new technologies, which would then become more competitive due to the energy price increase and the sociopolitical impact of the oil gap. In this case the fuel supply from indigenous sources (hydro, nuclear, oil from shales, etc.) would be reconsidered.

Due to the uncertainties in the fossil fuel supply and the lead times needed for new energies and advanced technologies (breeder), a somewhat flexible approach maintains an acceptable degree of dependence on foreign supply and underlines the French need to stress efforts to adapt the social and economic structure to take account of the relatively precious product "energy."

10.6 NEW ENERGIES

10.6.1 Introduction

New energies are considered to have the potential to contribute a significant percentage of the national energy supply. New energies include solar energy and geothermal heat as well as steam and hot water distribution (district heating, industrial uses) from combined electricity and heat generation.

Two alternatives have been investigated to illustrate the margin of uncertainty in the field of new energies.

- For the upper limit of all scenarios, new energies have a relatively modest share of the total primary demand (10 MTOE new energies in 2000).
- In scenarios C-1′ and D-7′ (lower limit), a greater potential for technical and economical developments was assumed, leading to a contribution of 30 MTOE from new energies in 2000.

10.6.2 Technological and Economic Investigations of Solar Energy[1]

Solar energy could be used in France both for space heating of houses and buildings and for decentralized production of electricity (with hot water or steam for industry or district heating).

Three major applications of solar space heating could be developed by 2000: (1) space heating and hot water for single-family houses, (2) hot water and base-load space heating for multifamily buildings, and (3) space heating for public buildings such as schools and military buildings.

1. This subsection has been prepared by experts of the Centre National de la Recherche Scientifique and reflects some preliminary investigations. It is therefore underlined that certain of the figures are considered to be soft since they are not confirmed by technological-economic experience.

Single-Family Houses

Two techniques in existence in France at the present time are the "Trombe" technique, which has been used for about fifteen years in the Pyrenees, and the new flat collector and hot water technique tested for the past two or three years in some parts of France.

In the Trombe technique, the south wall of the house, covered with a single or double glazing, collects and stores solar energy. When the sun is up, the air between the glass and the wall is heated by the sun and naturally circulated into the house by the "thermosyphon" process. When the sun goes down, the heat of the wall radiates into the house. This technique has low cost and medium efficiency. However, a complementary heating system is needed.

Depending on climatic conditions, capital costs (not including the complementary system) are 60-99 FF/m^2 (12-18 \$/m^2). An energy savings of 50–65 percent (space heating) is possible, and storage time is about 24 hours.

The flat collector/hot water technique is more sophisticated and diversified than the Trombe technique. At the present time, the cost of this technique is quite high, much higher than the cost of the Trombe technique, in spite of a much higher efficiency and the ability to produce both space heating and domestic hot water. But one must keep in mind that this technique is fairly new and that its present cost is not relevant to large-scale production.

Capital costs today (excluding the complementary system costs) are 250-400 FF/m^2 (floor) or 50-80 \$/m^2. For large-scale production capital costs are estimated to be 100–150 FF/m^2 (floor) or 20–30 \$/m^2. Energy savings of 70–80 percent (both space heating and hot water) are possible, and storage time can be as much as three days.

Multifamily Buildings

Some attempts to apply the Trombe technique to multifamily buildings have been made. It seems, however, that this technique is only applicable to small buildings in rural southern France.

Hot water for all types of residential buildings could be supplied by flat collectors set up on the flat roofs of these buildings. All the hot water needs of a building could be supplied by solar energy, at compensatory costs.

For multifamily dwellings with eight units of 80 m^2 each, capital costs today (prototype) would be 100-150 FF/m^2 (floor) or 20-30 \$/m^2. With large-scale production, estimated capital costs would drop to 35-55 FF/m^2 (floor) or 10-15 \$/m^2. For hot water, 100 percent energy savings are possible, and of the total heat requirements of the building, a 30 percent savings is possible.

Public Buildings

Solar energy could be utilized in schools (mainly rural schools) and in military buildings. Unfortunately, it is impossible to evaluate either the actual cost of the application or the possible energy savings. A rough estimate of this potential for energy savings is included in the overall estimation.

Some developments have already taken place in various parts of France. Capital costs are estimated at 70 FF/m^2 (floor) or 14 \$/m^2 (including the cost of a transport and distribution network), energy savings are 70 percent (30 percent coming from fuel oil, gas, or the heat pump), and the operating costs are 6 FF/m^2 (floor) per year or 1.2 \$/m^2 per year.

10.6.3 District Heating from Combined Electricity Generation

Power plant heat generation for district heating could be developed to provide the heat and hot water needs of large towns (more than 200,000 dwellings). However, it is necessary to supply both old and new multifamily buildings in order to be economically feasible (in this case all kinds of buildings could be supplied). Capital costs are quite high (160 FF/m^2 (floor) or 32 \$/m^2, including 30 km hot water transportation and seasonal storage), but environmental and energy savings are very large. Operating costs (lower efficiency of the plant) are about 8 FF/m^2 per year, with lead times of at least five years to implement district heating.

10.8 WORKSHEETS

On the following pages will be found an energy demand summary (1972-1985-2000) and eleven national input worksheets for supply-demand integration (one each for 1985 scenarios A-E and for 2000 scenarios C-1, C-2, D-7, D-8, C-1′, and D-7′).

WAES NATIONAL ENERGY DEMAND STUDIES, 1972-1985-2000
ENERGY DEMAND SUMMARY

Country: France

Scenario Assumptions		1972	1985					2000			
WAES CASE		Base Year	A	B	C	D	E	C-1' C-1	C-2	D-7' D-7	D-8
GNP 1976-85; 1985-2000			High	Low	High	Low	High	High	High	Low	Low
Oil/Energy Price*			11.50-17.25	11.50-17.25	11.50	11.50	11.50-7.66	11.50-17.25	11.50-17.25	11.50	11.50
National Policy Response			Vig	Vig	Vig	Res	Res	Vig	Vig	Vig	Vig
Principal Replacement Fuel								Coal	Nuclear	Coal	Nuclear
Primary Energy Demand	**Units**										
Oil		107	88	72	96	88	178	95–130	130	85–110	110
Natural Gas		14	37	35	37	34	35	70	70	57	54
Coal		31	27	17	30	27	17	54	20	33–46	20
Nuclear		1	53	46	60	55	40	85–105	145	80–88	121
Hydroelectric		11	14	14	14	13	14	15	15	15	15
Geothermal		–	3	4	3	3	–	30–10	10	10–30	10
Solar Thermal		–	3	4	3	3	–	30–10	10	10–30	10
Solar Electric											
TOTAL PRIMARY ENERGY DEMAND		164	222	188	240	220	284	349–384	390	300–326	330

*1975 US dollars per barrel of Arabian light crude oil, fob Persian Gulf.
All Figures are primary production prior to any processing or conversion.

Footnotes:

**Including heat recovery.

NATIONAL INPUT WORKSHEET FOR SUPPLY/DEMAND INTEGRATION

Country: France

Year: 1985

Case: A

Units: MTOE

	(a) Coal	(b) Petroleum	(c) (Syn. Liquids)	(d) Nat. Gas	(e) (Syn. Gas)	(f) (Heat)	(g) (Electricity)††	(h) Hydro Energy	(i) Nuclear Energy	(j) Geothermal energy and other	(k) Total
(1) Transport		42					2				44
(2) Industry	12	13		13			33				71
(3) Agric.											
(4) Commercial	3	24		20		1	37			3	88
(5) Public											
(6) Residential											
(7) Non-energy Uses											
(8) Final Energy Demand*	15	79		33		1	72			3	203
(9) Electricity**	12	4		2		-1	-72	14	53		
(10) Syn. Gas											
(11) Syn. Liquids											
(12) Heat											
(13) Energy Sector Self Consumption & conversion losses	-	5		2			12			-	19
(14) Primary Energy Input	27	88		37				14	53	3	222
(15) Indigenous Supply	13	-		7				14	53	3	90
(16 Imports***	14	88		30				-	-	-	132
(17) Exports											

*Includes non-energy uses.

**Blocks 9g, 10e 11c, and 12f must be negative to avoid double counting of energy from primary fuels.

***Includes imported products.

††Includes only electricity to be purchased by each sector. 1 TWh = 0.22 MTOE

NATIONAL INPUT WORKSHEET FOR SUPPLY/DEMAND INTEGRATION

Country: France

Year: 1985

Case: B

Units: MTOE

	(a)	(b)	(c)	(d)	(e)	(f)	(g)	(h)	(i)	(j)	(k)
	Coal	Petro-leum	(Syn. Liquids)	Nat. Gas	(Syn. Gas)	(Heat)	(Electri-city)††	Hydro Energy	Nuclear Energy	Geothermal energy and other	Total
(1) Transport		32					2				34
(2) Industry	10	12		9			26				57
(3) Agric.											
(4) Commercial	3	22		24		2	29			4	84
(5) Public											
(6) Residential											
(7) Non-energy Uses											
(8) Final Energy Demand*	13	66		33		2	57			4	175
(9) Electricity**	4	2				-2	-57	14	46		
(10) Syn. Gas											
(11) Syn. Liquids											
(12) Heat											
(13) Energy Sector Self Consumption & conversion losses	-	4		2		-	7	-	-	-	15
(14) Primary Energy Input	17	72		35				14	46	4	188
(15) Indigenous Supply	13	-		7				14	46	4	84
(16 Imports***	4	72		28							104
(17) Exports											

*Includes non-energy uses.

**Blocks 9g, 10e 11c, and 12f must be negative to avoid double counting of energy from primary fuels.

***Includes imported products.

††Includes only electricity to be purchased by each sector. 1 TWh = 0.22 MTOE

NATIONAL INPUT WORKSHEET FOR SUPPLY/DEMAND INTEGRATION

Country: France

Year: 1985

Case: C

Units: MTOE

	(a) Coal	(b) Petro-leum	(c) (Syn. Liquids)	(d) Nat. Gas	(e) (Syn. Gas)	(f) (Heat)	(g) (Electri-city)††	(h) Hydro Energy	(i) Nuclear Energy	(j) Geothermal energy and other	(k) Total
(1) Transport		42					2				44
(2) Industry	20	19		16		-	32				87
(3) Agric.											
(4) Commercial	3	27		17		-	38			3	88
(5) Public											
(6) Residential											
(7) Non-energy Uses											
(8) Final Energy Demand*	23	87		33		-	72			3	119
(9) Electricity**	7	3		2		-	-72	14	60	-	
(10) Syn. Gas											
(11) Syn. Liquids											
(12) Heat											
(13) Energy Sector Self Consumption & conversion losses		6		2			14				
(14) Primary Energy Input	30	96		37				14	60	3	240
(15) Indigenous Supply	13	-		7				14	60	3	97
(16 Imports***	17	96		30				-	-	-	143
(17) Exports											

*Includes non-energy uses.

**Blocks 9g, 10e 11c, and 12f must be negative to avoid double counting of energy from primary fuels.

***Includes imported products.

††Includes only electricity to be purchased by each sector. 1 TWh = 0.22 MTOE

NATIONAL INPUT WORKSHEET FOR SUPPLY/DEMAND INTEGRATION

Country: France

Year: 1985

Case: D

Units: MTOE

	(a) Coal	(b) Petroleum	(c) (Syn. Liquids)	(d) Nat. Gas	(e) (Syn. Gas)	(f) (Heat)	(g) (Electricity)[††]	(h) Hydro Energy	(i) Nuclear Energy	(j) Geothermal energy and other	(k) Total
(1) Transport		38					2				40
(2) Industry	18	16		14			29				77
(3) Agric.											
(4) Commercial	3	25		16			34			3	81
(5) Public											
(6) Residential											
(7) Non-energy Uses											
(8) Final Energy Demand*	21	79		30			65			3	198
(9) Electricity**	6	3		2			-65	13	55		
(10) Syn. Gas											
(11) Syn. Liquids											
(12) Heat											
(13) Energy Sector Self Consumption & conversion losses	-	6		2			14				22
(14) Primary Energy Input	27	88		34				13	55	3	220
(15) Indigenous Supply	13	-		7				13	55	3	91
(16 Imports***	14	88		27							139
(17) Exports											

*Includes non-energy uses.

**Blocks 9g, 10e 11c, and 12f must be negative to avoid double counting of energy from primary fuels.

***Includes imported products.

††Includes only electricity to be purchased by each sector. 1 TWh = 0.22 MTOE

NATIONAL INPUT WORKSHEET FOR SUPPLY/DEMAND INTEGRATION

Country: France

Year: 1985

Case: E

Units: MTOE

	(a)	(b)	(c)	(d)	(e)	(f)	(g)	(h)	(i)	(j)	(k)
	Coal	Petro- leum	(Syn. Liquids)	Nat. Gas	(Syn. Gas)	(Heat)	(Electri- city)††	Hydro Energy	Nuclear Energy	Geothermal energy and other	Total
(1) Transport		48		–			3				51
(2) Industry	14	37		16			35				102
(3) Agric.											
(4) Commercial	3	45		16			42				106
(5) Public											
(6) Residential											
(7) Non-energy Uses											
(8) Final Energy Demand*	17	130		32			80				259
(9) Electricity**	–	39		2			–80	14	40		
(10) Syn. Gas											
(11) Syn. Liquids											
(12) Heat											
(13) Energy Sector Self Consumption & conversion losses		9		1			15				
(14) Primary Energy Input	17	178		35				14	40		284
(15) Indigenous Supply	10			7				14	40		71
(16 Imports***	7	178		28							193
(17) Exports											

*Includes non-energy uses.

**Blocks 9g, 10e 11c, and 12f must be negative to avoid double counting of energy from primary fuels.

***Includes imported products.

††Includes only electricity to be purchased by each sector. 1 TWh = 0.22 MTOE

NATIONAL INPUT WORKSHEET FOR SUPPLY/DEMAND INTEGRATION

Country: France

Year: 2000

Case: C-1

Units: MTOE

	(a)	(b)	(c)	(d)	(e)	(f)	(g)	(h)	(i)	(j)	(k)
	Coal	Petro-leum	(Syn. Liquids)	Nat. Gas	(Syn. Gas)	(Heat)	(Electri-city)††	Hydro Energy	Nuclear Energy	Geothermal energy and other	Total
(1) Transport		74					3				77
(2) Industry	42	28		20			48				138
(3) Agric.											
(4) Commercial		21		45			63			10	139
(5) Public											
(6) Residential											
(7) Non-energy Uses											
(8) Final Energy Demand*	42	123		65			114			10	354
(9) Electricity**	12						–132	15	105		
(10) Syn. Gas											
(11) Syn. Liquids											
(12) Heat											
(13) Energy Sector Self Consumption & conversion losses		7		5			18				30
(14) Primary Energy Input	54	130		70				15	105	10	384
(15) Indigenous Supply	13			5				15	105	10	148
(16 Imports***	41	130		65							236
(17) Exports											

*Includes non-energy uses.

**Blocks 9g, 10e 11c, and 12f must be negative to avoid double counting of energy from primary fuels.

***Includes imported products.

††Includes only electricity to be purchased by each sector.

NATIONAL INPUT WORKSHEET FOR SUPPLY/DEMAND INTEGRATION

Country: France

Year: 2000

Case: C-2

Units: MTOE

	(a) Coal	(b) Petro-leum	(c) (Syn. Liquids)	(d) Nat. Gas	(e) (Syn. Gas)	(f) (Heat)	(g) (Electri-city)††	(h) Hydro Energy	(i) Nuclear Energy	(j) Geothermal energy and other	(k) Total
(1) Transport		73					4				77
(2) Industry	7	33		28			70				138
(3) Agric.											
(4) Commercial		12		37			80			10	139
(5) Public											
(6) Residential											
(7) Non-energy Uses											
(8) Final Energy Demand*	7	118		65			154			10	354
(9) Electricity**	13	5		–			–178	15	145		
(10) Syn. Gas											
(11) Syn. Liquids											
(12) Heat											
(13) Energy Sector Self Consumption & conversion losses	–	7		5			24				36
(14) Primary Energy Input	20	130		70				15	145	10	390
(15) Indigenous Supply	13	–		5				15	145	10	188
(16 Imports***	7	130		65							202
(17) Exports											

*Includes non-energy uses.

**Blocks 9g, 10e 11c, and 12f must be negative to avoid double counting of energy from primary fuels.

***Includes imported products.

††Includes only electricity to be purchased by each sector.

NATIONAL INPUT WORKSHEET FOR SUPPLY/DEMAND INTEGRATION

Country: France

Year: 2000

Case: D-7

Units: MTOE

	(a)	(b)	(c)	(d)	(e)	(f)	(g)	(h)	(i)	(j)	(k)
	Coal	Petro-leum	(Syn. Liquids)	Nat. Gas	(Syn. Gas)	(Heat)	(Electri-city)††	Hydro Energy	Nuclear Energy	Geothermal energy and other	Total
(1) Transport		63					3				66
(2) Industry	38	21		20			36				115
(3) Agric.											
(4) Commercial	–	18		32			60			10	120
(5) Public											
(6) Residential											
(7) Non-energy Uses											
(8) Final Energy Demand*	38	102		52			99				301
(9) Electricity**	8	2		–			−113	15	88		
(10) Syn. Gas											
(11) Syn. Liquids											
(12) Heat											
(13) Energy Sector Self Consumption & conversion losses		6		5			14				25
(14) Primary Energy Input	46	110		57				15	88	10	326
(15) Indigenous Supply	13			5				15	88	10	131
(16 Imports***	33	110		52							195
(17) Exports											

*Includes non-energy uses.

**Blocks 9g, 10e 11c, and 12f must be negative to avoid double counting of energy from primary fuels.

***Includes imported products.

††Includes only electricity to be purchased by each sector.

NATIONAL INPUT WORKSHEET FOR SUPPLY/DEMAND INTEGRATION

Country: France

Year: 2000

Case: D-8

Units: MTOE

	(a)	(b)	(c)	(d)	(e)	(f)	(g)	(h)	(i)	(j)	(k)
	Coal	Petro-leum	(Syn. Liquids)	Nat. Gas	(Syn. Gas)	(Heat)	(Electri-city)††	Hydro Energy	Nuclear Energy	Geothermal energy and other	Total
(1) Transport		63					3				66
(2) Industry	7	29		20			59				115
(3) Agric.											
(4) Commercial	-	10		30			70			10	120
(5) Public											
(6) Residential											
(7) Non-energy Uses											
(8) Final Energy Demand*	7	102		50			132				301
(9) Electricity**	13	2					-151	15	121		
(10) Syn. Gas											
(11) Syn. Liquids											
(12) Heat											
(13) Energy Sector Self Consumption & conversion losses		6		4			19				29
(14) Primary Energy Input	20	110		54				15	121	10	330
(15) Indigenous Supply	13			5				15	121	10	164
(16 Imports***	7	110		49							166
(17) Exports											

*Includes non-energy uses.

**Blocks 9g, 10e 11c, and 12f must be negative to avoid double counting of energy from primary fuels.

***Includes imported products.

††Includes only electricity to be purchased by each sector.

NATIONAL INPUT WORKSHEET FOR SUPPLY/DEMAND INTEGRATION

Country: France

Year: 2000

Case: C-1'

Units: MTOE

	(a) Coal	(b) Petro-leum	(c) (Syn. Liquids)	(d) Nat. Gas	(e) (Syn. Gas)	(f) (Heat)	(g) (Electri-city)††	(h) Hydro Energy	(i) Nuclear Energy	(j) Geothermal energy and other	(k) Total
(1) Transport		55					3				58
(2) Industry	24	23		32			48				127
(3) Agric.											
(4) Commercial		10		33			63			30	135
(5) Public											
(6) Residential											
(7) Non-energy Uses											
(8) Final Energy Demand*	24	88		65			114			30	320
(9) Electricity**	30	2		–			–132	15	85		
(10) Syn. Gas											
(11) Syn. Liquids											
(12) Heat											
(13) Energy Sector Self Consumption & conversion losses		6		5			18				29
(14) Primary Energy Input	54	95		70				15	85	30	349
(15) Indigenous Supply	13			5				15	85	30	148
(16 Imports***	41	95		65							201
(17) Exports											

*Includes non-energy uses.

**Blocks 9g, 10e 11c, and 12f must be negative to avoid double counting of energy from primary fuels.

***Includes imported products.

††Includes only electricity to be purchased by each sector.

NATIONAL INPUT WORKSHEET FOR SUPPLY/DEMAND INTEGRATION

Country: France

Year: 2000

Case: D-7'

Units: MTOE

	(a) Coal	(b) Petro-leum	(c) (Syn. Liquids)	(d) Nat. Gas	(e) (Syn. Gas)	(f) (Heat)	(g) (Electri-city)††	(h) Hydro Energy	(i) Nuclear Energy	(j) Geothermal energy and other	(k) Total
(1) Transport		47					3				50
(2) Industry	17	25		30			36				108
(3) Agric.											
(4) Commercial		6		22			60			30	118
(5) Public											
(6) Residential											
(7) Non-energy Uses											
(8) Final Energy Demand*	17	78		52			99			30	276
(9) Electricity**	16	2					-113	15	80		
(10) Syn. Gas											
(11) Syn. Liquids											
(12) Heat											
(13) Energy Sector Self Consumption & conversion losses		5		5			14				24
(14) Primary Energy Input	33	85		57				15	80	30	300
(15) Indigenous Supply	13			5				15	80	30	143
(16 Imports***	20	85		52							157
(17) Exports											

*Includes non-energy uses.

**Blocks 9g, 10e 11c, and 12f must be negative to avoid double counting of energy from primary fuels.

***Includes imported products.

††Includes only electricity to be purchased by each sector.

11 German Federal Republic

11.1 SUMMARY

Primary energy demand of the Federal Republic of Germany in 2000, given the explicit WAES scenario assumptions, is estimated to range from about 580 to 690 million tonnes of coal equivalent. Figure 11.1 shows the results of all Workshop scenarios for 2000. The energy consumption growth rates for 2000 do not in any of the cases achieve the rates observed before 1974. Even from the low base rate of 1975, average annual energy growth rates in the high growth cases do not exceed 2.8 percent per year.

The two economic growth rate assumptions of 3.4 percent and 2.2 percent for 1976 to 2000 cause a wide difference in the projected primary energy demand, as can be seen by a comparison between Case D-3 and the C cases, which are based on the same price assumption. The analysis reveals that the implicit growth elasticity of primary energy demand in the scenarios is between 0.82 and 0.9. The impact of the difference in price assumptions is estimated to be much less. Primary demand in Case D-3 is less than 4 percent lower than in Case D-8 where the energy price level is 50 percent lower. Significant energy savings by political measures (vigorous or restrained policy) are not assumed, in part because intervention beyond the actual restrained conservation policy is unlikely in the Federal Republic of Germany.

Since the electricity demand is increasing much more rapidly than the final fossil fuel demand, consumption in the energy sector (i.e., from conversion and distribution losses and self-consumption) is gaining more quantitative importance. The growth rates of final energy demand are, therefore, on the average much lower than the growth of primary demand. Final demand is increasing from 1972 to 2000 only by 1.4–1.7 percent per year.

In particular, after 1985 energy demand in transport and residential sectors will increase somewhat (Figure 11.2). Decreasing population, saturation in demand for appliances and cars, and higher levels of efficiency (i.e., substitution of energy by capital) are the major factors determining the increase. Consequently, other sectors such as industry, commercial and public, and nonenergy use will consume a larger share of final demand. Naturally, nonenergy use and industrial and commercial energy demand are highly sensitive to economic growth. In the residential sector, sensitivity of total energy demand to economic growth (income) seems to be clearly smaller than in the industrial and commercial sectors. Changes in price relations among fuels because of both higher oil prices and higher electricity prices have a comparably strong impact on household energy consumption patterns. The negative growth rates of fossil fuel consumption in the residential sector are partly due to the use of more efficient forms of energy such as district heat. This "fuel shift" causes a shift of the losses originally incurred in final demand to the energy sector.

Looking at primary energy demand, all scenarios show the overwhelming importance of the nuclear choice and the need for coal when the nuclear option is renounced for the Federal Republic of Germany. Under the given scenario circumstances, oil and gas cannot replace nuclear use as coal can (Figure 11.3).

The import requirements of conventional sources resulting from the integration of supply and demand scenarios do not indicate radical changes in absolute terms if the nuclear option is pursued according to governmental plans. A nuclear drawback would, under the given circumstances, have a major impact on coal and gas imports, but only a minor impact on oil imports (Figure 11.4).

This study is not supposed to predict the energy situation of the Federal Republic of Germany. However, aside from the description of plausible energy futures—and there are certainly others—WAES scenarios reveal a number of mechanisms built into the energy supply-demand system. Considering the options beyond the scenarios, it seems possible that total energy demand is rather inelastic to crude oil prices and energy prices in general, if other parameters, such as economic growth, are equal (a rather naive assumption). Also, in the long run the substitutability among fuels is considerable, and therefore the demand reaction to a relative price change for a single fuel is elastic. Indigenous energy supply is inelastic to changes in all scenario variables. Even with very ambitious policies, large

This study was prepared at the Institute of Energy Economics at the University of Cologne by Dieter Schmitt and Paul H. Suding.

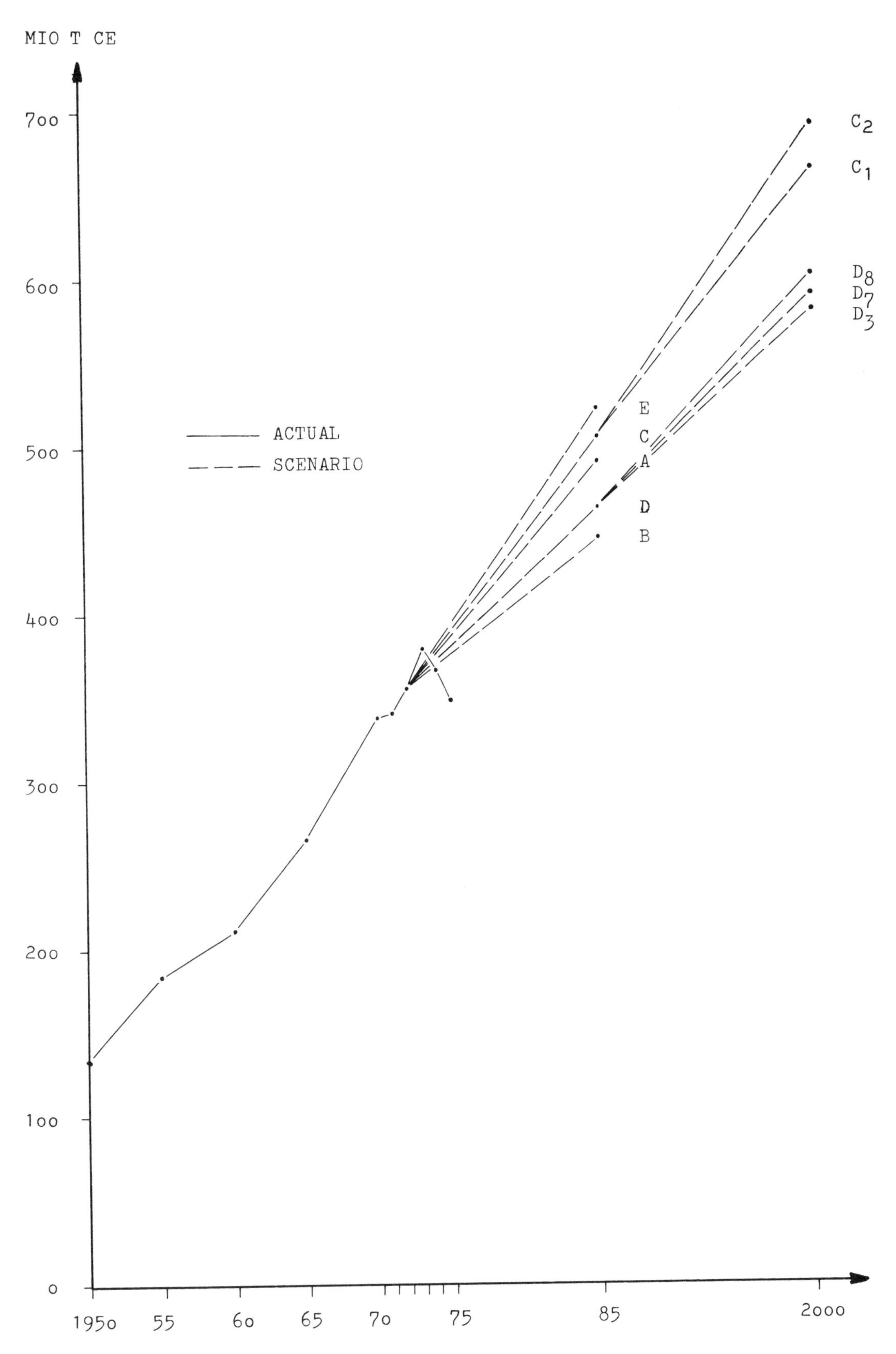

Figure 11.1
Primary energy input in the Federal Republic of Germany.

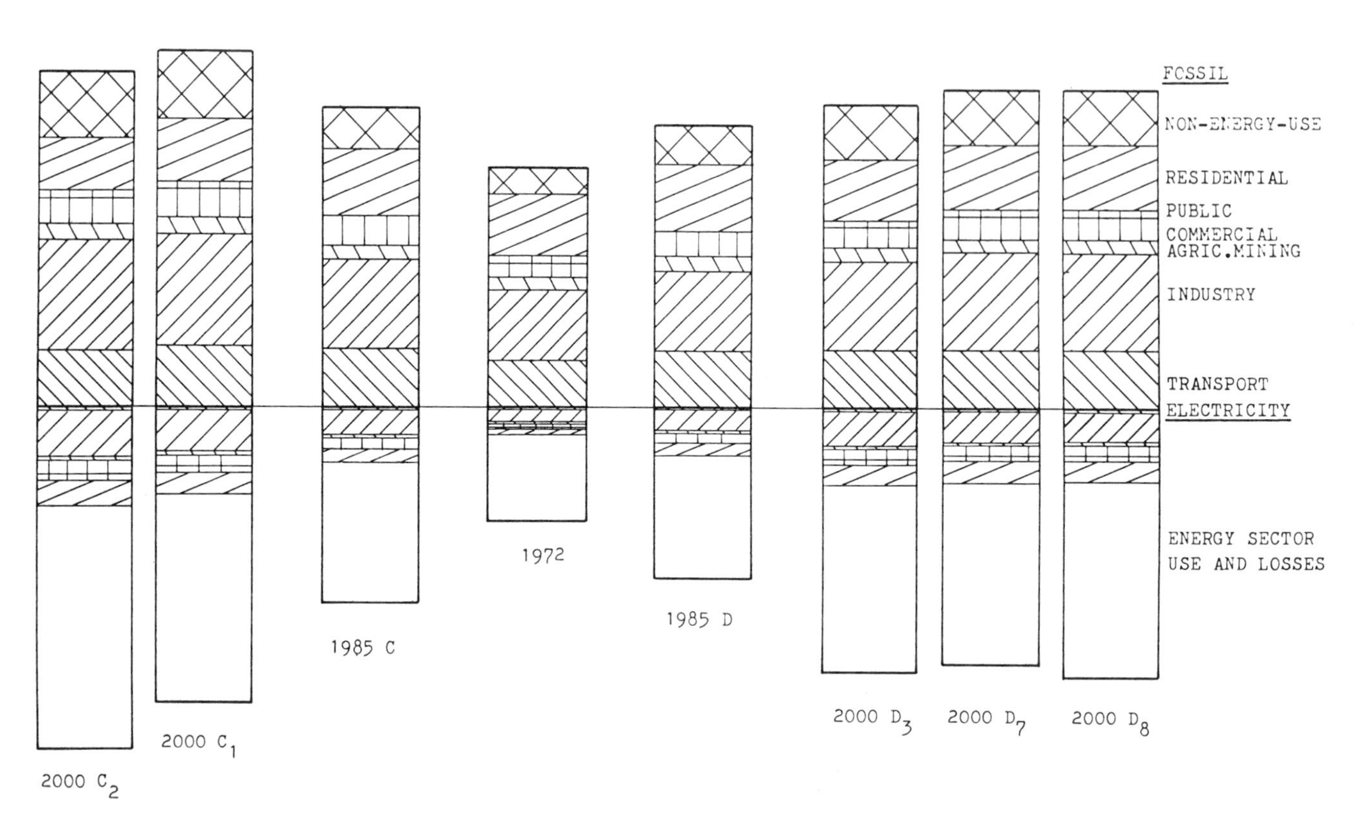

Figure 11.2
Final energy demand and energy sector use and losses.

additions to supply are hardly conceivable. Only nuclear energy, though not a genuine indigenous source, could be, in scarcity situations, an easy way out in economic terms. It is certainly not an easy way out in social and environmental security terms.

11.2 METHODOLOGY

The general methodology adopted by WAES for supply-demand integration studies for the period 1985 to 2000 is described in chapter 1 of this volume. This section discusses national scenario assumptions for the Federal Republic of Germany, the shape and implications of the scenarios, and the process by which the demand scenarios are combined with the supply scenarios to accomplish the supply-demand integration. Methods, techniques, prospects, and implications of the scenario approach applied are discussed in other Workshop publications (see chapter 7 in the WAES First Technical Report).

11.2.1 Economic Growth

Economic growth rates after 1985 are very uncertain. While the conditions for economic growth in the short and medium term are already partly determined, the conditions for long-run development will be determined in the future. Nevertheless, to be plausible, the scenario assumptions must take into account existing conditions in the Federal Republic of Germany, such as decreasing population, comparably low investment prospects, and lower investment productivity, all of which affect economic growth (Table 11.1). And these situations suggest a decreasing long-term growth rate for gross national product, which is used as an analogue for the world market price of primary energy, since it is expected that crude oil will lose its function as a price indicator.

Crude oil prices are not assumed to have a proportional impact on the internal energy price level in the Federal Republic of Germany (see the First Technical Report, p. 122). But the energy price assumption is assumed to cause the same relative development in both world and internal energy price levels.

For the WAES studies, a high economic growth rate of about 3.4 percent and a low growth rate of about 2.2 percent per year from 1976 to 2000 are assumed. These are the growth rates predicted by two of the very few available studies on economic growth to 2000, done by PROGNOS, Basel (3.4

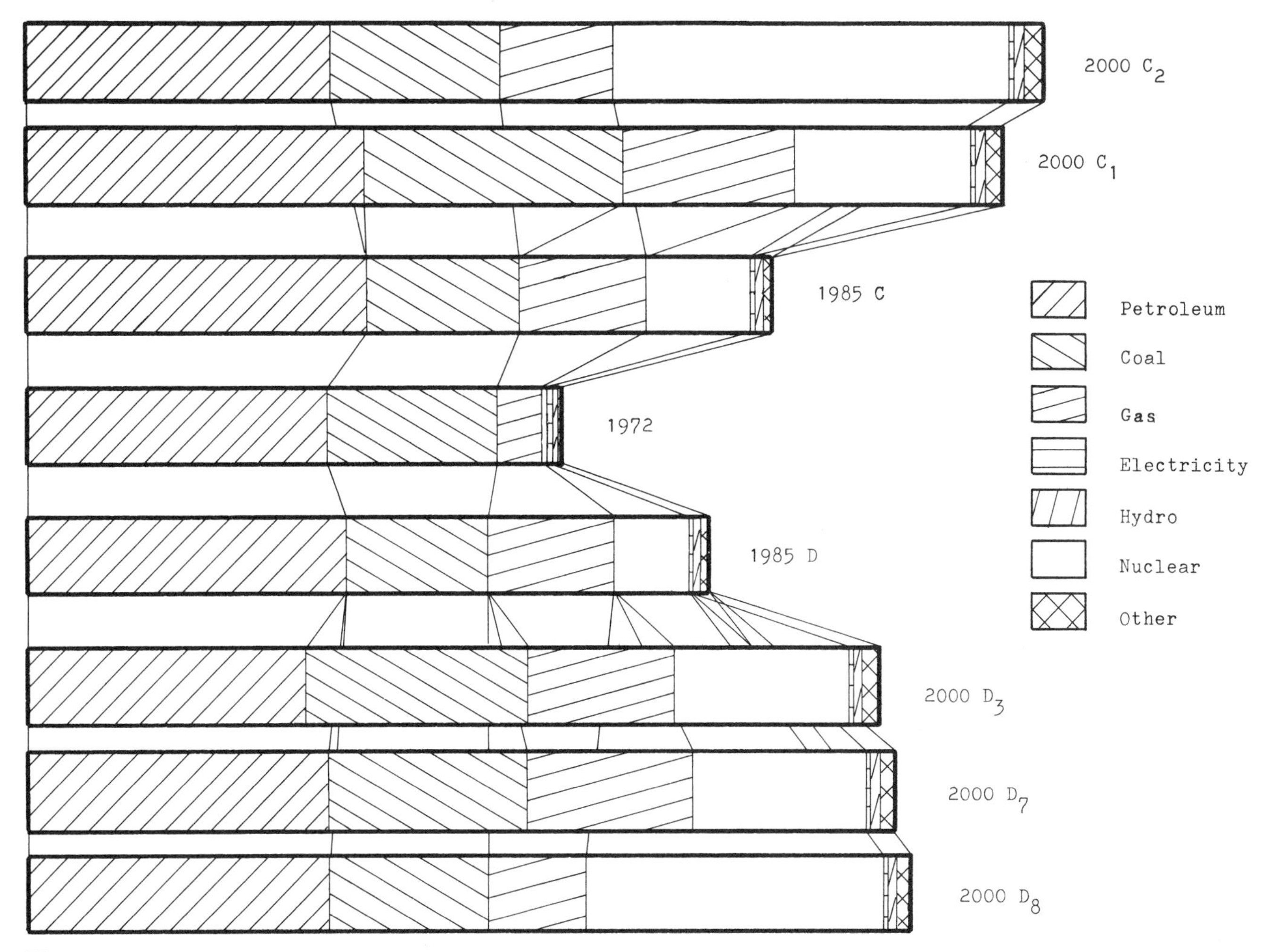

Figure 11.3
Primary energy demand by fuel.

percent) and by WSI, the economic research institute of the Federation of Trade Unions (2.2 percent). They result in a difference in the 2000 GDP assumptions of about 32 percent—sufficient alternatives of the state of the economy to provide opportunities for analysis with high learning value. The first period (1976 to 1985) is characterized by a growth in the productive working population and by the need to compensate for growth losses in the middle 1970s. Therefore, from 1976 to 1985 average GDP growth rates of 4.5 percent and 2.5 percent per year are assumed. This leaves for the second period (1985 to 2000) growth rates of 2.8 percent and 2.1 percent.

11.2.2 Energy Prices

The global price assumptions adopted by WAES have, for the two periods under study, slightly different characteristics. In the period to 1985 the Arabian light crude oil price FOB Persian Gulf means exactly what it says. For the second period, the internal energy price level is assumed to increase by about 50 percent from its 1976 level in the high price cases and to remain constant in relative terms in the medium price cases.

11.2.3 Energy Policy and Principal Replacement Fuel

The implicit assumption of a vigorous policy in the period to 2000 implies the continuation of the vigorous policy assumptions to 1985 (see the first technical report, pp. 123-125). Vigorous policy is characterized by the political objective of long-range energy cost rationality in the private sector.

The principal replacement fuel alternatives of low nuclear, high coal and high nuclear, low coal are largely politically determined. Nuclear development in the Federal Republic of Germany is per-

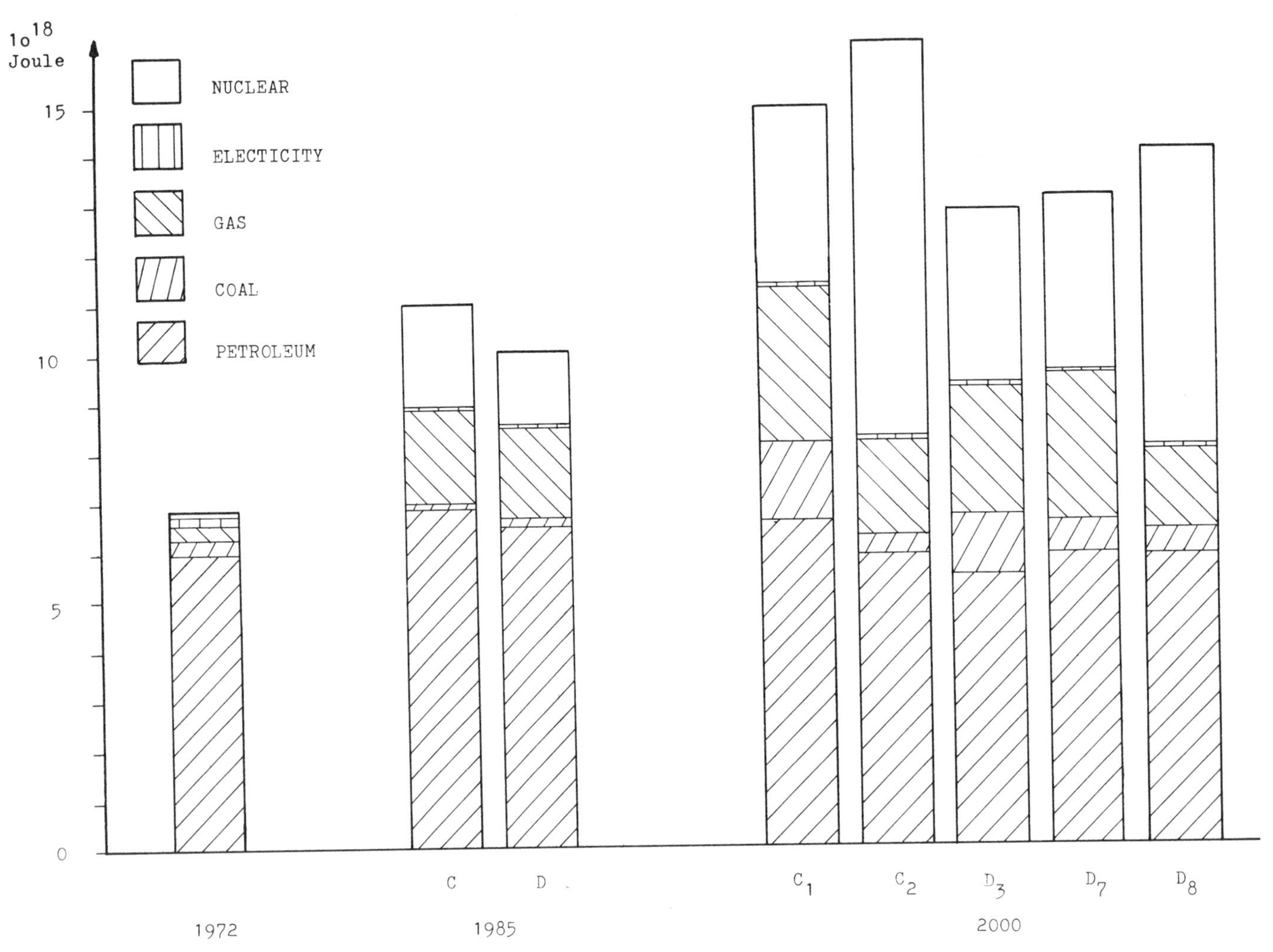

Figure 11.4
Import requirements.

Table 11.1
Economic growth, 1976-2000

	1960-1972	1972-1976[a]	1976-1985[b]		1985-2000[b]		1976-2000[b]	
			High	Low	High	Low	High	Low
World	5.0	2.5	5.1	3.5	4.0	2.7	4.4	3.0
Western Europe	4.7	2.0	4.5	2.8	3.5	2.2	3.9	2.4
Federal Republic of Germany	4.6	1.7	4.5	2.5	2.8	2.1	3.4	2.2

[a] 1976 early year estimates.
[b] WAES assumptions for the world and Western Europe are totals of national assumptions, not initial assumptions.

ceived as a function of technological, political, and social factors, all interdependent.

For the low nuclear variant, a quantitative assumption about installed nuclear capacity in the year 2000 was made (50 GWe). This figure was projected officially in the first "energy program" of the federal government for 1985. It represents a development in which the reprocessing and final waste disposal problems are not solved to the extent that sites and construction of plants for these purposes are socially accepted or politically agreed upon before the end of the 1970s. If siting for the nuclear power stations cannot follow existing plans, the nuclear expansion program will suffer a major setback during the 1980s and expansion during the 1990s will be slower than now projected. This low nuclear contribution is assumed to bring forward a political decision to open up three new hard coal mines to counteract rising coal imports. Low nuclear expansion also means a setback for all advanced nuclear technology until after 2000.

High nuclear expansion presumes that nuclear development is mainly determined by electricity consumption and a cost development favorable to nuclear energy. It assumes that 80 percent of electricity can be generated in base load and that this is almost entirely covered by nuclear energy. Also, in this case, nuclear energy is used in the 1990s to produce low- and high-temperature heat partly by advanced nuclear technology such as the high-temperature reactor (HTR).

11.3 THE SCENARIOS

WAES used several constellations of variables to develop the scenarios. This report presents results of five scenario cases to 1985 and five scenario cases to 2000 (Table 11.2).

The WAES approach of arbitrarily combining assumptions and writing scenarios around them does not necessarily produce predicted, desired, or feasible futures. However, simple a priori consideration of plausibility suggests that the scenarios have a high learning value and cannot be excluded as being economically unfeasible. Scenario D-3 is an example in which the implications are such that the scenario seems politically unacceptable; the assumption of worldwide low economic growth combined with high oil prices could result in a favorable energy supply-demand situation, but the worldwide economic situation would be regarded by many observers as critical. In such a case, one must be aware that exclusive focusing of the analysis on the energy sector would omit important implications for the whole of the economy and society.

Table 11.2
Economic growth rates for 1985 and 2000 scenarios (percent per year)

	1985 scenarios				
	A	B	C	D	E
1976-1985	4.5	2.5	4.5	2.5	4.5
	2000 scenarios				
	C-1	C-2	D-3	D-4	D-8
1976-1985	4.5	4.5	2.5	2.5	2.5
1985-2000	2.8	2.8	2.1	2.1	2.1

11.3.1 Scenario C-2

After a period of recovery (1976-1985) in which no deep economic depression occurs and average growth remains above 4.5 percent per year, a long-term trend of lower growth rates succeeds until the end of the century. The economy is able to cope with rising energy prices because of the demand effects of worldwide relative prosperity. The industrial and services sectors expand their GDP share to include the cost of the agriculture, mining, and construction sectors. Early technical solutions of the nuclear waste disposal problems as well as a sensitive siting strategy create public acceptance and political backing for rapid nuclear development. The present low-cost estimates for nuclear electricity are proved, and electricity prices are kept low in relation to the overall energy price level. Electricity is widely used.

11.3.2 Scenario D-8

Effective high economic growth rates are not achieved for a period of several years. Capital investments are not high enough to create a considerable extension of the growth potential. Domestically, saturation effects throttle demand expansion, and globally, low economic growth does not allow for continuity in export demand development. The economic situation is not capable of bearing an oil or energy price increase in real terms. Nuclear technology is favorable as in scenario C-2 but is more or less limited to the supply of the base load of electricity demand.

11.3.3 Scenario C-1

The economic growth assumption of this scenario is the same as that of scenario C-2. The difference lies in nuclear development, with only 50 GWe capacity in operation by 2000. The lower nuclear energy supply leads to higher costs for nuclear electricity, compared to the high nuclear energy scenarios, to overcome technological difficulties and to achieve public acceptance. By this and by the higher cost of electricity generation from other sources—primarily coal—electricity prices rise relative to fossil fuel prices, but not sufficiently to change the energy price level assumption. Reaction to higher prices leads to substitution among fuels. Electricity demand is considerably reduced in transportation, where electric cars do not succeed; in industry, where electricity cannot substitute for other fuels as rapidly; and in the residential sector, where—as in the case of electric car technology—the off-peak advantage given by high nuclear development fails to force electrical heat-storage systems.

11.3.4 Scenario D-7

The economic growth assumption of this scenario is the same as that of scenario D-8. As in scenario C-1 a capacity of only 50 GWe is installed by the year 2000. But while total electricity demand and production rise because of low economic growth coupled with a comparably low growth rate, electricity generating costs rise at a smaller margin because of the nuclear limitation. Also, with this development the demand price elasticity is assumed to be lower than in the high growth scenarios. Therefore, the electricity consumption is only marginally lower than in scenario D-8.

11.3.5 Scenario D-3

Scenario D-3 was adopted recently by WAES to analyze a boundary case in which the determining factors for energy supply and demand can establish more easily an equivalence or a calculated surplus of supply. In this case a rising energy price is assumed in order to force increased supply; a lower economic growth rate and high prices diminish demand. This case is critical: higher prices put a particularly heavy burden on low-growth economies. The international division of energy claims is shifted to the energy producers. The position of the low-income countries is weakened even more. With the same growth rates as in scenarios D-7 and D-8, the internal prices, employment, and international economic situation are less stable in D-3. This scenario does not seem to be economically feasible.

11.4 SUPPLY-DEMAND INTEGRATION APPROACH AND ASSUMPTIONS

11.4.1 Overview

After an appraisal of the final energy demand in each scenario case, the supply-demand integration procedure allocates final demand to fuels and adds calculated distribution and conversion losses and energy sector use. The result is the primary energy demand. By adding the exports and subtracting the indigenous production, one obtains, as a residual, import requirements. This demand-based procedure is rather strict and sometimes difficult to follow, since there are considerable impacts from the supply side. But by using the scenario approach, one can easily follow this procedure according to the Workshop's conventions.

11.4.2 Final Energy Demand

In the German study, the Workshop methodology for estimating activity levels and energy coefficients for each scenario was pursued further.

Contrary to the 1985 demand studies, only about 30 demand categories were distinguished. Estimates of the structural development of the German economy in the different cases were aided by long-term studies of PROGNOS and WSI. Energy coefficients are estimated according to trend, regression, and systems analysis techniques. Both economic growth (through investment activity) and energy price are presumed to have an impact on the development of energy coefficients.

For several demand sectors (e.g., transport and feedstocks) clear preferences for a single energy source are presumed to prevail. Also, there are certain political and market trends envisaged with regard to specific energy sources. Given the assumption that the price relations between oil, gas, and coal do not vary considerably, these hypotheses lead to useful directions for the fuel allocation process. It is assumed that coal can maintain only a small share of the residential and nonenergy use markets and a largely technically determined share in the industrial market. District heat is pushed politically, and the use of gas is forced by the gas industry during the 1970s and 1980s to reach a high market share for the already contracted gas and to sustain the use of the existing infrastructure.

11.4.3 Energy Sector

Electricity conversion is assumed to operate at an efficiency of 35 percent. The share of nuclear energy and coal for electricity and synthetic gas production is determined by the principal replacement fuel assumption. Hydro and urban waste power capacity are used to the peak of their economic potential. Oil is practically eliminated from electricity production by political measures. Some gas-fueled power stations apart from peak load are necessary when nuclear power fails to meet the optimistic forecasts. The residual share is provided by coal. Gas is produced as a by-product in refineries and coking processes and is therefore linked to the activity levels in these sectors. Synthetic gas is produced from lignite using high-temperature reactors in the high growth nuclear cases.

Heat is largely produced in combined purpose plants with electricity. Only about 20 percent is obtained by pure heat plants for peak loads and local compensations. Losses are determined technically. Small efficiency improvements in the transportation of electricity are assumed. The consumption in the energy sector is determined by the infrastructure and the techniques used.

11.4.4 Primary Energy Balance

From the final demand figure and the allocation procedure in the energy sector (including losses), primary energy demand can be calculated. The export is then estimated as a function of transport economies, the mismatchings of production outputs, product market demands, and also export demand for domestic sources (coking coal), which depends on global economic growth. This export estimate is added to the primary energy demand and the total is compared to the indigenous supply estimates of the different scenarios. The residual amount is the import requirement.

11.5 ENERGY DEMAND

Detailed numbers for the scenarios are given in the worksheets in section 11.9. Some remarks on major activity levels and energy coefficients are made below.

11.5.1 Major Activity Levels

The aggregation of line items does not allow for an immediate comparison between the activity levels for 2000 and those for 1985 and 1972, published in the WAES First Technical Report. Therefore, some explicit comparisons of the estimates for 2000, 1985, and 1972 are made in addition to the detailed numbers given in the worksheets.

Transport

The population-to-auto ratio is assumed to come down to about 2.4 by the year 2000 from about 3.8 in 1972 and 2.8 in 1985 (see Table 11.3). The yearly performance per car decreases considerably, partly because of the introduction of electric cars (7500 km/yr). This adds up to an even lower total of driven km in 2000 than in 1985.

The number of trucks is still increasing such that the figures for total driven km and ton-km transported continue to rise.

Considerable expansion of passenger transport is assumed to occur only for air travel, whereas freight transport shows expansion tendencies in all categories.

Production and Services

The growth rates in Table 11.4 disclose that the industrial sector, particularly the chemicals industry, is expected to carry the burden of overall growth. Nonenergy mining and agriculture contribute very little in all cases. Construction activity is rather low because of saturation effects. Likewise, the minerals industry is also growing slowly. The public GDP contribution is getting smaller because of smaller additional infrastructure needs. Also because of saturation, the growth of the commercial sector is assumed to be disproportionately slow.

Table 11.3
Major activity levels in transport

	2000 C-2	2000 D-8	1985 C	1972
Autos				
Fossil-fueled (10^6)	22.8	23.3	21.7	16.1
Electric (10^6)	2.0	0.4	0	0
Km/year/auto (10^3)	11.8	12.1	13.9	15.5
Total km/year (10^9)	293	287	302	249
Trucks				
Fossil-fueled (10^6)	1.75	1.65	1.4	1.05
Electric (10^6)	0.175	0.075	0	0
Km/year/truck (10^3)	22.6	22.8	23.6	23.9
Total km/year (10^9)	43.5	39.3	33.0	25.1
Public Road Transport				
Scheduled passenger-km (10^9)	51.0	51.25	50.8	48.3
Unscheduled passenger-km (10^9)	17.0	16.5	16.3	14.0
Rail				
Gross ton-km (10^9)	440	440	340	264.4
Freight ton-km (10^9)	112	112	86.6	66.7
Travel passenger-km (10^9)	52	52	49.4	39.6
Air				
Travel plus freight ton-km (10^9)	8	7	5.8	2.5

Table 11.4
Value added growth between 1972 and 2000 in production and service categories (percent per year)

	1972–1985		1985–2000	
	C	D	C-2	D-8
Industry				
Total	3.0	2.5	3.4	2.4
Chemicals	4.5	4.0	5.6	4.0
Iron and steel	1.0	1.0	1.6	1.1
Minerals	1.0	1.0	1.6	1.4
Commercial	2.5	1.5	2.8	2.0
Public (GDP contribution)	1.7	1.7	1.8	1.6
Agriculture	0.8	0.8	1.0	1.0
Mining	0.0	0.0	0.0	0.0
Construction	1.5	1.0	1.6	1.0

Residential

The assumed number of dwellings for 2000 varies from 25.54×10^6 for the high growth scenarios to 24.5×10^6 for the low economic growth scenarios (compared to 24.97–24.1×10^6 in 1985). This represents very little new housing. Economic growth (allowing for higher private income) is also assumed to have an impact on the average size of dwellings (88–90 m^2), so that the total floor space differs in the various scenarios.

The saturation levels of household appliances are assumed to be the same in all scenarios, although different in absolute terms. The shift to central space heating (including electric storage systems) is assumed to be almost completed by 2000.

11.5.2 Energy Coefficients

Transport

Major changes between 1985 and 2000 in efficiency as defined here are assumed only in individual transport. Higher fuel prices are assumed to push the development of electric cars and the fuel efficiency of air transport (Table 11.5). Electric cars have higher efficiencies. In year 2000 scenario C-2, the larger market share in the auto and truck categories results in lower overall specific consumption figures.

Production and Services

Continuing mechanization of agricultural production through 2000 results in higher energy consumption per value added, whereas in the mining and construction categories the specific consumption figures are expected to remain constant.

Small conservation gains in relation to the value added or GDP contribution are assumed in the commercial and public sector.

A slightly lower specific energy consumption with a shift to electricity when compared to 1985 is assumed for the industrial sectors by 2000, although the intensity of this development differs

Table 11.5
Energy coefficients in transport

	2000		1985	1972
	C-2	D-8	C	
Autos (10^6 J/km)				
Including electric cars	284	294	337	338
Excluding electric cars	292	297	337	338
Trucks (10^6 J/truck)				
Including electric trucks	208	223	206	197
Excluding electric trucks	225	232	206	197
Public Road Transport (10^6 J/passenger-km)				
Fossil	0.36	0.36	0.37	0.38
Electric	0.6	0.6	0.6	0.6
Rail (10^6 J/gross ton-km)				
Fossil	1	1	0.8	0.8
Electric	0.14	0.14	0.13	0.14
Air (10^6 J/ton-km)	27	30	31	32

by sector. This is due to changes or shifts in product mix (chemicals) and process mix (iron and steel). Changes in process mix particularly occur under the premises of changing factor cost relations through higher energy prices.

Residential

The distribution of heating systems by fuel type is influenced heavily by the development of the absolute and relative energy prices. Higher energy prices (scenarios C-1, C-2, and D-3) are assumed to force the development of district heat, electricity, and advanced supply systems such as heat pumps and solar energy.

Heat loss is reduced continually up to 2000 such that the average standards in 2000 are close to the top standards of 1975. In particular, single-family houses are expected to show a relatively large improvement. The same development is assumed for system efficiencies. Old systems are replaced and renovated. Thus, the efficiencies of modern systems in 1975 are the average efficiencies in 2000.

11.5.3 Nonenergy Use

About 65 percent of total energy consumption for nonenergy purposes in 1972 flowed into the chemicals industry. Another 26 percent was used in the form of bitumen for road construction, and 6 percent went for lubricants. While road construction and traffic are assumed to increase moderately, bitumen and lubricants consumption are not assumed to grow. But the comparatively extensive growth of the chemicals industry requires a steadily increasing supply of raw materials. It is assumed that the new capacity for production of petrochemicals will largely be installed domestically. This indicates a continuing expansion of nonenergy use in the chemicals industry (average growth 13.7 percent per year, 1960 to 1972) but with lower growth rates than in the past (4.4 percent compared to 3.4 percent per year from 1972 to 2000).

11.5.4 Final Energy Demand 2000

Tables 11.6–11.7 and Figure 11.5 show the development of final energy demand in the Federal Republic of Germany in four scenarios for 2000 as compared to 1985 and 1972 (between 4.7 and 3.7 percent average annual growth from 1972 to 2000).

Further analysis of final demand was not made before the supply-demand integration. Such a partial view could be misleading without the integration.

11.6 SUPPLY-DEMAND INTEGRATION

11.6.1 Final Demand 1985

The worksheets show the expected final demand by fuel and indicate a change in the percentages from 1972 to 1985. The share of coal and oil is decreasing strongly. Coal demand is even lower in absolute terms. Heat, electricity, and gas demand will almost double during this period.

Table 11.6
Final energy demand, 2000, 1985, and 1972, by sector (10^6 TCE)

	2000 C-1		2000 C-2		2000 D-7		2000 D-8		1985 C		1985 D		1972	
	Fossil	Electricity	Fossil	Electricity	Fossil	Electricity	Fossil	Electricity	Fossil	Electricity	Fossil	Electricity	Fossil	Electricity
Transport	62.1	2.4	60.3	3.4	61.6	2.3	61.1	2.5	61.3	1.56	59.2	1.56	48.6	1.05
Industry	115.5	42.5	110.2	47.8	97.6	32.7	97.6	32.7	90.6	28.1	80.3	22.6	73.4	14.7
Agriculture, mining, and construction	15.6	3.3	15.5	3.4	14.5	3.3	14.5	3.3	14.6	2.01	13.9	1.97	11.97	1.22
Commercial	27.5	12.4	26.7	13.4	21.9	10.7	21.9	10.7	27.6	9.97	26.6	9.36	21.7	4.15
Public	8.2	6.6	8.2	6.6	8.0	6.5	8.0	6.5						
Residential	64.9	22.4	52.2	27.0	64.7	21.0	64.7	21.0	67.1	14.4	67.4	12.65	61.7	6.87
Nonenergy use	66.6	—	66.6	—	55.5	—	55.5	—	41.9	—	38.4	—	26.0	—
Total	360.3	89.8	339.4	101.6	323.8	76.5	323.3	76.7	303.0	56.1	285.9	48.11	243.34	28.08

Table 11.7
Sectoral growth rates of final energy demand, 1972–2000 and 1985–2000 (percent per year)

	2000 C-2 (from 1985 C)		2000 D-8 (from 1985 D)		2000 C-1 (from 1972)		2000 C-2 (from 1972)			2000 D-8 (from 1972)		
	Fossil	Electricity	Fossil	Electricity	Fossil	Electricity	Fossil	Electricity	Total	Fossil	Electricity	Total
Transport	-0.1	5.3	0.2	3.2	0.9	3.0	0.8	4.3	0.9	0.8	3.1	0.9
Industry	1.3	3.6	1.3	2.5	1.6	3.9	1.5	4.3	2.1	1.0	2.9	1.4
Agriculture, mining, and construction	0.4	3.6	0.3	3.5	1.0	3.6	0.9	3.7	1.3	0.7	3.6	1.0
Commercial and public	1.6	4.7	0.8	4.1	1.8	5.6	1.7	5.8	2.7	1.2	5.2	2.2
Residential	-1.7	4.2	-0.3	3.4	0.2	4.3	-0.6	5.0	0.5	0.2	4.1	0.8
Nonergy use	3.1	—	2.5	—	3.4	—	3.4	—		2.7	—	
Overall growth rate	0.76	4.0	0.82	3.16	1.4	4.2	1.2	4.7		1.0	3.7	

Energy Sector 1985

The electricity generation estimate can largely be based on existing and planned capacity coupled with an electricity import assumption. Nuclear power is assumed to contribute between about 33 percent (in the low economic growth case) and 40 percent (in the high growth case) of the generated electricity. Low electricity demand growth is combined with low base-load demand growth. Nuclear growth cannot economically replace existing lignite and hydro power capacity. Therefore, the nuclear share is lower in the low growth case. The envisaged nuclear contribution can be generated by the capacity of 22 to 33 GW (according to the scenario and assuming high utilization).

Drawing heat from waste heat in power stations requires some additional electrical generating capacity, since the partial efficiency of electricity generation (ratio of electricity output to fuel input) in the double-purpose plants is smaller.

11.6.2 Primary Energy Demand 1985

The primary energy demand picture in the scenarios to 1985 is an immediate result of the described development in the final demand and energy sectors.

The ratios between primary energy and GDP growth for 1972 to 1985 range from 0.69 to 0.91 (Table 11.8). Ratios higher than in the past (0.86 in 1960 to 1972) only occur in cases of low economic growth, constant prices, and restrained policy, or with falling prices.

These figures are slightly different than the numbers given in the WAES First Technical Report because of the rough preliminary calculation of primary demand at that stage.

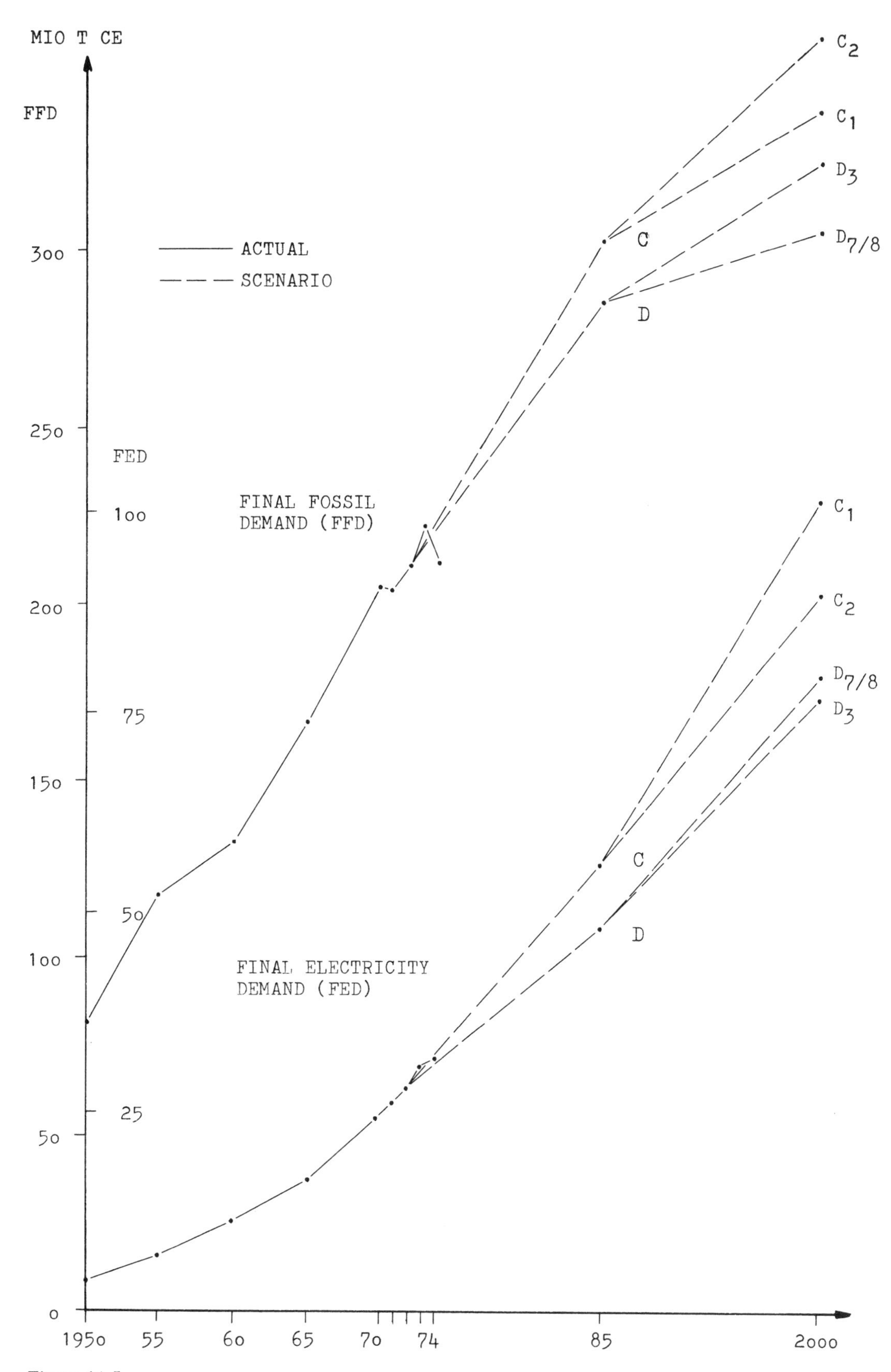

Figure 11.5
Energy demand in the Federal Republic of Germany. Scenarios include some 8.5×10^6 TCE.

The oil price and income elasticities assumed for 1985 are still valid (Table 11.9). An approximately 20 percent higher GDP results in about 10 percent higher energy demand. The oil price rise of 50 percent leads to about 3 percent lower primary energy demand in the period to 1985.

The impact of a vigorous conservation policy is minimal. Compared to the artificially constructed vigorous scenarios D* and E*, the restrained scenarios show only about 1 percent lower demand (Table 11.10).

11.6.3 Indigenous Supply and Exports 1985

The supply estimates are described in the WAES Second Technical Report (see also section 11.8).

With regard to exports in 1985, small quantities of oil products and gas are reasonable to expect because of regional transport advantages, while there is likely to be a considerable export of coking coal.

11.6.4 Imports 1985

The oil, gas, and coal import requirements of the Federal Republic of Germany are determined by adding the domestic and export demand and then comparing this total to the indigenous supply estimates.

Oil import requirements for 1985 are, except for the high price, low growth scenario, higher than oil imports in 1972. But since the total primary energy demand is increasing, the quota of oil imports and total primary energy demand are both lower (Table 11.11). The difference between the projected and actual "relative dependence" on oil import—defined as the percentage of demand provided by imports—is almost balanced by the higher gas imports. Coal import requirements increase total imports only marginally.

Although the uranium supply is entirely imported, the contribution obtained from nuclear energy is usually not regarded as import. Excluding nuclear power, the "import dependency" would be slightly lower in 1985 than in 1972, whereas the inclusion of nuclear power would increase "import dependency" to about 75 percent in 1985 from 65 percent in 1972.

Higher oil prices lower the oil import dependency from 46.2 percent to 43.5 percent in scenarios A to C. But this impact is not decisive. By forcing energy demand and import requirements to run almost in parallel, economic growth has almost no effect on the relative import dependence, though it clearly increases the absolute import dependence.

Table 11.8
GDP and primary energy demand growth, 1972-1985

	A	B	C	D	E
Average annual growth (%/yr)					
Primary energy	2.52	1.75	2.73	2.04	3.00
GDP	3.63	2.25	3.63	2.25	3.36
Ratio of growth rates (primary energy: GDP)	0.69	0.78	0.75	0.91	0.89

Table 11.9
Oil price[a] and income elasticities (a and γ) to 1985

	a[b]	γ[c]
Final energy demand		
Fossil fuels	-0.13	0.37
Electricity	0.03	0.91
Primary energy demand	-0.067	0.56

[a]The definition of price elasticities used in this analysis does not take into account a time factor. Later a suitable equation will be introduced.

[b]$(E_A/E_C) = (P_A/P_C)^a$; P = oil price in 1985.

[c]$(E_A/E_B) = (\mathrm{GDP}_A/\mathrm{GDP}_C)^\gamma$; GDP = GDP in 1985.

Table 11.10
Savings

	$(E_D - E_D^*)/E_D$	$(E_E - E_E^*)/E_E$
Final energy demand		
Fossil fuels	0.6	0.5
Electricity	0.5	0.1
Primary energy demand	1.0	0.7

Note: $E_D^* = E_C(\mathrm{GDP}_D/\mathrm{GDP}_C)^\gamma$.
$E_E^* = E_C(P_E/P_C)^a$.

Table 11.11
Import dependence (percentage of total final demand)

		1985					2000				
	1972	A	B	C	D	E	C-1	C-2	D-7	D-8	D-3
Oil	57.3	43.5	44.0	46.2	47.5	48.6	33.7	29.0	33.8	33.1	31.9
Oil and gas	60.4	56.6	58.0	58.8	61.2	60.8	49.9	38.6	51.2	42.3	47.1
Oil, gas, and coal	63.3	58.4	58.5	59.8	62.6	64.1	58.0	40.4	54.9	45.2	54.1
Oil, gas, coal, nuclear, and electricity	65.3	73.6	70.7	74.3	74.0	77.8	76.5	80.1	75.9	79.2	75.3

11.6.5 Final Demand 2000

Coal and oil are delivered to final consumers in the 2000 scenarios in almost the same quantities as in the 1985 scenarios C and D. Relatively high additions to district heat (partly through political measures), additions to industrial heat from double-purpose plants (particularly in the rising oil price cases), and an increase in electricity and gas consumption (both with decreasing growth rates) all contribute to the increased final energy demand.

11.6.6 Energy Sector 2000

About 80 percent of the electricity generating capacity in the high growth nuclear cases is assumed to be base load, consisting of hydro power, lignite, urban waste, and nuclear capacity. In the low nuclear growth scenarios, the assumed limit of 50 GW installed capacity requires the use of hard coal for base load.

In the low nuclear growth cases, there is no coal gasification assumed. Also, practically no heat is drawn from nuclear power stations. About 20 percent of the heat comes from local single-purpose plants, the rest from conventional double-purpose plants.

The growing share of electricity in final demand expands the quantitative importance of the energy sector. In 1972 about 77 percent of primary energy was used for final demand and bunkers, in the 1985 scenarios the percentage is 70–72 percent, and in the 2000 scenarios it is only 64–68 percent. This is based on an electricity conversion efficiency of 35 percent held constant over time—a Workshop convention.

11.6.7 Primary Energy Demand 2000

Primary energy demand in the Federal Republic of Germany is projected to grow with a slightly lower intensity compared to economic growth in the period after 1985 than in the period before (see Tables 11.8 and 11.12).

The elasticities (defined in the WAES First Technical Report) show the same signs for 2000 as for 1985 (see Tables 11.9 and 11.13). Higher oil prices are supposed to force a shift from fossil fuels to electricity in the final demand but to have only a small conservation effect on primary energy demand.

The relative contribution of oil continues to decrease through 2000. Natural gas has almost reached its maximum share in 1985. When gasification becomes competitive, it cannot adhere to its 1985 share. The nuclear power contribution is increasing relatively in all scenarios. The coal share depends on the size of the nuclear contribution.

11.6.8 Indigenous Supply and Exports 2000

A summary of the indigenous supply to 2000 can be found in section 11.9. For energy exports in the 2000 cases, only coking coal exports are considered.

11.6.9 Imports 2000

The import requirements of natural gas, coal, and under certain conditions, oil are strongly dependent on nuclear development.

In scenarios C-2, D-3, D-7, and D-8, oil import requirements are down slightly below 1972 levels. But high economic growth and a low nuclear contribution (C-1) make more fossil fuel imports necessary. Coal gasification and a high nuclear contribution can stabilize gas import requirements at the 1985 level. Low nuclear expansion leads to another increase in natural gas import requirements over and above the contracted quantities for 1990.

Table 11.12
GDP and primary energy demand growth, 1972-2000

	C-1	C-2	D-7	D-8	D-3
Average annual growth (%/yr)					
Primary energy	2.26	2.40	1.83	1.90	1.77
GDP	3.40	3.40	2.20	2.20	2.20
Ratio of growth rates (primary energy: GDP)	0.66	0.71	0.83	0.86	0.80

Table 11.13
Oil price and income elasticities (α and γ) to 2000

	α[a]	γ[b]	Deviation (%)[c]
Final demand			
Fossil fuels	-0.13	0.57	- 6
Electricity	0.10	0.43	11
Primary demand	-0.04	0.49	- 2

[a] $(E_3/E_7) = (P_3/P_7)^{\alpha}$.

[b] $(E_3/E_7) = (\text{GDP}_3/\text{GDP}_1)^{\gamma}$.

[c] $(E_8/E_8) - 1 = \text{Deviation}$

$E_8 = E_2 \times (P_8/P_2)^{\alpha} \times (\text{GDP}_8/\text{GDP}_2)^{\gamma}$

Here E = energy consumption; P = oil price in 2000; GDP = GDP in 2000. Since the specification of P assumes that all adaptation processes are brought to an end by 2000, the figure underestimates the real elasticity. It does not take into account long-term reactions on price changes in the late 1990s.

The low nuclear growth assumptions also create a high need for coal import. With high economic growth and low nuclear growth, even if all other sources are employed, a coal import of about 50 million coal tons must be expected.

11.7 NATIONAL RESPONSE

Global aggregation of the independent national supply-demand scenarios leads necessarily to surpluses or gaps between global supply and demand. Only by accident would prospective supply equal prospective demand exactly. If the prospective gap or surplus is larger than the permissible error range, then the assumed scenario is not feasible. This means that the particular set of assumptions of such a case does not represent a consistent and plausible state of reality.

The global integrations of the Workshop 2000 scenarios resulted in gaps between the demand desired and the supply provided in the original, fully studied scenarios C-1, C-2, D-1, and D-8. The "excursion" scenario D-3, though not completely analyzed, points to a more or less balanced global demand and supply situation. But this condition (low economic growth and a high energy price of $17.25 combined with vigorous government policy) shows only one possible future. There are several others (higher growth with prices above $17.25 and/or supervigorous policy, etc.).

To evaluate the global and national results of the scenarios and to reach a more complete understanding of the national options, it is appropriate to analyze the impacts of the scenario variables, in particular potential energy policies with respect to energy supply and demand. The analysis of the national potential and the global conditions together with sensible assumptions about energy policies of other nations form the basis for long-term success.

To study the impacts and options of the scenario assumptions, the use of a partial analysis is appropriate. Partial analysis of parametric variations can use the concept of price and income (GNP) elasticity as well as the concept of modules. National and global modules and elasticities form a set of tools for policy evaluation.

This partial analysis must not be applied arbitrarily. Elasticities should only be used for interpolation between the scenarios and marginal excursions of completely worked-out scenarios, since in reality, the variables are not independent from each other and the elasticities can vary when the variation exceeds a certain range. Modules must be envisaged not just as a quantitative potential but must be qualified by the description of their economic and political implications.

11.7.1 Elasticities of Demand and Supply

Demand

In addition to the demand function applied above and in the WAES First Technical Report, another specification can be used for the partial analysis, which the Workshop calls the "Bernardini function." It describes the demand for energy as a function of the base year demand, the economic growth, energy prices, and national policies. (Formula and detailed results may be found in subsection 11.7.4.)

With regard to economic growth, the elasticity of primary energy demand from 1972 to 2000 lies between about 0.8 and 0.9 in the Federal Republic of Germany.

Because of the rather autonomous development of some consumption categories such as households and individual transport, the elasticity coefficient is higher in the low economic growth cases (0.88 and 0.91) and lower in the high economic growth cases (0.76 and 0.83).

The elasticity of primary energy demand with regard to oil price variation is assumed to be efficient but rather low (-0.015 for the price change from \$2.90 to \$7.66, −0.059 from \$7.66 to \$11.50, and -0.046 from \$11.50 to \$17.25).

The percentage savings in primary energy demand gained by a switch from restrained to vigorous policy is assumed to be about 1 percent.

The impact of the low nuclear expansion assumption depends on the level of electricity demand determined by the other variables. Primary energy demand is decreasing, through substitution of electricity by fossil fuels by about 4 percent in the high economic growth cases and by about 2 percent in the low economic growth cases.

Indigenous Supply

The differences in indigenous supply of fossil fuels among the scenarios are almost entirely due to policies as stated in chapter 7 of the WAES Second Technical Report. That means that in the analyzed price range the elasticity of indigenous supply is close to zero. The supply of nuclear electricity is demand-restricted in the high nuclear expansion cases and dependent on economic growth.

Disregarding nuclear power, few additions to supply at competitive conditions in a conceivable price range above 17.25 U.S. \$/barrel and high economic growth can be expected in the Federal Republic of Germany. Oil and natural gas are almost depleted. The hydro power potential is assumed to be almost completely in use, and nonfossil nonnuclear sources either do not have any potential at all or offer only very small additional supply at the relevant price levels. Therefore, lignite and hard coal remain the only potential additional supplies. With these resources very high prices would be necessary to overcome the environmental and cost restrictions.

11.7.2 Modules

Modules are physically defined units of additional supply, substitution potential, and demand reduction and allow for partial policy analysis. Differing from elasticities, they are not based on the aspect of functional relationships between scenario variables and energy supply or demand quantities. Rather, the definition of modules requires first a technical description and an evaluation of the physical potential of a technology and then a consideration of the political and economic (price and growth) implications. By that, modules allow generally for variations of the policy assumptions. Since the policy assumptions are made explicit, the addition or aggravation of political measures means a variation in policy.

Tables 11.14–11.16 attempt to define, to qualify, and to quantify supervigorous measures in the Federal Republic of Germany. Only measures that create more supply or more conservation are relevant in this context. The quantities saved or added are given with respect to the lead time needed to implement such measures.

11.7.3 Evaluations and Strategies

The numerical national and global analysis forms a sound basis for the formulation of national strategies. For that purpose, it is necessary to discuss energy policy objectives and availability and impact of policy measures not only with regard to the energy situation but also in terms of other economic and social objectives, since the cost and benefits of energy policy measures do not occur only in the energy sector of the economy.

11.7.4 Application of the Bernardini Function

The Bernardini function, presented in chapter 12 of this volume, is

$$E = E_0 \left(\frac{G}{G_0}\right)^{\epsilon} \left[\prod_{i=1}^{3} \left(\frac{P_i}{P_{i-1}}\right)^{\eta_i} \right] (1\text{-}R\sigma)(1\text{-}N\tau),$$

where E, G = energy demand and GDP 2000

E_0, G_0, P_0 = energy demand, GDP, price 1972

P_1, P_2, P_3 = oil prices \$7.65, \$11.50, \$17.25

R = policy; σ = fractional savings

N = nuclear contribution; τ = fractional impact

N and R are dichotomic (0 or 1)

ϵ = income elasticity

η = price elasticity

The calculation of the elasticities for 2000 can only approximate the real elasticities, since there are too few different scenarios for too many unknowns. For that reason, some elasticities (η_1, η_2) and σ were taken from the analysis of 1985.

Assumptions and Results

1. σ = 1 percent (from the 1985 demand study)
2. σ = 2 percent (for D-7 and D-8), 4 percent (for C-1 and C-2)
3. η_1 = -0.015 (from the 1985 demand study)
   η_2 = 0.059 (from the 1985 demand study)
   η_3 = 0.0458 (for D-7 and D-3)
4. $\epsilon_{C\text{-}1}$ = 0.753
   $\epsilon_{C\text{-}2}$ = 0.827
   $\epsilon_{D\text{-}3}$ = 0.876
   $\epsilon_{D\text{-}7}$ = 0.876
   $\epsilon_{D\text{-}8}$ = 0.907

Table 11.14
National response. I. Demand reduction

No.	Target	Measure type	Measure definition	Lead time (years)	Saved (10^{15}J) Energy	Oil
1.	Residential	Mandatory	Reduction of room temperature by 2° C	1	170	50
2.	Residential	Mandatory	Higher insulation standards, 10% specific heat loss reduction per new dwelling	15	70	20
3.	Residential	Incentive	Heat loss reduction in old dwellings by 10%	5	100	30
4.	Residential	Incentive	Solar and heat pump promotion, substitution of oil in rural areas	10	70	100
5.	Residential	Mandatory	District heating obligation for higher consumer efficiency; heat substitutes other fuel	10	80	60
6.	Conservation	Mandatory	Production of additional district heat for measure 5			
			• from conventional power stations	10	-80	
			• from waste nuclear heat (net effect; recycled heat minus additional production)	10	220	
7.	Agriculture	Incentive	Solar and heat pump	10	150	100
8.	Commercial	Incentive/ Mandatory	Same as residential measures 5 and 6			
9.	Transport	Mandatory	Speed limit 90 km/hr			
			• 1st-order effect	1	50	50
			• 2nd-order effect	10	150	150
10.	Transport	Mandatory	Specific fuel consumption standards	10	200	200
11.	All categories	Incentive	Energy tax (effect comparable to price rise measurable in terms of price elasticity)			
12.	All categories	Incentive	Higher tax on oil products			
13.	All categories	Incentive	Import tariffs on energy imports			

Note: Measures 9 and 10 are partly alternatives. Since the measures 12 and 13 apply to only a part of the energy demand, the elasticity of the considered part of energy demand is probably higher because of additional interfuel substitution.

Table 11.15
National response. II. Supply expansion

No.	Target	Measure type	Measure definition	Lead time (years)	Added fuel (10^{15} J)
1.	Nuclear	Mandatory	Alleviation of nuclear restrictions in Case C-1	20	4300
2.	Nuclear	Incentive	Higher capacity of HTR and FBR	15	100
3.	Coal	Mandatory	Alleviation of environmental constraints on lignite production	15	300
4.	Coal	Incentive	Additional production of hard coal	10	200
5.	Gas	Incentive	Import of NGL	10	300
6.	Gas	Incentive	Import of LPG	5	200

Note: These measures can allow for more oil substitution. Alleviation of import quotas now in existence is already part of an assumed "vigorous policy."

Table 11.16
National response. III. Fuel substitution

No.	Target	Measure type	Measure definition	Lead time (years)	Saved oil (10^{15} J)	Additional (10^{15} J)		
						Oil	Gas	Other
1.	Oil-Gas/Coal	Incentive	Increase lignite gasification;	10			130	-200
			substitute oil by gasified lignite			130	-130	
2.	Oil-Gas	Incentive	Substitute oil products by gas from NGL imports	10		300	-300	
3.	Oil-Gas	Incentive	Substitute oil products by gas from LPG imports	5		300	-300	
4.	Oil-Coal	Incentive	Increase share of coal in the industrial sector	10	200			-200
5.	Oil-Coal	Incentive	Produce methanol from hard coal and substitute oil	10	200			-500
6.	Oil-Coal	Incentive	Lignite liquefaction	10		130		-200
7.	Oil-Coal	Incentive	Hard coal liquefaction	10		130		-200
8.	Oil-Nuclear	Incentive	Nonelectric use of nuclear (coal gasification)	15			55	-100
9.	Oil-Nuclear	Incentive	Electric auto on nuclear electricity	15	100			-100
10.	Oil-Nuclear	Incentive	Intrasource substitution leading to oil savings	5	50			

Note: The lead times include the times for R&D efforts based on today's states of the arts.

11.8 REFERENCES

The references and data sources for the German studies were listed in the first and second technical reports. Major sources for the demand study were the long-term prospects of economic and social developments in the Federal Republic of Germany projected by PROGNOS A.G. and by WSI, published as "Künftige Bedarf an electrisches Energie in Abhängigkeit von wirtschafts- und gesellschafts-politischer Entwicklung und dessen Deckung, insbesondere mit Hïlfe der Kernenergie" (report K 76-03 of the Ministry for Research and Technology, BMFT, of July 1976). This publication also contains electricity projections to the year 2000 prepared at the Institute of Energy Economics.

11.9 WORKSHEETS

This section contains summary tables and worksheets in the following order:

Energy demand summary (final demand) 1972–1985–2000

Energy demand summary (primary demand) 1972–1985–2000 (2 sheets)

Energy supply summary (indigenous supply) 1972–2000

National energy demand projections (2 sheets)

National input worksheets for supply/demand integration (11 sheets)

WAES NATIONAL ENERGY DEMAND STUDIES, 1972-1985-2000
ENERGY DEMAND SUMMARY FINAL DEMAND (Units: 10^{15} J)

Country: Federal Republic of Germany

Scenario Assumptions	1972	1985					2000				
WAES CASE	Base Year	A	B	C	D	E	C-1	C-2	D-7	D-8	D-3
GNP 1976-85; 1985-2000		High	Low	High	Low	High	High	High	Low	Low	Low
Oil/Energy Price*		11.50-17.25	11.50-17.25	11.50	11.50	11.50-7.66	11.50-17.25	11.50-17.25	11.50	11.50	11.50 17.25
National Policy Response		Vig	Vig	Vig	Res	Res	Vig	Vig	Vig	Vig	Vig
Principal Replacement Fuel							Coal	Nuclear	Coal	Nuclear	Coal
FINAL **Energy Demand** **Units**											
Oil	5027	5183	4793	5699	5339		5787	5442	5322	5308	4903
Natural Gas	1013	2272	2210	2252	2221	2253	3112	2891	2958	2958	2545
Coal	977	715	652	662	602	631	690	640	660	660	680
Heat	127	288	272	299	248	268	909	909	534	534	818
Electricity	827	1670	1425	1649	1415	1631	2640	2987	2250	2257	2340
Geothermal											
Solar Thermal											
Solar Electric											
Other	21						100	100	50	50	100
TOTAL PRIMARY ENERGY DEMAND	7992	10128	9352	1051	9815	11071	13238	12969	11774	11767	11386

*1975 US dollars per barrel of Arabian light crude oil, fob Persian Gulf.
All Figures are primary production prior to any processing or conversion.

WAES NATIONAL ENERGY DEMAND STUDIES, 1972-1985-2000
ENERGY DEMAND SUMMARY PRIMARY DEMAND (Percentage by Fuels)

Country: Federal Republic of Germany

Scenario Assumptions	1972	1985					2000				
WAES CASE	Base Year	A	B	C	D	E	C-1	C-2	D-7	D-8	D-3
GNP 1976-85; 1985-2000		High	Low	High	Low	High	High	High	Low	Low	Low
Oil/Energy Price*		11.50-17.25	11.50-17.25	11.50	11.50	11.50-7.66	11.50-17.25	11.50-17.25	11.50	11.50	11.50-17.25
National Policy Response		Vig	Vig	Vig	Res	Res	Vig	Vig	Vig	Vig	Vig
Principal Replacement Fuel							Coal	Nuclear	Coal	Nuclear	Coal
Primary Energy Demand **Units**											
Oil	55.4	42.5	43.0	45.3	46.6	47.7	34.3	29.6	34.2	33.5	32.3
Natural Gas	8.7	17.5	18.8	16.9	18.3	16.3	17.7	11.0	19.0	10.8	16.9
Coal	32.3	22.3	23.5	20.9	21.2	19.9	26.4	16.8	23.0	18.4	26.1
Nuclear	0.9	14.7	11.6	14.0	10.9	13.3	18.2	39.3	20.5	34.0	20.8
Hydroelectric	1.1	1.2	1.3	1.2	1.3	1.1	1.0	1.0	1.1	1.1	1.2
Geothermal											
Solar Thermal							0.5	0.5	0.3	0.3	0.6
Solar Electric											
	1.6	1.8	1.8	1.7	1.7	1.7	1.9	1.8	1.9	1.9	2.1
TOTAL PRIMARY ENERGY DEMAND	100	100	100	100	100	100	100	100	100	100	100

*1975 US dollars per barrel of Arabian light crude oil, fob Persian Gulf.
All Figures are primary production prior to any processing or conversion.

WAES NATIONAL ENERGY DEMAND STUDIES, 1972-1985-2000
ENERGY DEMAND SUMMARY PRIMARY DEMAND

Country: Fed. Rep. of Germany

Scenario Assumptions	1972	1985					2000				
WAES CASE	Base Year	A	B	C	D	E	C-1	C-2	D-7	D-8	D-3
GNP 1976-85; 1985-2000		High	Low	High	Low	High	High	High	Low	Low	Low
Oil/Energy Price*		11.50-17.25	11.50-17.25	11.50	11.50	11.50-7.66	11.50-17.25	11.50-17.25	11.50	11.50	11.50-17.25
National Policy Response		Vig	Vig	Vig	Res	Res	Vig	Vig	Vig	Vig	Vig
Principal Replacement Fuel							Coal	Nuclear	Coal	Nuclear	Coal
Primary Energy Demand **Units**	10^{15} Joules										
Oil	5770	6128	5613	6701	6311	7305	6687	5992	5922	5908	5503
Natural Gas	906	2516	2457	2493	2473	2493	3446	2229	3292	1908	2879
Coal	3364	3206	3064	3091	2879	3038	5141	3405	3974	3247	4445
Nuclear	91	2113	1513	2074	1480	2031	3545	7971	3545	5996	3545
Hydroelectric	120	175	175	175	175	175	200	200	200	200	200
Geothermal	-	-	-	-	-	-	-	-	-	-	-
Solar Thermal	-	-	-	-	-	-	100	100	50	50	100
Solar Electric	-	-	-	-	-	-	-	-	-	-	-
	169	260	235	260	235	260	375	375	325	325	345
TOTAL PRIMARY ENERGY DEMAND	10480	14398	13057	14794	13553	15302	19494	20373	17308	17634	17017

*1975 US dollars per barrel of Arabian light crude oil, fob Persian Gulf.
All Figures are primary production prior to any processing or conversion.

WAES NATIONAL ENERGY SUPPLY STUDIES, 1972-2000
ENERGY SUPPLY SUMMARY – INDIGENOUS SUPPLY 10^{15} Joules

Country: Federal Republic of Germany

Scenario Assumptions	1972	1985					2000				
WAES CASE	Base Year	A	B	C	D	E	C-1	C-2	D-7	D-8	D-3
GNP 1976-85; 1985-2000		High	Low	High	Low	High	High	High	Low	Low	Low
Oil/Energy Price*		11.50-17.25	11.50-17.25	11.50	11.50	11.50-7.66	11.50-17.25	11.50-17.25	11.50	11.50	11.50-17.25
National Policy Response		Vig	Vig	Vig	Res	Res	Vig	Vig	Vig	Vig	Vig
Principal Replacement Fuel							Coal	Nuclear	Coal	Nuclear	Coal
Energy Supply **Units**											
Crude Oil	299	168	168	168	168	168	112	122	67	67	67
Natural Gas	548	645	645	645	645	645	293	293	293	293	293
Coal	3883	3800	3800	3800	3390	3390	4371	3891	4018	3423	3968
Nuclear	90	2113	1513	2074	1480	2031	3545	7973	3545	5996	3545
Hydro-electric	120	175	175	175	175	175	200	200	200	200	200
Geothermal											
Solar Thermal											
Solar Electric											
	88	185	160	185	160	185	400	400	300	300	370
Total Energy Supply	5028	7086	6461	7047	6018	6594	8921	12817	8423	10285	8443

*1975 US dollars per barrel of Arabian light crude oil, fob Persian Gulf.
All Figures are primary production prior to any processing or conversion.

NATIONAL ENERGY DEMAND PROJECTIONS

Country: Fed. Rep. of Germany

Year: 2000

WAES Case: C 2

Population: 58.7 million

No. of Autos: 24.8 million

Number of Residential Dwellings: 25.45 million (24.5 heated centrally)

Sector	Activity Level (10^9)	Intensity (MJ/Activ.)		Energy Use (PJ/Yr)	
Transportation	Total Vehicle Distance/Yr. (%elec.)	Fossil/Elec Energy Efficiency		Fossil and Other Fuels	Electric[1]
Automobiles	293 km(5.1%)	2.92/	1.44	812	22
Air Travel & Freight	8 ton-km	27		216	
Other Passenger	68 pass-km(37%)	.36/	.6	14	15
Truck ~~& Air Freight~~	165 ton-km (4%)	2.43		394	7
Rail Freight & Pass.	440 gross-ton-km(92%)	.21		35	57
Shipping (Domestic & All Bunker)				303	

Industry	Value Added[x)] Output/Year	Total Energy/Output per $ Value Added	Elec		
Metals (Iron & St.)	5961	.17	.025	864	149
Chemicals	52210	.026	.011	783	574
Other Energy Intensive (Minerals)	5865	.074	.008	387	47
Non-Energy Intensive	167572	.011	.0038	1206	637
Petrochemical Feedstocks				1751	
Asphalt & Road Tar				207	
Agriculture	30025	.008		165	75
Mining (Non-Energy)	260	.090		18	5
Construction	39242	.0075		274	20

NATIONAL ENERGY DEMAND PROJECTIONS (con't) Case C 2

Commercial & Public	Value Added[x)]	Total Energy/VA	Elec/VA		
Commercial	82514	.0142	.0047	784	390
	GDP-Contribution				
Public	43672	.01	.0044	240	195

Residential	(10^6) Dwellings	kcal/m^2	m^2/Dwelling Heat Loss/ Area/Yr.	Syst. Eff.		
Foss. Fuel Space Heat / Distr.Heat	11.0/4.5	110	90	.73/.9	1065/ 353	
Electric Space Heat / Heat Pump	8.0/1.0	108	90	.98/3.0		283/ 24
Foss. Fuel Appliances[3]	25.54		4.3 GJ/Yr/Dw		111	
Electric Appliances[3]	25.54		19.1 GJ/Yr/Dw			487
Total Delivered Energy: Demand (Including Fuel used for Feedstocks)					9982	2987

[1] Purchased electricity

[2] The number for heat loss/area/year represents the total fossil fuel (electric) energy demand per unit area per year.

[3] The number for heat loss/area/year represents the total fossil fuel (electric) energy demand per dwelling.

x)Monetary Values in 1972-US $ ($=3.19 DM)

NATIONAL ENERGY DEMAND PROJECTIONS

Country: Fed. Rep. of Germany

Year: 2000

WAES Case: D-8

Population: 58.7 million

No. of Autos: 23.7 million

Number of Residential Dwellings: 24.5 million (23.5 heated centrally)

Sector	Activity Level (10^9)	Intensity (MJ/Act.)	Energy Use (PJ/Yr)	
Transportation	Total Vehicle Distance/Yr. (% elec.)	Fossil/Elec Energy Efficiency or Total	Fossil and Other Fuels	Electric[1]
Automobiles	287 km(1%)	2.97/1.44	844	4
Air Travel & Freight	7 ton-km	30	210	
Other Passenger	68 pass-km(30%)	.3 / .6	16	12
Truck	148 ton-km (2%)	2.6	383	3
Rail Freight & Pass	440 gross-ton-km(90%)	.23	44	55
Shipping (Domestic and All Bunker)			300	

Industry	Value Added x) Output/Year	Total Energy / Elec Energy/Output per Value Added			
Metals (Iron & St.)	5277	.173	.022	796	111
Chemicals	35160	.028	.01	632	352
Other Energy Intensive (Minerals)	5571	.076	.008	378	45
Non-Energy Intensive	131644	.0115	.0034	1066	454
Petrochemical Feedstocks				1430	
Asphalt & Road Tar				230	
Agriculture	30025	.0081		168	75
Mining	260	.093		19	5
Construction	33552	.0077		240	18

NATIONAL ENERGY DEMAND PROJECTIONS (con't) Case D–8

Commercial & Public	Value Added x) ~~Floor Area~~	Total Energy/VA	~~Heat Loss/ Area/Yr~~	Elec/VA ~~Syst. Eff.~~		
~~Foss. Fuel Space Heat~~ Commercial	66437	.0144		.0047	644	314
~~Electric Space Heat~~	GDP-Contribution					
~~Other Fossil Fuel²~~ Public	41605	.01		.0046	234	190
~~Other Electric²~~						

Residential	10^6 Dwellings	kcal/m^2	m^2/Dwelling ~~Heat Loss/ Area/Yr.~~	Syst. Eff.		
Foss. Fuel Space Heat / District Heat	16.0/3.0	110	90	.73/.9	1549/ 235	
Electric Space Heat/ Heat Pump	4.5/0	108	90	.98/3.0		159
Foss. Fuel Appliances[3]	24.5		4.86 GJ/Yr/Dw		119	
Electric Appliances[3]	24.5		18.8 GJ/Yr/Dw			460
Total Delivered Energy: Demand (Including Fuel used for Feedstocks)						

[1] Purchased electricity

[2] The number for heat loss/area/year represents the total fossil fuel (electric) energy demand per unit area per year.

[3] The number for heat loss/area/year represents the total fossil fuel (electric) energy demand per dwelling.

x) Monetary Values in 1972-US \$ (\$=3.19 DM)

NATIONAL INPUT WORKSHEET FOR SUPPLY/DEMAND INTEGRATION

Country: Fed. Rep. of Germany

Year: 1972

Case: Base Case

Units: 10^{15} Joules

	(a) Coal	(b) Petroleum	(c) (Syn. Liquids)	(d) Nat. Gas[6)]	(e) (Syn. Gas)	(f) (Heat)	(g) (Electricity)††	(h) Hydro Energy	(i) Nuclear Energy	(j) Geothermal energy and other	(k) Total
(1) Transport[1)]	34	1391		2			31				1459
(2) Industry	442	1061		620		34	435			1	2593
(3) Agric., Mining, Construction	1	323		28			36				388
(4) Commercial	19	352		44			73				488
(5) Public[2)]	67	109		40		8	49				272
(6) Residential	348	1159		204		85	202			21	2018
(7) Non-energy Uses											
(8) Final Energy Demand*	976	5021		1013		127	825			22	7983
(9) Electricity**	1780	345		322			-990	120	91	27	1695
(10) Syn. Gas[3)]	515	311		-790							36
(11) Syn. Liquids											
(12) Heat	88	59		32		-162					18
(13) Energy Sector[4)] Self Consumption & conversion losses	51	192		322		34	284				884
(14) Primary Energy Input	3411	5927		899			118	120	91	49	10616
(15) Indigenous Supply[5)]	3787	283		581				120	91	49	4912
(16) Imports***	308	5954		338			186				6788
(17) Exports	685	310		20			68				1083

*Includes non-energy uses.

**Blocks 9g, 10e 11c, and 12f must be negative to avoid double counting of energy from primary fuels.

***Includes imported products.

††Includes only electricity to be purchased by each sector.

Footnotes:
1) Includes Bunker
2) Includes Military
3) Byproduct in Coking & Refineries
4) Include Distribution Losses
5) Includes Stock Saldo
6) In Final Demand Categories::Natural & By- & Syn-Gas

NATIONAL INPUT WORKSHEET FOR SUPPLY/DEMAND INTEGRATION

Country: Fed. Rep. of Germany

Year: 1985

Case: A

Units: 10^{15} Joules

	(a)	(b)	(c)	(d)	(e)	(f)	(g)	(h)	(i)	(j)	(k)
	Coal	Petro- leum	(Syn. Liquids)	Nat. Gas 6)	(Syn. Gas)	(Heat)	(Electri- city)††	Hydro Energy	Nuclear Energy	Geothermal energy and other	Total
(1) Transport 1)		1727					45				1772
(2) Industry	494	723		1302		66	823				3408
(3) Agric., Mining, Construction		403					58				461
(4) Commercial											
(5) Public 2)		478		315			293				1086
(6) Residential	141	1010		425		222	451				2249
(7) Non-energy Uses	80	842		230							1152
(8) Final Energy Demand*	715	5183		2272		288	1670				10128
(9) Electricity**	1850	400		630			−1929	175	2113	185	3424
(10) Syn. Gas 3)	480	310		− 756							
(11) Syn. Liquids 7)											34
(12) Heat 8)	105	35		60		−363					
(13) Energy Sector 4)											−163
Self Consumption & conversion losses	56	200		310		75	334				975
(14) Primary Energy Input	3206	6128		2516			75	175	2113	185	14398
(15) Indigenous Supply 5)	3800	168		645				175	2113	185	7086
(16 Imports***	256	6260		1891		75					8482
(17) Exports	850	300		20							1170

*Includes non-energy uses.

**Blocks 9g, 10e 11c, and 12f must be negative to avoid double counting of energy from primary fuels.

***Includes imported products.

††Includes only electricity to be purchased by each sector.

Footnotes:

1) Includes Bunker
2) Includes Military
3) Byproduct in Coking % Refineries
4) Include Distribution Losses
5) Includes Stock Saldo
6) In Final Demand Categories: Natural & By- & Syn-Gas
7) Synthetic gas from lignite gasification by HTR
8) Accounting in 1985: input in single purpose plants in fuel columns heat from double purpose plants = negative quantity in totals column. Accounting in 2000: All inputs (single and double purpose plants) in fuel columns. Therefore, heat line contain also electricity production.

NATIONAL INPUT WORKSHEET FOR SUPPLY/DEMAND INTEGRATION

Country: Fed. Rep. of Germany

Year: 1985

Case: B

Units: 10^{15} Joules

	(a) Coal	(b) Petroleum	(c) (Syn. Liquids)	(d) 6) Nat. Gas	(e) (Syn. Gas)	(f) (Heat)	(g) (Electricity)††	(h) Hydro Energy	(i) Nuclear Energy	(j) Geothermal energy and other	(k) Total
(1) Transport 1)		1660					45				1705
(2) Industry	430	531		1250		65	659				2935
(3) Agric., Mining, Construction		385					57				442
(4) Commercial		460		292			275				1027
(5) Public 2)											
(6) Residential	152	946		470		207	389				2164
(7) Non-energy Uses	70	811		198							1079
(8) Final Energy Demand*	652	4793		2210		272	1425				9352
(9) Electricity**	1800	300		630			-1657	175	1513	160	2921
(10) Syn. Gas 3)	460	295		- 723							32
(11) Syn. Liquids 7)											
(12) Heat 8)	99	35		60		-343					- 149
(13) Energy Sector 4) Self Consumption & conversion losses	53	190		280		71	307				901
(14) Primary Energy Input	3064	5613		2457			75	175	1513	160	13057
(15) Indigenous Supply 5)	3800	168		645				175	1513	160	6461
(16 Imports***	64	5745		1832			75				7716
(17) Exports	700	300		20							1120
	+ 100										

*Includes non-energy uses.

**Blocks 9g, 10e 11c, and 12f must be negative to avoid double counting of energy from primary fuels.

***Includes imported products.

††Includes only electricity to be purchased by each sector.

Footnotes:

1) Includes Bunker
2) Includes Military
3) Byproduct in Coking % Refineries
4) Include Distribution Losses
5) Includes Stock Saldo
6) In Final Demand Categories: Natural & By- & Syn-Gas
7) Synthetic gas from lignite gasification by HTR
8) Accounting in 1985: input in single purpose plants in fuel columns heat from double purpose plants = negative quantity in totals column. Accounting in 2000: All inputs (single and double purpose plants) in fuel columns. Therefore, heat line contain also electricity production.

NATIONAL INPUT WORKSHEET FOR SUPPLY/DEMAND INTEGRATION

Country: Fed. Rep. of Germany

Year: 1985

Case: C

Units: 10^{15} Joules

	(a) Coal	(b) Petro-leum	(c) (Syn. Liquids)	(d) Nat. Gas 6)	(e) (Syn. Gas)	(f) (Heat)	(g) (Electri-city)††	(h) Hydro Energy	(i) Nuclear Energy	(j) Geothermal energy and other	(k) Total
(1) Transport 1)		18o2					46				1848
(2) Industry	441	886		1272		66	827				3492
(3) Agric., Mining, Construction		428					59				487
(4) Commercial		5o2		3o9			293				11o4
(5) Public 2)											
(6) Residential	141	1149		451		233	424				2398
(7) Non-energy Uses	8o	932		22o							1232
(8) Final Energy Demand*	662	5699		2252		299	1649				1o561
(9) Electricity**	18oo	4oo		65o			−19o4	175	2o74	185	338o
(10) Syn. Gas 3)	48o	35o		−794							36
(11) Syn. Liquids 7)											
(12) Heat 8)	94	47		65		−377					−171
(13) Energy Sector 4) Self Consumption & conversion losses	55	2o5		32o		78	33o				988
(14) Primary Energy Input	3o91	67o1		2493			75	175	2o74	185	14794
(15) Indigenous 5) Supply	38oo	168		645				175	2o74	185	7074
(16 Imports***	141	6833		1868			75				8917
(17) Exports	85o	3oo		2o							117o

*Includes non-energy uses.

**Blocks 9g, 10e 11c, and 12f must be negative to avoid double counting of energy from primary fuels.

***Includes imported products.

††Includes only electricity to be purchased by each sector.

Footnotes:

1) Includes Bunker
2) Includes Military
3) Byproduct in Coking % Refineries
4) Include Distribution Losses
5) Includes Stock Saldo
6) In Final Demand Categories: Natural & By- & Syn-Gas
7) Synthetic gas from lignite gasification by HTR
8) Accounting in 1985: input in single purpose plants in fuel columns heat from double purpose plants = negative quantity in totals column. Accounting in 2000: All inputs (single and double purpose plants) in fuel columns. Therefore, heat line contain also electricity production.

NATIONAL INPUT WORKSHEET FOR SUPPLY/DEMAND INTEGRATION

Country: Fed. Rep. of Germany

Year: 1985

Case: D

Units: 10^{15} Joules

	(a) Coal	(b) Petro- leum	(c) (Syn. Liquids)	(d) Nat. Gas 6)	(e) (Syn. Gas)	(f) (Heat)	(g) (Electri- city)††	(h) Hydro Energy	(i) Nuclear Energy	(j) Geothermal energy and other	(k) Total
(1) Transport 1)		1744					46				1790
(2) Industry	390	678		1230		65	664				3027
(3) Agric., Mining, Construction		410					58				468
(4) Commercial											
(5) Public 2)		490		293			275				1058
(6) Residential	152	1151		495		183	372				2353
(7) Non-energy Uses	60	866		203							1129
(8) Final Energy Demand*	602	5339		2221		248	1415				9825
(9) Electricity**	1700	400		630			÷1644	175	1480	160	2901
(10) Syn. Gas 3)	430	325		- 723							32
(11) Syn. Liquids 7)											
(12) Heat 8)	94	47		65		-313					- 107
(13) Energy Sector 4) Self Consumption & conversion losses	53	200		280		65	304				902
(14) Primary Energy Input	2879	6311		2473			75	175	1480	160	13553
(15) Indigenous Supply 5)	3390	168		645				175	1480	160	6018
(16 Imports***	189	6443		1848			73				8555
(17) Exports	700	300		20							1020

*Includes non-energy uses.

**Blocks 9g, 10e 11c, and 12f must be negative to avoid double counting of energy from primary fuels.

***Includes imported products.

††Includes only electricity to be purchased by each sector.

Footnotes:

1) Includes Bunker
2) Includes Military
3) Byproduct in Coking % Refineries
4) Include Distribution Losses
5) Includes Stock Saldo
6) In Final Demand Categories: Natural & By- & Syn-Gas
7) Synthetic gas from lignite gasification by HTR
8) Accounting in 1985: input in single purpose plants in fuel columns heat from double purpose plants = negative quantity in totals column. Accounting in 2000: All inputs (single and double purpose plants) in fuel columns. Therefore, heat line contain also electricity production.

NATIONAL INPUT WORKSHEET FOR SUPPLY/DEMAND INTEGRATION

Country: Fed. Rep. of Germany

Year: 1985

Case: E

Units: 10^{15} Joules

	(a)	(b)	(c)	(d)	(e)	(f)	(g)	(h)	(i)	(j)	(k)
	Coal	Petro-leum	(Syn. Liquids)	Nat. Gas 6)	(Syn. Gas)	(Heat)	(Electri-city)††	Hydro Energy	Nuclear Energy	Geothermal energy and other	Total
(1) Transport 1)		1839					45				1884
(2) Industry	420	1101		1260		65	838				3684
(3) Agric., Mining, Construction		430					58				488
(4) Commercial											
(5) Public 2)		549		303			293				1145
(6) Residential	141	1337		470		203	397				2548
(7) Non-energy Uses	70	1032		220							1322
(8) Final Energy Demand*	631	6288		2253		268	1631				11071
(9) Electricity**	1800	400		630			-1881	175	2031	185	3340
(10) Syn. Gas 3)	460	360		- 785							35
(11) Syn. Liquids 7)											
(12) Heat 8)	94	47		65		-338					-132
(13) Energy Sector 4) Self Consumption & conversion losses	53	210		330		70	325				988
(14) Primary Energy Input	3038	7305		2493			75	175	2031	185	15302
(15) Indigenous Supply 5)	3390	168		645				175	2031	185	6594
(16 Imports***	498	7437		1868			75				9878
(17) Exports	850	300		20							1170

*Includes non-energy uses.

**Blocks 9g, 10e 11c, and 12f must be negative to avoid double counting of energy from primary fuels.

***Includes imported products.

††Includes only electricity to be purchased by each sector.

Footnotes:

1) Includes Bunker
2) Includes Military
3) Byproduct in Coking % Refineries
4) Include Distribution Losses
5) Includes Stock Saldo
6) In Final Demand Categories: Natural & By- & Syn-Gas
7) Synthetic gas from lignite gasification by HTR
8) Accounting in 1985: input in single purpose plants in fuel columns heat from double purpose plants = negative quantity in totals column. Accounting in 2000: All inputs (single and double purpose plants) in fuel columns. Therefore, heat line contain also electricity production.

NATIONAL INPUT WORKSHEET FOR SUPPLY/DEMAND INTEGRATION

Country: Fed. Rep. of Germany

Year: 2000

Case: C1

Units: 10^{15} Joules

	(a) Coal	(b) Petroleum	(c) (Syn. Liquids)	(d) Nat. Gas[6]	(e) (Syn. Gas)	(f) (Heat)	(g) (Electricity)††	(h) Hydro Energy	(i) Nuclear Energy	(j) Geothermal energy and other	(k) Total
(1) Transport[1]		1826					72				1898
(2) Industry	500	850		1697		350	1250				4647
(3) Agric., Mining, Construction		459					98				557
(4) Commercial		384		325		100	365				1174
(5) Public[2]				140		100	195				435
(6) Residential	100	700		650		359	660			100	2569
(7) Non-energy Uses	90	1561		300							1958
(8) Final Energy Demand*	690	5787		3112		909	2650			100	13238
(9) Electricity**	1624	300		700			−2334	200	3545	300	4335
(10) Syn. Gas[3]	400	350		−716							34
(11) Syn. Liquids[7]											
(12) Heat[8]	2377					−1059	−601				717
(13) Energy Sector[4] Self Consumption & conversion losses	50	250		350		150	370				1170
(14) Primary Energy Input	5141	6687		3446			75	200	3545	400	19494
(15) Indigenous Supply[5]	4371	112		293				200	3545	400	8921
(16 Imports***	1570	6575		3153			75				11373
(17) Exports	800										800

*Includes non-energy uses.

**Blocks 9g, 10e 11c, and 12f must be negative to avoid double counting of energy from primary fuels.

***Includes imported products.

††Includes only electricity to be purchased by each sector.

Footnotes:

1) Includes Bunker
2) Includes Military
3) Byproduct in Coking % Refineries
4) Include Distribution Losses
5) Includes Stock Saldo
6) In Final Demand Categories: Natural & By- & Syn-Gas
7) Synthetic gas from lignite gasification by HTR
8) Accounting in 1985: input in single purpose plants in fuel columns heat from double purpose plants = negative quantity in totals column. Accounting in 2000: All inputs (single and double purpose plants) in fuel columns. Therefore, heat line contain also electricity production.

NATIONAL INPUT WORKSHEET FOR SUPPLY/DEMAND INTEGRATION

Country: Fed. Rep. of Germany

Year: 2000

Case: C 2

Units: 10^{15} Joules

	(a)	(b)	(c)	(d)	(e)	(f)	(g)	(h)	(i)	(j)	(k)
	Coal	Petroleum	(Syn. Liquids)	Nat.[6] Gas	(Syn. Gas)	(Heat)	(Electricity)††	Hydro Energy	Nuclear Energy	Geothermal energy and other	Total
(1) Transport[1]		1774					101				1875
(2) Industry	500	800		1590		350	1407				4647
(3) Agric., Mining, Construction		457					100				557
(4) Commercial		384		300		100	390				1174
(5) Public[2]				140		100	195				435
(6) Residential	50	459		561		359	794			100	2323
(7) Non-energy Uses	90	1569		300							1958
(8) Final Energy Demand*	640	5442		2891		909	2987			100	12969
(9) Electricity**	982			370			-2701	200	5865	300	5016
(10) Syn. Gas[3]	400	300		-668							32
(11) Syn. Liquids[4]	546			-684			- 20		516		358
(12) Heat[8]	787					-1059	-601		1590		717
(13) Energy Sector[4] Self Consumption & conversion losses	50	250		320		150	410				1180
(14) Primary Energy Input	3405	5992		2229			75	200	7971	400	20272
(15) Indigenous Supply[5]	3841	112		293				200	7971	400	12817
(16 Imports***	364	5880		1936			75				8255
(17) Exports	800										800

*Includes non-energy uses.

**Blocks 9g, 10e 11c, and 12f must be negative to avoid double counting of energy from primary fuels.

***Includes imported products.

††Includes only electricity to be purchased by each sector.

Footnotes:

1) Includes Bunker
2) Includes Military
3) Byproduct in Coking % Refineries
4) Include Distribution Losses
5) Includes Stock Saldo
6) In Final Demand Categories: Natural & By- & Syn-Gas
7) Synthetic gas from lignite gasification by HTR
8) Accounting in 1985: input in single purpose plants in fuel columns heat from double purpose plants = negative quantity in totals column. Accounting in 2000: All inputs (single and double purpose plants) in fuel columns. Therefore, heat line contain also electricity production.

NATIONAL INPUT WORKSHEET FOR SUPPLY/DEMAND INTEGRATION

Country: Fed. Rep. of Germany

Year: 2000

Case: D3

Units: 10^{15} Joules

	(a) Coal	(b) Petroleum	(c) (Syn. Liquids)	(d) Nat.6) Gas	(e) (Syn. Gas)	(f) (Heat)	(g) (Electricity)††	(h) Hydro Energy	(i) Nuclear Energy	(j) Geothermal energy and other	(k) Total
(1) Transport 1)		1757					67				1824
(2) Industry	480	624		1248		300	1059				3711
(3) Agric., Mining, Construction		417					98				515
(4) Commercial		300		256		80	320				956
(5) Public 2)				146		80	190				416
(6) Residential	100	552		615		358	606			100	2331
(7) Non-energy Uses	100	1253		280							1633
(8) Final Energy Demand*	680	4903		2545		818	2340			100	11386
(9) Electricity**	1296			700			-2069	200	3545	270	3842
(10) Syn. Gas 3)	400	350		-716							34
(11) Syn. Liquids 7)											
(12) Heat 8)	2119					-958	-531				630
(13) Energy Sector 4) Self Consumption & conversion losses	50	250		350		140	335				1125
(14) Primary Energy Input	4445	5503		2879			75	200	3545	370	17017
(15) Indigenous Supply 5)	3968	67		293				200	3545	370	8443
(16 Imports***	1177	5436		2586			75				9274
(17) Exports	700										700

*Includes non-energy uses.

**Blocks 9g, 10e 11c, and 12f must be negative to avoid double counting of energy from primary fuels.

***Includes imported products.

††Includes only electricity to be purchased by each sector.

Footnotes:

1) Includes Bunker
2) Includes Military
3) Byproduct in Coking % Refineries
4) Include Distribution Losses
5) Includes Stock Saldo
6) In Final Demand Categories: Natural & By- & Syn-Gas
7) Synthetic gas from lignite gasification by HTR
8) Accounting in 1985: input in single purpose plants in fuel columns heat from double purpose plants = negative quantity in totals column. Accounting in 2000: All inputs (single and double purpose plants) in fuel columns. Therefore, heat line contain also electricity production.

NATIONAL INPUT WORKSHEET FOR SUPPLY/DEMAND INTEGRATION

Country: Fed. Rep. of Germany

Year: 2000

Case: D7

Units: 10^{15} Joules

	(a) Coal	(b) Petroleum	(c) (Syn. Liquids)	(d) Nat. Gas [6)]	(e) (Syn. Gas)	(f) (Heat)	(g) (Electricity)††	(h) Hydro Energy	(i) Nuclear Energy	(j) Geothermal energy and other	(k) Total
(1) Transport [1)]		1811					67				1878
(2) Industry	480	731		1461		200	962				3834
(3) Agric., Mining, Construction		427					98				525
(4) Commercial		324		260		60	314				958
(5) Public [2)]				200		34	190				424
(6) Residential	100	756		757		240	619			50	2522
(7) Non-energy Uses	80	1273		280							1633
(8) Final Energy Demand*	660	5322		2958		534	2250			50	11774
(9) Electricity**	1414			700			-2138	200	3545	250	3971
(10) Syn. Gas [3)]	400	350		- 716							34
(11) Syn. Liquids [7)]											
(12) Heat [8)]	1450					-644	-367				439
(13) Energy Sector [4)] Self Consumption & conversion losses	50	250		350		110	330				1090
(14) Primary Energy Input	3974	5922		3292			75	200	3545	300	17308
(15) Indigenous Supply [5)]	4018	67		293				200	3545	300	8423
(16) Imports***	656	5855		2999			75				9585
(17) Exports	700										700

*Includes non-energy uses.

**Blocks 9g, 10e 11c, and 12f must be negative to avoid double counting of energy from primary fuels.

***Includes imported products.

††Includes only electricity to be purchased by each sector.

Footnotes:

1) Includes Bunker
2) Includes Military
3) Byproduct in Coking % Refineries
4) Include Distribution Losses
5) Includes Stock Saldo
6) In Final Demand Categories: Natural & By- & Syn-Gas
7) Synthetic gas from lignite gasification by HTR
8) Accounting in 1985: input in single purpose plants in fuel columns heat from double purpose plants = negative quantity in totals column. Accounting in 2000: All inputs (single and double purpose plants) in fuel columns. Therefore, heat line contain also electricity production.

NATIONAL INPUT WORKSHEET FOR SUPPLY/DEMAND INTEGRATION

Country: Fed. Rep. of Germany

Year: 2000

Case: D8

Units: 10^{15} Joules

	(a) Coal	(b) Petro-leum	(c) (Syn. Liquids)	(d) Nat. Gas[6]	(e) (Syn. Gas)	(f) (Heat)	(g) (Electri-city)††	(h) Hydro Energy	(i) Nuclear Energy	(j) Geothermal energy and other	(k) Total
(1) Transport[1]		1797					74				1871
(2) Industry	480	731		1461		200	962				834
(3) Agric., Mining, Construction		427					98				525
(4) Commercial		324		260		60	314				958
(5) Public[2]				200		34	190				424
(6) Residential	100	756		757		240	619			50	2522
(7) Non-energy Uses	80	1273		280							1633
(8) Final Energy Demand*	660	5308		2958		534	2257			50	11767
(9) Electricity**	1271						−2125	200	4350	250	3946
(10) Syn. Gas[3]	400	350	−	716							34
(11) Syn. Liquids[7]	546		−	684		−	20		516		358
(12) Heat[8]	320					−644	−367		1130		439
(13) Energy Sector[4] Self Consumption & conversion losses	50	250		350		110	330				1090
(14) Primary Energy Input	3247	5908		1908			75	200	5996	300	17634
(15) Indigenous Supply[5]	3429	67		293				200	5996	300	10385
(16 Imports***	518	5841		1615			75				8049
(17) Exports	700										700

*Includes non-energy uses.

**Blocks 9g, 10e 11c, and 12f must be negative to avoid double counting of energy from primary fuels.

***Includes imported products.

††Includes only electricity to be purchased by each sector.

Footnotes:

1) Includes Bunker
2) Includes Military
3) Byproduct in Coking % Refineries
4) Include Distribution Losses
5) Includes Stock Saldo
6) In Final Demand Categories: Natural & By- & Syn-Gas
7) Synthetic gas from lignite gasification by HTR
8) Accounting in 1985: input in single purpose plants in fuel columns heat from double purpose plants = negative quantity in totals column. Accounting in 2000: All inputs (single and double purpose plants) in fuel columns. Therefore, heat line contain also electricity production.

Italy

12.1 SUMMARY

12.1.1 National Structure and Reference Group

The Italian WAES study is being carried out under the supervision of two Participants: Prof. Umberto Colombo, Director, Research and Development Division, Montedison, and Prof. Sergio Vaccà, Director, Istituto di Economia delle Fonti di Energia, Università Bocconi. A team of three Associates—Dr. Riccardo Galli, Dr. Oliviero Bernardini, and Dr. William Mebane—organizes and coordinates activities in Italy related to WAES.

Key Italian organizations active in the energy context form the backbone of the support reference group. These include the main energy industries, financial holdings, and industrial firms and several institutes active in the socioeconomic and energy fields.

Contributions on specific problems requiring technical expertise also come from Ministries, industrial and other category associations, municipal authorities, and a large number of industries and companies.

12.1.2 Framework of the Energy Problem

The Energy Problem from Italy's Standpoint

The present study is an attempt to quantify the evolution of the Italian energy system over the last quarter of this century and provides a framework for policy action for the medium- and long-term solution of the energy problem in this country.

Italy has only a limited resource base in terms of traditional fuels. In 1972 domestic supply, principally hydropower and natural gas, covered barely 18.5 percent of primary energy for internal consumption. In any likely development scenario the dependence on imports must continue increasing at least up to the end of the century. In this period the Italian energy problem can be stated fairly accurately as a problem of identifying the optimal mix of imported fuels while at the same time decreasing overall consumption of energy by improvements in end use efficiency. The solution requires a long-term horizon stretching to well beyond the year 2000 in light of the expanding role that new end use technologies and renewable and unlimited energy resources will play in the future energy scene. The solution of the energy problem in the transition period will require major decisions which must be carefully balanced so as not to preclude the possibility of rapid expansion of new technologies as they arrive and even so as to favor their development.

There are few options open to Italy to solve its energy problem in the coming decades. Energy savings through improved efficiency of use is not an option in the sense that it cannot alone solve the problem. Compared to some countries the Italian economy is fairly efficient in its use of energy in most sectors, but there is still plenty of space for action. A strong energy savings policy can effectively alleviate the task of meeting the nation's energy demand in the transition period, and it is mostly in this sense that such a policy is considered in this study.

The major options involve the choice of replacement fuel or fuels necessary to fill the rapidly increasing energy demand that cannot be satisfied by oil and gas except at possibly very high costs to the nation in terms of balance of payments, low economic growth, and unemployment. Nuclear energy, by liberating oil otherwise used in electricity generation, is the most important of these replacement options because the technology is already available and because it can develop much of its potential impact during the period considered. Coal is also important, but its potential is limited in Italy in consideration of the major transportation, handling, and processing infrastructures that would be required in any major development program for this fuel. Moreover, if the future of coal lies in liquefaction and gasification technologies, then it may be inappropriate and uneconomic for Italy to embark on a high coal program requiring major infrastructure changes that could rapidly become obsolete as liquefaction and gasification reach a stage of competitiveness around the turn of the century.

Coal and nuclear are the major options for the transition period, but the choice and balance between them and the extent of the commitment must be moderated in consideration of the strategic role that renewable and unlimited energy resources could play beginning in the first decades of the next century. Solar and geothermal energy are singled out in this study as major replacement options in Italy for the next century, bearing in mind that they require a very early start both in current applications and in research and development if they are to have any significant impact on the other side of the transition period.

Development Hypotheses to the Year 2000

The demand for energy is closely linked to economic development, the price of energy, and other factors. Unless the problem is quantified, the solutions proposed tend to remain vague. The optimal solution and the size of the task cannot be assessed without taking into consideration the possible development paths. For this reason a number of different futures have been chosen for discussion. The variables included for analysis in this study are economic growth, oil price, government policy, and major replacement fuel.

The transition period to the year 2000 is divided in two parts: a 13-year period from 1972 to 1985 and a 15-year period from 1985 to 2000. A number of scenarios ranging widely over the variable values are selected for analysis in the first period in order to sound the system's sensitivity to economic growth, oil price, and government policy. These are described in detail in *Energy Demand Studies: Major Consuming Countries* and below. The scenarios analyzed to the year 2000 are outgrowths of two of these, namely, C* and D.

Scenario C-1: Economic growth is assumed to average 3.6 percent in the period 1972 to 1985 followed by 3.5 percent between 1985 and 2000. The price of oil is assumed to remain constant at \$11.50/bbl (1975 dollars) until 1985 and to slowly increase thereafter as a result of increased pressures on world trade, until it reaches \$17.25/bbl in the year 2000. Policy measures to save energy through rationalization of use and improvements in end use efficiency are assumed to be applied very early during the first period and to continue to be applied to the year 2000 and beyond. A strong diversification strategy is implemented beginning with the late seventies or early eighties favoring solid coal, geothermal, and solar energy in final uses. Although nuclear energy is allowed ample development, solid coal is given some prominence in electricity generation. Considerable importance is given to research and development of new technologies for the utilization of geothermal and solar energy.

Scenario C-2: Economic growth, oil price, and policy measures to save energy are assumed the same as in scenario C-1. The diversification of imports relies essentially on nuclear energy. A vigorous nuclear program is enacted beginning not later than 1977. Solar and geothermal energy are kept in low key both as regards research and development financing and as regards public policy.

Scenario D-7: Economic growth is assumed to average 2.2 percent in the period 1972 to 1985 followed by 2.4 percent in the period 1985 to 2000. The price of oil is assumed to remain constant at \$11.50/bbl all the way from 1972 to 2000 since the slower depletion of oil and gas allows pressures on international trade to build up much more slowly. Policy measures to save energy through rationalization and improvements in end use efficiency are applied only after 1985 but are carried on with effectiveness to the end of the century and beyond. A strong diversification strategy is implemented beginning with the late seventies or early eighties favoring solid coal, geothermal, and solar energy in final uses. Although nuclear energy is allowed ample development, solid coal is given some prominence in electricity generation. Considerable importance is given to research and development of new technologies for the utilization of geothermal and solar energy.

Scenario D-8: Economic growth, oil price, and policy measures to save energy are assumed the same as in scenario D-7. The diversification of imports relies essentially on nuclear energy. A vigorous nuclear program is enacted beginning not later than 1977. Solar and geothermal energy are kept in low key both as regards research and development financing and as regards public policy.

12.1.3 Major Findings of the Energy Study

The WAES methodology separates in a systematic and rigorous way the analysis and projection of final demand from the analysis and projection of energy conversion and distribution.

Energy Demand in Final Uses

Average annual growth rates in final energy demand projected over the period 1972 to 2000 vary between 1.9 and 3.0 percent in scenarios D-8 and C-1, respectively. This is a considerable decrease relative to the average annual growth rate of 8.6 percent between 1953 and 1972 but represents nonetheless roughly a further doubling (between 1.7 and 2.3 times) in final energy demand in the 28 years between 1972 and 2000.

The reasons for the much lower growth rates relative to the past can be attributed, in order of decreasing importance, to the lower economic growth assumed in these scenarios (between 2.3 and 3.5 percent average annual growth in gross national product in the period 1972 to 2000 compared to 5.4 percent in the period 1953 to 1972), to saturation effects and changes in structure, to increasing oil prices, and finally to vigorous government policies to save energy.

The magnitude of the effects of these last two factors in decreasing energy demand can be illustrated by reference to the net energy required to produce a unit of gross national product. This has steadily increased in the past starting from about 7.5 kcal/Lira (1972 prices) in 1953 to about 13.1 kcal/Lira in 1973. The rate of increase has been slowing down slightly since the late sixties as a result of saturation effects and changes in system structure.

This analysis indicates that if the price of a barrel of oil were to have maintained its 1972 value of about $2.90 (1975 $) all the way to the year 2000 and if vigorous government policies to save energy were never applied, then the net energy required to produce a unit of gross national product would increase to between 15.6 and 14.4 kcal/Lira, respectively, in the C and D scenarios in the year 2000. The increasing price of oil, by inducing slight changes in economic structure and by favoring more efficient use of energy, reduces this to about 12.5 and 12.8 kcal/Lira in the C and D scenarios, respectively. Finally, a vigorous government policy to save energy through rationalization of use and improvements in end use efficiency beyond what oil price increases alone can achieve, brings the net energy required to produce a unit of gross national product to 10.3 and 11.0 kcal/Lira, respectively, in scenarios C and D in 2000.

The ratio of net energy to economic growth rates in the period 1972 to 2000 implied in these results lies between 0.83 and 0.86, compared to an average of 1.48 in the period 1953 to 1972. Despite the impressive drop, the projected ratios are still substantially higher than those encountered in most industrialized countries (for example, ~0.5 in the WAES projections for North America in the period 1972 to 2000).

The share of electricity in final energy demand, which has in the past shown a steadily declining trend (11.5 percent in 1953 to 10.7 percent in 1972), is projected to increase to between 14.2 percent in scenario D-7 and 20.8 percent in scenario C-2, resulting in a significant increase in processing losses (from 20.4 percent of total primary energy input in 1972 to 27.6 percent in scenario C-2 in the year 2000). The implied average annual growth rates in electricity demand in scenarios D-7 and C-2 in the period 1972 to 2000 are 3.0 and 5.4, respectively, compared to 8.2 percent in the period 1953 to 1972. The ratio of electricity growth to economic growth varies between 1.30 in the high coal (diversification) scenario D-7 to 1.54 in high nuclear scenario C-2, compared to the 1.41 average over the period 1953 to 1972. Saturation effects in electricity demand only *begin* to appear before the end of the century. Moreover, increasing oil price and vigorous government policies have only minor impact on the demand for this preferred fuel.

The Primary Energy Input

The primary energy demand for internal consumption in the different scenarios studied to the year 2000 are shown in Figure 12.1. The important features of these results are the declining role of oil and the increasing role of nuclear energy, coal, and solar energy.

The assumptions about fuel substitution in final uses and in the energy conversion sector implicit in these results adhere closely to the scenario specifications on the replacement fuel strategy. They are, however, subject to a number of technical, economic, and financial constraints of a short- to medium-term nature and to a long-term strategic guidance relating to the role of new technologies in the twenty-first century. A somewhat different pattern of consumption would have resulted if the last of these constraints had been relaxed. For example, there would have been far less solar and geothermal energy and a bit more oil, coal, and nuclear energy, according to the scenario.

The percentage of oil in total primary energy input for internal consumption still remains quite high (between 46 and 53 percent in scenarios C-2 and D-7, respectively). This is considerably higher than the world average of about 35 percent indicated as the maximum possible by the WAES global supply-demand integrations to the year

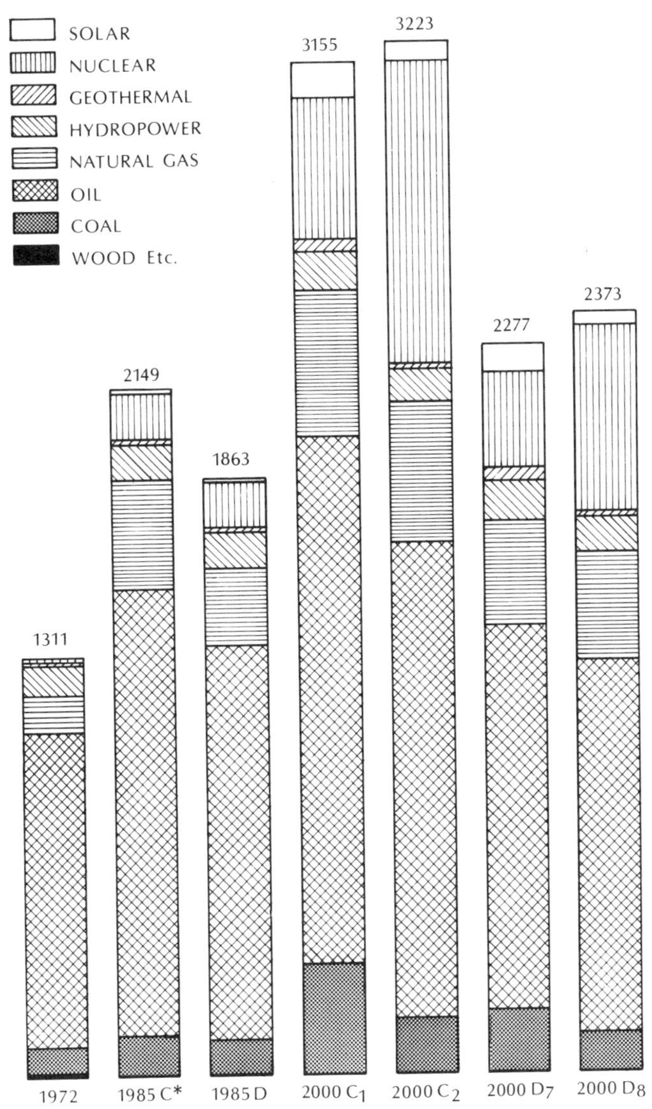

Figure 12.1
Primary energy demand for internal consumption (in units of 10^{12} kcal).

2000. However, the ratio of average world to Italian oil dependence remains essentially unchanged from the past: respectively, 54 percent on a global basis compared to 75 percent in Italy in 1972.

Natural gas increases its share from 10.4 percent in 1972 to between 13.0 and 15.6 percent in 1985 but thereafter follows the same declining fate as oil in response to depletion on a world scale as well as high transportation costs.

The growth in nuclear energy is substantial in all scenarios but close to its maximum potential only in scenario C-2, where it covers 29 percent of total primary energy input. The more contained growth of nuclear energy in the other high nuclear scenario D-8 reflects the limitations on expansion deriving from technical and economic factors circumstantial to the lower growth in electricity demand. For example, a higher growth in nuclear energy, which by the end of the century is not likely to play any significant role other than in electricity generation, would require early retirement of fossil fuel power plants already in existence.

Coal increases its share substantially in the high coal scenarios but continues the declining historic trend in the high nuclear scenarios. The amount of coal in the system in scenario C-1 (a fourfold increase over the 1972 levels) is expected to require only limited restructuring of distribution and handling infrastructures. Much of this coal is assumed to be used in power generation in plants located not too far from seaports where the coal could be shipped from abroad. Here again the lower penetration in the D scenarios is due to the much lower growth in electricity demand.

Among renewable and unlimited energy resources, hydro power continues losing share as its potential is progressively exhausted. Geothermal energy acquires renewed vigor as a result of expansion of old fields and production from new fields as well as increased utilization as heat.

Solar energy penetration before the end of the century is due almost entirely to applications in heating and cooling in the residential, commercial, and public sectors. In the high coal (diversification) scenarios the growth in solar energy can be considered the maximum compatible with technical and economic constraints such as turnover in housing stock and the cost of competing traditional systems. Whereas in the high nuclear scenarios solar energy develops at a rate compatible with the economic and oil price assumptions of the case, in the high coal scenarios its development is favored by precise national policies aiming at replacing a substantial part of the primary energy input with solar energy in the first decades of the twenty-first century. Solar penetration discussed in this study could therefore be considered as constituting a possible take-off phase.

The share of imports in total primary energy input increases substantially in all scenarios, reaching as much as 91 percent by the year 2000 in scenario C-2 compared to less than 82 percent in 1972. The results of the projections indicate appreciable diversification in the consumption of energy imports compared to the past, however, largely as a result of the increased role played by nuclear energy and coal and, to a minor extent, the substitution of oil and gas by solar energy. In scenario C-2, for example, almost 38 percent of the imports in the year 2000 comprise fuels other than oil and gas, compared to only 7.7 percent in 1972.

12.1.4 Implications for Action

Clearly no more information is available from the present analysis than was put in to start with. But the information in its original form was too disorganized and voluminous to assimilate and use for the purpose of decision making. This analysis has attempted to transform the raw input into a form that is more relevant to the problem at hand. It is in this spirit that it is thought possible to shed more light on strategies for action.

There is general consensus on the need for energy savings and on the need to replace traditional fuels such as oil and gas with nuclear and renewable or unlimited energy resources. In this sense the present analysis offers no new solution to the energy problem. A full comprehension of the problem, however, also requires an evaluation of the timing, intensity, and duration of the various possible actions as well as a careful characterization of their contents. These in themselves depend on the temporal development of the system in question. Finally, the choice among strategies depends on the overriding objectives.

The study indicates that all available energy options need to be mustered to some extent if the country is to experience satisfactory economic growth in the coming decades and into the next century. It is clear, for example, that the energy problem in Italy cannot be solved by nuclear energy alone. Nuclear energy can contribute significantly to alleviating the problem during the transition period but must be integrated with other options such as energy savings through rationalization and improvements in efficiency and possibly a partial return to coal. Moreover, the energy policy program should be chosen with a very long-term objective that gives ample space to solar and geothermal energy and also makes an imaginative effort in possible technological breakthroughs that could become important on a global scale in the more distant future.

The problem during the transition period is also one of careful introduction and balancing of options. Our present energy system in built on technologies using basically oil and gas. It does not seem possible to alter this radically in a period shorter than three to four decades, even if the new technologies were already available, without seriously hampering economic growth. It will take much longer than this unless adequate strategic decisions are taken immediately.

The Potential Impact of Energy Savings

Energy savings are incorporated in the present projections in the spirit of maintaining lifestyle and lifestyle expectations in keeping with the scenario specifications on economic development. Thus energy is considered to be saved only if this occurs through improvements in efficiency and not through decreases in activity.

This analysis has made a clean distinction between savings in energy resulting from the simple price mechanism and savings in energy that require specific government measures. These are not always easy to separate since domestic fuel prices in Italy usually contain a considerable policy component.

The assumption made in this study is that in the absence of a strong policy response the fiscal and tariff structure of fuel prices remains essentially unvaried relative to 1972, neither adhering entirely to standard principles reflecting the costs of production and distribution to the user nor particularly favoring the optimal allocation of available fuel resources. Under a strong policy response, on the other hand, the fiscal pricing structure is assumed to be adjusted so as to favor the best allocation of resources and so that no sector is protected from the effects of increasing oil prices unless for very particular social reasons or unless the sector is of vital importance to an overriding energy saving policy (for example, public transport).

This is, however, the minimum that can be expected of a vigorous government policy to save energy. Most of the energy savings incorporated in the scenarios are long-term effects that could not come about through the price mechanism alone and that require considerable government effort in legislation and financing. The time factor is particularly evident in considering the energy saving potential of public transport systems and thermal insulation in the civilian sector, which require at least 10 to 15 years to have any significant impact.

The particular spirit in which energy savings is considered in this study is again stressed. Limiting fuel deliveries for space heating, for example, effectively decreases energy consumption, but only at the expense of living standards (lower ambient temperatures) unless simultaneous measures are taken to decrease heat losses in buildings by a program to improve thermal insulation and burner efficiency. Increasing the fiscal component of gasoline price can save energy but can also lead to a decrease in travel activity unless adequate measures are taken to improve the quality and increase the capacity of public transport. Thus a program aiming to save energy in an enduring way without impacting negatively on the economy must carefully balance fiscal and regulatory measures with incentives to use energy more efficiently. It requires planning and massive capital outlays.

Energy saving policy conceived as an effort in technological restructuring can only be a long-term objective. Whereas the planning and scheduling of such a policy must take into account both short-term features and constraint characteristics of the system, the long-term policy must be given greater freedom of action. Thus the structure of refinery output, which is a fairly short-term constraint, should be considered only in modulating the development of an energy savings program and not in defining the upper limit of its potential. The upper limit can be achieved only over the course of decades, that is, in a period sufficiently long as to permit readjustment.

The energy savings, in the sense described, that can be accomplished by the year 2000 given a vigorous government policy enacted beginning in the very near future, is estimated in these scenarios as between 12 and 13 percent of total final demand. Most of this energy comes from oil and gas, so that in terms of primary energy input the policy measures considered save around 9 percent. In absolute terms this amounts to between 20 and 29 million tons of oil equivalent in scenarios D-8 and C-1, respectively, corresponding to a reduction in the imports bill of between 1.7 and 3.7 billion dollars at 1975 prices in the year 2000.

The Role of Fuel Substitution

One of the results emerging from this study is that in the period to the year 2000 Italy's dependence on oil can be reduced more rapidly by a program focusing on nuclear expansion than by concentrating on a diversification of primary sources (including coal, solar, and nuclear energy). The difference, however, is not great: respectively, 52 and 53 percent oil in scenarios C-1 and D-7 and 46 and 49 percent in scenarios C-2 and D-8. The opportunity for high nuclear development measured in these terms may, moreover, lose some of its thrust if the first quarter of the twenty-first century is brought into the picture. In this period solar energy could emerge to some importance, coal gasification and liquefaction technologies could contribute on a world scale to maintaining a substantial share of oil and gas in the total primary energy mix, and geothermal energy in the form of hot dry rocks could well contribute substantial quantities of energy.

The development of nuclear energy is often taken as tantamount to an increased role of electricity in final demand. Since nuclear energy replaces oil in electricity generation, the long-term objective of an intensive nuclear program is to substitute electricity in end uses to the maximum extent possible. In the present study an attempt has been made to examine the extent to which supply-push policies favoring the substitution of electricity in final demand are compatible with a strategic policy to decrease oil dependence in the transition period. Clearly, if electricity growth resulting from such a policy is faster than the capacity of the system to substitute nuclear energy for oil in electricity generation, then the result of such a strategy in the medium term is to increase rather than decrease oil dependence.

The success of a supply-push policy aiming to reduce the nation's oil dependence as quickly as possible will depend to a large extent on economic growth and on the timing and intensity of such a policy. If the policy is applied with vigor right away, then oil dependence in the period to 2000 increases substantially under any of the scenarios considered. If the policy is applied beginning around 1985, the effect depends on economic growth. In scenario C-2, for example, oil dependence integrated between 1985 and 2000 increases somewhat (about 3 to 4 percent), whereas in scenario D-8 it decreases slightly (about 2 percent). In order to achieve a decrease in scenario C-2 a supply-push policy for electricity should not be applied before the nineties.

The potential for coal is limited in Italy by the present infrastructure and by possible alternative uses of the capital that would be needed for infrastructure development. To the extent that it does not require major changes in barge, railway, and port systems, however, increased coal use, especially in electricity generation, can be an important strategic element in easing the transition out of the oil and gas era. Its role can be especially important in the event of slack in the development of a nuclear power program.

Research and Development in the Context of Long-Term Energy Strategies

The question of the intensity of a nuclear program depends in a fundamental way on the preferred mix of primary fuels in the first decades of the twenty-first century. Both solar energy and geothermal energy may play a considerable role in this period depending on technological development. These sources are particularly favored in a long-term strategy for Italy because they are domestic. In many cases these and other new technologies require substantial financing and research effort, which may not obtain if the major thrust of an energy policy is in the direction of nuclear power. Their importance in the first decades of the next century depends on the technical and economic space left for their development by a nuclear program as much as on the financing and research effort dedicated to their development. If this is not begun soon enough, it will take much longer for them to displace the energy systems that have been developed in the meanwhile to take the place of oil and gas.

The assumptions made about the vigor of government policy in favor of solar and geothermal energy are in line with this general concept. The solar energy penetration in residential, commercial, and public uses in the diversification scenarios C-1 and D-7 is close to the maximum that can be expected and is not likely to occur unless decisive policy actions on the part of the government, in the form of credit incentives and regulation, are taken in the near future.

12.1.5 Conclusion

Insofar as it is a long-term issue, the energy problem cannot be approached with short-term optimizations of a shortsighted nature, as was the rule in the past when oil and gas were abundant and in no immediate danger of depletion. This is not the first time that the nation has experienced a transition between fuel sources. Never in the past, however, has a transition been so constrained by time, capital, and the need for planning. The transitions of the past, from wood to coal and from coal to oil, took perhaps 30 to 40 years. They went almost unnoticed and required little or no long-term government planning. The amount of energy consumed per capita was very much less than it is today (in Italy: one-tenth in the last decade of the nineteenth century, one-fourth in the fourth decade of this century). The problem we face for the next 20 to 30 years is of a different nature. The demand for energy is much greater and continues to rise. The complexity of the energy conversion and distribution system has increased immensely. The primary components of this transition must be government policy and planning.

The projections made in this study are optimistic in the sense that they make favorable assumptions about the outcome of major decisions:

- rapid nuclear development giving adequate consideration to economic and technical constraints
- timely application of conservation policies aiming at increasing the efficiency of energy use rather than decreasing activity
- a rationalization of the fuel pricing system
- greater opening to coal in electricity generation to the extent permitted by infrastructural constraints
- an intensive solar and geothermal development program.

If these measures are not taken to some extent in the immediate future, it is almost certain that the Italian economy will not be in a position to develop at the rates postulated in this study. For example, if the nuclear programs specified are not enacted, then in the year 2000 the country will need to import between 29 and 94 million tons oil equivalent more (in scenarios D-7 and C-2, respectively) in the form of oil and coal in order to maintain the economic growth specified. If a major energy savings program is not enacted soon, then the nation will need to import between 20 and 29 million more tons of oil equivalent. If no effort is made to introduce solar energy into the system, then the additional oil equivalent demand ranges up to 18 million tons taking into account that over 30 percent of the solar energy substitutes electricity in final uses. The additional energy in the form of oil, coal, and gas that will be required if energy policy decisions of major importance are not taken in the near future and assuming the economic growth specified in the scenarios can somehow be maintained, determines an increase in the imports bill of up to 16 billion dollars (in 1975 prices) per year at the turn of the century.

There is no easy way out of this problem. Economic growth demands the consumption of increasing quantities of energy. Some have postulated reduced or zero energy growth as a panacea to the energy problem, but in the time available to the end of the century this amounts to condemning the nation to very low economic growth and almost a standstill in per capita income. The indications of this study are that if the nation were to accept zero energy growth between 1980 and 2000 unbacked by a strong energy saving program, then economic growth would average between 0.3 and 0.7 percent per year in this period according as the price of a barrel of oil remained constant at $11.50 or increased to $17.25. Per capita income in the year 2000 would actually decrease by 7 percent from its level in 1972 in the former case, whereas in the latter case it would only increase by 4 percent.

12.2 METHODOLOGY

12.2.1 Impact of the Energy Crisis on Modeling Techniques

Income, Price, and Policy before 1973

The early seventies witnessed a major change in the energy environment. Prior to 1971 the price of oil had been showing a slow but steady tendency to decline in real terms. In Italy the average price at import decreased some 70 percent between 1955 and 1970. Government policies to save energy by rationalizing its use and by improvements in efficiency were not known. While the price of oil remained generally low, the question of energy savings ranked lowest or not at all in the scale of government priorities. In Italy, which imported some 70 percent of its primary energy in the form of oil in 1970, the share of oil in total imports of goods and services amounted to barely 12 percent. Energy savings through rationalization of use and improvements in efficiency were limited to highly energy-intensive industries such as cement manufacture and occurred basically through the price mechanism.

It cannot be said that policies were not enacted that involved saving energy; however, they were never applied with this express purpose. For example, the high tax on gasoline (~70 percent) and the low price of public transport (buses and trains) tended to favor public over private transport. However, the high tax was applied principally to provide income for the government, and the low tariffs were a measure in favor of lower-income people. Quite frequently government policies during the sixties actually favored wasteful use of energy. Such, for example, was the relatively low price of electricity for domestic use.[1]

Inadequacy of Traditional Modeling Techniques

A number of traditional methods for projecting energy demand can be criticized as not responding to the new situation emerging after 1973.[2] The price of oil suddenly began rising, so that price coefficients known from the era of declining prices were no longer applicable. Price changes occurred far outside the range that had been experienced in the recent past, and there was apparently no quick way of telling how the energy system would react.

For the first time it became evident to many economies that the government would be required to play a major role in rationalizing the use of energy. Classical econometric demand functions contained only income and price components.[3] After 1973 government policy acquired importance as a fundamental variable to the extent that it plays a role in determining the rate of important technological advances in the energy field.

The Case of Italy

For the Italian economy in particular, the early seventies marked the beginning of the end of a period of exceptionally rapid economic develop-

1. Part of the rapid growth of electricity demand in the Italian household sector (11.6 percent per year over the period 1960 to 1972) is a result of underpricing. This was made possible by both government subsidizing and by overcharging the industrial sector. In 1976 kWh pricing to user and cost to producer (in Lire) were on average as follows:

	heavy industry	small/medium industry	households
price	19	26	29
cost	17	23	54

2. O. Bernardini, *A Forecast Study of Energy Demand in Italy*, lectures delivered at the 2nd Course of the International School of Energetics: "Energy Demand and Use," Ettore Majorana Center for Scientific Culture, Erice, October 1976.
3. See, for example, W. D. Nordhaus, *The Demand for Energy: An International Perspective*, in Proceedings of the Workshop on Energy Demand, May 22–23, 1975, IIASA, Laxenburg, Austria.

ment that had brought with it profound and irreversible changes in the structure and fabric of the energy system. This has been outlined in a number of publications.[4]

The overall characteristics of the change are summarized in Figure 12.2, which shows the income elasticity of energy demand reaching a maximum in the late sixties and beginning to decline with the early seventies. This elasticity is calculated from the expression

$$E_{t_1} = E_{t_2} \left(\frac{G_{t_1}}{G_{t_2}}\right)^{\alpha} \left(\frac{P_{t_1}}{P_{t_2}}\right)^{\beta},$$

where E, G, P are energy consumption in final uses, gross national product, and oil price, respectively, α and β are income and price elasticities, and (t_1, t_2) is an interval of time.) The main questions were:

1. How fast was the rapidly changing Italian energy system reaching a steady state? Did the dynamics of change of the post-World War II period contain all the germs of structural change in the foreseeable future, or were major changes in structure foreseen which might further postpone approach to a steady state?

2. How would the Italian energy system respond to the major changes in oil price experienced after 1973?

3. What was the scope of government policy in altering the energy picture?

In other words: What values would income elasticity take over the period to 1985 and beyond? What were the price elasticities? And what was the energy savings impact of government policy?

The WAES methodology provides an extremely useful tool for answering these questions.

12.2.2 The WAES Methodology as Applied to the Italian Energy System

Of the three basic components of the WAES methodology—demand, supply, and integration—only the first and last are discussed at length in this report. The methodology for projecting supply of domestic sources through the year 2000 is described in the Italian contribution to the second WAES technical report, *Energy Supply to the Year 2000.*

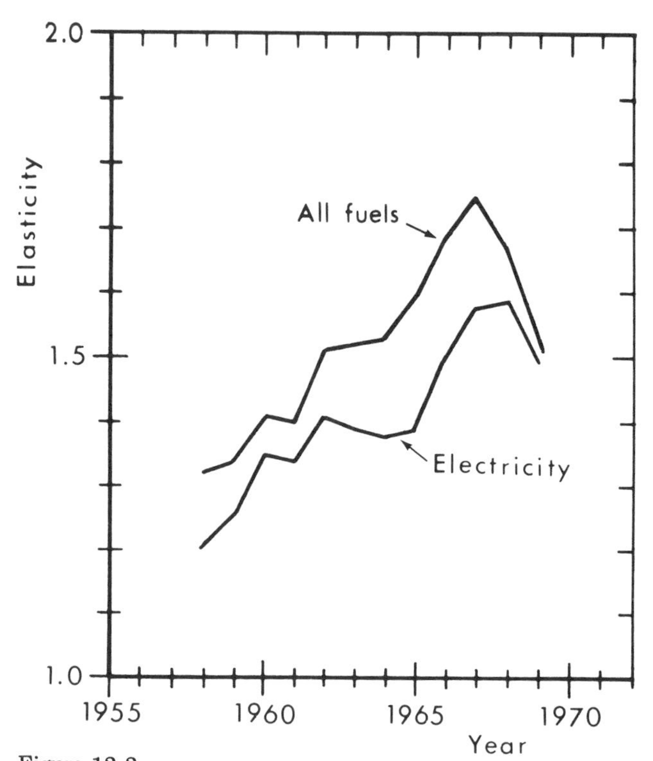

Figure 12.2
Income elasticity of energy demand in final uses (10-year moving average centered on midpoint).

A Time Frame for Application of the WAES Demand Methodology

A projection methodology is characterized by a time frame for application. For Italy it is believed that the WAES methodology as described in *Energy Demand Studies: Major Consuming Countries* cannot be applied for projecting forward more than 10 years or so.

The dynamics of change of the Italian energy system are such that after 1985 the system structure becomes ill-defined, so it is not possible to carry the system resolution and the techniques used in the 1985 projections to much beyond this date. For some sectors resolution becomes poor even before 1985. By the year 2000 resolution within most of the energy subsystems is extremely low. In manufacturing and mining, just to take one example, there is no reasonable way of telling how much primary aluminum, steel, and chemicals will be produced in Italy and how much will be imported. In other sectors substitution between processes is likely to remain in a high state of flux for

4. U. Colombo, O. Bernardini, R. Galli, and W. Mebane, *Previsioni sui Consumi di Energia*, Proceedings of the Round Table on: "E' possibile una politica dell'energia in Italia?" Camera di Commercio Industria Artigianato e Agricoltura di Pavia, October 1976. See also Bernardini, *Forecast Study of Energy Demand in Italy*.

some time to come. It appears futile to conjecture on how many passenger-km will be delivered by motorcycle and how many by bus. Motorcycles may be out of fashion by the year 2000. There is no way of knowing.

Looking ahead at the year 2000, fine details become fuzzy and the system can only be defined meaningfully in a more highly aggregated form. The disaggregation used in the 2000 projections and described below is judged to be the maximum possible consistent with the resolution available.

A Theoretical Basis for Projecting Energy Demand to the Year 2000

The approach used in projecting energy demand E to the year 2000 is structurally econometric. In the light of the profound changes in energy environment indicated in the previous section, the choice of demand function may pose some difficulties. The WAES methodology as set up for the period 1972 to 1985 is particularly well adapted to a specification in which income G, price P, and policy R appear as separate functions.

The separability hypothesis at the basis of the WAES methodology can be shown to be equivalent to specifying the energy projection problem as a number of independent equations:

$$E = k\, g(G)\, p(P)\, r(R),$$
$$G = a(P,R),$$
$$P = b(G,R),$$
$$R = c(G,P).$$

As regards the income and price variables this is in general accordance with standard econometrics as applied to historic periods. The separation of the policy component finds no historic validation, however, and must be taken as axiomatic until further justification is available. The specification of the income, price, and policy components of the energy demand function can only be conjectured since, unlike historic periods, no time series data are available for regression analysis.

On the basis of information available over the last two decades the income component can safely be approximated by the function

$$g(G) = G^{\alpha},$$

where the income elasticity α is a function of per capita income and population may thus be specified as a function of time.

On the basis of past experience the price component can also be approximated by an exponential of price. The possibility of highly nonlinear price elasticities over a range of price variations as great as that speculated, however, suggests introduction of a price-dependent price coefficient. By introducing a number of different prices and a price schedule $(P_o, P_1, \ldots, P_n)$, the WAES methodology suggests the following specification:

$$p(P) = \prod_{i=1}^{n} \left(\frac{P_i}{P_{i-1}}\right)^{\beta_i},$$

where β_i is the price elasticity of energy demand for price variations in the range (P_{i-1}, P_i).

The policy component is specified as

$$r(R) = 1 - R\sigma,$$

where R, the intensity of policy response, takes on values between 0 (restrained policy) and 1 (vigorous policy) and σ is the fractional energy savings impact of a vigorous policy, defined as a constant depending only on the period of application of the policy. This specification is thus equivalent to assuming that the income and price dependence of energy savings due to government policy is accounted for entirely in the intensity of policy response R:

$$R = f(G,P).$$

Thus, in the interval (t_o, t) during which income grows from G_o to G at the average annual rate $r_{t_o/t}$, during which price varies over the range (P_o, P) according to the schedule $(P_o, P_1, \ldots, P_n = P)$, and during which policy measures of intensity R are enacted, total final energy demand E^{τ} and final demand for electricity E^{ϵ} are specified as

$$E_t^{\tau} = E_{t_o}^{\tau} (1 + r_{t_o/t})^{\alpha_{t_o/t}^{\tau}(t - t_o)} \times \left[\prod_{i=1}^{n} \left(\frac{P_i}{P_{i-1}}\right)^{\beta_i^{\tau}}\right] (1 - R\sigma^{\tau}),$$

$$E_t^{\epsilon} = E_{t_o}^{\epsilon} (1 + r_{t_o/t})^{\alpha_{t_o/t}^{\epsilon}(t - t_o)} \times \left[\prod_{i=1}^{n} \left(\frac{P_i}{P_{i-1}}\right)^{\beta_i^{\epsilon}}\right] (1 - R\sigma^{\epsilon}).$$

This equation is assumed to be applicable at the general system level and also at the subsystem level as long as the partitioning criterion is chosen with a view to independence of subsystems. The time range for application is taken as the second half of the twentieth century and perhaps some years beyond.

No further justification can presently be given for this particular specification except that it seems reasonable and it does not stray too far from classical expressions. Standard statistical techniques are available that permit deciding whether or not this specification is the most suitable, but these can be applied only if a sufficiently large number of distinct data points (projections) are given. In fact, only five projections are presently available (scenarios A, B, C, D, and E to the year 1985), and these are just sufficient to determine the coefficient unknowns appearing in the expression and valid for the period 1972 to 1985. It is for this reason that the projection methodology cannot be considered strictly econometric.

Taking note of the fact that energy demand is the product of activity A and a technical coefficient T,

$$E = AT,$$

and that both activities and technical coefficients, like energy demand, are functions of income, price, and policy, a specification analogous to that for energy demand can be adopted:

$$A = A_o \left(\frac{G}{G_o}\right)^{\alpha'} \left[\prod_{i=1}^{3} \left(\frac{P_i}{P_{i-1}}\right)^{\beta'_i}\right] (1\text{-}R\sigma'),$$

$$T = T_o \left(\frac{G}{G_o}\right)^{\alpha''} \left[\prod_{i=1}^{3} \left(\frac{P_i}{P_{i-1}}\right)^{\beta''_i}\right] (1\text{-}R\sigma''),$$

where

$$E_o = A_o T_o,$$

$$\alpha = \alpha' + \alpha'',$$

$$\beta_i = \beta'_i + \beta''_i, \quad i = 1, 2, 3,$$

$$1 - \sigma = (1 - \sigma')(1 - \sigma'').$$

These expressions can be of use in gaining insight into the system in the year 2000 and in adjusting system parameters wherever it may be felt necessary. The extrapolation of income elasticities of energy demand, discussed below, is based partly on sectoral activity growths and changes in the technical coefficient projected by means of these expressions. Moreover, these expressions give enough information about the system to allow inferences to be made about the finer resolution, thus helping to give a more complete picture of the scenario projections. This is done in section 12.3.

Demographic and Economic Growth

The general demand equation can be specified in terms either of total system consumption or of per capita consumption. The latter form is preferred since it permits separation of the demographic component, which deviates appreciably from its past values because of a projected decrease in population growth rate over the period to the year 2000 (Figure 12.3).

Growth in per capita income over the projection time frame was chosen to bracket the long-term growth rate over the 70-year period from 1903 to 1973, namely 2.16 percent. In particular, the high and low growth rates between 1985 and 2000 were chosen as 2.86 and 1.76 percent, respectively, so that the median (2.31 percent) lies above the long-term trend (Figure 12.4).

The Parameters of Change

In the econometric approach used in energy demand projections to the year 2000, the WAES

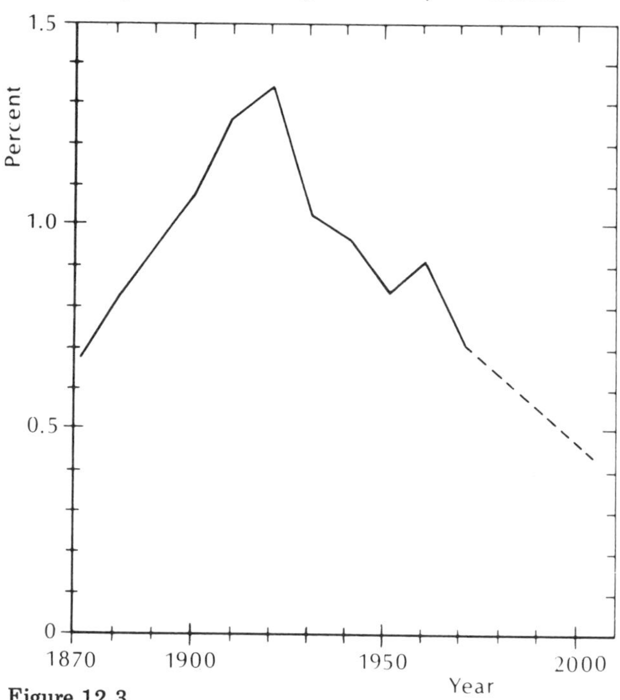

Figure 12.3
Population growth rate.

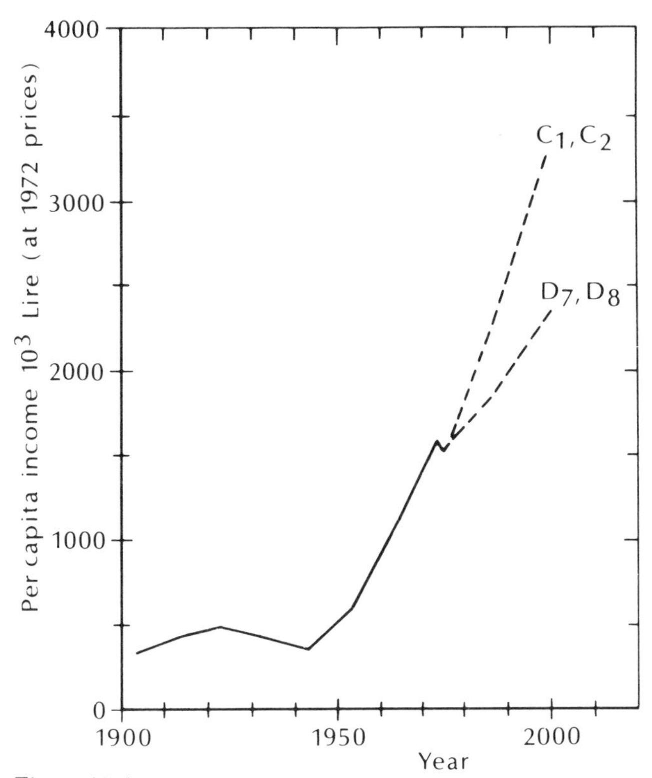

Figure 12.4
Per capita income.

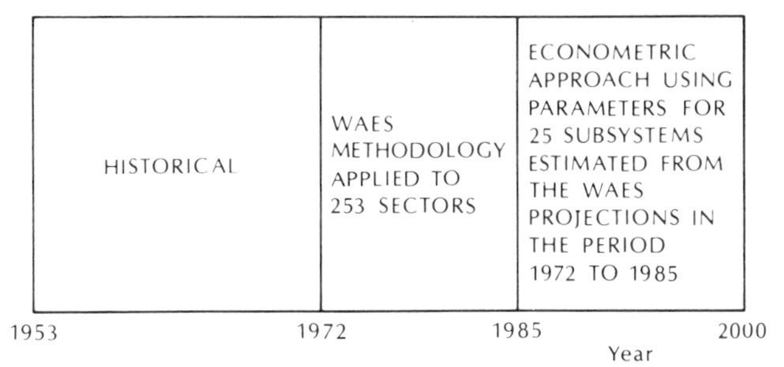

Figure 12.5
Origin of energy demand data.

methodology applied to the year 1985 serves the function of providing values for the basic parameters of change: income and price elasticities and the energy savings impact of a vigorous policy response (Figure 12.5).

The time (income) dependence of income elasticity is available up to 1972 from historic time series of energy consumption, income, and price. Income elasticities over the period 1972 to 1985 are obtained from the scenario projections to 1985, and for the period 1985 to 2000 they are obtained by extrapolation with adjustments for foreseen changes in system structure. (Historical and projected values of income elasticities are given in Table 12.1.) The extrapolability of income elasticities finds a general theoretical foundation in the theory of the stages of growth.[5] Income elasticity generally increases with time from low values, reaches a maximum, and then declines again. The detailed dynamics of change depend on per capita income growth. The extrapolations to the year 2000 take this into account by considering the income assumptions of the scenario projections.

The price dependence of price elasticities in the range $2.90 to $17.25/bbl is available from the projections for the period 1972 to 1985. The fundamental assumption is made that price elasticities for the period 1972 to 1985 carry over to the period 1985 to 2000. This is consistent with the assumption that they are not income-dependent. This assumption is not entirely satisfactory since in theory price elasticities should depend somewhat on income. In the specification used, however, income and price components are completely separated, so that price elasticities are only a function of price. In a more complete treatment of the subject an attempt could be made to specify the price elasticities as income-dependent. As stated before, however, any further refinements of the demand function would require extensive econometric analysis. Price elasticities used in the projections are given in Table 12.2.

A number of observations can be made. Increases in oil price bring about a decrease in energy demand. The incremental effect of price increases tends to increase as price rises, but only up to a point. Beyond this, further increases in price have only a small effect on decreasing energy demand. This is explained in the following way. Since by the separability hypothesis price is allowed to increase without any effect whatever on income, it follows that any depressing effect on energy demand must be attributable either to changes in the mix of goods and services in the economy or to improvements in the use of energy. The fact that price elasticity as a function of price goes through a minimum is a consequence of the fact that beyond certain prices the ability of the system to absorb increases in price by changing composition and by using energy more efficiently declines.

This, in turn, suggests limits on the extent to which price can increase without impacting on economic growth as long as the price increases are slow and not discontinuous. Within these limits the market system readjusts and economic measures can be implemented which alleviate the effect of

5. See W. W. Rostow, *Stages of Economic Growth,* 2nd ed. London: Cambridge University Press, 1971.

Table 12.1
Income elasticity of energy demand (per capita specification)

Subsystem	All fuels 1953/1960	1960/1966	1966/1972	1972/1985	1985/2000	Electricity 1953/1960	1960/1966	1966/1972	1972/1985	1985/2000
Urban passenger transport	1.34	2.85	2.24	1.09	0.58	0.82	-0.80	~0	-0.21	2.7
Long-distance passenger transport										
requiring sea crossing	1.15	1.42	2.68	3.61	3.3					
in competition with land travel	1.10	2.79	2.57	2.50	2.1	0.69	0.21	1.03	0.19	1.8
Long-distance freight transport										
requiring sea crossing	1.72	2.48	2.58	1.94	1.6					
in competition with land travel	0.48	0.81	1.21	0.69	0.45	1.50	0.15	1.45	0.47	0.80
Local freight transport	1.12	1.34	0.90	0.53	0.55					
Manufacturing and mining	1.50	1.58	1.59	1.37	1.00	1.20	1.43	1.23	1.35	1.38
Agriculture, livestock, forestry, and fishing	0.81	1.42	1.81	1.74	0.90	1.25	0.90	1.30	2.43	3.1
Construction	1.72	2.48	0.46	1.63	1.6	1.72	2.48	0.46	1.63	1.6
Cooling										
residential	—	4.70	4.69	3.88	3.0	—	4.70	4.69	3.88	3.0
commercial	1.72	4.03	3.94	3.46	2.5	1.72	4.03	3.94	3.46	2.5
public	—	4.41	2.46	2.39	1.6	—	4.41	2.46	2.39	1.6
Street lighting	0.69	0.94	1.57	1.46	1.2	0.69	0.94	1.57	1.46	1.2
Appliances										
cooking	0.18	0.11	0.06	0.06	0.04	1.72	2.48	0.46	0.63	0.45
dishwashing	1.56	1.64	2.23	1.88	1.8	2.66	3.20	3.27	2.92	2.0
clotheswashing	1.72	1.83	2.00	0.83	0.80	2.62	4.70	2.99	1.38	1.05
personal hygiene	0.66	0.60	0.54	0.72	0.64	1.25	2.84	3.60	1.95	1.40
lighting	1.39	1.34	1.31	0.94	0.90	1.39	1.34	1.31	0.94	0.90
refrigeration	5.31	3.67	2.92	2.28	1.90	5.31	3.67	2.92	2.28	1.90
entertainment	4.57	2.81	1.75	1.91	1.60	4.57	2.81	1.75	1.91	1.60
general services	1.72	1.71	1.53	0.43	0.58	1.72	1.71	1.53	0.43	0.58
other	1.72	2.24	2.58	1.44	1.30	1.72	2.24	2.58	1.44	1.30
Space heating										
residential	1.08	1.88	2.88	1.50	0.70	0.69	0.94	3.07	1.01	1.90
commercial	1.21	1.64	2.48	1.39	1.00	~0	1.42	1.63	1.37	1.50
public	1.30	1.72	2.86	1.29	0.90	~0	~0	1.87	-0.63	1.30
Total system[a]	1.24	1.64	1.90			1.29	1.57	1.56		

[a]Projections of total system elasticities depend on the scenario assumptions and thus are not included.

Table 12.2
Price elasticities of energy demand

Subsystem[a]	All fuels $ 2.90/7.66	All fuels $ 7.66/11.50	All fuels $ 11.50/17.25	Electricity $ 2.90/7.66	Electricity $ 7.66/11.50	Electricity $ 11.50/17.25
Urban passenger transport	-0.26	-0.52	-0.12	~0	~0	-0.05
Long-distance passenger transport requiring sea crossing	-0.28	-0.42	-0.40			
in competition with land travel	-0.28	-0.39	-0.08	~0	-0.03	-0.05
Long-distance freight transport requiring sea crossing	-0.14	-0.25	-0.16			
in competition with land travel	-0.05	-0.15	-0.04	~0	-0.03	-0.06
Local freight transport	-0.02	-0.04	-0.01			
Manufacturing and mining	-0.05	-0.137	-0.22	-0.015	-0.055	-0.13
Agriculture, livestock, forestry, and fishing	~0	-0.01	~0	~0	~0	~0
Construction	~0	~0	~0	~0	~0	~0
Cooling						
residential	-0.18	-0.38	-0.38	-0.18	-0.38	-0.38
commercial	-0.10	-0.17	-0.11	-0.10	-0.17	-0.11
public	~0	~0	-0.07	~0	~0	-0.07
Street lighting	-0.01	-0.03	-0.05	-0.01	-0.03	-0.05
Appliances cooking	~0	~0	~0	~0	~0	~0
dishwashing	-0.05	-0.14	-0.12	-0.04	-0.09	-0.08
clotheswashing	-0.01	-0.03	0.03	-0.01	-0.02	-0.04
personal hygiene	-0.10	-0.17	-0.18	-0.01	-0.06	-0.07
lighting	~0	~0	-0.03	~0	~0	-0.03
refrigeration	-0.02	-0.05	-0.05	-0.02	-0.05	-0.05
entertainment	~0	0.01	-0.02	~0	-0.01	-0.02
general services	0.06	0.10	~0	-0.06	-0.10	~0
other	0.01	0.02	-0.02	-0.01	-0.02	-0.02
Space heating residential	-0.05	-0.29	-0.76	-0.16	-0.36	-0.49
commercial	-0.07	-0.37	-0.78	-0.35	-0.70	-0.62
public	-0.03	-0.21	-0.69	~0	~0	-0.32

[a]Total system elasticities depend on the scenario assumptions and thus are not included.

price increases by distributing them differently within the economic system. Beyond these limits price increases inevitably impact on economic growth and the separability hypothesis breaks down. This study indicates that for Italy, prices in excess of $17.25 are still compatible with the application of the separability hypothesis.

In all projections for the period 1985 to 2000, the assumption made is that policy response is vigorous ($R = 1$). The energy savings impact in 1985 of a vigorous policy applied beginning in the late seventies (Table 12.3) is insufficient information to evaluate the impact of a continuing vigorous policy to the year 2000 (as in the C scenarios) or of a vigorous policy beginning in and around 1985 (the D scenarios). This is because the energy savings impact σ is a function of the time of introduction of the policy and of the projection period, and both of these change in passing from the 1977 to 1985 period to the 1985 to 2000 period. For example, the subsystem structure changes substantially between 1977 and 1985. The problem is not easily solved since, as explained above, methodology adopted for 1985 cannot generally be extended to 2000.

The treatment of energy saving through policy leaves space for considerable arbitrariness. It is, for example, unrealistic to assume that government policies could be applied across the board beginning in a single year. It is also unrealistic to assume that each and all of the policies adopted could be vigorous in the WAES sense of "maximum possible within technical, financial, and lifestyle constraints."

The approach adopted in quantifying the energy savings impact of vigorous policy in the year 2000 attempts to overcome both of these problems by toning down the implications of a vigorous policy as defined in the 1985 projections. This is not to suggest that a vigorous policy could not achieve the savings indicated in the 1985 projections, but rather that the likelihood of occurrence of such a policy is extremely low. In fact, vigorous in the WAES sense, though possible, is unlikely because no matter how great the organizational effort, events always arise which tend to reduce the vigor. In the approach adopted the estimates of energy savings through policy in 1985 obtained using the WAES methodology and listed in Table 12.3 are taken merely as benchmarks.

For this reason scenario C has been redefined as a sliding in time in the application of the vigorous policy measures in such a way that the full impact postulated for 1985 is felt only toward the end of the century. It is assumed that a vigorous policy enacted beginning in the late seventies is roughly 50 percent successful depending on the sector. The remaining energy savings are assumed to obtain from continued application or completion of the vigorous policy measures.

In the D scenarios the assumption is made that a vigorous policy applied beginning in 1985 has the same overall impact as it would have had had it been applied in the 1977 to 1985 period. This is felt not to be an overstatement of energy savings potential relative to the C scenarios since the time available is considerably greater—15 years as opposed to the 9 years of the 1985 projections—and the compulsion to save should be greater.

As a result of the revised definition of vigorous policy, two C scenarios are presented for 1985 in the supply-demand integrations of section 12.4: a scenario C assuming 100 percent success of vigorous policy measures by 1985 and a scenario C* assuming only 50 percent success by 1985; scenario C* is carried over as scenarios C-1 and C-2 to the year 2000.

Some excursion runs were tried to test the sensitivity of primary fuel mix by the year 2000 to electricity supply-push policies applied with the expressed purpose of speeding up substitution of nuclear energy for oil. In these projections the maximum feasible substitution in markets of electricity for other fuels was estimated at the subsystem level and assumed only slightly dependent on economic growth (Table 12.4).

The Integration of Supply with Demand

The WAES demand methodology refers to the projection of energy consumed in final uses and therefore includes energy not only purchased but also produced by the user (for example, self-produced electricity). This takes a variety of forms ranging from wood and coal to natural gas and electricity. Sometimes delivered energy is primary energy, as in the case of coal or natural gas. However, by far the greatest part of energy consumed by the user is secondary energy, implying it has been through a conversion process.

As described in the demand projection methodol-

Table 12.3
Energy savings impact of a vigorous policy in the period 1977-1985

Subsystem	Maximum energy savings due to policy response[a]	
	All fuels	Electricity[b]
Urban passenger transport	35	- 8
Long-distance passenger transport		
requiring sea crossing	~ 0	
in competition with land travel	26	-24
Long-distance freight transport		
requiring sea crossing	~ 0	
in competition with land travel	51	12
Local freight transport	59	
Manufacturing and mining	6	4
Agriculture, livestock, forestry, and fishing	~ 0	~ 0
Construction	~ 0	~ 0
Cooling		
residential	18	18
commercial	15	15
public	~ 0	~ 0
Street lighting	1	1
Appliances		
cooking	~ 0	~ 0
dishwashing	10	15
clotheswashing	7	10
personal hygiene	1	14
lighting	5	5
refrigeration	13	13
entertainment	14	14
general services	4	4
other	5	5
Space heating		
residential	23	15
commercial	23	15
public	24	16

[a]By 1985 assuming implementation in the late seventies.

[b]The negative savings are due to a vigorous energy policy inducing a shift from private (basically gasoline) to public (partly electric) transport.

Table 12.4
Maximum feasible penetration of electricity under high nuclear supply-push (%)[a]

Subsystem	1972	1985 C*	1985 D	2000 C-2 demand-pull	2000 C-2 supply-push	2000 D-8 demand-pull	2000 D-8 supply-push
Urban passenger transport	0.9	1.8	1.0	4.6	9.0	2.9	5.0
Long-distance passenger transport							
requiring sea crossing	—	—	—	—	—	—	—
in competition with land travel	2.6	4.0	2.0	3.6	3.6	3.0	3.0
Long-distance freight transport							
requiring sea crossing	—	—	—	—	—	—	—
in competition with land travel	4.8	7.6	4.5	8.6	8.6	8.7	8.7
Local freight transport	~0	~0	~0	~0	10.0	~0	10.0
Manufacturing and mining	13.1	14.1	14.1	18.1	19.0	16.4	17.0
Agriculture, livestock, forestry, and fishing	4.6	6.4	6.4	16.2	16.2	11.4	11.4
Construction	~100.0	100.0	100.0	~100.0	~100.0	~100.0	~100.0
Cooling							
residential	100.0	100.0	100.0	~100.0	~100.0	~100.0	~100.0
commercial	100.0	100.0	100.0	~100.0	~100.0	~100.0	~100.0
public	100.0	100.0	100.0	~100.0	~100.0	~100.0	~100.0
Street lighting	100.0	100.0	100.0	100.0	100.0	100.0	100.0
Appliances							
cooking	4.2	5.5	4.5	6.8	25.0	5.0	20.0
dishwashing	52.6	82.4	75.6	91.1	95.0	76.3	85.0
clotheswashing	65.6	82.4	79.8	91.2	95.0	82.4	90.0
personal hygiene	17.4	30.5	28.0	43.9	70.0	29.9	50.0
lighting	100.0	100.0	100.0	100.0	100.0	100.0	100.0
refrigeration	100.0	100.0	100.0	100.0	100.0	100.0	100.0
entertainment	100.0	100.0	100.0	100.0	100.0	100.0	100.0
general services	100.0	100.0	100.0	100.0	100.0	100.0	100.0
other	~100.0	~100.0	~100.0	~100.0	~100.0	~100.0	~100.0
Space heating							
residential	1.3	1.0	1.0	1.9	10.0	1.4	10.0
commercial	1.3	1.1	0.9	1.4	10.0	1.1	10.0
public	1.4	1.0	0.8	1.6	10.0	1.1	10.0
Total system	10.7	13.2	11.4	18.3	20.8	14.6	17.8

[a]Before solar penetration in the household sector.

ogy, electricity is singled out as a privileged fuel in that its demand is assumed to depend only on income, oil price, and policy. This is equivalent to assuming that its penetration potential in total final energy demand is independent of the supply availability of other fuels. This is clearly a simplification, though it is justified by the fact that electricity is a preferred fuel in practically all sectors and quite frequently unsubstitutable in end use.

The integration of supply with demand is interpreted as the problem of determining the primary energy fuel mix that satisfies projected energy demand for final uses at the least long-term cost to the nation.

Fuel substitution is the result of price differences and fuel performance, of technology and policy, of technical constraints such as the gas distribution network and economic constraints such as the useful lifetime of machines. The methodology used to determine the optimal fuel mix involves consideration of the substitution potential of the various fuels in each of the final demand systems and in the conversion processes that lead to the different forms of delivered energy. In the case of Italy, which has no significant coal deposits, only one such conversion process, electricity generation, is examined closely.

Both the delivery of energy to the user through the distribution system and the transformation of primary to secondary energy involve energy losses. Whereas it is possible to make approximate calculations of the energy lost in transformation and distribution by multiplying delivered energy by appropriate coefficients, in long-term projections it is more correct to consider the system in its basic constituent parts: the production and consumption of energy. These parts evolve independently subject to the constraint that energy delivered and consumed in any one year is equal to the energy produced/transformed/distributed in the same year. The equality may not be perfect because of the accumulation of producer's and consumer's stocks, but the difference is always slight.

The starting point for calculating the primary energy input to the total system is the energy consumed in end uses by the system (in its various fuel forms) and the structure of the transformation and distribution system. The procedure by means of which the calculation is performed is outlined in Figure 12.6. The calculation is complicated by the fact that secondary energy is practically always an input to the transformation and distribution of primary energy (Figure 12.7). For example, fuel oil is an output of refineries and an input to electricity generation. Because the flows of energy are not all in one direction (from primary to secondary), the calculation of primary energy input from delivered energy consumption requires inversion of a matrix T whose coefficients are related to the structure of the transformation and distribution system. In particular, the coefficients are established once the composition by fuel of electricity generation plants and the input/output ratios (direct efficiencies) of the various transformation processes are known.

A problem arises from the combined production of secondary fuels. This is the case with refinery output, for example, which takes the form of a large variety of products. Since the number of products of the transformation of energy is by far superior to the number of transformation processes,

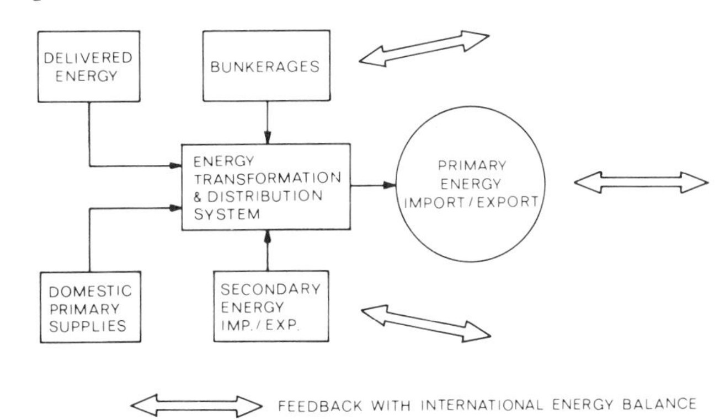

Figure 12.6
Schematic of the energy system.

	Charcoal production	Coking plants	Refineries	Town gas	Electricity generation
Wood	*				*
Coal I		*		*	*
Oil I			*		
Natural gas I				*	*
Hydraulic					*
Geothermal					*
Nuclear					*
Wood II					
Coal II		*		*	*
Oil II			*	*	*
Natural gas II				*	
Electricity		*	*	*	*

Figure 12.7
Principal energy flows in the energy conversion system (asterisks indicate where fuels enter into the process).

the entries to the matrix are greater than the exits, and the matrix cannot be inverted. A solution to this problem is provided by aggregating all the delivered energy issuing from a given transformation process since then the rows and columns are equal in number and the matrix can be inverted (Table 12.5 and Figure 12.8). This solution is not rigorous since it implies that the secondary energy outputs of a given transformation process (for example, refineries) are all input proportionally to other transformation processes (for example, electricity generation). In reality, the distribution of petroleum derivatives input to electricity generation is quite different from the product distribution of refinery output: fuel oil is the almost exclusive petroleum input to electricity generation. Fortunately, the error introduced in this approximation is negligible.

Once the delivered energy vector E_d is determined, calculation of the primary energy input vector E_p presents no particular problem:

$$E_p = T^{-1} E_d.$$

The integration is made in two steps:

1. The fuel mix in final uses and in energy conversion processes is determined as the preferred distribution. The resulting fuel demand is input to the global supply-demand integration.

2. Results from the international integration are used to modify the fuel distribution in order to comply with the global constraints on fuel availability. This is the integration presented in section 12.4.

Table 12.5
Identification of fuels in the supply-demand integration

Wood I:	wood
Coal I:	coal and lignites
Oil I:	oil
Gas I:	natural gas
Hydroelectric:	hydraulic energy
Geothermal:	geothermal energy
Nuclear:	nuclear energy
Wood II:	charcoal
Coal II:	coke from coal, coking gas, blast furnace gas, nonenergy coal products
Oil II:	liquid petroleum gas, refinery gas, light distillates, gasoline, jet fuel, kerosene, gas oil, fuel oil, petroleum coke, nonenergy petroleum products
Gas II:	coke from natural gas, town gas
Electricity:	secondary electricity

The most natural solution to a global supply-demand imbalance is an increase in the price of oil and a consequent decrease in energy demand and increase in supply. Despite the fact that such imbalances are forecast in 1985 for scenario E and in 2000 for scenarios C-1, C-2, D-7, and D-8, the price is not assumed to rise in these scenarios, and the fuel mix is assumed to adjust to the global constraints, even though this is not sufficient to bring supply into equilibrium with demand on a world scale. On account of the very long lead times involved in some forms of fuel substitution, especially in energy conversion processes, the second step is equivalent to assuming the application of a long-term strategy for fuel substitution well before strains in the international energy system become noticeable.

12.3 ENERGY DEMAND TO THE YEAR 2000

This section examines the demand for energy in final uses or net energy demand. Primary or gross energy demand, which is usually of greater interest to the planner and policy maker, is examined in the next section. The difference between gross and net energy demand represents the processing and distribution losses of the energy system. Whereas an examination of net energy demands may be misleading if not coupled to the processing and distribution losses, it does give a much more faithful representation of the dynamics of change of the

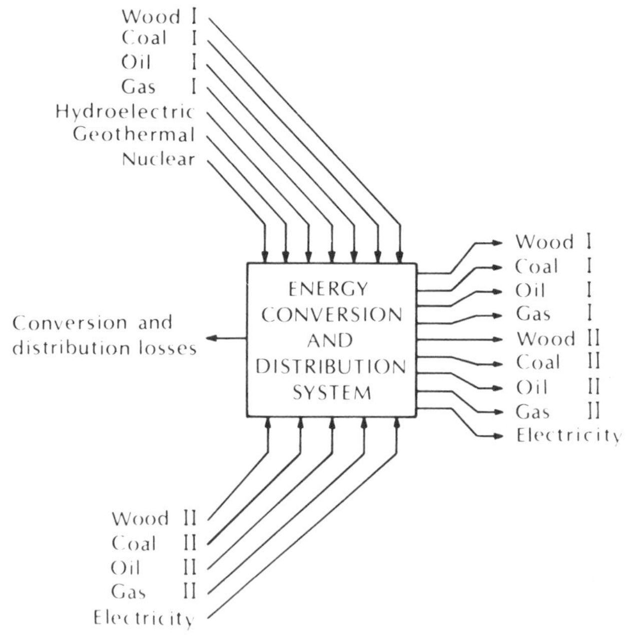

Figure 12.8
Schematic of the energy conversion system.

demand system. The same net energy demand determines different gross energy demands according to the choice of production and distribution system. Focusing on gross energy demand alone therefore conceals the diverse dynamics of two systems (the demand and processing systems) which are essentially different in their response to variations in income, price, and government policy.

Both total and electrical demands are discussed. The reasons for singling out electricity are related to the projection methodology described in the previous section and justified by the peculiar quality of electricity as a preferred fuel in most sectors and an unsubstitutable fuel in many end uses.

12.3.1 Summary

Overall Results

Total net energy demands projected to 1985 and to 2000 in the C and D scenarios and their breakdown by subsystem are compared with historic values in Table 12.6. The demand for electricity in end uses grows substantially above the 1972 levels in all scenarios as shown in Table 12.7. Part of this demand, mostly in the domestic sector, is satisfied by solar energy and therefore does not appear as electricity in the supply-demand balances.

The projected growth rates in total energy and electricity demand for final uses (Table 12.8) are substantially below the historic values. The slowdown is due basically to lower economic growth, to saturation effects accompanying economic growth, to increases in the price of oil, and to energy saving policies to be discussed below.

Sectoral Shifts

Historic and projected sector shares in total net energy demand are given in Table 12.9. Variations in sector shares occurring over the 50-year time span are largely the result of diffusion and then saturation effects in transport and civilian energy use. Intersectoral shifts between 1972 and 2000 are also due to decreased economic growth, to increases in the price of oil, and to vigorous energy policies acting with different emphasis in the different sectors.

Rapid diffusion of automobile ownership, modern domestic space heating systems, and appliances. occurring mainly during the sixties, was responsible for the net decrease in the share of energy going to industry between 1960 and 1972.

The drop in net energy share going to transport between 1972 and 1985 is due largely to increasing saturation in demand for urban transport, which accounted for 39 percent of all energy consumed for transport in 1972 as opposed to only 23 percent and 28 percent projected to 1985 in scenarios C* and D, respectively. The substantial differences in sector shares between scenarios C* and D in 1985 can be attributed largely to the difference in government energy policy, which is projected to act with increasing vigor in the direction industry→civilian uses→transport. Part of the difference is also due to the different degree to which industry, a major component of gross national product, and the domestic sector respond to economic growth.

The period 1985 to 2000 is characterized by a different order of events. Saturation effects begin to dominate in the domestic sector following the rapid diffusion phase of the sixties, seventies, and eighties. The substantial differences occurring between the C and D scenarios in respect to the weight of the civilian sector in total net energy demand is largely a result of the slower economic growth and consequent decrease in the rate of saturation in the D relative to the C scenarios. The share of net energy going to transport decreases in the D scenarios as vigorous government policies are enacted in this sector; it increases substantially in the C scenarios as a result of two basic factors. First, long-distance passenger transport, the fastest-growing component in this sector, rapidly gains share in total transport energy demand:

1972		29 percent
1985	C*	47 percent
1985	D	39 percent
2000	C-1, C-2	68 percent
2000	D-7, D-8	58 percent

Second, suburban living is expected to increase with economic growth, leading to renewed increase in energy demand for urban passenger transport as the average daily travel distance increases. Economic growth in the D scenarios is sufficiently slower that these two effects are not felt significantly before the next century.

Per Capita Energy Use

Historic and projected per capita net energy demand is shown in Figure 12.9. Although the increase over the period 1972 to 2000 is appreciable (between 48 and 95 percent), per capita net energy

Table 12.6
Breakdown of net energy demand by projection subsystem (10^{12} kcal)

				1985	2000		1985	2000	
	1953	1960	1972	C*	C-1	C-2	D	D-7	D-8
Urban passenger transport	9.2	16.6	68.6	47.0	51.7	51.7	60.9	50.4	50.4
Long-distance passenger transport requiring sea crossing	0.5	0.8	2.5	9.3	35.0	35.0	5.4	14.0	14.0
in competition with land travel	6.5	10.7	47.0	88.7	198.8	198.8	80.0	112.2	112.2
Long-distance freight transport requiring sea crossing	0.1	0.2	0.8	1.8	3.7	3.7	1.1	1.9	1.9
in competition with land travel	15.7	20.1	37.3	40.1	39.7	38.1	46.9	28.7	28.0
Local freight transport	5.2	8.6	17.2	14.1	14.1	13.5	23.0	12.1	11.7
Manufacturing and mining	96.8	186.1	470.5	833.6	1261.4	1209.1	632.9	856.6	834.4
Agriculture, livestock, forestry, and fishing	5.7	8.3	21.1	50.5	80.8	80.8	48.2	66.8	66.8
Construction	0.1	0.2	0.5	1.1	2.4	2.4	1.0	1.7	1.7
Cooling residential	~0	~0	0.2	0.9	2.8	2.8	0.6	1.2	1.2
commercial	~0	0.1	0.7	2.9	8.1	8.1	1.9	3.3	3.3
public	~0	~0	0.3	0.9	1.9	1.9	0.9	1.5	1.5
Street lighting	0.6	0.8	1.7	3.5	6.3	6.3	2.6	3.8	3.8
Appliances cooking	9.1	10.3	11.7	12.8	14.2	14.2	12.9	14.2	14.2
dishwashing	1.6	3.1	9.3	21.4	43.0	43.0	15.3	24.1	24.1
clotheswashing	1.3	2.8	8.4	12.9	18.2	18.2	11.0	13.8	13.8
personal hygiene	4.0	5.5	8.1	10.3	13.6	13.6	9.2	11.8	11.8
lighting	1.4	2.6	5.7	9.4	14.1	14.1	8.0	10.6	10.6
refrigeration	0.1	0.7	4.2	12.4	25.8	25.8	8.5	13.3	13.3
entertainment	0.1	0.6	2.0	5.2	9.6	9.6	3.8	5.4	5.4
general services	0.4	0.8	2.1	3.0	4.0	4.0	2.7	3.3	3.3
other	0.2	0.4	1.4	3.1	5.4	5.4	2.4	3.5	3.5
Space heating residential	30.0	48.9	181.2	285.1	272.4	272.4	312.6	316.2	316.2
commercial	6.9	11.9	37.7	54.8	59.0	59.0	59.5	65.1	65.1
public	2.1	3.7	13.0	19.4	20.6	20.6	21.7	22.8	22.8
Total system	197.6	343.8	953.3	1544.2	2206.6	2152.1	1373.0	1658.3	1635.0

Table 12.7
Percentage share of electricity in end uses[a]

				1985	2000		1985	2000	
	1953	1960	1972	C*	C-1	C-2	D	D-7	D-8
Urban passenger transport	5.2	4.3	0.9	1.4	4.6	9.1	1.0	3.0	5.0
Long-distance passenger transport									
requiring sea crossing	—	—	—	—	—	—	—	—	—
in competition with land travel	8.9	7.6	2.6	3.1	3.6	3.6	2.0	3.0	3.0
Long-distance freight transport									
requiring sea crossing	—	—	—	—	—	—	—	—	—
in competition with land travel	3.7	5.5	4.8	6.1	8.3	8.7	4.5	9.1	8.7
Local freight transport	~0	~0	~0	~0	~0	10.4	~0	~0	10.2
Manufacturing and mining	16.7	14.9	13.1	14.0	16.3	19.4	14.1	15.3	17.4
Agriculture, livestock, forestry, and fishing	5.1	6.1	4.7	6.3	16.2	16.7	6.4	11.4	11.7
Construction	~100	~100	~100	~100	~100	~100	~100	~100	~100
Cooling									
residential	100	100	100	100	100	100	100	100	100
commercial	100	100	100	100	100	100	100	100	100
public	100	100	100	100	100	100	100	100	100
Street lighting	100	100	100	100	100	100	100	100	100
Appliances									
cooking	1.1	2.0	4.2	5.5	7.0	25.4	4.5	4.9	19.7
dishwashing	18.2	27.4	52.6	80.3	91.1	95.1	75.6	76.1	85.1
clotheswashing	17.9	25.0	65.6	80.8	90.9	95.1	79.8	82.4	90.2
personal hygiene	3.6	4.6	17.4	32.4	54.9	69.9	28.0	44.3	59.2
lighting	100	100	100	100	100	100	100	100	100
refrigeration	100	100	100	100	100	100	100	100	100
entertainment	100	100	100	100	100	100	100	100	100
general services	100	100	100	100	100	100	100	100	100
other	~100	~100	~100	~100	~100	~100	~100	~100	~100
Space heating									
residential	1.9	1.6	1.3	1.0	1.9	9.1	1.0	1.4	9.3
commercial	2.8	1.7	1.3	1.0	1.4	9.0	0.9	1.1	9.2
public	4.7	2.7	1.4	1.1	1.5	9.2	0.8	1.3	9.2
Total system	11.5	11.7	10.7	13.2	17.3	20.8	11.4	14.2	17.8

[a]Before solar penetration.

Table 12.8
Historic and projected energy demand for final uses
(average annual percent growth rates)

	All fuels	Electricity
1953/1960	8.2	8.5
1960/1966	9.1	8.8
1966/1972	8.9	7.1
1972/1985		
scenario C*	4.0	5.7
scenario D	2.8	3.6
1985/2000		
scenario C-1	2.3	3.7
scenario C-2	2.1	5.2
scenario D-7	1.3	2.2
scenario D-8	1.2	4.0

Table 12.9
Sector shares in total net energy demand (%)

	Industry and agriculture	Transport	Civilian uses	Total
1953	52.7	19.1	28.2	100
1960	57.6	16.9	25.5	100
1972	52.7	18.6	28.8	100
1985				
scenario C*	56.7	13.6	29.8	100
scenario D	49.7	15.8	34.5	100
2000				
scenario C-1	60.9	15.5	23.5	100
scenario C-2	60.0	15.8	24.1	100
scenario D-7	55.8	13.2	31.0	100
scenario D-8	55.2	13.3	31.4	100

demand in Italy in the year 2000 (34.4 and 26.0 $\times$ 10^6 kcal/inhabitant in scenarios C and D, respectively) remains substantially lower or does not rise much above that of most advanced economies in the year 1972 (Japan, 25.9; Norway, 34.2; Sweden, 42.5; Canada, 57.7). No conclusions should be drawn from these comparisons, however, since differences between countries in per capita net energy consumption are a consequence not only of different living standards but also of the composition of the economy, industrial mix, climate, nature of terrain, and other factors.

The average annual rate of growth in per capita net energy demand projected for the period 1972 to 2000 (~2.4 and ~1.4 in the C and D scenarios, respectively) represents a considerable decrease when compared to the period 1953 to 1972 (7.9 percent average annual) and even to the period 1900 to 1972 (~3.1 percent average annual).

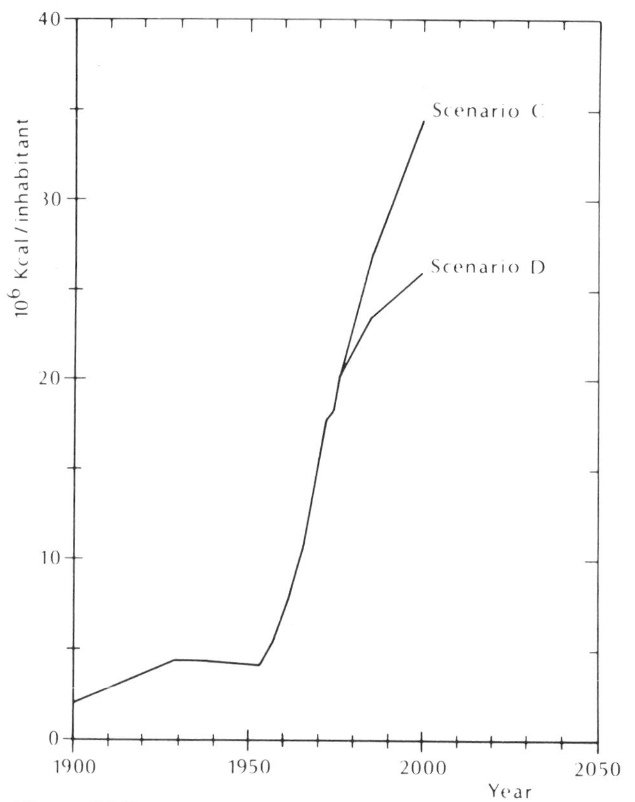

Figure 12.9
Per capita net energy consumption in Italy.

The decrease should not be considered as a forced or voluntary change in lifestyle or lifestyle expectation, at least to the extent that improvements in living standards are permitted by the lower economic growth projected in these scenarios. Whereas the projected slowdown in per capita economic growth relative to the 4.8 percent average of the period 1953 to 1972 is responsible for part of the decrease in per capita energy growth rates, a roughly equal part of the decrease is due to increases in oil price and to the introduction of vigorous energy saving policies, and also to some saturation effects that make their appearance in the latter part of the period 1972 to 2000 as a consequence of economic growth.

An attempt is made in Table 12.10 to assess the contributions of the different factors in decreasing the rate of growth in per capita net energy demand from the average of the period 1953 to 1972. (This assessment is based on the econometric formulation of the previous section and is made by examining the dynamics of energy growth, assuming in turn historic values of economic growth, income elasticity, oil price, and government policy.)

The decrease in energy growth rate is not so great when referred to gross energy demand. This

Table 12.10
Contributions of various factors in decreasing the rate of growth in per capita net energy demand in the period 1972-2000 (% annual average)

	Scenario C	Scenario D
Saturation and change in structure	-1.5	-1.5
Decreasing growth rate	-2.5	-3.9
Increasing price	-1.0	-0.6
Government policy	-0.5	-0.5
Total	-5.5	-6.5

is due to the increased role played by electricity and the consequently greater processing losses. This matter will be taken up in the next section.

Table 12.11
Net energy required to produce a unit of gross national product

	Net energy per unit of GNP (kcal/Lira (1972))
1960	8.2
1972	12.2
1985	
scenario C*	12.6
scenario D	13.1
2000	
scenario C-1	10.6
scenario C-2	10.3
scenario D-7	11.1
scenario D-8	10.9

Energy/GNP Ratios

Historical and projected values of net energy required to produce a unit of gross national product in Italy are shown in Table 12.11. These figures as they stand, however, mask a number of important trends and the effects of developments subsequent to 1972.

Historically the net energy required to produce a unit of gross national product in Italy has been substantially lower than in most other countries. In 1960, for example, it was 3.7 BOE/U.S.$1000 (1972), as compared to an average of about 8.7 for the world outside Communist areas (WOCA). Over time there has been a distinct tendency for the net energy/GNP ratio in Italy to increase, in contrast to an inverse trend in North America and on the average for Western Europe, so that in 1972 the ratios stood at 5.7 for Italy and 7.4 BOE/U.S.$1000 (1972) for WOCA. No attempt will be made in this context to explain these differences in any detail.

Figure 12.10 shows the historic trends together with the values implicit in the scenario projections. In order to make these comparable the projections have been corrected for the scenario assumptions on oil price and government policy (using the econometric formulation of the previous section applied to the standard scenario projections of energy demand in Italy and in the different world regions) so that they represent net energy/GNP ratios in a scenario in which oil retains its 1972 international price and in which no government policies are enacted to save energy. The results of these projections indicate a continuation of the historic trends, so that by the year 2000 the net energy/GNP ratios in Italy and in the main regions of the world are about equal.

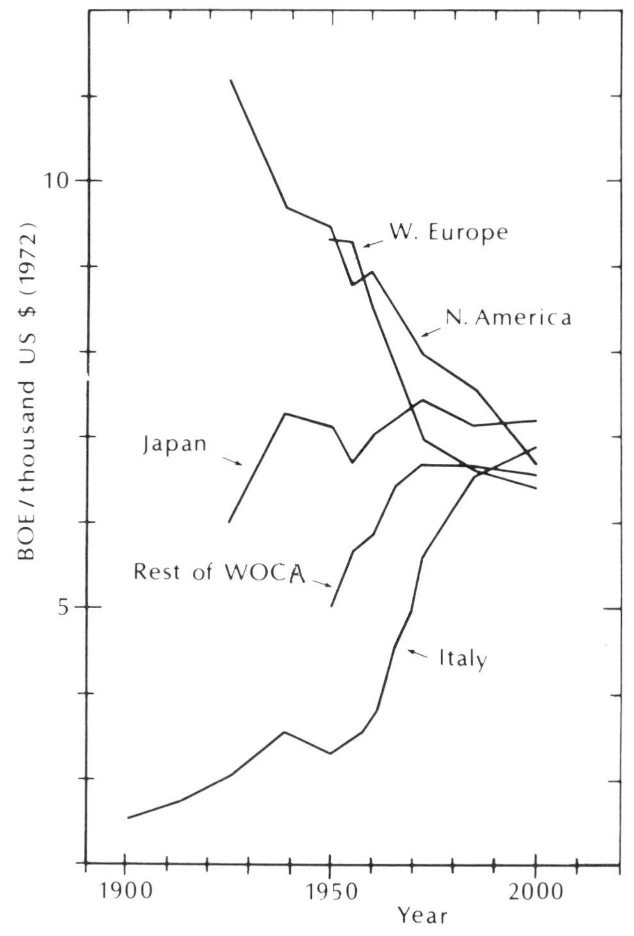

Figure 12.10
Net energy required to produce a unit of gross national product assuming constant 1972 oil price and no policy response (all data based on WAES projections).

The strong increase in the price of oil and the assumptions about government policy alter the picture considerably. Figure 12.11 shows the extent to which first increases in the price of oil and then vigorous government policies are projected to decrease the energy intensity of gross national product in Italy to the values of Table 12.11. Similar effects are projected to occur in the other regions of the world, and the converging trend of Figure 12.10 repeats itself at reduced net energy/GNP ratios.

Tracing the Impact of Income, Price, and Policy
Each energy system reacts differently to changes in income, oil price, and energy policy. This is shown in Figure 12.12 for four Italian energy systems: urban passenger transport, long-distance passenger transport, manufacturing and mining, and space heating. In these figures an attempt has been made, based on the general econometric formulation outlined previously, to separate the independent effects of income, price, and policy in the determination of final energy demand. The projections refer in order to situations in which: oil price and government policy retain their 1972 values (upper curve); oil price increases according to the scenario specification but government policy retains its 1972 value (middle curve); and both oil price and government policy assume their scenario values (lower curve).

This approach will be used in the sections below to illustrate what is in the scenarios in relation to the separate effects on energy demand of income, price, and policy. It is noted that, in keeping with the lifestyle constraint, only government policies that do not affect lifestyle and lifestyle expectations are considered. The Italian energy system structure in terms of 25 basic noninteracting systems (see, for example, Table 12.1) has been chosen partly with a view to reflecting lifestyle de-

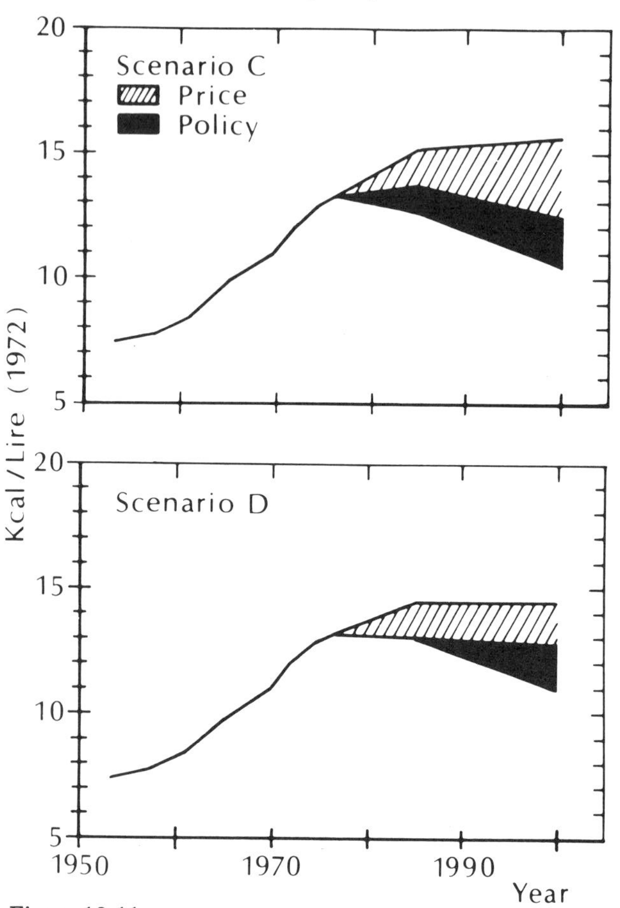

Figure 12.11
The impact of income price and policy on the net energy required to produce a unit of gross national product.

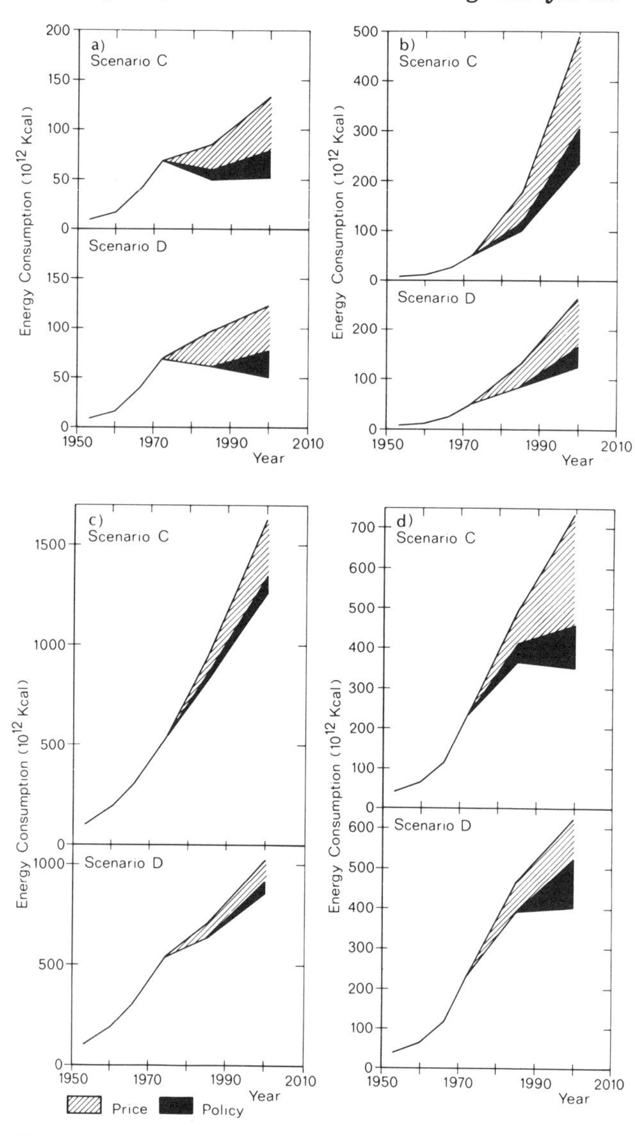

Figure 12.12
Impact of income price and policy on net energy consumption in (a) urban passenger transport, (b) long-distance passenger transport, (c) manufacturing and mining, and (d) space heating.

scriptors. The activities related to the basic energy systems are essentially unaffected by government policies enacted to save energy (for example, demand for urban transport). Nor are price variations in the range studied expected to have any substantial effect on the basic system activities.

In the following discussion basic system activities are therefore essentially only a function of income. The same cannot be said for subsystem activities (for example, the balance between private and public urban transport), which are a function of oil price and government policy as well as economic growth.

12.3.2 Energy Demand in Transport

Urban Passenger Transport

Per capita demand for urban passenger transport has increased over the period 1953 to 1972 basically as a consequence of economic growth acting principally through migrations from rural to urban areas (in the order of 7 million inhabitants in the period 1951 to 1961 and 3 million inhabitants in the period 1961 to 1971), and through increased private spending in cultural and commercial activities:

1953	880 km/yr
1960	1507 km/yr
1966	2270 km/yr
1972	2691 km/yr

The increase has been slowing down substantially and is expected to continue slowing down as a result of a decrease in rural-to-urban migrations and improvements in the layout of new metropolitan areas related to decentralization of hospitals, schools, commercial and cultural centers, and industry.

Per capita demand in 1985 is actually projected to decrease slightly, principally as a consequence of this last effect, which is more pronounced with greater economic growth (2470 and 2630 km/yr in scenarios C* and D, respectively).

In the period 1985 to 2000 a new component of growth appears in the higher-growth scenario as residential quarters begin to transfer into rural districts. In scenario D this effect does not appear in its own right before the end of the century, and per capita urban travel demand continues the decline of the 1972 to 1985 period (2730 and 2540 km/yr in scenarios C and D, respectively, in the year 2000).

The energy intensity of urban passenger transport more than doubled in the period 1953 to 1972, basically as a result of the income-induced shift from public to private forms of transport occurring after 1960 (private means delivered 31.4 percent of all urban transport in 1953 as compared to 69.6 percent in 1972):

1953	217 kcal/passenger-km
1960	218 kcal/passenger-km
1966	320 kcal/passenger-km
1972	467 kcal/passenger-km

Part of the increase was also due to increasing traffic congestion in cities, to the shift from two-wheel to four-wheel private transport, and to the diffusion of larger automobiles as private income increased.

Economic growth is projected to have only minor effects in increasing the energy intensity of urban passenger transport. The principal mechanism that functioned in the past is not likely to continue into the future since the quality of private urban transport is rapidly deteriorating as congestion increases.

The increase in oil price projected in scenarios C and D is expected to provoke a reversal of trend in favor of public transport. This has already been observed in Milan and other large cities in 1975 and 1976. The drop in energy intensity projected to 1985 in scenario D (395 kcal/passenger-km) is essentially the effect of increases in the price of oil (59 percent of urban transport delivered by private means).

A sizable shift back to public transport will come about only as a result of strong government policies enacted principally to increase the quantity and improve the quality of this form of transport. The assumptions embodied in the projections to 1985 and 2000 include large devolvements of public expenditure in mass transit to be used in increasing the stock of buses and trams, in developing rapid electrified transit, and in introducing advanced computer systems in traffic regulation. The time required to bring these measures to fruition is likely to be quite long, and only partial effects can be expected by 1985. (In scenario C*, which assumes a vigorous policy beginning in the late seventies, private means are projected to account for 55 percent of urban travel in 1985 and 45 percent in the year 2000.)

In scenario C* the energy intensity is projected

to decrease to 324 kcal/passenger-km by 1985, which includes the effects of both increases in the price of oil and a vigorous government policy. Further increases in the price of oil assumed in this scenario over the period 1985 to 2000 and a full development of government policies pursued with vigor are expected to bring the energy intensity to 295 kcal/passenger-km. The decrease is not so strong in scenario D (309 kcal/passenger-km by 2000) because the price of oil does not increase further in this scenario and because implementation of government policies, although well under way, is not fully developed by the year 2000.

The drop in energy intensity is, of course, smaller if processing losses are taken into account since a shift to public transport in the period considered implies greater electrification. The penetration of electricity to the year 2000 in this sector has been projected as being quite strong in relation to the growth of electrified urban rapid transit systems (Table 12.2). Electric cars, however, are assumed to be of importance only in limited applications.

Long-Distance Passenger Transport

Long-distance passenger transport represents a vigorously growing component of energy demand. Over the period 1953 to 1972 rapid economic growth produced a remarkable increase in per capita passenger travel over long distances:

1953	680 km/yr
1960	1121 km/yr
1966	1965 km/yr
1972	3572 km/yr

The growth was due basically to weekend pleasure trips and holiday travel and is likely to continue for many years to come considering that even as late as 1972 diffusion of regular holiday travel had spread to less than 50 percent of the populace. Per capita demand for travel over long distances embodied in the projections assumes implicitly that as income increases the share of personal expenditures going to travel activities also tends to increase, as has been the case in the past both in Italy and in other industrialized countries. The projections to 1985 and 2000 show the weight of economic growth in this system:

1985	scenario C*	7630 km/yr
	scenario D	5890 km/yr
2000	scenario C	13100 km/yr
	scenario D	7850 km/yr

The energy intensity of long-distance passenger travel did not increase substantially over the period 1953 to 1972 (from 213 to 254 kcal/passenger-km) despite a strong shift from public to private transport.

Increases in the price of oil are expected to have only minor effects on energy intensity, basically by slowing down the growth of air and sea travel relative to less energy-intensive terrestrial modes. The decrease in energy intensity projected in 1985 scenario D (247 kcal/passenger-km) is essentially due to this mechanism. Except in the short term, increases in the price of oil alone have little effect in slowing down the shift from public to private transport. In scenario D the percentage of long-distance travel delivered by private means actually increases to 79.8 by 1985, compared to 78.1 in 1972.

The scope for government policy is limited in this system. Policies envisaged involve greater development of railway and bus travel, but even the most optimistic public expenditure scenarios cannot effect a truly significant shift from private to public travel because of the very rapid growth in demand for travel. The decrease to 235 kcal/passenger-km projected to 1985 in scenario C* is due partly to a relatively slower growth of air and sea travel (respectively 1530 and 1310 kcal/passenger-km in 1972) induced by higher oil prices but also includes the effect of restrictions on speed and of fiscal measures enacted to favor diesel over gasoline engines (to have a significant savings effect the price of diesel per kcal should be no greater than one-third that of gasoline) as well as a small reduction in the share of private transport (to 76 percent).

Economic growth over the period 1972 to 2000 favors increasing energy intensity. This occurs partly through a move to larger automobiles and growth in air travel.The most important effect of economic growth, however, is to increase the demand for travel from the mainland to the islands, which in 1972 was almost six times more intensive than travel over land. From 5.1 percent of all long-distance travel activity in 1972, this develops to 10.8 and 6.3 percent in 1985 scenarios C* and D and to 15.0 and 11.1 percent in 2000 scenarios C and D. It is for these reasons that by the year 2000 evergy intensity increases above the 1972 level in scenario C (278 kcal/passenger-km) notwithstanding the further increase in the price of oil and the effect of vigorous government policies. Even in scenario D, where economic growth is much more contained, neither price nor policy is successful in decreasing the energy intensity substantially below the 1972 level (251 kcal/passenger-km in 2000, scenario D, against 254 in 1972).

Local and Long-Distance Freight Transport

Local and long-distance freight transport demand are intimately related with industrial and agricultural activity and with industrial siting. Increasing density of industrial activities and a more even distribution over the country have been largely responsible for the decline over time in the average distance traveled by a ton of freight:

1953 380 km

1960 318 km

1966 297 km

1972 274 km

The effect has prevailed despite the general decrease in the share of local freight transport (<100 km) in total freight transport (15.5 percent of the activity in 1953 against 10.8 percent in 1972) and is expected to continue as industrialization spreads to the southern regions of Italy (to 250 km in 1985 and 230 km in 2000).

The effect described is a major component of the slowdown in the growth in energy demand for freight transport projected to 1985 and 2000. Another major component, to be discussed below, is the slowdown in the growth of industry and agriculture as other activities increase their share in gross national product. Other factors leading to a slower growth in freight transport include more complete coverage and integration of commercial distribution systems, which has important containment effects in local transport.

Slight differences are introduced in the projections to 2000 between the high nuclear and high coal scenarios to take account of the increased freight transport requirements in the latter case (Table 12.12).

The energy intensities of local and long-distance freight transport over time show opposite trends, the former increasing and the latter decreasing. Local freight transport is structurally inefficient. It tends to involve small carriers traveling, on average, only 30 percent full. The efficiency has fallen over time as larger carriers have been introduced without exploiting their full carrying potential. Efficiency in this sector is essentially a managerial problem, which can only be solved by maximizing the average capacity utilization of the fleet. This is very difficult to achieve if ownership of carrying capacity continues to be highly dispersed (only very slightly over one truck per company in 1972).

Table 12.12
Historic and projected demand for freight transport (10^9 ton-km)

	Local freight	Long-distance freight
1953	6.7	36.4
1960	9.7	55.5
1966	11.3	75.5
1972	13.5	111.3
1985		
scenario C*	16.3	179
scenario D	14.8	144
2000		
scenario C	19.2 (18.3)[a]	240 (231)[a]
scenario D	17.0 (16.4)[a]	164 (160)[a]

[a]Numbers in parentheses refer to the high nuclear scenarios.

The increase in end use efficiency of long-distance freight transport is a consequence of two basic factors. During the fifties a substantial part of the increase in efficiency was due to the replacement of steam trains (1410 kcal/ton-km in 1959) by electric trains (84 kcal/ton-km in 1972). Throughout the period 1953 to 1972 sea transport (220 kcal/ton-km in 1972) was consistently increasing its share in long-distance freight activity (from 7.9 to 16.2 percent).

In road transport economies of scale could have been achieved as larger and larger trucks came into use (average carrying capacity increased from 9.5 to 13.6 tons in the period 1953 to 1972). As in the case of local freight transport, however, economies of scale in road travel have been counteracted by poor capacity utilization traceable directly to the very low average number of trucks per company (1.5 in the late sixties). In fact, the efficiency of long-distance freight transport by road actually decreased between 1953 and 1972 (from 390 to 511 kcal/ton-km).

Economic growth is expected to continue inducing a shift to larger road carriers and to sea transport, which is projected to account for as much as 23 percent of all long-distance freight activity in 1985 and 30 percent in 2000.

The influence of increases in the price of oil specified in the scenarios on the energy intensity of freight transport is expected to be minor since the cost of fuel, which is only a small part of the total cost, is simply passed on as a service charge.

The impact of government policy on the efficiency of freight transport can be great. The policy measures included in the scenario projections involve development of inland waterways, such as the Po River network, and of rail transport, which should gain considerable momentum once automatic coupling and uncoupling systems begin diffusing in the late eighties. Policy measures to increase energy efficiency of freight transport by road are expected to double capacity utilization in the long term from the 1972 average of 30 percent. This can be achieved by abolishing legislation prohibiting issuance of new truck licenses to third-customer trucking companies and by creating consortia providing computer services for optimal location of transport facilities.

The combined effects of income, price, and policy on the efficiency of freight transport between 1972 and 2000 are shown in Table 12.13.

Table 12.13
Historic and projected energy intensity of freight transport (kcal/ton-km)

	Local freight	Long-distance freight
1953	780	430
1960	890	370
1966	1190	360
1972	1280	340
1985		
scenario C*	873	242
scenario D	1554	334
2000		
scenario C	736	181
scenario D	712	187

It must be emphasized that some of the projected increase is due to developments driven essentially by economic growth and expected to occur even independently of a vigorous energy policy. In scenario C, for example, the 29 percent increase in efficiency of long-distance freight transport between 1972 and 1985 can be ascribed to changes in modal mix (5 percent), economies of scale (4 percent), development of inland waterways (6 percent), dieselization of trucks (3 percent), and restructuring of the trucking sector (11 percent). Policy actions are essential only to the last three items, corresponding to about two-thirds of the improvement in efficiency.

12.3.3 Energy Demand for Industry and Agriculture

The changing composition of Italian gross national product over time is shown in Table 12.14. Essential elements of this change are the strong decline in total share of agriculture during the fifties and sixties, the decline in share of the construction industries during the sixties, and the strong increase in the share of manufacturing and mining industries and energy products industries.

The trends from the past are not generally expected to continue into the future. Embodied in or consistent with the scenario projections are the following developments. Agriculture increases its share slightly in total value added up to 1985 and beyond. (This presumes a strong development policy in agriculture.) Industry reaches a maximum in share around 1985. Energy products industries lose share slightly or stabilize after 1985 as growth in energy demand slows down relative to economic growth. Construction industries continue to lose share up to 1985 and then stabilize to 2000 partly

Table 12.14
Composition of gross national product (% value added at market prices)

	1954	1964	1974	1985		2000	
				C*	D	C	D
Agriculture	14.1	11.1	7.9	8.9	9.6	9	10
Industry							
energy products	3.7	5.5	6.2	6.7	6.6	6	6
manufacturing and mining	20.9	24.9	29.9	30.9	28.5	29	27
construction	9.4	10.4	7.5	5.5	6.2	6	6
Services	51.9	48.1	48.5	48.0	49.1	50	51
Total	100	100	100	100	100	100	100

as a result of the increased contribution from the construction of nuclear power plants. Manufacturing and mining begin losing share after 1985. During the whole second half of the twentieth century services maintain their share in total value added within about two percentage points around an average of 50 percent.

Manufacturing and Mining

Embodied in or consistent with the projections to 1985 and 2000 is the following increase in value added (at factor cost in 1972 prices) in manufacturing and mining (energy products industries are excluded from consideration here since their energy consumption is treated under processing losses in the next section):

1972		16.6×10^{12} Lire
1985	C*	28.5×10^{12} Lire
	D	22.5×10^{12} Lire
2000	C	50×10^{12} Lire
	D	33×10^{12} Lire

Table 12.15 shows the changing composition of value added in manufacturing and mining over the period from 1953 to 2000 in terms of four relevant energy-intensity sectors. The most conspicuous feature is the decline in share of the nonintensive sectors and the increase in the chemicals and metal sectors. This trend, however, runs out its course and essentially stops in the latter part of the period. In the high nuclear scenarios, furthermore, there is a slight inversion in trend as a result of the faster development of the nuclear power industry. Stabilization of the composition of manufacturing and mining projected to the end of the century can be related to Italy's development toward a mature industrialized society.

Increases in the price of oil are projected to have appreciable effects in changing the structure of this sector in favor of less energy-intensive industries. This effect is due partly to the decreased competitiveness of energy-intensive products in substitution markets.

Energy is only one, and usually not even a major, component of the industrial picture. For this reason no assumptions have been made on the possibility of altering the "natural" trend in industrial composition as a primary objective of a government policy to save energy.

The average energy intensity of manufacturing and mining increased quite substantially in the period 1953 to 1972, basically as a result of the change in composition in favor of the metals and chemicals sectors. The projections to 1985 and 2000 indicate a leveling off and then decline in energy intensity (Table 12.16). This is due partly to the stabilization in structure projected to the latter part of the century and to changes in mix within the basic sectors, and partly to improvements in efficiency resulting from application of new technologies and replacement of old plants and from economies of scale accompanying changes in distribution by company size class.

In the metals sector the continuing decline in energy intensity through the period 1953 to 2000 is due almost entirely to technological improvements in the steel industry and only partly to changes in mix. The slowdown in decrease between 1963 and 1972 can, however, be attributed in part to this latter effect in the form of strong growth in primary aluminum production during this period.

The increasing price of energy is an important component of improvements in technology in the metals sector and is responsible for about 50 per-

Table 12.15
Composition of manufacturing and mining (% value added at factor cost)

				1985		2000[a]	
	1953	1963	1972	C*	D	C	D
Metals	4.1	5.7	6.2	7.4	7.1	7.6 (7.3)	7.4 (7.1)
Chemicals	5.3	9.6	11.4	12.0	11.3	12.4 (12.0)	11.9 (11.5)
Other energy-intensive industries (>40 kcal/Lira)	7.0	8.6	8.0	7.4	7.7	6.9 (6.7)	7.3 (7.1)
Non-energy-intensive industries (≤40 kcal/Lira)	83.6	76.1	74.4	73.2	73.9	73.1 (73.4)	74.0 (74.3)
Total	100	100	100	100	100	100 (100)	100 (100)

[a]High nuclear scenarios in parentheses.

Table 12.16
Energy intensity in manufacturing and mining (kcal/Lira (1972))

				1985		2000	
	1953	1963	1972	C*	D	C	D
Metals	99.5	73.4	73.8	60.1	65.5	53	59
Chemicals	98.8	85.4	103.6	114.1	112.5	100	105
Other energy-intensive industries (>40 kcal/Lira)	79.7	80.2	73.0	68.2	69.0	55	59
Non-energy-intensive industries (⩽40 kcal/Lira)	7.9	7.7	8.4	7.2	7.3	6.5	6.7
Total[a]	21.5	25.1	28.5	28.5	28.1	25.0(24.4)	26.1(25.4)

[a]Figures in parentheses refer to high nuclear.

cent of the projected decrease in intensity between 1972 and 1985 in scenario C* and practically all of the decrease in scenario D. In scenario C* government policy in the form of credit incentives, especially in favor of the smaller steel companies which are still abundant in Northern Italy and which are generally less efficient in energy use, is projected to contribute some 40 percent to the decrease in energy intensity of this sector. A small part of the decrease is due to changes in mix in the steel industry. The continuing decrease in intensity over the last 15 years of the century is due partly to changes in mix in favor of steel products and away from basic iron and steel. The increasing price of oil induces still further savings in scenario C, whereas the drop in energy intensity in scenario D is due almost entirely to government policy measures of the type specified above, applied beginning around 1985.

The average energy intensity of the chemicals industry has increased substantially since the early fifties, primarily as a result of growth in heavy petrochemicals (307 kcal/Lira in 1972). During the first decade of this period the effect of changes in mix in increasing the energy intensity of the sector is masked by the displacement of old inefficient plants and outdated electrochemical production techniques. Were it not for this effect the average energy intensity of the chemicals industry in 1953 would have been close to 80 kcal/Lira, some 30 percent lower than its 1972 value.

Faster-than-average growth of heavy petrochemicals is projected to increase the energy intensity of the chemicals industry substantially above the 1972 value at least until 1985. The increase would be substantially greater were it not for the negative impacts of increasing oil price and vigorous government policy. To 1985 these two factors are estimated to decrease energy consumption per unit of value added by 9 and 4 percent, respectively. Thus, in the absence of an increase in oil price and without application of a vigorous government policy, the average energy intensity of the chemicals industry in 1985 in a high economic growth scenario would be about 129 kcal/Lira.

The development of heavy petrochemicals is related to the present fairly low average consumption of plastics and synthetic fiber in Italy as compared to other industrialized countries. This effect is projected to be partly compensated by more vigorous growth in fine and speciality chemicals in the more distant future. Part of the decline in energy intensity after 1985 is due to this effect. The further increase in the price of oil in scenario C is also responsible for an additional 4 percent decline in energy intensity by the year 2000 through technological improvements. In scenario D energy intensity in 2000 is somewhat higher than in scenario C, despite the application of vigorous government policies beginning in 1985, partly because the price of oil remains constant at $11.50 and partly because of the lower economic growth.

Other energy-intensive industries include mineral products and paper and paperboard. The historic change in the energy intensity of this sector is due largely to the change in composition over time and to substantial improvements in the technology of cement production. The projected decrease to 1985 is due almost entirely (~80 percent) to the effect of increasing oil prices. Space for policy action in this sector is minor since the industry, on average a low value product/high energy content industry, is greatly sensitive to the price of energy. The further decrease in intensity to the year 2000 is due partly to the decreased role played by cement production (386 kcal/Lira in 1972) in this sector and partly to technological improvements.

The non-energy-intensive sector comprises essentially the textile, mechanical, transport, and food industries. In 1953 they accounted for 31 percent of the energy consumed in manufacturing and mining, compared to 22 percent in 1972 and about 19 percent in 1985 and 2000. They are usually high value product/low energy content industries, so that the incentive to save energy is low. The decrease in energy intensity projected to the year 2000 includes the effect of increasing average company size. Many of these industries are presently small. In 1971, 27 percent of the employment in these industries was in firms with fewer than ten employees, down from 37 percent in 1951. The historic trend in the direction of increasing average company size will lead to a substantial decrease in energy intensity.

Agriculture, Livestock, Forestry, and Fishing

Value added (at factor cost in 1972 prices) in agriculture, livestock, forestry, and fishing is projected to increase to the end of the century as follows:

1972		5.9×10^{12} Lire
1985	C*	8.2×10^{12} Lire
	D	7.2×10^{12} Lire
2000	C	15.0×10^{12} Lire
	D	12.0×10^{12} Lire

The average energy intensity increased almost fourfold between 1953 and 1972, reaching 3.6 kcal/Lira in the latter year. The increase, due largely to mechanization of the agricultural sector, is expected to continue. The energy intensities projected to 1985 are 6.2 and 6.7 kcal/Lira in scenarios C* and D, respectively, and include a further doubling of the energy intensity of the agricultural sector. During this period and beyond, the livestock sector, a low energy intensity sector (0.8 kcal/Lira in 1972), is projected to gain progressively greater share, accounting for the projected decrease in energy intensity by the year 2000 (5.4 and 5.6 kcal/Lira in scenarios C and D, respectively).

In this sector increases in oil prices and vigorous energy policies are not envisaged as having any substantial effect.

12.3.4 Energy Demand for Civilian Uses

Residential uses account for most of the energy consumed in this sector. The distribution, which has stayed roughly constant since the late fifties, was in 1972:

residential	77 percent
commercial	17 percent
public	6 percent

It is not expected to change substantially before the end of the century. Most of the following discussion will therefore concentrate on this sector.

Embodied in or consistent with the scenario projections is a steady decrease in the number of persons per household (4.3 in 1953, 4.2 in 1960, 3.5 in 1972, 3.1 and 3.2 in 1985 and 2.8 and 3.0 in 2000 in scenarios C and D, respectively), due to a decrease both in the number of families per household and in the number of persons per family. Population projections of 58.6 million in 1985 and 64.1 million in 2000 lead to the following numbers of households:

1972		15.6 million
1985	C*	19.2 million
	D	18.2 million
2000	C	22.9 million
	D	21.0 million

Appliances

Energy demand for appliances (excluding space heating and cooling uses) is concentrated in the residential sector: 81 percent in 1960, 75 percent in 1972. The increase in share of the commercial and public sectors is expected to continue slowly until it reaches about 30 percent in the year 2000. The change in distribution is explained by a shift from the residential sector of uses such as cooking, dishwashing, and clotheswashing (which together account for over 50 percent of the energy use) with increasing economic growth.

In the residential sector average consumption per household for appliances almost doubled between 1953 and 1972. Growth to the end of the century is projected to slow down substantially as a result of both lower economic growth and saturation effects:

1953		1.3×10^6 kcal/household
1960		1.7×10^6 kcal/household
1972		2.5×10^6 kcal/household
1985	C*	3.6×10^6 kcal/household
	D	3.0×10^6 kcal/household
2000	C	4.4×10^6 kcal/household
	D	3.6×10^6 kcal/household

A breakdown of the increase in energy demand in scenario C indicates that growth in the number of households is responsible for 39 percent of the increment between 1972 and 1985 and 26 percent of that between 1985 and 2000. Continued diffusion of energy-consuming appliances in place of more traditional low-intensity technology is responsible for 34 percent of the increase in both periods. The remaining increase can be attributed to greater intensity of use, to the diffusion of multiple appliances, and to larger models. A similar breakdown is obtained for scenario D but with some differences on account of the slower economic development. Growth in households continues to play the most important role up to 2000 (43 percent of the increase between 1985 and 2000), whereas increased use, the diffusion of multiple appliances, and larger models are less important (25 percent of the increase between 1985 and 2000).

Between 1953 and 1972 the composition by use changed substantially as increasing economic growth led to the displacement of low-intensity uses by energy-consuming appliances. The composition is expected to continue changing but at reduced pace until by the end of the century it has almost stabilized (Table 12.17).

Implicit in the projections to 1985 and 2000 are also the effects of increasing oil prices, which are minor, and of vigorous government policies. The latter are envisaged as taking the form of an energy purchase tax on appliances proportional to the power output. Whereas the effect of these measures is not expected to be important in directly decreasing energy demand in this sector, the income from the energy tax ($\sim 100 \times 10^9$ Lire (1972)/yr) could be usefully directed to developing more energy-efficient appliances.

Space Heating

Space heating accounts for most of the energy consumed in the residential, commercial, and public sectors. During the fifties and sixties energy demand for space heating actually grew faster than demand for other uses, so that space heating increased its share in total civilian energy use from 67 percent in 1953 to 70 percent in 1970 and 81 percent in 1972. The part played by space heating is projected to decline somewhat as a consequence of saturation effects accompanying economic growth, higher oil prices, and vigorous energy policies. The share going to space heating in scenario C diminishes to 78 percent in 1985 and 68 percent in 2000. In scenario D the decrease is delayed by slower economic growth and by later application

Table 12.17
Energy consumption by appliances in the residential sector: distribution by type of use (%)[a]

				1985		2000	
	1953	1960	1972	C*	D	C	D
Cooking	52	41	21	14	15	10	12
Dishwashing	7	11	18	23	19	27	23
Clotheswashing	8	11	17	15	15	14	14
Personal hygiene	26	25	18	14	14	12	13
Lighting	4	6	8	8	9	9	9
Refrigeration	<1	2	9	14	16	15	16
Entertainment	<1	1	5	6	7	6	7
Other	2	3	4	6	5	7	6
Total	100	100	100	100	100	100	100
Hot sanitary water	41	46	49	49	46	49	47
Heating and cooling[b]	54	45	34	31	33	29	31
Mechanical force	1	2	4	6	5	7	6
Light[c]	4	7	13	14	16	15	16

[a]For absolute values of energy consumption see Table 12.6.

[b]Other than hot water, space heating, and space cooling.

[c]Includes television.

of vigorous policies; in fact, the share continues to increase to 83 percent in 1985 but decreases thereafter, reaching 79 percent in the year 2000.

Almost 80 percent of the demand for space heating is concentrated in the residential sector. The share of this sector remains stable over time, varying at most by three percentage points between 1953 and 2000.

Average household demand for space heating grew very rapidly during the fifties and sixties, from 2.7×10^6 kcal/household in 1953 to 11.6×10^6 kcal/household in 1972. A number of factors contributed to this effect. Economic growth (greater wellbeing) was responsible for an increase in the average household volume and in degree-days as houses were heated for longer periods and to higher average internal temperatures. This effect was minor, however, compared to the impact of migrations from rural to urban areas and from south to north which accompanied rapid economic development. A major effect of these migrations was a change in heating habits as families passed from single-family households in rural settings to multifamily dwellings in urban and suburban areas. The former were largely warmed by room heating reduced to essentials using stoves, chimneys, and kitchen fires, the latter by centralized heating functioning for prolonged periods to maintain "comfortable" temperatures throughout the house. The substantial change in distribution by household class observed between 1953 and 1972 (Table 12.18) is principally the result of urbanization. (The effect of changes in the distribution by household class on energy intensity would actually have been negative if it had not been accompanied by changes in heating system. This is due to the increasing number of shared surfaces as the number of households per building increased: from about 13 in 1951 to 15 in 1961 and to 16 in 1971.)

Displacement of traditional heating systems also occurred independently of changes in the type of dwelling as a direct consequence of economic growth. The period between 1953 and 1972 saw an almost complete revolution in the way Italians heated their houses as stoves and other traditional methods were replaced by single-household radiators and central heating. An indicator of the changes in heating systems is the distribution by fuel of the energy consumed, which leaned heavily on wood and coal products (~76 percent) in 1953 but had moved essentially to oil products (84 percent) by 1972.

The different factors contributing to the increase in household demand for space heating are closely interrelated and difficult to disentangle. As a consequence the breakdown given in Table 12.19 is only indicative. The displacement of traditional heating systems (chimneys, braziers, kitchen fires) will proceed rapidly to completion by the end of the century. Another factor that will contribute to containing the increase in energy demand for space heating is the continued increase in the number of households per building projected as a consequence of continued urbanization, unaccompanied this time by major changes in heating system.

Projections to 1985 and 2000 also include the negative effect on energy demand of increases in the price of oil and of vigorous energy policies, which are envisaged to include application of severe building codes for new constructions, incentives for retrofitting, and control measures to improve the efficiency of burners. In scenario C the household space heating demand continues to increase to 1985 (14.8×10^6 kcal/household), principally as a consequence of continued urbanization and as the substitution of traditional heating systems continues to exhaustion. The effect of a vigorous policy can be roughly estimated by com-

Table 12.18
Distribution of dwellings by household class in the period 1953–2000

	1953	1971	1985	2000
1-family units	12.7	11.3	11.6	11.4
2-family units	13.0	11.1	9.7	8.5
3- to 15-family apartment buildings	45.5	42.3	41.8	41.1
16- to 30-family apartment buildings	18.2	19.4	18.4	17.9
> 30-family apartment buildings	10.6	15.9	18.6	21.0
Total	100	100	100	100

Table 12.19
Contributions of various factors to the increase in energy intensity for space heating during the period 1953-1972 (10^6 kcal household)

1953 intensity	2.7
Increase in average household volume	0.4
Migrations from Southern to Northern Italy	0.5
Higher room temperature and longer heating periods	0.9
Rural to urban migrations	2.5
Replacement of traditional heating systems	4.6
1972 intensity	11.6

parison with scenario D in 1985 (17.2×10^6 kcal/household) since the effect of differences in economic growth is only marginal.

The impact of vigorous energy policies becomes apparent with the increasing turnover in housing stock in the period between 1985 and 2000. In scenario C average consumption per household drops back down to 11.9×10^6 kcal/yr in 2000 as a result of a further increase in oil price. The effect of delaying vigorous policy measures in this sector to 1985 can be estimated from a comparison with the energy demand in scenario D in the year 2000 (15.1×10^6 kcal/household).

12.4 SUPPLY-DEMAND INTEGRATION

Up to 1985 the allocation of fuels to the various consumption and production sectors of the Italian economy can be taken as roughly equivalent to the preferred distribution (in the sense that both "constrained" and "unconstrained" supply-demand integration give similar results on a global basis). In the period between 1985 and 2000 the allocation of resources differs substantially from the preferred distribution on account of insufficient availability (at the scenario energy price) of the major fuels, oil and gas.

In line with the general WAES methodology two major replacement options, coal and nuclear, have been considered as alternatives to oil and gas in this period.

Italy has only limited coal resources and would therefore need to import increasing quantities of coal in a high coal scenario. If the extra amount of coal introduced into the system were sufficiently great, this could involve major restructuring of freight transport and handling systems to make space for coal movement over land and major construction work to improve and increase port facilities for coal imports from overseas. The financial requirements would not in themselves constitute an impediment to the reintroduction of coal in Italy if distributed over a sufficiently long period of time. It is doubtful, however, that it would be possible to carry forward at the same time both an ambitious coal program and a sustained nuclear program such as would be required even in a high coal scenario. The basic argument against considering a major effort to reintroduce coal use in Italy is related, however, to the longer-term aspect of such a policy. Given sufficient lead time for research and development, the future of coal lies in liquefaction and gasification. The particular attractiveness of coal conversion processes lies in the fact that the resulting gaseous and liquefied fuels require no change in end use technology and utilize the same or analogous distribution systems as presently exist.[6] It therefore seems inappropriate for a country like Italy to concentrate major capital outlays in creating transportation and handling systems that would find economic use only for a short transition period. A moderately high coal policy of the type presented here would not require major infrastructure changes and would thus allow more flexibility.

The first part of the present section deals principally with the allocation of fuels to final uses in

6. The role played by coal synthetics is likely to be of only minor importance under the scenario price schedule assumptions to the end of the century: constant at \$11.50/bbl or *slowly* rising to \$17.25/bbl. An alternative price path could bring on more coal synthetics in the period under discussion. This is one in which the price of oil increased rapidly during the early eighties to values sufficiently high as to bring on substantial coal liquefaction and gasification. Within 15 to 20 years, research and development and economies of scale could bring the price of coal synthetics down to the scenario prices. This scenario, however, has not been considered.

Whatever coal synthetics may be in use by the end of the century, they are not distinguished from oil and gas in the present study for two reasons. First, they are very similar physically, and second, the Italian energy system is indifferent to the extent of world coal liquefaction and gasification. Since Italy has no important coal resources, strategies involving Italian participation in coal liquefaction and gasification could be taken into serious consideration only as joint international ventures. Moreover, extensive coal synthetics production abroad would, from Italy's point of view, result basically only in a reduction of the pressure on the world demand for oil and gas. World trade would continue to concentrate in crude oil and natural gas, with synthetics consumed largely in the producing countries.

For these reasons direct participation of coal synthetics in the Italian energy economy is likely to be minimal by the end of the century, justifying reference to "oil" and "gas" throughout the period to 2000 in this report.

the period 1985 to 2000 as crude oil and natural gas become increasingly scarce. The reasoning behind fuel allocation in final use sectors in the period 1972 to 1985 is outlined in some detail in *Energy Demand Studies: Major Consuming Countries* and will not be discussed in the present context. The second part of the section discusses in detail the allocation of primary fuels to conversion processes, electricity generation in particular, during both periods 1972 to 1985 and 1985 to 2000. In the third and last part, the primary energy demands resulting from the integration are examined for all scenario combinations studied, in particular as they relate to the fuel mix and conversion losses evolving over the second half of the twentieth century.

12.4.1 Allocation of Fuels in Final Uses

Among alternative fuels to oil and gas competing for final use sectors, only solid coal, solar, and geothermal energy assume some importance in the period 1985 to 2000.

Nuclear heat, although an important alternative, is not expected to contribute significantly before the end of the century. While the technology for combined heat and electricity production and utilization in industry is likely to be well developed by the 1990s through the high-temperature gas reactor, the contribution of this source of energy will be minimal before the end of the century because of the time required to make any impact and because of possible siting problems. Siting problems will also hold back the utilization of low-temperature nuclear heat already available from current technology for space heating and hot sanitary water in the civilian sector.

A whole series of uncertainties about hot dry rocks technology makes estimates on the potential for geothermal heat in final uses uncertain to the end of the century. The projections presented in this study do not include possible contributions from hot dry rocks and are assumed to result from conventional geothermal fields through new finds and better use of existing resources. Most of this energy (1.7 and 12.0×10^{12} kcal by the year 2000 in the high nuclear and high coal scenarios, respectively) is assumed to be employed in agriculture and smaller amounts in space heating and industry.

The penetration of electricity in end uses has been determined as part of the demand projection methodology and described in section 12.2. The allocation of nonelectrical demand among fuels in final uses is estimated sector by sector following simple rules. The approach used in these projections is to allocate oil and gas (including possible synthetics) to the demand residue after the maximum feasible penetration of other fuels has been estimated in line with the scenario assumptions. Given the ready substitutability of oil and gas in most final use sectors and the relative indifference from a long-term strategic standpoint of oil and gas as alternative fuels, the problem of allocating these resources to final uses in the period 1985 to 2000 becomes generally indeterminate. For this reason no particular criticality has been assumed in relation to the way oil and gas distribute in the energy system, and the very simple hypothesis is made that in each sector the distribution in the year 2000 maintains the 1985 partial shares.

The following subsections examine the penetration potential of solid coal and solar energy in final use sectors up to the year 2000.

The Potential for Coal

Solid fuels were important in all sectors of the Italian economy in the early fifties. In 1953 they accounted for some 44 percent of the energy consumed in final uses (35 percent in industry, 31 percent in transport, 52 percent in civilian uses). Wood and wood products such as charcoal made up a substantial portion of the solid fuel: 36 percent in all final uses and 67 percent in civilian uses. During the fifties and sixties solid fuels declined very rapidly in importance. In 1963 they covered barely 17 percent of all final consumption (11.6 percent of all end uses in industry). In 1972 solid fuels, by then mostly coal and coke (76 percent), had lost favor and practically vanished in all end use sectors of the economy except industry, where they accounted for 7.5 percent of all final consumption concentrated almost entirely (74 percent) in the steel industry.

Prospects for a return of solid fuels in Italy in final uses other than industrial are considered to be very poor. A return of solid coal in transport end uses is equivalent to assuming a comeback of the steam locomotive and is therefore not taken into consideration. A return of solid coal in civilian uses has been examined but found to be unattractive and essentially uneconomical. In the appliances sector a reintroduction of coal has practically no impact since most of the demand is for electricity (~85 percent). In space heating it is undesirable for a number of reasons: pollution, low efficiency, high handling costs. Improvements in the technology of coal combustion (fluidized bed, etc.) can change the picture somewhat, but the handling costs remain high and are expected to be a major argument against a return of coal in this sector even if the cost per kcal is substantially lower than that of other fuels (which is unlikely in a country that has limited coal resources). For these reasons only marginal use of solid fuels has been assumed in the residential, commercial, and public sectors even under a high coal scenario. Much of this fuel (2.9 and 3.4×10^{12} kcal in scenarios C and D, respectively, in 2000) is wood and charcoal used for special cooking and open fires in holiday houses and tourist resorts.

In all industrial end uses except the metals sector (above all iron and steel), solid fuels and products were of only marginal importance in 1972 and are projected to decline even further over the period to 1985 and 2000 as antiquated plants and technology are replaced. Even in the metals sector solid fuels and derived products such as blast furnace gas are showing an overall tendency to decline in importance. In this sector the share covered by coal and coal products, which was 42 percent in 1972, is projected to reach 36 and 38 percent in 1985 in the high and low growth scenarios, respectively, as a result of substitution in end uses and changes in product mix.

Table 12.20 gives the percentage share of coal and coal products projected to 1985 and 2000 in the basic industrial sectors in the absence of a deliberate policy to reintroduce coal (high nuclear scenario). In absolute terms the growth of coal use in industry implicit in these projections is actually faster over the period 1972 to 2000 than it was in the period 1953 to 1972 when coal was displaced by fuel oil in many end uses (Table 12.21).

Table 12.20
Percentage share of solid fuels in industrial end use up to the year 2000: high nuclear scenarios[a]

		1985		2000	
	1972	C*	D	C	D
Metals	42.3	36.4	38.4	31	35
Chemicals	2.7	0.9	0.8	0.5	0.5
Other intensive	3.5	1.5	1.3	1	1
Nonintensive	4.5	3.7	4.2	3	3.5
Total	9.3	7.8	8.7	6.0	7.2

[a]Including all coal products in current use: coke, coking gas, blast furnace gas, and nonenergy coal products.

Relative to the high nuclear scenarios, the high coal projections assume a doubling in the growth rate of coal use in industry over the period 1985 to 2000. The amount of coal and coal products in use in industry in the year 2000 resulting from these assumptions is 90×10^{12} kcal in scenario C-1 and 71×10^{12} kcal in scenario D-7 compared to 44×10^{12} kcal in 1972, and therefore does not represent any particular burden in terms of transportation and handling costs to the manufacturing industries. The country would be transporting between 4 and 7 million tons/yr more coal than in 1972 for direct use in industry.

Most of the incremental coal would be used in place of fuel oil for process heat and other low-grade uses. If it were to be used only in the steel industry, which is largely endowed with the necessary structures for coal handling and transportation, it would determine a far smaller change in the fuel structure than if it were to be all used by other intensive industries such as mineral products. In scenario C-1 such extreme cases of allocation would determine an increase in the share of all end uses covered by coal from 31 to 38 percent in the metals sector and from about 1 to 8 percent in the other intensive industries. In practice, some intermediate situation would be expected to result in which all sectors consumed more coal.

The assumptions embodied in the projections to 1985 and 2000 imply a continuing decline in the share of solid fuels in overall final end uses:

1972		6.4 percent
1985	C*	4.7 percent
	D	4.6 percent
2000	C-1	4.2 percent
	C-2	3.5 percent
	D-7	4.5 percent
	D-8	3.9 percent

The Potential for Solar

In 1976 solar energy had only marginal application in Italy. Considerable research and development is under way, however, both in industry and in the universities, and the indications are that by the early eighties solar energy in some uses could become competitive with other technologies and be produced on a commercial scale.

Table 12.21
Average annual growth in the use of coal in industry to the year 2000

	Historical	Scenario C	Scenario D
1953-1960	0.2		
1960-1972	1.6		
1972-1985		3.0	1.6
1985-2000			
high coal		2.2	1.8
high nuclear		1.1	0.9

The potential for solar energy up to the year 2000 is found largely in the civilian sector in uses such as space heating, sanitary water heating, and cooling. The estimates of solar penetration to 1985 and 2000 presented in this study assume that solar energy will be used only marginally in industry and agriculture, mostly in experimental and pilot applications of solar thermal power and solar ovens for high-purity metallurgical processes and special glasses.

This section deals basically with the scenario estimates of the penetration potential of solar energy in civilian uses and in particular in the residential sector. These were made through an assessment of solar technology taking into account both the structure and turnover of the housing stock.

Italy receives a fairly high solar energy flux. The daily extraatmospheric equivalent solar energy flux varies between 4450 and 5250 kcal/m^2 according to the latitude in winter and between 8640 and 9120 kcal/m^2 in summer. Average radiation loss in passing through the atmosphere is between 50 and 60 percent according to the latitude and the season. In northern Italy the winter season lasts about 210 days and in southern Italy about 180 days.

The amount of solar energy available to households depends in a critical way on the assumptions made regarding the coverage of solar collectors. In the present estimates it is assumed that most of the energy is collected on roofs but that some collection (~20 percent of the total) also occurs on walls. Efficiencies of solar collectors vary greatly with the water temperature and the type and inclination of the collector. The efficiency incorporated in the projections, averaged over summer and winter, amounts to about 50 percent of the incident flux and is therefore essentially equivalent to present efficiencies.

The maximum amount of energy available to households is very much a function of the type of dwelling, the climatic zone, and the season. The very large differences encountered in Table 12.22 between different types of dwelling can be explained largely by the rapidly increasing average number of dwellings under one roof in passing from 1-family units to >30-family apartment buildings. Clustering of buildings in urban areas is another important factor built into the estimates of Table 12.22 to take into account shading of one building by another and also by trees. This is most important in the 3- to 15-family apartment buildings class and is negligible for 1- and 2-family units.

The solar energy available with the assumed technology is generally lower than the energy required for heating and cooling loads, as can be seen from a comparison with Table 12.23, implying that some integration is necessary with traditional technologies. As regards space heating, backup systems (oil or gas burners and electric stoves) are probably necessary in any event to cover the peaks in demand in the most cost-effective way. The optimal fraction of space heating demand satisfied by solar energy depends also on the price of oil and on the cost of storage systems. In this study the fraction is estimated as 50 percent at an oil price of $11.50/bbl and 70 percent at $17.25/bbl.

Solar energy is expected to displace traditional technology most effectively in new construction if the credit incentives are available to take care of the rather high front-end costs. The role of the government is therefore of paramount importance for the development of solar energy in Italy, as regards both financing research and facilitating the marketing process. The potential for solar energy in Italy in the medium term is severely restrained by the structure of the housing stock. The diffusion of solar technology in the present projections is assumed to be quite high, probably higher than can be expected to occur in practice in the time period under consideration. The optimism expressed is thought to be of use in giving a measure of the task that must be faced by government if solar energy is to make any significant headway in the period to the year 2000.

In the diversification (high coal) scenarios, government is assumed to take strong action in favor of a solar program in households beginning as soon as possible. Retrofitting of buildings existing in

Table 12.22
Solar energy available to households in the year 2000[a] (10^6 kcal/household)

	North		Center		South	
	Winter	Summer	Winter	Summer	Winter	Summer
1-family units	37	58	33	47	37	51
2-family units	24	36	24	34	21	37
3- to 15-family apartment buildings	4	7	5	6	4	4
16- to 30-family apartment buildings	5	7	4	6	4	6
> 30-family apartment buildings	3	3	3	3	3	3

[a]Assuming the technology specified.

Table 12.23
Average yearly heating and cooling demands in the residential sector in the year 2000: Scenario C (10^6 kcal/household)

	North	Center	South
Space heating			
1-family units	53	34	22
2-family units	46	33	20
3- to 15-family apartment buildings	19	14	9
16- to 30-family apartment buildings	13	10	6
> 30-family apartment buildings	7	6	4
Sanitary water heating	3.3	2.9	2.5
Cooling	0.1	0.1	0.1

1985 is given only moderate importance (<20 percent), but the technology is assumed to be applied to about 60 percent of new single-family housing and to almost as great an extent in multifamily apartment buildings due to the attractiveness of solar water heating, which requires a relatively small collection surface. (The figures cited refer to the number of households equipped with solar panels to one extent or another.) Even in large apartment buildings, sufficient solar energy is available to cover most of the demand for hot sanitary water that would otherwise be met largely by electricity (compare Tables 12.22 and 12.23).

Although the capital costs can be quite high, the operating costs are usually sufficiently low that the technology is assumed to diffuse on its own even without credit incentives. In the high nuclear scenarios a certain governmental neglect is assumed in regard to providing incentives for the diffusion of solar energy in households. In these projections the diffusion is assumed to be some 50 percent lower than in the high coal scenarios and to depend more markedly on the household class because of the difficulty in getting homeowners of multiapartment dwellings to agree on installing high-capital-cost units without credit facilities.

Allocation of available solar energy within individual household classes for space heating, sanitary water heating, and cooling is assumed to depend on the season and on the climatic zone. Preference is given to water heating in both summer and winter. Any residual solar energy is allotted to space heating in winter and to cooling in summer.

Notwithstanding a certain optimism in some of the assumptions described, solar energy penetration to 1985 and 2000 in the civilian sectors is not very great. This is largely due to the distribution by household class, which in Italy is heavily weighted on the multiapartment dwellings side: on average 16 dwellings per building in 1971 and 18 in 2000.

In these estimates solar energy covers as much as 20 percent of the total civilian heating and cooling loads by the year 2000 (20, 10, 16, and 8 percent in scenarios C-1, C-2, D-7, D-8, respectively). Most of the solar energy goes to space heating (69 and 77 percent in scenarios C-1 and D-7, respectively), reflecting the high space heating demands and solar energy availability of 1- and 2-family units. Between 16 percent (scenario D-7) and 19 percent (scenario C-1) of the solar energy is used for sanitary water heating. The remainder is used for cooling.

As an indication of the distance that has to be covered before the solar alternative can satisfy a substantial portion of the energy demand in the civilian sectors, it is noted that in these projections solar energy takes only 15 to 19 percent of the space heating load, about 33 percent of the sanitary water heating load, and about 31 percent of the cooling load by the year 2000.

12.4.2 Allocation of Electricity Supply

Electricity Supply Allocation for 1985

Electricity generation is based on complex technological systems requiring relatively long periods for their completion. As a consequence a certain caution in planning for future capacity is required, particularly in the short term for which the construction time nearly equals the planning horizon, in order to avoid the possibility of insufficient capacity which could limit economic development. For this reason the supply is allocated for maximum economic growth (4.5 percent increase in gross national product per year) and the least amount of energy savings (scenario E). The additional capacity required at the winter peak is calculated starting from the corresponding electricity demand.

In the last decade the ratio of the annual consumption of electric energy, before distribution, to the power required at the winter peak has been 5780 hours. Applying this coefficient, the capacity required at the winter peak in this scenario is 47,500 MW. To this the necessary reserve capacity must be added. In the years preceding 1972 the ratio of reserve capacity to the capacity required at the winter peak did not exceed 25 percent. In other countries, such as Germany and the United States, the fraction required for reserves is even lower. The increase in this ratio after 1972 is due to slower-than-anticipated growth in demand. In any case, a high percentage of reserve capacity should be avoided, particularly given the fact that as electricity systems expand, the demand normally tends to be more evenly distributed, thus requiring lower reserves. Assuming a 25 percent reserve, the total capacity requirement calculated for 1985 is 59,400 MW. Given the capacity available presently (33,500 MW), the maximum amount of new capacity required is 25,900 MW.

With additional hydroelectric and geothermal capacity of 2000 MW at the winter peak, and 4500 MW capacity from pumping stations and gas turbines, there remain 19,400 MW capacity of thermoelectric power to be installed by 1985. Approximately 9400 MW of conventional thermoelectric plants are presently under construction, resulting in a difference of 10,000 MW maximum additional capacity that may be provided by nuclear power.[7]

Adding in nuclear capacity already in existence, the total nuclear capacity in 1985 would be 10,600 MW, which if supplied by conventional thermoelectric plants would require an annual input of approximately 144×10^{12} kcal or 14.4×10^{6} TOE.

The objective of 10,000 MW additional capacity can be achieved through the 800 MW nuclear plant at Caorso, presently almost in operation, two 1000 MW plants already ordered that should enter into operation by 1983, plus a total additional capacity of 7200 MW by 1984 or 1985. This implies that the the necessary power plants must be ordered by the end of 1977.

Such a nuclear program is consonant to the objective of diversification of supply, technological development of important industrial sectors, and creation of export opportunities for nuclear equipment and plants, and it should therefore be actively pursued despite the well-known problems concerning the acceptability of nuclear power and the substantial capital investments involved.

In the longer term a nuclear program could be more flexible, allowing responses to changes in economic growth, the impact of energy saving, and possible new technological developments. For example, if gross national product increases at a rate in excess of 4.5 percent per year (which is unlikely), there would always be time to realize an adequate number of conventional thermoelectric plants. In fact, there are 4000 MW of such plants presently planned but not under construction that could be utilized for this purpose. These plants could also be constructed if the nuclear program should be delayed.

In Table 12.24 the electricity supply-demand balance is indicated for all the 1985 scenarios.

Electricity Supply Allocation for 2000

The maximum nuclear capacity assumed to be achieved by the year 2000 is 65,000 MW in

7. This nuclear capacity does not include the power that would be required for the Coredif plant if this were sited in Italy.

Table 12.24
Electricity supply-demand balance in 1985 (all figures in GW capacity at winter peak)

	A	B	C	C*	D	E
Supply						
Additional capacity						
nuclear	10.0	10.0	10.0	10.0	10.0	10.0
hydroelectric	1.7	1.7	1.7	1.7	1.7	1.7
modulation power and storage[a]	3.8	2.2	4.0	4.3	2.7	4.5
geothermal	0.7	0.7	0.7	0.7	0.3	0.3
thermoelectric conventional	2.4	-9.0	4.4	6.2	-4.9	9.4
1975 capacity	33.5	33.5	33.5	33.5	33.5	33.5
1985 capacity	52.1	39.1	54.3	56.4	43.3	59.4
Reserve[b]	10.4	7.8	10.9	11.3	8.7	11.9
Net available 1985 supply (before transmission)	41.7	31.3	43.4	45.1	34.6	47.5
Demand (before transmission)[c]	41.7	31.3	43.4	45.1	34.6	47.5
Balance[d]	0.0	0.0	0.0	0.0	0.0	0.0

[a]Includes pumped storage and gas turbines for supplying electricity during periods of modulation of the demand. Gas turbines are excluded from thermoelectric conventional. Modulation power and storage is set equal to 12 percent of the total capacity, with 2.5 GW present capacity.

[b]Using a 1:4 ratio of reserve to demand.

[c]The structure of the demand, in terms of average ratio of annual demand for electrical energy to power required at the winter peak, has been 5780 hr/yr from 1964 to 1974.

[d]Supply minus demand.

scenario C-2.[8] If 25,000 MW are operational by 1990, this policy implies putting into service an average of four 1000 MW nuclear plants each year from 1990 to 2000. This is considered to be near the productive limit of the Italian nuclear power industry. In this scenario 56 percent of the total capacity is nuclear. In the low economic growth/ high nuclear and high economic growth/high coal scenarios (D-8 and C-1) the nuclear program is kept at substantial levels (40,000 and 30,000 MW), ensuring more than enough nuclear construction for industrial "take-off" and possible exportation. With a vigorous diversification (high coal) strategy and low economic growth (scenario D-7) nuclear capacity is 20,000 MW, approximately equal to the number of orders presently being considered for the medium term and to the minimum capacity that would permit sufficient experience for export of related equipment.

Maximum coal-fired capacity (scenario C-1) is assumed to be 18,000 MW, as compared to the approximately 1000 MW capacity, equivalent to 4.75×10^9 kWh, produced by coal in 1976. This represents an average annual growth rate of 13.4 percent from 1977 to 2000. However, approximately 8000 MW of existing conventional thermoelectric capacity can be converted to coal-fired plants,[9] and all of the new thermoelectric plants (presently 9400 MW under construction) are dual purpose and can be fired by both coal and liquid fuels.

Coal use in electricity generation represents a diversification of the source of supply and can be of some importance in the event of slack in the development of a nuclear power program. Its cost per kWh is likely to remain competitive with or below that of oil-fired plants. Major problems are pollution from high-sulfur coal, the difficulty in obtaining local permits, the necessary expansion of transportation and handling facilities, and ash disposal.

As indicated in the supply study, geothermal and hydropower resources are more fully developed in the diversification scenarios C-1 and D-7. The complete electricity supply allocation for the year 2000 is given in Table 12.25.

8. In calculating capacity requirements from the forecast of annual demand the same basic methodology employed for 1985 is used, with a reserve/required power ratio of 25 percent, a level of modulation power and storage of 12 percent of total capacity, and a ratio of annual demand to power required at the winter peak of 5900 hours.

9. *Programma Energetico Nazionale,* Italian Ministry of Industry, 1975, p. 55.

Table 12.25
Electricity supply allocation in the year 2000

	High economic growth		Low economic growth	
	Coal C-1	Nuclear C-2	Coal D-7	Nuclear D-8
Delivered energy (10^9 kWh)	405	500	250	325
Demand before transmission[a] (10^9 kWh)	445	549	278	357
Demand[b]				
Power required (without reserves)[c]	75	93	47	61
Reserve[d]	19	23	12	15
Total power required at winter peak	94	116	59	76
Supply[b]				
Nuclear plants	30	65	20	40
Hydroelectric natural	13	12	13	12
Hydroelectric pumped storage[e]	11	14	7	9
Geothermal power	2	1	2	1
Coal-fired plants	18	6	8	3
Oil-fired plants	17	15	7	9
Gas-fired plants	3	3	2	2
Total capacity	94	116	59	76

[a]Transmission losses equal to 9 percent of demand before transmission.

[b]Figures given reflect power in GW at winter peak.

[c]The structure of the demand in terms of annual demand divided by power required at the winter peak increases slightly to 5900 hr/yr from the 1964–1974 average of 5780 hr/yr.

[d]Reserve capacity is equal to 25 percent of the requested power at the winter peak.

[e]Pumped storage capacity is set equal to 12 percent of total capacity.

12.4.3 The Primary Energy Demands in the WAES Scenarios

Fueling of the Energy System

The composition by fuel of the primary input of the Italian energy system changed substantially over the period 1953 to 1972 as traditional sources such as wood, coal, and hydropower were displaced by oil.

The present projections indicate continuing substantial changes over the period 1972 to 2000. The essential traits of these changes depend on the scenario and in particular on the choice, which must be made in the very near future, between a diversification (high coal) strategy and a nuclear strategy (Table 12.26).

The dependence on oil and gas, which was 84.4 percent in 1972, drops to 80.8 and 79.1 percent in 1985 scenarios C* and D, respectively. By the year 2000 it drops further to about 67 percent in the high coal scenarios and 60 percent in the high nuclear scenarios. The maximum development of nuclear energy projected to the end of the century captures as much as 29 percent of the total primary energy input (scenario C-2). In the high coal scenarios, coal increases its share substantially above the 1972 levels due largely to increased coal use in electricity generation. The implied coal expansion is close to the financial and economic limits that can be tolerated by the system and corresponds to an average annual growth rate in primary coal use over the period 1972 to 2000 of 5.4 and 3.3 percent in scenarios C-1 and D-7, respectively. Solar energy in these scenarios covers as much as 3.7 percent of total primary energy input (scenario D-7). Geothermal energy increases its share considerably from 0.5 percent in 1972 to between 1.3 and 1.8 percent in scenarios C-1 and D-7. Much of this energy is used for heating purposes rather than for power generation.

Processing and Distribution Losses

Over the period 1953 to 1972 energy processing and distribution losses in Italy decreased from 25.8

Table 12.26
Composition of the primary energy input for internal consumption (%)

				1985		2000			
	1953	1960	1972	C*	D	C-1	C-2	D-7	D-8
Solar				0.5	0.4	3.3	1.6	3.7	1.8
Nuclear			0.6	6.7	7.7	13.8	29.1	12.7	24.4
Geothermal	1.3	0.8	0.4	0.7	0.6	1.3	0.5	1.8	0.7
Hydropower	18.6	17.1	7.2	5.2	6.0	3.9	3.5	5.4	4.7
Natural gas	5.8	9.1	9.2	15.7	13.1	14.4	13.5	14.6	13.5
Oil	40.1	54.9	75.4	65.2	66.0	52.3	46.3	53.1	49.6
Coal	24.0	13.6	6.1	5.8	5.8	11.0	5.4	8.6	5.2
Wood	10.2	4.4	1.1	0.2	0.3	<0.1	<0.1	0.1	0.1
Total	100	100	100	100	100	100	100	100	100

Note: Nuclear, geothermal, and hydro power are accounted using conventional thermal power plant efficiency. Figures include bunkerages.

to 21.3 percent of the total primary energy input for internal consumption (assuming conventional thermal power plant efficiency for hydropower, geothermal electricity generation, and nuclear power). This relatively anomalous behavior can be correlated with the decreasing share of electricity in final energy demand (11.7 percent in 1953 and 10.3 percent in 1972), which can in turn be explained by the abundance of hydropower in early postwar Italy and the very rapid growth of oil products in final uses during the fifties and sixties.

The decreasing trend slowed down and essentially stopped in the early seventies as electricity demand began growing more rapidly than the demand for fossil fuels in final uses. The results of the projections to 1985 and 2000 indicate that processing and distribution losses tend to increase again, bringing Italy back in line with the trend prevalent in most other countries:

1972		20.4 percent
1985	C*	21.4 percent
	D	19.5 percent
2000	C-1	24.2 percent
	C-2	27.6 percent
	D-7	21.0 percent
	D-8	25.3 percent

Imports of Primary Energy

Italy shows a growing dependence on imports for its primary energy sources. Imports have risen from 62.6 percent of the total primary energy input for internal consumption in 1953 to 65.8 percent in 1960 and 81.5 percent in 1972.

The projections to 1985 and 2000 obtained from total primary energy demand by subtracting domestic production show a steadily growing dependence on imports:

1985	C*	85.1 percent
	D	83.3 percent
2000	C-1	88.1 percent
	C-2	91.3 percent
	D-7	84.4 percent
	D-8	88.6 percent

The major difference between 1972 and 2000 is an increasing diversification of the imported sources. Whereas oil remains the principal import to the end of the century, its share in total imports declines appreciably (Table 12.27).

12.5 SUPPLY-DEMAND INTEGRATION WORKSHEETS

On the following pages will be found eleven supply-demand integration worksheets (one each for the base year, 1985 scenarios A, B, C, C*, D, E, and 2000 scenarios C-1, C-2, D-7, and D-8). The reader should note the following points:

• The supply-demand integrations consider only energy used for internal consumption including bunkerage. They exclude the net import-export balance of secondary energy. In order to maintain consistency this accounting convention is also ap-

Table 12.27
Composition of energy imports for internal consumption (%)

		1985		2000			
	1972	C*	D	C-1	C-2	D-7	D-8
Coal	6.9	6.4	6.6	11.8	5.6	9.3	5.4
Oil	91.4	74.9	77.4	58.7	50.1	62.1	55.3
Natural gas	1.0	10.8	6.7	13.8	12.4	13.5	11.8
Nuclear fuel	0.7	7.9	9.3	15.7	31.9	15.1	27.5
Total	100	100	100	100	100	100	100

Note: Nuclear fuel is accounted using conventional thermal power plant efficiency. Internal consumption includes bunkerage. Coal includes wood.

plied to the year 1972, so that the balance sheet presented for this year does not correspond to the Bilanci Energetici, which do include the imports and exports of secondary energy.

• The primary energy input to hydroelectric, geothermal electric, and nuclear power plants is calculated assuming the equivalent conversion efficiency of conventional thermoelectric power plants (2360 kcal/kWh).

• Solar energy and nonelectric uses of geothermal energy are assumed conventionally to have no processing and distribution losses.

• Contributions from urban solid waste are not included.

SUPPLY/DEMAND INTEGRATION WORKSHEETS

Country : Italy Case : 1972

Units in 10^{12} Kcal

	Primary Energy Input									Processing and Distribution Losses	Bunkerage and changes in stocks	Final Demand			
	Wood	Coal	Oil	Natural Gas	Hydro-power	Geo-thermal	Nuclear	Solar	Total			Total	Transp.	Ind. & Agric.	Resid. Comm. & Public
Secondary Energy Conversion, Distribution and Final Demand															
Charcoal Production															
Coking Plants		59							59	16		43		41	2
Refineries			815		1				816	45	89	682	168	302	212
Town Gas		1	2	5					8	2		6			6
Electricity Generation	2	15	169	9	94	6	8		303	205		98	3	63	32
Total	2	75	986	14	95	6	8		1186	268	89	829	171	406	252
Primary Energy Distribution and Final Demand															
Wood	13								13			13			13
Coal		5							5			5		3	2
Natural Gas				107					107	2		105	1	83	21
Geothermal															
Solar															
Total	13	5		107					125	2		123	1	86	36
Total Primary Demand	15	80	986	121	95	6	8		1311						
Domestic Supply	14	7	11	110	95	6			243						
Imports	1	73	975	11			8		1068						

SUPPLY/DEMAND INTEGRATION WORKSHEETS

Country : Italy Case : 1985 A

Units in 10^{12} Kcal

	Primary Energy Input									Processing and Distribution Losses	Bunkerage and changes in stocks	Final Demand			
	Wood	Coal	Oil	Natural Gas	Hydro-power	Geo-thermal	Nuclear	Solar	Total			Total	Transp.	Ind. & Agric.	Resid. Comm. & Public
Secondary Energy Conversion, Distribution and Final Demand															
Charcoal Production															
Coking Plants		78							78	22		56		54	2
Refineries			872						872	28	112	732	157	433	142
Town Gas		1	2	5					8	2		6			6
Electricity Generation	3	27	252	5	112	16	144		559	371		188	6	113	69
Total	3	106	1126	10	112	16	144		1517	423	112	982	163	600	219
Primary Energy Distribution and Final Demand															
Wood	2								2			2			2
Coal		4							4			4		3	1
Natural Gas				283					283	5		278		188	90
Geothermal															
Solar								15	15			15			15
Total	2	4		283				15	304	5		299		191	108
Total Primary Demand	5	110	1126	293	112	16	144	15	1821						
Domestic Supply	4	9	29	140	112	16		15	325						
Imports	1	101	1097	153			144		1496						

SUPPLY/DEMAND INTEGRATION WORKSHEETS

Country : Italy Case : 1985 B

Units in 10^{12} Kcal

	Primary Energy Input									Processing and Distribution Losses	Bunkerage and changes in stocks	Final Demand			
	Wood	Coal	Oil	Natural Gas	Hydro-power	Geo-thermal	Nuclear	Solar	Total			Total	Transp.	Ind. & Agric.	Resid. Comm. & Public
Secondary Energy Conversion, Distribution and Final Demand															
Charcoal Production															
Coking Plants		62							62	17		45		44	1
Refineries			732						732	24	92	616	131	327	158
Town Gas			1	4					5	1		4			4
Electricity Generation	3	27	113	5	112	16	144		420	278		142	4	86	52
Total	3	89	846	9	112	16	144		1219	320	92	807	135	457	215
Primary Energy Distribution and Final Demand															
Wood	2								2			2			2
Coal		3							3			3		2	1
Natural Gas				207					207	4		203		137	66
Geothermal															
Solar								12	12			12			12
Total	2	3		207				12	224	4		220		139	81
Total Primary Demand	5	92	846	216	112	16	144	12	1443						
Domestic Supply	4	9	29	140	112	16		12	322						
Imports	1	83	817	76			144		1121						

SUPPLY/DEMAND INTEGRATION WORKSHEETS

Country : Italy Case : 1985 C

Units in 10^{12} Kcal

	Primary Energy Input									Processing and Distribution Losses	Bunkerage and changes in stocks	Final Demand			
	Wood	Coal	Oil	Natural Gas	Hydro-power	Geo-thermal	Nuclear	Solar	Total			Total	Transp.	Ind. & Agric.	Resid. Comm. & Public
Secondary Energy Conversion, Distribution and Final Demand															
Charcoal Production															
Coking Plants		87							87	25		62		60	2
Refineries			1014						1014	32	127	855	161	471	223
Town Gas		1	2	5					8	2		6			6
Electricity Generation	3	27	275	5	112	16	144		582	386		196	6	119	71
Total	3	115	1291	10	112	16	144		1691	445	127	1119	167	650	302
Primary Energy Distribution and Final Demand															
Wood	2								2						
Coal		5							5						
Natural Gas				308					308						
Geothermal															
Solar								10	10						
Total	2	5		308				10	325						
Total Primary Demand	5	120	1291	318	112	16	144	10	2016						
Domestic Supply	4	9	29	140	112	16		10	320						
Imports	1	111	1262	178			144		1696						

S U P P L Y / D E M A N D   I N T E G R A T I O N   W O R K S H E E T S

Country : Italy   Case : 1985 C*

Units in 10^{12} Kcal

	Primary Energy Input									Processing and Distribution Losses	Bunkerage and changes in stocks	Final Demand			
	Wood	Coal	Oil	Natural Gas	Hydro-power	Geo-thermal	Nuclear	Solar	Total			Total	Transp.	Ind. & Agric.	Resid. Comm. & Public
Secondary Energy Conversion, Distribution and Final Demand															
Charcoal Production															
Coking Plants		94							94	27		67		63	4
Refineries			1105						1105	36	137	932	195	485	252
Town Gas		1	2	5					8	1		7			7
Electricity Generation	3	27	293	5	112	16	144		600	398		202	6	121	75
Total	3	122	1400	10	112	16	144		1807	462	137	1208	201	669	338
Primary Energy Distribution and Final Demand															
Wood	2								2			2			2
Coal		3							3			3		2	1
Natural Gas				327					327	5		322		214	108
Geothermal															
Solar								10	10			10			10
Total	2	3		327				10	342	5		337		216	121
Total Primary Demand	5	125	1400	337	112	16	144	10	2149						
Domestic Supply	4	9	29	140	112	16		10	320						
Imports	1	116	1371	197			144		1829						

SUPPLY/DEMAND INTEGRATION WORKSHEETS

Country : Italy Case : 1985 D

Units in 10^{12} Kcal

	Primary Energy Input									Processing and Distribution Losses	Bunkerage and changes in stocks	Final Demand			
	Wood	Coal	Oil	Natural Gas	Hydro-power	Geo-thermal	Nuclear	Solar	Total			Total	Transp.	Ind. & Agric.	Resid. Comm. & Public
Secondary Energy Conversion, Distribution and Final Demand															
Charcoal Production															
Coking Plants		78							78	23		55		52	3
Refineries			1067						1067	31	123	913	212	378	323
Town Gas			2	4					6	1		5			5
Electricity Generation	3	27	161	5	112	11	144		463	308		155	4	93	58
Total	3	105	1230	9	112	11	144		1614	363	123	1128	216	523	389
Primary Energy Distribution and Final Demand															
Wood	3								3						3
Coal		4							4			4		2	2
Natural Gas				235					235	3		232	1	157	74
Geothermal															
Solar								7	7			7			7
Total	3	4		235				7	249	3		246	1	159	86
Total Primary Demand	6	109	1230	244	112	11	144	7	1863						
Domestic Supply	4	9	29	140	112	11		7	312						
Imports	2	100	1201	104			144		1551						

SUPPLY/DEMAND INTEGRATION WORKSHEETS

Country : Italy   Case : 1985 E

Units in 10^{12} Kcal

	Primary Energy Input									Processing and Distribution Losses	Bunkerage and changes in stocks	Final Demand			
	Wood	Coal	Oil	Natural Gas	Hydro-power	Geo-thermal	Nuclear	Solar	Total			Total	Transp.	Ind. & Agric.	Resid. Comm. & Public
Secondary Energy Conversion, Distribution and Final Demand															
Charcoal Production															
Coking Plants		109							109	32		77		74	3
Refineries			1394						1394	45	164	1185	291	527	367
Town Gas		1	2	5					8	1		7			7
Electricity Generation	3	27	327	5	112	11	144		629	418		211	4	125	81
Total	3	137	1723	10	112	11	144		2140	496	164	1480	295	726	458
Primary Energy Distribution and Final Demand															
Wood	3								3			3			3
Coal		6							6			6		4	2
Natural Gas				339					339	5		334	1	229	104
Geothermal															
Solar								2	2			2			2
Total								2	350	5					
Total Primary Demand	6	143	1723	349	112	11	144	2	2490						
Domestic Supply	4	9	29	140	112	11		2	307						
Imports	2	134	1694	209			144		2183						

SUPPLY/DEMAND INTEGRATION WORKSHEETS

Country : Italy Case : 2000 C_1

Units in 10^{12} Kcal

	Primary Energy Input									Processing and Distribution Losses	Bunkerage and changes in stocks	Final Demand			
	Wood	Coal	Oil	Natural Gas	Hydro-power	Geo-thermal	Nuclear	Solar	Total			Total	Transp.	Ind. & Agric.	Resid. Comm. & Public
Secondary Energy Conversion, Distribution and Final Demand															
Charcoal Production															
Coking Plants		122							122	35		87		86	1
Refineries		1	1431				2		1434	49	177	1208	330	708	170
Town Gas		1	3	7					11	2		9			9
Electricity Generation		219	215	8	122	29	433		1026	678		348	13	210	125
Total		343	1649	15	122	29	435		2593	764	177	1652	343	1004	305
Primary Energy Distribution and Final Demand															
Wood	2								2			2			2
Coal		4							4			4		4	
Natural Gas				439					439	8		431		315	116
Geothermal						12			12			12		10	2
Solar								105	105			105		11	94
Total	2	4		439		12		105	562			554		340	214
Total Primary Demand	2	347	1649	454	122	41	435	105	3155						
Domestic Supply	2	18	17	72	122	41		105	377						
Imports		329	1632	382			435		2778						

SUPPLY/DEMAND INTEGRATION WORKSHEETS

Country : **Italy** Case : 2000 C_2

Units in 10^{12} Kcal

	Primary Energy Input									Processing and Distribution Losses	Bunkerage and changes in stocks	Final Demand			
	Wood	Coal	Oil	Natural Gas	Hydro-power	Geo-thermal	Nuclear	Solar	Total			Total	Transp.	Ind. & Agric.	Resid. Comm. & Public
Secondary Energy Conversion, Distribution and Final Demand															
Charcoal Production															
Coking Plants		98							98	27		71		70	1
Refineries			1387				3		1390	49	172	1169	324	671	174
Town Gas		1	3	7	.				11	2		9			9
Electricity Generation		73	100	7	112	14	936		1242	811		431	17	245	169
Total		172	1490	14	112	14	939		2741	889	172	1680	341	986	353
Primary Energy Distribution and Final Demand															
Wood	2								2			2			2
Coal		3							3			3		3	
Natural Gas				422					422	8		414		297	117
Geothermal						2			2			2		2	
Solar								53	53			53		6	47
Total	2	3		422		2		53	482	8		474		308	166
Total Primary Demand	2	175	1490	436	112	16	939	53	3223						
Domestic Supply	2	9	17	72	112	16		53	281						
Imports		166	1473	364			939		2942						

SUPPLY/DEMAND INTEGRATION WORKSHEETS

Country : Italy Case : 2000 D_7

Units in 10^{12} Kcal

	Primary Energy Input									Processing and Distribution Losses	Bunkerage and changes in stocks	Final Demand			
	Wood	Coal	Oil	Natural Gas	Hydro-power	Geo-thermal	Nuclear	Solar	Total			Total	Transp.	Ind. & Agric.	Resid. Comm. & Public
Secondary Energy Conversion, Distribution and Final Demand															
Charcoal Production															
Coking Plants		95							95	25		70		69	1
Refineries			1121				1		1122	38	133	951	212	496	243
Town Gas		1	2	5					8	2		6			6
Electricity Generation		97	87	5	122	29	289		629	413		216	8	134	74
Total		193	1210	10	122	29	290		1854	478	133	1243	220	699	324
Primary Energy Distribution and Final Demand															
Wood	2								2			2			2
Coal		3							3			3		3	
Natural Gas				322					322	6		316		208	108
Geothermal						12			12			12		10	2
Solar								84	84			84		7	77
Total	2	3		322		12		84	423	6		417		228	189
Total Primary Demand	2	196	1210	332	122	41	290	84	2277						
Domestic Supply	2	18	17	72	122	41		84	356						
Imports		178	1193	260			290		1921						

SUPPLY/DEMAND INTEGRATION WORKSHEETS

Country : Italy Case : 2000 D_8

Units in 10^{12} Kcal

	Primary Energy Input									Processing and Distribution Losses	Bunkerage and changes in stocks	Final Demand			
	Wood	Coal	Oil	Natural Gas	Hydro-power	Geo-thermal	Nuclear	Solar	Total			Total	Transp.	Ind. & Agric.	Resid. Comm. & Public
Secondary Energy Conversion, Distribution and Final Demand															
Charcoal Production															
Coking Plants		83							83	24		59		58	1
Refineries			1102				2		1104	37	131	936	209	485	242
Town Gas		1	2	5					8	2		6			6
Electricity Generation		37	73	5	112	14	577		818	538		280	10	151	119
Total		121	1177	10	112	14	579		2013	601	131	1281	219	694	368
Primary Energy Distribution and Final Demand															
Wood	2								2			2			2
Coal		2							2			2			2
Natural Gas				311					311	4		307		202	105
Geothermal						2			2			2		2	
Solar								43	43			43		4	39
Total	2	2		311		2		43	360	4		356		208	148
Total Primary Demand	2	123	1177	321	112	16	579	43	2373						
Domestic Supply	2	9	17	72	112	16		43	271						
Imports		114	1160	249			579		2102						

13 Japan

13.1 OVERVIEW AND SUMMARY

According to the WAES boundary scenario approach, the estimated volume of energy demand and supply ranges from about 550×10^{13} kcal to 750×10^{13} kcal in 1985 and from 850×10^{13} kcal to 1300×10^{13} kcal in 2000 due to scenario assumptions.

Since Japan's domestic fossil fuel and hydroelectricity supply is very limited and future supplies will be more limited whatever the scenario assumptions are, the increased demand would mean a greater requirement for fossil fuel import and continued expansion of nuclear power plant capacity.

Since only one figure for the installed capacity of nuclear power plants (30 GW) is assumed for all 1985 cases, fossil fuel import dependency will be greatest in the high economic growth (and thus high demand) cases. Thus, in 1985 Case B (the lowest demand case) the import share of total energy supply is 82.5 percent, while in Case E (the largest demand case) the import share is 87.0 percent.

Different nuclear power plant installed capacities are assumed for the 2000 cases. The low demand and high nuclear capacity case (D-8) shows the smallest import dependency. The high demand cases, C-1 and C-2, are estimated to have a fossil fuel import dependency of 85 percent and 82 percent, respectively, and the low demand cases, D-7 and D-8, 77 percent and 72 percent, respectively.

Oil will remain the most important fossil fuel import, but the volumes of gas and steam coal imports are expected to increase rapidly. The desired oil import is estimated to range from 350×10^{13} kcal to 530×10^{13} kcal in 1985 and from 470×10^{13} kcal to 900×10^{13} kcal in 2000, according to the WAES boundary approach.

13.2 METHODOLOGY

We have developed an energy demand forecasting model by which supply-demand integration is carried out. The schema of this model is shown in chapter 9 of the WAES first technical report.

In this model the estimated final energy demand is divided into two categories, substitutable and nonsubstitutable.

The energy sector is assumed to supply the nonsubstitutable final energy by converting primary energy. In this conversion process, oil by-products such as fuel oil are necessarily produced. The yield of by-products depends on the structure of the conversion processes. The produced by-products are used to fill the demand for substitutable energy. If there is a shortage, it will be covered by import products; if there is a surplus, energy will be exported. The current yield pattern of oil refineries is assumed to remain substantially the same.

In the electricity-producing sector, a thermal efficiency of 38 percent is assumed for fossil fuel thermal power plants and 35 percent for nuclear, geothermal, and hydro power plants.

Projected indigenous energy supplies in Japan are relatively small. Domestic supplies of fossil fuels, hydro power, and geothermal power are the same in all cases. Figures for liquefied natural gas (LNG) are also given exogenously. Figures for nuclear energy and coal import in 2000 differ according to scenario assumptions. The oil import is calculated as the difference between the estimated demand and the exogenously given primary energy.

13.3 ENERGY DEMAND 2000

The gross national product of Japan has grown with an elasticity of 1.9 to the world GNP growth rate in the last 15 years. The Japanese economy will probably grow at a slower rate in the future, with the elasticity dropping to around 1.3 in the next decade, and then to 0.9. Accordingly, the Japanese GNP is assumed to grow by 8.0 percent per year between 1977 and 1985, which corresponds to the WAES high growth case of 6.0 percent per year. And for the WAES low growth case of 3.5 percent, a growth rate of 4.5 percent is assumed for Japan. Between 1985 and 2000, it is assumed that Japan will grow by 4.5 percent and 2.7 percent per year corresponding to the WAES high growth case of 5.0 percent and the low growth case of 3.0 percent. The assumed GNP and its components are summarized in Table 13.1.

Based on this economic forecast, an energy demand projection was made. The relations between GNP, price, and national response and their complex effect on energy demand are difficult to sepa-

Table 13.1
GNP assumptions (1970 price, 10^9 Yen)

	1970	1975 (estimate)	1977 (estimate)	1985 A, C	1985 E	1985 B, D	2000 C-2	2000 D-8
GNP	70,634.5	90,183.0	104,200	192,870	192,870	148,170	373,200	220,920
Private consumption	36,258.9	48,515.8	53,660	96,030	97,390	82,240	216,460	134,760
Government consumption	5,796.1	7,603.0	8,330	12,150	12,540	11,850	29,860	22,090
Housing investment	4,760.9	6,392.0	8,330	15,820	16,390	10,370	24,260	12,150
Private investment	14,195.3	14,928.3	17,710	36,650	36,650	20,740	51,130	24,310
Government investment	5,811.2	8,841.0	10,420	19,290	20,250	12,590	26,120	15,470
Stock investment	3,031.1	616.0	3,130	5,210	4,820	4,450	6,720	5,520
Export	8,271.7	14,032.3	16,670	36,650	36,650	26,670	82,090	41,970
Import	7,490.7	10,745.0	14,050	28,930	31,820	20,740	63,440	35,350

rate theoretically and incorporate into the projection model. (See chapter 9 of the WAES first technical report for further details on this point as well as for an explanation of the definition of national policy response.) The estimated volume of energy demand ranges from about 550×10^{13} kcal to 750×10^{13} kcal in 1985 and from 850×10^{13} to 1300×10^{13} kcal in 2000 according to the scenario assumptions.

The results of the energy demand projection as well as the details by sectors for 2000 are summarized in the worksheets (section 13.6). For 1972 and 1985 the information is given in the first technical report.

13.4 SUPPLY-DEMAND INTEGRATIONS

Estimates for energy supply and supply-demand integrations for all cases are summarized in the worksheets (section 13.6).

13.4.1 Effects of Scenario Assumptions

An installed nuclear power plant capacity of 75 GW is assumed in the low nuclear growth case and 120 GW in the high nuclear growth case. For Cases C-1 and C-2 (high energy demand) this assumed nuclear capacity will bring a reasonable balance to the electricity supply system.

For Cases D-7 and D-8 (low demand) the projected nuclear capacity gives too high a share to nuclear power plants, especially for Case D-8. The nuclear figure for Case D-8 has therefore been changed to 100 GW (i.e., 20 GW less than Case C-2).

The installed capacity of coal power plants is assumed to be 40 GW for the high economic growth, high coal case (C-1) and 20 GW for the high economic growth, low coal case (C-2). The assumption for Case D-7 (low economic growth, high coal) is 20 GW and that for D-8 (low economic growth, low coal) is zero because demand is lower in this case. This might be unrealistic, and the supply-demand integration figures in Cases D-7 and D-8 are therefore somewhat artificial.

13.4.2 Implications

The implications of the supply-demand integration figures are quite different for cases of high economic growth and for cases of low economic growth because the estimated energy demand differs greatly under these two assumptions. In the high economic growth cases (C-1 and C-2), demand is estimated to be around 1300×10^{13} kcal (1300 MTOE), which is about 50 percent higher than that of the low economic growth cases (D-7 and D-8) (850×10^{13} kcal).

In the low economic growth cases the vigorous nuclear or vigorous coal and supervigorous conservation policies of Cases C-1 and C-2 might not be necessary. Rather, moderate development of nuclear and coal power plants would be able to cover electricity demand, and oil imports at about the same level as 1985 would be sufficient to cover the projected oil demand. In short, Japan would not face serious supply problems in the low economic growth cases.

In cases of high economic growth, even with a vigorous nuclear policy accompanied by both vigorous coal and vigorous conservation policies, Japan will face a difficult situation in terms of the availability of desired oil imports.

The desired oil import in the high economic growth cases (C-1, C-2) is estimated at approximately 900×10^{13} kcal (900 MTOE = 18 MBD including bunker demand). If oil imports fall short of demand by between 200 and 300×10^{13} kcal, a more stringent conservation effort might be effective. However, if the oil import shortage should exceed this range, even a combination of vigorous nuclear, vigorous coal, and supervigorous conservation policies would not be able to close the gap. In this situation, one realistic way to cope with the shortage would be to assume a lower economic growth rate than the originally assumed high rate of 4.5 percent between 1985 and 2000.

13.5 NATIONAL RESPONSE

The WAES global supply-demand integration analysis to 2000 shows significant prospective shortages of oil. Nations can respond with different measures to reduce such prospective shortages. The Japanese national response is presented here.

Considering this picture of future world oil supply, it is estimated that a GNP of 302×10^{13} Yen (1970 prices) for Japan (about 20 percent less than that of the originally projected 373×10^{12} Yen) would be reasonable for the high economic growth case. This reduces the annual economic growth rate in Japan to approximately 3 percent between 1985 and 2000.

In addition to a slowing down of economic growth, a more vigorous conservation effort will also be needed as part of the national response. A primary energy demand reduction of about 15 percent in place of the original estimate of about 8 percent is assumed to occur through additional conservation efforts and structural changes.

In the low economic growth cases, no adjustment to the economic growth assumption is required, but a more vigorous conservation effort is needed. Therefore a supervigorous (rather than vigorous) conservation policy is assumed in Cases D-7 and D-8.

With this change in assumptions, total primary energy demand is estimated to decrease to 938×10^{13} kcal by 2000 in Revised Cases C-1 and C-2 in place of the original 1300×10^{13} kcal, and to 780×10^{13} kcal in Revised Cases D-7 and D-8 in place of the original 850×10^{13} kcal. Some of the specifics of these changes are shown in the Revised Case worksheets in section 13.6.

It should be emphasized that for Japan the industrial-structural change to a less energy-intensive type will bring about a significant reduction in energy demand.

In the transportation sector, conservation is assumed to occur only through a change in transportation modes. Any improvement in energy efficiency in this sector will be very limited and will be offset by antipollution measures, which require more energy.

Because of the total energy demand reduction, especially the oil demand reduction, it is estimated that oil imports in the year 2000 will be about 9.9 MBD in Revised Cases C-1 and C-2 and 8.3 MBD in Revised Cases D-7 and D-8. The desired oil import in the high growth cases, including the responses described above, is about 6 MBD lower than desired level of import excluding national policy response and remains close to the level of 1985 Case C.

13.6 WORKSHEETS

This section consists of summary tables and worksheets presented in the following order:

Energy demand summary 1972–1985–2000

Energy supply summary 1972–2000

Energy supply/demand integration summary—Imports and exports of primary energy

National input worksheets for supply/demand integrations (14 sheets, 1 each for 1972, for 1985 Cases A–E, and for 2000 Cases C-1, C-2, D-7, D-8, and Revised Cases C-1, C-2, D-7, and D-8)

National energy demand projections (2 sheets)

National response worksheets (2 sheets, 1 each for Cases C-1 and D-8)

Demand reduction modules efficiency (2 sheets, 1 each for Cases C and D)

WAES NATIONAL ENERGY DEMAND STUDIES, 1972-1985-2000
ENERGY DEMAND SUMMARY

Country: Japan

Scenario Assumptions	1972	1985					2000			
WAES CASE	Base Year	A	B	C	D	E	C-1	C-2	D-7	D-8
GNP 1976-85; 1985-2000		High	Low	High	Low	High	High	High	Low	Low
Oil/Energy Price*		11.50-17.25	11.50-17.25	11.50	11.50	11.50-7.66	11.50-17.25	11.50-17.25	11.50	11.50
National Policy Response		Vig	Vig	Vig	Res	Res	Vig	Vig	Vig	Vig
Principal Replacement Fuel							Coal	Nuclear	Coal	Nuclear
Primary Energy Demand **Units** 10^{13} Kcal										
Oil	261.6	467.2	351.4	492.3	372.8	527.4	924.9	890.9	502.6	490.7
Natural Gas	4.0	48.4	45.6	48.4	45.6	48.4	84.0	84.0	84.0	84.0
Coal	57.2	91.2	73.6	91.2	73.6	91.2	136.4	113.2	106.9	83.6
Nuclear	2.3	45.1	45.1	45.1	45.1	45.1	105.0	167.0	105.0	150.2
Hydroelectric	21.3	29.2	28.4	29.1	28.4	29.1	44.4	44.4	42.1	42.1
Geothermal	-	1.0	1.0	1.0	1.0	1.0	4.8	4.8	4.8	4.8
Solar Thermal										
Solar Electric										
TOTAL PRIMARY ENERGY DEMAND	346.4	682.0	545.1	707.1	566.5	742.2	1299.5	1304.3	845.4	855.4

*1975 US dollars per barrel of Arabian light crude oil, fob Persian Gulf.
All Figures are primary production prior to any processing or conversion.

WAES NATIONAL ENERGY SUPPLY STUDIES, 1972-2000
ENERGY SUPPLY SUMMARY

Country: Japan

Scenario Assumptions	1972	1985					2000			
WAES CASE	Base Year	A	B	C	D	E	C-1	C-2	D-7	D-8
GNP 1976-85; 1985-2000		High	Low	High	Low	High	High	High	Low	Low
Oil/Energy Price*		11.50-17.25	11.50-17.25	11.50	11.50	11.50-7.66	11.50-17.25	11.50-17.25	11.50	11.50
National Policy Response		Vig	Vig	Vig	Res	Res	Vig	Vig	Vig	Vig
Principal Replacement Fuel							Coal	Nuclear	Coal	Nuclear
Energy Supply **Units** 10^{13} Kcal										
Crude Oil (includes imports of products)	261.6	467.2	351.4	492.3	372.8	527.4	924.9	890.9	502.6	490.7
Natural Gas	4.0	48.4	45.6	48.4	45.6	48.4	84.0	84.0	84.0	84.0
Coal	57.2	91.2	73.6	91.2	73.6	91.2	136.4	113.2	106.9	83.6
Nuclear	2.3	45.1	45.1	45.1	45.1	45.1	105.0	167.0	105.0	150.2
Hydro-electric	21.3	29.1	28.4	29.1	28.4	29.1	44.4	44.4	42.1	42.1
Geothermal		1.0	1.0	1.0	1.0	1.0	4.8	4.8	4.8	4.8
Solar Thermal										
Solar Electric										
Total Energy Supply	346.4	682.0	545.1	707.1	566.5	742.2	1299.5	1304.3	845.4	855.4

*1975 US dollars per barrel of Arabian light crude oil, fob Persian Gulf.
All Figures are primary production prior to any processing or conversion.

WAES NATIONAL ENERGY SUPPLY/DEMAND INTEGRATION STUDIES, 1972-1985-2000
ENERGY SUPPLY/DEMAND INTEGRATION SUMMARY—IMPORTS AND EXPORTS OF PRIMARY ENERGY

Country: JAPAN

Scenario Assumptions	1972	1985					2000			
WAES CASE	Base Year	A	B	C	D	E	C-1	C-2	D-7	D-8
GNP 1976-85; 1985-2000		High	Low	High	Low	High	High	High	Low	Low
Oil/Energy Price*		11.50-17.25	11.50-17.25	11.50	11.50	11.50-7.66	11.50 17.25	11.50 17.25	11.50	11.50
National Policy Response		Vig	Vig	Vig	Res	Res	Vig	Vig	Vig	Vig
Principal Replacement Fuel							Coal	Nuclear	Coal	Nuclear
(Import) & Export **Units** 10^{13}Kcal		"	"	"	"	"	"	"	"	"
Oil	(260.8)	(465.8)	(350.0)	(490.9)	(371.4)	(526.0)	(904.9)	(870.9)	(482.6)	(470.7)
Natural Gas	(1.3)	(44.6)	(41.9)	(44.6)	(41.9)	(44.6)	(80.0)	(80.0)	(80.0)	(80.0)
Coal	(38.9)	(75.5)	(57.8)	(75.5)	(57.8)	(75.4)	(I20.6)	(97.4)	(91.1)	(67.8)
Nuclear										
Hydroelectric										
Total Net (Imports) or Exports	(301.0)	(585.9)	(449.7)	(611.0)	(471.1)	(646.0)	(1105.5)	(1048.3)	(653.7)	(618.5)

*1975 US dollars per barrel of Arabian light crude oil, fob Persian Gulf.
All Figures are primary production prior to any processing or conversion.

NATIONAL INPUT WORKSHEET FOR SUPPLY/DEMAND INTEGRATION

Country: JAPAN

Year: 1972

Case:

Units: 10^{13}Kcal

	(a)	(b)	(c)	(d)	(e)	(f)	(g)	(h)	(i)	(j)	(k)
	Coal	Petro-leum	(Syn. Liquids)	Nat. Gas	(Syn. Gas)	(Heat)	(Electri-city)[††]	Hydro Energy	Nuclear Energy	Geothermal energy and other	Total
(1) Transport		42.0					1.0				43.0
1' Bunker		12.1									12.1
(2) Industry	43.4	74.6		1.7	0.4		16.8				136.9
(3) Agric., Mining, Construction	0.6	11.5			0.1		0.7				12.9
(4) Commercial) (5) Public	0.9	16.7			1.0		5.0				23.6
(6) Residential	0.6	15.0			3.9		3.9				23.4
(7) Non-energy Uses	-	3.5									3.5
7' Chemical feed-stocks	-	23.8									23.8
(8) Final Energy Demand*	45.5	199.2		1.7	5.4		27.4				279.2
(9) Electricity**	5.7	51.1		1.4			(-30.4)	21.3	2.3		51.4
(10) Syn. Gas	2.5	2.6		0.9	(-6.0)						-
(11) Syn. Liquids											
(12) Heat											
(13) Energy Sector Self Consumption & conversion losses	3.5	8.7			0.6		3.0				15.8
(14) Primary Energy Input	57.2	261.6		4.0				21.3	2.3		346.4
(15) Indigenous Supply	18.3	0.8		2.7				21.3	2.3		45.4
(16 Imports***	38.9	260.8		1.3							301.0
(17) Exports											

*Includes non-energy uses.

**Blocks 9g, 10e 11c, and 12f must be negative to avoid double counting of energy from primary fuels.

***Includes imported products.

††Includes only electricity to be purchased by each sector.

NATIONAL INPUT WORKSHEET FOR SUPPLY/DEMAND INTEGRATION

Country: JAPAN

Year: 1985

Case: A

Units: 10^{13}Kcal

	(a)	(b)	(c)	(d)	(e)	(f)	(g)	(h)	(i)	(j)	(k)
	Coal	Petro-leum	(Syn. Liquids)	Nat. Gas	(Syn. Gas)	(Heat)	(Electri-city)††	Hydro Energy	Nuclear Energy	Geothermal energy and other	Total
(1) Transport		71.9					2.3				74.2
1' Bunker		41.7									41.7
(2) Industry	75.5	157.7		2.8	0.7		34.9				271.6
(3) Agric., Mining, Construction		17.0					1.0				18.0
(4) Commercial) (5) Public		39.9			2.3		7.3				49.5
(6) Residential		26.8			8.3		9.2				44.3
(7) Non-energy Uses		11.1									11.1
~~7' Chemical feed-stocks~~		~~38.2~~									~~38.2~~
(8) Final Energy Demand*	75.5	404.3		2.8	11.3		54.7				548.6
(9) Electricity**	11.0	42.9		36.6			(-60.8)	29.1	45.1	1.0	104.9
(10) Syn. Gas	1.2	3.4		9.0	(-12.6)						1.0
(11) Syn. Liquids											
(12) Heat											
(13) Energy Sector Self Consumption & conversion losses	3.5	16.6			1.3		6.1				27.5
(14) Primary Energy Input	91.2	467.2		48.4				29.1	45.1	1.0	682.0
(15) Indigenous Supply	15.7	1.4		3.8				29.1	45.1	1.0	96.1
(16 Imports***	75.5	465.8		44.6							585.9
(17) Exports											

*Includes non-energy uses.

**Blocks 9g, 10e 11c, and 12f must be negative to avoid double counting of energy from primary fuels.

***Includes imported products.

††Includes only electricity to be purchased by each sector.

NATIONAL INPUT WORKSHEET FOR SUPPLY/DEMAND INTEGRATION

Country: JAPAN

Year: 1985

Case: B

Units: 10^{13}Kcal

	(a) Coal	(b) Petro- leum	(c) (Syn. Liquids)	(d) Nat. Gas	(e) (Syn. Gas)	(f) (Heat)	(g) (Electri- city)††	(h) Hydro Energy	(i) Nuclear Energy	(j) Geothermal energy and other	(k) Total
(1) Transport		68.2					1.9				70.1
1' Bunker		30.1									30.1
(2) Industry	61.0	111.1		2.7	0.5		25.6				200.9
(3) Agric., Mining, Construction		14.8					0.8				15.6
(4) Commercial) (5) Public		38.2			1.8		5.7				45.7
(6) Residential		23.4			7.0		8.0				38.4
(7) Non-energy Uses		9.0									9.0
7' Chemical feed-stock		30.6									30.6
(8) Final Energy Demand*	61.0	325.4		2.7	9.3		42.0				440.4
(9) Electricity**	8.8	8.7		36.6			(-46.7)	28.4	45.1	1.0	81.9
(10) Syn. Gas	1.2	3.9		6.3	(-10.3)						1.1
(11) Syn. Liquids											
(12) Heat											
(13) Energy Sector Self Consumption & conversion losses	2.6	13.4			1.0		4.7				21.7
(14) Primary Energy Input	73.6	351.4		45.6				28.4	45.1	1.0	545.1
(15) Indigenous Supply	15.8	1.4		3.7				28.4	45.1	1.0	95.4
(16 Imports***	57.8	350.0		41.9							449.7
(17) Exports											

*Includes non-energy uses.

**Blocks 9g, 10e 11c, and 12f must be negative to avoid double counting of energy from primary fuels.

***Includes imported products.

††Includes only electricity to be purchased by each sector.

NATIONAL INPUT WORKSHEET FOR SUPPLY/DEMAND INTEGRATION

Country: JAPAN

Year: 1985

Case: C

Units: 10^{13}Kcal

	(a)	(b)	(c)	(d)	(e)	(f)	(g)	(h)	(i)	(j)	(k)
	Coal	Petro-leum	(Syn. Liquids)	Nat. Gas	(Syn. Gas)	(Heat)	(Electri-city)††	Hydro Energy	Nuclear Energy	Geothermal energy and other	Total
(1) Transport		84.1					2.2				86.3
1' Bunker		41.7									41.7
(2) Industry	75.5	164.0		2.8	0.7		35.7				278.7
(3) Agric., Mining, Construction		17.3					1.0				18.3
(4) Commercial) (5) Public)		41.3			2.3		7.6				51.2
(6) Residential		28.0			8.3		9.5				45.8
(7) Non-energy Uses		11.1									11.1
7' Chemical feed-stock		38.2									38.2
(8) Final Energy Demand*	75.5	425.7		2.8	11.3		56.0				571.3
(9) Electricity**	11.0	46.6		36.6			(-62.2)	29.1	45.1	1.0	107.2
(10) Syn. Gas	1.2	3.4		9.0	(-12.6)						1.0
(11) Syn. Liquids											
(12) Heat											
(13) Energy Sector Self Consumption & conversion losses	3.5	16.6			1.3		6.2				27.6
(14) Primary Energy Input	91.2	492.3		48.4				29.1	45.1	1.0	707.1
(15) Indigenous Supply	15.7	1.4		3.8				29.1	45.1	1.0	96.1
(16 Imports***	75.5	490.9		44.6							611.0
(17) Exports											

*Includes non-energy uses.

**Blocks 9g, 10e 11c, and 12f must be negative to avoid double counting of energy from primary fuels.

***Includes imported products.

††Includes only electricity to be purchased by each sector.

NATIONAL INPUT WORKSHEET FOR SUPPLY/DEMAND INTEGRATION

Country: JAPAN

Year: 1985

Case: D

Units: 10^{13}Kcal

	(a) Coal	(b) Petro-leum	(c) (Syn. Liquids)	(d) Nat. Gas	(e) (Syn. Gas)	(f) (Heat)	(g) (Electri-city)††	(h) Hydro Energy	(i) Nuclear Energy	(j) Geothermal energy and other	(k) Total
(1) Transport		74.4					1.7				76.1
1' Bunker		30.1									30.1
(2) Industry	61.0	118.0		2.7	0.5		26.4				208.6
(3) Agric., Mining, Construction		15.2					0.9				16.1
(4) Commercial (5) Public		40.5			1.8		6.0				48.3
(6) Residential		25.1			7.0		8.4				40.5
(7) Non-energy Uses		9.0									9.0
7' Chemical feed-stock		30.6									30.6
(8) Final Energy Demand*	61.0	342.9		2.7	9.3		43.4				459.3
(9) Electricity**	8.8	12.6		36.6			(-48.2)	28.4	45.1	1.0	84.3
(10) Syn. Gas	1.2	3.9		6.3	(-10.3)						1.1
(11) Syn. Liquids											
(12) Heat											
(13) Energy Sector Self Consumption & conversion losses	2.6	13.4			1.0		4.8				21.8
(14) Primary Energy Input	73.6	372.8		45.6				28.4	45.1	1.0	566.5
(15) Indigenous Supply	15.8	1.4		3.7				28.4	45.1	1.0	95.4
(16 Imports***	57.8	371.4		41.9							471.1
(17) Exports											

*Includes non-energy uses.

**Blocks 9g, 10e 11c, and 12f must be negative to avoid double counting of energy from primary fuels.

***Includes imported products.

††Includes only electricity to be purchased by each sector.

NATIONAL INPUT WORKSHEET FOR SUPPLY/DEMAND INTEGRATION

Country: JAPAN

Year: 1985

Case: E

Units: 10^{13}Kcal

	(a) Coal	(b) Petroleum	(c) (Syn. Liquids)	(d) Nat. Gas	(e) (Syn. Gas)	(f) (Heat)	(g) (Electricity)††	(h) Hydro Energy	(i) Nuclear Energy	(j) Geothermal energy and other	(k) Total
(1) Transport		92.3					2.0				94.3
1' Bunker		43.5									43.5
(2) Industry	75.5	175.9		2.8	0.7		37.3				292.2
(3) Agric., Mining, Construction		17.8					1.0				18.8
(4) Commercial) (5) Public		43.8			2.2		8.0				54.0
(6) Residential		30.6			8.5		10.2				49.3
(7) Non-energy Uses		11.1									11.1
7' Chemical feed-stock		38.2									38.2
(8) Final Energy Demand*	75.5	453.2		2.8	11.4		58.5				601.4
(9) Electricity**	11.0	54.0		36.6			(-65.0)	29.1	45.1	1.0	111.8
(10) Syn. Gas	1.2	3.6		9.0	(-12.7)						1.1
(11) Syn. Liquids											
(12) Heat											
(13) Energy Sector Self Consumption & conversion losses	3.5	16.6			1.3		6.5				27.9
(14) Primary Energy Input	91.2	527.4		48.4				29.1	45.1	1.0	742.2
(15) Indigenous Supply	15.8	1.4		3.8							96.2
(16 Imports***	75.4	526.0		44.6							646.0
(17) Exports											

*Includes non-energy uses.

**Blocks 9g, 10e 11c, and 12f must be negative to avoid double counting of energy from primary fuels.

***Includes imported products.

††Includes only electricity to be purchased by each sector.

NATIONAL INPUT WORKSHEET FOR SUPPLY/DEMAND INTEGRATION

Country: JAPAN

Year: 2000

Case: C-1

Units: 10^{13}Kcal

	(a) Coal	(b) Petro-leum	(c) (Syn. Liquids)	(d) Nat. Gas	(e) (Syn. Gas)	(f) (Heat)	(g) (Electri-city)††	(h) Hydro Energy	(i) Nuclear Energy	(j) Geothermal energy and other	(k) Total
(1) Transport		171.1					4.8				175.9
1' Bunker		92.3									92.3
(2) Industry	85.1	348.9		3.0	0.4		58.9				496.3
(3) Agric., Mining, Construction		25.1					1.5				26.6
(4) Commercial, (5) Public		84.6			4.4		22.5				111.5
(6) Residential		62.0			19.6		15.3				96.9
(7) Non-energy Uses		12.8									12.8
7' Chemical feed-stock		41.8									41.8
(8) Final Energy Demand*	85.1	838.6		3.0	24.4		103.0				1054.1
(9) Electricity**	46.5	50.9		60.1			(-114.4)	44.4	105.0	4.8	197.3
(10) Syn. Gas	1.2	6.5		20.9	(-27.1)						1.5
(11) Syn. Liquids											
(12) Heat											
(13) Energy Sector Self Consumption & conversion losses	3.6	28.9			2.7		11.4				46.6
(14) Primary Energy Input	136.4	924.9		84.0				44.4	105.0	4.8	1299.5
(15) Indigenous Supply	15.8	20.0		4.0				44.4	105.0	4.8	194.0
(16 Imports***	120.6	904.9		80.0							1105.5
(17) Exports											

*Includes non-energy uses.

**Blocks 9g, 10e 11c, and 12f must be negative to avoid double counting of energy from primary fuels.

***Includes imported products.

††Includes only electricity to be purchased by each sector.

NATIONAL INPUT WORKSHEET FOR SUPPLY/DEMAND INTEGRATION

Country: JAPAN

Year: 2000

Case: C-2

Units: 10^{13}Kcal

	(a)	(b)	(c)	(d)	(e)	(f)	(g)	(h)	(i)	(j)	(k)
	Coal	Petro-leum	(Syn. Liquids)	Nat. Gas	(Syn. Gas)	(Heat)	(Electri-city)††	Hydro Energy	Nuclear Energy	Geothermal energy and other	Total
(1) Transport		171.1					4.8				175.9
1' Bunker		92.3									92.3
(2) Industry	85.1	348.9		3.0	0.4		58.9				496.3
(3) Agric., Mining, Construction		25.1					1.5				26.6
(4) Commercial) (5) Public		84.6			4.4		22.5				111.5
(6) Residential		62.0			19.6		15.3				96.9
(7) Non-energy Uses		12.8									12.8
7' Chemical feed-stock		41.8									41.8
(8) Final Energy Demand*	85.1	838.6		3.0	24.4		103.0				1054.1
(9) Electricity**	23.3	16.9		60.1			(-114.4)	44.4	167.0	4.8	202.1
(10) Syn. Gas	1.2	6.5		20.9	(-27.1)						1.5
(11) Syn. Liquids											
(12) Heat											
(13) Energy Sector Self Consumption & conversion losses	3.6	28.9			2.7		11.4				46.6
(14) Primary Energy Input	113.2	890.9		84.0				44.4	167.0	4.8	1304.3
(15) Indigenous Supply	15.8	20.0		4.0				44.4	167.0	4.8	256.0
(16 Imports***	97.4	870.9		80.0							1048.3
(17) Exports											

*Includes non-energy uses.

**Blocks 9g, 10e 11c, and 12f must be negative to avoid double counting of energy from primary fuels.

***Includes imported products.

††Includes only electricity to be purchased by each sector.

NATIONAL INPUT WORKSHEET FOR SUPPLY/DEMAND INTEGRATION

Country: JAPAN

Year: 2000

Case: D-7

Units: 10^{13}Kcal

	(a) Coal	(b) Petro-leum	(c) (Syn. Liquids)	(d) Nat. Gas	(e) (Syn. Gas)	(f) (Heat)	(g) (Electri-city)[††]	(h) Hydro Energy	(i) Nuclear Energy	(j) Geothermal energy and other	(k) Total
(1) Transport		114.3					2.7				117.0
1' Bunker		49.1									49.1
(2) Industry	80.8	136.2		46.6	0.2		36.4				300.2
(3) Agric., Mining, Construction		18.3					1.0				19.3
(4) Commercial, (5) Public		70.3			2.6		9.3				82.2
(6) Residential		43.9			12.0		14.4				70.3
(7) Non-energy Uses		9.5									9.5
7' Chemical feed-stock		39.2									39.2
(8) Final Energy Demand*	80.8	480.8		46.6	14.8		63.8				686.8
(9) Electricity**	23.3	0		21.8			(-70.9)	42.1	105.0	4.8	126.1
(10) Syn. Gas	1.2	0.2		15.6	(-16.4)						0.6
(11) Syn. Liquids											
(12) Heat											
(13) Energy Sector Self Consumption & conversion losses	1.6	21.6			1.6		7.1				31.9
(14) Primary Energy Input	106.9	502.6		84.0				42.1	105.0	4.8	845.4
(15) Indigenous Supply	15.8	20.0		4.0				42.1	105.0	4.8	191.7
(16 Imports***	91.1	482.6		80.0							653.7
(17) Exports											

*Includes non-energy uses.

**Blocks 9g, 10e 11c, and 12f must be negative to avoid double counting of energy from primary fuels.

***Includes imported products.

††Includes only electricity to be purchased by each sector.

NATIONAL INPUT WORKSHEET FOR SUPPLY/DEMAND INTEGRATION

Country: JAPAN

Year: 2000

Case: D-8

Units: 10^{13}Kcal

	(a)	(b)	(c)	(d)	(e)	(f)	(g)	(h)	(i)	(j)	(k)
	Coal	Petro-leum	(Syn. Liquids)	Nat. Gas	(Syn. Gas)	(Heat)	(Electri-city)††	Hydro Energy	Nuclear Energy	Geothermal energy and other	Total
(1) Transport		114.3					2.7				117.0
1' Bunker		49.1									49.1
(2) Industry	80.8	144.3		38.5	0.2		36.4				300.2
(3) Agric., Mining, Construction		18.3					1.0				19.3
(4) Commercial (5) Public		70.3			2.6		9.3				82.2
(6) Residential		23.9			32.0		14.4				70.3
(7) Non-energy Uses		9.5									9.5
7' Chemical feed-stock		39.2									39.2
(8) Final Energy Demand*	80.8	468.9		38.5	34.8		63.8				686.8
(9) Electricity**	0	0		4.5			(-70.9)	42.1	150.2	4.8	130.7
(10) Syn. Gas	1.2	0.2		41.0	(-38.6)						3.8
(11) Syn. Liquids											
(12) Heat											
(13) Energy Sector Self Consumption & conversion losses	1.6	21.6			3.8		7.1				34.1
(14) Primary Energy Input	83.6	490.7		84.0							855.4
(15) Indigenous Supply	15.8	20.0		4.0							236.9
(16 Imports***	67.8	470.7		80.0							618.5
(17) Exports											

*Includes non-energy uses.

**Blocks 9g, 10e 11c, and 12f must be negative to avoid double counting of energy from primary fuels.

***Includes imported products.

††Includes only electricity to be purchased by each sector.

NATIONAL INPUT WORKSHEET FOR SUPPLY/DEMAND INTEGRATION

Country: JAPAN

Year: 2000

Case: Revised C-1

Units: 10^{13}Kcal

	(a) Coal	(b) Petroleum	(c) (Syn. Liquids)	(d) Nat. Gas	(e) (Syn. Gas)	(f) (Heat)	(g) (Electricity)††	(h) Hydro Energy	(i) Nuclear Energy	(j) Geothermal energy and other	(k) Total
(1) Transport		149.1					1.2				150.3
1' Bunker		50.8									50.8
(2) Industry	82.9	134.0		23.0	0.4		58.3				298.6
(3) Agric., Mining, Construction		21.3					1.0				22.3
(4) Commercial) (5) Public		42.6			3.5		13.9				60.0
(6) Residential		33.0			18.5		23.0				74.5
(7) Non-energy Uses		11.9									11.9
7' Chemical feed-stock		40.8									40.8
(8) Final Energy Demand*	82.9	483.5		23.0	22.4		97.4				709.2
(9) Electricity**	50.6	29.7		39.9			(△ 108.2)	38.7	139.5	4.8	195.0
(10) Syn. Gas	1.2	3.8		20.9	(△24.9)						1.0
(11) Syn. Liquids											
(12) Heat											
(13) Energy Sector Self Consumption & conversion losses	2.3	16.7			2.5		10.8				32.3
(14) Primary Energy Input	137.0	533.7		83.8				38.7	139.5	4.8	937.5
(15) Indigenous Supply	14.0	8.7		3.9				38.7	139.5	4.8	209.6
(16 Imports***	123.0	525.0		79.9							727.9
(17) Exports											

*Includes non-energy uses.

**Blocks 9g, 10e 11c, and 12f must be negative to avoid double counting of energy from primary fuels.

***Includes imported products.

††Includes only electricity to be purchased by each sector.

NATIONAL INPUT WORKSHEET FOR SUPPLY/DEMAND INTEGRATION

Country: JAPAN

Year: 2000

Case: Revised C-2

Units: 10^{13}Kcal

	(a)	(b)	(c)	(d)	(e)	(f)	(g)	(h)	(i)	(j)	(k)
	Coal	Petro-leum	(Syn. Liquids)	Nat. Gas	(Syn. Gas)	(Heat)	(Electri-city)††	Hydro Energy	Nuclear Energy	Geothermal energy and other	Total
(1) Transport		149.1					1.2				150.3
1' Bunker		50.8									50.8
(2) Industry	82.9	134.0		23.0	0.4		58.3				298.6
(3) Agric., Mining, Construction		21.3					1.0				22.3
(4) Commercial) (5) Public)		42.6			3.5		13.9				60.0
(6) Residential		33.0			18.5		23.0				74.5
(7) Non-energy Uses		11.9									11.9
7' Chemical feed-stock		40.8									40.8
(8) Final Energy Demand*	82.9	483.5		23.0	22.4		97.4				709.2
(9) Electricity**	21.0	29.7		39.9			(△ 108.2)	38.7	169.1	4.8	195.0
(10) Syn. Gas	1.2	3.8		20.9	(△ 24.9)						1.0
(11) Syn. Liquids											
(12) Heat											
(13) Energy Sector Self Consumption & conversion losses	2.3	16.7			2.5		10.8				32.3
(14) Primary Energy Input	107.4	533.7		83.8				38.7	169.1	4.8	937.5
(15) Indigenous Supply	14.0	8.7		3.9				38.7	169.1	4.8	239.2
(16 Imports***	93.4	525.0		79.9							698.3
(17) Exports											

*Includes non-energy uses.

**Blocks 9g, 10e 11c, and 12f must be negative to avoid double counting of energy from primary fuels.

***Includes imported products.

††Includes only electricity to be purchased by each sector.

NATIONAL INPUT WORKSHEET FOR SUPPLY/DEMAND INTEGRATION

Country: JAPAN

Year: 2000

Case: Revised D-7

Units: 10^{13}Kcal

	(a)	(b)	(c)	(d)	(e)	(f)	(g)	(h)	(i)	(j)	(k)
	Coal	Petro-leum	(Syn. Liquids)	Nat. Gas	(Syn. Gas)	(Heat)	(Electri-city)††	Hydro Energy	Nuclear Energy	Geothermal energy and other	Total
(1) Transport		107.3					1.2				108.5
1' Bunker		45.0									45.0
(2) Industry	80.8	118.3		23.0	0.4		38.0				260.5
(3) Agric., Mining, Construction		18.0					0.9				18.9
(4) Commercial) (5) Public		41.0			2.6		11.3				54.9
(6) Residential		31.5			17.4		20.4				69.3
(7) Non-energy Uses		9.5									9.5
7' Chemical feed-stock		39.2									39.2
(8) Final Energy Demand*	80.8	409.8		23.0	20.4		71.8				605.8
(9) Electricity**	35.0	19.8		39.9			(△79.8)	35.0	91.0	4.8	145.7
(10) Syn. Gas	1.2	1.6		20.9	(△22.7)						1.0
(11) Syn. Liquids											
(12) Heat											
(13) Energy Sector Self Consumption & conversion losses	1.6	15.6			2.3		8.0				27.5
(14) Primary Energy Input	118.6	446.8		83.8				35.0	91.0	4.8	780.0
(15) Indigenous Supply	14.0	8.7		3.9				35.0	91.0	4.8	157.4
(16 Imports***	104.6	438.1		79.9							622.6
(17) Exports											

*Includes non-energy uses.

**Blocks 9g, 10e 11c, and 12f must be negative to avoid double counting of energy from primary fuels.

***Includes imported products.

††Includes only electricity to be purchased by each sector.

NATIONAL INPUT WORKSHEET FOR SUPPLY/DEMAND INTEGRATION

Country: JAPAN

Year: 2000

Case: Revised D-8

Units: 10^{13}Kcal

	(a)	(b)	(c)	(d)	(e)	(f)	(g)	(h)	(i)	(j)	(k)
	Coal	Petroleum	(Syn. Liquids)	Nat. Gas	(Syn. Gas)	(Heat)	(Electricity)††	Hydro Energy	Nuclear Energy	Geothermal energy and other	Total
(1) Transport		107.3					1.2				108.5
1' Bunker		45.0									45.0
(2) Industry	80.8	118.3		23.0	0.4		38.0				260.5
(3) Agric., Mining, Construction		18.0					0.9				18.9
(4) Commercial (5) Public)		41.0			2.6		11.3				54.9
(6) Residential		31.5			17.4		20.4				69.3
(7) Non-energy Uses		9.5									9.5
7' Chemical feed-stock		39.2									39.2
(8) Final Energy Demand*	80.8	409.8		23.0	20.4		71.8				605.8
(9) Electricity**	21.0	19.8		39.9			(△ 79.8)	35.0	105.0	4.8	145.7
(10) Syn. Gas	1.2	1.6		20.9	(△22.7)						1.0
(11) Syn. Liquids											
(12) Heat											
(13) Energy Sector Self Consumption & conversion losses	1.6	15.6			2.3		8.0				27.5
(14) Primary Energy Input	104.6	446.8		83.8				35.0	105.0	4.8	780.0
(15) Indigenous Supply	14.0	8.7		3.9				35.0	105.0	4.8	171.4
(16 Imports***	90.6	438.1		79.9							608.6
(17) Exports											

*Includes non-energy uses.

**Blocks 9g, 10e 11c, and 12f must be negative to avoid double counting of energy from primary fuels.

***Includes imported products.

††Includes only electricity to be purchased by each sector.

NATIONAL ENERGY DEMAND PROJECTIONS

Country: JAPAN

Year: 2000

WAES Case: C-1, C-2

Population: 135×10^6

No. of Autos:

Number of Residential Dwellings: 45.3×10^6

Sector	Activity Level	Intensity	Energy Use	
Transportation	Total Vehicle Distance/Yr.	Energy Efficiency	Fossil and Other Fuels	Electric[1]
Automobiles	646.9×10^6 Pass.Km	612 Kcal/Pass. Km	$395{,}903 \times 10^6$ Kcal	
Air Travel	47.9 "	842 "	40,332 "	
Other Passenger	1679.5 "	85.3 "	135,741 "	7500×10^6 Kcal
Truck & Air Freight	530.4×10^6 Ton.Km	1218 Kcal/Ton.Km	646,027 "	
Rail Freight	343.3 "	180 "	57,294 "	4500 "
Shipping	533.2 "	406 "	216,479 "	

Industry	Output/Year[1]	Energy/Output (Kcal/Yen)	Energy Use (10^9 Kcal)
Metals (Iron & Steel)	3197.4	278.1	889,109
Chemicals	9623.3	121.7	1171,164
Other Energy Intensive[2]	5532.7	119.8	663,077
Non-Energy Intensive[3]	57023.2	11.8	670,754
Petrochemical Feedstocks[4]			
Asphalt & Road Tar			
Agriculture	5403.2	20.0	107,805
Mining	304.5	47.0	14,304
Construction	14227.9	7.1	101,476

1. Given in the value added amount in price of 1965, 10^9 Yen
2. Non ferrous metals, Paper & Allied, Mineral products
3. Food & Related, Fabric, Metal products, Machinery, Textile, Misc.
4. Included in Chemicals.

NATIONAL ENERGY DEMAND PROJECTIONS (con't)

Commercial & Public	Floor Area	Heat Loss/ Area/Yr.	Syst. Eff.	Energy Use 10^9Kcal
Foss. Fuel Space Heat				599,500
Electric Space Heat				
Other Fossil Fuel[2]				
Other Electric[2]				

Residential	Dwellings	Heat Loss/ Area/Yr.	Syst. Eff.	Energy Use 10^9Kcal
Foss. Fuel Space Heat				745,000
Electric Space Heat				
Foss. Fuel Appliances[3]				
Electric Appliances[3]				
Total Delivered Energy: Demand (Including Fuel used for Feedstocks)				

[1] Purchased electricity

[2] The number for heat loss/area/year represents the total fossil fuel (electric) energy demand per unit area per year.

[3] The number for heat loss/area/year represents the total fossil fuel (electric) energy demand per dwelling.

NATIONAL ENERGY DEMAND PROJECTIONS

Country: JAPAN

Year: 2000

WAES Case: D-7, D-8

Population: 135×10^6

No. of Autos:

Number of Residential Dwellings: 45.3×10^6

Sector	Activity Level	Intensity	Energy Use	
Transportation	Total Vehicle Distance/Yr.	Energy Efficiency	Fossil and Other Fuels	Electric[1]
Automobiles	547.7×10^6 Pass.Km	612 Kcal/Pass.Km	$335{,}192 \times 10^6$ Kcal	
Air Travel	40.6 "	842 "	34,185 "	
Other Passenger	1422.1 "	85.3 "	113,778 "	7500×10^6 Kcal
Truck & Air Freight	338.4 "	1218 Kcal/Ton. Km	412,171 "	
Rail Freight	219.0 "	180 "	34,920 "	4500 "
Shipping	340.2 "	406 "	138,121 "	

Industry	Output/Year[1]	Energy/Output (Kcal/Yen)	Energy Use (10^9 Kcal)
Metals (Iron & Steel)	3,034.0	296.0	898,125
Chemicals	6,700.5	142.8	957,001
Other Energy Intensive[2]	4,091.6	129.3	529,206
Non-Energy Intensive[3]	44,770.5	13.7	613,068
Petrochemical Feedstocks[4]			
Asphalt & Road Tar			
Agriculture	4,827.6	20.7	99,746
Mining	304.5	48.5	14,761
Construction	10,223.9	7.3	75,025

1. Given in the value added amount in price of 1965, 10^9 Yen
2. Non ferrous metals, Paper & Allied, Mineral products
3. Food & Related, Fabric, Metal products, Machinery, Textile, Misc.
4. Included in Chemicals

NATIONAL ENERGY DEMAND PROJECTIONS (con't)

Commercial & Public	Floor Area	Heat Loss/ Area/Yr.	Syst. Eff.	Energy Use 10^9 Kcal
Foss. Fuel Space Heat				
Electric Space Heat				549,000
Other Fossil Fuel[2]				
Other Electric[2]				

Residential	Dwellings	Heat Loss/ Area/Yr.	Syst. Eff.	Energy Use 10^9 Kcal
Foss. Fuel Space Heat				
Electric Space Heat				693,000
Foss. Fuel Appliances[3]				
Electric Appliances[3]				
Total Delivered Energy: Demand (Including Fuel used for Feedstocks)				

[1] Purchased electricity

[2] The number for heat loss/area/year represents the total fossil fuel (electric) energy demand per unit area per year.

[3] The number for heat loss/area/year represents the total fossil fuel (electric) energy demand per dwelling.

NATIONAL RESPONSE WORKSHEET

Country: JAPAN

WAES SCENARIO CASE: C-1

Type of Response (Demand Reduction, Supply Expansion, or Fuel Substitution)	Examples of Response Action[1]	Lead Time (Years)[2]	% of Max. Potential Above Scenario Case[3]	Savings Achieved or Supply Added over Scenario Case[4] Oil Saved[5]	Oil Added	Gas Added	Other Supply Added
Demand Reduction	Slowing down of Economic growth			(3.77)			
Demand Reduction (Industry)	Slowing down of Activity and Improved efficiency			3.73			
Demand Reduction (Transport)	Slowing down of Activity and Improved efficiency			0.49			
Demand Reduction (Residential & Commercial)	Slowing down of Activity and Improved efficiency			1.39			
Demand Reduction (Energy Sector)	Slowing down of Activity and Improved efficiency			0.32			
Demand Reduction (Bunker)	Slowing down of Activity			0.73			
Supply Expansion (Nuclear)	Increase from 75 Mil. KW to 100 Mil. KW						
Supply Expansion (Coal)	Increase Steam Coal Import from 25 Milt. to 75 Milt.						

[1] Responses should reflect "supervigorous" policies; energy prices high enough to draw forth alternative supplies, and demand reductions and fuel substitutions resulting from conservation actions or improved efficiencies even at prices substantially higher than scenario prices.

[2] Number of years from start up to full implementation.

[3] Percent used of total maximum potential for each class of module.

[4] These figures are not necessarily additive; they represent components of alternative energy strategies. Units—Million Barrels a Day Oil Equivalent.

[5] All demand reduction and fuel substitution savings should be noted in "oil saved" column.

NATIONAL RESPONSE WORKSHEET

Country: JAPAN

WAES SCENARIO CASE: D-8

Type of Response (Demand Reduction, Supply Expansion, or Fuel Substitution)	Examples of Response Action[1]	Lead Time (Years)[2]	% of Max. Potential Above Scenario Case[3]	Savings Achieved or Supply Added over Scenario Case[4] Oil Saved[5]	Oil Added	Gas Added	Other Supply Added
Demand Reduction (Industry)	Improved Efficiency			0.75			
Demand Reduction (Transport)	Change of transportation module			0.16			
Demand Reduction (Residential & Commercial)	Improved Efficiency			0.54			
Supply Reduction (Nuclear)	Reduce from 100 Mil. KW to 75 Mil. KW						0.77
Supply Expansion (Coal)	Increase Steam Coal Import from 25 Milt. to 50 Milt.						0.36

[1] Responses should reflect "supervigorous" policies; energy prices high enough to draw forth alternative supplies, and demand reductions and fuel substitutions resulting from conservation actions or improved efficiencies even at prices substantially higher than scenario prices.

[2] Number of years from start up to full implementation.

[3] Percent used of total maximum potential for each class of module.

[4] These figures are not necessarily additive; they represent components of alternative energy strategies. Units—Million Barrels a Day Oil Equivalent.

[5] All demand reduction and fuel substitution savings should be noted in "oil saved" column.

DEMAND REDUCTION MODULES EFFICIENCY

CASE C JAPAN

Demand Line Items and Units	(Units)	User Performance 1972	1985	2000	"What If?" National Efficiency Response[1]	National Response Savings over Year 2000 Scenario Case	Lead Time for National Response Years[2]
Autos	(Module Share %)	34.0	29.0	27.0		Transportation as a whole	
Air Travel	(Module Share %)	2.0	2.0	2.0		0.49	
Truck Freight	(Module Share %)	44.7	39.7	37.7			
Rail Freight	(Module Share %)	17.4	22.4	24.4			
Commercial/Public Heat Loss	Kcal/GNP	2.8	2.7	2.0		Residential & Commercial as a whole 1.39	
System Efficiency (FF)							
System Efficiency (E)							
Primary Metals	(Kcal/Yen V.A.) (Iron & Steel)	328.2	296.0	278.1		Industry as a whole 3.73	
Chemicals	(Kcal/Yen V.A.) (Incld. Naptha)	201.2	180.9	121.7			
Energy Intensive Ind.	Iron & Steel, nonfer.Metal Paper,Chemical,Metal Prod.	224.7	189.5	148.3			
Non-Energy Uses (Petrochemical, Asphalt)							
Residential Heat Loss	(10^6Kcal/dwelling)	8.0	11.5	16.5			
System Efficiency (FF)							
System Efficiency (E)							

[1] Given supervigorous policy and substantially higher than case energy price. These figures are not necessarily additive: they represent components of alternative energy strategies.

[2] Number of years from start-up of national response to full implementation.

DEMAND REDUCTION MODULES EFFICIENCY

CASE D JAPAN

Demand Line Items and Units	(Units)	User Performance 1972	1985	2000	"What If?" National Efficiency Response[1]	National Response Savings over Year 2000 Scenario Case	Lead Time for National Response Years[2]
Autos	(Module Share %)	34.0	34.0	27.0		Transportation as a whole 0.16	
Air Travel	(Module Share%)	2.0	2.0	2.0			
Truck Freight	(Module Share %)	44.7	44.7	37.7			
Rail Freight	(Module Share %)	17.4	17.4	24.4			
Commercial/Public Heat Loss	(Kcal/Yen)	2.8	3.3	2.5		Residential & Commercial as a whole 0.54	
System Efficiency (FF)							
System Efficiency (E)							
Primary Metals	(Kcal/Yen V.A.) (Iron & Steel)	328.2	305.9	296.0		Industry as a whole 0.75	
Chemicals	(Kcal/Yen) (incld. Naptha)	201.2	186.9	142.8			
Energy Intensive Ind.	Iron & Steel, Nonf. Metal Paper, Chemic. Mineral Prod.	224.7	197.9	172.5			
Non-Energy Uses (Petrochemical, Asphalt)							
Residential Heat Loss	10^6 Kcal/dwelling	8.0	10.1	15.3			
System Efficiency (FF)							
System Efficiency (E)							

[1] Given supervigorous policy and substantially higher than case energy price. These figures are not necessarily additive: they represent components of alternative energy strategies.

[2] Number of years from start-up of national response to full implementation.

Mexico

14.1 OVERVIEW AND SUMMARY

14.1.1 Major Shifts among Fuels

Mexico's total energy demand rose at an annual rate of 8.2 percent between 1964 and 1974; Figure 14.1 shows the values for each year during that period. Our projection of demand growth indicates that the rate will be between 6.7 and 7.7 percent between 1985 and 2000, considering energy conservation in the industrial and transportation sectors. By 2000 our energy needs will be between 2993 and 3875 $\times$ 10^{12} kcal, five to seven times more than the 570 $\times$ 10^{12} kcal consumed in 1975 (see Figure 14.2).

The investment required to satisfy our energy needs during the period from 1976 to 2000 is estimated at 1.6 $\times$ 10^{12} pesos at 1975 prices; this figure includes the costs of exploring for and exploiting resources and installing production, transport, and distribution systems.

Because this growth rate for Mexico's energy sector is high, our strategy is to maintain economic growth while directing our efforts toward distributing wealth more equitably.

Our national energy sector is highly dependent on hydrocarbons, the best-known fuel source (see Figure 14.3). We do not expect any major changes in this composition before 2000. Our basic reserves are estimated at 54 $\times$ 10^9 barrels, and we project that hydrocarbons will provide at least 70 percent of the total supply in 2000.

14.1.2 Projections under Varying Assumptions

The C-1 and C-2 projections correspond to coal and nuclear energy, respectively, under the assumptions of high economic growth and high oil prices. The main difference between the projections in these two cases is a major shift toward consumption of electric energy in Case C-2, which assumes an acceptance of nuclear plants. Similarly, the main difference between Cases D-7 and D-8 is an increase in consumption of electricity.

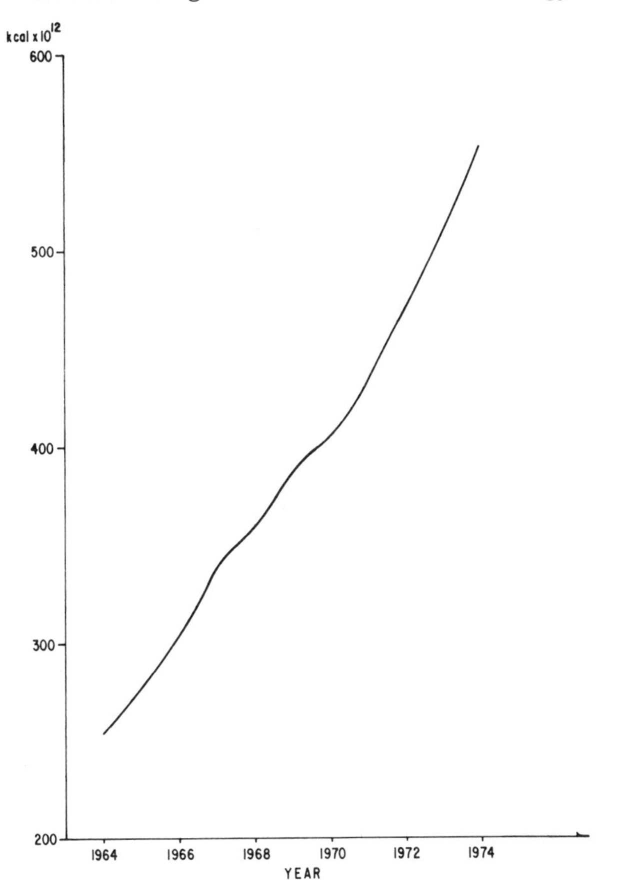

Figure 14.1
Mexican energy consumption, 1964–1974.

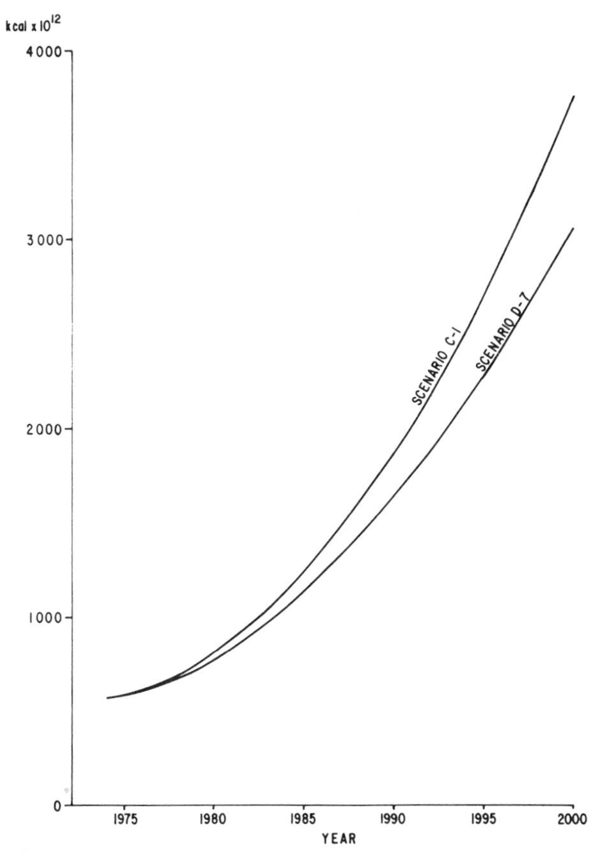

Figure 14.2
Mexican energy demand, 1975–2000.

14.1.3 National Goals

Certain national goals are implicit in these projections. Mexico has 60 million people, a very unequal distribution of wealth, and a high rate of development. Therefore, a high growth rate in energy development is essential to economic growth and industrialization.

It is plausible to assume that the energy supply structure will not change greatly in the future. Also, no important changes are expected in the composition of the consumption demand sector. Figure 14.4 shows Mexico's energy balance for 1975.

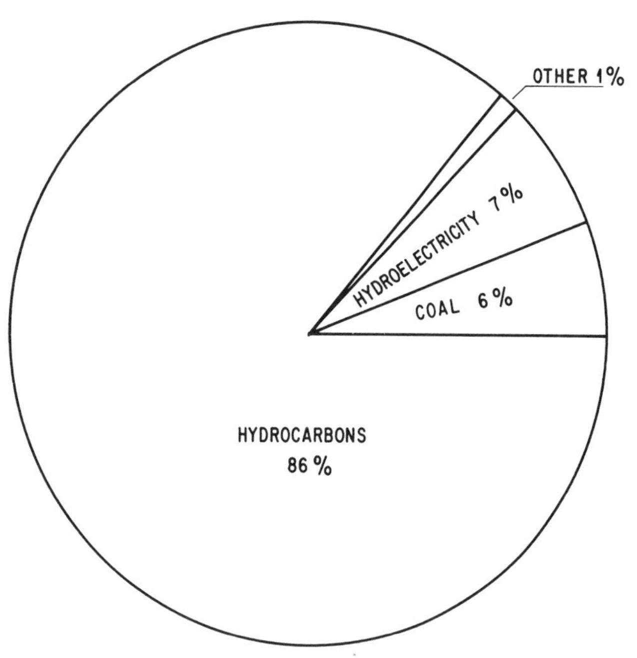

Figure 14.3
Actual present-day primary energy composition in Mexico.

14.2 METHODOLOGY

For both high and low GWP (5 and 3 percent per year, respectively) during the period 1985 to 2000, the selected growth objectives in GNP were 6.5 and 5.5 percent (Table 14.1). These estimates are consistent with those made by groups responsible for projecting the economic future of the country.

The data that form the basis of our projections are from Petróleos Mexicanos, Comisión Federal de Electricidad (CFE), Instituto de Energía Nuclear, and Comisión de Energéticos. The data were integrated through a subjective process to obtain the WAES projections.

In conducting this study, we frequently encountered deficient information, and all estimates of demand are on the aggregate level. Assuming a stable structure of the consumption sector is rea-

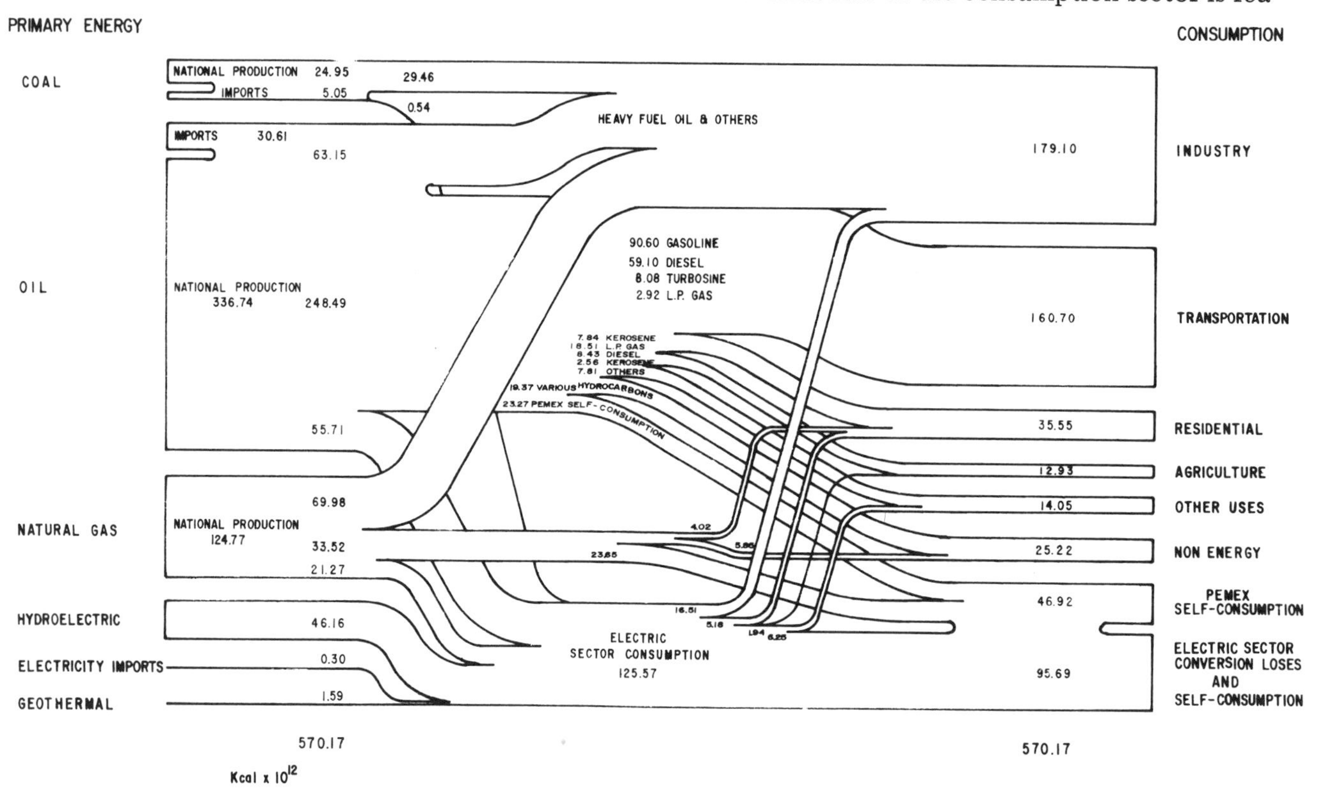

Figure 14.4
Mexican energy balance, 1975.

sonable for the relative short term but could distort long-term estimates. To more accurately estimate future needs, a major effort is needed to disaggregate current information so that the probable consumption patterns of individual units can be analyzed.

14.3 ENERGY DEMAND TO THE YEAR 2000

Our projections for the year 2000 have provided us with both indirect and direct results.

The main indirect result is that to satisfy the minimal needs of Mexico's growing population and to meet the national objective of raising the standard of living, energy consumption will have to grow at least 6.5 percent per year between 1985 and 2000.

A direct result is that hydrocarbons, the main source of energy in Mexico, will continue to provide a major share of energy demand to the year 2000, when demand is expected to vary between 4.6 and 5.6 MBD. Yet the foreseen volumes of needed energy are of such a magnitude that hydrocarbon reserves, large as they are, will not be sufficient to satisfy the needs of even a few decades of the next century. By that time the total hydroelectric potential, estimated at 60,000 GWh—equivalent to 150×10^{12} kcal (at 2500 kcal/kWh)—will have been totally developed. This represents approximately 20,000 MW of installed capacity, compared to the 3900 currently in existence, a fivefold increase over the next 24 years.

The electric energy required in the year 2000 will be between 320,000 and 470,000 GWh, eight to eleven times the 42,000 GWh generated in 1975.

Coal has provided only a small share of Mexico's energy needs because of the low prices of petroleum products. To satisfy the foreseen demand for coal in the steel as well as in the electric industries, an output of 50-70 million tons will be needed. Coal reserves are high compared to production, which may permit it to play a significant role in the future national energy market. The same is true for uranium. Both sources could substantially modify the energy supply structure.

The distribution of energy by sectors in the year 2000 is not expected to differ from that of 1975 (see Figure 14.5).

14.4 ALTERNATIVES FOR THE GENERATION OF ELECTRIC ENERGY UNDER HISTORICAL SCENARIOS

The expansion programs of the electrical sector are based on the economic analysis of feasible solutions. We assume that the sector will utilize all the hydroelectric energy that can be developed; this level is estimated by the Plan Nacional Hidráulico

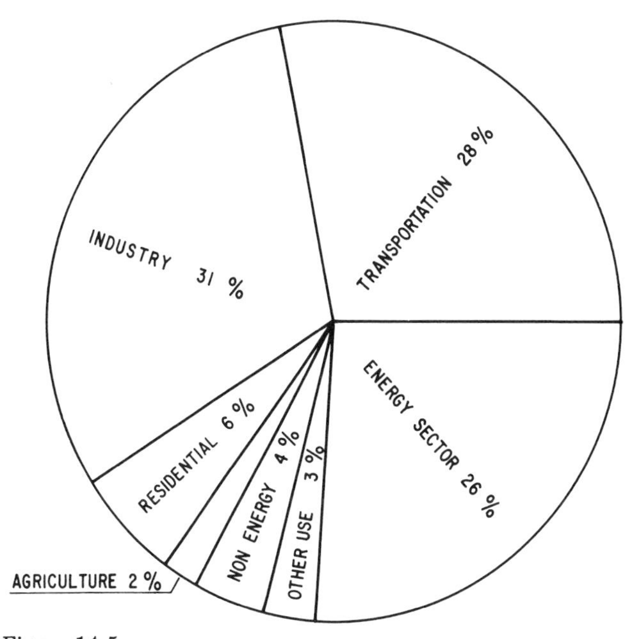

Figure 14.5
Historical distribution of energy consumption in Mexico by sectors.

Table 14.1
Scenario assumptions, 1985-2000

	World		Mexico					
Scenario	Economic growth rate (%/yr)	Energy price	Principal replacement fuel	Government response	Economic growth rate (%/yr)	Population growth rate (%/yr)	Energy growth rate (%/yr)	Electricity growth rate (%)/yr)
C-1	5	high	coal	vigorous	6.5	3.5	7.6	9.6
C-2	5	high	nuclear	vigorous	6.5	3.5	7.8	10.4
D-7	3	medium	coal	restrained	5	3.5	6.6	8.4
D-8	3	medium	nuclear	restrained	5	3.5	6.8	9

and the CFE to be 83 TWh per year. The CFE estimates that 60 TWh will be generated in the year 2000, assuming the construction of 43 additional plants. Thermal sources will probably supply most of the rest of the electric energy, with contributions from hydrocarbons, coal, and uranium. New sources, such as solar and sea power, are not expected to contribute much because they will not become economical before the end of the century.

If only carboelectric power plants were used to produce electricity, the electric sector would consume 1044 million tons of coal between 1985 and 2000, and consumption would be 134 million tons in the year 2000.

If only nuclear power plants were used, the corresponding values would be 67,000 tons of U_3O_8 by the year 2000, and 11,000 tons would be consumed in that year.

Table 14.2 shows for three alternatives the consumption of primary energy (amounts of hydrocarbons, coal, and uranium needed to produce electricity being added to the amounts needed in the nonelectric sector).

14.5 SUPPLY-DEMAND INTEGRATION

From the demand and supply projections we conclude that Mexico can satisfy its energy needs with its own resources. Mexico may be able to increase its supply of natural gas sufficiently to export marginal quantities of hydrocarbons by the year 2000.

The policy guidelines for the development of the energy sector have been and will continue to be

- Maintain self-sufficiency.
- Nationalize the service of energy supply.
- Increase supply to satisfy the needs of the rural sector as well.
- Satisfy, without restriction and at the least cost possible, the demand for energy.

14.6 NATIONAL RESPONSE

In Cases C-1, C-2, D-7, and D-8 Mexico expects no gap between supply and demand.

The energy sector is characterized primarily by the dependence of the national economy on hydrocarbons, the high economic growth rate (about 8 percent per year over several decades), and the fact that the energy supply agencies are government-owned.

With more than 60 million people, uneven wealth distribution, a large industrial infrastructure in absolute terms, and one of the highest rates of population increase, Mexico has the characteristics of both a developed and of a developing country. Slowing the country's growth is an unthinkable alternative at this time. The only feasible strategy seems to be to sustain the rate of economic growth, to improve the efficiency of energy use, and to promote better income distribution.

A comparison of demand and supply estimates shows that an energy economy dominated by hydrocarbons is feasible until the end of the century. However, the expected energy needs are of such magnitude that efforts to incorporate new energy sources should begin as early as possible.

Table 14.2
Primary energy consumption under three alternatives for electricity production

	No carboelectric or nuclear plants	Only carboelectric plants	Only nuclear plants
Accumulated consumption 1976-2000			
Hydrocarbons (10^6 BPCE)[a]	27,751	23,327	23,279
Coal (10^6 tons)[b]	540	1,507	540
Uranium (tons)[c]	5,213	5,213	67,600
Share of energy supply (%)			
Hydrocarbon	8.6	63.7	63.7
Coal	7	31.9	7.1
Hydroelectricity	4	4	4
Uranium	0.4	0.4	25.2

[a]BPCE = barrels of petroleum crude equivalent.

[b]Including Río Escondido.

[c]Consumption of Laguna Verde.

Although the best-known alternative energy source is hydroelectricity, it could supply only about 15 percent of the expected electricity demand in 2000. In addition to this source, Mexico has coal, uranium, and a potential for geothermal energy that could be important; in the longer run, the favorable geographical conditions for solar energy could make this source significant.

Coal and uranium have been considered for the "high coal" and "high nuclear" cases up to their maximum participation for electricity production. The two cases are to a certain extent mutually exclusive because electricity contributes only about 15 percent to total demand.

Diversifying energy sources will require considerable efforts, and since the alternative sources require higher investments, their selection must be analyzed with extreme care.

It may be possible to increase the supply of primary energy while reducing demand. However, considerable efforts are still needed to incorporate lower-income people into the mainstream of society, and this can be achieved only with the consumption of increasing amounts of energy.

14.7 WORKSHEETS

This section contains summary tables and worksheets in the following order:

Energy demand summary, 1972-1985-2000

1972 national input worksheet

1985 national input worksheets (5 sheets)

2000 national input worksheets (4 sheets)

National response worksheets (2 sheets)

WAES NATIONAL ENERGY DEMAND STUDIES, 1972-1985-2000
ENERGY DEMAND SUMMARY

Country: MEXICO

Scenario Assumptions		1972	1985					2000			
WAES CASE		Base Year	A	B	C	D	E	C-1	C-2	D-7	D-8
GNP 1976-85; 1985-2000			High	Low	High	Low	High	High	High	Low	Low
Oil/Energy Price*			11.50-17.25	11.50-17.25	11.50	11.50	11.50-7.66	11.50-17.25	11.50-17.25	11.50	11.50
National Policy Response			Vig	Vig	Vig	Res	Res	Vig	Vig	Vig	Vig
Principal Replacement Fuel								Coal	Nuclear	Coal	Nuclear
Primary Energy Demand	**Units** Kcal x 10^{12}										
Oil		296	913	708	817	744	730	1901	1901	1511	1511
Natural Gas		122	297	231	279	255	255	745	745	637	637
Coal		30	87	70	82	70	70	340	260	340	260
Nuclear			30	20	20	20	20	622	815	351	502
Hydroelectric		37	51	51	51	51	51	150	150	150	150
Geothermal			2	2	2	2	2	4	4	4	4
Solar Thermal											
Solar Electric											
TOTAL PRIMARY ENERGY DEMAND		485	1380	1082	1252	1142	1128	3762	3875	2993	3064

*1975 US dollars per barrel of Arabian light crude oil, fob Persian Gulf.
All Figures are primary production prior to any processing or conversion.

NATIONAL INPUT WORKSHEET FOR SUPPLY/DEMAND INTEGRATION

Country: MEXICO

Year: 1972

Case: Base year

Units: Kcal x 10^{12}

	(a)	(b)	(c)	(d)	(e)	(f)	(g)	(h)	(i)	(j)	(k)
	Coal	Petro-leum	(Syn. Liquids)	Nat. Gas	(Syn. Gas)	(Heat)	(Electri-city)[††]	Hydro Energy	Nuclear Energy	Geothermal energy and other	Total
(1) Transport		132									132
(2) Industry	29	34		71			13				147
(3) Agric., Mining, Construction		1					2				3
(4) Commercial							8				8
(5) Public							1				1
(6) Residential		42					4				46
(7) Non-energy Uses		14		4							18
(8) Final Energy Demand*	29	223		75			28				355
(9) Electricity**	1	34		16			(33)	37			55
(10) Syn. Gas											
(11) Syn. Liquids											
(12) Heat											
(13) Energy Sector Self Consumption & conversion losses		38		32			5	0	0	0	75
(14) Primary Energy Input	30	295		123			0	37	0	0	485
(15) Indigenous Supply	18	268		125			0	37	0	0	448
(16 Imports***	12	42									54
(17) Exports		15		2							17

*Includes non-energy uses.

**Blocks 9g, 10e 11c, and 12f must be negative to avoid double counting of energy from primary fuels.

***Includes imported products.

††Includes only electricity to be purchased by each sector.

NATIONAL INPUT WORKSHEET FOR SUPPLY/DEMAND INTEGRATION

Country: MEXICO

Year: 1985

Case: A

Units: Kcal x 10^{12}

	(a) Coal	(b) Petroleum	(c) (Syn. Liquids)	(d) Nat. Gas	(e) (Syn. Gas)	(f) (Heat)	(g) (Electricity)††	(h) Hydro Energy	(i) Nuclear Energy	(j) Geothermal energy and other	(k) Total
(1) Transport		355									355
(2) Industry	52	106		200			46				404
(3) Agric., Mining, Construction		3					7				10
(4) Commercial							30				30
(5) Public							3				3
(6) Residential		130					14				144
(7) Non-energy Uses		43		11							54
(8) Final Energy Demand*	52	637		211			100				1000
(9) Electricity**	35	158		24			(117)	51	30	2	181
(10) Syn. Gas											
(11) Syn. Liquids											
(12) Heat											
(13) Energy Sector Self Consumption & conversion losses		118		62			17	0	0	0	197
(14) Primary Energy Input	87	913		297			0	51	30	2	1380
(15) Indigenous Supply	70	1006		297			0	51	30	2	1456
(16 Imports***	17										17
(17) Exports		93									93

*Includes non-energy uses.

**Blocks 9g, 10e 11c, and 12f must be negative to avoid double counting of energy from primary fuels.

***Includes imported products.

††Includes only electricity to be purchased by each sector.

NATIONAL INPUT WORKSHEET FOR SUPPLY/DEMAND INTEGRATION

Country: MEXICO

Year: 1985

Case: B

Units: Kcal x 10^{12}

	(a) Coal	(b) Petroleum	(c) (Syn. Liquids)	(d) Nat. Gas	(e) (Syn. Gas)	(f) (Heat)	(g) (Electricity)††	(h) Hydro Energy	(i) Nuclear Energy	(j) Geothermal energy and other	(k) Total
(1) Transport		290									290
(2) Industry	52	64		154			37				307
(3) Agric., Mining, Construction		2					6				8
(4) Commercial							25				25
(5) Public							2				2
(6) Residential		100					12				112
(7) Non-energy Uses		34		8							42
(8) Final Energy Demand*	52	490		162			82				786
(9) Electricity**	18	133		24			(96)	51	20	2	152
(10) Syn. Gas											
(11) Syn. Liquids											
(12) Heat											
(13) Energy Sector Self Consumption & conversion losses		85		45			14	0	0	0	144
(14) Primary Energy Input	70	708		231			0	51	20	2	1082
(15) Indigenous Supply	70	720		231			0	51	20	2	1094
(16 Imports***											
(17) Exports		12									12

*Includes non-energy uses.

**Blocks 9g, 10e 11c, and 12f must be negative to avoid double counting of energy from primary fuels.

***Includes imported products.

††Includes only electricity to be purchased by each sector.

NATIONAL INPUT WORKSHEET FOR SUPPLY/DEMAND INTEGRATION

Country: MEXICO

Year: 1985

Case: C

Units: Kcal x 10^{12}

	(a) Coal	(b) Petro-leum	(c) (Syn. Liquids)	(d) Nat. Gas	(e) (Syn. Gas)	(f) (Heat)	(g) (Electri-city)††	(h) Hydro Energy	(i) Nuclear Energy	(j) Geothermal energy and other	(k) Total
(1) Transport		325									325
(2) Industry	52	86		187			42				367
(3) Agric., Mining, Construction		3					7				10
(4) Commercial							28				28
(5) Public							3				3
(6) Residential		127					13				140
(7) Non-energy Uses		41		10							51
(8) Final Energy Demand*	52	582		197			93				924
(9) Electricity**	30	135		24			(109)	51	20	2	153
(10) Syn. Gas											
(11) Syn. Liquids											
(12) Heat											
(13) Energy Sector Self Consumption & conversion losses		100		59			16				175
(14) Primary Energy Input	82	817		280			0	51	20	2	1252
(15) Indigenous Supply	70	827		280			0	51	20	2	1250
(16 Imports***	12										12
(17) Exports		10									10

*Includes non-energy uses.

**Blocks 9g, 10e 11c, and 12f must be negative to avoid double counting of energy from primary fuels.

***Includes imported products.

††Includes only electricity to be purchased by each sector.

NATIONAL INPUT WORKSHEET FOR SUPPLY/DEMAND INTEGRATION

Country: MEXICO

Year: 1985

Case: D

Units: Kcal x 10^{12}

	(a) Coal	(b) Petroleum	(c) (Syn. Liquids)	(d) Nat. Gas	(e) (Syn. Gas)	(f) (Heat)	(g) (Electricity)[††]	(h) Hydro Energy	(i) Nuclear Energy	(j) Geothermal energy and other	(k) Total
(1) Transport		315									315
(2) Industry	52	80		171			37				340
(3) Agric., Mining, Construction		2					6				8
(4) Commercial							25				25
(5) Public							2				2
(6) Residential		104					12				116
(7) Non-energy Uses		34		8							42
(8) Final Energy Demand*	52	535		179			82				848
(9) Electricity**	18	119		24			(96)	51	20	2	138
(10) Syn. Gas											
(11) Syn. Liquids											
(12) Heat											
(13) Energy Sector Self Consumption & conversion losses		90		52			14	0	0	0	156
(14) Primary Energy Input	70	744		255			0	51	20	2	1142
(15) Indigenous Supply	70	744		255			0	51	20	2	1142
(16 Imports***											
(17) Exports											

*Includes non-energy uses.

**Blocks 9g, 10e 11c, and 12f must be negative to avoid double counting of energy from primary fuels.

***Includes imported products.

††Includes only electricity to be purchased by each sector.

NATIONAL INPUT WORKSHEET FOR SUPPLY/DEMAND INTEGRATION

Country: MEXICO

Year: 1985

Case: E

Units: Kcal x 10^{12}

	(a)	(b)	(c)	(d)	(e)	(f)	(g)	(h)	(i)	(j)	(k)
	Coal	Petro-leum	(Syn. Liquids)	Nat. Gas	(Syn. Gas)	(Heat)	(Electri-city)††	Hydro Energy	Nuclear Energy	Geothermal energy and other	Total
(1) Transport		306									306
(2) Industry	52	77		171			37				337
(3) Agric., Mining, Construction		2					6				8
(4) Commercial							25				25
(5) Public							2				2
(6) Residential		104					12				116
(7) Non-energy Uses		34		8							42
(8) Final Energy Demand*	52	523		179			82				836
(9) Electricity**	18	119		24			(96)	51	20	2	138
(10) Syn. Gas											
(11) Syn. Liquids											
(12) Heat											
(13) Energy Sector Self Consumption & conversion losses		88		52			14	0	0	0	154
(14) Primary Energy Input	70	730		255			0	51	20	2	1128
(15) Indigenous Supply	70	730		255			0	51	20	2	1128
(16 Imports***											
(17) Exports											

*Includes non-energy uses.

**Blocks 9g, 10e 11c, and 12f must be negative to avoid double counting of energy from primary fuels.

***Includes imported products.

††Includes only electricity to be purchased by each sector.

NATIONAL INPUT WORKSHEET FOR SUPPLY/DEMAND INTEGRATION

Country: MEXICO

Year: 2000

Case: C-1

Units: Kcal x 10^{12}

	(a) Coal	(b) Petroleum	(c) (Syn. Liquids)	(d) Nat. Gas	(e) (Syn. Gas)	(f) (Heat)	(g) (Electricity)††	(h) Hydro Energy	(i) Nuclear Energy	(j) Geothermal energy and other	(k) Total
(1) Transport		1000									1000
(2) Industry	160	202		488			173				1023
(3) Agric., Mining, Construction		7					25				32
(4) Commercial							107				107
(5) Public							8				8
(6) Residential		258		60			57				375
(7) Non-energy Uses		128		32							160
(8) Final Energy Demand*	160	1595		580			370				2705
(9) Electricity**	180	80		16			(435)	150	622	4	617
(10) Syn. Gas											
(11) Syn. Liquids											
(12) Heat											
(13) Energy Sector Self Consumption & conversion losses		226		149			65	0	0		440
(14) Primary Energy Input	340	1901		745			0	150	622	4	3762
(15) Indigenous Supply	340	1901		745			0	150	622	4	3762
(16 Imports***											
(17) Exports											

*Includes non-energy uses.

**Blocks 9g, 10e 11c, and 12f must be negative to avoid double counting of energy from primary fuels.

***Includes imported products.

††Includes only electricity to be purchased by each sector.

NATIONAL INPUT WORKSHEET FOR SUPPLY/DEMAND INTEGRATION

Country: MEXICO

Year: 2000

Case: C-2

Units: Kcal x 10^{12}

	(a) Coal	(b) Petroleum	(c) (Syn. Liquids)	(d) Nat. Gas	(e) (Syn. Gas)	(f) (Heat)	(g) (Electricity)††	(h) Hydro Energy	(i) Nuclear Energy	(j) Geothermal energy and other	(k) Total
(1) Transport		1000									1000
(2) Industry	160	202		488			192				1042
(3) Agric., Mining, Construction		7					28				35
(4) Commercial							119				119
(5) Public							9				9
(6) Residential		258		60			63				381
(7) Non-energy Uses		128		32							160
(8) Final Energy Demand*	160	1595		580			411				2746
(9) Electricity**	100	80		16			(483)	150	815	4	682
(10) Syn. Gas											
(11) Syn. Liquids											
(12) Heat											
(13) Energy Sector Self Consumption & conversion losses		226		149			72	0	0	0	447
(14) Primary Energy Input	260	1901		745			0	150	815	4	3875
(15) Indigenous Supply	260	1901		745			0	150	815	4	3875
(16 Imports***											
(17) Exports											

*Includes non-energy uses.

**Blocks 9g, 10e 11c, and 12f must be negative to avoid double counting of energy from primary fuels.

***Includes imported products.

††Includes only electricity to be purchased by each sector.

NATIONAL INPUT WORKSHEET FOR SUPPLY/DEMAND INTEGRATION

Country: MEXICO

Year: 2000

Case: D-7

Units: Kcal x 10^{12}

	(a) Coal	(b) Petroleum	(c) (Syn. Liquids)	(d) Nat. Gas	(e) (Syn. Gas)	(f) (Heat)	(g) (Electricity)††	(h) Hydro Energy	(i) Nuclear Energy	(j) Geothermal energy and other	(k) Total
(1) Transport		822									822
(2) Industry	160	138		430			128				856
(3) Agric., Mining, Construction		5					19				24
(4) Commercial							80				80
(5) Public							8				8
(6) Residential		199		40			42				281
(7) Non-energy Uses		89		21							110
(8) Final Energy Demand*	160	1253		491			277				2181
(9) Electricity**	180	80		16			(326)	150	351	4	455
(10) Syn. Gas											
(11) Syn. Liquids											
(12) Heat											
(13) Energy Sector Self Consumption & conversion losses		178		130			49	0	0	0	357
(14) Primary Energy Input	340	1511		637			0	150	351	4	2993
(15) Indigenous Supply	340	1511		637			0	150	351	4	2993
(16 Imports***											
(17) Exports											

*Includes non-energy uses.

**Blocks 9g, 10e 11c, and 12f must be negative to avoid double counting of energy from primary fuels.

***Includes imported products.

††Includes only electricity to be purchased by each sector.

NATIONAL INPUT WORKSHEET FOR SUPPLY/DEMAND INTEGRATION

Country: MEXICO

Year: 2000

Case: D-8

Units: Kcal x 10^{12}

	(a) Coal	(b) Petro-leum	(c) (Syn. Liquids)	(d) Nat. Gas	(e) (Syn. Gas)	(f) (Heat)	(g) (Electri-city)††	(h) Hydro Energy	(i) Nuclear Energy	(j) Geothermal energy and other	(k) Total
(1) Transport		822									822
(2) Industry	160	138		430			140				868
(3) Agric., Mining, Construction		5					19				24
(4) Commercial							86				86
(5) Public							10				10
(6) Residential		199		40			46				285
(7) Non-energy Uses		89		21							110
(8) Final Energy Demand*	160	1253		491			301				2205
(9) Electricity**	100	80		16			(354)	150	502	4	498
(10) Syn. Gas											
(11) Syn. Liquids											
(12) Heat											
(13) Energy Sector Self Consumption & conversion losses		178		130			53	0	0	0	361
(14) Primary Energy Input	260	1511		637			0	150	502	4	3064
(15) Indigenous Supply	260	1511		637			0	150	502	4	3064
(16 Imports***											
(17) Exports											

*Includes non-energy uses.

**Blocks 9g, 10e 11c, and 12f must be negative to avoid double counting of energy from primary fuels.

***Includes imported products.

††Includes only electricity to be purchased by each sector.

NATIONAL RESPONSE WORKSHEET

Country: MEXICO

WAES SCENARIO CASE: C 1

Type of Response (Demand Reduction, Supply Expansion, or Fuel Substitution)	Examples of Response Action[1]	Lead Time (Years)[2]	% of Max. Potential Above Scenario Case[3]	Savings Achieved or Supply Added over Scenario Case[4] Oil Saved[5]	Oil Added	Gas Added	Other Supply Added
S.E.(Geoth)	Develop all fields (9000MW)	10-15					0.2
S.E.(Coal)	Increase coal production	4-6					0.3
S.E.(Solar)	60% of domestic sector with solar	5-10					0.4
S.E.(Nuclear)	Increase nuclear from 40 to 50 GWe	10					0.3

[1] Responses should reflect "supervigorous" policies; energy prices high enough to draw forth alternative supplies, and demand reductions and fuel substitutions resulting from conservation actions or improved efficiencies even at prices substantially higher than scenario prices.

[2] Number of years from start up to full implementation.

[3] Percent used of total maximum potential for each class of module.

[4] These figures are not necessarily additive; they represent components of alternative energy strategies. Units—Million Barrels a Day Oil Equivalent.

[5] All demand reduction and fuel substitution savings should be noted in "oil saved" column.

NATIONAL RESPONSE WORKSHEET

Country: MEXICO

WAES SCENARIO CASE: D 8

Type of Response (Demand Reduction, Supply Expansion, or Fuel Substitution)	Examples of Response Action[1]	Lead Time (Years)[2]	% of Max. Potential Above Scenario Case[3]	Savings Achieved or Supply Added over Scenario Case[4] Oil Saved[5]	Oil Added	Gas Added	Other Supply Added
D.R.(Ind)	Improve eff. of industrial sector 5%	6		0.1			
D.R.(Trans)	Reduce auto consumption by 10%	6-8		0.2			
S.E.(Nuc)	Increase nuclear up to C case level	10-15					0.6
S.E.(Solar)	60% of domestic sector with solar	5-10					0.3

[1] Responses should reflect "supervigorous" policies; energy prices high enough to draw forth alternative supplies, and demand reductions and fuel substitutions resulting from conservation actions or improved efficiencies even at prices substantially higher than scenario prices.

[2] Number of years from start up to full implementation.

[3] Percent used of total maximum potential for each class of module.

[4] These figures are not necessarily additive; they represent components of alternative energy strategies. Units—Million Barrels a Day Oil Equivalent.

[5] All demand reduction and fuel substitution savings should be noted in "oil saved" column.

15.1 ORGANIZATION

Participation in WAES was made possible by cooperation between the Ministry of Economic Affairs and the Future Shape of Technology Foundation. This report has been drawn up by a working party consisting of A.G. Melman, ESTEL Hoesch-Hoogovens; A.C. Sjoerdsma, Future Shape of Technology Foundation; and M.J. Stoffers, Centraal Planbureau.

The data and supporting assumptions represent the professional opinion of the working party and do not necessarily reflect official views. Data have been derived from various sources and have been supplemented by personal information from a number of authorities and specialists.

15.2 METHODOLOGY

15.2.1 General Outline

In conformity with the WAES approach, the development of energy consumption up to 2000 in The Netherlands has been investigated with respect to three scenario variables—economic growth, energy price, and national policy response.

The activity levels used for the calculation of demand are based on an economic model to be dealt with below. The technical coefficients reflect the effect of further improvement in efficiency. The product of activity levels and technical coefficients gives the energy demand estimate for each item. The projections have been based on a single set of demographic trends (Table 15.1).

Table 15.1
Demographic trends

	1972	1985	2000
Population (millions)	13.3	13.9	15.2
Dwellings (millions)	4.0	4.95	5.55

Demand and supply have been integrated manually, taking into account fuel preferences on the demand side and principal policy lines with respect to the allocation of energy sources on the supply side.

15.2.2 Economic Growth

The data on economic growth are governed (partially) by the national economic model calculations of the Centraal Planbureau. In principle, the economic scenario is founded on an assumed development of labor productivity and labor potential. In the high growth variant, it has been assumed that at full employment increases in labor productivity will be largely converted into production. For the low growth variant, increases in productivity are considered to be converted into more leisure time, so that a lower growth rate is attained. Aside from the 1985 calculations, the consistency of the Dutch growth variants with the assumed growth of the gross world product in the WAES approach has not been further investigated.

As with the 1985 WAES studies, it has been assumed that oil price changes have no direct influence on economic growth.

Activity levels are presented in Table 15.2 for the high and the low growth variants. No effort was made to specify further the production of the industrial sector in 2000 by means of the sector activity model. Adherence to the existing structural relations seem to have little value. However, it has been assumed that the macroelasticity of the growth of industrial energy consumption ($\epsilon = dE/dP$) declines from approximately 1.3 to 1 from 1985 to 2000.

15.2.3 Policy and Energy Price

The most important factor influencing energy policy in the WAES cases for the period 1985 to 2000 is energy supply. It must be decided whether to expand nuclear capacity or to expand coal imports to offset projected declines in world oil and gas production.

The WAES approach for 2000 does not explicitly mention a national policy response to demand (for instance, by means of conservation measures). It has been assumed that further improvement of efficiency will take place, especially if energy prices continue to rise (C-1 and C-2). The government response will anticipate these price effects. Thus, a vigorous policy is assumed in both the high and the low growth cases for the period 1985 to 2000. The assessment of the applied actions has been based to an important extent on the work done by the Stichting Toekomstbeeld der Techniek (Publication 19: *Energy Saving, Ways and Means*).

Table 15.2
Gross national product (in 1972 prices)

	1972	1975	1985		2000	
			High	Low	High	Low
Gross value added by sector (10^9 Dfl)						
Agriculture	8.0	9.3	13.0	12.8	19.1	11.4
Industry	49.5	51.1	86.5	65.5	138.3	73.6
Construction	13.0	11.7	12.2	11.9	14.5	9.8
Services	55.6	59.9	92.1	82.9	156.5	101.8
Gross national product (market prices)	148	156	228	197	343	248
Growth rate (%/year)						
1972–1985			3.4	2.2		
1985–2000					2.8	1.5

Note: U.S. $1 = 3.21 Dfl (1972).

Apart from the above-mentioned "normal" (i.e., vigorous) policies, a set of supervigorous policies on top of the scenario variables has been introduced in conformity with the WAES approach for the year 2000 cases. Energy costs are assumed to be substantially higher. The options (modules) are aimed at reduction of demand, substitution for oil and gas, and expansion of supply. By lowering the import requirements, these national measures contribute to closing the prospective gap between global energy demand and supply.

It should be noted that the influence of both normal and supervigorous policies on economic development has been disregarded whenever energy demand is affected indirectly. It is advisable to pursue investigations of such macroeconomic effects in such areas as investments, employment, and general price level in order to ascertain the consistency of governmental actions. In this respect, much work remains to be done as a follow-up of the WAES studies.

15.3 ANALYSIS OF THE DEMAND PROJECTIONS

15.3.1 General Outline

The projection of energy demand in 2000 has been related to two demand scenarios: (1) high economic growth of 2.8 percent per year (1985 to 2000) coupled with an energy price rising from $11.50 to $17.25 per barrel of oil equivalent, and (2) low economic growth of 1.5 percent per year coupled with a constant energy price of $11.50 per barrel of oil equivalent.

Taking into account the way in which the demand has been covered (coal or nuclear energy as the principal replacement fuel), four couplings between 1985 and 2000 cases have been made.

In Table 15.3 total primary energy input in the year 2000 cases is compared to the corresponding 1985 cases and to the base year 1972. It appears that energy consumption in Cases C-1 and C-2 is over 2.3 times as high as in the year 1972, and in Cases D-7 and D-8, approximately 1.7 times as high. The influence of the economic growth rate on energy consumption is decisive. At the high growth rate the energy demand is approximately 1.4 times as high as at the low growth rate. The small differences between Cases C-1 and C-2 and between D-7 and D-8 are connected with the differing fuel mixes of these cases:

Table 15.3 also shows that energy consumption between 1985 and 2000 increases less than the GNP. The elasticities of energy growth for 1985 to 2000 are 0.96 for the high growth case and 1.0 for the low growth case, as compared to an elasticity of 1.0 for both cases for 1972 to 1985. The trend toward electricity consumption increasing faster than energy demand as a whole is continued in the period 1972 to 2000. The increase in the share of electricity means that conversion losses in the electricity sector increase. Total processing losses in 1985, however, have a smaller share of primary energy input (as borne out by the final line in Table 15.3). This can be ascribed to the conversion losses in oil refining, which are included in domestic consumption. In 1972, 50 percent of refinery output was exported, but in 1985 and 2000 export of refinery products is assumed to be 20 percent (Case C), 40 percent (Case D), 5 percent (Cases C-1 and

Table 15.3
1972-2000 demand summary (10^{15} J)

Year			1985				2000							
WAES case			C		D		C-1		C-2		D-7		D-8	
GWP growth			High		Low		High		High		Low		Low	
Energy price[a]	1972		11.50		11.50		11.50→17.25		11.50→17.25		11.50		11.50	
Energy type	Fuel	Elec.	Fuel	Elec.	Fuel	Elec.	Fuel	Elec.	Fuel	Elec.	Fuel	Elec.	Fuel	Elec.
Transportation	802	3	977	5	977	5	1147	11	1147	11	1048	9	1048	9
Agriculture, etc.	183	10	260	12	252	11	317	13	317	13	212	10	212	10
Commercial and public	155	39	187	51	172	47	166	60	166	60	191	58	191	58
Industry	738	64	1751	134	1092	88	2873	507	2873	507	1501	265	1501	265
Residential	525	36	745	74	723	74	677	117	677	117	770	109	770	109
Final energy demand	2403	152	3919	275	3216	225	5180	707	5180	707	3723	452	3723	452
Processing losses	171	295	180	412	175	324	216	947	216	933	178	733	178	718
Primary energy input	3021		4786		3950		7050		7036		5086		5071	
Growth rate (index)	100		158		131		233		233		168		167	
Electricity														
Share of final energy demand (%)	5.9		6.6		6.5		11.2		11.2		10.8		10.8	
Share of primary energy input (%)	14.8		14.4		13.9		23.5		23.4		23.3		23.0	
Losses[b]														
Share of primary energy input (%)	15.4		12.4		12.6		16.5		16.3		17.9		17.6	

[a] Oil or energy price in constant 1975 U.S. dollars per barrel of light Arabian crude FOB Persian Gulf or equivalent.
[b] Determined by the supply-demand integration.

C-2), and 15 percent (Cases D-7 and D-8). Thus, the processing losses will have a relatively smaller share of total Dutch energy demand than before.

15.3.2 Energy Demand by Sector

Transportation

From many studies it appears that the number of motorcars, hitherto dependent solely on income growth, will reach saturation before the year 2000 because of a stabilization of population and a projected maximum car density of approximately 450 cars per 1000 inhabitants. A total of 6.7 million cars has been assumed for the high growth alternative, and 6.0 million for the low growth case. One reason for the lower figure in the low growth case is a lag in the number of second cars. (The influence of the second car is reflected in the average number of kilometers driven, so that the number of passenger-kilometers is more nearly equal for the two alternatives.)

By application of a number of very specific measures through supervigorous policies, the growth of the number of cars can be further reduced. Possibilities are a shift to public transportation and a reduction in the general need for transport (see also section 15.5).

The decrease in average automotive fuel consumption per kilometer continues on a modest scale. Consumption will decrease through technical modifications of transportation modes (*decreasing effects:* stratified charge, maximum speed, driving behavior, streamline, weight reduction; *increasing effects:* putting environmental exigencies and safety requirements into effect). A 40 percent increase in the share of diesel motors will also contribute to a consumption decrease (see Table 15.4).

Public Transportation

The maintenance of sufficient public transportation to cover transport needs is an important component of government policy. For this study it was assumed that promotion of public transportation will lead to a strengthening of its position (particularly rail traffic), but that its share of total transportation will slightly decrease. A small improvement in energy efficiency has been assumed for 1972 to 2000.

Table 15.4
Scenario assumptions: private cars

		1985		2000	
	1972	High	Low	High	Low
Cars (millions)	2.9	5.6	4.9	6.7	6.0
Driven km/car/year	16,000	14,300	15,400	14,300	14,500
Diesel (%)	1	40	10	40	40

Air Travel and Freight
Recent studies indicate that the development of air travel over the period 1985 to 2000 will be less significant than estimated earlier. (Amsterdam Airport will handle 40–60 million instead of 60–80 million passengers.) The number of take-offs, which is used as a measure in the determination of energy consumption, was projected at 227,000 for the high growth variant and 190,000 for the low growth variant.

Freight Traffic
Freight traffic by road as well as by inland shipping will develop according to existing trends, in proportion with the growth of the industrial and the construction sectors. Truck freight will continue to gain a larger proportion of the total freight traffic. Rail traffic, particularly international rail traffic, is assumed to increase little in an absolute sense.

An important improvement in road transport efficiency is expected through the growing penetration of the diesel engine (from 35 percent in 1972 to 45 percent in 1985 to approximately 60 percent in 2000).

Ship Freight International
In this study the consumption of bunkers is included in domestic energy demand. The figures used are approximations because insufficient data are available on the future of refining in The Netherlands. It has been assumed that consumption will increase little after 1985.

Nonenergy use refers to the consumption of lubricants in the transport sector. It is assumed that approximately 50 percent is recycled.

Commercial and Public
As in the demand projection for 1985, the 2000 cases utilize the number of employees in the services sector as a measure of energy consumption.

Besides a further improvement of the thermal insulation of buildings due to higher energy costs and government policy incentives, an amelioration of the average boiler efficiency has been assumed for the high growth variant (C-1). Electricity consumption per employee is at the 1972 level, and improvement is not very likely because of the increasing use of air conditioners.

Industry
As stated in section 15.2, we did not break down the development of industrial production to 2000 by industry. Taking the actual situation as a basis will certainly not provide realistic propositions. Using ideas expressed in some policy projects, we calculated total industrial energy consumption on the basis of the supposition that energy growth elasticity will be reduced gradually to 1.0 by 2000. For C-1 and C-2 we assumed an average elasticity of 1.10; for D-7 and D-8, 1.21. An ongoing improvement in efficiency has been assumed in this approach.

Residential
For the calculation of energy consumption in the residential sector, it has been assumed that the number of dwellings will increase to 5.55 million, of which almost 100 percent will be centrally heated in the high growth cases (C-1 and C-2) and 90 percent in the low growth cases (D-7 and D-8).

The period 1975 to 2000 will be characterized by improvements in the quality of dwellings, for example, by the application of improved insulation standards. The savings are independent of whether a dwelling was built before 1976 or after (a fuel saving of 20 percent at high growth and 30 percent at low growth for single-family houses, and 15 percent at high growth and 20 percent at low growth for apartments as compared with the 1972 situation). Because of technical improvements in heating installations and hot water generation, heating efficiency will be improved in the high growth case.

As to electricity consumption, a further penetration of electrical appliances in households has been taken into account. The estimate of average power consumption for dwellings amounts to 5700 kWh annually in Cases D-7 and D-8 and 6000 kWh annually (higher because of air conditioning) in Cases C-1 and C-2. Important and economically feasible savings can be realized by more drastic technical measures that create the possibility of a "thrifty" household consuming 3100 kWh annually. This option will be dealt with in section 15.5 as an example of what can be achieved by supervigorous policies.

In the period under consideration, solar heating, combined with the heat pump, will ease the market for room heating. By the application of low-temperature collectors, a savings of approximately 35.2×10^9 J per dwelling (1000 Nm^3 natural gas equivalent) can be achieved under Dutch conditions. In combination with the heat pump, approximately 90×10^9 J will fulfill the heat requirements of the average well-insulated house, although at the expense of approximately 5500 kWh needed for the activation of the heat pump. It has been assumed that solar heating will be used in 20 percent (at high growth) and 10 percent (at low growth) of buildings built after 1985. Further expansion as an option of supervigorous policy is elaborated in section 15.5. Table 15.5 summarizes a number of important parameters determining residential energy consumption.

15.4 SUPPLY-DEMAND INTEGRATION

15.4.1 General Outline

The Dutch supply-demand data have been integrated manually. The results for the 2000 cases have been compared with Cases C and D for 1985. The allocation of energy sources by sector and energy type (determination of the fuel mix) and import requirements and export possibilities have all been calculated.

For the allocation of fuels, the fuel preferences of each sector have been taken into account, as well as the possible use of domestic fuels (natural gas, electricity based on nuclear energy) and the policies used with respect to the input of solid fuel or nuclear energy as the principal replacement fuel. Oil has been considered as a balance item.

The fuel mixes of final users and of power stations have been determined, yielding the total primary energy input after addition of processing losses from oil refining, electricity generation, and coke production. Determination of the energy supply then allows the import requirements to be calculated.

The following trends are anticipated for 1972–2000 (see Table 15.6):

- The primary energy input, resulting from demand, grows in the period by a factor of 2.3 in the high growth alternative and 1.7 in the low growth alternative.
- Expansion of coal and nuclear energy as replacements for oil brings their share of primary energy input from approximately 4 percent in 1972 to 20 to 25 percent in 2000.
- Because indigenous supply decreases in 2000, natural gas contributes 15 percent of the primary energy input in the high growth alternative and 20 percent in the low growth alternative (compared to 40 percent in 1972).
- Natural gas will no longer be exported; on the contrary, an additional quantity of 10 billion Nm^3 (Slochteren equivalents) will be imported.
- Total domestic supply will be less than the level of 1972 at the end of the century.
- To cope with the increasing supply shortage, The

Table 15.5
Scenario assumptions: residential

		1985		2000	
	1972	C	D	C-1, C-2	D-7, D-8
Dwellings (millions)	4	4.95	4.95	5.55	5.55
Central heating (%)	32	85	75	100	90
Boiler efficiency	0.55	0.55	0.55	0.65	0.55
Solar dwellings (thousands)	—	—	—	240	120

Table 15.6
Composition of primary energy input and import quota (%)

		1985		2000			
	1972	C	D	C-1	C-2	D-7	D-8
Oil	55.4	57.0	51.7	57.9	61.1	55.0	58.1
Natural gas	40.3	34.3	41.5	15.8	15.8	21.9	21.9
Coal	4.3	5.1	4.1	16.4	7.2	14.2	6.2
Nuclear	0	3.6	2.6	9.8	15.7	8.8	13.6
Solar				0.2	0.2	0.2	0.2
Total	100	100	100	100	100	100	100
Import quota (net)	27.0	33.3	19.0	70.0	64.1	63.4	58.5

Netherlands will depend on the world market for coal, oil, and natural gas. Therefore the import quota increases strongly.

• The consumption of oil grows in absolute terms to about 90 million tonnes, i.e., 2.5 times the 1972 level, for the high growth rate variant. In the low growth case, oil imports hardly surpass the 1985 level.

• Oil dependency in 2000 surpasses the 55 percent level of 1972; in both the high and low cases, the oil share of primary energy input will be about 60 percent.

• Coal imports grow from approximately 3 million tonnes in 1972 to over 7 million tonnes in 1985 (Case C) to almost 40 million tonnes in the highest growth case (C-1). This is an increase of over 10 percent annually after 1985.

15.4.2 Scenario Assumptions

Fuel preferences are inherent in the various final-user processes. Transport, fishing, farm machinery, and the construction industry, for instance, are almost completely dependent on oil, whereas pig-iron production requires a minimum quantity of coke.

Indigenous natural gas resources are limited. The policy for natural gas, taken as a guiding rule for the supply-demand integrations, implies that (1) the supply to power stations will be limited to existing contracts, (2) there will be no extension or prolongation of export contracts, and (3) there will be only selective sales for underfiring in large (industrial) boiler installations. Priority will be given to public distribution and applications such as chemical feedstocks.

For the distribution of the natural gas that is assumed to be available in 2000, government policy has been translated as follows. The residential fuel requirements are completely covered by natural gas for those dwellings that have been connected to the natural gas grid. This holds true for horticulture also. Eighty percent of commercial and public energy use consists of natural gas; the remainder is intended for industry.

In the WAES scenario approach, nuclear energy is the principal replacement fuel in the C-2 and D-8 cases, coal in the C-1 and D-7 cases. Up to 2000, nuclear energy is assumed to be used only for electric power production. The maximum potential is applicable at high growth (C-2), whereas at low growth (D-8), a 30 percent lower potential has been assumed. The Netherlands is completely dependent on imports for its fossil fuel supply in the electricity sector. It is probable that from 1985 to 2000 coal will be used only for firing of industrial boilers and for power generation. The practical value of substantially higher coal consumption is doubtful. For the WAES cases, it has been assumed that coal will cover only 20 percent of the fuel requirements of industry, including iron and steel production, in Case C-1 and 10 percent in Case D-7.

Table 15.7 shows the fuel mix for public electricity generation. Comparable figures for 1972 and 1985 have been included. After determination of the fuel mix for final users and for electricity generation, the processing losses from energy conversion and the energy consumption in the energy sector itself were determined. The conversion factors used are given in Table 15.8. For comparison, the figures for 1972 and 1985 have been included.

Table 15.7
Scenario assumptions: public electricity generation

		1985		2000			
	1972	C	D	C-1	C-2	D-7	D-8
Nuclear							
Share (%)	1	24	17	41	66	37	57
Power (GW)	0.06	2.5	1.5	10	16	6.5	10
Coal							
Share (%)	—	10	10	30	10	30	10
Input (10^6 tonnes)	0.5	3	2.2	18.3	6.1	13.1	4.4

Table 15.8
Conversion factors[a]

		Oil	Gas	Coal	Nuclear
Refineries	1972	94.5			
	1985	95.25			
	2000	95.5			
Power stations	1972		34		
	1985	38		32	34
	2000	38		32	34
Coke batteries	1972				
	1985			88.4	
	2000			88.7	

[a]Self-consumption included.

15.5 NATIONAL RESPONSE

In consideration of the "prospective gap," particularly that for oil, alternative energy strategies must be developed on a national level. These strategies are aimed at the reduction of energy demand, substitution of oil and gas, and expansion of the supply of energy. They are assumed to become effective at prices over $17.25 per barrel of oil equivalent and with supervigorous policies on top of the normal scenario policy measures. The results of the options elaborated for The Netherlands are presented in the demand reduction and national response worksheets (section 15.6).

It should be noted that this analysis does not pretend to be complete: the list of measures is meant to be exemplary, and its consistency with economic and environmental policies has not been investigated.

15.5.1 Demand Reduction

Important savings, which are already part of the scenario approach, can be achieved by a reduction of energy demand. We estimate that the energy demand, if extrapolated without change, could be approximately 11 percent higher in the C-1 and C-2 cases.

Additional savings are also possible. First, in the transportation sector, the expansion of the number of cars and the resulting number of kilometers driven annually can be restrained by measures that diminish the transport needs of the population and/or favor public transportation. These policies should be able to absorb any detrimental effects. Efficiency can be further improved by the introduction of the electrocar in urban areas. In terms of primary energy input, the savings will be small, but it means in any case a substitution of oil.

Second, additional savings will accrue in the residential sector from space heating if the maximum potential for heat loss reduction indicated in the study *Energy Conservation, Ways and Means* is realized. Residential electricity consumption can also be decreased further by the introduction of energy-saving appliances, in which the electricity used for heating is replaced, if possible, by natural gas, heat losses are reduced by better insulation, and other efficiency improvements are effected. The thrifty household will consume only 3100 kWh per year.

Considerable savings are possible with the changeover to district heating (a system in which individual heating of each house is replaced by central heat generation combined with electricity generation in medium-sized units). This means, in fact, a reduction of conversion losses in power production by utilization of residual heat. In The Netherlands, where district heating is already applied in two places, major possibilities exist both in future town districts and in older districts that are scheduled for renewal. From calculations for a randomly chosen residential agglomeration of 100,000 dwell-

Table 15.9
Data for demand reduction options

	Normal policy		Supervigorous policy	
	High growth	Low growth	High growth	Low growth
Transportation				
Number of cars (millions)	6.7	6.0	5.3	4.8
Driven km/car/year	14,300	14,500	13,600	13,800
Public traffic (10^6 passenger-km)	34,200	30,200	45,200	42,200
Electrocars (millions)	—	—	0.5 (10%)	0.5 (10%)
Dwellings				
Savings (% vs. 1972)	20–30	20–30	20–30	20–30
Power consumption (kWh/dwelling)	6000	5700	3100	

ings, a savings of approximately 1300 Nm^3 natural gas equivalent per well-insulated dwelling can be obtained. This estimate is based on a module of ten projects of approximately the same size (1 million dwellings, or 80 percent of dwellings built after 1985).

Table 15.9 summarizes the most important parameters forming the basis of several energy-saving options.

15.5.2 Expansion of Supply

The possibilities for expanding energy supply are in fact limited until a further penetration of solar heating systems takes place. For the attainable savings per dwelling, see section 15.3. The module used is based on the application of solar heating in 1 million dwellings. This figure could be reached by 2000 if solar heating equipment is installed in 100 percent of dwellings built after 1985 and if approximately 10 percent of the existing dwellings are converted to solar heating.

It should be noted that district heating and solar heating systems are not completely complementary. Scattered houses and smaller residential centers may offer better opportunities for solar heating, while urban living areas should be considered part of the natural supply area for district heating.

15.5.3 Coal Gasification

A real possibility for direct substitution, other than expansion of nuclear energy to its maximum potential of 16 GWe, is offered by coal gasification. The option in question is based on the production of 10 billion Nm^3 per year of natural gas equivalent, although the economic potential of such substitution is negatively affected by the rather high conversion loss of approximately 30 percent.

15.6 WORKSHEETS

This section contains summary tables and worksheets in the following order:

Energy demand summary 1972–1985–2000

Aggregated demand worksheets (5 sheets, one each for 1972, for 1985 Cases C and D, and for 2000 Cases C-1/C-2 and D-7/D-8)

Energy supply summary, 1972–2000

Supply/demand integration summary—Imports and exports of primary energy

National input worksheets for supply/demand integration (6 sheets)

Demand reduction modules efficiency

National response worksheet

WAES NATIONAL ENERGY DEMAND STUDIES, 1972-1985-2000
ENERGY DEMAND SUMMARY

Country: The Netherlands

Scenario Assumptions	1972	1985					2000			
WAES CASE	Base Year	A	B	C	D	E	C-1	C-2	D-7	D-8
GNP 1976-85; 1985-2000		High	Low	High	Low	High	High	High	Low	Low
Oil/Energy Price*		11.50-17.25	11.50-17.25	11.50	11.50	11.50-7.66	11.50-17.25	11.50-17.25	11.50	11.50
National Policy Response		Vig	Vig	Vig	Res	Res	Vig	Vig	Vig	Vig
Principal Replacement Fuel							Coal	Nuclear	Coal	Nuclear
Primary Energy Demand — **Units** 10^{15}J										
Oil	1679			2727	2043		4082	4301	2797	2947
Natural Gas	1209[1)]	1641	1641	1641	1641	1641	1112	1112	1112	1112
Coal	129			246	163		1153	508	722	316
Nuclear	4	172	103	172	103	172	688	1100	447	688
Hydroelectric										
Geothermal										
Solar Thermal							15	15	8	8
Solar Electric										
Oil + coal		2693	1822			3491				
TOTAL PRIMARY ENERGY DEMAND	3021	4506	3566	4786	3950	5304	7050	7036	5086	5071

*1975 US dollars per barrel of Arabian light crude oil, fob Persian Gulf.
All Figures are primary production prior to any processing or conversion.

Footnotes:
1) Losses and not accounted for: 6 x 10^{15}J.

WAES AGGREGATED DEMAND WORKSHEET

Country: The Netherlands

Year: 1972

WAES Case: Base year

GNP: 148×10^9 Dfl 1)

Population: 13.3×10^6

Sector	Activity Level	Energy Intensity	Other PJ/y	Electric PJ/y	Foss. Fuel PJ/y	Feed-stocks PJ/y
Transportation						
1. Automobiles	2.9×10^6 autos x 16,000 km/auto/y	3.241 MJ/km			150.4	
2. Airtravel & freight	100 2)	MJ/pass-km			37.0	
3. Other passengers	$17,502 \times 10^6$ pass-km/y	0.491 MJ/pass-km		3.0	5.6	
4. Truck	$13,973 \times 10^6$ ton-km/y	4.0 MJ/ton-km			55.9	
5. Rail freight	$3,070 \times 10^6$ ton-km/y	0.521 MJ/ton-km		0.3	1.3	
6. Shipping 3)	$29,334 \times 10^6$ ton-km/y 4)	1.4 MJ/ton-km 4)			545.3 5)	
1-6 Total				3.3	795.5	6.7
Agricult., Mining, Construct.						
7. Agriculture	$2,504 \times 10^6$ \$/y output	4.2 MJ/\$		2.6	105.4	
8. Mining	855×10^6 \$/y output	4.3 MJ/\$		2.1	35.6	
9. Construction	$4,054 \times 10^6$ \$/y output	1.1 MJ/\$		5.1	11.9	29.6
7-9 Total				9.8	152.9	29.6

1) 1\$ = 3.21 Dfl (1972 – prices). 2) index. 3) inland and international shipping (bunkers). 4) inland shipping. 5) of which bunkers 504.0×10^{15}J.

Commercial, Public							
10 Space Heat (Fossil Fuel)	____ m²		____ MJ/m²/y	____ Effic			
11. – – (Electric)	____ m²		____ MJ/m²/y	____ Effic			
12. Other	____ m²						
10-12 Total	2.4×10^6 my [6]		81 MJ/my/y	0.65 Effic.	39.1	155.1	
Industry							
13. Metals	567×10^6 \$/y	5.85×10^6 (ton/y)	3.0 MJ/\$ (elec)	19.6 MJ/\$ (fos)	17.3	114.8	
14. Chemicals	$1,645 \times 10^6$ \$/y	____ (ton/y)	1.2 MJ/\$ (–)	24.6 MJ/\$ (–) [7]	21.3	177.2	240.0
15. Other Energy Industries ____ \$/y 16. Non-Energy Industries ____ \$/y	$11,219 \times 10^6$ \$/y		0.2 MJ/\$ (–)	1.8 MJ/\$ (–)	25.3	206.3	
13-16 Total					63.9	498.3	240.0
Residential							
17. Space Heat (Fossil Fuel) [8]	4.042×10^6 Dwellings (fossil)		105.4 MJ/dw/y	0.55 Effic		452.9	
18. – – (Electric)	– Dwellings (elect)		– MJ/dw/y	– Effic	1.8 [9]		
			24.4 MJ fossil/dwelling			98.8	
19. Appliances	4.042×10^6 Dwellings		8.5 MJ elect/dwelling		34.2		
17-19 Total					36.0	524.7	
1-19 Total					152.1	2,126.5	276.3

6) employment (man-years). 7) excl. feedstock: 10.2 MJ/\$. 8) central heating penetration 32%. 9) additional heating.

WAES AGGREGATED DEMAND WORKSHEET

Country: The Netherlands

Year: 1985

WAES Case: C

GNP: 228 x 10^9 Dfl 1)

Population: 13.9 x 10^6

Sector Transportation	Activity Level	Energy Intensity	Other PJ/y	Electric PJ/y	Foss. Fuel PJ/y	Feed-stocks PJ/y
1. Automobiles	5.6 x 10^6 autos x 14,300 km/auto/y	2.606 MJ/km			208.7	
2. Airtravel & freight	131 2)	MJ/pass-km			51.4	
3. Other passengers	21,900 x 10^6 pass-km/y	0.489 MJ/pass-km		4.6	6.1	
4. Truck	20,350 x 10^6 ton-km/y	3.8 MJ/ton-km			77.4	
5. Rail freight	3,000 x 10^6 ton-km/y	0.533 MJ/ton-km		0.3	1.3	
6. Shipping 3)	38,400 x 10^6 ton-km/y 4)	1.4 MJ/ton-km 4)			624.0 5)	
1-6 Total				4.9	968.9	8.0
Agricult., Mining, Construct.						
7. Agriculture	4,058 x 10^6 \$/y output	4.2 MJ/\$		4.2	170.6	
8. Mining	1,206 x 10^6 \$/y output	4.3 MJ/\$		3.0	50.2	
9. Construction	3,796 x 10^6 \$/y output	1.1 MJ/\$		4.7	11.1	27.7
7-9 Total				11.9	231.9	27.7

1) 1\$ = 3.21 Dfl (1972-prices). 2) index. 3) inland and international shipping (bunkers).
4) inland shipping. 5) of which bunkers 570 x 10^{15}J.

Commercial, Public							
10 Space Heat (Fossil Fuel)	____ m^2		____ $MJ/m^2/y$	____ Effic			
11. – – (Electric)	____ m^2		____ $MJ/m^2/y$	____ Effic			
12. Other	____ m^2						
10-12 Total	3.2×10^6 my 6)		74 MJ/my/y	0.65 Effic.	51.2	186.5	
Industry							
13. Metals	$1,183 \times 10^6$ $/y	12.5×10^6 (ton/y)	2.8 MJ/$ (elec)	18.7 MJ/$ (fos)	33.9	227.4	
14. Chemicals	$4,952 \times 10^6$ $/y	____ (ton/y)	1.2 MJ/$ (–)	24.5 MJ/$ (–) 7)	63.4	527.8	722.6
15. Other Energy Industries	____ $/y } $17,557 \times 10^6$ $/y		____ MJ/$ (–) 0.2	____ MJ/$ (–) 1.5	36.3	273.2	
16. Non-Energy Industries	____ $/y }		____ MJ/$ (–)	____ MJ/$ (–)			
13-16 Total					133.6	1,028.4	722.6
Residential							
17. Space Heat (Fossil Fuel) 8)	4.95×10^6 Dwellings (fossil)		119.5 MJ/dw/y	0.55 Effic		591.3	
18. – – (Electric)	– Dwellings (elect)		– MJ/dw/y	– Effic	0.9 9)		
			31.0 MJ fossil/dwelling			153.4	
19. Appliances	4.95×10^6 Dwellings		14.7 MJ elect/dwelling		72.8		
17-19 Total					73.7	744.7	
1-19 Total					275.3	3,160.4	758.3

6) employment (man-years). 7) excl. feedstock: 10.0 MJ/$. 8) central heating penetration 85%.
9) additional heating.

WAES AGGREGATED DEMAND WORKSHEET

Country: The Netherlands

Year: 1985

WAES Case: D

GNP: 197 x 10^9 Dfl 1)

Population: 13.9 x 10^6

Sector	Activity Level	Energy Intensity	Other PJ/y	Electric PJ/y	Foss. Fuel PJ/y	Feed-stocks PJ/y
Transportation						
1. Automobiles	4.9 x 10^6 autos x 15,400 km/auto/y	2.998 MJ/km			226.2	
2. Airtravel & freight	131 ~~pass-km/y~~ 2)	____ MJ/pass-km			51.4	
3. Other passengers	21,900 x 10^6 pass-km/y	0.489 MJ/pass-km		4.6	6.1	
4. Truck ~~& Air freight~~	17,300 x 10^6 ton-km/y	3.8 MJ/ton-km			65.6	
5. Rail freight	3,000 x 10^6 ton-km/y	0.533 MJ/ton-km		0.3	1.3	
6. Shipping 3)	34,800 x 10^6 ton-km/y 4)	1.4 MJ/ton-km 4)			619.0 5)	
1-6 Total				4.9	969.6	7.5
Agricult., Mining, Construct.						
7. Agriculture	4,000 x 10^6 $/y output	4.2 MJ/$		4.1	168.2	
8. Mining	1,117 x 10^6 $/y output	4.3 MJ/$		2.7	46.5	
9. Construction	3,702 x 10^6 $/y output	1.1 MJ/$		4.6	10.7	27.0
7-9 Total				11.4	225.4	27.0

1) 1$ = 2.21 Dfl (1972-prices). 2) index. 3) inland and international shipping (bunkers). 4) inland shipping. 5) of which bunkers 570 x 10^{15}J.

Commercial, Public							
10 Space Heat (Fossil Fuel)	___ m^2		___ $MJ/m^2/y$	___ Effic			
11. – – (Electric)	___ m^2		___ $MJ/m^2/y$	___ Effic			
12. Other	___ m^2						
10-12 Total	2.9×10^6 my 6)		74 MJ/my/y	0.65 Effic	47.3	172.0	
Industry							
13. Metals	757×10^6 $/y	7.8×10^6 (ton/y)	2.8 MJ/$ (elec)	18.6 MJ/$ (fos)	21.9	145.4	
14. Chemicals	$2,801 \times 10^6$ $/y	___ (ton/y)	1.2 MJ/$ (–)	24.5 MJ/$ (–) 7)	35.9	298.4	408.6
15. Other Energy Industries ___ $/y; 16. Non-Energy Industries ___ $/y	$13,885 \times 10^6$ $/y		0.2 MJ/$ (–)	1.7 MJ/$ (–)	30.2	239.8	
13-16 Total					88.0	683.6	408.6
Residential							
17. Space Heat (Fossil Fuel) 8)	4.95×10^6 Dwellings (fossil)		115.0 MJ/dw/y	0.55 Effic		569.3	
18. – – (Electric)	– Dwellings (elect)		– MJ/dw/y	– Effic	0.9 9)		
			31.0 MJ fossil/dwelling			153.4	
19. Appliances	4.95×10^6 Dwellings		14.7 MJ elect/dwelling		72.8		
17-19 Total					73.7	722.7	
1-19 Total					225.3	2,773.3	443.1

6) employment (man-years). 7) excl. feedstock: 10.3 MJ/$. 8) central heating penetration 75%. 9) additional heating.

WAES AGGREGATED DEMAND WORKSHEET

Country: The Netherlands

Year: 2000

WAES Case: C-1/C-2

GNP: 343×10^9 Dfl 1)

Population: 15.2×10^6

Sector **Transportation**	Activity Level	Energy Intensity	Other PJ/y	Electric PJ/y	Foss. Fuel PJ/y	Feed-stocks PJ/y
1. Automobiles	6.7×10^6 autos x 14,300 km/auto/y	2.596 MJ/km			249.2	
2. Airtravel & freight	257 ~~pass-km/y~~ 2)	____ MJ/pass-km			101.0	
3. Other passengers	$34{,}200 \times 10^6$ pass-km/y	0.474 MJ/pass-km		10.0	6.3	
4. Truck ~~& Air freight~~	$32{,}600 \times 10^6$ ton-km/y	3.3 MJ/ton-km			107.6	
5. Rail freight	$4{,}000 \times 10^6$ ton-km/y	0.450 MJ/ton-km		0.4	1.4	
6. Shipping 3)	$53{,}600 \times 10^6$ ton-km/y 4)	1.4 MJ/ton-km 4)			675.0 5)	
1-6 Total				10.4	1140.5	6.0

Agricult., Mining, Construct.	Activity Level	Energy Intensity	Other PJ/y	Electric PJ/y	Foss. Fuel PJ/y	Feed-stocks PJ/y
7. Agriculture	$5{,}950 \times 10^6$ \$/y output	4.2 MJ/\$		6.3	250.8	
8. Mining	872×10^6 \$/y output	2.2 MJ/\$		1.1	19.5	
9. Construction	$4{,}517 \times 10^6$ \$/y output	1.1 MJ/\$		5.7	13.2	33.0
7-9 Total				13.1	283.5	33.0

1) 1\$ = 3.21 Dfl (1972-prices). 2) index. 3) inland and international shipping (bunkers). 4) inland shipping. 5) of which bunkers 600.0×10^{15} J.

Commercial, Public								
10 Space Heat (Fossil Fuel)		___ m^2	___ $MJ/m^2/y$	___ Effic				
11. – – (Electric)		___ m^2	___ $MJ/m^2/y$	___ Effic				
12. Other		___ m^2						
10-12 Total	3.7×10^6 my [6]		61 MJ/my/y	0.75 Effic		60.3	165.8	
Industry								
13. Metals	___ $/y	___ (ton/y)	___ MJ/$ (elec)	___ MJ/$ (fos)				
14. Chemicals	___ $/y	___ (ton/y)	___ MJ/$ (–)	___ MJ/$ (–)				
15. Other Energy Industries	___ $/y		___ MJ/$ (–)	___ MJ/$ (–)				
16. Non-Energy Industries	___ $/y		___ MJ/$ (–)	___ MJ/$ (–)				
13-16 Total	$38,598 \times 10^6$ $/y		1.3 MJ/$	7.2 MJ/$		507.0	1,723.0	1,150
Residential								
17. Space Heat (Fossil Fuel) [7]		5.55×10^6 Dwellings (fossil)	92.4 MJ/dw/y	0.65 Effic	15.0[8]	2.4	496.8	
18. – – (Electric)		– Dwellings (elect)	– MJ/dw/y	– Effic				
			29.8 MJ fossil/dwelling				165.3	
19. Appliances		5.55×10^6 Dwellings	20.5 MJ elect/dwelling			113.9		
17-19 Total					15.0	116.3	662.1	
1-19 Total					15.0	707.1	3,974.9	1,189

6) employment (man-years). 7) central heating penetration 100%. 8) solar energy.

WAES AGGREGATED DEMAND WORKSHEET

Country: The Netherlands

Year: 2000

WAES Case: D-7/D-8

GNP: 237×10^9 Dfl [1)]

Population: 15.2×10^6

Sector Transportation	Activity Level	Energy Intensity	Other PJ/y	Electric PJ/y	Foss. Fuel PJ/y	Feed-stocks PJ/y
1. Automobiles	6.0×10^6 autos x 14,500 km/auto/y	2.600 MJ/km			225.9	
2. Airtravel & freight	215 [2)]	____ MJ/pass-km			84.5	
3. Other passengers	$30,200 \times 10^6$ pass-km/y	0.474 MJ/pass-km		8.8	5.4	
4. Truck	$20,730 \times 10^6$ ton-km/y	3.3 MJ/ton-km			68.4	
5. Rail freight	$4,000 \times 10^6$ ton-km/y	0.450 MJ/ton-km		0.4	1.4	
6. Shipping [3)]	$44,740 \times 10^6$ ton-km/y [4)]	1.4 MJ/ton-km [4)]			656.8 [5)]	
1-6 Total				9.2	1,042.5	5.0
Agricult., Mining, Construct.						
7. Agriculture	$3,548 \times 10^6$ \$/y output	4.2 MJ/\$		3.8	149.2	
8. Mining	872×10^6 \$/y output	2.3 MJ/\$		1.1	19.5	
9. Construction	$2,586 \times 10^6$ \$/y output	1.1 MJ/\$		5.3	12.3	30.7
7-9 Total				10.2	181.0	30.7

1) 1\$ = 3.21 Dfl (1972-prices). 2) index. 3) inland and international shipping (bunkers).
4) inland shipping. 5) of which bunkers 600.0×10^{15} J.

Commercial, Public									
10 Space Heat (Fossil Fuel)		___ m^2	___ $MJ/m^2/y$	___ Effic					
11. – – (Electric)		___ m^2	___ $MJ/m^2/y$	___ Effic					
12. Other		___ m^2							
10-12 Total		3.7×10^6 my 6)	68 MJ/my/y	0.65 Effic			57.7	191.2	
Industry									
13. Metals	___ $/y	___ (ton/y)	___ MJ/$ (elec)	___ MJ/$ (fos)					
14. Chemicals	___ $/y	___ (ton/y)	___ MJ/$ (–)	___ MJ/$ (–)					
15. Other Energy Industries	___ $/y		___ MJ/$ (–)	___ MJ/$ (–)					
16. Non-Energy Industries	___ $/y		___ MJ/$ (–)	___ MJ/$ (–)					
13-16 Total	$21,620 \times 10^6$ $/y		1.2 MJ/$	6.7 MJ/$			265.0	910.0	591.0
Residential									
17. Space Heat (Fossil Fuel) 7)		5.55×10^6 Dwellings (fossil)	___ MJ/dw/y	0.55 Effic		7.5 8)	1.2	597.5	
18. – – (Electric)		- Dwellings (elect)	- MJ/dw/y	- Effic					
			29.8 MJ fossil/dwelling					165.3	
19. Appliances		5.55×10^6 Dwellings	19.5 MJ elect/dwelling				108.2		
17-19 Total						7.5	109.4	762.8	
1-19 Total						7.5	451.5	3,087.5	626.7

6) employment (man-years). 7) central heating penetration 90%. 8) solar energy.

WAES NATIONAL ENERGY SUPPLY STUDIES, 1972-2000
ENERGY SUPPLY SUMMARY

Country: The Netherlands

Scenario Assumptions	1972	1985					2000			
WAES CASE	Base Year	A	B	C	D	E	C-1	C-2	D-7	D-8
GNP 1976-85; 1985-2000		High	Low	High	Low	High	High	High	Low	Low
Oil/Energy Price*		11.50-17.25	11.50-17.25	11.50	11.50	11.50-7.66	11.50-17.25	11.50-17.25	11.50	11.50
National Policy Response		Vig	Vig	Vig	Res	Res	Vig	Vig	Vig	Vig
Principal Replacement Fuel							Coal	Nuclear	Coal	Nuclear
Energy Supply **Units** 10^{15} Joules										
Crude Oil	67			80	80		80	80	80	80
Natural Gas 1)	2058			3318	3318		760	760	760	760
Coal	82									
Nuclear 2)	4			172	103		688	1100	447	688
Hydro-electric										
Geothermal										
Solar Thermal							15	15	8	8
Solar Electric										
Total Energy Supply	2211			3570	3501		1543	1955	1295	1536

*1975 US dollars per barrel of Arabian light crude oil, fob Persian Gulf.
All Figures are primary production prior to any processing or conversion.

Footnotes:

1) Onshore gas has been calculated as 1 m^3 = 35.2 x 10^6J. Offshore gas has been calculated as 1 m^3 = 39.4 x 10^6J.

2) All reactors are LWR's with an operating time of 6,500 hours per year, and a steam cycle efficiency of 34 percent. Capacities represent the maximum potential under WAES assumptions, not necessarily the likely capacities.

WAES NATIONAL ENERGY SUPPLY/DEMAND INTEGRATION STUDIES, 1972-1985-2000
ENERGY SUPPLY/DEMAND INTEGRATION SUMMARY—IMPORTS AND EXPORTS OF PRIMARY ENERGY

Country: The Netherlands

Scenario Assumptions		1972	1985					2000			
WAES CASE		Base Year	A	B	C	D	E	C-1	C-2	D-7	D-8
GNP 1976-85; 1985-2000			High	Low	High	Low	High	High	High	Low	Low
Oil/Energy Price*			11.50-17.25	11.50-17.25	11.50	11.50	11.50-7.66	11.50 17.25	11.50 17.25	11.50	11.50
National Policy Response			Vig	Vig	Vig	Res	Res	Vig	Vig	Vig	Vig
Principal Replacement Fuel								Coal	Nuclear	Coal	Nuclear
(Import) & Export	**Units**										
Oil		(1612)			(2647)	(1963)		(4002)	(4221)	(2771)	(2947)
Natural Gas		855	1677	1677	1677	1677	1677	(352)	(352)	(352)	(352)
Coal		(47)			(246)	(163)		(1153)	(508)	(722)	(316)
Nuclear											
Hydroelectric											
Oil + coal			(2613)	(1742)			(3411)				
Total Net	(Imports) or Exports	804	(936)	(65)	(1216)	(449)	(1734)	(5507)	(5081)	(3791)	(3535)

*1975 US dollars per barrel of Arabian light crude oil, fob Persian Gulf.
All Figures are primary production prior to any processing or conversion.

NATIONAL INPUT WORKSHEET FOR SUPPLY/DEMAND INTEGRATION

Country: The Netherlands

Year: 1985

Case: C

Units: 10^{15} Joules

	(a) Coal	(b) Petroleum	(c) (Syn. Liquids)	(d) Nat. Gas	(e) (Syn. Gas)	(f) (Heat)	(g) (Electricity)††	(h) Hydro Energy	(i) Nuclear Energy	(j) Geothermal energy and other	(k) Total
(1) Transport		977					5				982
(2) Industry	116	1152	4	456	23		134				1885
(3) Agric., Mining, Construction		117		143			12				272
(4) Commercial, (5) Public		40		146			51				238
(6) Residential		47		698			74				818
(7) Non-energy Uses 1)											(759)
(8) Final Energy Demand*	116	2333	4	1443	23		275				4194
(9) Electricity**	88	234		197			-279		172 2)		412
(10) Syn. Gas	23				-23						
(11) Syn. Liquids	4		-4								
(12) Heat											
(13) Energy Sector Self Consumption & conversion losses	15	160 3)		1			4				180
(14) Primary Energy Input	246	2727		1641					172		4786
(15) Indigenous Supply		80		3318					172		3570
(16 Imports***	246	2647									2893
(17) Exports				1677							1677

*Includes non-energy uses.

**Blocks 9g, 10e 11c, and 12f must be negative to avoid double counting of energy from primary fuels.

***Includes imported products.

††Includes only electricity to be purchased by each sector.

Footnotes:

1) Already included in line-items (1), (2), and (3)

2) 2.5 GWe

3) Input refineries 75 x 10^6t (45.4 x 10^6J/t)

NATIONAL INPUT WORKSHEET FOR SUPPLY/DEMAND INTEGRATION

Country: The Netherlands

Year: 1985

Case: D

Units: 10^{15} Joules

	(a)	(b)	(c)	(d)	(e)	(f)	(g)	(h)	(i)	(j)	(k)
	Coal	Petroleum	(Syn. Liquids)	Nat. Gas	(Syn. Gas)	(Heat)	(Electricity)††	Hydro Energy	Nuclear Energy	Geothermal energy and other	Total
(1) Transport		977					5				982
(2) Industry	73	522	3	480			88				1180
(3) Agric., Mining, Construction		114		138			11				264
(4) Commercial, (5) Public		24		148			47				219
(6) Residential		46		677			74				796
(7) Non-energy Uses 1)											(445)
(8) Final Energy Demand*	73	1683	3	1443	14		225				3441
(9) Electricity**	63	200		197			-229		103 2)		334
(10) Syn. Gas	14				-14						
(11) Syn. Liquids	3		-3								
(12) Heat											
(13) Energy Sector Self Consumption & conversion losses	10	160		1			4				175
(14) Primary Energy Input	163	2043		1641					103		3950
(15) Indigenous Supply		80		3318					103		3501
(16 Imports***	163	1963									2126
(17) Exports				1677							1677

*Includes non-energy uses.

**Blocks 9g, 10e 11c, and 12f must be negative to avoid double counting of energy from primary fuels.

***Includes imported products.

††Includes only electricity to be purchased by each sector.

Footnotes:

1) Already included in line-items (1), (2), and (3)

2) 1.5 GWe

NATIONAL INPUT WORKSHEET FOR SUPPLY/DEMAND INTEGRATION

Country: The Netherlands

Year: 2000

Case: C-1

Units: 10^{15} Joules

	(a) Coal	(b) Petroleum	(c) (Syn. Liquids)	(d) Nat. Gas	(e) (Syn. Gas)	(f) (Heat)	(g) (Electricity)††	(h) Hydro Energy	(i) Nuclear Energy	(j) Geothermal energy and other	(k) Total
(1) Transport		1147					11				1158
(2) Industry	575	2167	5	103	23		507				3380
(3) Agric., Mining, Construction		107		210			13				330
(4) Commercial } (5) Public }		16		150			60				226
(6) Residential		13		649			116			15 1)	793
(7) Non-energy Uses	2)										1189
(8) Final Energy Demand*	575	3450	5	1112	23		707			15	5887
(9) Electricity**	535	437					-713		688 3)		947
(10) Syn. Gas	23				-23						
(11) Syn. Liquids	5		-5								
(12) Heat											
(13) Energy Sector Self Consumption & conversion losses	15	195	4)				6				216
(14) Primary Energy Input	1153	4082		1112					688	15	7050
(15) Indigenous Supply		80		760					688	15	1543
(16 Imports***	1153	4002		352							5507
(17) Exports											

*Includes non-energy uses.

**Blocks 9g 10e 11c; and 12f must be negative to avoid double counting of energy from primary fuels.

***Includes imported products.

††Includes only electricity to be purchased by each sector.

Footnotes:

1) Solar Energy
2) Already included in line-items (1), (2), and (3)
3) 10 GWe
4) Input refineries 95 x 10^6 tonnes (45.4 x 10^6J/t)

NATIONAL INPUT WORKSHEET FOR SUPPLY/DEMAND INTEGRATION

Country: The Netherlands

Year: 2000

Case: C-2

Units: 10^{15} Joules

	(a) Coal	(b) Petro-leum	(c) (Syn. Liquids)	(d) Nat. Gas	(e) (Syn. Gas)	(f) (Heat)	(g) (Electri-city)††	(h) Hydro Energy	(i) Nuclear Energy	(j) Geothermal energy and other	(k) Total
(1) Transport		1147					11				1158
(2) Industry	287	2455	5	103	23		507				3380
(3) Agric., Mining, Construction		107		210			13				330
(4) Commercial } (5) Public }		16		150			60				226
(6) Residential		13		649			116			15 [1]	793
(7) Non-energy Uses	[2]										(1189)
(8) Final Energy Demand*	287	3738	5	1112	23		707			15	5887
(9) Electricity**	178	368					-713		1100 [3]		933
(10) Syn. Gas	23				-23						
(11) Syn. Liquids	5		-5								
(12) Heat											
(13) Energy Sector Self Consumption & conversion losses	15	195					6				216
(14) Primary Energy Input	508	4301		1112					1100	15	7036
(15) Indigenous Supply		80		760					1100	15	1955
(16) Imports***	508	4221		352							5081
(17) Exports											

*Includes non-energy uses.

**Blocks 9g, 10e 11c, and 12f must be negative to avoid double counting of energy from primary fuels.

***Includes imported products.

††Includes only electricity to be purchased by each sector.

Footnotes:

1) Solar energy

2) Already included in line-items (1), (2), and (3)

3) 16 GWe

NATIONAL INPUT WORKSHEET FOR SUPPLY/DEMAND INTEGRATION

Country: The Netherlands

Year: 2000

Case: D-7

Units: 10^{15} Joules

	(a) Coal	(b) Petro- leum	(c) (Syn. Liquids)	(d) Nat. Gas	(e) (Syn. Gas)	(f) (Heat)	(g) (Electri- city)††	(h) Hydro Energy	(i) Nuclear Energy	(j) Geothermal energy and other	(k) Total
(1) Transport		1048					9				1057
(2) Industry	300	1095	4	82	20		265				1766
(3) Agric., Mining, Construction		80		132			10				222
(4) Commercial (5) Public		41		150			58				249
(6) Residential		15		748			109			8 1)	880
(7) Non-energy Uses	2)										(591)
(8) Final Energy Demand*	300	2279	4	1112	20		452			8	4175
(9) Electricity**	384	358					-456		447 3)		733
(10) Syn. Gas	20				-20						
(11) Syn. Liquids	4		-4								
(12) Heat											
(13) Energy Sector Self Consumption & conversion losses	14	160					4				178
(14) Primary Energy Input	722	2797		1112					447	8	5086
(15) Indigenous Supply		80		760					447	8	1295
(16 Imports*** (17) Exports	722	2717		352							3791

*Includes non-energy uses.

**Blocks 9g, 10e 11c, and 12f must be negative to avoid double counting of energy from primary fuels.

***Includes imported products.

††Includes only electricity to be purchased by each sector.

Footnotes:

1) Solar Energy

2) Already included in line-items (1), (2), and (3)

3) 6.5 GWe

4) Input refineries 75 x 10^6 tonnes (45.4 x 10^6 J/t)

NATIONAL INPUT WORKSHEET FOR SUPPLY/DEMAND INTEGRATION

Country: The Netherlands

Year: 2000

Case: D-8

Units: 10^{15} Joules

	(a) Coal	(b) Petroleum	(c) (Syn. Liquids)	(d) Nat. Gas	(e) (Syn. Gas)	(f) (Heat)	(g) (Electricity)††	(h) Hydro Energy	(i) Nuclear Energy	(j) Geothermal energy and other	(k) Total
(1) Transport		1048					9				1057
(2) Industry	150	1245	4	82	20		265				1766
(3) Agric., Mining, Construction		80		132			10				222
(4) Commercial, (5) Public		41		150			58				249
(6) Residential		15		748			109			8 1)	880
(7) Non-energy Uses	2)										(591)
(8) Final Energy Demand*	150	2429	4	1112	20		452			8	4175
(9) Electricity**	128	358					-456		688 3)		718
(10) Syn. Gas	20				-20						
(11) Syn. Liquids	4		-4								
(12) Heat											
(13) Energy Sector Self Consumption & conversion losses	14	160					4				178
(14) Primary Energy Input	316	2947		1112					688	8	5071
(15) Indigenous Supply		80		760					688	8	1536
(16 Imports***	316	2867		352							3535
(17) Exports											

*Includes non-energy uses.

**Blocks 9g, 10e 11c, and 12f must be negative to avoid double counting of energy from primary fuels.

***Includes imported products.

††Includes only electricity to be purchased by each sector.

Footnotes:

1) Solar energy

2) Already included in line-items (1),(2), and (3)

3) 10 GWe

DEMAND REDUCTION MODULES EFFICIENCY

CASE C

Demand Line Items and Units	(Units)	User Performance 1972	1985	2000	"What If?" National Efficiency Response[1]	National Response Savings over Year 2000 Scenario Case[3)]	Lead Time for National Response Years[2]
Autos	10^6J/km	3241	2606	2596	2452	0.006	
	km/a	16900	14300	14300	13600	0.026	
Air Travel	.	.	.	.	-		
Truck Freight	10^6J/tkm	4.0	3.8	3.3	-		
Rail Freight	10^6J/tkm	0.7	0.7	0.7	-		
Commercial/Public							
Heat Loss	.	.	.	.	-		
System Efficiency (FF)	-	0.65	0.65	0.75	-		
System Efficiency (E)	-	-	-	-	-		
Primary Metals	10^6J/$	100	95	.	-		
Chemicals	10^6J/$	100	99	.	-		
Energy Intensive Ind.	10^6J/$	.	.	.	-		
Non-Energy Uses (Petrochemical, Asphalt)	10^6J/$	100	100	.	-		
Residential							
Heat Loss	10^9J/dw/y	63.4	56.6	38.8	34.3	0.011	
System Efficiency (FF)	-	0.55	0.55	0.65	-		
System Efficiency (E)	-	-	-	-	-		

[1] Given supervigorous policy and substantially higher than case energy price. These figures are not necessarily additive: they represent components of alternative energy strategies.

[2] Number of years from start-up of national response to full implementation.

3) Units = million barrels a day Oil Equivalent.

NATIONAL RESPONSE WORKSHEET

Country: The Netherlands

WAES SCENARIO CASE: C-1/C-2

Type of Response (Demand Reduction, Supply Expansion, or Fuel Substitution)	Examples of Response Action[1]	Lead Time (Years)[2]	% of Max. Potential Above Scenario Case[3]	Savings Achieved or Supply Added over Scenario Case[4] Oil Saved[5]	Oil Added	Gas Added	Other Supply Added
Demand Reduction autos	Decrease of auto km's (-25%) (shift to public traffic)	25		0.027			0.001 6)
	Improvement of effic. (e.g. by introduction el. car)	10		0.009			0.002 6)
Demand Reduction residential use	Further improvement of insulation (max. potential)	25		0.024			
	Electr.Appl. of energy saving type	10		0.023			
	Expansion of district heating	25		0.022			
Supply expansion residential use	Further expansion of solar heating/heat pumps	25		0.023			0.004 6)
Fuel substitution	Coal gassification	15		0.160			0.234 7)

[1] Responses should reflect "supervigorous" policies; energy prices high enough to draw forth alternative supplies, and demand reductions and fuel substitutions resulting from conservation actions or improved efficiencies even at prices substantially higher than scenario prices.

[2] Number of years from start up to full implementation.

[3] Percent used of total maximum potential for each class of module.

[4] These figures are not necessarily additive; they represent components of alternative energy strategies. Units—Million Barrels a Day Oil Equivalent.

[5] All demand reduction and fuel substitution savings should be noted in "oil saved" column.

6) Additional electricity demand.

7) Additional coal demand.

16.1 SUMMARY

16.1.1 Demand to 2000

The scenarios studied indicate that the primary energy demand in 2000 will amount to 2.0–2.2 million TJ (0.9–1.0 MBDOE) as compared to the 1972 demand of 1.1 million TJ (0.5 MBDOE). This is an increase by a factor of about 1.8 to 2.0 of the primary energy demand, corresponding to an economic growth rate for Norway of around 3.4 percent up to 1985 and 1.6–2.4 percent for 1985 to 2000.

This occurs despite the fact that the relationship between growth in energy demand and economic growth is expected to change in such a way that a given economic growth rate can be sustained with a lower growth in energy demand than indicated by the historical relationship. The reasons for this change are (1) consumer reaction to higher energy prices, (2) a different economic growth pattern, and (3) government policies aiming at improved energy efficiencies.

16.1.2 Supply-Demand Integration

The indigenous supply estimates for 2000 indicate that the total supply potential for Norway is in the range of 4.1–5.1 million TJ (1.9–2.3 MBDOE). About 75 percent of the total supply is in the form of oil and gas from the Norwegian North Sea. The remaining supply will mainly be electricity, of which which around 80–90 percent is expected to be produced by hydro power.

The abundance of North Sea oil and gas thus makes Norway more than self-sufficient for these resources even after 2000. All desired demands for these preferred fuels can therefore be easily satisfied.

All electricity is presently produced from hydro power. By early 1990, however, expected economic and environmental restrictions will limit any further expansion of hydro power. Although oil and gas from the North Sea would also be available for electricity production, it has been assumed that the global scarcity of oil at that time will influence the choice of primary energy resources to be used for this purpose, so that new demands for electricity will most likely be covered mainly by nuclear power.

This chapter was written by Kai Killerud and Pål Kristensen.

16.2 METHODOLOGY

16.2.1 Supply-Demand Integration Procedure

Due to the very favorable energy supply situation in Norway to 2000, the supply-demand integration procedure has been fairly straightforward. Basically it has been assumed that all fuels will be available to the consumer in the desired quantities. Thus, once the preferred demands have been calculated, the fuel mix in the supply-demand integration is also given.

To complete the supply-demand integration process, however, it is necessary to make assumptions about which primary fuels will be used for electricity production. It has been assumed that by the time thermal power plants are required to supplement electricity production from hydro power, the global scarcity of oil will probably lead to the use of nuclear power to meet these new demands. This has been assumed for all scenarios.

16.2.2 Economic Assumptions

The latest available official document that provides basic figures for alternative economic developments up to 1985 and beyond is Stortingsmelding No. 50, 1974–1975, "Natural Resources and Economic Development," a proposition to the parliament. This paper deals with natural resources, international perspectives, population growth, welfare, policies concerning resources and environment, and economic development. In this paper, an attempt was made to take into consideration all of the important factors and constraints that will have long-term influence on the development of Norwegian society. The economic trends outlined in this document should therefore, to the extent possible, include the most significant social, political, and resource-oriented factors. It has been possible, with the help of economists, to match the alternatives presented in this paper to the WAES global scenarios.

A summary of the economic assumptions that

have served as a basis for the Norwegian demand analyses follows. Table 16.1 shows the data used by sector. The data have been taken from the source mentioned above. Small adjustments have been made only for the high growth alternative in the period 1974–1980.

The distribution of the industrial sector's total economic growth across different industrial groups is of great importance for the energy demand. Therefore, as the paper mentioned did not present this kind of detailed data, it was necessary to determine the distribution to be used in the analysis. This distribution was made in consultation with Ministry economists. Table 16.2 presents the rates of growth used for different industrial groups. For transport, the growth rates for the various transportation modes are mainly based upon work done by Transportokonomisk Institutt (Institute of Transport Economics). These figures provided the required basis for determination of growth rates consistent with the data presented in Table 16.1. The resulting basic data for the transportation sector are presented in Table 16.3.

16.2.3 National Energy Policies

As was the case for economic development, official political documents have been sought to obtain signals about future energy policies. Stortingsmelding No. 100, 1973–1974, "Future Supply of Energy in Norway," is the latest official document dealing with energy questions. This paper deals primarily with energy supply, while this chapter focuses on supply-demand integrations. However, certain points in Stortingsmelding No. 100 have influenced the demand analysis, for example the assumption that long-term marginal costs will, in principle, determine electricity prices, unless special conditions point toward a different policy. It has been assumed in this chapter that prices corresponding to long-term marginal costs will be in effect by 1985 for the scenarios with a vigorous policy response. For scenarios with restrained policy response, it has been assumed that this level will be reached somewhat later. Partly based upon indications in Stortingsmelding No. 50, the growth in electricity supply for the metal-producing industries is assumed to be approximately 2 percent per year. It is assumed that there is a good correspondence between this rate of growth, which allows for the improvement of energy efficiency and increases in energy prices, and the economic growth rates given in Table 16.2.

16.2.4 The Effect of Different Oil Prices

Due to the production in the North Sea, Norway has become a net exporter of oil, and in twenty

Table 16.1
Economic growth by sector for the low and high growth scenarios (% per year)

	1974–1980		1980–2000	
	Low growth[a]	High growth[b]	Low growth	High growth
Agriculture, forestry, fishing	1.3	1.3	0.8	0.7
Mining, industry, power production	3.2	3.4	4.4	4.7
Construction	4.8	5.0	4.0	4.4
Commercial	5.4	5.4	4.0	4.2
International shipping	9.3	9.3	1.2	1.8
Other transport	4.5	4.7	4.1	4.9
Miscellaneous public and private services	4.3	4.4	4.2	4.4
Subtotal	4.6	4.7	3.8	4.1
Drilling, extraction, and transport of oil and gas	87.5	87.5	1.1	4.2
Total	7.8	7.9	3.4	4.2

[a]This alternative is directly based on "St. meld. no 50, 1974-75."

[b]Based on "St. meld. no 50," but with small adjustments for certain activities.

Table 16.2
Growth rates by industry for low and high growth scenarios (% per year)

	1974–1980		1980–2000	
	Low growth	High growth	Low growth	High growth
Iron and steel	3.0	3.0	3.0	3.0
Nonferrous metals	3.0	3.0	3.0	3.0
Paper and allied	3.0	3.0	2.0	2.0
Chemicals and allied	6.0	6.0	6.0	7.0
Mineral products	3.0	3.0	4.0	4.0
Food and related	1.5	1.5	1.5	1.5
Transport equipment	2.0	3.0	3.0	4.0
Manufacture of other metal products	4.0	5.0	6.0	6.5
Remaining industry	1.5	1.5	2.0	2.0
Total (excluding refining)	3.0	3.2	4.3	4.6
Total industry	3.2	3.4	4.4	4.7

Table 16.3
Activity growth (% per year) for different modes of transportation for the low and high growth scenarios

	1972-1985		1985-2000	
	Low growth	High growth	Low growth	High growth
Passenger travel				
Automobile (number at end of period)	1.6×10^6	1.7×10^6	2.5×10^6	2.5×10^6
Bus	1.1	1.1	1.1	1.1
Rail	2.4	2.6	2.3	2.6
Airplane, domestic	6.8	8.3	6.2	8.0
Airplane, international	8.2	9.2	7.4	8.6
Ship	1.9	2.0	1.7	1.2
Freight				
Rail	2.5	2.7	2.4	2.7
Airplane, domestic	8.0	10.1	7.2	9.7
Truck	6.4	7.6	5.9	7.5
Ship, domestic	5.7	7.0	5.4	7.0
Ship, international	7.5	7.8	4.3	4.7

years production is expected to grow to about 90 MTOE per year, or about 10 times Norway's present demand. The economic activity related to exploration for and production of oil and gas represents a significant part of the total economic activity in Norway. It is assumed that this situation will make it possible to use the activity in the oil sector as a lever to control the total economic activity and thus to keep economic growth within desired limits. This will most probably be the case with oil prices at or above existing levels, and it has therefore been assumed that oil price will have no influence on economic activity. Oil price will, to a certain extent, influence energy efficiency in some sectors: in oil-consuming processes, higher oil prices will lead to improved specific consumptions. In the transportation sector, on the other hand, the oil price will have only minor influence on energy efficiency, because, with the added refinery costs and and heavy taxation on the fuel itself, the crude oil price will have only negligible impact on the price of gasoline, diesel fuel, etc.

16.2.5 National Policy Response Studies

As a part of the demand analysis, an evaluation has been made of national policies that can help to improve energy efficiency. The evaluation has concentrated on the three sectors having the largest energy demands: transport, industry, and residential. The policies that have been evaluated may be divided into categories according to type of government action required to obtain improved energy efficiencies and energy savings. The categories are financial support, tax incentives, price incentives, mandatory measures, convenience incentives, information and education, research and development, and finally, removal of barriers to energy conservation.

16.3 ENERGY DEMAND TO 2000

16.3.1 Introduction

Compared to most other industrialized countries, the energy situation in Norway is favorable. Even so, the country is facing a new situation since the abundance of potential hydro power enjoyed by Norway for many decades is nearing its end. Some two-thirds of the hydro power resources that can be economically utilized are already developed. In addition, the last one-third will be more expensive to develop, and environmental concerns weigh more heavily. On the other hand, oil and gas discoveries on the Norwegian continental shelf represent new, nearly unlimited energy sources. Nuclear power, now becoming widespread in industrialized countries, may also become another new energy source.

The objective of Norway's energy policies, up until quite recently, has been to ensure an abundant supply of energy, especially electricity. In the last few years, this has been changing, and in general, the future supply of electricity to the electrometallurgical industries will most likely be limited, and strict government allocation will be practiced. The government stresses, however, the importance of safeguarding the many industrial communities built around this type of industry, by assuring these industries' long-term electricity supply. It is not yet clear whether oil and gas will be used to any large extent for production of electricity in Norway.

16.3.2 Summary of Analysis

All the scenarios analyzed for Norway assume high economic growth. The analysis shows that under such conditions the growth of energy demand will be of the same character as economic growth; that is, if the growth rate for the economy is constant, then the growth rate of energy demand will be fairly constant. The analysis shows, however, that the growth rate of energy demand will be substantially lower than that of the economy in the period up to year 2000. This is a significant change from earlier experience.

During the period 1950 to 1972, the ratio between the rate of energy growth and the rate of GNP growth was approximately 1.2, and in the period 1960 to 1972 the ratio was even higher. The scenarios analyzed have similar ratios or elasticities, averaging 0.65–0.8 for the period up to year 2000. The analysis, therefore, shows that under the chosen assumptions, this ratio will be reduced substantially compared to what it was during the last two decades.

Figure 16.1 shows curves for energy demand for two of the most important scenarios, C-1 and D-3, that have been analyzed. Net final energy demand is estimated to increase from about 570,000 TJ (0.26 MBDOE) in 1972 to about 1.4–1.2 million TJ (0.64–0.55 MBDOE) in year 2000, or by a factor of about 2.4 for scenario C-1 and 2.1 for scenario D-3.

Figure 16.1 also shows a traditional development curve based upon the same GNP growth as scenario C-1 but assuming an elasticity of about 1.2. The elasticity is based upon energy delivered at the consumer end and GNP, excluding extraction, refining, and primary transporation of oil and gas, making this development consistent with that before 1972. The difference between this curve and the one for scenario C-1 has long-term significance. The reduction of the energy demand corresponding to the difference between traditional development and

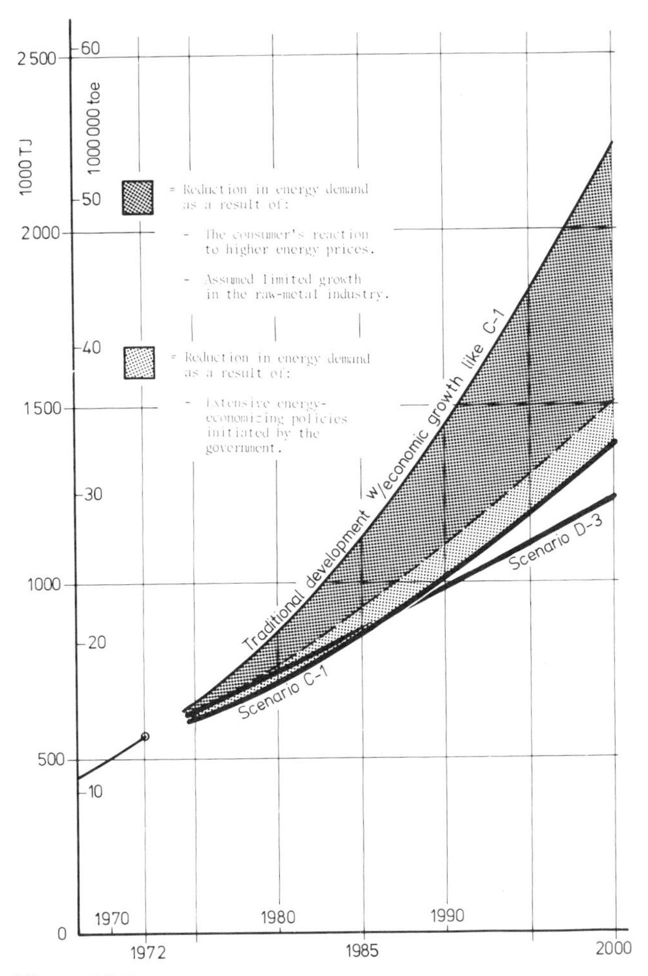

Figure 16.1
Development of energy demand at consumer end, 1972–2000, for scenarios C-1 and D-3.

scenario C-1, which is reflected in the reduction of elasticity from 1.2 to 0.65–0.8, is due to the following:

1. Consumer reaction to the higher energy prices which are assumed in the period up to year 2000.

2. A different growth pattern, particularly in industry, where reduced growth in the metal-producing industries has been assumed.

3. Government policies leading to improvement of energy efficiencies beyond what is automatically attained due to consumer reactions.

It is important to note that in the period up to year 2000, rising oil prices have been assumed, whereas in the period 1950 to 1972, the energy prices were low and falling in real terms.

The dotted curve in Figure 16.1 depicts the assumed development of energy demand if no government policies are put into operation. The other assumptions for this curve are the same as for scenario C-1.

Some of the assumed conservation measures can only be realized by means of research and development. Effects from such research are, of course, also reflected in the reduction of energy consumption due to consumer reaction to higher energy prices. Important areas of research and development are described below.

Table 16.4 presents some of the main results in the form of key figures for scenarios C-1, D-3, and D-7. In these scenarios, a shift from electricity to oil has been assumed. The nuclear scenarios C-2 and D-8 are not described here since they differ very little from scenarios C-1 and D-7. For Norway, these scenarios can be characterized as "low electricity" rather than "coal" scenarios.

Where not otherwise stated, the following figures and conclusions for energy demand will apply to energy delivered to the consumer. Table 16.4 shows that the growth rate for total final energy demand is between 2.4 and 3.4 percent per year throughout the period, about 1 percent per year lower than the economic growth rates. Table 16.4 also shows that the growth rate of total demand for fossil fuels is higher than for electricity; that is, there is a shift toward oil and gas. This shift is not a result of an expected shift away from electricity in traditional activities and processes. Only in one area, residential heating, is an extensive shift expected, and here the shift will be in the opposite direction. The shift that is evident from the figures, therefore, is due to the fact that sectors that mainly utilize fossil fuels will expand the most. These sectors are transport and the chemicals and allied industries.

Energy demand within the transportation sector is estimated to increase much faster than total demand because international trade increases faster than production as a result of the continuous trend toward increased international independence and specialization.

Growth within the residential sector is much slower than that of total energy demand, in spite of assumed higher public standards and greatly in-

Table 16.4
Key figures characterizing the growth of energy demand for the WAES scenarios

	1972–1985		1985–2000		
	C	D	C-1	D-3	D-7
GNP growth (% per year)[a]	4.4	4.2	4.1	3.8	3.8
Net growth in energy demand by fuel (% per year)[b]					
Total	3.4	3.4	3.3	2.4	2.5
Fossil fuels	3.7	3.6	3.7	2.8	3.0
Electricity	2.9	3.0	2.2	1.6	1.6
Net growth in energy demand by sector (% per year)[b]					
Transport	5.6	5.3	4.9	4.0	4.0
Industry	3.4	3.3	3.3	2.3	2.4
Residential	0.8	1.4	-0.1	-0.6	-0.1
Net energy growth/GNP growth	0.77	0.81	0.79	0.64	0.67
Gross growth in total energy demand (% per year)[c]	3.4	3.4	2.4	1.6	1.7
Gross energy growth/GNP growth	0.75	0.80	0.57	0.43	0.45
Energy demand 1985/ Energy demand 1972					
Final	1.55	1.55			
Primary	1.53	1.53			
Energy demand 2000/ Energy demand 1985					
Final			1.62	1.43	1.45
Primary			1.42	1.28	1.29

[a]Growth rates from Table 16.1 subtotal recalculated for the actual periods.

[b]Delivered to the consumer. Includes demand for oil and gas as feedstock.

[c]Includes losses during transportation, production, and distribution.

creased saturation of all presently known household appliances by 2000. The reasons for this are the low population growth and, for most of the scenarios, extensive measures aiming at improved energy efficiency. Furthermore, a significant shift toward electric heating is assumed for most scenarios. Since an efficiency of 100 percent is being used for electric heating, compared to barely 70 percent on average for heating with fossil fuels, such a shift leads to a growth in energy demand at the consumer end that is lower than the growth of utilized energy (i.e., the energy that is really contributing to residential heating).

In the period 1985 to 2000 primary energy growth is significantly lower than final energy growth. This is partially due to the shift away from electricity and partially due to the assumed improvement in efficiency of electricity production (from 35 percent in 1985 to 40 percent in 2000), both factors reducing conversion losses in relative terms.

16.3.3 Effect of Different Scenario Parameters

Figure 16.2 gives the energy demand for the 2000 scenarios. The low economic growth of scenarios D-3 and D-7 gives reductions in energy demand of about 11 percent and 10 percent, respectively, compared to scenario C-1 in the year 2000. The difference in energy price for D-3 and D-7 in the period 1985 to 2000 is of less importance.

Figure 16.3 shows the effect of vigorous national policy response on energy demand for scenarios C and C-1 and the distribution of the total savings by sector. Based on Figure 16.3 the following conclusions can be drawn.

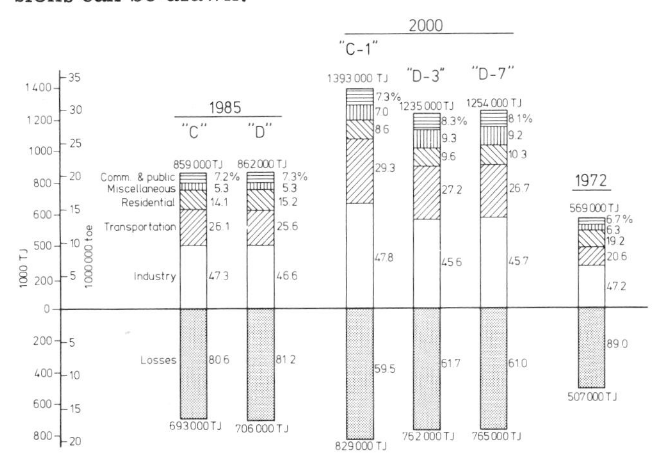

Figure 16.2
Energy demand for scenarios C and D (1985), C-1, D-3, and D-7 (2000), with distribution by sector.

• The assumed vigorous policy response results in a reduction of the energy demand by about 9 percent for scenario C and 8.4 percent for scenario C-1. The total effect is greater for scenario C due to larger savings in energy losses.

• The greatest energy savings are obtained within the three most energy-intensive sectors: transportation, industry, and residential. Within the transportation sector, the main savings are due to a reduction in automobile activity. In industry, the greatest savings are expected within the following groups: iron and steel, nonferrous metals, chemicals and allied, and paper and allied. Significant savings are also obtained because the energy losses are reduced in proportion to the reduction of final energy demand. The most important energy saving policies are described and an indication of their effect is given in section 16.4.

16.3.4 Energy Demand by Sector

Each sector's share of the total energy demand in the year 2000 appears in Figure 16.2. The following conclusions can be drawn with regard to the development up to year 2000.

• Industry, the largest sector, retains a share of about 47 percent throughout the period, in spite of the restrained growth in metal-producing industries, which is balanced by the rapid expansion of the petrochemicals industry.

• Transportation increases its share significantly from 19 percent in 1972 to about 28 percent in the year 2000. This is due to the continuous development toward international independence and specialization leading to a more rapid increase in freight than in production.

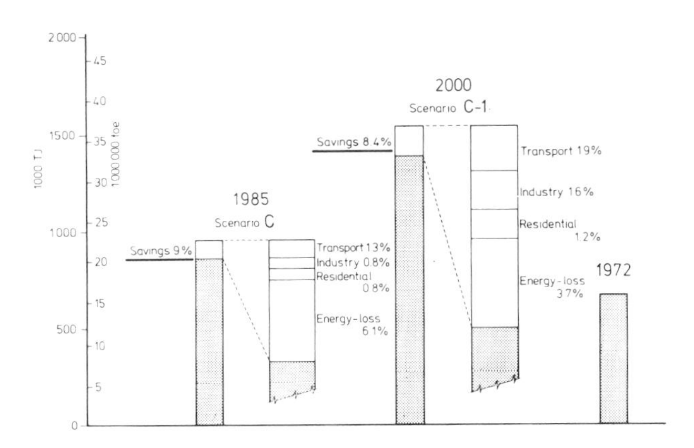

Figure 16.3
The effect of vigorous national policy response for scenarios C and C-1.

• Residential has a significantly reduced share (from 19 percent to about 9 percent in the year 2000). One reason for this is the strong shift toward more effective electric heating and extensive measures aiming at improved energy efficiency. The most fundamental cause is, however, that some of the most important demands within this sector are related to the size of the population rather than the economic activity level, since saturation effects are becoming apparent.

• The shares of other sectors change very little during the period.

16.3.5 Energy Demand by Fuel

Figure 16.4 shows the fuel mix of energy demand for the 2000 scenarios. The following conclusions can be drawn regarding the development of the fuel mix during the period.

• Oil, which is the most important fuel, retains a fairly constant share of about 57 percent from 1972 to 2000.

• The share of electricity decreases from 36 percent in 1972 to around 30 percent in 2000. This is basically due to the rapid expansion of oil- and gas-consuming activities, such as transport and the petrochemicals industry, and the restrained growth of the metal-producing industries.

• The share of gas increases from a negligible 0.2 percent in 1972 to about 7 percent in 2000. This is due to the use of gas in the rapidly expanding petrochemicals industry.

• The share of coal and coke decreases from barely 7 percent in 1972 to around 6 percent in 2000 as a result of the restrained growth assumed in the metal-producing industries.

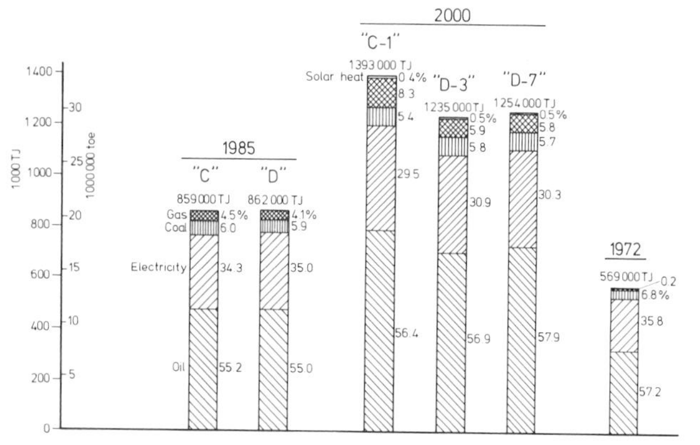

Figure 16.4
Energy demand by fuel for scenarios C and D (1985), C-1, D-3, and D-7 (2000).

• On the whole, the share of fossil fuels increases from around 64 percent in 1972 to about 70 percent in 2000.

• The shift from electricity to fossil fuels is most evident in scenarios with high growth since transportation and the chemicals industry are relatively more dominant under these circumstances.

16.3.6 Energy Demand in the Transportation Sector

Figures 16.5 and 16.6 show the energy demand for 2000 within the transportation sector for passenger travel and freight, respectively. In general, it appears that freight's share of the total demand in the transportation sector increases from 58 percent in 1972 to around 71 percent for scenario C-1 and 65 percent for scenarios D-3 and D-7.

Another general conclusion from this analysis is that the demand for electricity within the transportation sector will have an insignificant share of total demand in 2000. In the scenarios analyzed, it makes up less than 1 percent of the total. Electric cars are not expected to have a significant share of the automobile activity in 2000 and therefore are not included in the demand analysis.

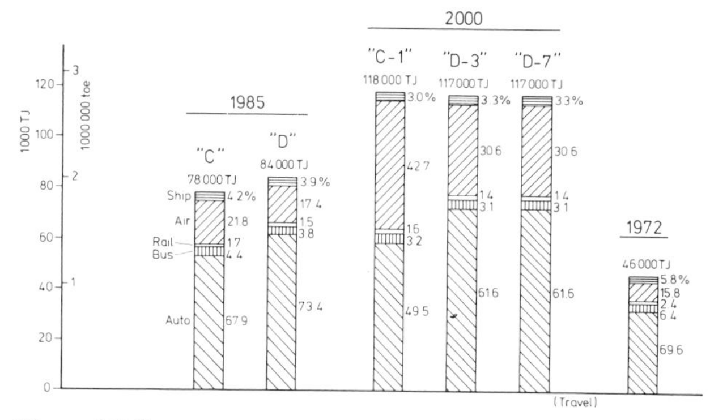

Figure 16.5
Energy demand for passenger travel for scenarios C and D (1985), C-1, D-3, and D-7 (2000).

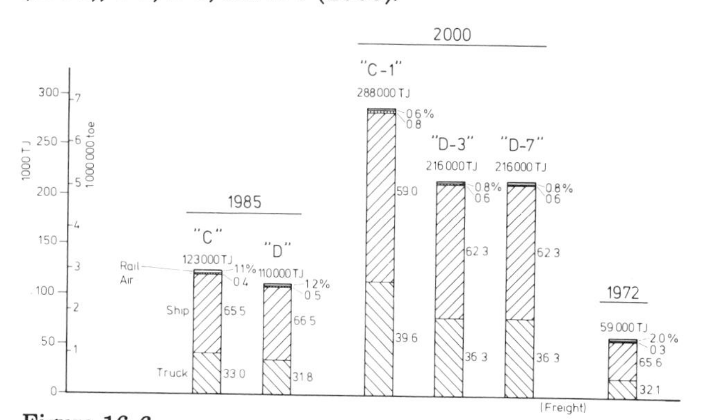

Figure 16.6
Energy demand for freight transport for scenarios C and D (1985), C-1, D-3, and D-7 (2000).

Passenger Travel

The following conclusions on passenger travel development can be drawn for the period 1972 to 2000.

- The energy demand in 2000 increases by a factor of around 2.6 compared to 1972.
- Scenario C-1, which assumes a higher economic growth rate than scenarios D-3 and D-7, has only a slightly higher energy demand than the D-3 and D-7 scenarios because auto activity in the C scenarios (which have a vigorous national policy) is less than in the D scenarios.
- Scenarios D-3 and D-7 show the same energy demand in spite of the 50 percent difference in crude oil price in the period 1985 to 2000. The price difference has very little influence on the price of the final fuel products.
- The share of air travel increases from barely 16 percent in 1972 to about 43 percent for scenario C-1 and about 31 percent for scenarios D-3 and D-7 in 2000. There is a particularly strong increase in the period 1985 to 2000. The saturation effect for autos is the main reason for this expansion.
- Due to the same saturation effect, the share of total energy demand required by automobiles decreases, particularly after 1985, from about 70 percent in 1972 to between 50 and 62 percent in 2000, depending on the scenario. This is true in spite of the fact that the number of autos triples during the period, giving a ratio of two persons per auto.
- Bus and rail transport (urban and intercity) are reduced in their shares of the total in spite of a relative reduction in automobile commuting that transfers activity toward bus and rail.

Average distance traveled per auto per year is reduced by 9–14 percent, depending on the scenario (12,500 km in 1972). The specific fuel consumption for autos is reduced by around 33 percent from around 10 km/liter (24 miles per U.S. gallon) in the base year.

Freight

For development within freight, the following conclusions can be drawn (see Figure 16.6).

- The energy demand increases by a factor of 3.4 to 4.5, depending on the scenario, in the period 1972 to 2000.
- Scenario C-1 has 33 percent higher energy demand than the D-3 and D-7 scenarios due to higher economic activity.
- Scenarios D-3 and D-7 are almost identical with regard to both activity levels and energy demand due to identical growth in GNP. The differences in policy response and crude oil prices have very little influence.
- Truck (local plus intercity) and ship freight dominate energy demand in 2000 as in 1972. Rail and air freight still account for a very small share of the total energy demand.

The specific energy consumptions have been reduced on average between 20 and 40 percent as a result of improved technology and better utilization of capacity.

16.3.7 Energy Demand in the Industrial Sector

Figure 16.7 shows the energy demand in the industrial sector for the 2000 scenarios. From this figure, one can conclude the following.

- The energy demand increases by a factor of 2.1 to 2.5 in the period 1972 to 2000.
- The lower economic growth of scenarios D-3 and D-7 results in a reduction in total energy demand by about 15 percent compared to scenario C-1.
- Scenarios D-3 and D-7 are almost identical in the energy shares of different industries, but D-3 shows somewhat lower energy demand due to higher energy price in the period 1985 to 2000.
- The chemicals and allied industries show the most rapid increase of energy demand. This is caused by the planned and expected expansion of the petrochemicals industry with the energy-intensive production of primary plastics.

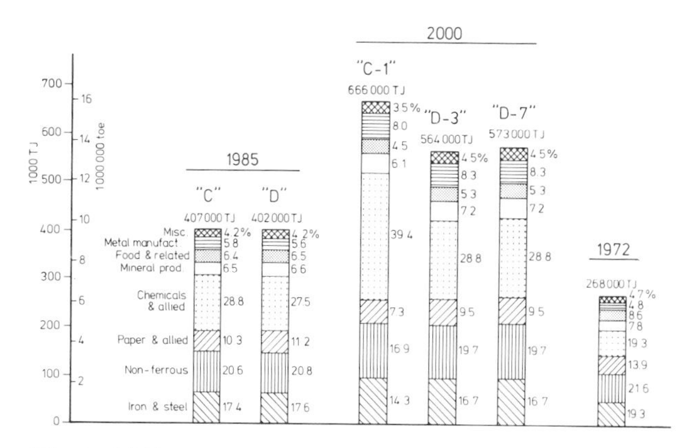

Figure 16.7
Energy demand within the industrial sector for scenarios C and D (1985), C-1, D-3, and D-7 (2000).

- The share of iron and steel and nonferrous metals is reduced since it is assumed that these industries will be allocated approximately 2 percent energy growth per year.
- The share of paper and allied is reduced due to low growth and significant savings in specific energy consumption.
- Metal manufacturing (transport equipment and manufacturing of other metal products) is the only group other than chemicals that has an increasing share of the energy demand.

It appears from the above points that oil price has no major influence on energy demand in Norwegian industry. This is caused by the relatively high share of hydroelectric energy and the utilization of a substantial portion of fossil fuels as feedstock.

For different processes within the iron and steel industry, the assumed improvements in specific energy consumption are in the area of 5–20 percent, compared to 1972. Utilization of waste heat for internal production of electricity has been assumed to affect these improvements, particularly after 1985. Improvement of the Hall process for aluminum production is expected to reduce specific energy consumption by approximately 20 percent to an average of 15,800 kWh/ton in the year 2000. Within the paper and allied industries, a 25 percent reduction in average specific energy consumption is anticipated. Energy savings are mainly a result of better utilization of internal fuels and structural changes leading to larger production units. The substantial expansion in the production of primary plastics in the petrochemicals industry causes an increase in average specific energy consumption in the chemicals and allied industries. This is because after 1977 plastics will mainly be produced directly from natural gas, a process that consumes seven to eight times more energy than production based on imported semiproducts, which have thus far formed the raw material for Norwegian production.

16.3.8 Energy Demand in the Residential Sector

Figure 16.8 shows the energy demand in the residential sector for the 2000 scenarios. From Figure 16.8 the following conclusions about development of energy demand in the residential sector up to year 2000 can be drawn.

- The total energy demand for the 2000 scenarios increases very little (9–18 percent) above the 1972 level.

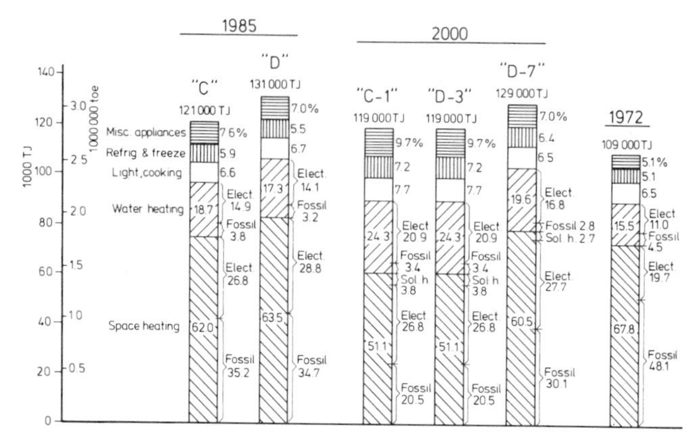

Figure 16.8
Energy demand within the residential sector for scenarios C and D (1985), C-1, D-3, and D-7 (2000).

- Scenarios C-1 and D-3, both of which have a high energy price, show significantly lower energy demand than scenario D-7, which has a lower energy price.
- The share of total energy demand for household appliances, including water heating, increases at the expense of space heating for all scenarios. The reasons are better insulation of houses, more widespread use of electric household appliances, and an approximately 50 percent contribution of waste heat from these appliances to space heating. The introduction of heat pumps after 1985 also increases average efficiency of space heating.
- Solar energy starts contributing to space heating after 1985 with a share of 4 percent for scenarios C-1 and D-3 and 3 percent for scenario D-7.
- Over the whole period there is a significant shift toward electric space heating.

A total of 1.95 million dwellings is anticipated in the year 2000 compared to 1.3 million in 1972. An increase of around 10–15 percent in average floor area per resident is also assumed as well as almost complete saturation for all important appliances. The relatively modest increase in total energy demand for the most realistic scenarios is the result of vigorous policies and higher electricity prices, which lead to a high insulation standard, and the contribution to space heating from household appliances.

16.4 CONCLUSIONS

The change in the energy situation created by the

strong rise in the oil price, quadrupling during the fall of 1973 and spring of 1974, means that many processes and pieces of equipment are no longer at a technical and economic optimum. Significant energy savings can thus be obtained in a sound, economic way by improvement of existing processes and systems or by the use of new competitive processes and technology.

This price change has also influenced prices for other types of fuels and the price of electricity on the world (WOCA) market. In an indirect way, the new situation influences electricity-consuming processes and equipment in Norway.

Government policy actions in the energy field can be utilized to accelerate a development toward processes and equipment that are at a technical optimum in this new situation. Such a development might, in many cases, be combined with fulfillment of the more stringent environmental policies now being enforced.

16.4.1 National Strategies and Policies

Energy conservation possible through policy actions has been analyzed within the framework of the demand study. Such actions may have a significant effect, equivalent to around 8 percent reduction in total energy demand in the year 2000. Energy policy actions include research and development.

The evaluation of possible policy actions has included only the most energy-intensive sectors: transportation, industry, and residential. The objectives of the policy actions are to influence consumers in such a way that the primary energy sources are utilized as efficiently as possible in order to save energy, considering the three constraints given below.

1. *Financial and economic:* As capital usually is a limiting factor, only those possibilities for energy savings that give a return comparable to that of other investment possibilities are realistic.

2. *Technological:* Only technology that is commercially available can be relied upon. Lead times for entirely new technologies are long, usually one or more decades, even if large resources are used for research and development.

3. *Social and political:* Only energy-saving policies that are socially and politically acceptable can be utilized. Policies giving effects that are otherwise unacceptable socially can be accepted if they are combined with countermeasures that render the combined effects acceptable.

The following strategies are the most important within the different sectors.

Transport

- Shifting passengers and goods to less energy-intensive means of transport, which will usually mean more transit time, thus exacting an economic and a social cost.
- Increased occupancy and loading (balanced against reduced schedules and reduced competition).
- Regulation and information aimed at improving the energy efficiency of automobiles.

Industry

- Increased energy prices, reducing energy demand through the effect of price elasticity.
- Support of changeover to more energy-efficient equipment.
- Encouraging better utilization of waste heat.
- Supporting recycling of waste products.

Residential

- Reducing heat loss of new buildings through regulation.
- Increased prices of electricity to reduce energy demand through the effect of price elasticity.
- Supporting improvements that reduce heat loss in old buildings.
- Providing knowledge about more efficient energy use.
- Encouraging the use of the most efficient appliances and heating systems.
- Supporting development of new energy-efficient systems for space heating, water heating, and ventilation.

16.4.2 Research and Development

Research and development must be regarded as a significant part of the policy actions within energy-consuming sectors. Many of the assumed energy savings are dependent upon the necessary technological basis being established through research and development.

Industry

The most important possibilities for energy savings within industry in Norway lie in the few dominant groups with the highest energy demand: iron and steel, nonferrous metals, chemicals and allied, and paper and allied. The possibilities for energy savings, which are regarded as most promising for research and development, are better control and improvement of existing smelting processes within the metal-producing industries; utilization of waste heat, particularly within the metal-producing industries but also within the chemicals and paper industries, and better utilization of waste-type fuels within the paper industry.

It is possible to identify a relatively small number of high-energy-consumption processes being used in many plants. As far as it is possible to judge today, it is expected that most of these processes will not be replaced by new ones for a long time. Concentration of research and development on these processes should, therefore, make significant energy savings possible.

Residential

The best possibilities for energy savings within the residential sector are related to space heating, but significant amounts of energy may also be saved in connection with water heating.

The possibilities for energy savings regarded as most promising in relation to research and development efforts are utilization of heat pump and solar energy systems for space and water heating and various measures for reduction of heat loss from buildings, such as better ventilation and insulation. At present, residential heating in Norway accounts for about 15 percent of the total energy demand. By implementing these energy-saving measures, the total energy demand within the residential sector can be stabilized at about its present level. Achievement of such an effect will require considerable research and development effort. But development with regard to residential heating might also lead to energy savings in connection with space heating in other sectors.

Transport

Transport has an increasing share of the total energy demand and is assumed to make up about one-third of the total energy demand in Norway in year 2000. Ship freight represents the largest consumption of energy within this sector, even when only bunkers filled in Norway are included. Since Norway has a large merchant fleet, improved energy efficiency of ships would be of even greater importance than would appear from the national energy balance.

The following tasks, which all aim at improved energy efficiency of ships, stand out as important areas for Norwegian research and development.

- Improvement of hull configuration.
- Improvement of propulsion systems, especially propeller systems and possible alternatives to such systems.
- Improvement of cleaning systems for hulls and the study of other possible methods for reduction of friction.

16.4.3 Effects of Different Policy Actions

The results of the evaluation of policy actions for the 2000 scenarios are given in summary form below. The effects given relate to scenario C-1, which assumes vigorous policy response.

Transport

Several policies have usually been applied to each of the different modes of transportation. The most significant energy-saving possibilities are related to passenger travel. The fact that policies which influence one mode of transportation often result in secondary effects on other modes of transportation has been taken into account.

The auto activity (passenger-kilometers) is almost five times higher than for bus and rail activity together in 2000. The policies that apply to automobiles directly, therefore, result in larger energy savings than those feasible for all other transportation modes. For autos, the following policies have been evaluated:

- Registration taxes that increase with weight and engine power.
- A 50 percent increase in the price of gasoline.
- Minimum requirements for fuel economy.
- Miscellaneous policies, such as higher parking fees.

In total, these policies are estimated to give a reduction in the average distance traveled per auto per year of about 11 percent and a reduction in the average specific fuel consumption of around 15 percent.

For bus and rail, the following policies have been evaluated:

- Subsidies on ticket prices to such an extent that prices are kept at today's level in real terms.
- Increased information on and better coordination of routes and schedules.
- Special lanes for buses.
- Expansion of rail transport systems.
- "Park and ride" arrangements at suburban train stations.

In total these policies are estimated to increase the activity level (passenger-kilometers) by about 11–17 percent for bus and rail and to reduce the specific energy consumption by about 8–14 percent. This last reduction is basically due to better utilization of capacity.

Industry

The evaluation has been concentrated on the most energy-intensive industrial groups. These groups are iron and steel, nonferrous metals, paper and allied, and chemicals and allied. The following policies have been evaluated:

- Low-cost loans for installation of energy-saving equipment.
- Tax credit for installation of energy-saving equipment and for recycling.
- Increased electricity prices.
- Financial support for research and development, especially for existing energy-intensive processes.

In total, these policies are estimated to reduce the average specific energy consumptions as follows: iron and steel, 5 percent; nonferrous metals, 7 percent; paper and allied, 16 percent; and chemicals and allied, 6 percent.

By far the greatest reduction of specific energy consumption is obtained for paper and allied. For the metal-producing industries as a whole, the energy savings achieved are also significant in absolute terms, since this part of the industry accounts for about 35 percent of the total demand.

Residential

Within the residential sector, the following policies have been evaluated:

- Low-cost loans for improvement of old buildings.
- More stringent requirements in building codes.
- Information about means for energy saving and selection of more energy-efficient equipment.
- Increased electricity prices.
- Research and development.

In total, these policies are assumed to give a reduction in the average demand for heating per unit of floor area of about 25 percent in the year 2000. Some of the above-cited policies also have the effect of reducing energy demand per unit of household appliances by around 14 percent.

16.5 SUPPLY-DEMAND INTEGRATION

The primary energy demand estimate for the year 2000 varies between 2.0 and 2.2 TJ (0.9 and 1.0 MBDOE), depending upon the scenario. The total demands of the two high growth scenarios (C-1 and C-2) are almost identical, as are those of the three low growth scenarios (D-3, D-7, and D-8).

The indigenous supply estimate for year 2000 varies between approximately 5.1 and 4.1 million TJ (2.3 and 1.9 MBDOE), depending upon the scenario. Of this, between 4.0 and 3.0 million TJ (1.9 and 1.4 MBDOE) represents oil and gas from the Norwegian sector of the North Sea. Most of this supply will be exported, that is, around two-thirds of the oil and around 90 percent of the gas. The potential for increased production is much larger but will most probably be limited due to the inflationary effect this production would have on the Norwegian economy. The supply of electricity will balance with the demand; a small amount of coal will be imported.

Thus, the abundance of North Sea oil and gas makes Norway more than self-sufficient for these resources in both 1985 and 2000. All desired demands for these preferred fuels can be easily satisfied. Oil and gas will cover almost all demand for fossil fuels for thermal uses.

Basically, it has been assumed that the desired demand for electricity will be met. The only exception is related to the industries producing raw metals, such as aluminum and ferrosilicon. For these industries some limitations on the supply have been assumed.

All electricity is presently produced from hydro power. By the early 1990s, hydro power development will have reached a limit beyond which further expansion will be unlikely, given the expected economic and environmental limitations.

Electricity produced by some kind of thermal power plants will therefore be required in the 1990s. Oil and gas from the North Sea would be available for this potential demand. However, it has been assumed that the global scarcity of oil at that time will influence the choice of primary energy source for this purpose, which means that new demands on electricity will most likely be covered mainly by nuclear power. This is assumed for all scenarios. The nuclear scenarios (C-2 and D-8), however, assume somewhat higher electricity demands than the coal scenarios (C-1, D-3, and D-7). It is unlikely that coal will be used in any significant quantities for electricity production due to low indigenous production capacity.

Basically, the pattern of the supply-demand integrations for Norway does not vary much among the 1985 or 2000 scenarios. This is due to the very satisfactory supply situation in Norway. The most noticeable difference between the 1985 and 2000 supply-demand integrations is the utilization of uranium and oil for electricity production in the year 2000. The greatest difference occurring in the 2000 supply-demand integrations is found on the supply side. Production of oil and gas in the North Sea is 25 percent less in the low economic growth than in the high economic growth scenarios. This difference, as well as the production levels themselves, are assumed to be unaffected by whether coal or nuclear is assumed as the principal replacement fuel. Regarding the 1985 supply-demand integrations, oil and gas production is assumed the same in both high and low economic growth cases. On the demand side, the differences between high coal and high nuclear scenarios are minor, causing a changeover from the use of fossil fuel to electricity. Total energy demand remains almost unchanged.

The different production levels for electricity in the year 2000 are assumed to affect only the capacity development of nuclear power plants, varying from around 10 to 20 percent of total electricity production depending upon the scenario. Hydro power is assumed to be fully developed by the early 1990s. It has been assumed that only one oil- or gas-fueled power plant (~500 MW) will be built during the transition period before nuclear power is accepted.

16.6 NATIONAL RESPONSE

The global supply-demand integrations show that world oil is becoming scarce. In contrast, Norway has and will continue to have abundant resources of oil and gas in the North Sea. This will most likely have a dampening effect on the national response to the global energy situation. However, for demand reduction, chosen scenarios up to year 2000 have built-in, extensive energy-economizing policies. In addition, a strong reduction in the supply of electricity to the raw metals industry (approximately 2 percent growth per year) is assumed. On this basis, it is considered highly unlikely that further expansion of energy-saving measures will take place on the demand side, even with a serious global scarcity of oil.

Since hydro power will be fully utilized by the early 1990s, thermal power plants will have to be introduced for production of electricity, with oil, gas, and uranium as potential fuels. Given a serious deficit of energy on a global basis, nuclear power plants will likely be preferred to conventional power plants to a somewhat greater extent than assumed in the scenarios. The share of the total production of electricity that comes from thermal power plants is assumed to account for about 10–20 percent of the total, or 2–4 percent of the primary energy demand for Norway in the year 2000. The savings of fossil fuels obtained by wider use of nuclear power plants will have mainly a moral effect from a global perspective, though such an effect could prove to be important enough in itself.

16.7 WORKSHEETS

Summary tables and worksheets appear in this section in the following order:

Energy demand summary, 1972–1985–2000

National input worksheets for supply/demand integrations (7 sheets)

National energy demand projections (5 sheets)

National response worksheets (2 sheets)

Demand reduction modules efficiency (2 sheets)

WAES NATIONAL ENERGY DEMAND STUDIES, 1972-1985-2000
ENERGY DEMAND SUMMARY

Country: NORWAY

Scenario Assumptions	1972	1985					2000			
WAES CASE	Base Year	A	B	C	D	E	C-1	C-2	D-7	D-8
GNP 1976-85; 1985-2000		High	Low	High	Low	High	High	High	Low	Low
Oil/Energy Price*		11.50-17.25	11.50-17.25	11.50	11.50	11.50-7.66	11.50-17.25	11.50-17.25	11.50	11.50
National Policy Response		Vig	Vig	Vig	Res	Res	Vig	Vig	Vig	Vig
Principal Replacement Fuel							Coal	Nuclear	Coal	Nuclear
Primary Energy Demand — **Units** 10^3TJ										
Oil	366	516	492	535	534	582	903	896	836	817
Natural Gas	1	42	39	42	40	43	127	126	80	80
Coal	39	51	51	51	51	51	75	74	72	72
**Nuclear	0	0	0	0	0	0	193	218	109	150
**Hydroelectric	650	920	900	924	944	949	919	919	919	919
Geothermal										
Solar Thermal							5	5	3	3
Solar Electric										
TOTAL PRIMARY ENERGY DEMAND	1076	1529	1482	1552	1568	1624	2222	2238	2019	2041

*1975 US dollars per barrel of Arabian light crude oil, fob Persian Gulf.
All Figures are primary production prior to any processing or conversion.

Footnotes:

**All types of electric power plants: η = 35% for 1972 and 1985
η = 40% for 2000

NATIONAL INPUT WORKSHEET FOR SUPPLY/DEMAND INTEGRATION

Country: NORWAY

Year: 1985

Case: C

Units: 10^3 TJ/y

	(a) Coal	(b) Petroleum	(c) (Syn. Liquids)	(d) Nat. Gas	(e) (Syn. Gas)	(f) (Heat)	(g) (Electricity)††	(h) Hydro Energy	(i) Nuclear Energy	(j) Geothermal energy and other	(k) Total
(1) Transport		221.4					2.9				224.3
(2) Industry	51.0	111.6		18.3			184.3				365.2
(3) Agric., Mining, Construction		35.3					10.5				45.8
(4) Commercial		17.0		.1			13.0				30.0
(5) Public		20.1					11.5				31.6
(6) Residential		48.0					73.0				121.0
(7) Non-energy Uses		21.5		20.0							41.5
(8) Final Energy Demand*	51.0	474.9		38.4			295.1				859.5
(9) Electricity**							−323.4	924.0			600.6
(10) Syn. Gas											
(11) Syn. Liquids											
(12) Heat											
(13) Energy Sector Self Consumption & conversion losses		60.1		3.7			28.3				92.1
(14) Primary Energy Input	51.0	535.0		42.1			0	924.0			1552.1
(15) Indigenous Supply	42	1740		860				1045			3687
(16 Imports***	9										9
(17) Exports		1205		818				121 (=42e)			2144

*Includes non-energy uses.

**Blocks 9g, 10e 11c, and 12f must be negative to avoid double counting of energy from primary fuels.

***Includes imported products.

††Includes only electricity to be purchased by each sector.

Footnotes:

**For all types of electric power plants: $\eta = 35\%$

(3) Includes forestry and fishing

NATIONAL INPUT WORKSHEET FOR SUPPLY/DEMAND INTEGRATION

Country: NORWAY

Year: 1985

Case: D

Units: 10^3 TJ/y

	(a) Coal	(b) Petroleum	(c) (Syn. Liquids)	(d) Nat. Gas	(e) (Syn. Gas)	(f) (Heat)	(g) (Electricity)††	(h) Hydro Energy	(i) Nuclear Energy	(j) Geothermal energy and other	(k) Total
(1) Transport		218.2					2.8				221.0
(2) Industry	50.9	112.3		17.5			182.9				363.6
(3) Agric., Mining, Construction		35.4					10.6				46.1
(4) Commercial		17.1		.1			13.1				30.3
(5) Public		20.3					11.6				31.9
(6) Residential		50.5					80.5				131.0
(7) Non-energy Uses		20.0		18.5							38.6
(8) Final Energy Demand*	50.9	473.9		36.1			301.5				862.5
(9) Electricity**							-330.4	943.9			613.5
(10) Syn. Gas											
(11) Syn. Liquids											
(12) Heat											
(13) Energy Sector Self Consumption & conversion losses		60.0		3.5			28.9				155.9
(14) Primary Energy Input	50.9	533.9		39.6			0	943.9			1568.3
(15) Indigenous Supply	42	1740		860				1045			3687
(16 Imports***	9										9
(17) Exports		1206		820				101 (=35e)			2127

*Includes non-energy uses.

**Blocks 9g, 10e 11c, and 12f must be negative to avoid double counting of energy from primary fuels.

***Includes imported products.

††Includes only electricity to be purchased by each sector.

Footnotes:

**For all types of electric power plants: η = 35%

(3) Includes forestry and fishing

NATIONAL INPUT WORKSHEET FOR SUPPLY/DEMAND INTEGRATION

Country: NORWAY

Year: 2000

Case: C-1

Units: 10^3 TJ/y

	(a) Coal	(b) Petroleum	(c) (Syn. Liquids)	(d) Nat. Gas	(e) (Syn. Gas)	(f) (Heat)	(g) (Electricity)††	(h) Hydro Energy	(i) Nuclear Energy	(j) Geothermal energy and other	(k) Total
(1) Transport		451					5				456
(2) Industry	75	182		51			271				579
(3) Agric., Mining, Construction		38					12				50
(4) Commercial		29					20				49
(5) Public		35					18				53
(6) Residential		29					85			5	119
(7) Non-energy Uses		22		65							87
(8) Final Energy Demand*	75	786		116			411			5	1393
(9) Electricity**		18					−452	919	193		678
(10) Syn. Gas											
(11) Syn. Liquids											
(12) Heat											
(13) Energy Sector Self Consumption & conversion losses		99		11			41				151
(14) Primary Energy Input	75	903		127			0	919	193	5	2222
(15) Indigenous Supply	42	2662		1320				919	193⊕	5	5141
(16 Imports***	33										33
(17) Exports		1759		1193							2952

*Includes non-energy uses.

**Blocks 9g, 10e 11c, and 12f must be negative to avoid double counting of energy from primary fuels.

***Includes imported products.

††Includes only electricity to be purchased by each sector.

Footnotes:

** For all types of electric power plants: $\eta \doteq 40\%$

(3) Includes forestry anf fishing

⊕ Uranium imported

NATIONAL INPUT WORKSHEET FOR SUPPLY/DEMAND INTEGRATION

Country: NORWAY

Year: 2000

Case: C-2

Units: 10^3 TJ/y

	(a)	(b)	(c)	(d)	(e)	(f)	(g)	(h)	(i)	(j)	(k)
	Coal	Petro-leum	(Syn. Liquids)	Nat. Gas	(Syn. Gas)	(Heat)	(Electricity)††	Hydro Energy	Nuclear Energy	Geothermal energy and other	Total
(1) Transport		450					5				455
(2) Industry	74	177		51			278				580
(3) Agric., Mining, Construction		38					13				51
(4) Commercial		28					22				50
(5) Public		34					20				54
(6) Residential		30					82			5	117
(7) Non-energy Uses		22		64							86
(8) Final Energy Demand*	74	779		115			420			5	1393
(9) Electricity**		18					-462	919	218		693
(10) Syn. Gas											
(11) Syn. Liquids											
(12) Heat											
(13) Energy Sector Self Consumption & conversion losses		99		11			42				152
(14) Primary Energy Input	74	896		126			0	919	218	5	2238
(15) Indigenous Supply	42	2662		1320				919	193⊕	5	5162
(16 Imports***	32										32
(17) Exports		1766		1194							2960

*Includes non-energy uses.

**Blocks 9g, 10e 11c, and 12f must be negative to avoid double counting of energy from primary fuels.

***Includes imported products.

††Includes only electricity to be purchased by each sector.

Footnotes:

** For all types of electric power plants: η = 40%

(3) Includes forestry and fishing

⊕ Uranium imported

NATIONAL INPUT WORKSHEET FOR SUPPLY/DEMAND INTEGRATION

Country: NORWAY

Year: 2000

Case: D-3

Units: 10^3 TJ/y

	(a)	(b)	(c)	(d)	(e)	(f)	(g)	(h)	(i)	(j)	(k)
	Coal	Petro-leum	(Syn. Liquids)	Nat. Gas	(Syn. Gas)	(Heat)	(Electri-city)††	Hydro Energy	Nuclear Energy	Geothermal energy and other	Total
(1) Transport		393					4				397
(2) Industry	72			33			242				511
(3) Agric., Mining, Construction		40					13				53
(4) Commercial		29					20				49
(5) Public		35					18				53
(6) Residential		29					85			5	119
(7) Non-energy Uses		13		40							53
(8) Final Energy Demand*	72	703		73			382			5	1235
(9) Electricity**		18					−420	919	114		631
(10) Syn. Gas											
(11) Syn. Liquids											
(12) Heat											
(13) Energy Sector Self Consumption & conversion losses		89		7			38				134
(14) Primary Energy Input	72	810		80			0	919	114	5	2000
(15) Indigenous Supply	42	2662		1320				919	114⊕	5	5052
(16 Imports***	30										30
(17) Exports		1852		1240							3082

*Includes non-energy uses.

**Blocks 9g, 10e 11c, and 12f must be negative to avoid double counting of energy from primary fuels.

***Includes imported products.

††Includes only electricity to be purchased by each sector.

Footnotes:

** For all types of electric power plants: η = 40%

(3)Includes forestry anf fishing

⊕ Uranium imported

NATIONAL INPUT WORKSHEET FOR SUPPLY/DEMAND INTEGRATION

Country: NORWAY

Year: 2000

Case: D-7

Units: 10^3 TJ/y

	(a) Coal	(b) Petroleum	(c) (Syn. Liquids)	(d) Nat. Gas	(e) (Syn. Gas)	(f) (Heat)	(g) (Electricity)††	(h) Hydro Energy	(i) Nuclear Energy	(j) Geothermal energy and other	(k) Total
(1) Transport		393					4				397
(2) Industry	72	173		33			242				520
(3) Agric., Mining, Construction		40					13				53
(4) Commercial		29					20				49
(5) Public		35					18				53
(6) Residential		43					83			3	129
(7) Non-energy Uses		13		40							53
(8) Final Energy Demand*	72	726		73			380			3	1254
(9) Electricity**		18					-418	919	109		628
(10) Syn. Gas											
(11) Syn. Liquids											
(12) Heat											
(13) Energy Sector Self Consumption & conversion losses		92		7			38				137
(14) Primary Energy Input	72	836		80			0	919	109	3	2019
(15) Indigenous Supply	42	2055		1006				919	109⊕	3	4134
(16 Imports***	30										30
(17) Exports		1219		926							2145

*Includes non-energy uses.

**Blocks 9g, 10e 11c, and 12f must be negative to avoid double counting of energy from primary fuels.

***Includes imported products.

††Includes only electricity to be purchased by each sector.

Footnotes:

** For all types of electric power plants: η = 40%

(3) Includes forestry and fishing

⊕ Uranium imported

NATIONAL INPUT WORKSHEET FOR SUPPLY/DEMAND INTEGRATION

Country: NORWAY

Year: 2000

Case: D-8

Units: 10^3 TJ/y

	(a) Coal	(b) Petro-leum	(c) (Syn. Liquids)	(d) Nat. Gas	(e) (Syn. Gas)	(f) (Heat)	(g) (Electri-city)††	(h) Hydro Energy	(i) Nuclear Energy	(j) Geothermal energy and other	(k) Total
(1) Transport		393					4				397
(2) Industry	72	167		33			248				520
(3) Agric., Mining, Construction		40					13				53
(4) Commercial		28					21				49
(5) Public		34					20				54
(6) Residential		34					89			3	126
(7) Non-energy Uses		13		40							53
(8) Final Energy Demand*	72	709		73			395			3	1252
(9) Electricity**		18					−434	919	150		653
(10) Syn. Gas											
(11) Syn. Liquids											
(12) Heat											
(13) Energy Sector Self Consumption & conversion losses		90		7			39				136
(14) Primary Energy Input	72	817		80			0	919	150	3	2041
(15) Indigenous Supply	42	2055		1006				919	150⊕		4175
(16 Imports***	30										30
(17) Exports		1238		926							2164

*Includes non-energy uses.

**Blocks 9g, 10e 11c, and 12f must be negative to avoid double counting of energy from primary fuels.

***Includes imported products.

††Includes only electricity to be purchased by each sector.

Footnotes:

** For all types of electric power plants: η = 40%

(3) Includes forestry and fishing

⊕ Uranium imported

NATIONAL ENERGY DEMAND PROJECTIONS

Country: NORWAY

Year: 1972

WAES Case: Base Year

Population: $3.9x10^6$

No. of Autos: $780x10^3$

Number of
Residential Dwellings: $1,297x10^3$

Sector	Activity Level	Intensity	Energy Use	
Transportation	Total Vehicle Distance/Yr.(1)	Energy Efficiency(2)	(TJ) Fossil and Other Fuels	(TJ) Electric[1]
Automobiles	12500 km/auto	$3.253x10^6$J/km	32000	
Air Travel (domestic)	.92	4.19	3850	
(3)Other Passenger	3.27	.73	2380	
(4)Truck & Air Freight	4.7	4.02	18973	
Rail Freight	2.48	.73/.34	597	565
(5)Shipping	156.06	.14	21848	

Industry	Output/Year	Energy/Output		
(6)Metals	(10^3ton) 3098	(10^6J/ton) 16674	26248	25416
Chemicals	2407	13360	11177	20981
(7)Other Energy Intensive	844	68604	13820	44089
(8)Non-Energy Intensive	2924	7859	15920	7067
Petrochemical Feedstocks			19670	
Asphalt & Road Tar				
(9)Agriculture			22378	5690
Mining	29346	149	1950	2420
Construction			3790	

NATIONAL ENERGY DEMAND PROJECTIONS (con't)

Commercial & Public	Floor Area(10) (14)	Heat Loss/(15) Area/Yr.	Syst. Eff.		
Foss. Fuel Space Heat	702000	$\frac{22918}{702\text{x}10^3}$		22918 (11)	
Electric Space Heat	702000	$\frac{15001}{702\text{x}10^3}$			15001 (12)
Other Fossil Fuel[2]	702000	$\frac{46}{702\text{x}10^3}$		46 (13)	
Other Electric[2]					

Residential	Dwellings	Heat Loss/ kWhArea/Yr.(16)	Syst. Eff.		
Foss. Fuel Space Heat	803994	15700 − 3175(17)	.66x.95 (18)	52540	
Electric Space Heat	492770	15700 − 3175(17)	1.00x.95 (18)		21510
Foss. Fuel Appliances[3]				5430	
Electric Appliances[3]		7760			29730
Total Delivered Energy: Demand (Including Fuel used for Feedstocks)				370768	204127

[1] Purchased electricity

[2] The number for heat loss/area/year represents the total fossil fuel (electric) energy demand per unit area per year.

[3] The number for heat loss/area/year represents the total fossil fuel (electric) energy demand per dwelling.

Footnotes:

(1) 10^9pass.km. or 10^9ton.km.
(2) 10^6J/pass.km. or 10^6J/ton.km.
(3) Intercity bus.
(4) Truck only.
(5) International, fuel filled in Norway, all operators.
(6) Iron & Steel.
(7) Aluminium & Allied.
(8) Food & Related.
(9) Includes forestry & fishing.
(10) Floor area is not available. The building statistics are based on yearly changes.
(11) Includes energy for hot water supply and some unspecified uses.
(12) Total value for the category, includes lighting, water heating, etc.
(13) Gasworks, gas mainly for cooking and heating (restaurants, etc.).
(14) Represented by number of employed.
(15) TJ/person.
(16) Per dwelling.
(17) Indirect heating from appliances (50% of total consumption of electricity to such equipment).
(18) $\frac{\text{Climate factor 1972}}{\text{Climate factor normal year}}$.

NATIONAL ENERGY DEMAND PROJECTIONS

Country: NORWAY

Year: 1985

WAES Case: C

Population: $4.4x10^6$

No. of Autos: $1.68x10^6$

Number of Residential Dwellings: $1.64x10^6$

Sector	Activity Level	Intensity	Energy Use	
Transportation	Total Vehicle Distance/Yr.(1)	Energy Efficiency(2)	(TJ) Fossil and Other Fuels	(TJ) Electric[1]
Automobiles	11375km/auto	$2.667x10^6$J/km	52630	
Air Travel (domestic)	2.53	3.53	8483	
(3) Other Passenger	4.82	.59	2841	
(4) Truck & Air Freight	11.81	3.43	40508	
Rail Freight	3.97	.52/.24	680	637
(5) Shipping	341.16	.13	44351	

Industry	Output/Year	Energy/Output		
	(10^3ton)	(10^6J/ton)		
(6) Metals	3851	18425	36046	34910
Chemicals	5009	15215	35219	40996
(7) Other Energy Intensive	1365	61401	20006	63822
(8) Non-Energy Intensive	3549	7318	18256	7718
Petrochemical Feedstocks			40960	
Asphalt & Road Tar				
(9) Agriculture			25456	6718
Mining	47062	146	3066	3805
Construction			6772	

NATIONAL ENERGY DEMAND PROJECTIONS (con't)

Commercial & Public	Floor Area (10)	Heat Loss/ Area/Yr (11)	Syst. Eff.		
Foss. Fuel Space Heat	1.8	$\frac{22918}{1.00}$x.90		37125	
Electric Space Heat	1.8	$\frac{15001}{1.00}$x.90			24410
Other Fossil Fuel[2]	1.86	$\frac{46}{100}$		86 (12)	
Other Electric[2]					

Residential	Dwellings	Heat Loss/ (kWh)Area/Yr. (13)	Syst. Eff.		
Foss. Fuel Space Heat	$.738 \times 10^6$	14494 -3440(14)	.69	42570	
Electric Space Heat	$.902 \times 10^6$	13423 -3440(14)	1.00		32420
Foss. Fuel Appliances[3]				5420	
Electric Appliances[3]		7760			40610
Total Delivered Energy: Demand (Including Fuel used for Feedstocks)				564327	295128

[1] Purchased electricity

[2] The number for heat loss/area/year represents the total fossil fuel (electric) energy demand per unit area per year.

[3] The number for heat loss/area/year represents the total fossil fuel (electric) energy demand per dwelling.

(1) 10^9pass-km or 10^9ton-km
(2) 10^6J/pass-km or 10^6J/ton-km
(3) Intercity bus
(4) Truck only
(5) International: fuel filled in Norway, all operators
(6) Iron and steel
(7) Aluminum and allied
(8) Food and related
(9) Includes forestry and fishing
(10) Production index, 1972 = 1.00
(11) Unit: TJ/index
(12) Gas
(13) Per dwelling
(14) Indirect heating from appliances (50% of total consumption of electricity by such equipment)

NATIONAL ENERGY DEMAND PROJECTIONS

Country: NORWAY

Year: 1985

WAES Case: D

Population: 4.4×10^6

No. of Autos: 1.77×10^6

Number of Residential Dwellings: 1.64×10^6

Sector	Activity Level	Intensity	Energy Use (TJ) Fossil and Other Fuels	(TJ) Electric[1]
Transportation	Total Vehicle Distance/Yr.(1)	Energy Efficiency(2)		
Automobiles	12125km/auto	2.873×10^6 J/km	61554	
Air Travel (domestic)	2.16	3.31	7160	
(3) Other Passenger	4.22	.61	2572	
(4) Truck & Air Freight	10.21	3.43	35034	
Rail Freight	3.86	.52/.24	661	620
(5) Shipping	329.40	.13	42822	

Industry	Output/Year (10^3 ton)	Energy/Output (10^6 J/ton)	Fossil and Other Fuels	Electric
(6) Metals	3800	18675	36046	34910
Chemicals	4652	15573	33478	38970
(7) Other Energy Intensive	1328	63116	20006	63822
(8) Non-Energy Intensive	3549	7318	18256	7718
Petrochemical Feedstocks			38041	
Asphalt & Road Tar				
(9) Agriculture			25668	6788
Mining	47656	146	3105	3853
Construction			6673	

NATIONAL ENERGY DEMAND PROJECTIONS (con't)

Commercial & Public	Floor Area(10)	Heat Loss/ Area/Yr(11)	Syst. Eff.		
Foss. Fuel Space Heat	1.79	22918 / 1.00 x.92		37780	
Electric Space Heat	1.79	15001 / 1.00 x.92			24658
Other Fossil Fuel[2]	1.84	46 / 1.00		85(12)	
Other Electric[2]					

Residential	Dwellings	Heat Loss/ (kWh) Area/Yr. (13)	Syst. Eff.		
Foss. Fuel Space Heat	$.738 \times 10^6$	15244 −3620(14)	.68	45420	
Electric Space Heat	$.902 \times 10^6$	15244 −3620(14)	1.00		37750
Foss. Fuel Appliances[3]				5077	
Electric Appliances[3]		7760			42755
Total Delivered Energy: Demand (Including Fuel used for Feedstocks)				560979	301494

[1] Purchased electricity

[2] The number for heat loss/area/year represents the total fossil fuel (electric) energy demand per unit area per year.

[3] The number for heat loss/area/year represents the total fossil fuel (electric) energy demand per dwelling.

(1) 10^9 pass-km or 10^9 ton-km
(2) 10^6 J/pass-km or 10^6 J/ton-km
(3) Intercity bus
(4) Truck only
(5) International: fuel filled in Norway, all operators
(6) Iron and steel
(7) Aluminum and allied
(8) Food and related
(9) Includes forestry and fishing
(10) Production index, 1972 = 1.00
(11) Unit: TJ/index
(12) Gas
(13) Per dwelling
(14) Indirect heating from appliances (50% of total consumption of electricity by such equipment)

NATIONAL ENERGY DEMAND PROJECTIONS

Country: NORWAY

Year: 2000

WAES Case: C-1

Population: $5.1x10^6$

No. of Autos: $2.5x10^6$

Number of Residential Dwellings: $1.95x10^6$

	Sector	Activity Level	Intensity	Energy Use (TJ) Fossil and Other Fuels	(TJ) Electric[1]
	Transportation	Total Vehicle Distance/Yr.(1)	Energy Efficiency(2)		
	Automobiles	10750km/auto	$2.183x10^6$J/km	58665	
	Air Travel (domestic)	7.81	3.10	24197	
(3)	Other Passenger	5.72	.54	3088	
(4)	Truck & Air Freight	35.17	3.24	113963	
	Rail Freight	6.28	.43/.20	891	841
(5)	Shipping	679.45	.12	81535	

	Industry	Output/Year (10^3ton)	Energy/Output (10^6J/ton)		
(6)	Metals	5632	16957	48512	46985
	Chemicals	10414	16956	102379	74199
(7)	Other Energy Intensive	2045	55182	26952	85896
(8)	Non-Energy Intensive	4438	6733	21358	8493
	Petrochemical Feedstocks			85883	
	Asphalt & Road Tar				
(9)	Agriculture			22420	7684
	Mining	61397	146	4000	4964
	Construction			11979	

NATIONAL ENERGY DEMAND PROJECTIONS (con't)

Commercial & Public	Floor Area(10)	Heat Loss/ Area/Yr. (11)	Syst. Eff.		
Foss. Fuel Space Heat	1.92	$\frac{37125}{1.00}$x.90		64000	
Electric Space Heat	1.73	$\frac{24410}{1.00}$x.90			38000
Other Fossil Fuel[2]	1.85			159(12)	
Other Electric[2]					

Residential	Dwellings	(kWh) Heat Loss/ Area/Yr(13)	Syst. Eff.		
Foss. Fuel Space Heat	$.585 \times 10^6$	12619 -3772(14)	.70	24341	(1381) (15)
Electric Space Heat	1.365×10^6	12225 -3772(14)	1.21	(3115) (15)	31752
Foss. Fuel Appliances[3]				5090	
Electric Appliances[3]		7710			52960
Total Delivered Energy: Demand (Including Fuel used for Feedstocks)				977000	410900 (4496)(15)

[1] Purchased electricity

[2] The number for heat loss/area/year represents the total fossil fuel (electric) energy demand per unit area per year.

[3] The number for heat loss/area/year represents the total fossil fuel (electric) energy demand per dwelling.

(1) 10^9 pass-km or 10^9 ton-km
(2) 10^6J/pass-km or 10^6J/ton-km
(3) Intercity bus
(4) Truck only
(5) International: fuel filled in Norway, all operators
(6) Iron and steel
(7) Aluminum and allied
(8) Food and related
(9) Includes forestry and fishing
(10) Production index, 1972 = 1.00
(11) Unit: TJ/index
(12) Gas
(13) Per dwelling
(14) Indirect heating from appliances (50% of total consumption of electricity by such equipment)
(15) 5% contribution from solar heat

NATIONAL ENERGY DEMAND PROJECTIONS

Country: NORWAY

Year: 2000

WAES Case: D-7

Population: 5.1×10^6

No. of Autos: 2.8×10^6

Number of Residential Dwellings: 1.95×10^6

Sector	Activity Level	Intensity	Energy Use	
Transportation	Total Vehicle Distance/Yr. (1)	Energy Efficiency (2)	(TJ) Fossil and Other Fuels	(TJ) Electric[1]
Automobiles	11375km/auto	2.26×10^6 J/km	72137	
Air Travel (domestic)	5.33	3.01	16038	
(3) Other Passenger	5.32	.56	2979	
(4) Truck & Air Freignt	24.13	3.24	78192	
Rail Freight	5.86	.43/.20	831	785
(5) Shipping	619.43	.12	74332	

Industry	Output/Year	Energy/Output		
(6) Metals	(10^3 ton) 5620	(10^6 J/ton) 16993	49736	45761
Chemicals	6354	17833	65706	47606
(7) Other Energy Intensive	1970	57258	26925	85896
(8) Non-Energy Intensive	4438	6852	22098	8308
Petrochemical Feedstocks			51925	
Asphalt & Road Tar				
(9) Agriculture			23382	7528
Mining	65226		4249	5274
Construction			12369	

NATIONAL ENERGY DEMAND PROJECTIONS (con't)

Commercial & Public	Floor Area (10)	Heat Loss/ Area/Yr. (11)	Syst. Eff.		
Foss. Fuel Space Heat	1.86	$\frac{37780}{1.00}$x.91		64000	
Electric Space Heat	1.69	$\frac{24658}{1.00}$x.91			38000
Other Fossil Fuel[2]	1.80	$\frac{85}{1.00}$		153 (12)	
Other Electric[2]					

Residential	Dwellings	Heat Loss/ (kWh) Area/Yr. (13)	Syst. Eff.		
Foss. Fuel Space Heat	$.78x10^6$	13433 -3340(14)	.69	39018	(1416) (15)
Electric Space Heat	$1.17x10^6$	13014 -3340(14)	1.08	(2037) (15)	35843
Foss. Fuel Appliances[3]				4340	
Electric Appliances[3]		7760			
Total Delivered Energy: Demand (Including Fuel used for Feedstocks)				871000 (3453) (15)	380400

[1] Purchased electricity

[2] The number for heat loss/area/year represents the total fossil fuel (electric) energy demand per unit area per year.

[3] The number for heat loss/area/year represents the total fossil fuel (electric) energy demand per dwelling.

(1) 10^9 pass-km or 10^9 ton-km
(2) 10^6J/pass-km or 10^6J/ton-km
(3) Intercity bus
(4) Truck only
(5) International: fuel filled in Norway, all operators
(6) Iron and steel
(7) Aluminum and allied
(8) Food and related
(9) Includes forestry and fishing
(10) Production index, 1972 = 1.00
(11) Unit: TJ/index
(12) Gas
(13) Per dwelling
(14) Indirect heating from appliances (50% of total consumption of electricity by such equipment)
(15) 5% contribution from solar heat

NATIONAL RESPONSE WORKSHEET

Country: NORWAY

WAES SCENARIO CASE: C-1

Type of Response (Demand Reduction, Supply Expansion, or Fuel Substitution)	Examples of Response Action[1]	Lead Time (Years)[2]	% of Max. Potential Above Scenario Case[3]	Savings Achieved or Supply Added over Scenario Case[4] Oil Saved[5]	Oil Added	Gas Added	Other Supply Added
Supply Expansion (Nuclear)	Increase nuclear from 3.44GWe to approx.4.0 GWe	10	33	.01			.01

[1] Responses should reflect "supervigorous" policies; energy prices high enough to draw forth alternative supplies, and demand reductions and fuel substitutions resulting from conservation actions or improved efficiencies even at prices substantially higher than scenario prices.

[2] Number of years from start up to full implementation.

[3] Percent used of total maximum potential for each class of module.

[4] These figures are not necessarily additive; they represent components of alternative energy strategies. Units—Million Barrels a Day Oil Equivalent.

[5] All demand reduction and fuel substitution savings should be noted in "oil saved" column.

NATIONAL RESPONSE WORKSHEET

Country: NORWAY

WAES SCENARIO CASE: D-7

Type of Response (Demand Reduction, Supply Expansion, or Fuel Substitution)	Examples of Response Action[1]	Lead Time (Years)[2]	% of Max. Potential Above Scenario Case[3]	Savings Achieved or Supply Added over Scenario Case[4] Oil Saved[5]	Oil Added	Gas Added	Other Supply Added
Supply Expansion (Nuclear)	Increase nuclear from 2.0GWe - 2.8 GWe	10	27	.01			.01

[1] Responses should reflect "supervigorous" policies; energy prices high enough to draw forth alternative supplies, and demand reductions and fuel substitutions resulting from conservation actions or improved efficiencies even at prices substantially higher than scenario prices.

[2] Number of years from start up to full implementation.

[3] Percent used of total maximum potential for each class of module.

[4] These figures are not necessarily additive; they represent components of alternative energy strategies. Units—Million Barrels a Day Oil Equivalent.

[5] All demand reduction and fuel substitution savings should be noted in "oil saved" column.

DEMAND REDUCTION MODULES EFFICIENCY

CASE C,C-1,C-2

Demand Line Items and Units	(Units)	User Performance 1972	1985	2000	"What If?" National Efficiency Response[1]	National Response Savings over Year 2000 Scenario Case	Lead Time for National Response Years[2]
Autos	(10^6J/km)	3.25	2.67	2.18	NONE		
(Domestic) Air Travel	(10^6J/pass.km)	4.19	3.35	3.10			
Truck Freight	(10^6J/tonkm)	4.02	3.43	3.24			
(Electric) Rail Freight	(10^6J/tonkm)	.34	.24	.20			
Commercial/Public							
Heat Loss							
System Efficiency (FF)							
System Efficiency (E)							
Primary Metals (Iron & Steel)	(GJ/ton)	16.7	18.4	17.0			
Chemicals	(GJ/ton)	13.4	15.2	16.9			
(Aluminium) Energy Intensive Ind.	(GJ/ton)	68.6	61.4	55.2			
Non-Energy Uses (Petrochemical, Asphalt)							
Residential (single family)							
Heat Loss	(kWh)	15000	13000	10500			
System Efficiency (FF)		.66	.69	.70			
System Efficiency (E)		1.00	1.00	1.21			

[1] Given supervigorous policy and substantially higher than case energy price. These figures are not necessarily additive: they represent components of alternative energy strategies.

[2] Number of years from start-up of national response to full implementation.

DEMAND REDUCTION MODULES EFFICIENCY

CASE D,D-7,D-8

Demand Line Items and Units	(Units)	User Performance 1972	1985	2000	"What If?" National Efficiency Response[1]	National Response Savings over Year 2000 Scenario Case	Lead Time for National Response Years[2]
Autos	(10^6J/km)	3.25	2.87	2.26	NONE		
(Domestic) Air Travel	(10^6J/pass.km)	4.19	3.31	3.01			
Truck Freight	(10^6J/tonkm)	4.02	3.43	3.24			
(Electric) Rail Freight	(10^6J/tonkm)	.34	.24	.20			
Commercial/Public Heat Loss							
System Efficiency (FF)							
System Efficiency (E)							
Primary Metals (Iron & Steel)	(GJ/ton)	16.7	18.7	17.0			
Chemicals	(GJ/ton)	13.4	15.6	17.8			
(Aluminium) Energy Intensive Ind.	(GJ/ton)	68.6	63.1	57.3			
Non-Energy Uses (Petrochemical. Asphalt)							
Residential (single family) Heat Loss	(kWh)	15000	13750	11750			
System Efficiency (FF)		.66	.68	.69			
System Efficiency (E)		1.00	1.00	1.08			

[1] Given supervigorous policy and substantially higher than case energy price. These figures are not necessarily additive: they represent components of alternative energy strategies.

[2] Number of years from start-up of national response to full implementation.

17 Sweden

17.1 OVERVIEW AND SUMMARY

The demand projections to 2000 are mainly based on the forecasts made by EPU (the Government Energy Forecast Commission) in 1974, adjusted to conform to WAES scenario parameters through discussions with the Swedish Reference Group, other experts in the energy field, and review of preliminary results in ongoing energy studies. However, we are responsible for the results.

The selected supply-demand integrations are mainly based on plans or proposals formulated in the spring of 1976 regarding probable expansion of hydro power, nuclear power, and district heating. For WAES coal scenario cases for 1985 to 2000, nuclear power after 1985 is assumed to be almost eliminated, upper limits for existing plans or proposals to expand hydro power and district heating have been used, and the introduction of coal, wind power, and solar heat to balance the projected energy demand is assumed.

Sweden has no known oil or gas reserves but does have large resources of peat, uranium, and shale oil and a small amount of coal. It is assumed that there will be no introduction of natural gas and no domestic production of oil, gas, coal, peat, or shale oil, but that there will be production of uranium to match Sweden's domestic demand in 1985 to 2000. The uncertainties around this last assumption will probably be resolved in 1978. The losses due to distribution, conversion, and energy sector self-consumption are calculated according to WAES conventions, using, for instance, 35 percent efficiency for hydro power.

These demand projections and supply-demand integrations should be interpreted as soft figures under certain defined assumptions. They will, however, be revised and further developed during the spring of 1977. The effects of eliminating nuclear power in the 1980s will then be taken into account. In spite of projected energy savings, fossil fuel per dwelling for residential heating and specific fuel consumption (TWh per unit of valued added) in industry are projected to be about 50 percent lower in 2000 than in 1972. The growth rate of electricity per dwelling for appliances and lighting is about three to five times lower in 1985–2000 than in 1972–1985, the final energy demand (excluding distribution and conversion losses) is 65–90 percent higher in 2000 than in 1972, and net electricity to consumers is about 180–320 percent higher (see Table 17.1). These projections imply the growth rates shown in Table 17.2. Electricity for space heating in residential, commercial, and public sectors is projected to increase from 6 TWh in 1972 to 21–37 TWh in WAES Cases C-1 and C-2 in 2000.

The importance of oil in the Swedish energy system is projected to gradually decrease under the WAES assumptions, but in 2000 Sweden will still be dependent on oil for around 50 percent of final energy demand in spite of the projected vigorous response and 50 percent higher oil prices. This compares to an oil dependency of 69 percent of final energy demand for 1972.

We tried to estimate the impact of WAES parameters on the demand projections to 2000. We then found that these projections probably are correct within a margin of ±10–15 percent. The activity levels of certain sectors, such as energy-intensive industries and truck freight, could change the demand significantly, and much more detailed studies are needed for these and other sectors in order to arrive at accurate projections. Other assumptions, such as calculation of losses and energy supply assumptions to 2000, also need further study.

An Energy Commission appointed by the government will study all relevant issues regarding energy and publish their findings in 1978. Parliament will then define Sweden's future energy policy. Thus, it is possible that nuclear power will be eliminated in the 1980s, peat and shale oil may be domestically produced, and natural gas may be introduced in Sweden before 2000.

A methodology has been developed which defines the consequences of various national policies, such as energy independence, low cost, or environmental protection, the capital and operating costs for incremental increases of electricity and heat from 1985 to 2000, and the potential to develop domestic energy resources without increasing the cost to consumers by more than 10 percent above the prices assumed in the selected WAES cases.

This study has been discussed and approved by the Swedish Reference Group. They have also re-

Table 17.1
Projected energy supply and demand, 1972–2000 (TWh)

Year	1972	1985		2000			
WAES case Economic growth National policy response World energy price Principal replacement fuel		C high vig rising	D low restrained constant	C-1 high vig high coal	C-2 high vig high nuc	D-7 low vig medium coal	D-8 low vig medium nuc
Final energy demand							
Electricity	64	139	125	222	266	179	197
Fossil fuels	343	439	420	540	504	506	471
Total	407	578	545	762	770	685	668
Primary energy input (including losses)	580	911	847	1257	1372	1103	1106
Oil imports	374	454	413	566	471	484	449
Coal imports	16	44	38	159	60	100	50
Hydro power (net)	54	66	66	85	75	85	75
Nuclear power (net)	1	62	57	64	193	62	112
Wind power (net)	—	—	—	15	—	5	1
Solar heating (net)	—	—	—	7	1	5	1
Combined production of electricity and heat for district heating (net)	7	42	23	87	74	87	87

Table 17.2
Projected growth rates of energy demand and oil imports (% per year)

	1972–1985		1985–2000			
	C	D	C-1	C-2	D-7	D-8
Net electricity	6.1	5.3	3.2	4.4	2.4	3.1
Final energy demand	2.7	2.3	1.9	1.9	1.5	1.4
Primary energy input	3.5	3.0	2.2	2.8	1.8	1.8
Oil import	1.5	0.7	1.5	0.3	1.1	0.6

vised all the cost figures and suggested another approach to calculate environmental costs and district-heating prices. The revisions are quite small and do not change the results significantly.

Fuel supply mixes for each WAES case result from an interaction between national policies and the specific characteristics of energy resources (see Figure 17.1).

For each case, a potential supply-demand integration with a corresponding figure for fossil fuel imports and their impact on balance of payments has been developed. A least-cost energy policy for 1985–2000 in Sweden will result in the highest oil import of all the energy policies (Table 17.3). Oil imports can most readily be reduced in the higher oil price cases since the cost differences between oil and other alternatives are then less.

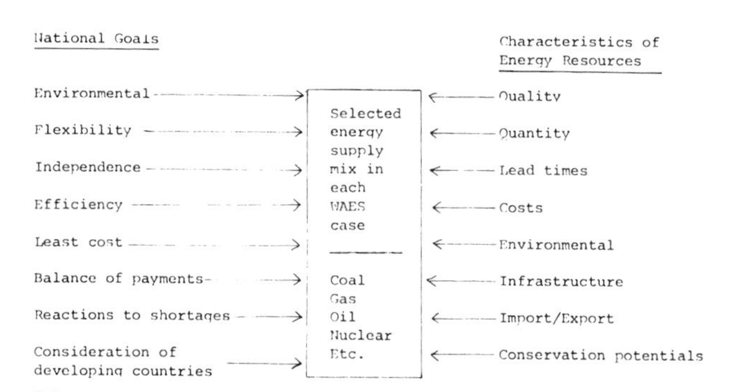

Figure 17.1
Interaction between national goals and the characteristics of energy sources.

Projected final energy demand, net electricity, and district heat demand are fixed in the WAES cases. However, the supply mix changes according to national policy, with the constraint that in 2000 the total costs for fuel, the additional yearly capi-

Table 17.3
Percentage change of oil imports from the selected case due to different national goals

	Oil price	Oil imports in 2000 Selected case (TWh)	Least cost (% change)	Independence (% change)	Environment (% change)
C-1 (coal)	rising	517	+13	-46	-33
C-2 (nuclear)	rising	421	+ 4	-36	-21
D-7 (coal)	constant	443	+ 6	-21	-23
D-8 (nuclear)	constant	402	+ 2	-26	-15

Table 17.4
Percentage change of costs in 2000 from the selected case due to different national goals[a]

	Selected case (billions of 1975 U.S. dollars)	Least cost (% change)	Independence (% change)	Environment (% change)
C-1 (coal)	10.4	-5	+8	+8
C-2 (nuclear)	11.3	-2	+7	+6
D-7 (coal)	6.3	0	+8	+8
D-8 (nuclear)	6.7	-3	+9	+9

[a]Costs for total fuel and annuity for additional electricity, heat generation, and distribution.

Table 17.5
Percentage change of consumer cost for additional electricity from the selected case due to different national goals, 1985-2000

	Selected case (1975 U.S. ¢/kWh)	Least cost (% change)	Independence (% change)	Environment (% change)
C-1 (coal)	5.7	-10	0	+9
C-2 (nuclear)	4.6	- 3	+5	+7
D-7 (coal)	4.4	-10	0	-1
D-8 (nuclear)	4.4	- 5	+7	+7

tal costs for electricity, heat generation, and distribution, and consumer cost for electricity and heat should be no more than 10 percent above the corresponding figures for the selected WAES case. Such price increases would of course affect the demand. This, however, has not been taken into account but will be resolved during the spring of 1977.

As can be seen from the cost comparisons, shown in Tables 17.4 and 17.5, potential reductions of oil imports can be achieved by relatively small increases in costs. The costs for the selected cases are higher than least cost due to a mix of national policy goals.

The main policies that would reduce fossil fuel import are to increase hydro power, wind power, solar heat, and domestic production of shale oil, methanol, and peat. The degree to which this can be done depends on the availability, acceptability, and costs of these energy resources. Knowledge of these costs are limited. The following prices have been assumed:

	U.S.$/BOE	U.S.$($\times 10^6$)/TWh
Wind power	93-107	57-66
Solar heat	39-50	24-31
Peat, shale oil	24-39	11-18
Methanol	24	11

All costs for capital and fuel are expressed in millions of 1975 U.S. dollars per TWh. Thus, the low usage time for wind power and solar heat, which greatly increases the cost per TWh, and the costs for environmental protection, which increase with large-scale use of wind power and domestic production of peat and shale oil, have been added. In this way, it is easy to see the consequences of

Table 17.6
Impact of oil imports on balance of payments (billions of U.S. dollars)

			2000			
	1972	1985	Selected case	Least cost	Independence	Environment
C-1 (coal)	-0.9	-3.5	-8.0	-8.3	-4.6	-7.6
C-2 (nuclear)[a]	-0.9	-3.5	-5.4	-5.6	-3.8	-4.7
D-7 (coal)	-0.9	-3.2	-4.5	-4.5	-3.5	-4.1
D-8 (nuclear)	-0.9	-3.2	-4.1	-4.1	-3.3	-3.8

[a]Domestic production and enrichment of uranium assumed.

different energy strategies—for instance, the greater impact on balance of payments of a coal rather than a nuclear policy (Table 17.6).

Tables 17.7–17.10 give a summary for each of the WAES cases to 2000. It should be noted that the cost (U.S. $15 million per TWh) for local distribution of electricity is included in these four tables.

The results of this analysis indicate that the coal cases require less investment in electricity and heat generation and distribution than the nuclear cases but result in increases in oil imports, balance-of-payments deterioration, and average fuel, electricity, and heat prices, especially if an independence policy is adopted.

An independence policy, as opposed to a least-cost policy, would increase the investment in electricity generation, keep electricity prices fairly constant, and greatly increase oil imports and balance-of-payments deterioration.

If the same price relationships exist between oil and other fossil fuels, a further price increase for oil (110 percent above today's price) would not change the level of oil imports or the infrastructure in electricity and heat generation but would increase prices and balance-of-payments deterioration. The consequences of such a price development would trigger a change of policy toward independence.

In the sensitivity analysis, the projected electricity demand is probably 8–16 percent higher. If this is correct, a 16 percent (36 TWhe) reduction in electricity demand would reduce the cost per year for electricity generation and distribution by U.S. $1 billion, the balance-of-payments deterioration by U.S. $0.8 billion, and the price of electricity by 3 percent. But how much would it cost to reduce electricity demand by 16 percent? What actions would be needed and what would the consequences be?

An independence policy for Case C-1 (coal) would increase the electricity investment per year and total fuel costs by 8 percent, the price to consumers for additional 1985 to 2000 electricity by 12 percent (assuming that domestic consumers pay the subsidy for the extra energy costs to the export industry), and cause environmental problems for shale oil production, but it would reduce the balance-of-payments deterioration by 39 percent, reduce some environmental pollution, and increase domestic employment.

A rough analysis indicates that if Sweden foresees high economic growth and wants independence, the time to act is today. If low economic growth is assumed, Sweden can afford to wait.

Important Questions in Strategy Formulation for 1985–2000

Is domestic development of peat, shale oil, methanol, wind power, and solar heat a realistic alternative? Should more research and development be focused on these alternatives to get more realistic figures for costs, lead times, environmental impact, actions needed, etc.? If not, how should Sweden compensate for the decrease in oil supply? Reduce energy demand further? Increase coal imports? What are the actions needed, consequences of lead times, costs, and impacts on balance of payments and environment? With no additional nuclear or domestic development of peat, shale oil, wind power, or solar heat, Sweden remains heavily dependent on oil or coal after 2000; what are the consequences?

17.2 ENERGY DEMAND TO 2000

17.2.1 Methodology

The energy demand studies for 1985 to 2000 have

Table 17.7
Case C-1 summary

	Selected case	Least cost	Independence	Environment
Costs (billions of U.S. $)				
Annuity per year, 1985-2000, for additional electricity and heat	2.5	1.6	4.4	3.6
Total fuel costs	8.5	8.8	7.6	8.4
Effect on balance of payments	-8.0	-8.3	-4.6	-7.6
Price (millions of U.S. $ per TWh)				
Average fuel	9.9	10.2	10.7	10.6
Additional electricity	42.0	36.3	42.2	47.1
Additional heat	15.7	14.8	21.2	16.5
Fuel (TWh)				
Refuse	14	10	16	3
Shale oil	—	—	84	—
Peat	—	—	26	—
Methanol	—	—	30	30
Oil imports	517	584	280	348
Coal imports	147	60	97	60
Gas imports	—	32	—	168
Uranium imports	183	183	183	183
Solar heat (TWh)	7	—	50	50
Wind power (TWh)	15	—	25	15
Hydro power (TWh)	85	96	96	85
Nuclear power (TWh)	64	64	64	64
Oil import share of final energy demand (%)	68	77	37	46
Final energy demand (TWh)	762	762	762	762
Net electricity (TWh)	222	222	222	222
Net district heat (TWh)	67	67	67	67
Processing losses in electricity and heat generation (TWh)	224	214	173	211

Table 17.8
Case C-2 summary

	Selected case	Least cost	Independence	Environment
Costs (billions of U.S. $)				
Annuity per year, 1985-2000, for additional electricity and heat	3.3	3.1	4.2	4.6
Total fuel costs	7.4	7.4	7.4	6.9
Effect on balance of payments	-5.4	-5.6	-3.8	-4.7
Price (millions of U.S. $ per TWh)				
Average fuel	6.7	7.0	7.3	7.0
Additional electricity	31.4	29.8	33.6	34.5
Additional heat	7.0	6.1	7.7	7.7
Fuel (TWh)				
Refuse	5	6	8	4
Shale oil	—	—	75	—
Peat	—	—	—	—
Methanol	—	—	30	30
Uranium	620	538	582	534
Oil imports	421	437	270	332
Coal imports	60	60	60	60
Gas imports	—	—	—	20
Solar heat (TWh)	1	—	30	30
Wind power (TWh)	—	—	7	8
Hydro power (TWh)	75	96	96	76
Nuclear power (TWh)	193	170	180	163
Oil import share of final energy demand (%)	55	57	35	43
Final energy demand (TWh)	770	770	770	770
Net electricity (TWh)	266	266	266	266
Net district heat (TWh)	59	59	59	59
Processing losses in electricity and heat generation (TWh)	423	375	388	371

Table 17.9
Case D-7 summary

	Selected case	Least cost	Independence	Environment
Costs (billions of U.S. $)				
Annuity per year, 1985-2000, for additional electricity and heat	1.3	1.3	2.1	2.1
Total fuel costs	4.9	4.9	4.7	4.7
Effect on balance of payments	-4.5	-4.5	-3.5	-4.1
Price (millions of U.S. $ per TWh)				
Average fuel	6.9	6.9	6.9	6.9
Additional electricity	31.0	26.2	31.0	30.6
Additional heat	13.0	12.9	14.5	13.5
Fuel (TWh)				
Refuse	9	6	14	3
Shale oil	—	—	25	—
Peat	—	—	—	—
Methanol	—	—	30	20
Oil imports	443	468	348	342
Coal imports	88	50	50	50
Gas imports	—	—	—	71
Uranium imports	177	180	180	163
Solar heat (TWh)	5	—	14	30
Wind power (TWh)	5	—	7	5
Hydro power (TWh)	85	96	96	85
Nuclear power (TWh)	63	62	62	63
Oil import share of final energy demand (%)	65	68	51	50
Final energy demand (TWh)	685	685	685	685
Net electricity (TWh)	179	179	179	179
Net district heat (TWh)	66	66	66	66
Processing losses in electricity and heat generation (TWh)	153	136	134	129

Table 17.10
Case D-8 summary

	Selected case	Least cost	Independence	Environment
Costs (billions of U.S. $)				
Annuity per year, 1985-2000, for additional electricity and heat	1.9	1.8	2.6	2.6
Total fuel costs	4.6	4.5	4.6	4.6
Effect on balance of payments	-4.1	-4.1	-3.3	-3.8
Price (millions of U.S. $ per TWh)				
Average fuel	5.8	6.1	6.8	6.0
Additional electricity	29.3	27.0	32.2	32.2
Additional heat	9.5	8.6	10.6	10.3
Fuel (TWh)				
Refuse	5	4	6	4
Shale oil	—	—	40	—
Peat	—	—	—	—
Methanol	—	—	30	30
Oil imports	402	410	298	340
Coal imports	50	50	50	50
Gas imports	—	—	—	9
Uranium imports	335	279	284	329
Solar heat (TWh)	1	—	30	30
Wind power (TWh)	1	—	10	5
Hydro power (TWh)	75	96	96	75
Nuclear power (TWh)	112	94	94	110
Oil import share of final energy demand (%)	60	61	45	51
Final energy demand (TWh)	668	668	668	668
Net electricity (TWh)	197	197	197	197
Net district heat (TWh)	65	65	65	65
Processing losses in electricity and heat generation (TWh)	220	187	172	214

been done in three steps.

1. Estimates were made of the energy demand in 2000 by sector according to EPU and WAES parameters. WAES Cases D-7 and D-8 correspond to EPU alternatives 3 and 4. However, WAES Cases C-1 and C-2 differ from EPU alternatives 1 and 2 in that a vigorous instead of a restrained government policy is used and energy price is 50 percent higher instead of constant (in real terms). Adjustments have been made for these differences. The results of these projections were used as demand inputs in the supply-demand integration worksheets and give a first estimate of potential energy imports and exports.

2. The sensitivity of the soft demand figures was studied to get an estimate of high, medium, or low impact of WAES parameters for economic growth, energy price, and national response. The projected electricity demand probably results from a high impact of WAES parameters, but for fossil fuel demand the projected results probably represent medium impact with a margin of ±7 percent between low or high impact.

3. The industrial sector was further broken down into high or low energy-intensive industries, and the transport sector into autos, trucks, air freight, etc. This sectoral breakdown underlined the importance of making detailed studies of energy-intensive industries, especially their potential activity levels; the need to break down each sector as much as possible to get more realistic estimates; and the need for much more detailed studies, especially for international travel and truck freights. The sensitivity of the figures are probably in the order of ±5-20 percent.

17.2.2 WAES Cases D-7 and D-8

The Government Energy Forecast Commission (EPU) has made forecasts for 1985-2000 for energy demand in industry, transportation, and other sectors. EPU alternatives 3 and 4 have the same parameters as WAES Cases D-7 and D-8, namely, low economic growth, vigorous government response, and constant oil price. Therefore, EPU growth rates for 1985 to 2000 are used for these cases and are applied to the demand figures for 1985 Case D.

It should be noted that EPU's definitions of industry, transportation, and other sectors, which differ slightly from those of WAES, have been used. These growth rates as well as the demand estimates are given in Table 17.11. The energy demand in 2000 for D-7 and D-8 is lower than for EPU alternatives 3 and 4 mainly because of actual low energy use in 1974-1975.

17.2.3 WAES Cases C-1 and C-2

These two cases differ quite substantially (vigorous rather than restrained policy and 50 percent higher rather than constant energy price) from the corresponding EPU alternatives 1 and 2, but they correspond in the high economic growth parameter.

Energy price increases by 50 percent from 1985 to 2000 in Cases C-1 and C-2 and in 1985 Case A. An indication of the price effect can be obtained

Table 17.11
Energy demand estimates and growth rates for Cases D, D-7, and D-8

	1985 Case D				2000 Case D-7 (coal)				2000 Case D-8 (nuclear)			
	Fossils		Electricity		Fossils		Electricity		Fossils		Electricity	
	(TWh)	(% yr, 1972-1985)	(TWh)	(% yr, 1972-1985)	(TWh)	(% yr, 1985-2000)	(TWh)	(% yr, 1985-2000)	(TWh)	(% yr, 1985-2000)	(TWh)	(% yr, 1985-2000)
Industry including mining	166	+2.4	66	+4.8	223	+2.0	98	+2.7	217	+1.8	106	+3.2
Transportation including agriculture	97	+2.1	3	+3.2	116	+1.2	5	+3.5	116	+1.2	5	+3.5
Other	117	-0.9	56	+6.1	117	± 0	76	+2.1	88	-1.9	86	+2.9
Final energy demand	380	+1.1	125	+5.3	456	+1.3	179	+2.5	421	+0.9	197	+3.1

by comparison of Case A (rising oil price) with Case C (constant oil price). For the period 1985-2000, the results of 1985 Case A have, to some extent, been used. The development of new techniques and increased energy savings in 1985-2000 have been considered.

Industry (Including Mining)
Especially in Case C-1 (coal) conservation policies for electricity will have to take place. The growth rates shown in Table 17.12 are assumed for specific energy consumption from 1985 to 2000.

The value added increases by 6 percent from 1976 to 1985. For 1985-2000, a 5 percent growth rate is projected. It should be stressed that these projections represent WAES boundary high economic growth and not probable actual developments. Application of these growth rates gives the figures shown in Table 17.13. Multiplying the value added by the specific energy demand gives the energy demand as 265 TWh for fossil fuels and 137 TWh for electricity in Case C-1 and 253 TWh for fossil fuels and 161 TWh for electricity in Case C-2.

Transportation Sector (Including Agriculture)
According to EPU, the transportation demand for fossil fuel in 2000 is 140 TWh, and that for electricity is 5 TWh. The corresponding estimates for Case C in 1985 were 105 TWh for fossil fuels and 3 TWh for electricity.

In Case A (50 percent higher oil price), the demand for fossil fuel in 1985 was reduced by 8 percent compared to Case C. As a result of WAES parameters, fossil fuel demand is estimated to be 10 percent less in 2000 than EPU estimates. The EPU demand for fossil fuel increased by 32 TWh from 1985 to 2000. Due to actual low energy use in 1974-1975, figures for 1985 Case C are slightly lower than EPU 1985 figures. The corresponding energy demand for WAES Cases C-1 and C-2 can then be estimated as 0.90(105 + 32) = 123 TWh.

Commercial and Public Sectors
The same growth rate of building volume is used for 2000 as for 1985, and the same total energy per cubic meter is used in 2000 as is used in Case A for 1985. This gives the total year 2000 energy demand for these sectors for Cases C-1 and C-2 (Table 17.14).

Specific electricity demand (excluding electricity for space heat) increases from 1972 to 1985. A slower growth rate from 1985 to 2000 is estimated due to higher energy price, saturation tendencies, etc., especially in Case C-1 (see Table 17.14). Re-

Table 17.12
Growth rate for energy intensity in the industrial sector (% per year growth in TWh per unit of value added)

	1972-1985			1985-2000		
	Fossil fuels	Electricity	Total	Fossil fuels	Electricity	Total
Case A	-3.1	-0.3				
Case C	-2.7	+0.3	-2.0			
Case C-1				-2.8	-1.0	-2.3
Case C-2				-3.1	0	-2.1

Table 17.13
Value added and energy intensity in the industrial sector

	Value added (10^9 U.S. $)	Intensity (TWh/dollar of value added)		
		Fossil fuels	Electricity	Total
1972	12.4	9.8	2.9	12.7
1985, Case C	26.4	6.8	3.0	9.8
2000, Case C-1	58.9	4.5	2.4	6.9
2000, Case C-2	58.9	4.3	2.8	7.1

Table 17.14
Energy demand for the commercial and public sectors

	Commercial	Public	Total
1972			
Total energy demand (TWh/m³)	141	143	
Space heat electricity (TWh)	0.4	0.5	0.9
Other electricity (TWh/m³)	20	24	
1985			
Total energy demand (TWh/m³)			
Case A	130	130	
Case C	145	145	
Space heat electricity (TWh)			
Case A	4.0	5.0	9.0
Case C	2.0	2.5	4.5
Other electricity (TWh/m³)			
Case A	35	48	
Case C	36	50	
2000			
Total energy demand for Cases C-1 and C-2			
TWh/m³	130	130	
TWh	26	36	62
Space heat electricity (TWh)			
Case C-1	2	3	5
Case C-2	6	8	14
Other electricity (TWh/m³)			
Case C-1	45	65	
Case C-2	50	70	
Other electricity (TWh)			
Case C-1	9	18	27
Case C-2	10	19	29
Case C-1			
Fossil	15	15	30
Electricity	11	21	32
Case C-2			
Fossil	10	9	19
Electricity	16	27	43

garding electricity for space heating, a higher growth rate is projected in Case C-2 than in Case C-1. An end to electric space heating has been projected in Case C-1, an assumption that could be modified. In Table 17.14 total energy less electricity and space heat electricity gives fossil fuel demand.

Residential Sector

Estimates of dwellings heated by electricity and fossil fuels from 1972 to 2000 are made according to EPU and "Kraftvärme-75" (a study on district-heating potentials). Recent studies show that the population growth rate to 2000 may be lower than previously estimated. The number of dwellings, however, is a result of factors that could counteract a lower population growth rate. We shall therefore use these projections, keeping in mind the uncertainty of the numbers (Table 17.15).

Our 1985 studies for Case C indicate a 21–33 percent reduction of fossil fuel used for heat per dwelling in the three types of dwellings shown in Table 17.15 as compared to 1972. A series of ongoing studies regarding this sector as well as concrete government policies to stimulate energy savings exist. There are, however, different opinions regarding the impact of conservation policies. As most of the policies concern new houses, a 20 percent further reduction of fossil fuel use per dwelling has been projected from 1985 to 2000. This will result in a situation in 2000 in which fossil fuel use per dwelling is about 40–50 percent lower than in 1972.

Regarding electricity for space heating, a 4–6 percent reduction of electricity per dwelling has been projected for Case C from 1972 to 1985. For Cases C-1 and C-2 a further reduction of 4 percent has been projected for electricity (heating) per dwelling in apartments as compared to 8 percent for single-family houses.

Electricity used for appliances, ventilation, lighting, etc., is projected to increase much more slowly from 1985 to 2000. Electricity use was 3.5 MWh per dwelling in 1972 and is estimated to be 5.5 MWh in 1985 for Case C; 50 MWh for Case A; 6.5 MWh in 2000 for Cases C-2 and D-8; and 6.0 MWh for Cases C-1 and D-7. It should be noted that the figures for the year 2000 cases are significantly below the EPU estimates because these estimates include the impact of higher energy prices and vigorous government response coupled with saturation effects.

From the 2000 figures an estimate of the residential demand for energy can be made (Table 17.16). Recent Swedish studies have pointed out that families in higher-income groups use much more energy than families in lower-income groups. A movement of families from lower-income to higher-income brackets is highly probable in the next 10–25 years, so that the estimates may be too conservative. However, we foresee a high degree of government response, which would make the projections more probable.

For agriculture (excluding transportation) and construction, projections are (in TWh) 0.2 for 1972; 0.3 for 1985 Case C; 0.5 for 2000 Case C-1; and 0.4 for 2000 Case C-2. Electricity use projec-

Table 17.15
Increase in number of dwellings, 1972-2000

Type of dwelling	Heating source	Number of dwellings (10^6)			
		1972	1985 Case C	2000 Case C-1	2000 Case C-2
Single family	Fossil fuels	1.17	1.25	0.9	1.2
	Electricity	0.20	0.60	1.1	0.8
Apartments	Fossil fuels	1.87	1.97	1.9	2.1
	Electricity	0.03	0.08	0.3	0.1
Vacation houses	Fossil fuels	0.2	0.2	0.1	0.3
	Electricity	0.3	0.6	0.8	0.6

Table 17.16
Residential demand for energy (TWh)

	1972	1985 Case C	2000 Case C-1	2000 Case C-2
Fossil fuels	88	70	59	48
Space heat electricity	5	11	16	23
Other electricity	12	20	25	27
Total	105	101	100	98

tions are (in TWh) 1.5 for 1972; 2.5 for 1985 Case C; 3.0 for 2000 Case C-1; and 3.5 for 2000 Case C-2.

These projections result in final energy demand for 2000 as shown in Table 17.17. These estimates of final energy demand are used in the supply-demand integration worksheets.

17.2.4 Consumer Preferences and Nonenergy Demand

Except for use of coal, coke, and wood (mainly for the iron and steel and pulp and paper industries), Sweden is almost wholly dependent on oil for its fossil fuel supply. To a very small extent gas is used in residential areas in big cities, but it is produced by oil. A few power plants using coal exist. Very few houses use wood for heating. This means that Sweden today has almost no infrastructure for using coal or gas for heating or electricity generation. Public resistance to the use of coal for heating and electricity is strong. Such use would require investments in harbors, domestic transportation, storage, and environmental protection. The introduction of gas would require building an infrastructure, such as pipe lines; this has been studied, and it is possible that gas use will increase in Sweden in the future.

The figures shown in Table 17.18 have been projected for nonenergy demand (plastics, etc.) and for the use of coal, coke, and wood, which to a great extent depends on the development of the chemical, iron and steel, and pulp and paper industries.

17.2.5 Sensitivity Analysis for Energy Demand

To make an accurate projection of energy demand to 2000 is impossible. It is possible, however, to estimate the effects of a high, medium, or low impact of the WAES parameters of vigorous response and rising energy price. In order to arrive at these estimates, the development from 1972 to 1985 has been compared with 1985 to 2000 and the probable sensitivity of our estimates has been discussed with experts.

We define "sensitivity" as an indication of the probable percentage variance between high or low impact of a parameter on the demand estimate in 2000 and the estimate resulting from a medium impact of the parameter.

We shall here mainly use the relations (TWh in 1985)/(TWh in 1972) and (TWh in 2000)/(TWh in 1985), or the factor increase. One should note that the period 1972-1985 is thirteen years as compared to fifteen years from 1985 to 2000. The same factor increase in both these periods implies, therefore, a somewhat slower increase of demand from 1985 to 2000.

Electricity

In spite of higher economic growth in Case C as compared to Case D, the factor increase for elec-

Table 17.17
Projected and revised changes in electricity demand and specific electricity consumption, 1972–2000

	1972–1985		1985–2000			
	C	D	C-1	C-2	D-7	D-8
Principal electricity demand						
Industry (factor increase)	2.2	1.8	1.8	2.1	1.5	1.6
Transportation (factor increase)	1.5	1.5	1.7	1.7	1.7	1.7
Other sectors (factor increase)	2.2	2.2	1.2	1.7	1.4	1.5
Total net electricity (factor increase)	2.2	2.0	1.6	1.9	1.4	1.6
Space heat electricity						
Commercial (%/year)	+13.2		0	+ 7.6		
Public (%/year)	+13.2		+ 1.2	+ 8.1		
Residential (%/year)	+ 6.3		+ 2.5	+ 5.0		
Projected change in specific energy consumption						
Industrial (%/year change in TWh/value added)	+ 0.3		- 1.5	- 0.5		
Commercial[a] (%/year change in kWh/m³)	+ 4.6		+ 1.5	+ 2.2		
Public[a] (%/year change in kWh/m³)	+ 5.8		+ 1.8	+ 2.3		
Residential[a] (%/year change in kWh/m³)	+ 3.5		+ 0.6	+ 1.1		
Revised figures for "other sectors" TWh (factor increase)	2.5	2.2	2.0	2.2	1.9	2.0
Change from projected figures at end of period (TWh)	+ 8	0	+51	+43	+30	+26
Change from projected figures at end of period (%)	+ 6	0	+23	+18	+17	+13
Revised figures for total net electricity demand (TWh)	142	125	258	293	194	210
Change from projected figures (TWh)	+ 4	0	+36	+27	+15	+13
Sensitivity change (%)	± 3	0	±14	± 9	± 8	± 6

[a]Excluding space heat electricity.

Table 17.18
Projections for nonenergy use of oil, coal, and wood (TWh)

		1985		2000	
	1972	C	D	C-1, C-2	D-7, D-8
Oil	15	45	40	60	50
Coal, coke	16	44	38	60	50
Wood	31	39	37	40	40

tricity demand is only slightly higher due to vigorous policy, as shown in Table 17.17.

It is possible that we have underestimated the electricity growth in other sectors or overestimated the effect of vigorous policy.

A low impact of WAES parameters would probably result in a factor increase of 2.5 for other sectors in Case C instead of 2.2. This would imply that the total demand for electricity in Case C (1985) would increase by about 8 TWh (from 138 to 146 TWh). The demand for electricity in the transportation sector (1985–2000) is so small that it is not considered. As compared to the development from 1972 to 1985, two parameters change, namely a 50 percent higher oil (energy) price in Cases C-1 and C-2 and a vigorous policy response in Cases D-7 and D-8. Also, it is possible that we have overestimated the impact of a 50 percent higher energy price and vigorous policies, especially for other sectors.

The factor increases shown in Table 17.17 represent the probable low impact of vigorous response and higher energy price. The demand for electricity would then increase. The electricity demand increases seem, however, very high. In the projected electricity demand, the figures shown in Table 17.17 have been used for development of specific electricity consumption. Strong conservation measures have been assumed in Cases C-1 and D-7.

It should be noted that a continued increase of specific electricity consumption is projected in residential, commercial, and public sectors. However, a strong reduction in the growth rate, compared to the development in 1972–1985, is projected. One could argue that these projections represent a high impact of WAES parameters.

Cost studies indicate that the price of electricity in Cases C-1 and D-7 will increase by about 10–20 percent above C-2 and D-8. One could, however, discuss whether such a cost increase would result in a large difference between the coal and nuclear cases for specific energy consumption in industry and for space heat by electricity.

It could also be discussed whether industry can reduce specific electricity consumption as indicated in Table 17.17. A change of -0.5 percent per year implies a reduction of electricity demand in 2000 by about 11 TWh.

A low impact of WAES parameters would probably result in a reduction of the specific electricity consumption for industry from +0.3 percent per year for 1972–1985 to no increase for 1985–2000 in Case C-2 and to -0.5 percent per year in Case C-1. This would increase the electricity demand by 22 TWh in Case C-1 and 11 TWh in Case C-2.

Preliminary cost studies indicate that the price per TWh to consumers for additional electricity in 1985–2000 would increase by about 20 percent even if the electricity price increase were lower. (About one-third of the electricity in 2000 will come from existing electricity generation plants with low costs.) The question is the extent to which such a price increase coupled with vigorous policy would affect the consumption of electricity. More detailed analysis is needed to answer this question and to arrive at more exact projections of electricity demand.

Projected total electricity demand represents the probable high impact of supervigorous conservation policies and high market response to price increases. A further decrease in electricity demand without change in activity levels is not very probable.

The main exception is space heating by electricity. In reality, the electricity damand in 2000 will probably be between the projected figures and the above calculated increases. The average of these ranges represents a medium impact of WAES parameters. The revised electricity demand and sensitivity between high or low impact are shown in Table 17.17.

Fuel (Excluding Nonenergy Demand)

The factor increase is about the same for total fuel demand in Cases C and D in spite of higher economic growth in Case C. This is mainly due to vigorous policy, especially in sectors such as residential, commercial, and public.

It appears that we have underestimated the effect of a 50 percent higher energy price in Cases C-1 and C-2 and a change to vigorous policy in Cases D-7 and D-8 in industry and other sectors. For industry, however, we have projected reductions of specific fuel consumption as shown in Table 17.19, which result overall in about a 50 percent reduction of the specific fuel consumption in 2000 as compared to 1972. We do not think it possible to reduce this further in spite of vigorous policy and 50 percent higher oil (energy) price, and we regard this as a high impact of WAES parameters. More detailed studies are needed to arrive at a probable low impact of WAES parameters. For transportation a more detailed sensitivity analysis has been made. For other sectors the projected development implies a continued decrease of fuel consumption.

The projected increase in number of dwellings (50 percent increase in vacation houses included) and volume (cubic meters) of commercial and public space is quite small except for dwellings heated by electricity. As pointed out earlier, however, these projections of "activity levels" could be discussed and require more in-depth studies. It is quite possible that the commercial and public sectors especially will increase their space volume above the estimates.

The projections of specific fossil fuel consumption indicate that we have probably underestimated the effect of a 50 percent higher energy price and vigorous policy in Cases C-1 and C-2 for residential dwellings. At the same time it is possible that we have overestimated the energy savings in 1985 Case C in this sector. The factor increases will be revised to represent high impact from WAES parameters as shown in Table 17.19.

We believe that these changes represent a high or low impact of WAES parameters. We have therefore revised our old projections to represent medium impact of WAES parameters.

Table 17.19
Projected and revised changes in fossil fuel demand and specific fossil fuel consumption, 1972-2000

	1972-1985		1985-2000			
	C	D	C-1	C-2	D-7	D-8
Projected fossil fuel demand						
Industry (factor increase)	1.4	1.3	1.5	1.4	1.3	1.3
Transportation (factor increase)	1.5	1.4	1.2	1.2	1.2	1.2
Other sectors (factor increase)	0.8	0.9	0.8	0.6	1.0	0.8
Total, TWh (factor increase)	1.3	1.2	1.2	1.1	1.2	1.1
Projected change in activity levels in buildings						
Commercial and public (factor increase)	1.1	1.1	1.1	1.1	1.1	1.1
Residential heated by fossil fuels (factor increase)	1.1	1.1	1.0	0.8	1.0	0.8
Residential heated by electricity (factor increase)	2.7	2.7	1.5	2.3	1.5	2.3
Projected change in specific energy consumption						
Industry (%/year change in TWh/value added)	-2.8		-2.8	-3.1		
Single-family houses (factor increase in MWh/dwelling)	0.75		0.80	0.80		
Apartments (factor increase in MWh/dwelling)	0.79		0.80	0.80		
Vacation houses (factor increase in MWh/dwelling)	0.67		0.80	0.80		
Revised projections for residential dwellings heated by fossil fuels						
MWh/dwelling (factor increase)	0.8		0.6	0.6		
Revised fuel demand at end of time period (TWh)	74		53	44		
Sensitivity change (TWh)	±6		±11	±9		

17.2.6 Sensitivity Analysis for Electricity Demand for Industry in Cases C-1 and C-2

WAES parameters to 1985 have resulted in the projections stated in Table 17.20 for electricity use in industry from 1972 to 1985 in Case C. It should be noted that the figures for 1985 are still soft.

We have divided the industrial sector into four groups, A through D, in which we now consider the level of specific electricity consumption and the trend of its development.

Groups A and B have a very high kWh to dollar value added ratio as compared to groups C and D. The percent per year 1972-1985 change in kWh per dollar value added is positive for groups A and C and negative for groups B and D.

The industries in groups A and B dominate the electricity consumption and increase their share from 73 to 78 percent from 1972 to 1985, partly due to a projected higher increase in their activity levels.

For the period 1985-2000, a series of factors will probably decrease the activity levels in these electricity-intensive branches, such as higher electricity costs and more limited possibilities for mining and forestry. For iron and steel it is doubtful that the high increase in activity levels will be realistic even if WAES assumes high economic growth in Cases C, C-1, and C-2. Increasing electricity costs, especially in the coal cases, could force some of the electricity-intensive nonferrous metals industry out of Sweden and cause a lower demand for these energy-intensive products. More integrated production in the paper and allied industries would further decrease the specific energy consumption, but a move toward mechanical pulp to save wood would counteract this decrease. Another factor is that, compared to most other countries, the advantage of low electricity prices will probably disappear in Sweden from 1985 to 2000, and this will mainly affect industries in groups A and B.

There are many unknown factors regarding the development to 2000, but we believe that the reduced activity levels and trends in specific electricity consumption shown in Table 17.21 are probable.

The 4.4-4.6 percent per year increase in activity level for total industry from 1985 to 2000 may seem low. If the real development were 5 percent for 1985-2000, this would result in an increase of

Table 17.20
Projected changes in value added (VA, in 10^9 U.S. dollars) and specific electricity consumption (SEC, in kWh/dollar of VA) in industry, Case C, 1972-1985

		1972 VA	1985 VA	Growth rate of VA (%/year, 1972-1985)	1972 SEC	1985 SEC	Growth rate of SEC (%/year, 1972-1985)
A	Mining	0.4	0.8	5.6	4.5	10.6	6.8
	Nonferrous metals	0.2	0.5	7.7	16.3	17.8	0.7
	Chemicals	1.2	3.0	7.3	4.0	4.5	0.9
	Overall A	1.8	4.3	6.9	5.4	7.2	2.2
B	Iron and steel	0.8	2.3	8.5	5.6	5.5	-0.1
	Pulp and paper	0.9	1.7	5.0	13.3	10.4[a]	-1.9
	Overall B	1.7	4.0	6.8	9.7	7.6	-1.9
A + B (all electricity-intensive)		3.5	8.3	6.9	7.5	7.4	-0.1
C	Food and related	1.2	1.7	2.7	0.92	1.1	1.1
	M'fg and transport	5.3	12.5	6.8	0.72	0.8	0.8
	Overall C	6.5	14.2	6.2	0.75	0.83	0.8
D	Minerals	0.5	0.6	1.4	2.6	2.0	-2.0
	Miscellaneous	1.9	3.3	4.3	1.7	1.4	-1.5
	Overall D	2.4	3.9	3.8	1.9	1.5	-1.7
C + D		8.9	18.1	5.6	1.1	1.0	-0.7
Total industry (A + B + C + D)		12.4	26.4	6.0	2.9	3.0	0.3

[a]Probably too high; the National Board of Industry has recently suggested savings that would imply a 1985 SEC of 9.0 kWh/dollar of value added.

Table 17.21
Revised projections for activity levels, value added (VA), and specific electricity consumption (SEC) in industry, 1985-2000, Cases C-1 and C-2

	Activity				Value added				SEC			
	Growth rate (%/yr)		Level (TWh)		Growth rate (%/yr)		Level (10^9 U.S. $)		Growth rate (%/yr)		Level (kWh/$VA)	
	C-1	C-2	C-1	C-2	C-1	C-2	C-1	C-2	C-1	C-2	C-1	C-2
Revised projections												
Group A	5.1	6.1	64.7	74.7	4.0	4.5	7.7	8.3	1.0	1.5	8.4	9.0
Group B	1.9	3.0	40.3	47.0	4.0	4.5	7.2	7.7	-2.0	-1.5	5.6	6.1
Group C	4.7	5.6	23.6	26.0	5.0	5.0	29.5	29.5	0	0.5	0.8	0.9
Group D	0.5	1.1	6.1	6.1	3.0	3.0	6.1	6.1	-2.5	-2.0	1.0	1.1
Total industry	3.7	4.6	134.7	155.0	4.4	4.6	50.5	51.6	-0.7	0	2.7	3.0
Old projections for total industry			141	164	5.5	5.5			-1.5	-0.5		
Alternative projections for total industry			120 159	145 176	4.4 5.5	4.6 5.5			-1.5 -0.7	-0.5 0		
Average sensitivity change (%)			±14	± 9								

3-4 TWhe if group C (mainly the manufacturing sector) counted for activity level increases. This would not be impossible since it would imply an increase of about 6 percent per year in activity levels from 1985 to 2000 for the manufacturing sector, as compared to 6.2 percent from 1976 through 1985.

However, if this activity level increase for total industry from 4.4-4.6 percent per year to 5 percent per year for 1985-2000 were caused by groups A and B, this would imply an increase in their activity levels from 4.0 to 4.5 to 5.8 to 6.2 percent per year, which would not be impossible as compared to the projected 6.8-6.9 percent increases for 1976 to 1985. But the effect on total TWhe demand would be an increase of 25-32 TWhe.

This illustrates how sensitive the results are to our assumptions of activity levels in the electricity-intensive industries and points to the need for much more detailed sutdies of the probable future developments of infrastructure not only in these branch groups but in each electricity-intensive branch. The factors that determine specific electricity consumption, such as more integrated mills, continuous casting, new techniques to save electricity, and substitution possibilities, also need further study. The consequences (and to some extent the sensitivity) of different activity levels and specific energy consumption for total industry are shown in Table 17.21.

17.2.7 Sensitivity Analysis for Transportation Energy Demand

In the demand projections for 1985 to 2000, the methodology mainly used the EPU estimates. For Cases C-1 and C-2 the results from WAES cases to 1985 were also applied as an indication of the impact of energy price (the oil price is 50 percent higher in Case A than in Case C).

The following more detailed analysis illustrates the difficulties in making estimates of large sectors, such as transportation, without breaking them down as much as possible to get more realistic estimates of probable developments in energy demand.

For 1985-2000, revised projected figures for fossil fuels for transportation are given in Table 17.22. This analysis also indicates that a detailed study of international air travel and ship freight, as well as a comparison between freight by truck and by railway, could be useful, since future energy demand could be greatly affected by developments in these sectors. A shift from truck freight to railway would significantly reduce energy demand.

The increase in energy demand between 1985 and 2000 comes mainly from truck freight and is due to the projected high increases in activity levels.

Autos account for about 50 percent of the energy demand in the transportation sector. Although an increase in energy demand for autos to 2000 was previously presented, revised figures show no increase from 1985 to 2000.

Autos

There are, of course, many different factors that influence the demand for energy for autos. Here we consider mainly the factors shown in Table 17.23. Load factor increases resulting from policy actions could change the distance traveled per auto per year and thus the total demand for energy. If the distance traveled could be held constant, this would result in a 4-15 percent decrease in the projected energy demand for autos, depending upon scenario.

The projected increase in distance traveled per auto per year for Case C represents a very small increase of 0.8 percent per year. The above change could therefore be regarded as high impact. Other factors such as changing infrastructure of residence and shopping locations as well as increases in the

Table 17.22
Revised projections for transportation sector fossil fuel demand

		1985		2000	
	1972	C	D	C-1, C-2	D-7, D-8
Old projection (TWh)	69.6	97.9	90.5	123.0	116.0
Revised projection (TWh)	69.6	97.9	90.5	135.0	106.5
Sensitivity (%)				±15	±10

number of second cars in a household, with lower distance per auto, make predictions difficult.

For Cases C-1 and C-2, the predicted increase of distance traveled per auto per year represents an increase of 0.4 percent per year. The high impact of higher energy prices and vigorous actions to increase the load factor could result in no increase in distance traveled per auto per year from 1985 to 2000. This would result in a demand of 41.2 TWh instead of 44.0 TWh (a decrease of 6 percent).

For cases D-7 and D-8, the predicted increase in distance traveled per auto per year is 0.27 percent per year, as compared to 0.44 percent for Case D (1985). Vigorous policy actions could eliminate any increase in 1985 to 2000. The sensitivity would then be -4 percent.

The energy per distance traveled could be reduced by different methods, such as lighter cars. We have predicted a 22 percent increase in efficiency (0.07 liter/km in 2000 Cases C-1 and C-2, as compared to 0.09 liter/km in 1985 Case C) for Cases C-1 and C-2 as compared to 11 percent for Cases D-7 and D-8. The difference is due to a 50 percent higher energy price in Cases C-1 and C-2.

A high impact of WAES parameters might increase the efficiency by 33 percent (0.06 liter/km) for Cases C-1 and C-2, which would result in an energy demand of 37.7 TWh (a decrease of 14 percent). If the distance traveled per auto per year were constant between 1985 and 2000, the energy demand would be 35.3 TWh (a decrease of 20 percent from the predicted demand).

It is, however, also possible that a low impact of WAES parameters could result in an energy per distance traveled of 0.08 liter/km in Cases C-1 and C-2. This would result in an energy demand of 50 TWh (14 percent higher than predicted).

A high impact of WAES parameters might increase the efficiency for Cases D-7 and D-8 by 22 percent instead of the predicted 11 percent (0.075 liter/km). This would result in an energy demand of 37.3 TWh (a decrease of 12 percent). If distance traveled per auto per year were constant, the demand would be 35.8 TWh (a decrease of 15 percent from predicted demand).

The possible variations in the predicted number of autos will not be considered. The figures used are the same as those of EPU.

Instead of an increase in energy demand from 1985 to 2000, there would be a decrease if the load factor and efficiency increased as above, especially for Cases C-1 and C-2, due to the high impact of a 50 percent energy price increase and vigorous policy response. We believe that our figures represent medium impact of WAES parameters. The above preliminary analysis indicates the magnitude of changes with a high or low impact. The sensitivity from the previous projected figures is 14–20 percent. The study, "Motor Fuel Consumption in Sweden towards 1990, Passenger Cars," made by Transportforskningsdelegationen in May 1976 arrives at about the same estimates of fuel demand: 38–43 TWh in 1985 and 36–44 TWh in 2000.

In accordance with the fuel consumption study and in order to reduce the sensitivity, we would like to revise our projected demand figures and predict constant use of energy between 1985 and 2000.

Domestic Air Travel

The figures shown in Table 17.24 depend on whether we have overestimated the increase of efficiency (kWh/passenger-km) in Case C. At the same

Table 17.23
Projected changes of activity levels for automobiles, 1972–2000

		1985		2000	
	1972	C	D	C-1, C-2	D-7, D-8
Population (10^6)	8.2	8.6	8.6	8.9	8.9
Population/auto	3.25	2.46	2.64	1.98	2.34
Autos (10^6)	2.52	3.50	3.26	4.5	3.8
Distance/auto/year (10^4 km)	1.36	1.50	1.44	1.60	1.50
Energy/distance (liters/10 km)	1.0	0.9	0.95	0.7	0.85
Energy/distance (mpg)	23.0	25.5	24.2	32.9	27.1
Energy consumed (TWh)	29.9	41.3	38.9	44.0	42.3

time, the percentage increase per year of activity level may be too high, although not according to historical trends. Since this sector requires relatively small amounts of energy as compared to autos and trucks, however, we shall not do a special sensitivity analysis but shall regard the above figures as representative of the probable magnitude of energy demand.

International Air Travel

It should be noted that calculations are based on air petrol tanked in Sweden. As most of Swedish international travel goes by way of Copenhagen, the kWh/passenger-km figures in Table 17.24 do not represent the real specific energy consumption for international travel.

To what extent the trends in activity levels will continue is unknown. Since this sector is quite small, however, a special sensitivity analysis will not be made. Such an analysis is needed, though, for overall international travel because the historical increase of activity levels has been quite high and global international air travel may require a large proportion of transportation energy by 2000.

Truck Freight

The figures shown in Table 17.24 for activity levels and efficiency in Cases C, C-1, and C-2 represent the probable high impact of WAES parameters.

Low impact would probably result in an increase of specific energy consumption (kWh/ton-km) to 0.50 in Cases C-1 and C-2 and 0.40 in Case C. This would result in an energy demand of 24 TWh for Case C and 46 TWh for C-1 and C-2, or a margin of 25 percent between low and high impact.

Energy demand for truck freight is quite sensitive to the activity levels. A change from 7.8 to 7.0 percent per year for Case C and from 6 to 5 percent per year for Cases C-1 and C-2 would decrease energy demand to 17.6 TWh for Case C and 29.2 TWh for Cases C-1 and C-2, or a sensitivity of –8 percent compared to –21 percent for C-1 and C-2. These large sensitivity values indicate that a special study would be needed to show the probable devel-

Table 17.24
Projected and revised demand estimates for transportation, 1972–2000

		1985		2000	
	1972	C	D	C-1, C-2	D-7, D-8
Domestic Air Travel					
Activity level (10^9 passenger-km)	0.68	2.35	1.35	7.45	2.43
kWh/passenger-km	0.88	0.54	0.76	0.44	0.50
Increase per year of activity level (%)		10.0	5.4	8.0	4.0
International Air Travel					
Activity level (10^9 passenger-km)	8.2	25.7	14.8	70.9	24.8
kWh/passenger-km	0.28	0.23	0.27	0.16	0.22
Increase per year of activity level (%)		9.2	4.7	7.0	3.5
Truck Freight					
Activity level (10^9 ton-km)	18.2	48.0	31.7	115.0	49.4
kWh/ton-km	0.65	0.40	0.60	0.32	0.50
Increase per year of activity level (%)		7.8	4.4	6.0	3.0
Energy Demand					
Autos	29.9	41.3	38.9	41.3	38.9
Air travel, domestic	0.6	1.3	1.0	3.3	1.5
Air travel, international	2.3	5.9	3.9	11.3	5.5
Truck freight	11.8	19.2	19.0	36.8	24.7
Agriculture transportation	2.9	4.8	4.1	8.0	6.0
Ship freight international	14.9	16.3	16.3	17.6	17.6
Other sectors	10.1	13.9	11.4	4.7	21.8
Projected Total Demand (TWh)	69.6	97.9	90.5	123.0	116.0
Revised Total Demand (TWh)	69.6	97.9	90.5	135.0	106.5
Sensitivity (%)				±15	±10

opments in this sector.

Rail Freight
This sector requires only about 1-3 TWh electricity, so no special sensitivity study has been made. The figures indicate, however, that the specific energy consumption for rail freight is only about 30 percent of that for truck freight. Large energy savings would be possible if domestic freight switched from trucks to railways. There would be many other effects to be considered if such a change were forced through, however, and a special study would be needed before such proposed actions could be made.

Other Sectors
Other transportation sectors include bus, rail, air freight, and miscellaneous. The figures for the year 2000 in Table 17.24 indicate a grave error in the earlier analysis of other sectors, if other projections are correct. It illustrates the difficulties of estimating large sectors and points to the need to break down each sector as much as possible.

If the figures to 2000 are corrected for other sectors and an increase of 2 percent per year for Cases C-1 and C-2 and 0.5 percent for Cases D-7 and D-8 is assumed, the energy demand becomes 18.7 and 12.3 TWh (a change from the projected figures of +12 and -9.5 TWh, respectively).

If all figures to 2000 were revised according to this analysis, the result would be an increase of demand in Cases C-1 and C-2 and a decrease in Cases D-7 and D-8. Sensitivity analyses indicate that these figures are quite insecure. A sensitivity of 10–15 percent is therefore indicated.

17.3 SUPPLY-DEMAND INTEGRATION

17.3.1 Analysis

This section describes the methodology and worksheets used to project energy imports and exports with due regard to losses through conversion, distribution, and self-consumption of fuels by the energy sector.

In order to arrive at the different supply mixes, existing plans or proposals to expand hydro power, district heating, etc., have been used. No introduction of gas and no production of peat or shale oil has been assumed, but uranium production is assumed to match domestic demand. (This last assumption can be discussed.) Existing plans (as of spring 1976) to increase nuclear power to 1985 have been used. For WAES coal cases to 2000, it is assumed that nuclear power will be almost eliminated, and upper limits for existing plans or proposals to expand hydro power and district heating have been used. We have assumed that coal as well as solar heat and wind power will be introduced to balance the energy demand. Estimates for production of electricity and heat are summarized in Table 17.25.

In spite of a projected slowing down of the growth rate for demand, especially for 1985–2000, due to vigorous government response and a 50 percent higher energy price, the growth rate of oil imports is projected to continue or even increase in the coal cases to 2000, as shown in Table 17.26. In the nuclear cases to 2000 the growth rate of oil imports decreases, especially in C-2. The main factor behind the drastically reduced growth rate for oil imports in C-2 is the great increase in space heating by electricity.

The importance of oil in the Swedish energy system is projected to gradually diminish under WAES assumptions, but Sweden will still be quite heavily dependent on oil in 2000 (around 40–50 percent) in spite of vigorous government response and rising oil prices.

17.3.2 Methodology

The main WAES methodology is described in chapter 2. There are, however, several points peculiar to the Swedish studies that need to be further explained and discussed.

In the worksheets (section 17.5) lines 1–8 present projected consumer final energy demand. Lines 9–13 present distribution, conversion losses, and energy sector self-consumption.

In estimating the conversion losses, WAES has adopted a convention of 35 percent efficiency for nuclear power, oil condenser power, and hydro power and 85 percent efficiency for district heating and hot water centrals. The reasoning behind this assumption is set out in the WAES report, *ENERGY: Global Prospects 1985–2000* (McGraw-Hill, 1977). A wind power efficiency of 35 percent is used to be consistent with the WAES convention.

The following conventions for estimating losses

Table 17.25
Projected production source of electricity and heat, 1972–2000 (TWh)

		1985		2000			
	1972	C	D	C-1	C-2	D-7	D-8
Production Source of Electricity							
Hydro power	54	66	66	85	75	85	75
Nuclear	1	62	57	64	193	62	112
Oil	14	22	11	50	20	30	25
Industrial backpressure	3	6	6	10	10	10	8
Coal	—	—	—	25	—	8	—
Wind power	—	—	—	15	—	5	1
Total	72	156	140	249	298	200	221
Percent used for district heating	3%	15%	8%	28%	23%	28%	28%
Production Source of Heat							
Industrial waste heat	—	2	2	5	3	4	3
Oil	12.5	35	35	25	26	28	32
Coal	—	—	—	24	—	24	—
Nuclear	—	—	—	—	24	—	24
Refuse, etc.	0.5	3	3	9	4	7	4
Solar	—	—	—	7	1	5	1
Total	13	40	40	70	58	68	64
Combined Production of Electricity and Heat for District Heating	7	42	23	87	74	87	87

Table 17.26
Projected growth rates for energy demand, electricity, and oil imports

	1972–1985		1985–2000			
	C	D	C-1	C-2	D-7	D-8
Final energy demand (%/year)	+2.1	+2.3	+1.9	+1.9	+1.5	+1.4
Primary energy input (%/year)	+3.5	+3.0	+2.2	+2.8	+1.8	+1.8
Projected net electricity (%/year)	+6.1	+5.3	+3.2	+4.4	+2.4	+3.1
Revised net electricity (%/year)	+6.3	+5.3	+4.0	+4.9	+3.0	+3.5
Oil imports ($/year)	+1.5	+0.7	+1.5	+0.3	+1.1	+0.6
Oil imports share of final energy demand (%)	69	54	56	47	49	47
Oil imports share of primary energy input (%)	64	50	49	45	44	41

have been adopted:

Distribution

Electricity: 12 percent of net electricity delivered to consumers (line 8,g).

Heat: 1 percent of net heat delivered to consumers (line 8,f).

Energy Sector Self-Consumption

Nuclear Power: 6 percent of net electricity delivered to consumers (line 9,i).

Hydro Power: 1 percent of net electricity delivered to consumers (line 9,h).

Oil Condensers, District heat, Hot Water Centrals: 6 percent of net electricity or heat delivered to consumers (lines 9,b; 12,b; and 12,j).

Refineries: By 1985, 4.75 percent and by 2000, 5 percent of net oil demand to consumers, exclusive of oil used in district heating, hot water centrals, and oil condensers (line 8,b).

One could raise the point that the convention for conversion losses is so rough that energy self-consumption ought to be included in the energy sector. One could also discuss the percentage figures used, especially for refineries, and the method of calculating the refinery losses.

Sweden has no oil production but does export some refinery products. Arbitrary estimates of the export figures for 1985 and 2000 have been used. These oil export figures are included in line 13,b.

The supply-demand integration worksheets do not have either a line or a row for industrial backpressure or industrial waste heat. These energy supply resources are estimated in lines 9,k and 12,k, respectively.

Regarding indigenous supply, Sweden has no known oil or gas resources but has potential resources of peat, shale oil, uranium, and hydro power, most of which have already been developed. A small amount of coal resources exist. It is tentatively assumed that from 1985 to 2000 Sweden will produce uranium to match the domestic demand. This assumption could be discussed and will probably be resolved in 1978.

In the selected supply-demand integrations used by WAES to calculate global integrations, we have not assumed any domestic production of peat or shale oil. It is assumed that such production, as well as the introduction of natural gas, is a possibility. These alternatives are considered in section 17.4.

In order to make it easier to see how the figures in lines 9-13 in the national input worksheets for supply-demand integration were arrived at, an additional worksheet, attached to each integration worksheet, was developed.

The methodology used to get figures for the supply of electricity and heat was the following: We started with existing plans or proposals according to "Kraftvärme-75," "CDL 1975," and other sources to increase district heat, hydro and nuclear power, industrial backpressure, and industrial waste heat. The balance to 1985 according to our final energy demand projections was met by oil condensers and hot water centrals.

For the 2000 Cases C-2 and D-8 (nuclear), we used the projections made by "Kraftvärme-75" and "CDL 1975" for district heat, hot water centrals, industrial back pressure, hydro power, and oil condensers, and let nuclear power be the backup, closing the gap of projected figures for final energy demand for electricity and heat.

For the 2000 Cases C-1 and D-7 (coal), we more or less stopped further nuclear power expansion, introduced the upper limits according to "Kraftvärme-75" and "CDL 1975" for district heating and hydro power, and assumed that coal, to some extent, had to be used both in district heat and in "coal condenser power." We also projected that wind power and solar heat in these cases would be available to some extent in order to arrive at the projected demand figures for electricity and heat.

It should be noted that nuclear power estimates were made according to the situation before the Swedish election in the autumn of 1976. The situation has since changed, and it is not unlikely that nuclear power will be eliminated after 1985.

17.3.3 References for Demand Projections and Supply-Demand Integration

Konsumentverket: Hushållens energikonsumtion, Dec 1976, Nov 1976:1.

Direktiv till energikommissionen, Dec 1976

Villkorsproposition, Jan 1977.

Solenergidagen SI-IVA, Nov 1976.

Lag om allmänna fjärrvärmeanläggningar samt propositionen 1975:76:149.

Beredskapslargring av olja, kol, uran, SOU 1976:67.

Naturgas i Sverige, SOU 1972:25.

Nordiske Naturgasutredningen 1976:1.

Kraftvärme-75.

Program från Nämnden för energiproduktionsforskning.

Program från Delegationen för energiforskning.

Vattenkraft och miljö, SOU 1976:28.

Petroindustrin i Sverige, SOU 1976:59.

Kommunalenergiplanering, SOU 1976:55

Transportforskningsdelegationen, "Motor fuel consumption in Sweden towards 1990," May 1976.

Studies made by Statens Råd för byggnadsforskning, Statens Planverk, IVA, IUI, Statens Industriverk and CDL.

"Energy conservation in the built environment," Höglund m.fl., KTH, Stockholm.

Svensk Gasförsörjning, Föredrag vid Sveriges Industriförbunds styrelsesammanträde den 17.3.76 av direktör Claes Lindgren, Svenska Gasförsörjningen.

Sekretariatet för framtidsstudier: "Energi och ekonomisk tillväxt," "Energi och inkomstfördelning," "Energi och sysselsättning," "Energi och handlingsfrihet," 1976.

Bergman m.fl. "An energy forecasting model for Sweden," SIND PM 1976:6, Statens Industriverk.

See also the references in chapter 13 of *Energy Demand Studies: Major Consuming Countries* (MIT Press, 1976).

17.4 ENERGY POLICY ANALYSIS

17.4.1 Methodology

Strategy Formulation

The margins of error in our WAES case projections to 2000 are approximately ±10 percent. Within this range and under the costs of each WAES case supply mix (with fixed projected final energy demand for electricity, district heat, and fuel), we can, by consideration of national goals and costs and characteristics of energy supply sectors, change the mix of energy supply, capital costs for electricity and heat generation, import costs and dependence, and fuel needed as more efficient electricity and heat systems are implemented. From these changes, one can arrive at the selected WAES cases and show the national policy behind the selected demand-supply balance, the total, and for each type of supply, the resulting capital costs for electricity and heat generation, fuel and import costs, potential areas for increased supply, the options available and their consequences, and the background for national strategies if shortages occur.

Some Alternative National Goals

- *Independence:* Develop domestic resources and reduce strains on balance of payments as much as possible with due regard to potential reserves, realistic uses, and least costs, but with no regard to environmental impact except that costs will correspond to plausible environmental concerns. (Constraint: The sum of total fuel costs plus capital costs for electricity, heat generation, and distribution and the cost of electricity and heat to consumers should not be more than 10 percent above the corresponding figures for the selected WAES case.)
- *Environment:* Focus on the development of fuel resources with the least environmental impact according to both political and plausible environmental requirements with due regard to potential reserves, realistic uses, and least costs. (Constraint: The sum of total fuel costs plus capital costs for electricity, heat generation, and distribution and the cost of electricity and heat to consumers should not be more than 10 percent above the corresponding figures for the selected WAES case.)
- *Least cost:* Minimize the cost of energy supply for each selected WAES case by combining the above and other national goals to arrive at a plausible supply mix. (Constraint: The sum of total fuel costs plus capital costs for electricity, heat generation, and distribution and the cost of electricity and heat to consumers should not be more than 10 percent above the corresponding figures for the least-cost national goal.)

Costs

There are many different types of costs involved in energy systems, such as exploration for and transportation of primary energy, conversion of primary energy to electricity and heat, distribution of electricity and heat, and end-use equipment. Considered here are (1) additional capital costs per year for 1985–2000 (expressed as the annuity with 10 percent interest) for conversion of primary energy to and regional distribution of electricity and heat; (2) fuel and operating costs for electricity and heat generation; and (3) total fuel cost per year in 2000 for domestic and imported fuel.

An environmental cost figure for electricity and heat plants and for domestic production of primary fuel has been added. Such a cost for fuel used outside centralized electricity and heat generation plants has not, however, been added. The cost figures should include any additional infrastructure costs related to the introduction of new supply systems such as, for Sweden, the use of peat or gas, and also additional costs related to environmental protection. The costs could increase if a fuel or electricity sector is utilized on a larger scale.

To make costs more easily comparable, they are all expressed in millions of constant 1975 U.S. dollars per TWh. It should be noted that the capital costs for electricity and heat generation are difficult to specify exactly and to relate to TWh output since the heat and/or power plants differ according to their usage of base or peak load, size, percentage of time in actual use, etc. One can, however, make average estimates and include in the capital cost figures per TWh the consequences of, for instance, low usage. For wind power and solar heat, the low usage time greatly increases the capital costs per TWh.

Table 17.27
Projected prices for different fuels in 2000 (millions of U.S. dollars per TWh)

	Compared to oil price	D-7, D-8	C-1, C-2
Coal	-10%	7.8	11.1
Gas	+20%	10.4	14.8
Uranium[a]		2.0	2.5
Methanol		15	15
Refuse		6	6
Peat and shale oil[b]		15-30	15-30

[a]Includes enrichment and waste treatment.

[b]The price increases if these fuels are utilized on a larger scale.

Fuel Costs

According to WAES price parameters, the energy price for 1985-2000 should correspond to an oil price of $11.50 in Cases D-7 and D-8 and $17.25 in Cases C-1 and C-2 (U.S. dollars per barrel light Arabian crude oil) or U.S. $7.1 million and U.S. $10.7 million per TWh, respectively. If transport costs to Sweden and oil refining costs are included, the prices for oil would be U.S. $8.7 million and U.S. $12.7 million per TWh. The costs in 2000 of other fuels should be linked to the oil price. The prices used are shown in Table 17.27.

For electricity, heat generation, and distribution capital costs, the annuity (calculated with 10 percent interest and the economic life length defined below) can be found in Tables 17.28 and 17.29, which give a summary for the price to consumers of additional electricity and heat from 1985 to 2000. It should be noted that the price (about U.S. $15 million per TWh) for local distribution of electricity is not included in these tables.

Table 17.28
Projected economic life length, annuity, and yearly operating costs for additional electricity and heat generation plants, 1985-2000

	Economic life (years)	Annuity (10^6 U.S. $)	Yearly operating costs other than fuel (% of capital costs)
Hydro power plants	40	10.23	0.5
Nuclear power plants	25	11.02	2.0
Wind power plants	20	11.75	0.25
Industrial backpressure or waste heat	10	16.27	1.0
Condenser power plants, district heating plants, and hot water centrals using:			
oil and gas	25	11.02	1.0-2.0
other fuels	25	11.02	2.0-2.5
Solar heat	15	13.15	0.5
Distribution of electricity and heat	30	10.61	

Table 17.29
Costs to consumers for additional electricity and heat, 1985-2000

	Increase of TWh, 1985-2000	Annuity/TWh (10^6 U.S. $)			Operating costs/TWh (10^6 U.S. $)			Total costs/TWh	
		Plant	Environmental	Regional distribution	Fuel D-7, D-8	Fuel C-1, C-2	Other	D-7, D-8	C-1, C-2
Electricity									
Industrial backpressure	0-4	8.1	—	1.1	7.7	10.7	0.5	17.4	20.4
Hydro power	0-20	15.4	2.0	2.7	—	—	0.8	20.9	20.9
District heat, gas		8.8	—	1.1	9.2	12.8	1.6	20.7	24.3
Industrial backpressure	4-15	16.3	—	1.1	7.7	10.7	1.0	26.1	29.1
Hydro power	20-30	21.5	4.0	2.7	—	—	1.1	29.3	29.3
Nuclear	0-60	15.4	2.2	3.2	5.7	7.1	2.8	29.3	30.7
District heat, oil		14.3	2.2	1.1	7.7	10.7	2.6	27.9	30.9
Nuclear	60-120	15.4	4.4	3.2	5.7	7.1	2.8	31.5	32.7
District heat nuclear		23.1	2.2	1.1	2.4	2.9	5.3	34.1	34.6
coal		18.1	3.3	1.1	6.9	9.7	4.4	34.3	37.1
refuse		20.9	3.3	1.1	7.1	7.1	4.8	37.2	37.2
shale oil		14.3	2.2	1.1	12.9	—	2.6	33.1	33.1
Condense oil		8.8	2.2	2.1	18.6	26.0	0.8	32.5	39.9
gas		7.2	—	2.1	22.3	31.1	0.7	32.3	41.1
coal		11.6	3.3	2.1	16.9	23.4	2.1	36.0	42.5
District heat, peat	0-25	20.9	3.3	1.1	12.9	—	4.8	43.0	43.0
Condense, shale oil	0-25	8.8	2.2	2.1	31.4	—	0.8	45.3	45.3
Wind power	0-15	52.9	1.8	1.6	—	—	1.1	57.4	57.4
Condense, peat	0-25	13.8	3.3	2.1	31.4	—	2.5	53.1	—
Wind power	15-25	58.9	3.5	2.1	—	—	1.3	65.8	65.8
Heat									
District heat nuclear				3.2	2.4	2.9		5.6	6.1
refuse				3.2	7.1	7.1		10.3	10.3
coal				3.2	6.9	9.7		10.1	12.9
oil				3.2	7.7	10.7		10.9	13.7
gas				3.2	9.2	12.8		12.4	16.0
Hot water centrals, refuse		2.2	2.2	4.4	7.1	7.1	1.4	17.3	17.3
Industrial waste heat	0-10	14.6	—	2.1	—	—	0.9	17.6	17.6

Table 17.29 (continued)
Costs to consumers for additional electricity and heat, 1985–2000

	Increase of TWh, 1985–2000	Annuity/TWh (10^6 U.S. $)			Operating costs/TWh (10^6 U.S. $)				
					Fuel			Total costs/TWh	
		Plant	Environmental	Regional distribution	D-7, D-8	C-1, C-2	Other	D-7, D-8	C-1, C-2
Hot water centrals									
oil		1.7	1.1	4.4	7.7	10.7	0.8	15.7	18.7
gas		1.1	—	4.4	9.2	12.8	0.5	15.2	18.8
coal		2.2	2.2	4.4	6.9	9.7	1.4	17.1	19.9
District heat									
peat		—	—	3.2	12.9	12.9	—	16.1	16.1
shale oil		—	—	3.2	12.9	12.9	—	—	—
Industrial waste heat	10–20	17.3	—	3.2	—	—	1.3	21.8	21.8
Solar heat	0–30	23.0	—	—	—	—	0.9	23.9	23.9
Hot water centrals									
shale oil		1.7	1.1	4.4	12.9	12.9	0.8	20.9	20.9
peat		2.2	2.2	4.4	12.9	12.9	1.4	23.1	23.1
Solar heat	30–50	29.6	—	—	—	—	1.1	30.7	30.7

For consumer costs for wind power and solar heat, the sum of annuity for investment, environment, distribution, and operating costs have been calculated as

wind power U.S. $57–66 million per TWh

solar heat U.S. $24–31 million per TWh

The price increases if these fuels are utilized on a larger scale.

For the cost of electricity and heat to consumers, the figures in Table 17.28 for the economic life length of power plants, heat plants, etc., will be used. With 10 percent interest, they give the corresponding annuity for an investment of U.S.$100 million. Interest costs during construction are included in the capital costs. The yearly operating costs other than fuel costs are assumed to be a percentage of the capital costs. The costs to consumers for additional electricity and heat from 1985 to 2000 can then be ranked as shown in Table 17.29.

Manual for Worksheets

Tables 17.30–17.32 for Case C-1 indicate how the final results were arrived at for the four WAES cases. The figures for "environment" will be revised. These tables illustrate the increase of TWhe and TWh thermal from different types of generation plants, when the consumer price limits of possible TWh increases and the price constraint due to different national policies have been considered.

For each national goal, a table has been done in which the annuity according to the projected increase of TWh and the fuel needed for electricity and heat generation have been calculated.

By subtracting from the final energy demand in 2000 the net electricity, net heat, and certain unchangeable requirements such as coal and wood needed by the iron and steel and pulp and paper industries, oil demand other than for electricity and heat generation is obtained. By multiplying the amounts of fuel needed by the cost per unit, the total costs for fuel are obtained. Within the cost constraint, oil demand can be reduced by increasing domestic shale oil, methanol, or peat production.

17.4.2 Advantages and Disadvantages of an Independence Policy for Case C-1

In Case C-1 a Swedish policy of independence has the following consequences if the export industry is compensated for the price increase in electricity and fuel costs (U.S. $500 million) and if this subsidy is paid by a price increase for electricity and fuel to domestic consumers (Tables 17.33 and 17.34).

The effect on balance of payments is calculated as the cost of imported fuel, where the refinery cost for oil is assumed to be the domestic cost.

One should try to consider that a higher energy price in Sweden as a result of an independence policy would have an impact on exports and thereby affect the balance of payments. The price for electricity increases by 3.6 (56.5 – 52.9) and average fuel price by 0.5 (8.0 – 7.5)as compared to Case C-1.

Table 17.30
Case C-1, summary for heat

	1985 contribution (TWh)	Price to consumers in 2000 (10^6 U.S.$/TWh)	Increase in TWh, 1985-2000			
			Selected case	Least cost	Independence	Environment
Industrial waste heat	2	17.6	+ 3		+10	+10
		21.8			+ 3	+ 6
District heat						
Oil	25	17.5			-25	- 9
Coal		16.3	+24			
Gas		14.7		+16		+25
Nuclear						
Refuse		10.3	+ 4	+ 2	+ 4	
Peat		20.9			+14	
Shale Oil		20.9			+25	
Hot water centrals						
Refuse	3			+ 4	+ 1	
Oil	10		- 9		-10	-10
Gas						
Coal						
Total district heat	40		+22	+22	+22	+22
Solar heat			+ 7		+30	+30
					+20	+20
Cost to consumers for additional heat (10^6 U.S.$/TWh)			+15.7	14.8	21.2	16.5

Table 17.31
Case C-1, summary for electricity

	1985 contribution (TWh)	Price to consumers in 2000 (10^6 U.S. $/TWh)	Increase in TWh, 1985–2000			
			Selected case	Least cost	Independence	Environment
Hydro power	66	20.9	+19	+20	+20	+19
		29.3		+10	+10	
Nuclear	62	30.7	+ 2	+ 2	+ 2	+ 2
Industrial backpressure	6	24.2	+ 4	+ 4	+ 4	+ 4
		32.9		+ 6	+ 6	+ 6
Condense power						
Oil	7	49.0	+23	+38	- 7[a]	- 7
Coal		50.8	+17		+13	
Gas		52.3				+41
Shale oil		56.8			+ 7[a]	
District heat/power						
Oil	15	34.7			-15[a]	- 5
Coal		40.5	+ 8			
Gas		34.7[b]		+11		+18
Nuclear		34.6				
Peat		47.8			+ 8	
Shale oil		37.9			+15[a]	
Refuse		37.2	+ 5	+ 2	+ 5	
Wind power		57.4	+15		+15	+15
		65.8			+10	
Total electricity	156		+93	+93	+93	+93
Cost to consumers for additional electricity (10^6 U.S.$/TWh)			42.0	36.3	42.2	47.1

[a] Shale oil instead of oil in existing plants.

[b] Should be 28.9. Increased to correspond to the cost for district heat by oil to get a more realistic price for heat.

Table 17.32
Case C-1, summary for "independence"

	1985 contribution (TWh)	Increase in TWh, 1985-2000	Annuity (10^6 U.S.$/year) Plant	Environment	Distribution	Fuel needs, 2000 (TWh) Net	Industrial processing losses
Hydro power	66	+20	308	40	54	—	—
		+10	215	40	27	—	—
Nuclear	62	+ 2	31	4	6	64	183
Industrial backpressure	6	+ 4	32	—	4	16	19
		+ 6	98	—	7		
Condense							
Oil	7	- 7	—	—	—	—	—
Coal		+13	151	43	27	13	37
Shale oil		+ 7				7	20
District heat/power							
Oil	40	-40	—	—	—	—	—
Peat		+22	167	26	54	22	26
Refuse		+ 9	105	17	18	9	11
Shale oil		+40				40	47
Wind power		+15	794	27	24	—	—
		+10	589	35	21	—	—
Industrial waste heat	2	+10	146	—	21	—	—
		+ 6	104	—	19	—	—
Hot water centrals							
Oil	10	-10	—	—	—	—	—
Refuse	3	+ 1	2	1	4	4	5
Solar heat		+30	690	—	—	—	—
		+20	592	—	—	—	—
Total			4022	233	229	—	—

Final energy demand in 2000	762 TWh
Electricity	-222 TWh
Coal, wood	-100 TWh
Net heat	-112 TWh
Shale oil	- 37 TWh
Methanol	- 30 TWh
Oil demand	261 TWh

Fuel	TWh	Price/TWh (10^6 U.S.$)	Cost (10^6 U.S.$)
Oil	280	12.3	3444
Coal	97	11.1	1077
Refuse	16	6	96
Uranium	183	2.5	457
Peat	26	15	390
Shale	84	20.5	1722
Methanol	30	15	450
Total	716	10.7	7636

Table 17.33
Disadvantages and advantages of an independence policy as compared to the selected case, C-1

		Effect of policy
Disadvantages	Increase in the annuity/year for additional electricity, heat, 1985-2000	+8 percent
	Increase in total fuel costs in 2000	+U.S. $800 million
	Increase in the price/TWh to domestic consumers in 2000 for additional electricity, 1985-2000	+12 percent (+3 ore/kWh)
	average fuel cost	+11 percent (+0.4 ore/kWh)
	Environmental problems involved in producing shale oil, peat, etc.	
Advantages	Reduction in the impact on balance of payments in 2000	39 percent (U.S. $2.4 billion)
	Reduction in oil imports in 2000	47 percent (242 TWh)
	Reduction in environmental pollution, since less fuel is consumed due to solar heat implementation	-95 TWh
	Export industry compensated for price increases	
	Increased domestic employment to produce wind power and solar heat equipment and peat, shale oil, methanol	

Table 17.34
Comparison in costs, prices, and oil imports between "independence" and the selected case, C-1, in 2000

	Case C-1	Independence
Costs (10^6 U.S. $)		
Annuity/year 1985-2000, for additional electricity and heat	3.9	5.0
Total fuel costs	6.4	6.2
Effect on balance of payments	- 6.1	- 3.8
Prices (10^6 U.S. $/TWh)		
Average fuel, 2000	7.5	8.0
Additional electricity, 1985-2000	52.9	56.5
Additional heat, 1985-2000	13.0	14.3
Oil imports, 2000 (TWh)	517	275

According to SOU 1974:64, page 127, about 50 percent of the energy used by industry in 1970 was incorporated in exported articles. This percentage has been steadily increasing since 1955, when the corresponding figure was about 35 percent. Calculations utilize 70 percent in 2000 as shown in Table 17.35.

If industry were compensated for these extra energy costs due to price increases with an independence policy, the total costs of annuity and fuel would increase by U.S. $1.4 billion (11.2 - 10.4 + 0.5) or by 13 percent above the costs for Case C-1. But the impact on balance of payments with an independence policy is reduced by U.S. $2.4 billion (6.1 - 3.8)

If the export industry were compensated for the price increases, it can be shown that the cost of electricity to domestic consumers would be increased by 12 percent above the projection for Case C-1. If the industry were not compensated, the price of additional electricity would increase by 7 percent if an independence policy were adopted.

For fuel in 2000, domestic demand is 354 TWh (total demand of 540 TWh minus 186 TWh of energy in exports). The price increase of fuel would then be 93/354 or U.S.$0.26 million per TWh. The average fuel price to domestic consumers would be U.S.$8.3 million (8.0 + 0.3) per TWh, or 11 percent more than the average fuel price for Case C-1. If the industry were not compensated, the average fuel price would increase by 7 percent if an independence policy were adopted.

The overall effects on prices of increases in fuel and electricity prices have not been considered in this analysis. This line of analysis could, however, be expanded as needed.

17.4.3 Effects of a Change in Demand

In section 17.2 we found by a sensitivity analysis that electricity demand probably shows a minimum increase due to strong reactions to vigorous policy and a 50 percent higher energy price. The

Table 17.35
Energy in exported industrial articles and increased exported energy costs with an independence policy, 2000

	Total industrial consumption (TWh)	Energy in exports (TWh)[a]	Price increase due to independence policy (10^6 U.S. $)	Energy export costs (10^6 U.S. $)
Electricity	144	101	3.6	364
Fossil fuels	265	186	0.5	93

[a]Calculated as 70 percent of total industrial consumption.

likely development would probably be about 10 percent (7-16 percent in the various WAES cases) higher electricity demand in 2000.

This could be interpreted in such a way that our results for prices, capital costs, oil import, etc., represent the situation after the electricity demand has reacted further with higher prices, etc.

If electricity demand growth follows the pattern described above, the following results might occur:

- Oil imports and the balance-of-payments deficit would increase in the coal cases with about the same percentages as the electricity demand increase, but in the nuclear cases no such change would occur.
- The price of additional 1985-2000 electricity and average fuel costs would increase more in the coal cases than in the nuclear ones.

This analysis could also be used in the following way for Case C-1, least cost policy, assuming that the development of electricity demand outlined above is the real one.

What actions would be needed, how much would it cost, etc., if the electricity demand in 2000 should be cut by 16 percent or 36 TWhe? The consequences of such a reduction would be:

- Smaller annuity/year -U.S. $1.0 billion
- Smaller impact on balance of payments -U.S. $800 million
- Lower electricity price for additional electricity 1985-2000 -3 percent
- Lower oil import -96 TWh

17.4.4 Sensitivity Analysis for the Impact of Different Assumptions

Costs for wind power, solar heat, peat, and shale oil have been calculated on the high side. If these costs are much lower, they certainly would have an impact on the resulting electricity, fuel costs, and prices in the independence policy cases.

Another possibility is to go further in the analysis by including the costs of end-use equipment for electricity and fossil fuel to see how electricity may compete with fossil fuels, especially in heating. In the WAES First Technical Report, we have already tried to consider how WAES parameters would influence space heating by electricity or fuel.

It should be noted that we have tried to consider the effects of higher prices and vigorous policy in the demand projections. It would be possible to use the results from these studies to arrive at some estimates of price elasticity for energy demand.

Estimates could then be made of the change in demand with the price changes due to, for instance, independence or least-cost policies.

A study should also be made of the impact of changes in capital costs for electricity generation and distribution and of possible different price relations between oil, coal, gas, and uranium.

The availability of some alternative energy resources is shown in Table 17.36. To use refuse to the highest possible extent would be in the public interest. The constraint is the cost of collecting it. Methods such as pyrolysis to utilize this energy resource instead of burning it are also possible alternatives.

For utilizing peat or shale oil, the environmental concern, and thereby public resistance, must be regarded as strong. We have tried to implement in the cost figures costs for environmental protection, such as landscaping. If Sweden should choose an independence policy, these domestic resources, as well as uranium, would be taken into serious consideration.

For wind power and solar heat, no public resistance or environmental concern can be visualized, except if wind power is utilized on a large scale,

Table 17.36
Availability of Swedish reserves according to EPU and other sources (TWh)

	Reserves available
Peat	
Total	46,000
Southern Sweden	6,600
Yearly increase	23
Shale oil, Southern Sweden	9,300
Refuse per year	
Residential and commercial	8 (in 2000)
Agriculture[a]	26
Industrial[b]	?
Forestry[c]	?
Wind power	10–30
Solar heat[d]	?
Wood[e]	?
Uranium	large reserves

[a]Straw contributes 21 TWh and animal refuse 5 TWh; most of this is, however, needed for cultivation.

[b]To what extent is this refuse recyclable?

[c]Mostly needed for earth cultivation.

[d]Limited by number of sunshine hours, need for storage, current costs, and technology availability.

[e]Experiments to cultivate fast-growing plants and trees are under way.

and we have tried to consider this in the cost figures. The problem with solar heat is mainly one of implementation.

17.4.5 Constraints and Choices

A rough estimate indicates that about 60 percent of the electricity and 80 percent of the fuel in 2000 is already more or less predetermined as compared to about 92 percent in 1985. If these estimates are correct, possibilities exist to influence about 25 percent of the energy needed in 2000, or about 200 TWh. After 1985, the possibilities may have decreased to about 8 percent or about 60 TWh.

An independence policy requires, according to the assumptions and methodology of this chapter (for instance, the costs for the energy system should not vary by more than ±10 percent), the increases in energy shown in Table 17.37. These figures imply, if this method is correct, that if Sweden believes in low economic growth and continued nuclear expansion, we could wait until 1985 to decide upon an independence policy. If there is to be low economic growth and no nuclear expansion after 1985, we must probably start soon to develop alternative resources if independence is wanted.

Table 17.37
Domestic energy needed for independence policy (TWh)

	C-1	C-2	D-7	D-8
Shale oil	150	150	70	40
Peat	26	—	—	—
Methanol	20	20	20	20
Solar heat	30	41	14	10
Wind power	26	7	9	5
Total	252	218	113	75
Uranium	182	525	185	284

If Sweden believes in high economic growth and wants independence, however, the time for action is today.

This method is, of course, very rough and should be refined to determine lead times, costs, etc. It would be especially interesting to see the costs and other consequences if Sweden tried to rush through an independence policy in 1985 with high economic growth.

17.4.6 Alternative Energy Strategies

WAES global studies imply a stagnation or decline in oil production some time after 1985.

In summary, it looks as if we are going through a transition period in which nuclear, oil, and other fossil fuels will help us to enter a new period, where renewable energy sources and breeders, fusion, and coal will probably make up the energy supply. At the same time, oil is an excellent source for the chemicals industry and difficult to replace in the transportation sector.

Sweden will be quite heavily dependent on oil as a primary energy resource even in 2000, especially in the WAES coal cases, unless an independence policy is chosen and implemented. But even then, oil will account for about 30–50 percent of the final energy demand.

In this chapter we have only dealt with a few national goals. As indicated above, a number of such goals exist, such as flexibility, consideration of developing countries, preserving oil for special purposes, etc. This analysis has been an attempt to show some of the consequences of a few national goals. In a further development during the spring of 1977 we shall try to consider the consequences of an elimination of nuclear power in the 1980s.

17.5 WORKSHEETS

This section presents summary tables and worksheets in the following order:

Energy demand summary, 1972–1985–2000

Energy supply/demand integration summary—imports and exports of primary energy.

National input worksheets for supply/demand integration (10 sheets, one each for the base year, 1985 Cases A–E, and 2000 Cases C-1, C-2, D-7, and D-8)

WAES NATIONAL ENERGY DEMAND STUDIES, 1972-1985-2000
ENERGY DEMAND SUMMARY

Country: Sweden

Scenario Assumptions	1972	1985					2000			
WAES CASE	Base Year	A	B	C	D	E	C-1	C-2	D-7	D-8
GNP 1976-85; 1985-2000		High	Low	High	Low	High	High	High	Low	Low
Oil/Energy Price*		11.50-17.25	11.50-17.25	11.50	11.50	11.50-7.66	11.50-17.25	11.50-17.25	11.50	11.50
National Policy Response		Vig	Vig	Vig	Res	Res	Vig	Vig	Vig	Vig
Principal Replacement Fuel							Coal	Nuclear	Coal	Nuclear
Primary Energy Demand **Units**	TWh									
Oil 1)	374	401	340	454	413	569	566	471	484	449
Natural Gas 6)										
Coal 2)	16	40	36	44	38	50	159	60	100	50
Nuclear 3)	4	181	181	181	166	181	187	580	208	343
Hydroelectric 4)	154	190	190	190	190	190	244	215	244	215
Geothermal										
Solar Thermal							7	1	5	1
Solar Electric										
Wood, Refuse etc.	31	40	39	42	40	38	93 5)	46	62 5)	48
TOTAL PRIMARY ENERGY DEMAND	580	852	786	911	847	1,028	1,257	1,372	1,103	1,106

*1975 US dollars per barrel of Arabian light crude oil, fob Persian Gulf.
All Figures are primary production prior to any processing or conversion.

Footnotes:

1) of which less than 10 % is projected to be exported in form of refined products.

2) The figures for 1985 to 2000 might be on the high side, as the activity levels in iron & steel sector might be smaller than we have projected and the extent to which coal might be used for electricity and heat generation could be discussed.

3) 35 % efficiency for electricity generation and 85 % for district heating.

4) 35 % efficiency. A WAES convention that could be discussed.

5) including windpower, with 35 % efficiency, a convention that could be discussed.

6) It is not impossible that Sweden will introduce natural gas in the 80's.

WAES NATIONAL ENERGY SUPPLY/DEMAND INTEGRATION STUDIES, 1972-1985-2000
ENERGY SUPPLY/DEMAND INTEGRATION SUMMARY—IMPORTS AND EXPORTS OF PRIMARY ENERGY

Country: Sweden

Scenario Assumptions	1972	1985					2000			
WAES CASE	Base Year	A	B	C	D	E	C-1	C-2	D-7	D-8
GNP 1976-85; 1985-2000		High	Low	High	Low	High	High	High	Low	Low
Oil/Energy Price*		11.50-17.25	11.50-17.25	11.50	11.50	11.50-7.66	11.50 17.25	11.50 17.25	11.50	11.50
National Policy Response		Vig	Vig	Vig	Res	Res	Vig	Vig	Vig	Vig
Principal Replacement Fuel							Coal	Nuclear	Coal	Nuclear
(Import) & Export Units TWh										
Oil	(374)	(401)	(340)	(454)	(413)	(569)	(566)	(471)	(484)	(449)
Natural Gas 4)										
Coal 5)	(16)	(40)	(36)	(44)	(38)	(50)	(159)	(60)	(100)	(50)
Nuclear	(4)	2)	2)	2)	2)	2)	2)	2)	2)	2)
Hydroelectric 3)										
Total Net 1) (Imports) or Exports	(368)	(431)	(346)	(458)	(422)	(579)	(685)	(491)	(549)	(464)

*1975 US dollars per barrel of Arabian light crude oil, fob Persian Gulf.
All Figures are primary production prior to any processing or conversion.

Footnotes:

1) Sweden exports a small amount of refined oil products. That is why this figure is less than the sum of the above import figures.

2) Domestic production of uranium to match domestic demand of uranium assumed. This assumption could be discussed and will not be resolved until 1978 when the Parliament will decide upon Sweden's future energy policy. It is not then unlikely that nuclear will be eliminated in the 80's.

3) Small amount of import or export exists between the Scandinavian Countries. We have not considered these in these integrations.

4) It is not impossible that Sweden will introduce natural gas in the 80's.

5) The figures for 1985 to 2000 might be on the high side, as the activity levels in iron & steel sector might be smaller than we have projected and the extent to which coal might be used for electricity and heat generation could be discussed.

NATIONAL INPUT WORKSHEET FOR SUPPLY/DEMAND INTEGRATION

Country: Sweden

Year: 1972

Case: Base year

Units: TWh

	(a) Coal	(b) Petro- leum	(c) (Syn. Liquids)	(d) Nat. Gas	(e) (Syn. Gas)	(f) (Heat)	(g) (Electri- city)††	(h) Hydro Energy	(i) Nuclear Energy	(j) Geothermal energy and other	(k) Total
(1) Transport		69.6					2.0				71.6
(2) Industry	15.9	72.1					33.9			30.5 (wood etc.)	152.4
(3) Agric., Mining, Construction		7.4					3.3				10.7
(4) Commercial		15.0				2.0	3.3				20.3
(5) Public		24.4				3.0	6.0				33.4
(6) Residential		78.6			1.0	8.0	15.9				103.5
(7) Non-energy Uses		15.0									15.0
(8) Final Energy Demand*	15.9	282.1			1.0	13.0	64.4			30.5	406.9
(9) Electricity**		13.7					−71.7	53.7	1.4		−2.9[1)]
(10) Syn. Gas		1.0			-1.0						-
(11) Syn. Liquids											-
(12) Heat		12.5				−13.0				0.5 (refuse)	-
(13) Energy Sector Self Consumption & conversion losses		65.1					7.3	100.2	3.6		176.2
(14) Primary Energy Input	15.9	374.4						153.9	4.0	31.0	580.2
(15) Indigenous Supply								153.9		31.0	185.9
(16 Imports***	15.9	374.4							4.0		394.3
(17) Exports		26.2									26.2

*Includes non-energy uses.

**Blocks 9g, 10e 11c, and 12f must be negative to avoid double counting of energy from primary fuels.

***Includes imported products.

††Includes only electricity to be purchased by each sector.

NATIONAL INPUT WORKSHEET FOR SUPPLY/DEMAND INTEGRATION

Country: Sweden

Year: 1985

Case: A

Units: TWh

	(a)	(b)	(c)	(d)	(e)	(f)	(g)	(h)	(i)	(j)	(k)
	Coal	Petroleum	(Syn. Liquids)	Nat. Gas	(Syn. Gas)	(Heat)	(Electricity)††	Hydro Energy	Nuclear Energy	Geothermal energy and other	Total
(1) Transport		90					3				93
(2) Industry	40	85					65			37 (wood etc.)	254
(3) Agric., Mining, Construction		17					10				27
(4) Commercial) (5) Public)		14				14	27				55
(6) Residential		5			1	33	44				83
(7) Non-energy Uses		40									40
(8) Final Energy Demand*	40	251			1	47	149			37	525
(9) Electricity**		31					167	66	62		-8 [1)]
(10) Syn. Gas		1			-1						–
(11) Syn. Liquids										3	-2 [2)]
(12) Heat		42				-47				(refuse)	
(13) Energy Sector Self Consumption & conversion losses		76					18	124	119		347
(14) Primary Energy Input	40	401						190	181	40	852
(15) Indigenous Supply								190	181	40	411
(16 Imports***	40	401									441
(17) Exports		30									30

*Includes non-energy uses.

**Blocks 9g, 10e 11c, and 12f must be negative to avoid double counting of energy from primary fuels.

***Includes imported products.

††Includes only electricity to be purchased by each sector.

NATIONAL INPUT WORKSHEET FOR SUPPLY/DEMAND INTEGRATION

Country: Sweden

Year: 1985

Case: B

Units: TWh

	(a)	(b)	(c)	(d)	(e)	(f)	(g)	(h)	(i)	(j)	(k)
	Coal	Petro-leum	(Syn. Liquids)	Nat. Gas	(Syn. Gas)	(Heat)	(Electri-city)††	Hydro Energy	Nuclear Energy	Geothermal energy and other	Total
(1) Transport		78					3				81
(2) Industry	36	72					57			36 (wood etc.)	201
(3) Agric., Mining, Construction		14					7				21
(4) Commercial) (5) Public)		15				20	22				57
(6) Residential		20			1	20	42				83
(7) Non-energy Uses		35									35
(8) Final Energy Demand*	36	234			1	40	131			36	478
(9) Electricity**		11					−147	66	62		−8[1)]
(10) Syn. Gas		1			−1						-
(11) Syn. Liquids											
(12) Heat		35				−40				3 (refuse)	−2[2)]
(13) Energy Sector Self Consumption & conversion losses		59					16	124	119		318
(14) Primary Energy Input	36	340						190	181	39	786
(15) Indigenous Supply								190	181	39	410
(16 Imports***	36	340									376
(17) Exports		+30									30

*Includes non-energy uses.

**Blocks 9g, 10e 11c, and 12f must be negative to avoid double counting of energy from primary fuels.

***Includes imported products.

††Includes only electricity to be purchased by each sector.

NATIONAL INPUT WORKSHEET FOR SUPPLY/DEMAND INTEGRATION

Country: Sweden

Year: 1985

Case: C

Units: TWh

	(a) Coal	(b) Petro-leum	(c) (Syn. Liquids)	(d) Nat. Gas	(e) (Syn. Gas)	(f) (Heat)	(g) (Electri-city)††	(h) Hydro Energy	(i) Nuclear Energy	(j) Geothermal energy and other	(k) Total
(1) Transport		98					3				101
(2) Industry	44	86					70			39 (wood etc.)	239
(3) Agric., Mining, Construction		19					11				30
(4) Commercial) (5) Public)		18				20	23				61
(6) Residential		49			1	20	31				101
(7) Non-energy Uses		45									45
(8) Final Energy Demand*	44	315			1	40	139			39	578
(9) Electricity**		22					-156	66	62		-8[1)]
(10) Syn. Gas		1			-1						-
(11) Syn. Liquids											
(12) Heat		35				-40				3 (refuse)	-2[2)]
(13) Energy Sector Self Consumption & conversion losses		81					17	124	119		341
(14) Primary Energy Input	44	454						190	181	42	911
(15) Indigenous Supply								190	181	42	413
(16 Imports***	44	454									498
(17) Exports		40									40

*Includes non-energy uses.

**Blocks 9g, 10e 11c, and 12f must be negative to avoid double counting of energy from primary fuels.

***Includes imported products.

††Includes only electricity to be purchased by each sector.

NATIONAL INPUT WORKSHEET FOR SUPPLY/DEMAND INTEGRATION

Country: Sweden

Year: 1985

Case: D

Units: TWh

	(a)	(b)	(c)	(d)	(e)	(f)	(g)	(h)	(i)	(j)	(k)
	Coal	Petroleum	(Syn. Liquids)	Nat. Gas	(Syn. Gas)	(Heat)	(Electricity)††	Hydro Energy	Nuclear Energy	Geothermal energy and other	Total
(1) Transport		91					3				94
(2) Industry	38	82					61			37 (wood etc.)	254
(3) Agric., Mining, Construction		15					7				22
(4) Commercial) (5) Public)		25				20	22				67
(6) Residential		51			1	20	32				104
(7) Non-energy Uses		40									40
(8) Final Energy Demand*	38	304			1	40	125			37	545
(9) Electricity**		11					–140	66	57		–6[1)]
(10) Syn. Gas		1			–1						–
(11) Syn. Liquids											
(12) Heat		35				–40				3 (refuse)	–2[2)]
(13) Energy Sector Self Consumption & conversion losses		62					15	124	109		310
(14) Primary Energy Input	38	413			–	–	–	190	166	40	847
(15) Indigenous Supply	–	–						190	166	40	396
(16 Imports***	38	413						–	–	–	451
(17) Exports		30						–	–	–	30

*Includes non-energy uses.

**Blocks 9g, 10e 11c, and 12f must be negative to avoid double counting of energy from primary fuels.

***Includes imported products.

††Includes only electricity to be purchased by each sector.

NATIONAL INPUT WORKSHEET FOR SUPPLY/DEMAND INTEGRATION

Country: Sweden

Year: 1985

Case: E

Units: TWh

	(a)	(b)	(c)	(d)	(e)	(f)	(g)	(h)	(i)	(j)	(k)
	Coal	Petroleum	(Syn. Liquids)	Nat. Gas	(Syn. Gas)	(Heat)	(Electricity)††	Hydro Energy	Nuclear Energy	Geothermal energy and other	Total
(1) Transport		115					5				120
(2) Industry	50	104					69			37 (wood etc.)	266
(3) Agric., Mining, Construction		19					13				32
(4) Commercial)											
(5) Public)		25				20	27				72
(6) Residential		67			1	20	33				121
(7) Non-energy Uses		50									50
(8) Final Energy Demand*	50	380			1	40	153			37	661
(9) Electricity**		37					−171	66	62		−6[1)]
(10) Syn. Gas		1			−1						-
(11) Syn. Liquids											
(12) Heat		39				−40				1 (refuse)	-
(13) Energy Sector Self Consumption & conversion losses		112					18	124	119		373
(14) Primary Energy Input	50	569						190	181	38	1,028
(15) Indigenous Supply								190	181	38	409
(16 Imports***	50	569									619
(17) Exports		40									40

*Includes non-energy uses.

**Blocks 9g, 10e 11c, and 12f must be negative to avoid double counting of energy from primary fuels.

***Includes imported products.

††Includes only electricity to be purchased by each sector.

NATIONAL INPUT WORKSHEET FOR SUPPLY/DEMAND INTEGRATION

Country: Sweden

Year: 2000

Case: C-1, Coal

Units: TWh

	(a) Coal	(b) Petroleum	(c) (Syn. Liquids)	(d) Nat. Gas	(e) (Syn. Gas)	(f) (Heat)	(g) (Electricity)††	(h) Hydro Energy	(i) Nuclear Energy	(j) Geothermal energy and other	(k) Total
(1) Transport 1) 2)		123					5				128
(2) Industry	70	155					141			40 (wood etc.)	406
(3) Agric., Mining, Construction 1) 2)		1					3				4
(4) Commercial) (5) Public)		13				19	32				64
(6) Residential		8			1	43	41			7 (solar)	100
(7) Non-energy Uses		60									60
(8) Final Energy Demand*	70	360			1	62	222			47	762
(9) Electricity**	25	50					−249	85	64	15 (wind power)	−10 3)
(10) Syn. Gas		1			−1						
(11) Syn. Liquids											
(12) Heat	24	25				−63				9 (refuse etc.)	−5 4)
(13) Energy Sector Self Consumption & conversion losses	40	130				1	27	159	123	30	510
(14) Primary Energy Input	159	566						244	187	101	1,257
(15) Indigenous Supply								244	187	101	532
(16 Imports***	159	566									725
(17) Exports		40									40

*Includes non-energy uses.

**Blocks 9g, 10e 11c, and 12f must be negative to avoid double counting of energy from primary fuels.

***Includes imported products.

††Includes only electricity to be purchased by each sector.

NATIONAL INPUT WORKSHEET FOR SUPPLY/DEMAND INTEGRATION

Country: Sweden

Year: 2000

Case: C-2, Nuclear

Units: TWh

	(a)	(b)	(c)	(d)	(e)	(f)	(g)	(h)	(i)	(j)	(k)
	Coal	Petro- leum	(Syn. Liquids)	Nat. Gas	(Syn. Gas)	(Heat)	(Electri- city)††	Hydro Energy	Nuclear Energy	Geothermal energy and other	Total
(1) Transport [1]		123					5				128
(2) Industry + mining	60	153					164			40 (wood etc.)	417
(3) Agric., ~~Mining~~ Construction		1					4				5
(4) Commercial) (5) Public)		5				14	43				62
(6) Residential		4			1	42	50			1 (solar)	98
(7) Non-energy Uses		60									60
(8) Final Energy Demand*	60	346			1	56	266			41	770
(9) Electricity**		20					−298	75	193		−10 [2]
(10) Syn. Gas		1			−1						
(11) Syn. Liquids											
(12) Heat		26				−57			24	4(refuse)	−3 [3]
(13) Energy Sector Self Consumption & conversion losses		78				1	32	140	363	1	615
(14) Primary Energy Input	60	471						215	580	46	1,372
(15) Indigenous Supply								215	580	46	841
(16 Imports***	60	471									531
(17) Exports		40									40

*Includes non-energy uses.

**Blocks 9g, 10e 11c, and 12f must be negative to avoid double counting of energy from primary fuels.

***Includes imported products.

††Includes only electricity to be purchased by each sector.

NATIONAL INPUT WORKSHEET FOR SUPPLY/DEMAND INTEGRATION

Country: Sweden

Year: 2000

Case: D-7, Coal

Units: TWh

	(a) Coal	(b) Petro-leum	(c) (Syn. Liquids)	(d) Nat. Gas	(e) (Syn. Gas)	(f) (Heat)	(g) (Electri-city)††	(h) Hydro Energy	(i) Nuclear Energy	(j) Geothermal energy and other	(k) Total
(1) Transport [1)]		116					5				121
(2) Industry + mining	60	123					98			40 (wood etc.)	321
(3) Agric., ~~Mining~~) Construction)											
(4) Commercial)		49			1	62	76			5 (solar)	193
(5) Public)											
(6) Residential)											
(7) Non-energy Uses		50									50
(8) Final Energy Demand*	60	338			1	62	179			45	685
(9) Electricity**	8	30					−200	85	62	5 (wind power)	−10 [2)]
(10) Syn. Gas		1			−1						
(11) Syn. Liquids											
(12) Heat	24	28				−63				7 (refuse etc.)	−4 [3)]
(13) Energy Sector Self Consumption & conversion losses	8	87				1	21	159	146	10	432
(14) Primary Energy Input	100	484						244	208	67	1,103
(15) Indigenous Supply								244	208	67	519
(16 Imports***	100	484									584
(17) Exports		35									35

*Includes non-energy uses.

**Blocks 9g, 10e 11c, and 12f must be negative to avoid double counting of energy from primary fuels.

***Includes imported products.

††Includes only electricity to be purchased by each sector.

NATIONAL INPUT WORKSHEET FOR SUPPLY/DEMAND INTEGRATION

Country: Sweden

Year: 2000

Case: D-8, Nuclear

Units: TWh

	(a)	(b)	(c)	(d)	(e)	(f)	(g)	(h)	(i)	(j)	(k)
	Coal	Petroleum	(Syn. Liquids)	Nat. Gas	(Syn. Gas)	(Heat)	(Electricity)††	Hydro Energy	Nuclear Energy	Geothermal energy and other	Total
(1) Transport		116					5				121
(2) Industry + mining	50	127					106			40 (wood etc.)	323
(3) Agric., ~~Mining~~, Construction)											
(4) Commercial) (5) Public) (6) Residential)		24			1	62	86			1 (solar)	174
(7) Non-energy Uses		50									50
(8) Final Energy Demand*	50	317			1	62	197			41	668
(9) Electricity**		25					−221	75	112	1 (wind power)	−8[1)]
(10) Syn. Gas		1			−1						
(11) Syn. Liquids											
(12) Heat		32				−63			24	4 (refuse etc.)	−3[2)]
(13) Energy Sector Self Consumption & conversion losses		74				1	24	140	207	3	449
(14) Primary Energy Input	50	449						215	343	49	1,106
(15) Indigenous Supply								215	343	49	607
(16 Imports***	50	449									499
(17) Exports		35									35

*Includes non-energy uses.

**Blocks 9g, 10e 11c, and 12f must be negative to avoid double counting of energy from primary fuels.

***Includes imported products.

††Includes only electricity to be purchased by each sector.

18.1 SUMMARY

The estimated energy demand of the U.K. over the period to 2000 in the various WAES scenario cases has been previously set out in the WAES first technical report. In addition, in both that report and the WAES second technical report, the energy supplies estimated to be available in the U.K. in the same period are given. This chapter includes summaries of both these studies in the form of supply-demand integration worksheets and goes on to give a plausible picture of the U.K. response to a foreseen world situation in which all energy supplies are strained by year 2000, insofar as such response is not already implicit in the earlier studies.

In section 18.2, we describe the guidelines of U.K. energy policy and emphasize the conclusion of the demand and supply studies that imports of energy to the U.K. will be rising rapidly around 2000. We suggest scenarios designed to minimize the impact of such imports by developing all indigenous supplies to their maximum extent, while imposing conservation measures. The new scenarios are described in section 18.3, the developments required in supply industries in section 18.4, and the conservation possibilities in section 18.5. Finally, section 18.6 reiterates a few of the problem areas.

The possible developments under the WAES scenario assumptions for the U.K. are summarized in Figures 18.1 and 18.2. These show how the total requirements for U.K. energy might be met under the high growth assumptions (scenarios C, C-1, and C-2) illustrated in Figure 18.1, or the low growth assumptions (scenarios D, D-7, and D-8) illustrated in Figure 18.2. All scenarios involve a period of energy surplus (oil exports) in the 1980s followed by a period of increasing imports of energy. In the following paragraphs policies for the U.K. alternative to those implied by the WAES scenarios are discussed with a view to a reduction in the impact of rising energy imports toward the year 2000.

18.2 U.K. ENERGY POLICY

The objectives of U.K. energy policy have traditionally been formulated in such terms as "The energy needs of the country should be met at minimum cost in real resources over time, while paying due regard to security of supply, to public safety, to the protection of the environment, and to the social consequences of change."

The secretary of state for energy has said, in reply to a parliamentary question on 16 February 1976,

> The governing assumptions and methods by which energy policy has developed over the past years now need to be reexamined in the light of changing circumstances. New criteria, for example, the need for long-term conservation measures and the need to develop all the fuel industries on a secure basis to meet the expected energy gap in the 1990s and beyond, are required. The key role of government in developing an integrated energy policy in

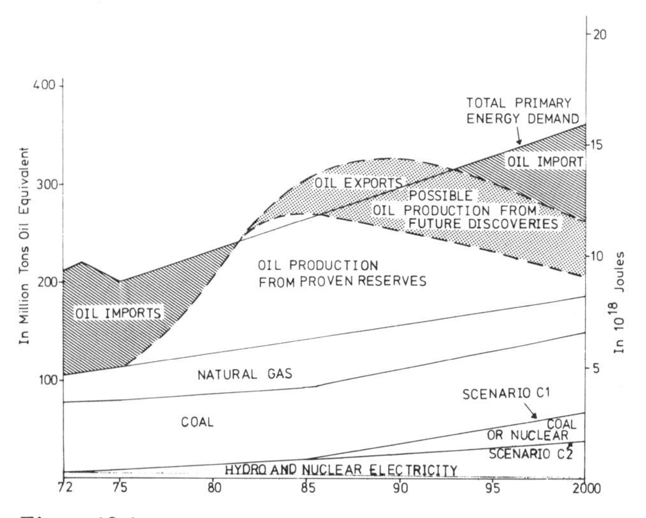

Figure 18.1
U.K. primary energy supply and demand for scenarios C-1 and C-2 (high economic growth).

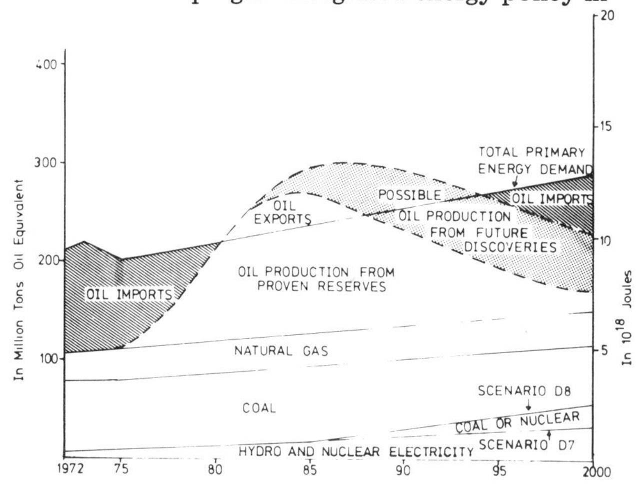

Figure 18.2
U.K. primary energy supply and demand for scenarios D-7 and D-8 (low economic growth).

conjunction with the industries and those that work in them also requires further consideration. This should be done as publicly as possible so that the interests of domestic and industrial consumers, long term as well as short term, can be safe-guarded, along with the interests of supplying industries and the community as a whole. I intend to proceed on this basis.

We are in general agreement with these statements, but would explicitly emphasize that the reason for the "new criteria" is the prospects for the world as a whole, since these long-term criteria are liable to lead to conflicts with short-term policies arising naturally from the traditional statements quoted above.

The series of WAES papers is intended to illustrate some of the problems and alternative strategies that must be considered for a risk/benefit analysis of policy objectives. The first WAES study indicated rapidly increasing energy import requirements for the U.K. in the 1990s, following a period of net exports in the 1980s. This increase in imports, taking place against an expected background of rising world energy prices, suggests that a reduction in U.K. energy imports in the year 2000 should be included as a policy objective, although the optimum level of imports cannot be decided without assumptions about the price and availability of world traded energy.

This report describes two scenarios for the period 1985–2000 in which, through high indigenous energy production and rigorous conservation, energy imports are reduced. Indigenous production is limited mainly by physical constraints while energy conservation is stimulated by high internal energy prices or by policies that simulate such prices.

18.3 SCENARIO ASSUMPTIONS

The two new scenarios for the U.K. in the period 1985–2000 are denoted herein by C and D. They are continuations of the 1972–1985 scenarios C and D described in the demand study, and differ from the WAES scenarios in the supply and conservation assumptions, for the supply is taken as high, while the conservation assumptions are replaced by assumptions about the U.K. internal energy prices. These internal prices may arise from higher world prices or from policy decisions (e.g., via taxes), or, alternatively, they may be an artefact used for calculation of the effects of changed criteria for energy conservation (see section 18.5). The major scenario assumptions are listed in Table 18.1, together with the previously defined scenario assumptions for comparison.

18.4 INDIGENOUS ENERGY PRODUCTION

During the period 1985–2000, the U.K. will be producing coal, oil, gas, and nuclear and hydro electricity from indigenous sources. (The small contributions from other sources, such as solar, are accounted for in the level of energy demand.) The export potential for electricity is negligible, as is that for large quantities of steam coal from U.K. mines, at least before 1985–1990. Thus, production of coal and nuclear electricity is essentially

Table 18.1
Scenario definitions, 1972–1985 and 1985–2000

Scenario label	Average annual economic growth rate (%)	World oil price (fixed 1975 U.S. dollars per barrel)	Effective average U.K. internal energy price (1972 = 1.0)
1972–1985			
C	2.3	11.50	1.40
D	1.5	11.50	1.25
1985–2000			
C-1	3.0	17.25	2.40
C-2	3.0	17.25	2.40
D-7	1.6	11.50	1.60
D-8	1.6	11.50	1.60
C	3.0	(see text)	2.50
D	1.6	(see text)	2.00

Notes: (1) Scenarios C-1 and D-7 differ from C-2 and D-8 through assumptions on indigenous supply policy, the former being high coal, low nuclear, while the latter correspond to low coal, high nuclear. (2) Scenarios C and D for 1985–2000 assume a policy for high coal and high nuclear production.

controlled by U.K. demand, as is production of gas as long as no facilities exist for its export. This leaves oil as the indigenous energy source which may be exported during the period that the U.K. has a net energy surplus. It follows that the best use of installed capacity in the supply industries requires flexibility in the demand pattern (achieved possibly by different relative fuel pricing or taxation) so that total energy requirements are balanced by oil with the other energy production industries operating at their economic optimum during the period of surplus, and thereafter so that indigenous sources can meet the largest share possible of demand. We examine below the implications for supply strategies.

It is planned that current annual coal production of 120 megatonnes should be increased to 130 megatonnes by 1985. Maintaining this production until 2000 would consume 3.2 gigatonnes of the 4.0 gigatonnes of the current and potential classified reserves, and hence would require substantial investment in new and existing mines, since without such investment annual production would fall below 80 megatonnes. Since U.K. coal resources are estimated at more than 100 gigatonnes, there is unlikely to be a resource problem, but since future reserves may be outside traditional mining areas, there may be manpower or environmental problems. The economic resources required would exceed the current Plan for Coal, since four or five new large mines (10 megatonnes per year) would be needed in the period 1985–2000. Such mines have a lead time of about ten years, so a radical policy might achieve a higher production, 140–150 megatonnes being about the plausible upper limit.

The current overcapacity in the electricity supply industry is likely to result in a low ordering rate for new power stations for the next ten years. The estimates in the supply and demand studies of a maximum of 58 GWe nuclear capacity in 2000 assumed that the nuclear program would be based on the steam-generating heavy water reactor (SGHWR). With the current delay in this program, that figure is now optimistic. It is possible that an alternative program, based on the advanced gas reactor (AGR) or pressurized water reactor (PWR), may be followed. Using either of these technologies would allow expansion at the rate previously assumed, while relaxation of the capital expenditure constraint that we have assumed on the production rate could lead to an installed capacity of 70 GWe in the high growth scenario.

Our estimates in the earlier WAES reports for oil production were based on a somewhat optimistic assumption of 4 gigatonnes of reserves. We may split this figure into four components:

1. Proved reserves with committed production facilities. These will give a peak annual production of about 100 megatonnes in 1980–1981, declining thereafter. Production delay would be economically disadvantageous.

2. Proved reserves with no committed production facilities. Production delay would be economic if a large enough oil price rise occurred in the 1980s. Current government policy is not to delay production, but this remains as a commercial option.

3. Probable reserves (making the total up to 3 gigatonnes). Government policy could delay production, but again economic justification is contingent on a substantial oil price rise.

4. Possible reserves (about 1 gigatonne). Undiscovered and speculative, depending on discoveries in, e.g., areas west of the U.K.. The possibility of government intervention in depletion rates may discourage exploration for such reserves. Production profiles cannot be estimated until 1980 at the earliest.

Production figures for natural gas in earlier WAES reports assumed reserves of 1.3 gigatonnes oil equivalent (GTOE), a figure which excluded the possible gas associated with undiscovered oil. Including this associated gas, we now would estimate reserves at 1.6 GTOE allowing production to be 10 MTOE higher through 1980–2000 than our previous estimates, with a peak of 50–60 MTOE. Gas production from the dry gas wells of the southern North Sea has lower cost than other energy sources, associated gas production depends on oil production rates (although reinjection is possible at extra cost), while in certain markets (e.g., domestic) gas is the most efficient fuel, all of which mitigates against any reduction in the production rate, which should be sufficient to supply the main premium markets (domestic and some industrial) until 2000. Thereafter, with total production already falling, continued supply to these markets will require LNG imports or gas manufacture from coal or oil. As with other fuels there is uncertainty about potential reserves and production. Although reserves may be higher than those indicated, it would seem prudent to follow an energy policy that takes a median rather than an optimistic view.

18.5 ENERGY DEMAND

Table 18.2 summarizes the aggregate primary energy demand for the U.K. given in the demand study.

In order to reanalyze plausible perturbations on the figures for primary energy demand given in Table 18.2, we first derive the implied income (GDP) elasticity γ and price elasticity β. These quantities define the energy demand E in terms of the GDP Y and the energy price P through the formula

$$(E/E_0) = (Y/Y_0)^{\gamma} (P/P_0)^{\beta}.$$

If we assume the price variable is the crude oil price, then the figures in Table 18.2 for 1972, 1977, and 1985 C imply an income elasticity of +0.88 and a price elasticity of -0.031. These elasticities, however, extrapolate to an energy demand in 2000 C-1 which is 6 percent higher than that given in the table, the magnitude of the price elasticity needing to be five times larger (-0.172) during the period 1985–2000 to obtain the tabulated value. These elasticities also do not fit the 1985 D, 2000 D-7 figures very well.

It is more instructive to discuss elasticities with respect to the internal U.K. average energy price, i.e., the price the consumer finally meets, rather than the crude oil price. In the U.K. the average (real) price of energy to final consumers rose by about 20 percent between 1972 and 1976. For the high growth scenario, C, this price may be expected to rise between now and 1985 as the price of gas, in particular, rises toward that of oil (with coal following the lead of oil products). The limit is likely to be an average energy price 40–50 percent above the 1972 figure. In the low growth scenario, D, with less pressure on resources to encourage early policy responses, the internal price is unlikely to rise so much. The figures in Table 18.2 are consistent with those in Table 18.3 (see also Table 18.1).

Beyond 1985, the internal energy price seems likely to rise in all the scenarios. In scenarios D-7 and D-8, a figure of about 60 percent above the 1972 value (i.e., slightly higher than the 40–50 percent limit mentioned above) is a plausible policy objective, considering the worsening energy component in the U.K. balance of payments in the 1990s. In scenarios C-1 and C-2, the scenario definitions require an energy price 50 percent higher than for D-7 and D-8, implying an internal energy price 140 percent above the 1972 value. Thus, the internal price of energy (1972 = 1.00) implied by the results in Table 18.2 and the figures in Table 18.3 are P(2000 C-1, C-2) = 2.4, and P(2000 D-7, D-8) = 1.6. The total primary energy demands given in the demand study and summarized in Table 18.2 are consistent with these internal prices and the income and price elasticities (constant in time) given above.

We investigate the effect on energy consumption of raising the internal price of energy. This may arise from actual price increases or it could be a notional change to simulate the effect of new economic criteria applied to energy conservation measures. The internal price of energy in scenarios C-1 and C-2 has been taken at 2.4 times the 1972 price in real terms. This implies significant

Table 18.3
U.K. internal energy price, 1972–1985 (1972 = 1.00)

			1985	
	1972	1977	Case C	Case D
Internal price	1.00	1.20	1.40	1.25

Note: Based on an income elasticity of 1.07 and a price elasticity (with respect to the internal price) of -0.29.

Table 18.2
U.K. primary energy demand, GDP, and crude oil price assumptions (1972 = 1.00)

			1985		2000	
	1972	1977	Case C	Case D	Case C-1	Case D-7
Primary energy demand	1.00	1.00	1.24	1.15	1.71	1.38
GDP	1.00	1.05	1.34	1.21	2.09	1.54
Crude oil price	1.00	4.00	4.00	4.00	6.00	4.00

increases in taxation and other overheads for secondary energy supply. This pricing policy is already "vigorous" in relation to measures for energy conservation, and for the new scenario C we limit additional measures to a further 4 percent increase to 2.5 times the 1972 price. This would give a reduction in primary energy demand of 1 percent. For the new D scenario we perturb the D-7 and D-8 scenarios by increasing the internal price of energy from 1.6 to 2.0 times the 1972 price. This reduces demand by 6 percent.

A decision, politically difficult to implement during a period of apparently plentiful energy, to raise the internal price as quickly as possible toward the long-run marginal cost of energy may be worthwhile. This would multiply the 1972 internal price of energy by 1.9 by 1985 (the average delivered price of oil products has been multiplied by 1.9 since 1972). If the factor rose to 2.5 by 2000, as before, the total cumulative saving due to this retiming of the price response would be about 250 MTOE. Following this policy in scenario C gives a demand similar to that of Case B (high oil price), while the 2000 demand pattern is unchanged. The difference lies in the increased reserves remaining (or extra balance-of-payments advantage from the extra exports).

18.6 PROBLEM AREAS

Up to about 1980 the supply situation of the U.K. will improve, as North Sea oil is developed. Thereafter, for 10–20 years supply is expected to exceed demand and exports should be possible. After this, however, the U.K. will move rapidly into deficit, at a time when world energy supplies are strained. Thus, the major policy problems for the U.K. for the next two decades will be to adjust the supply industries and demand patterns so as to meet the later deficits, while also making best use of the good supply position in the short term.

A substantial fraction of the coal accessible from existing mines will have been used by the year 2000. Many of these mines are old, and they will be closed when it becomes uneconomic to work them after full account is taken of the social costs involved in closing mines. However, physical coal resources are not a constraint on future production, though this will be limited by the rate at which new mines and major extensions to existing mines can be developed. Coal production will be the sum of the declining output from older and higher cost workings and the increasing output from newer and more efficient workings. The former dominate output now and will continue to do so in the 1980s, but the latter could dominate by the year 2000 provided there is adequate investment and a suitable workforce is available in the right areas. If production from the older mines is maintained at a high level, there may be difficulty in finding markets in the 1980s, but unless new workings and new mines are developed on a substantial scale, there will be a shortage of coal by the year 2000.

The electricity industry is the largest consumer of coal, but burns a large part of it in older, less efficient, lower merit order plants. The current overcapacity, likely to last several years, will delay ordering of new plants, giving problems in the heavy electrical supply industries, especially the nuclear industry. The implied low ordering rate for nuclear plants may lead to a situation where, when a large expansion in nuclear capacity is likely to be needed, the industry may be unable to provide it.

The gas supply industry is growing rapidly, and the low cost of production makes gas competitive with other fuels in most markets. Present policy aims for the rapid penetration of premium markets such as the domestic sector where gas has special advantages in the efficiency with which it can be used. This will contribute toward a reduction in the the future growth of electricity demand and will prolong the problems of overcapacity noted above. The production of large quantities of gas associated with North Sea oil will require natural gas also to be consumed for nonpremium uses such as underboiler fuel during periods of low total demand for gas, but such uses, which could also be met by coal, will obviously be kept to a minimum. It is likely that indigenous natural gas can supply premium markets until around 2000. If oil production is delayed, this will also delay the use of associated gas. The economics of such delays would depend on the timing and extent of a rise in the world price of oil.

If the world price of oil were to double in the period 1985–1990, it might be advantageous to delay some of the development of North Sea oil. In practice we would not know in advance when and by how much the price of oil might change. If the price of oil remained constant, economic arguments would appear to favor a policy of rapid depletion. The results of the WAES global projections suggest that the price of oil may rise within the period 1985–1995. On that basis, it might seem prudent to delay the development of some of the North Sea oil. Unfortunately such delay would also probably reduce exploration and would therefore introduce more uncertainties about potential reserves. It should be noted also that from the viewpoint of world oil supplies, restrictions on North Sea production would have an adverse effect, which if followed by other producers, would lead to a higher price for oil. However, if consumer countries (including the U.K.) follow vigorous measures for energy conservation and for the production of both indigenous oil and other forms of energy, this could lead to a balance of world energy supply and demand without the "aid" of a substantial rise in the world price of oil. This result is more likely to be achieved under conditions of low world economic growth, as in WAES scenario D, than for high economic growth, as in scenario C, primarily because of the long lead times for the development of alternative energy sources.

From the viewpoint of long-term planning and energy conservation, it would seem prudent for the internal price of energy in the U.K. to anticipate further increases in the world oil price. However, it is equally important, if industrial planning is to approach optimum, that both the absolute energy price and the relative prices of fuels should have a reasonable measure of stability and predictability. The possible increases in the internal price of energy could be reduced if some of the response toward energy conservation were stimulated by other measures, for example, new economic criteria for assessing the benefits of energy conservation, the further improvement of regulations and standards, and subsidies for appropriate conservation measures.

18.7 CONVERSION FACTORS AND CONVENTIONS

The conversion factors used in this report are shown below. It should be noted that the thermal content of U.K. coal is somewhat lower than the conventional figure of 28.8 GJ per tonne that is often used internationally. These conversion factors are conventions used to obtain the quoted figures in the text, the actual thermal contents of fuels varying with the specific type of fuel.

1 tonne oil equivalent (TOE) =
44 GJ = 44×10^9 J

1 tonne coal equivalent (TCE) =
25.5 GJ = 25.5×10^9 J

18.8 WORKSHEETS

Summary tables and worksheets appear in this section in the following order:

Energy demand summary, 1972–1985–2000
National input worksheets for supply/demand integration (10 sheets)

WAES NATIONAL ENERGY DEMAND STUDIES, 1972-1985-2000
ENERGY DEMAND SUMMARY

Country: UNITED KINGDOM

Scenario Assumptions	1972	1985					2000			
WAES CASE	Base Year	A	B	C	D	E	C-1	C-2	D-7	D-8
GNP 1976-85; 1985-2000		High	Low	High	Low	High	High	High	Low	Low
Oil/Energy Price*		11.50-17.25	11.50-17.25	11.50	11.50	11.50-7.66	11.50-17.25	11.50-17.25	11.50	11.50
National Policy Response		Vig	Vig	Vig	Res	Res	Vig	Vig	Vig	Vig
Principal Replacement Fuel							Coal	Nuclear	Coal	Nuclear
Primary Energy Demand **Units**	10^{18} JOULES PER ANNUM, PRIMARY ENERY EQUIVALENT									
Oil	4.68	5.05	4.48	5.31	4.62	5.45	7.66	7.57	6.22	6.28
Natural Gas	1.19	2.08	1.95	2.10	1.95	2.12	1.63	1.63	1.50	1.50
Coal	3.15	3.30	3.30	3.30	3.30	3.30	4.88	3.62	3.68	2.48
Nuclear	0.25	0.81	0.76	0.81	0.77	0.81	1.67	2.93	1.47	2.51
Hydroelectric	0.04	0.04	0.04	0.04	0.04	0.04	0.04	0.04	0.04	0.04
Geothermal	–	–	–	–	–	–	–	–	–	–
Solar Thermal	–	–	–	–	–	–	–	–	–	–
Solar Electric	–	–	–	–	–	–	–	–	–	–
TOTAL PRIMARY ENERGY DEMAND	9.31	11.28	10.53	11.55	10.68	11.71	15.88	15.79	12.70	12.81

*1975 US dollars per barrel of Arabian light crude oil, fob Persian Gulf.
All Figures are primary production prior to any processing or conversion.

NATIONAL INPUT WORKSHEET FOR SUPPLY/DEMAND INTEGRATION

Country: U.K.

Year: 1972

Case: BASE YEAR

Units: EXAJOULES (10^{18}J)

	(a)	(b)	(c)	(d)	(e)	(f)	(g)	(h)	(i)	(j)	(k)
	Coal	Petro-leum	(Syn. Liquids)	Nat. Gas	(Syn. Gas)	(Heat)	(Electri-city)††	Hydro Energy	Nuclear Energy	Geothermal energy and other	Total
(1) Transport	0	1.24					0.01				1.25
(2) Industry	0.70	1.17		0.28			0.26				2.42
(3) Agric., Mining, Construction (4) Commercial	0.04	0.19		0.07			0.11				0.40
(5) Public	0.08	0.21		0.04			0.04				0.37
(6) Residential	0.57	0.16		0.48			0.31				1.52
(7) Non-energy Uses		0.52		0.11							0.62
(8) Final Energy Demand*	1.39	3.49		0.97			0.74				6.58
(9) Electricity**	1.60	0.85		0.07			-0.87	0.04	0.25		1.94
(10) Syn. Gas											
(11) Syn. Liquids											
(12) Heat											
(13) Energy Sector Self Consumption & conversion losses	0.16	0.34		0.15			0.13				0.78
(14) Primary Energy Input	3.15	4.68		1.19			0.04	0.25			9.31
(15) Indigenous Supply	3.02	0		1.15			0.04	0.25			4.47
(16) Imports***	0.13	4.68		0.03							4.84
(17) Exports											

*Includes non-energy uses.

**Blocks 9g, 10e 11c, and 12f must be negative to avoid double counting of energy from primary fuels.

***Includes imported products.

††Includes only electricity to be purchased by each sector.

NATIONAL INPUT WORKSHEET FOR SUPPLY/DEMAND INTEGRATION

Country: U.K

Year: 1985

Case: A

Units: EXAJOULES (10^{18} J)

	(a) Coal	(b) Petro-leum	(c) (Syn. Liquids)	(d) Nat. Gas	(e) (Syn. Gas)	(f) (Heat)	(g) (Electri-city)[††]	(h) Hydro Energy	(i) Nuclear Energy	(j) Geothermal energy and other	(k) Total
(1) Transport		1.58					0.01				1.58
(2) Industry	0.73	1.08		0.59			0.40				2.80
(3) Agric., Mining, Construction; (4) Commercial	0.01	0.10		0.13			0.17				0.41
(5) Public	0.06	0.18		0.08			0.08				0.40
(6) Residential	0.26	0.12		0.94			0.46				1.77
(7) Non-energy Uses		0.86		0.11							0.97
(8) Final Energy Demand*	1.05	3.92		1.86			1.11				7.94
(9) Electricity**	2.12	0.76		0.02			-1.31	0.04	0.81		2.43
(10) Syn. Gas											
(11) Syn. Liquids											
(12) Heat											
(13) Energy Sector Self Consumption & conversion losses	0.13	0.37		0.21			0.20				0.91
(14) Primary Energy Input	3.30	5.05		2.08				0.04	0.81		11.28
(15) Indigenous Supply	3.30	7.00		1.68				0.04	0.81		12.83
(16) Imports***				0.40							
(17) Exports		1.95									1.55

*Includes non-energy uses.

**Blocks 9g, 10e 11c, and 12f must be negative to avoid double counting of energy from primary fuels.

***Includes imported products.

††Includes only electricity to be purchased by each sector.

NATIONAL INPUT WORKSHEET FOR SUPPLY/DEMAND INTEGRATION

Country: U.K.

Year: 1985

Case: B

Units: EXAJOULES (10^{18}J)

	(a)	(b)	(c)	(d)	(e)	(f)	(g)	(h)	(i)	(j)	(k)
	Coal	Petro-leum	(Syn. Liquids)	Nat. Gas	(Syn. Gas)	(Heat)	(Electri-city)††	Hydro Energy	Nuclear Energy	Geothermal energy and other	Total
(1) Transport		1.46					0.01				1.47
(2) Industry	0.73	1.12		0.53			0.37				2.75
(3) Agric., Mining, Construction } (4) Commercial	0.01	0.17		0.10			0.14				0.42
(5) Public	0.04	0.13		0.14			0.06				0.36
(6) Residential	0.26	0.12		0.86			0.43				1.67
(7) Non-energy Uses		0.73		0.11							0.84
(8) Final Energy Demand*	1.03	3.73		1.74			1.01				7.51
(9) Electricity**	2.14	0.43		0.02			-1.18	0.04	0.76		2.20
(10) Syn. Gas											
(11) Syn. Liquids											
(12) Heat											
(13) Energy Sector Self Consumption & conversion losses	0.13	0.33		0.19			0.18				0.83
(14) Primary Energy Input	3.30	4.48		1.95				0.04	0.76		10.53
(15) Indigenous Supply	3.30	7.00		1.55				0.04	0.76		12.64
(16) Imports***				0.40							
(17) Exports		2.52									2.12

*Includes non-energy uses.

**Blocks 9g, 10e 11c, and 12f must be negative to avoid double counting of energy from primary fuels.

***Includes imported products.

††Includes only electricity to be purchased by each sector.

NATIONAL INPUT WORKSHEET FOR SUPPLY/DEMAND INTEGRATION

Country: U.K.

Year: 1985

Case: C

Units: EXAJOULES (10^{18}J)

	(a)	(b)	(c)	(d)	(e)	(f)	(g)	(h)	(i)	(j)	(k)
	Coal	Petro-leum	(Syn. Liquids)	Nat. Gas	(Syn. Gas)	(Heat)	(Electri-city)††	Hydro Energy	Nuclear Energy	Geothermal energy and other	Total
(1) Transport		1.59					0.01				1.60
(2) Industry	0.74	1.11		0.59			0.41				2.86
(3) Agric., Mining, Construction } (4) Commercial	0.01	0.10		0.13			0.18				0.42
(5) Public	0.06	0.18		0.08			0.08				0.40
(6) Residential	0.26	0.12		0.96			0.48				1.81
(7) Non-energy Uses		0.89		0.11							1.00
(8) Final Energy Demand*	1.06	4.00		1.87			1.16				8.09
(9) Electricity**	2.11	0.92		0.02			-1.36	0.04	0.81		2.53
(10) Syn. Gas											
(11) Syn. Liquids											
(12) Heat											
(13) Energy Sector Self Consumption & conversion losses	0.13	0.39		0.21			0.20				0.93
(14) Primary Energy Input	3.30	5.31		2.10				0.04	0.81		11.55
(15) Indigenous Supply	3.30	7.00		1.70				0.04	0.81		12.85
(16) Imports***				0.40							
(17) Exports		1.69									1.30

*Includes non-energy uses.

**Blocks 9g, 10e 11c, and 12f must be negative to avoid double counting of energy from primary fuels.

***Includes imported products.

††Includes only electricity to be purchased by each sector.

NATIONAL INPUT WORKSHEET FOR SUPPLY/DEMAND INTEGRATION

Country: U.K.

Year: 1985

Case: D

Units: EXAJOULES (10^{18} J)

	(a) Coal	(b) Petroleum	(c) (Syn. Liquids)	(d) Nat. Gas	(e) (Syn. Gas)	(f) (Heat)	(g) (Electricity)[††]	(h) Hydro Energy	(i) Nuclear Energy	(j) Geothermal energy and other	(k) Total
(1) Transport		1.49					0.01				1.50
(2) Industry	0.73	1.11		0.53			0.38				2.74
(3) Agric., Mining, Construction; (4) Commercial	0.01	0.17		0.11			0.14				0.43
(5) Public	0.04	0.14		0.15			0.06				0.37
(6) Residential	0.26	0.13		0.85			0.44				1.68
(7) Non-energy Uses		0.77		0.11							0.87
(8) Final Energy Demand*	1.03	3.81		1.73			1.03				7.60
(9) Electricity**	2.14	0.48		0.02			-1.21	0.04	0.77		2.24
(10) Syn. Gas											
(11) Syn. Liquids											
(12) Heat											
(13) Energy Sector Self Consumption & conversion losses	0.13	0.34		0.19			0.18				0.84
(14) Primary Energy Input	3.30	4.62		1.95				0.04	0.77		10.68
(15) Indigenous Supply	3.30	7.00		1.55				0.04	0.77		12.66
(16) Imports***				0.40							
(17) Exports		2.38									1.98

*Includes non-energy uses.

**Blocks 9g, 10e 11c, and 12f must be negative to avoid double counting of energy from primary fuels.

***Includes imported products.

[††]Includes only electricity to be purchased by each sector.

NATIONAL INPUT WORKSHEET FOR SUPPLY/DEMAND INTEGRATION

Country: U.K.

Year: 1985

Case: E

Units: EXAJOULES (10^{18}J)

	(a)	(b)	(c)	(d)	(e)	(f)	(g)	(h)	(i)	(j)	(k)
	Coal	Petro-leum	(Syn. Liquids)	Nat. Gas	(Syn. Gas)	(Heat)	(Electri-city)††	Hydro Energy	Nuclear Energy	Geothermal energy and other	Total
(1) Transport		1.62					0.01				1.62
(2) Industry	0.73	1.15		0.59			0.43				2.90
(3) Agric., Mining, Construction (4) Commercial	0.01	0.10		0.13			0.19				0.43
(5) Public	0.06	0.18		0.09			0.08				0.40
(6) Residential	0.26	0.12		0.97			0.49				1.83
(7) Non-energy Uses		0.87		0.11							0.98
(8) Final Energy Demand*	1.05	4.04		1.89			1.19				8.17
(9) Electricity**	2.11	1.01		0.02			-1.40	0.04	0.81		2.59
(10) Syn. Gas											
(11) Syn. Liquids											
(12) Heat											
(13) Energy Sector Self Consumption & conversion losses	0.13	0.40		0.21			0.21				0.95
(14) Primary Energy Input	3.30	5.45		2.12				0.04	0.81		11.71
(15) Indigenous Supply	3.30	7.00		1.72				0.04	0.81		12.86
(16) Imports***				0.40							
(17) Exports		1.55									1.15

*Includes non-energy uses.

**Blocks 9g, 10e 11c, and 12f must be negative to avoid double counting of energy from primary fuels.

***Includes imported products.

††Includes only electricity to be purchased by each sector.

NATIONAL INPUT WORKSHEET FOR SUPPLY/DEMAND INTEGRATION

Country: U.K.

Year: 2000

Case: C1

Units: EXAJOULES (10^{18}J)

	(a) Coal	(b) Petroleum	(c) (Syn. Liquids)	(d) Nat. Gas	(e) (Syn. Gas)	(f) (Heat)	(g) (Electricity)††	(h) Hydro Energy	(i) Nuclear Energy	(j) Geothermal energy and other	(k) Total
(1) Transport		2.11					0.01				2.13
(2) Industry	1.23	1.76		0.20			0.65				3.84
(3) Agric., Mining, Construction } (4) Commercial		0.12		0.15			0.28				0.55
(5) Public	0.03	0.30		0.10			0.12				0.54
(6) Residential	0.14	0.36		1.03			0.74				2.27
(7) Non-energy Uses	0.28	1.39									1.67
(8) Final Energy Demand*	1.68	6.05		1.47			1.80				11.00
(9) Electricity**	3.11	0.84					-2.12	0.04	1.67		3.53
(10) Syn. Gas											
(11) Syn. Liquids											
(12) Heat											
(13) Energy Sector Self Consumption & conversion losses	0.10	0.77		0.16			0.32				1.34
(14) Primary Energy Input	4.88	7.66		1.63				0.04	1.67		15.88
(15) Indigenous Supply	3.56	3.39		0.97				0.04	1.67		9.63
(16) Imports***	1.33	4.26		0.66							6.26
(17) Exports											

*Includes non-energy uses.

**Blocks 9g, 10e 11c, and 12f must be negative to avoid double counting of energy from primary fuels

***Includes imported products.

††Includes only electricity to be purchased by each sector.

NATIONAL INPUT WORKSHEET FOR SUPPLY/DEMAND INTEGRATION

Country: U.K.

Year: 2000

Case: C2

Units: EXAJOULES (10^{18}J)

	(a) Coal	(b) Petroleum	(c) (Syn. Liquids)	(d) Nat. Gas	(e) (Syn. Gas)	(f) (Heat)	(g) (Electricity)[††]	(h) Hydro Energy	(i) Nuclear Energy	(j) Geothermal energy and other	(k) Total
(1) Transport		2.11					0.01				2.13
(2) Industry	0.99	1.80		0.20			0.70				3.68
(3) Agric., Mining, Construction }											
(4) Commercial }		0.12		0.15			0.28				0.55
(5) Public	0.03	0.30		0.09			0.12				0.54
(6) Residential	0.14	0.36		1.03			0.74				2.27
(7) Non-energy Uses	0.17	1.50									1.67
(8) Final Energy Demand*	1.34	6.19		1.47			1.85				10.84
(9) Electricity**	2.21	0.63					−2.18	0.04	2.93		3.63
(10) Syn. Gas											
(11) Syn. Liquids											
(12) Heat											
(13) Energy Sector Self Consumption & conversion losses	0.07	0.76		0.16			0.33				1.32
(14) Primary Energy Input	3.62	7.57		1.63				0.04	2.93		15.79
(15) Indigenous Supply	2.30	3.39		0.97				0.04	2.93		9.63
(16) Imports***	1.32	4.18		0.66							6.16
(17) Exports											

*Includes non-energy uses.

**Blocks 9g, 10e 11c, and 12f must be negative to avoid double counting of energy from primary fuels.

***Includes imported products.

††Includes only electricity to be purchased by each sector.

NATIONAL INPUT WORKSHEET FOR SUPPLY/DEMAND INTEGRATION

Country: U.K.

Year: 2000

Case: D7

Units: EXAJOULES (10^{18}J)

	(a) Coal	(b) Petro-leum	(c) (Syn. Liquids)	(d) Nat. Gas	(e) (Syn. Gas)	(f) (Heat)	(g) (Electri-city)††	(h) Hydro Energy	(i) Nuclear Energy	(j) Geothermal energy and other	(k) Total
(1) Transport		1.80					0.01				1.80
(2) Industry	1.08	1.53		0.18			0.48				3.27
(3) Agric., Mining, Construction } (4) Commercial		0.22		0.11			0.18				0.51
(5) Public	0.02	0.22		0.15			0.07				0.46
(6) Residential	0.14	0.31		0.90			0.57				1.92
(7) Non-energy Uses	0.15	1.11									1.26
(8) Final Energy Demand*	1.40	5.18		1.35			1.31				9.23
(9) Electricity**	2.21	0.42					-1.55	0.04	1.47		2.58
(10) Syn. Gas											
(11) Syn. Liquids											
(12) Heat											
(13) Energy Sector Self Consumption & conversion losses	0.08	0.62		0.15			0.24				1.08
(14) Primary Energy Input	3.68	6.22		1.50				0.04	1.47		12.90
(15) Indigenous Supply	3.14	3.39		1.10				0.04	1.47		9.13
(16) Imports***	0.54	2.83		0.40							3.77
(17) Exports											

*Includes non-energy uses.

**Blocks 9g, 10e 11c, and 12f must be negative to avoid double counting of energy from primary fuels.

***Includes imported products.

††Includes only electricity to be purchased by each sector.

NATIONAL INPUT WORKSHEET FOR SUPPLY/DEMAND INTEGRATION

Country: U.K.

Year: 2000

Case: D8

Units: EXAJOULES (10^{18}J)

	(a)	(b)	(c)	(d)	(e)	(f)	(g)	(h)	(i)	(j)	(k)
	Coal	Petro-leum	(Syn. Liquids)	Nat. Gas	(Syn. Gas)	(Heat)	(Electri-city)††	Hydro Energy	Nuclear Energy	Geothermal energy and other	Total
(1) Transport		1.80					0.01				1.80
(2) Industry	0.91	1.54		0.18			0.52				3.14
(3) Agric., Mining, Construction }											
(4) Commercial }		0.22		0.11			0.18				0.51
(5) Public	0.02	0.22		0.15			0.07				0.46
(6) Residential	0.14	0.31		0.90			0.57				1.92
(7) Non-energy Uses	0.10	1.16									1.26
(8) Final Energy Demand*	1.17	5.24		1.35			1.35				9.10
(9) Electricity**	1.25	0.42					−1.58	0.04	2.51		2.64
(10) Syn. Gas											
(11) Syn. Liquids											
(12) Heat											
(13) Energy Sector Self Consumption & conversion losses	0.05	0.63		0.15		0.24					1.07
(14) Primary Energy Input	2.48	6.28		1.50				0.04	2.51		12.81
(15) Indigenous Supply	2.09	3.39		1.10				0.04	2.51		9.13
(16) Imports***	0.39	2.89		0.40							3.68
(17) Exports											

*Includes non-energy uses.

**Blocks 9g, 10e 11c, and 12f must be negative to avoid double counting of energy from primary fuels.

***Includes imported products.

††Includes only electricity to be purchased by each sector.

United States

19.1 SUMMARY AND CONCLUSIONS

19.1.1 Introduction

The intent of this report is to present a range of plausible energy futures for the United States for 1972–2000 within the overall WAES framework, and to highlight the important and major policy issues revealed by the analysis. The aim of this section is to set this report in its proper perspective.

U.S. national studies done within the WAES framework (described in chapter 1) include energy demand studies, energy supply studies, and national balancings of energy demand and supply—national supply-demand integrations. The analysis for 1972 and projections to 1985 of energy demand, and the projections of energy supply for the period 1972 to year 2000, are included in other WAES technical reports (1,2). This chapter presents projections of energy demand for the year 2000, based on the WAES scenario assumptions, and U.S. national supply-demand integrations for 1985 and the year 2000.

The integrations are performed for each of the WAES scenario cases (see Figure 1.2 of chapter 1). *Desired energy demands*—demands for end-use energy for several economic activities—are estimated for each of the various WAES scenarios and are taken as fixed. Desired energy demands for the United States to 2000 are presented in section 19.3, demands for 1972 and 1985 are summarized in section 19.2.3. *Potential energy supplies*—maximum U.S. production levels for fuels—are taken as physical upper limits on the availability of indigenous fuels consistent with the given assumptions in each scenario. The projections of U.S. potential energy supplies are presented in chapter 20 of the WAES Second Technical Report and are summarized in section 19.2.4 below. With these demand and supply inputs, the supply-demand integration process allocates potentially available indigenous fuels to meet end-use demands. The allocation results in required imports and potential exports for each fuel for each set of WAES scenario assumptions regarding energy price, economic growth, and national policy response.

The U.S. integrations, like other integrations, serve as inputs to the WAES global supply-demand integration process. Import desires and export potentials from each country are summed into regions to reveal imbalances—prospective shortages or surpluses—in the *unconstrained* global integrations. Demand preferences and supply potentials from each country combine in the *constrained* global integrations to show how fuels might be allocated to reduce potential shortages or surpluses and meet desired energy demands at the least total cost to the world's energy consumers. The procedures followed in the unconstrained and constrained global integrations are described in Part I of this report. There, they are placed in the general context of the overall WAES methodology for studying alternative energy futures. Chapters 4 and 5 of Part II present the results of these two global integration approaches.

19.1.2 Summary of Results

The result of an unconstrained national supply-demand integration is a particular pattern of supply and use of fuels—a mix of fuels that meets energy needs. In a nationally based global analysis such as the WAES analysis, one useful summary of the national integration results is the level of desired imports and/or potential exports of fuels. Such projected preferences are developed, initially, without regard to the broader global energy environment; they are "unconstrained" by global supply availabilities.

Table 19.1 presents a summary of the imports and exports of the United States, developed in this way, for the WAES cases to 1985 and 2000. Given the WAES assumptions, U.S. desired oil imports shown in Table 19.1 for the unconstrained cases range from 20 to 32 Quads (1 Quad = 10^{15} Btu)—or from about 9.5 to about 15 million barrels per day (MBD) in the year 2000. Gas imports are 3.2 to 5.0 Quads in 2000. These oil and gas figures indicate a continued and growing demand for imports of these fuels to the end of the century, given the range of assumptions of these cases.

The United States has the *potential* to export significant quantities of coal in the year 2000, as Table 19.1 shows. Potential coal exports are the excess of potential U.S. coal production over desired internal coal demand given the scenario as-

Table 19.1
U.S. energy supply-demand integration summary—(imports) and exports of primary energy (10^{15} Btu)

	1972	1985					2000			
WAES case	Base year	A	B	C	D	E	C-1	C-2	D-7	D-8
GNP 1976-85; 1985-2000		High	Low	High	Low	High	High	High	Low	Low
Oil/energy price[a]		11.50-17.25	11.50-17.25	11.50	11.50	11.50-7.66	11.50-17.25	11.50-17.25	11.50	11.50
National policy response		Vig	Vig	Vig	Res	Res	Vig	Vig	Vig	Vig
Principal replacement fuel							Coal	Nuclear	Coal	Nuclear
Oil	(10.1)	(11.6)	(8.6)	(16.1)	(25.9)	(39.8)	(19.9)	(20.5)	(32.3)	(32.3)
Natural gas	(1.0)	(1.6)	—	(2.7)	(2.1)	(6.0)	(3.2)	(3.6)	(5.0)	(5.0)
Coal	1.5	11.3	12.3	7.1	—	—	18.1	7.2	14.7	8.6
Nuclear	—	—	—	—	—	—	—	—	—	—
Hydroelectric	—	—	—	—	—	—	—	—	—	—
Total net (imports) or exports	(9.7)	(1.9)	3.7	(11.7)	(28.0)	(45.8)	(5.0)	(17.0)	(22.6)	(28.7)

All figures are primary production prior to any processing or conversion.
[a]1975 U.S. dollars per barrel of Arabian light crude oil FOB Persian Gulf.

sumptions. These potential exports range from 7 to 18 Quads in 2000, about 300 to 800 million tons of coal.

These basic results of the unconstrained U.S. cases are presented in more detail in section 19.4.

19.1.3 Chapter Overview

This chapter presents U.S. national studies done using the WAES methodology and WAES scenario conditions (see the discussion in chapter 1). It should be viewed within that framework.

The next section of this chapter describes the U.S. national supply-demand integration methodology. It explicitly states the main assumptions, summarizes the major demand and supply inputs, and describes the process of analysis that was used.

Section 19.3 of this chapter presents one of the major *inputs* to the integrations—the study of U.S. energy demand for the year 2000. It is an extension of the U.S. energy demand projections to 1985, reported as chapter 15 of the First Technical Report of WAES (1).

Section 19.4 reports the main findings of the U.S. unconstrained integration case analyses. "Excursion" cases, variations on the unconstrained integrations, are also presented. Some of the major issues raised by these studies are outlined, including the implications of this work for future directions in energy policy.

This chapter represents the work of the U.S. Associates in WAES. The demand study to 2000 was coordinated by Richard Cheston and Paul Basile of the MIT staff, with the assistance of Steven Carhart of Brookhaven National Laboratory. Anthony Finizza of the Atlantic Richfield Company played a major role in developing the 1985-2000 economic activity level projections. The bulk of the unconstrained integration analysis was performed for the WAES-U.S. team using WAES demand and supply studies by Steven Carhart and Kenneth Hoffman of the National Center for the Analysis of Energy Systems at Brookhaven National Laboratory, with the assistance of Paul Basile.

19.2 U.S. SUPPLY-DEMAND INTEGRATION METHODOLOGY

19.2.1 Major Assumptions

The analyses described here were performed under the WAES scenario assumptions (see Figure 1.2 of chapter 1), which were converted into national values for the U.S. projections of energy supply and demand. The specific values for these variables are presented in the separate U.S. demand and supply reports. The WAES scenarios call for different combinations of values: a *high* and *low* economic growth rate; a *rising* and *constant* real oil or energy price to 1985 and 2000 (with also a *falling* price case to 1985); a *vigorous* and *restrained* national energy policy response in relation to both supply and demand; and *coal* and *nuclear* options as principal replacement fuels for oil in the 1985 to 2000 period.

The U.S. national supply-demand integrations described here treat the supply estimates for each fuel as physical, upper limits that cannot be exceeded. They also assume that end-use demands must be met. These demands and supplies are *inputs* to the integration studies; the inputs are not changed by the integration.

End-use demands from the WAES U.S. demand studies are not specified as demands for specific fuels. Rather, *ranges of preference* are selected for each fuel for each demand category. These ranges mean that the integration results must ensure that, for example, at least 30 percent of domestic sector demands are met by electricity, or that no more than 50 percent of industrial demand is met by coal in 2000. These preference ranges are given, for all cases, in the summary of U.S. energy demand (section 19.2.3).

The U.S. national unconstrained integrations assume that there are unlimited oil imports available and that imports of natural gas are within realistic limits. (This is spelled out more explicitly in section 19.4.)

The assumption of unlimited oil imports—at the oil price assumptions of the particular scenario case—is the assumption that defines an "unconstrained" national integration in the WAES terminology; unconstrained integrations are those in which nations can obtain all the oil imports they desire. Some possible implications for the United States of global constraints on oil availability—of global "prospective shortages" of oil—are presented in section 19.4.3.

19.2.2 The Brookhaven Energy System Optimization Model

The allocation of primary fuel supplies to meet end-use demands requires many calculations. As each fuel is extracted, transported, processed, converted, and ultimately delivered to the consumer, there are losses and efficiencies and costs. A particular fuel might take any number of different "paths" from primary to delivered form.

To perform these many calculations, and to make the necessary tradeoffs among alternative paths, the Brookhaven Energy System Optimization Model (BESOM) was used.

BESOM is a linear programming model of the U.S. energy system, which is documented in detail elsewhere (3). It has been used extensively by the U.S. Energy Research and Development Administration for national research, development, and demonstration planning, technology assessment, and other purposes (4,5).

The model consists of a mathematical formulation of the flow of energy from primary resources in the ground, through various conversion and processing stages, to the satisfaction of final demand. Each stage is characterized by amortized capital and operating cost, efficiency, and environmental impact data. The model calculates the lowest-cost energy system under the assumption of constraints on maximum supply availability, demand levels that must be met, mix of fuels to be delivered to the consumer, and maximum capacity in various processing stages.

To perform supply-demand integrations, the model begins with WAES supply estimates for U.S. maximum potential production of fuels. These are taken as maximum constraints in the model. BESOM does not use more of any individual indigenous energy source than allowed.

The model also uses WAES U.S. demand estimates as inputs. In all integrations, the delivered energy must be the amount specified by the U.S. demand study. The types of energy delivered to the final consumers—oil products, gas, electricity, coal, solar energy, and so forth—must be within the preference ranges specified for each sector.

The costs and efficiency data for energy supply and delivery technology used in these integrations are based on extensive data and U.S. experience in this area.

In all WAES cases analyzed, the combination of U.S. demand projections, domestic supply estimates, and types of fuels required does not produce an energy balance without requiring imported oil. In the WAES unconstrained cases developed here, unlimited amounts of imported oil are allowed. This ensures that a balance can be achieved, since oil is one fuel that can be used in all sectors to offset any shortages.

In addition to a supply-demand balance, the model calculates the marginal costs (shadow prices) associated with demand, supply, and fuel mix assumptions. These must be evaluated for consistency with original cost inputs.

The results of using BESOM to perform such integrations can be illustrated schematically by the Reference Energy System flow chart shown in Figure 19.1. This chart shows the flow of energy from primary sources through various processing stages to the points of final consumption. Primary fuel production, and import and export, is shown at the left. Each fuel can then be traced from left to right through various stages of refining, transport, conversion, transmission, distribution, and utilization. Final demand categories are set out at the right.

19.2.3 Summary of U.S. Energy Demand, 1972–2000

The demand inputs to the U.S. supply-demand integrations for 1985 and 2000 are summarized in Table 19.2. These demand estimates, for final or end-use energy, result from detailed calculations in the 69 individual demand sectors in the 1985 projections and the 23 sectors in the 2000 projections, given the specific scenario assumptions about economic growth, oil or energy price, and government policy. They represent a substantial degree of energy conservation action—improvements in the technical efficiencies of energy use.

As can be seen in Table 19.2, the demand projections do not suggest any major shifts of the demand mix among sectors. The annual average growth rates do, however, vary by sector because of the combinations of economic activity projections and anticipated efficiency improvements. The total final or delivered energy demand is estimated to grow at an average annual rate of 1.6 and 1.8 percent for Cases C-1 and C-2, respectively, and 1.2 percent for Cases D-7 and D-8 between 1985 and 2000. Chemical feedstocks and the industrial sector show higher energy growth rates, but the transportation and domestic (residential, commercial, and public) sectors show growth rates of less than 1 percent per year with the residential/commercial total actually declining in some cases.

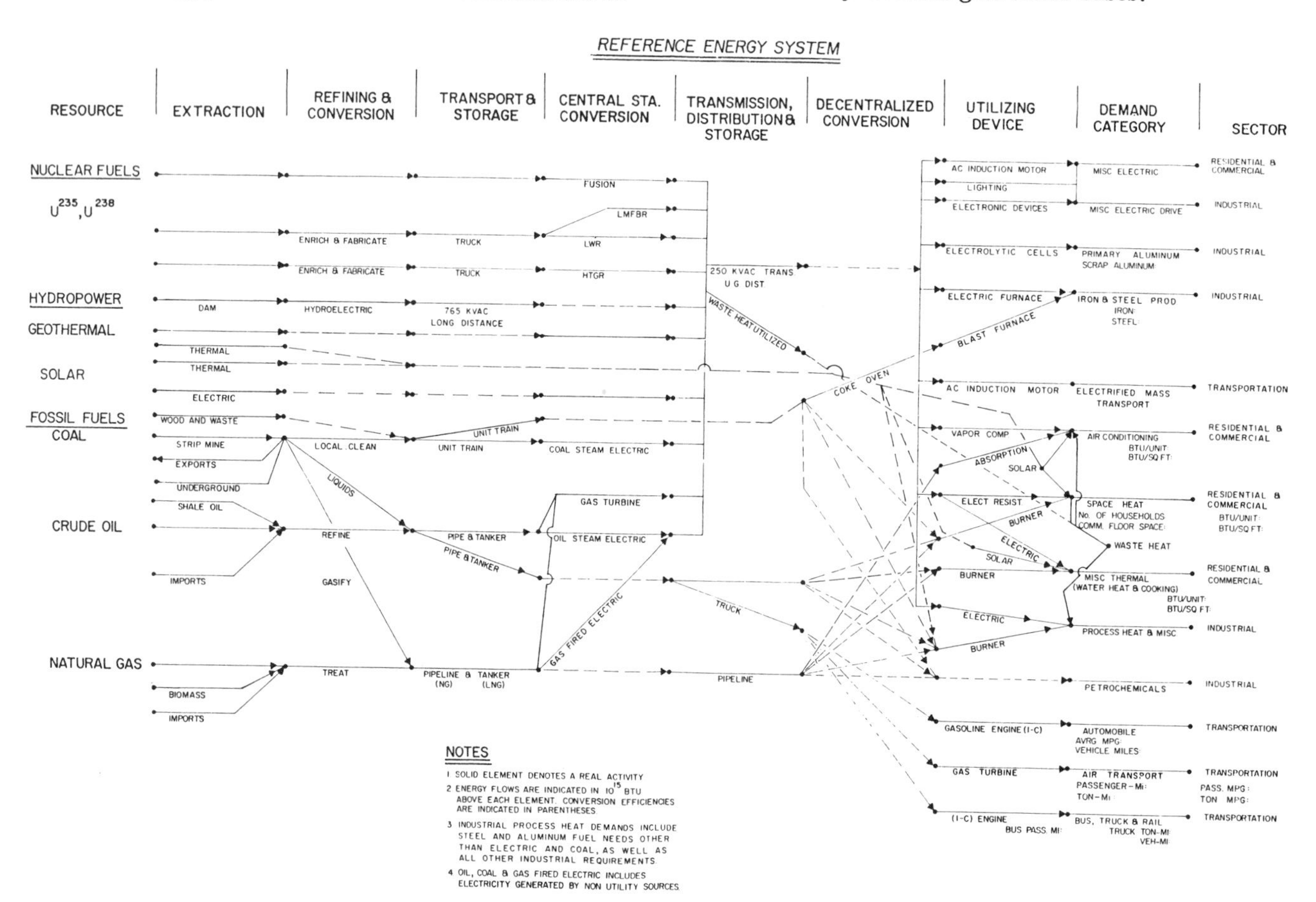

Figure 19.1
Reference energy system flow chart.

Table 19.2
U.S. energy demand summary (10^{15} Btu)

	1972	1985					2000			
WAES case	Base year	A	B	C	D	E	C-1	C-2	D-7	D-8
GNP 1976-85; 1985-2000		High	Low	High	Low	High	High	High	Low	Low
Oil/energy price[a]		11.50-17.25	11.50-17.25	11.50	11.50	11.50-7.66	11.50-17.25	11.50-17.25	11.50	11.50
National policy response		Vig	Vig	Vig	Res	Res	Vig	Vig	Vig	Vig
Principal replacement fuel							Coal	Nuclear	Coal	Nuclear
Transportation[b]	17.0	18.5	17.7	18.8	19.8	21.2	21.4	21.4	21.9	21.9
				(0.8)	(1.2)		(0.1)	(0.9)	(0.7)	(0.7)
Industry[b]	14.3	23.4	19.9	24.2	21.6	26.3	35.8	35.7	28.9	29.0
				(4.1)	(3.2)		(2.6)	(2.6)	(2.0)	(2.0)
Residential and commercial[b]	17.6	15.0	14.7	16.5	(19.6)	20.9	16.1	17.6)	19.7	20.3
				(-0.5)	(0.8)		(-0.2)	(0.4)	(0)	(0.2)
Feedstocks[b]	4.2	8.3	6.9	8.2	7.1	8.2	13.3	13.3	10.5	10.5
				(5.3)	(4.1)		(3.8)	(3.3)	(2.6)	(2.6)
Total delivered energy demand[b]	53.0	65.1	59.2	67.8	68.1	76.6	86.6	88.0	80.9	81.7
				(1.9)	(1.9)		(1.6)	(1.8)	(1.2)	(1.2)

[a] 1975 U.S. dollars per barrel of Arabian light crude oil FOB Persian Gulf or equivalent.

[b] Numbers in parentheses are average annual growth rates of delivered energy, 1972-1985 for the 1985 cases and 1985-2000 for the 2000 cases.

For each demand category, consumer preference ranges for each fuel are specified. These values, shown in Table 19.3, are based on judgments about the plausible *maximum* saturation levels that could be reached in the time period of interest by certain fuels in certain markets, and the plausible *minimum* amounts of fuel that any market might be expected to use. Table 19.3 contrasts these maxima and minima with the base year (1972) actual values. In the U.S. national integration, fuels are allocated to demands so that all minima are met, and so that no maxima are exceeded.

Known and expected trends are reflected in the preference constraints of Table 19.3 to the extent possible. For example, minimum preferences for electricity in the domestic and industrial sectors in 2000 are 6 to 20 percent higher than the actual 1972 percentage; oil and gas preference minima in all sectors are below historic levels. These minima *allow* a reduced share of these fuels in markets; of course, the actual integrated case results could show higher percentages. Preferences for coal, a relatively low-priced fuel, are set at levels that reflect an estimate of the maximum possible saturation levels of this fuel by 2000.

In the electricity generation sector of Table 19.3, the nuclear, hydro, geothermal, and solar maxima are supply study potentials.

19.2.4 Summary of U.S. Energy Supply, 1972-1985-2000

The U.S. supply study, which is presented as chapter 20 of the WAES Second Technical Report, includes projections of potential supply by fuel type for 1972 to 2000.

Potential domestic energy production of all fuels under the WAES scenario assumptions is summarized in Table 19.4. In 1985, the estimated production ranges from 61 to 86 Quads compared to 63 Quads in 1975. In the year 2000, the range is from 89 to 119 Quads. Fossil fuel resource estimates were developed for each WAES scenario, and a resulting production potential was calculated via the WAES fossil fuel worksheet, taking account of energy price and national policy response impacts on

Table 19.3
Maximum or (minimum) preferences for fuels[a] (percent of total sector demand)

	1972 (actual)	1985 (all cases)	2000			
			C-1	C-2	D-7	D-8
Industry						
Oil	19	(10)	(10)	(10)	(10)	(10)
Gas	45	40	(25)	(25)	(15)	(15)
Coal	24	40	50	50	50	50
Electricity	12	12	(18)	(20)	(18)	(20)
Residential						
Oil	30	(20)	(15)	(10)	(15)	(10)
Gas	47	45	(20)	(20)	(20)	(20)
Coal	1	(0)	5	5	5	5
Electricity	21	(25)	(30)	(40)	(30)	(40)
Electricity generation						
Oil	18	(10)	(8)	(8)	(8)	(8)
Gas	23	(5)	(3)	(3)	(3)	(3)
Coal	42	60	60	60	60	60

[a]When unspecified, minima are assumed to be zero and maxima are assumed to be 100 percent of demand for each category.

Table 19.4
U.S. energy supply summary, 1972-2000 (10^{15} Btu)

	1972	1985					2000			
WAES case	Base year	A	B	C	D	E	C-1	C-2	D-7	D-8
GNP 1976-85; 1985-2000		High	Low	High	Low	High	High	High	Low	Low
Oil/energy price[a]		11.50–17.25	11.50–17.25	11.50	11.50	11.50–7.66	11.50–17.25	11.50–17.25	11.50	11.50
National policy response		Vig	Vig	Vig	Res	Res	Vig	Vig	Vig	Vig
Principal replacement fuel							Coal	Nuclear	Coal	Nuclear
Crude oil	23.7	25.4	25.4	23.3	20.1	17.0	14.8	14.8	12.7	12.7
Natural gas[b]	22.1	20.0	20.0	19.0	16.0	14.0	15.0	15.0	11.5	11.5
Coal	14.1	26.2	24.0	24.0	20.2	20.2	48.0	31.7	33.1	24.8
Nuclear	0.5	8.5	8.5	8.5	6.6	6.6	26.0	42.4	26.0	42.4
Hydroelectric	2.7	4.6	4.6	4.6	3.4	3.4	4.6	4.6	3.6	3.6
Geothermal	0.0	0.5	0.5	0.5	0.2	0.2	3.0	3.0	1.0	1.0
Solar thermal	0.0	0.2	0.2	0.2	0.0	0.0	2.3	2.3	1.3	1.3
Solar electric	0.0	0.0	0.0	0.0	0.0	0.0	1.5	1.5	0.0	0.0
Shale oil	0.0	0.8	0.8	0.4	0.0	0.0	4.2	4.2	0.0	0.0
Total energy supply	63.1	86.2	84.0	80.5	66.5	61.4	119.4	119.5	89.2	97.3

All figures are primary production prior to any processing or conversion.

[a]1975 U.S. dollars per barrel of Arabian light crude oil FOB Persian Gulf or equivalent.

[b]Includes 1.0×10^{15} Btu biomass in Cases C-1 and C-2.

the likely development rates of the estimated potential reserves. Consideration was also given to infrastructure development, lead times, and environmental constraints. The results were then checked against the range of published estimates for reasonableness and also reviewed in detail with knowledgeable individuals.

The production estimates for nonfossil fuels and conversion processes were developed in relation to available published estimates and their related assumptions and have also been reviewed with knowledgeable individuals

The supply study indicates that the key determinant of the level of U.S. energy production in 1985 is national policy response, with domestic energy prices (influenced strongly by national policy response) playing a lesser but still important role.

In the WAES scenarios for 2000, vigorous policy response is assumed in all cases beyond 1985, with the national energy policy variable being the relative emphasis on coal or nuclear as the principal replacement fuel for declining oil supply. The difference in total energy supply between cases is primarily a function of energy prices. Coal and nuclear both increase in all cases, but the mix is a result of the policy assumption. In all cases, oil and gas decline to 30 percent or less of primary energy supply compared to 66 percent in 1975. The big increases are in coal production and nuclear capacity. Solar energy begins to make an important contribution by 2000 in the rising energy price cases (C-1 and C-2), at about 4 Quads or 3 percent of primary energy. While solar will very likely become a major energy source in the twenty-first century, its development is not anticipated to exceed the levels shown in Table 19.4 until after the year 2000.

19.3 U.S. ENERGY DEMAND, 1985-2000

The energy demand projections presented here represent an extension of the previously published WAES demand estimates through 1985 (1). As with all WAES demand studies, these projections cover only the final energy delivered to consumers. Total energy consumption is the sum of these final demands and processing losses in energy conversion and distribution. Numerous sources were used for the figures here, but several were particularly useful (6-11).

U.S. demand projections are derived on the basis of assumptions about national economic growth, energy prices, and national energy policy, consistent with the global variables described in chapter 1. For the U.S. studies of the WAES 2000 cases, real U.S. economic growth rates, consistent with the global rates, were estimated as 3.5 percent per year from 1985 to 2000 for Cases C-1 and C-2, and 2.5 percent per year for Cases D-7 and D-8. These rates are lower than the global averages of 5 and 3 percent because the U.S. is expected to have a lower growth rate than the world as a whole in this period.

19.3.1 Economic Activity Levels

Table 19.5 summarizes the projections of major economic levels for the period 1985-2000, based on the 1972-1985 projections of WAES Cases C and D. The main energy demand activities are shown here—population, number and distance traveled of autos, industrial output, and residential and commercial buildings—for 1972 with projections to 1985 and 2000.

The 69 disaggregated economic sectors in the demand projections through 1985 have been aggregated into 23 energy demand sectors for these longer-range projections. An extension of the 1985 economic activity projections was developed to form a basis for projecting these activity levels for the 1985 to 2000 period.

Gross national product (GNP) for 2000 was first projected on the basis of the growth rates for each WAES case and the sectoral growth trends from the 1980-1985 period. Then these GNP projections, in real terms, were modified to capture the effect of the transition from an oil and gas based economy to a coal or nuclear emphasis. This change assumed a shift in demand from consumption and government spending to business investment, with increases in nonresidential construction for transportation infrastructure and mining.

Real GNP projections for these cases were transformed into real output measured for 81 sectors using 1963 input-output tables. These sectoral estimates were then aggregated into the 23 categories used for the economic activity components of the WAES demand analysis.

The activity levels developed in this way assume a fairly straightforward relationship to estimated population trends. The population growth rate, which the U.S. Census Bureau projects to increase

Table 19.5
Economic activity levels

	1972	1985		2000	
		C	D	C-1, C-2	D-7, D-8
Population (10^6)	208.8	234.1	234.1	262.5	262.5
Total vehicle miles (10^9)	986.4	1444.6	1392.3	1660.0	1589.0
Autos (10^6)[a]	96.9	138.0	133.0	166.0	158.9
Air travel (10^9 passenger-miles)	169.0	410.8	382.1	654.1	561.5
Truck freight (10^9 ton-miles)	461.5	644.2	727.0		1068.7
Rail freight (10^9 ton-miles)	784.0	1163.2	1002.2	262.5	1473.2
Ship freight (10^9 ton-miles)	631.0	737.0	714.9	1589.0	1050.9
Primary metals[b]		2.026	1.622	158.9	1.535
Chemicals[c]		3.094	3.915	561.5	1.502
Energy-intensive industry[b]		1.454	1.373	1.651	1.427
Non-energy-intensive industry[b]		1.613	1.407	1.762	1.502
Commercial floor area (10^9 ft^2)	23.4	36.1	33.9	56.4	48.1
% fossil fuel space heat	95	88	89	80	80
% electric space heat	5	12	11	20	20
Total occupied dwellings (10^6)	66.7	86.6	85.6	107.8	104.4
% fossil fuel space heat	94	80	81	72	72
% electric space heat	6	20	19	28	28

[a]For 1985 Cases C and D, the automobile activity level was calculated according to total vehicle miles.

[b]Units for 1985 cases: ratio of value added, 1985 and 1972. Units for 2000 cases: ratio of industrial output, 2000 and 1985.

[c]Units for 1985 cases: FRB industrial production indices. Units for 2000 cases: ratio of industrial output, 2000 and 1985.

to 1 percent per year in the early 1980s, is projected to decline to 0.6 percent per year in the late 1990s and then to continue to decline toward zero population growth. This reduced population growth directly affects economic activity levels such as auto production and new housing starts.

The same basic formula was used to project both the auto stock and the housing stock in 2000. For example, to project the housing stock over the 15-year period from 1985 to 2000, the formula is:

$$\begin{aligned}&(\text{Housing Stock, 2000})\\&= (\text{Housing Stock, 1985})(1 - \text{RR})^{15}\\&\quad + (\text{Average Housing Sales})\left(\frac{1 - (1 - \text{RR})^{15}}{\text{RR}}\right),\end{aligned}$$

where RR is the replacement rate expressed in decimal fraction form.

For Cases C-1 and C-2, the projected housing stock replacement rate is just over 1.1 percent. Housing starts (including mobile homes), based on the economic assumptions and the population trends, average 2.5 million units per year. For Cases D-7 and D-8, the replacement rate is 1.0 percent and the annual housing starts are 2.2 million units. The housing mix in 2000 is assumed to follow the trends indicated by the Wharton econometric model[1] through 1985; that is, there will continue to be a small increase in the percentage of apartments, so that by 2000 they will comprise approximately one-third of all housing units.

Over the 1985-2000 period, autos, dwellings, and commercial floor space all grow at a much faster rate than the population, reflecting income growth effects and changes in age distribution. This leads to a continued decline in the ratio of people per dwelling and people per auto and a corresponding increase in the amount of commercial and public floor space per person. However, the growth rate of net additions to the housing stock, of commercial construction, and of autos on the road declines in 1985-2000 as compared to 1972-1985. In the earlier period, the average net increase in housing units per year is 1.5 million units. For 1985-2000, it drops to 1.3 million units per year. The

1. The use of the Wharton Annual and Industry Forecasting Model for the 1985 U.S. demand study is described in the first WAES technical report.

average net increase in autos on the road per year drops from about 3.0 million for 1972–1985 to about 1.9 million for 1985–2000.

19.3.2 Efficiency Coefficients

Table 19.6 shows the energy efficiencies for major line items in 1972, 1985, and 2000. These efficiencies are affected by the price and energy policy scenario variables for Cases C-1 and C-2 (rising price cases) and primarily by the energy policy variable for Cases D-7 and D-8 (constant price cases) for the period 1985 to 2000.

In the transportation sector, energy efficiency improvements are expected due to changing demands for more fuel-efficient vehicles and as the result of previous government action. For instance, almost all of the auto stock in the year 2000 will have been manufactured after 1985. Thus, they will have been affected by a federal law setting a new car average fuel economy standard of from 20.0 to 27.5 miles per gallon for the 1980 to 1985 period (based on 1976 emission standards). Further auto fuel efficiency improvements for new cars are assumed beyond 1985 due to reductions in auto weights, technology improvements, an increasing proportion of diesel engines, etc. One potential technological development that is not reflected in these studies because of the uncertainties that now exist is the widespread adoption of the electric car.

Government policy actions and energy prices are also expected to affect other transportation activities. Federal agencies are expected to review existing regulations to emphasize energy conservation. Examples include raising the average load factor

Table 19.6
Energy demand efficiency

		1985		2000	
	1972	C	D	C-1, C-2	D-7, D-8
Autos (Btu/mile)	9260	5774	6365	4310	4808
(mpg)	13.5	21.7	19.6	29.0	26.0
Air travel (Btu/passenger-mile)	7750	5266	6244	4525	5628
Truck freight (Btu/ton-mile)	3284	2964	2837	2666	2555
Rail freight (Btu/ton-mile)	710	630	630	566	563
Primary metals[a]		1.75	1.3	0.8	0.5
Chemicals[b]		1.7	1.5	1.4	1.0
Energy-intensive industry[a]		1.2	1.0	1.5	1.2
Non-energy-intensive industry[a]		0.6	0.6	1.0	0.6
Commercial space heat heat loss (10^4 Btu/ft^2)	7.10	5.68	6.70	3.68	5.10
fossil fuel system efficiency	0.70	0.75	0.72	0.78	0.74
electric system efficiency	1.0	1.0	1.0	1.3	1.2
Other commercial fossil fuel (10^3 Btu/ft^2)[c]	21.3	20.8	23.6	11.9	20.8
electric (10^3 Btu/ft^2)	57.0	51.2	65.5	45.4	55.7
Residential space heat heat loss (10^5 Btu/dwelling)	764	607	708	443	580
fossil fuel system efficiency	0.63	0.68	0.65	0.75	0.70
electric system efficiency	1.0	1.2	1.0	1.6	1.2
Residential appliances fossil fuel (10^6 Btu/dwelling)[c]	27.2	18.5	23.8	15.8	29.0
electric (10^6 Btu/dwelling)	24.2	26.9	29.7	35.1	35.0

[a] Percent per year change in energy use per dollar of value added for 1972–1985; percent per year change in energy use per unit of output for 1985–2000.

[b] Percent per year change in energy per unit of output for 1972–1985 and 1985–2000.

[c] The fossil fuel ratio is lower for 2000 than for 1985 because of the switch to solar energy for heating and cooling.

for commercial flights to 80 percent and stimulating the wider use of railraods for long-distance freight hauling.

In the industrial sector, energy use per dollar of output is projected to decrease between 1985 and 2000 at a rate of 0.8 percent per year for primary metals and 1.5 percent per year for other energy-intensive industries. Energy coefficient improvements are also expected from the normal process of replacing old plants with new facilities using the latest technologies. There is evidence that the average life of major plant equipment is 25 years, so that many facilities will replace old equipment between 1972 and 2000. For the aluminum industry, adoption of more energy-efficient processes is expected to lead to substantial energy usage reductions. For the cement industry, dry process technology will improve energy efficiency. For many industries, the reduction of heat loss from manufacturing processes and the recovery of waste heat from electricity generation hold large potentials for energy efficiency. Other energy measures include a vigorous recycling program to minimize potential shortages in the supply of scrap metal and the adoption of solar space heating and air conditioning in industrial facilities. Estimates of solar energy potentials were developed as part of the U.S. supply study and are based on stimulation from national policy and decontrol of energy prices.

Energy efficiencies in the residential, commercial, and public sectors correspond to the Energy Policy and Conservation Act of 1975. This law authorizes the Department of Housing and Urban Development to establish performance standards for all new residential, commercial, and public buildings. It also authorizes the Federal Energy Administration to make available up to $2 billion in loan guarantees for retrofitting and for solar energy equipment. Other federal and state insulation and building energy efficiency standards are assumed to be effective in reducing energy demand.

19.3.3 Sectoral Energy Demand

The results of multiplying the economic activity levels and the efficiency coefficients of the WAES cases to derive energy demand totals are shown in Tables 19.7 and 19.8. Note that these figures are

Table 19.7
Energy use and energy growth rates for major activities

Sector	1972 Quads	1972–1985 Case C		1985–2000 Cases C-1, C-2		1972–1985 Case D		1985–2000 Cases D-7, D-8	
		Quads	%/yr	Quads	%/yr	Quads	%/yr	Quads	%/yr
Transportation	17.0	18.8	0.8	21.3	0.8	19.8	1.2	21.8	0.6
Autos	9.1	8.3	(0.7)	7.2	(0.9)	8.9	(0.2)	7.6	(1.0)
Air travel	1.3	2.2	4.1	3.0	2.1	2.5	5.2	3.2	1.7
Truck freight	1.5	1.9	1.8	2.9	2.9	2.1	2.6	2.7	1.7
Rail freight	0.6	0.7	1.2	1.1	3.1	0.6	0.0	0.8	1.9
Industrial	14.3	21.6	3.2	31.7	2.6	19.5	2.4	25.8	1.9
Metals	4.4	7.1	3.7	11.6	3.3	6.0	2.4	8.5	2.3
Chemicals	3.0	5.4	4.6	7.4	2.1	5.2	4.2	6.7	1.7
Energy-intensive industry	4.0	4.9	1.6	6.5	1.9	4.7	1.2	5.6	1.2
Non-energy-intensive industry	2.8	4.1	3.0	6.3	2.9	3.6	2.0	4.9	2.1
Feedstocks	4.2	8.2	5.3	13.3	3.3	7.1	4.1	10.5	2.6
Agriculture, mining, and construction	2.0	2.7	2.3	4.2	3.0	2.1	0.4	2.8	1.9
Commercial and public	4.2	5.3	1.8	5.7	0.5	6.1	2.9	6.7	0.6
Space heat	2.3	2.7	1.2	2.5	(0.5)	3.1	2.3	3.1	0
Residential	11.4	11.2	(0.1)	11.1	(0.1)	13.5	1.3	14.4	0.4
Space heat	7.9	7.3	(0.6)	5.5	(1.9)	8.9	0.9	7.7	(1.0)
Total delivered energy	53.0	67.8	1.9	87.2	1.7	68.1	1.9	81.3	1.2

( ) indicates negative growth.

Table 19.8
Energy use and sectoral percentages of final energy demand

Sector	1972 Total Quads	1972 Sectoral %	1985 Case C Total Quads	1985 Case C Sectoral %	2000 Cases C-1, C-2 Total Quads	2000 Cases C-1, C-2 Sectoral %	1985 Case D Total Quads	1985 Case D Sectoral %	2000 Cases D-7, D-8 Total Quads	2000 Cases D-7, D-8 Sectoral %
Transportation	17.0	32.1	18.8	27.7	21.3	24.4	19.8	29.1	21.8	26.9
Industry (excluding feedstocks)	14.3	27.0	21.6	31.9	31.7	36.4	19.5	28.6	27.2	33.5
Feedstocks	4.2	7.9	8.2	12.1	13.3	15.3	7.1	10.4	9.1	11.1
Agriculture, mining, and construction	2.0	3.8	2.7	4.0	4.2	4.8	2.1	3.1	2.1	2.6
Commercial and public	4.2	7.9	5.3	7.8	5.7	6.5	6.1	9.0	6.7	8.3
Residential	11.4	21.5	11.2	16.5	11.1	12.7	13.5	19.8	14.4	17.7
Final energy demand (excluding processing losses)	53.0	100.0	67.8	100.0	87.2	100.0	68.1	100.0	81.3	100.0

for final or delivered energy only. The total resource consumption and the losses resulting from processing those resources between the point of extraction and the point of delivery to the final consumer are calculated in the course of the supply-demand integrations.

Total delivered energy demand is projected to grow slowly from 1985 to 2000 as shown in Table 19.7. For Cases C-1 and C-2, the delivered energy growth rate is 1.7 percent per year (1985–2000) in contrast to the assumed 3.5 percent annual GNP growth rate. For D-7 and D-8, the energy growth rate is only 1.2 percent per year compared to the assumed GNP growth rate of 2.5 percent. While all major sectors have positive energy growth rates, those for transportation, commercial and public, and residential are less than 1 percent per year.

The categories growing most rapidly in delivered energy use are freight transport, metals industries, non-energy-intensive industries, feedstocks, mining, and construction. All of these activities are tied to the production sectors of the economy. Output for these activities is expected to grow rapidly, while projected energy efficiency improvements are modest at best, reflecting relatively efficient existing systems with low energy consumption per unit of output.

The data in Table 19.8 indicate a continuing shift away from energy usage by individual consumers and toward energy use by industry. In 1972, the transportation, commercial and public, and residential sectors accounted for over 61 percent of the total end-use energy demand. For Case C in 1985, they are estimated to account for 52.0 percent of the total end-use energy demand, and for Cases C-1 and C-2 in 2000, they are projected to account for 43.6 percent (see Table 19.8).

19.3.4 Fuel Mix

Although the detailed fuel mix in each case results from the supply-demand integrations, the demand studies include an initial partition of the sector projections between electricity and fossil fuels.

Consistent with recent trends, WAES demand projections include a significant evolution toward greater electrification. Most of this growth represents the substitution of electric power for fossil fuels in the residential and commercial sectors. Over the period from 1972 to 1985, an estimated 45 percent of new residential construction will have electric space heating systems. By 2000, approximately 28 percent of all dwellings and 20 percent of all commercial floor area are projected to use electric space heating.

Electricity demand is also expected to expand because of the growing use of air conditioning, appliances, and office equipment. These trends plus the consistent growth of electricity demand in other sectors lead to a 75 percent increase in the delivered demand between 1985 and 2000.

In contrast, fossil fuel delivered demand expands only slightly between 1985 and 2000 because of improved energy efficiencies and the substitution of other energy sources. In transportation and industry, the main factor is improvements in energy efficiencies. For instance, improved automobile efficiencies (increases in miles per gallon figures) lead to actual declines in energy usage by automobiles, in spite of a larger vehicle population. In the residential and commercial sectors, the decline in fossil fuel demand is caused mainly by the substitution of electricity and solar energy.

In 2000, solar energy will be used mainly to provide hot water and space heating and cooling for buildings and to generate small amounts of electricity. Solar energy provides nearly 4 Quads of energy in Cases C-1 and C-2, compared to less than 0.2 Quads in 1985. However, widespread large-scale solar energy use is not expected to occur until after the year 2000. Solar water heating is likely to become cost-effective in the early 1980s, so that a significant percentage of all water heating is expected to use solar energy by 2000. Solar space heating and air conditioning is expected to reach the "take-off" point by around 1985. Because of the high first costs and the long payback period, federal government loan or subsidy programs may be necessary to encourage users to adopt solar energy systems in the early stages of development.

The detailed 23-category worksheets used for the projection of U.S. energy demand to 2000 for WAES Cases C-1, C-2, D-7, and D-8 are contained in section 19.7. When these demand estimates are matched with the maximum fuel supply, the result is supply-demand integration, described in the next section.

19.4 SUMMARY OF U.S. SUPPLY-DEMAND INTEGRATION RESULTS

19.4.1 Unconstrained Cases

National *unconstrained* supply-demand integrations are cases not constrained by any limits on oil import availability. The U.S. unconstrained cases for 1985 and 2000 are described here. Together with other inputs to the global unconstrained supply-demand integrations, they provide the basis for the global "prospective shortages" described in the WAES final report (12) and in Parts I and II of this book.

Unconstrained cases represent explorations of the the maximum physical limits of supply as well as the plausible ranges of consumer preferences. They are cases in which any potentially unfilled demand is met simply by increasing oil imports; major fuel switching is generally not required to balance supply and demand. This set of conditions can potentially lead to some unrealistic results, especially when viewed in a global context. Some of these issues are noted, and dealt with, in section 19.4.3.

In doing the integrations of the unconstrained cases, BESOM attempts to allocate primary supplies of fuels through various processing stages to specific end-use demand categories. For the unconstrained cases, this task was relatively simple because of the unlimited oil imports available at the case price. The model performed primarily a calculating function, allocating available energy supplies by least cost, including oil imports at the case price plus transportation.

The process of matching the primary energy supply to delivered energy demand produces calculations of the processing losses—the energy used to transport, refine, process, convert, and distribute energy. Table 19.9 summarizes the results of the unconstrained case calculations. Total primary energy demand ranges from 115 to 132 Quads in 2000. Of this, as much as 43 Quads (or 33 percent) is processing losses—compared to 27 percent in 1972. The detailed results of the national unconstrained integrations are contained in the worksheets given for all cases in section 19.7.

Primary Energy Totals

The primary energy totals in these cases represent the total amount of physical resources used—including imports but excluding any portion of U.S. indigenous production that is exported. This primary energy demand continues to increase with economic growth, but at lower than historic rates.

Primary energy consumption, 72.73 Quads in 1972, increases to over 90 Quads by 1985 and to 115–132 Quads by 2000. The two high nuclear cases, C-2 and D-8, have higher resource totals than the two high coal cases, C-1 and D-7, because (other things being equal) of the greater fraction of final demand met by electricity, with accompanying greater processing losses in electricity generation.

A summary of the energy sources that provide the total primary energy of the United States in these cases is shown in Table 19.10. The projected decline in domestic oil production and growth in oil imports can be seen in this table. Oil continues

Table 19.9
Energy demand totals (10^{15} Btu)

	1972	1985		2000			
		C	D	C-1	C-2	D-7	D-8
Economic growth		High	Low	High	High	Low	Low
Energy price		Const.	Const.	Rising	Rising	Const.	Const.
Policy response		Vig.	Vig.	Vig.	Vig.	Vig.	Vig.
Principal replacement fuel				Coal	Nuclear	Coal	Nuclear
Delivered (end-use) energy demand	53.0	67.8	68.1	86.6	88.0	80.9	81.7
Processing losses	19.7	24.4	26.3	37.9	44.1	34.2	38.6
Primary energy demand	72.7	92.2	94.4	124.5	132.1	115.1	120.3

Table 19.10
Energy sources (10^{15} Btu)

	1972	2000			
		C-1	C-2	D-7	D-8
Oil					
Domestic	23.71	14.80	14.80	12.70	12.70
Imports	10.11	19.90	20.52	32.25	32.29
Shale	0	4.30	4.30	0	0
From coal[a]	0	4.20	4.20	0	0
Gas					
Domestic[b]	22.12	15.00	15.00	11.50	11.50
Imported	1.00	3.24	3.62	5.00	5.00
From coal	0	0	0	0	0
Coal					
Internal consumption	12.60	29.84	27.13	21.15	19.09
Potential exports[c]	1.45	18.13	7.15	14.65	8.62
Solar thermal	—	2.35	2.35	1.18	1.18
Solar electric	—	1.50	1.50	.50	.50
Nuclear	0.53	25.96	35.24[d]	25.96	33.12[d]
Hydro	2.66	4.60	4.60	3.69	3.69
Geothermal	0	3.00	3.00	1.20	1.20
Total	72.73	124.49	132.06	115.13	120.27

All sources are given as primary energy—for example, in the case of sources of electricity, grossed up at an assumed 35 percent efficiency to input equivalence.

[a]Energy content included in coal figures.

[b]Cases C-1 and C-2 include approximately 1 Quad of methane from biomass.

[c]For 1985 and 2000, exportable coal, after U.S. uses (equals potential production minus U.S. internal consumption). For 1972, actual exports.

[d]Maximum potential is 42.36 Quads (620 GWe).

to be a fuel of major importance through 2000 in the unconstrained cases—ranging from 33 to 39 percent of total resource consumption.

Domestic gas production declines from 1972 to 2000, with imports failing to offset the declines in domestic production.

Coal and nuclear use is increased in all cases, with the relative importance of each depending on whether coal or nuclear is the principal replacement fuel scenario assumption.

Nonconventional alternative energy sources—synthetic fuels, solar energy, and geothermal energy—all assume significant but not dominant roles in the fuel mix. Their contribution is most notable in Cases C-1 and C-2, in which it is assumed that oil shale and synthetic crude oil from coal become competitive at or near the case price of $17.25 per barrel[2] and their early development is aided by some form of national policy support.

No synthetic fuels are anticipated to be available in Cases D-7 and D-8 in view of the $11.50 per barrel world oil price.

Each of the fuel sources in Table 19.10 can be used in different ways to meet different end-use energy demands. This allocation of fuels, one of the central purposes of supply-demand integration, is described below.

Fuel Mix by Sectors

Changes in the fuels used to meet sectoral demands occur gradually. This is particularly true for these unconstrained cases, since the assumption of unlimited oil import availability ensures that past fuel mix preferences—in which oil is the dominant and preferred fuel—can be continued.

A few major uses of energy—most notably transportation and petrochemical feedstocks—have depended, and will probably continue to depend, almost totally on oil. This situation would change only following the development of major technological improvements.

Other energy-using activities are more amenable to changes in fuel supply. Residential space and water heating and industrial process heat are examples. These demands could be met by a variety of fuels. Recently, for these uses, there has been a trend away from oil and gas toward electricity, coal, and solar energy.

2. All oil prices are given as constant 1975 U.S. dollars per barrel of Arabian light crude oil, FOB Persian Gulf.

The fuel-use trends projected for the industrial and residential sectors for the WAES cases to 2000 are illustrated in Figure 19.2. This figure shows substantial increases in the use of coal in industry and electricity in the residential sector, and reductions in the use of oil and gas in both sectors (particularly the industrial). Coal use in industry reaches 50 percent of the sector's demand in 2000, or about 18 Quads, compared to 28 percent (3.9 Quads) in 1972. Such a growth in coal use represents a maximum technically feasible level, and one that may be warranted on the basis of dwindling U.S. oil and gas supplies. As coal and, to a lesser extent, electricity form a greater portion of industrial demand, oil and gas use declines but does not disappear. They account for 29–32 percent of industrial energy demand in 2000 (8.6–11.5 Quads), compared to 60 percent (10 Quads) in 1972.

Electrification of the residential sector ranges from 35 to 47 percent in 2000 compared to 21 percent in 1972. At the same time, solar energy begins to provide up to 13 percent of residential energy needs (mostly hot water and some space heating and cooling) while oil and gas use declines to 39–54 percent of sectoral demand, compared to 77 percent in 1972.

The detailed fuel mix in four major sectors for the four WAES 2000 cases is shown in Table 19.11. It can be seen that coal use is concentrated in industry in all cases (coal use in electricity generation is considered separately below). Oil and gas use in all sectors in all cases is at or near the minimum preferred amounts (see Table 19.3).

The oil used in transportation and for petrochemical feedstocks can also be seen in Table 19.11. These two sectors consume large proportions of the available oil—from 67 to 75 percent of the oil in the 2000 cases—compared to about 60 percent in 1972. This, of course, is a direct consequence of the gradual shift away from oil and toward other fuels, wherever possible, observed during the period to 2000.

Processing Losses

Conventional fossil fuels—oil, gas, and coal—have relatively low transport and processing losses between extraction and final consumption, some 2 to 12 percent of primary energy content. Generating electricity involves much larger losses, about

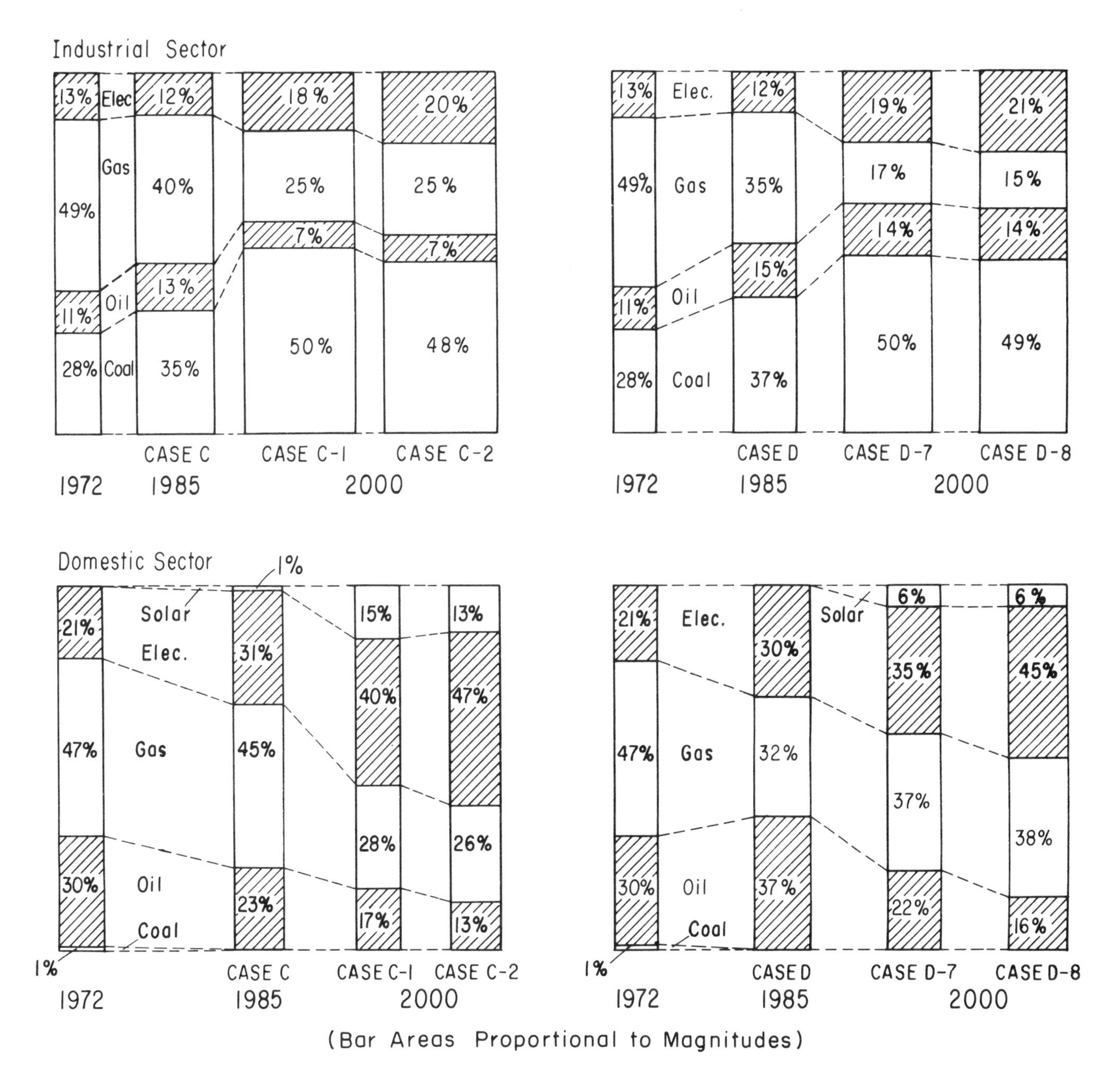

Figure 19.2
Fuel mix in two sectors

Table 19.11
End-use energy by sector and fuel (10^{15} Btu)

	Oil	Gas	Coal	Electricity	Solar thermal
Case C-1					
Residential	2.73	4.55	0	6.43	2.35
Industrial	2.48	8.99	17.99	6.34	—
Feedstocks	10.79	2.13	0.40	—	—
Transport	21.35	—	—	0.03	—
Case C-2					
Residential	2.32	4.55	0	8.33	2.35
Industrial	2.51	8.97	17.20	7.02	—
Feedstocks	10.79	2.13	0.40	—	—
Transport	21.35	—	—	0.03	—
Case D-7					
Residential	4.28	7.37	0	6.89	1.18
Industrial	3.98	5.04	14.35	5.49	—
Feedstocks	8.49	1.68	0.30	—	—
Transport	21.85	—	—	0.03	—
Case D-8					
Residential	3.21	7.61	0	8.28	1.18
Industrial	4.16	4.44	14.35	6.09	—
Feedstocks	8.49	1.68	0.30	—	—
Transport	21.85	—	—	0.03	—

70 percent of primary energy content.[3] Synthetic gas and oil are produced from coal with losses amounting to around 40 percent of the initial energy content of the coal. These differences are partly offset by the higher efficiency of electricity at the point of end-use. This factor is taken into account by end-use demand efficiencies that vary among fuels in the BESOM integration model.

In the cases examined here, there is a strong trend toward increases in processing losses in relation to total energy consumed. This is due to a gradual trend of substitution of coal (either as electricity or as synthetics) and nuclear (as electricity) for oil and gas. This trend is reflected in the figures for processing losses as a percentage of primary energy demand—growing from 27 percent in 1972 to 28–31 percent in 1985 to 30–33 percent in 2000.

The high nuclear year 2000 cases (C-2 and D-8) have slightly higher losses than the high coal cases because a greater fraction of total resources is used to generate electricity.

Table 19.12 summarizes the mix of fuels used for electric power generation. Oil and gas are expected to provide about the same level of electricity in 2000 as in the 1985 cases in our projection, with no new capacity added after 1985.

Nuclear power at the U.S. supply study maximum likely and minimum likely amounts of 620 GWe (42.4 Quads input) and 380 GWe (26.0 Quads input), respectively, supplies a large fraction of all electricity growth to 2000. All of the permitted capacity is used in the low nuclear cases, and a significant fraction of it in the high cases. This outcome is based on the apparent favorable economics of nuclear energy and the potential use of large amounts of coal in industry and for export. If the nuclear contribution were less, then the coal inputs to power plants would increase accordingly. This possibility—which may be more plausible (in view of some utility planning) than the figures in Table 19.12, especially given the relatively low overall energy growth in these cases—is discussed briefly in section 19.4.3.

Table 19.12 also shows that hydroelectric energy and electricity from geothermal and solar sources amount to only about 10–20 percent of electricity generation in 2000, although these sources are used at the maximum estimates for the U.S. supply study.

3. The 70 percent loss figure is based on an assumed 35 percent efficiency in electricity generation and includes losses in transmission and distribution. The efficiency figure has been adopted by WAES as a *convention* for the generating efficiency of all fuels.

Table 19.12
Electricity generation (10^{15} Btu)

		2000			
	1972	C-1	C-2	D-7	D-8
Oil	3.25	3.05	4.09	2.98	4.01
Gas	4.10	1.00	1.40	1.00	1.40
Coal	7.68	4.56	2.56	6.10	4.05
Solar	0	1.50	1.50	0.50	0.50
Nuclear	0.53	25.96	35.24	25.96	33.12
Hydro	2.66	4.60	4.60	3.69	3.69
Geothermal	0	3.00	3.00	1.20	1.20
Total electric inputs	18.22	43.67	52.39	41.33	47.97
Electricity growth rate, 1972-2000[a]		3.2%/yr	3.8%/yr	3.0%/yr	3.5%/yr
Total energy growth rate, 1972-2000[b]		1.9%/yr	2.2%/yr	1.7%/yr	1.8%/yr
Ratio of electric growth to energy growth, 1972-2000[c]		1.7	1.7	1.8	1.9

[a]Growth rate for 1950-1972 was about 7.0%/yr.

[b]Growth rate for 1950-1972 was about 3.0%/yr.

[c]Ratio for 1950--1972 was 2.3.

Total electricity growth rates from 1972 to 2000 range from 3 to nearly 4 percent per year for the WAES 2000 cases. These growth rates, significantly lower than historical rates, result from the low overall energy growth (because of assumed efficiency improvements) and from assumptions about the level of penetration of electricity into various end-use markets. Although electricity is in principle usable for all energy demands, it is thermodynamically inefficient in terms of primary energy use and very expensive to substitute for fossil fuels where the high-quality energy of electricity is not required (e.g., in applications where the demand is for heat).

Imports and Exports

One of the most important and relevant elements of these unconstrained cases is the resulting oil imports. In 1972, oil imports into the United States totaled 10.1 Quads, or about 5 MBD. The WAES integrations show oil imports increasing gradually to about 20 Quads or 10 MBD in Cases C-1 and C-2, and to over 30 Quads (about 15 MBD) in D-7 and D-8.

The gradual increase in oil imports is the central manifestation of the unconstrained nature of these cases. That is, with no restrictions on oil import availability at the case price assumptions, the United States will import increasing amounts of oil to 2000—even with rising prices and "vigorous" energy demand and supply policies. Oil will continue to be a preferred, and essential (for some purposes), fuel.

Oil import projections result, of course, from both oil demand and indigenous oil production projections. For Cases C-1 and C-2, the high economic growth assumption tends to increase oil demand, while the rising price assumption acts to moderate oil demand growth and stimulate domestic oil production. In Cases D-7 and D-8, the low economic growth assumption tends to moderate oil demand growth, while constant prices tend to encourage demand growth and discourage expansion of oil production. (Also, Case D to 1985, as a restrained policy case, has higher demand and lower supply than Case C, a vigorous policy case to 1985.) As a result, imports in Cases D-7 and D-8 are significantly higher than in Cases C-1 and C-2 in 2000.

Table 19.13 shows the trends from 1972 to 2000 for oil demand, production, and imports resulting from the U.S. integrated cases. The two 1985 cases

Table 19.13
U.S. oil: Demand, production, and imports (10^{15} Btu)

		1985		2000			
	1972	C	D	C-1	C-2	D-7	D-8
Oil production[a]	23.7	23.7	20.1	19.1	19.1	12.7	12.7
Oil imports	10.1	16.1	25.9	19.9	20.5	32.3	32.3
Share of oil demand	30%	40%	56%	51%	52%	72%	72%
Oil demand	33.8	39.8	46.0	39.0	39.6	45.0	45.0
Share of total primary energy demand	47%	43%	49%	31%	30%	39%	37%

[a]Includes 0.4 Quad of oil shale in the 1985 cases and 4.2 Quads of oil shale in 2000 Cases C-1 and C-2.

shown illustrate the combined impact of the low production and high demand character of Case D: oil imports are large—about 56 percent of total oil demand. With slightly higher production levels (due to vigorous policies) in Case C, oil imports represent 40 percent of total oil demand.

In year 2000 Cases C-1 and C-2, with rising energy prices and high economic growth, oil imports are substantial, accounting for some 50 percent of total oil demand. In Cases D-7 and D-8, with constant prices and significantly lower maximum oil production estimates, oil imports are much higher—over 70 percent of total oil demand.

Natural gas imports increase at a relatively moderate rate to 2000 in these cases, owing to large uncertainties concerning the possibility and acceptability of large-scale LNG imports to the United States. Imports reach about 3.5 Quads in 2000 in Cases C-1 and C-2, and 5.0 Quads in D-7 and D-8 (see Table 19.1).

Potential coal exports in these cases in 2000 range from 7 to 18 Quads compared to about 1.5 Quads in 1972. These potential exports result from maximum production estimates of 28–48 Quads (1.3–2.2 billion tons) and coal demand levels of 19–30 Quads, which is primarily coal used in industry (up to 50 percent of industrial demand). Coal use in power plants is relatively low in these cases, as noted earlier. A test case that examines (among other things) increased coal use in electricity generation is discussed below.

Conclusion

So far, the discussion has been of the unconstrained U.S. integrations, with emphasis on the 2000 cases where potential global oil shortages emerge. The detailed results of these cases are given on the worksheets in section 19.7. These cases are not projections, but are developed for illustrative purposes from explicit WAES assumptions. The learning value of these cases is in the messages they convey about the consequences of a range of plausible U.S. demand and supply estimates coupled with unlimited oil imports availability. Having looked at these cases, it may be instructive to reassess a few of the assumptions, postulate some different and perhaps more "plausible" conditions, and examine the results. The following section describes such "excursion" cases.

19.4.2 Excursion Cases

Although the demands for imported oil in the U.S. unconstrained integrations are fairly substantial, the overall impact of U.S. demand on world energy markets is mitigated by two important considerations: (1) it is assumed that some 4 Quads each of shale oil and coal-based synthetic oil will be produced under the conditions of Cases C-1 and C-2; and (2) it is assumed that nuclear power can be used for a large proportion of electricity generation —up to the maximum estimates for nuclear power for these cases.

The first assumption can be questioned on the grounds that synthetics may be expected to cost more than the case price of $17.25 (in 1975 dollars). Working from this assumption, "excursion" cases were developed—revisions of the C-1, C-2, D-7, and D-8 unconstrained cases. In these modified cases, labeled C-1*, C-2*, D-7*, and D-8*, a total of 1.0 Quad of each of the synthetic fuels was introduced on a subsidized basis. This amount is assumed to represent a demonstration program.

The nuclear share of the total in the unconstrained cases resulted from an assumed small economic advantage of nuclear over coal in electricity generation. However, this is a tenuous basis for the strong shift to nuclear that occurred, resulting in fact in the early closing down of some coal power plants that were operating in the 1985 cases. In fact, many utilities are currently ordering coal and nuclear on a 50/50 basis. In view of the uncertainty surrounding public acceptance of nuclear and the limited accuracy of long-term cost projections, a 50/50 mix seems to be a reasonable assumption. Thus, in the excursion cases, coal is assumed to be 50 percent of the coal/nuclear total in electric power generation, up to the limit of the available domestic coal supply.

The consequences of these modified assumptions are summarized in Tables 19.14, 19.15, and 19.16. These tables can readily be compared with Tables 19.10, 19.11, and 19.12, respectively, to see the specific differences between the unconstrained and excursion cases. In Tables 19.14–19.16, only the *italicized* numbers are different from the unconstrained cases results. Numbers that are not italicized are essentially identical in both unconstrained and excursion cases.

Table 19.14 shows the overall changes in energy sources that result from the synthetic fuels and coal/nuclear mix changes described above. Table 19.15 shows the resulting fuel mix changes in end-use markets—changes that are relatively minor. Table 19.16 shows the mix of electricity generation sources in the excursion cases, with the specified 50/50 coal/nuclear mix being the only change from the unconstrained case numbers.

Table 19.14
Energy sources: Excursion cases (10^{15} Btu)

		2000			
	1972	C-1*	C-2*	D-7*	D-8*
Oil					
Domestic	23.71	14.80	14.80	12.70	12.70
Imported	10.11	*27.1*	*27.7*	*31.0*	*31.7*
Shale	0	*1.0*	*1.0*	0	0
From coal[a]	0	*1.0*	*1.0*	0	0
Gas					
Domestic[b]	22.12	15.00	15.00	11.50	11.50
Imported	1.00	*2.1*	*2.5*	5.00	5.00
From coal[a]	0	*1.0*	*1.0*	0	0
Coal					
Internal consumption	12.60	*36.5*	*34.3*	*31.7*	*27.7*
Potential exports[c]	1.45	*11.5*	*0*	*4.1*	*0.0*
Solar thermal	—	2.35	*1.90*	1.18	1.18
Solar electric	—	1.50	1.50	0.50	0.50
Nuclear	0.53	*14.8*[d]	*23.5*[e]	*16.7*[d]	*24.7*[e]
Hydro	2.66	4.60	4.60	3.69	3.69
Geothermal	0	3.00	3.00	1.20	1.20
Total	72.73	*122.8*	*129.8*	*115.2*	*119.9*

Numbers in italics are those which differ from corresponding data for the unconstrained cases (see Table 19.10). All sources are given as primary energy—for example, in the case of sources of electricity, grossed up at an assumed 35 percent efficiency to input equivalence.

[a]Energy content included in coal figures.

[b]Cases C-1 and C-2 include approximately 1 Quad of methane from biomass.

[c]For 1985 and 2000, exportable coal, after U.S. uses (equals potential production minus U.S. internal consumption). For 1972, actual imports.

[d]Maximum potential is 25.96 Quads (380 GWe).

[e]Maximum potential is 42.36 Quads (620 GWe).

Table 19.15
End-use energy by sector and fuel: Excursion cases (10^{15} Btu)

	Oil	Gas	Coal	Electricity	Solar thermal
Case C-1*					
Residential	2.73	4.55	0	6.43	2.35
Industrial	*2.84*	8.97	*17.59*	6.34	—
Feedstocks	10.79	2.13	0.40	—	—
Transport	21.35	—	—	0.03	—
Case C-2*					
Residential	2.32	4.55	0	8.33	2.35
Industrial	*2.87*	8.98	*16.89*	7.05	—
Feedstocks	10.79	2.13	0.40	—	—
Transport	21.35	—	—	0.03	—
Case D-7*					
Residential	4.28	*6.13*	0	*7.59*	1.18
Industrial	*2.82*	*6.28*	14.35	5.52	—
Feedstocks	8.49	1.68	0.30	—	—
Transport	21.85	—	—	0.03	—
Case D-8*					
Residential	3.21	*7.08*	0	*8.53*	1.18
Industrial	*3.66*	*4.97*	14.35	6.04	—
Feedstocks	8.49	1.68	0.30	—	—
Transport	21.85	—	—	0.03	—

Numbers in italics are those which *differ* from corresponding data for the unconstrained cases (see Table 19.11).

Table 19.16
Electricity generation: Excursion cases (10^{15} Btu)

		2000			
	1972	C-1*	C-2*	D-7*	D-8*
Oil	3.25	3.05	4.09	2.98	4.01
Gas	4.10	1.00	1.40	1.00	1.40
Coal	7.68	*14.83*	*13.39*	*16.69*	*12.65*
Solar	0	1.50	1.50	0.50	0.50
Nuclear	0.53	*14.83*	*23.50*	*16.69*	*24.70*
Hydro	2.66	4.60	4.60	3.69	3.69
Geothermal	0	3.00	3.00	1.20	1.20
Total electric inputs	*18.22*	*42.81*	*50.48*	*42.75*	*48.15*

Numbers in italics are those which *differ* from corresponding data for the unconstrained cases (see Table 19.12).

These changes have significant effects on the desired imports of oil and potential exports of coal. Oil imports are higher, owing to the reduced estimates for synthetic fuel production. Potential exports are lower, due to the increased use of coal domestically in power plants. In a global context, these results represent a less favorable situation for world energy markets than the unconstrained cases because they include maximum U.S. oil imports and minimum potential U.S. coal exports. These important differences are compared with the basic unconstrained cases in Table 19.17.

This table shows that the oil imports are somewhat higher in the excursion C-1*, C-2* cases than in the unconstrained C-1, C-2 cases, while oil imports are nearly the same in the D-7*, D-8* cases and the D-7, D-8 cases.

The coal export potential is reduced substantially in each of the excursion cases, but particularly in the high nuclear (C-2 and D-8) cases, where the export potential is reduced to zero. The lower coal production of cases C-2 and D-8 is all required for industrial uses and power generation within the United States in the excursion cases C-2* and D-8*.

Table 19.17 highlights the two central observations of this excursion exercise: (1) if total synthetic fuel production is minimal (2 Quads or 1 MBDOE total, in these cases), oil imports will increase, and (2) as more indigenous coal is used (instead of nuclear) in electricity generation, potential exports of coal from the United States decline. Each of these observations could have important impacts on the global picture.

19.4.3 Issues Raised by Unconstrained Supply-Demand Integrations

Global Prospective Oil Shortages

Section 2.3 of chapter 2 of this volume summarizes the major findings of the global integrations of the WAES cases to 2000. The results show large prospective shortages of oil by 2000, given the range of WAES assumptions. Desired national imports of oil, aggregated globally, exceed total WOCA production of oil by 15–20 MBD in the WAES year 2000 cases. The magnitude of the prospective global shortages is approximately the same as the level of desired U.S. oil imports in unconstrained Cases C-1 and C-2, and is about half the size of the desired U.S. oil imports in D-7 and D-8. Clearly, the size of U.S. oil imports has important bearing on the character of the global energy futures.

The WAES cases to 2000 also exhibit potential global surpluses of coal—an excess of potential coal supply over desired demand.

In short, the world will have less oil than it desires under the WAES scenario conditions; and it will want less coal than it could have (although the prospective coal surplus is smaller than the prospective oil gap). In such a situation, what would be a viable U.S. "response"? What actions by the

Table 19.17
Oil imports and coal exports, 2000: Unconstrained and excursion cases (10^{15} Btu)

Unconstrained case	Oil imports	Excursion case	Oil imports	Change in oil imports[a]
C-1	19.9	C-1*	27.1	+ 7.2
C-2	20.5	C-2*	27.7	+ 7.2
D-7	32.3	D-7*	31.0	- 1.3
D-8	32.3	D-8*	31.7	- 0.6
Unconstrained case	**Coal exports**	**Excursion case**	**Coal exports**	**Change in coal exports[a]**
C-1	18.1	C-1*	11.5	- 6.6
C-2	7.2	C-2*	0	- 7.2
D-7	14.7	D-7*	4.1	-10.6
D-8	8.6	D-8*	0	- 8.6

[a]Plus sign indicates an *increase* in the excursion case over the unconstrained case; minus sign indicates a *reduction* in the excursion case from the unconstrained case.

United States could contribute to reducing the scale of the prospective global energy imbalance?

Substitutes for Oil

One major course of action for adapting to these prospective shortages is the substitution of other fuels for oil in sectors where this is feasible. On examination of the demand sector, one can conclude that of the various uses for oil, three would be very difficult to replace by other fuels. These are petrochemical feedstocks, transportation, and peak electric power generation. For other major uses—such as domestic appliances (principally space and water heating), industrial uses (process heat and steam raising), and electric power generation—technologies are available for the use of fuels other than oil. If oil were replaced in these sectors in the United States by other fuels—electricity, coal, solar, or natural gas—both U.S. oil imports and the prospective global shortages of oil would be reduced.

Table 19.18 illustrates the approximate magnitudes involved. These figures are not intended as predictions or recommendations of the size of the substitution task. They are intended to illustrate and explore the flexibility of response possible to the anticipated environment of declining world oil availability. They represent one measure of the potential substitution that might be achieved.

The substantial reduction in use of oil in the residential, industrial, and electrical power generation markets could contribute greatly to alleviating the oil deficits. However, further steps may be needed. Among the measures that may be available are increasing efficiency and electrification in transportation, and substitution for oil or efficiency improvements in feedstock use.

Substitution for natural gas, to reduce the need for LNG imports in the WAES cases, is generally not as difficult as substitution for oil, because of the nature of the sectors where gas is in use. The substitution could be made with electricity (albeit at a higher cost), solar heating (at an uncertain but potentially competitive lifetime cost), or with coal, especially in the industrial sector.

Table 19.18
Oil substitutability, 2000 (10^{15} Btu)

Case	Total oil demand	Oil used in essentially nonsubstitutable sectors[a]	Oil used in potentially substitutable sectors
C-1	36.2	29.4	6.8
C-2	36.9	31.4	5.5
D-7	41.6	32.3	9.3
D-8	41.7	31.8	9.9

[a]Here, these are taken to be transportation, petrochemical feedstocks, and some peak electric power generation.

For applications requiring heat (rather than, for example, electric drive), where oil and gas currently serve, electricity from coal, nuclear, or other sources is usually too expensive to compete with other potential substitutes. Exceptions to this are situations in which the electricity is available off-peak at low cost, or in which high-efficiency heat pumps can be used.

For other thermal applications currently using conventional petroleum or gas, there are other possible substitutes. Coal is comparatively inexpensive, but the equipment needed to handle it and burn it in an environmentally acceptable manner often is not. In contrast, synthetic oil and gas are more expensive but allow continued use of existing oil and gas distribution infrastructure and existing end-use devices. Individual consumers may need to assess which approach is best in their individual circumstances.

The exact levels of synthetic fuel production in the unconstrained integrations are entirely dependent upon the assumptions made concerning relative prices of synthetics and imports. The production figures presented in Table 19.10 for Cases C-1 and C-2 are based on the assumption that synthetic oil can be produced at a price of $18.25 per barrel (which is the case price for landed oil imports—including $1 for shipping). This is a highly optimistic assumption under current estimates of costs, but such production levels could be achieved if there were government assistance for large-scale demonstrations of promising technologies and measures by government to stimulate investment in synthetics plants, which cost $1 billion each, at a time when the estimated cost of the product is well above world oil prices.

These figures are based on the assumption that environmentally acceptable methods of production can be proven, and that water demands for production in the West can be met.

Coal Exports

One of the most significant issues raised by these integrated cases is the potential for U.S. coal exports. The maximum potential U.S. coal production indicated by the U.S. WAES supply study in some cases far exceeds the amount required for domestic consumption. The difference between the potential production and the amount actually used in the United States represents a quantity that is *potentially* available for export—from the viewpoint of the resource base lead times and infrastructure—provided foreign demand for coal exports materializes and the necessary policy decisions are made.

In view of the extensive investment in mines, transportation equipment, and other infrastructure that is required for coal development, the potential exports indicated in this analysis will not be produced in the absence of long-term contracts for delivery.

Would the United States accept the environmental consequences and the depletion of a nonrenewable national resource to provide energy for export to other countries? While there is little doubt that the United States is capable of producing great quantities of coal, it is not certain at this time that it would be prepared to undertake extensive coal mining and related transportation and infrastructure expansion to provide coal for export on long-term contracts with assurance of continuing delivery.

Yet there are a number of reasons to believe that such a policy would be adopted if the need for the coal were clearly established. Coal in the United States is not a scarce resource. Exporting it to meet the energy needs of others and to offset the cost of imported oil would be fully consistent with traditional U.S. foreign policy goals. Such coal production and export would have a positive effect on economic development, employment, and the balance of payments. But large-scale coal exports will not occur unless: (1) demand for coal exports is explicit in the form of firm long-term contracts, based on buyers' expectations that the coal will be available for export; (2) the environmental and social acceptability of western surface mining is demonstrated; and (3) there is sufficient advanced planning and commitments so that necessary transportation systems, including deep water port facilities, are built on a schedule that would allow coal production and export to increase at the rates suggested in the WAES cases.

In an energy-deficient world where oil production is declining in the face of growing demand, and in which U.S. coal is available at low cost relative to other alternatives, it seems reasonable to expect support for a U.S. policy of coal exports.

Coal/Nuclear Balance in Electricity Generation
Coal and nuclear power are both subjects of intense political debate concerning their future role in the energy supply system. Detailed analysis of these debates is beyond the scope of this study. Here, it is assumed that environmental and safety questions surrounding both technologies will be resolved through improved technology and planning processes.

Economic considerations for coal or nuclear as sources of electricity differ from region to region within the United States. Oil and gas are expected to remain constant as sources of electric power. Consequently, most of the new electric capacity *under these unconstrained assumptions* is projected to be nuclear.

Such an assumption results in a large potential for exports of coal. If nuclear is not developed at the rate anticipated, some of the potential U.S. coal exports would be used in electricity generation in the United States, leaving less for export.

In the high coal cases (C-1 and D-7), all of the maximum estimate of potential nuclear capacity, 380 GWe, is used in the energy mix. In the high nuclear cases, an amount somewhat less than the potential 620 GWe capacity is required, because of limits on the uses for electricity noted earlier. In all cases, nuclear energy represents a large proportion of the total electricity generated.

Effect of the Principal Replacement Fuel Variable
The principal replacement fuel scenario variable (nuclear in Cases C-2 and D-8; coal in Cases C-1 and D-7) had its greatest effects in the unconstrained cases in the electric sector and in the degree of electrification.

The results of the unconstrained cases, with unlimited oil imports, tended toward increased oil and gas imports rather than increased consumption of domestic coal and nuclear. This was because the marginal coal and nuclear was available to consumers only in the form of electricity, which was more expensive than imported oil products for most thermal applications.

19.5 CONCLUSION

This chapter has outlined the procedures and results of U.S. supply-demand integrations to the year 2000 performed using U.S. supply and demand estimates developed within a consistent international framework. Comparing these energy scenarios with world availability of fuels in trade, particularly oil, reveals major imbalances. The timing and magnitude of such potential imbalances, and the character of actions that will be required to avoid severe economic dislocations, have been examined in general terms by WAES. The types of actions needed include: switching from oil to other fuels in demand sectors wherever possible; development of replacement fuels for oil such as coal and nuclear; additional conservation measures beyond those in the scenario projections, especially in oil-dependent demand sectors; and development of renewables and other unconventional energy sources.

In the longer term, there will be some combination of reduced demand for liquid fuels and increased availability of alternatives at a reliable price that will provide the transition from the era of oil and gas to the energy economy of the future. The point that is emphasized by the results presented in this chapter is the need to make the necessary policy choices soon enough to overcome the long lead times inherent in developing our alternatives.

The United States has many energy options, all of which are characterized by political, environmental, and economic uncertainties which may not be quickly overcome. Policies and actions must be directed toward setting priorities and resolving uncertainties. These important challenges need further, more detailed attention. It is hoped that the results presented in this chapter can contribute to clarifying the issues and focusing further work.

19.6 REFERENCES

1. *Energy Demand Studies: Major Consuming Countries,* the First Technical Report of WAES, MIT Press, 1976.

2. *Energy Supply to the Year 2000: Global and National Studies,* the Second Technical Report of WAES, MIT Press, 1977.

3. Chemiavsky, C.A., Brookhaven Energy System Optimization Model, Topical Report, BNL—19569, Brookhaven National Laboratory, December, 1974.

4. *A National Plan for Energy Research, Development and Demonstration,* ERDA-48, Washington, Government Printing Office, 1975.

5. *A National Plan for Energy Research, Development and Demonstration,* ERDA-76-1, Washington, Government Printing Office, 1976

6. Edward Allen et al., "U.S. Energy and Economic Growth 1975-2000 (Draft)," Institute for Energy Analysis, Oak Ridge, TN, September 1976.

7. Arthur D. Little, Inc. "An Assessment of ASHRAE Standard 90-75, Energy Conservation in New Building Design." Report #C-78309, December 1975.

8. John G. Myers et al., *Energy Consumption in Manufacturing,* Cambridge, Massachusetts: Ballinger Publishing Company, 1974.

9. "26th Annual Electric Industry Forecast." *Electrical World,* New York, McGraw-Hill, September 15, 1975.

10. U.S. Department of Commerce, Bureau of the Census, *Population Estimates and Projections: Projections of the Population of the United States: 1975 to 2050,* P-25, No. 601, October 1975.

11. U.S. Department of Transportation, et. al., *The Report by the Federal Task Force on Motor Vehicle Goals Beyond 1980* (300-Day Study), Vol. 2: Task Force Report (Draft), September 2, 1976.

12. *Energy: Global Prospects 1985-2000,* McGraw-Hill, 1977.

19.7 WORKSHEETS

The pages that follow present the detailed results of the unconstrained U.S. supply-demand integrations for 1972, 1985, and 2000 for each of the WAES cases. The worksheets are presented in the following order:

National energy demand projections (Cases C-1, C-2 and D-7, D-8)

National input worksheets for supply/demand integrations (1972, 1985 Cases A–E, 2000 Cases C-1, C-2, D-7, and D-8)

NATIONAL ENERGY DEMAND PROJECTIONS

Country: United States

Year: 2000

WAES Case: C-1 and C-2

Population: 262.5×10^6

No. of Autos: 166.0×10^6

Number of Residential Dwellings: 107.8×10^6

Sector	Activity Level	Intensity	Energy Use (10^{15} Btu)	
Transportation	Total Vehicle Distance/Yr.	Energy Efficiency	Fossil and Other Fuels	Electric[1]
Automobiles	1660×10^9 mi.	4310 Btu/mi.	7.16	
Air Travel	654.1×10^9 pass-mi.	4525 Btu/pass-mi.	2.96	
Other Passenger			5.21	0.03
Truck & Air Freight	1114.2×10^9 ton-mi.		3.65	
Rail Freight	1977.2×10^9 ton-mi.	566 Btu/ton-mi.	1.12	
Shipping	1252.7×10^9 ton-mi.	966 Btu/ton-mi.	1.21	
Industry	Output/Year	Energy/Output		
Metals			9.63	1.92
Chemicals			6.02	1.36
Other Energy Intensive			5.68	0.82
Non-Energy Intensive			4.16	2.10
Petrochemical Feedstocks			10.55	
Asphalt & Road Tar			2.76	
Agriculture			1.91	0.19
Mining			0.50	0.10
Construction			1.48	

NATIONAL ENERGY DEMAND PROJECTIONS (con't) Cases C-1 and C-2

Commercial & Public	Floor Area	Heat Loss/ Area/Yr.	Syst.Eff.	Fossil fuels	Electric
Foss. Fuel Space Heat	45.11×10^9 sq.ft.	3.68×10^4 Btu/sq.ft.	0.78	2.13	
Electric Space Heat	11.28×10^9 sq.ft.	3.68×10^4 Btu/sq.ft.	1.3		0.32
Other Fossil Fuel[2]	56.39×10^9 sq.ft.	1.19×10^4 Btu/sq.ft.		0.67	
Other Electric[2]	56.39×10^9 sq.ft.	4.54×10^4 Btu/sq.ft.			2.56

Residential	Dwellings	Heat Loss/ Area/Yr.	Syst.Eff.	Fossil fuels	Electric
Foss. Fuel Space Heat	78.9×10^6	443×10^5 Btu/dwelling	0.75	4.66	
Electric Space Heat	28.9×10^6	443×10^5 Btu/dwelling	1.6		0.80
Foss. Fuel Appliances[3]	107.8×10^6	158×10^5 Btu/dwelling	1.70		
Electric Appliances[3]	107.8×10^6	351×10^5 Btu/dwelling			3.89
Total Delivered Energy: Demand (Including Fuel used for Feedstocks)				73.17	14.09

[1] Purchased electricity

[2] The number for heat loss/area/year represents the total fossil fuel (electric) energy demand per unit area per year.

[3] The number for heat loss/area/year represents the total fossil fuel (electric) energy demand per dwelling.

NATIONAL ENERGY DEMAND PROJECTIONS

Country: United States

Year: 2000

WAES Case: D-7 and D-8

Population: 262.5×10^6

No. of Autos: 166.0×10^6

Number of Residential Dwellings: 107.8×10^6

Sector	**Activity Level**	**Intensity**	**Energy Use** 10^{15} BTU	
Transportation	Total Vehicle Distance/Yr.	Energy Efficiency	Fossil and Other Fuels	Electric[1]
Automobiles	1580×10^9 mi.	4808 Btu/mi.	7.64	
Air Travel	561.5×10^9 pass-mi.	5628 Btu/pass-mi.	3.16	
Other Passenger			5.83	0.03
Truck & Air Freight	1084.1×10^9 ton-mi.		3.33	
Rail Freight	1473.2×10^9 ton-mi.	566 Btu/ton-mi.	0.83	
Shipping	1050.9×10^9 ton-mi.	971 Btu/ton-mi.	1.02	

Industry	Output/Year	Energy/Output		
Metals			7.08	1.45
Chemicals			5.47	1.24
Other Energy Intensive			4.90	0.73
Non-Energy Intensive			3.27	1.65
Petrochemical Feedstocks			9.06	
Asphalt & Road Tar			1.41	
Agriculture			1.58	0.15
Mining			0.31	0.06
Construction				

NATIONAL ENERGY DEMAND PROJECTIONS (con't) Cases D-7 and D-8

Commercial & Public	Floor Area	Heat Loss/ Area/Yr.	Syst. Eff.	Fossil fuels	Electric
Foss. Fuel Space Heat	38.51×10^9 sq.ft.	5.10×10^4 Btu/sq.ft.	0.74	2.65	
Electric Space Heat	9.63×10^9 sq.ft.	5.10×10^4 Btu/sq.ft.			0.41
Other Fossil Fuel[2]	48.14×10^9 sq.ft.	2.08×10^4 Btu/sq.ft.		1.00	
Other Electric[2]	48.14×10^9 sq.ft.	5.51×10^4 Btu/sq.ft.			2.68

Residential	Dwellings	Heat Loss/ Area/Yr.	Syst. Eff.	Fossil fuels	Electric
Foss. Fuel Space Heat	76.40×10^6	580×10^5 Btu/dwelling	0.70	6.33	
Electric Space Heat	28.00×10^6	580×10^5 Btu/dwelling	1.2		1.35
Foss. Fuel Appliances[3]	104.4×10^6	290×10^5 Btu/dwelling		3.03	
Electric Appliances[3]	104.4×10^6	350×10^5 Btu/dwelling			3.65
Total Delivered Energy: Demand (Including Fuel used for Feedstocks)				67.90	13.49

[1] Purchased electricity

[2] The number for heat loss/area/year represents the total fossil fuel (electric) energy demand per unit area per year.

[3] The number for heat loss/area/year represents the total fossil fuel (electric) energy demand per dwelling.

NATIONAL INPUT WORKSHEET FOR SUPPLY/DEMAND INTEGRATION

Country: U.S.A.

Year:

Case: 1972

Units: 10^{15} Btu

	(a) Coal	(b) Petroleum	(c) (Syn. Liquids)	(d) Nat. Gas	(e) (Syn. Gas)	(f) (Heat)	(g) (Electricity)[††]	(h) Hydro Energy	(i) Nuclear Energy	(j) Geothermal energy and other	(k) Total
(1) Transport	0	16.98		0			0.01				16.99
(2) Industry	3.93	1.56		6.96			1.81				14.26
(3) Agric., Mining, Construction	0	1.60		.30			.13				2.03
(4) Commercial / (5) Public	.19	.90		1.67			1.42				4.17
(6) Residential	0	3.77		5.70			1.88				11.35
(7) Non-energy Uses	.12	3.40		.67			0				4.19
(8) Final Energy Demand*	4.23	28.21		15.30			5.25				52.99
(9) Electricity**	7.68	3.25		4.10			-6.04	2.66	0.53		12.18
(10) Syn. Gas											
(11) Syn. Liquids											
(12) Heat											
(13) Energy Sector Self Consumption & conversion losses	.67	2.36		3.72			0.79				7.76
(14) Primary Energy Input	12.60	33.82		23.12			-	2.66	0.53		72.73
(15) Indigenous Supply	14.05	23.71		22.12			-	2.66	0.53		63.07
(16 Imports***		10.11		1.00							11.11
(17) Exports	1.45										1.45

*Includes non-energy uses.

**Blocks 9g, 10e 11c, and 12f must be negative to avoid double counting of energy from primary fuels.

***Includes imported products.

††Includes only electricity to be purchased by each sector.

NATIONAL INPUT WORKSHEET FOR SUPPLY/DEMAND INTEGRATION

Country: United States

Year: 1985

Case: A

Units: 10^{15} Btu

	(a)	(b)	(c)	(d)	(e)	(f)	(g)	(h)	(i)	(j)	(k)
	Coal	Petro-leum	(Syn. Liquids)	Nat. Gas	(Syn. Gas)	Solar	(Electri-city)††	Hydro Energy	Nuclear Energy	Geothermal energy and other	Total
(1) Transport		18.47					.02				18.49
(2) Industry (3) Agric., Mining, Construction	7.91	3.6		9.60			2.84				23.41
(4) Commercial (5) Public (6) Residential	0	3.44		6.63		0.20	4.69				14.96
(7) Non-energy Uses	.25	6.70		1.32							8.28
(8) Final Energy Demand*	8.16	31.67		17.55		0.20	7.55				65.14
(9) Electricity**	6.59	3.00		2.23			-8.89	4.6	8.48	.50	16.51
(10) Syn. Gas											
(11) Syn. Liquids											
(12) Heat											
(13) Energy Sector Self Consumption & conversion losses	.23	3.09		1.84			1.33				6.50
(14) Primary Energy Input	14.88	37.76		21.62		0.20		4.6	8.48	.50	88.04
(15) Indigenous Supply	26.20	26.20(1)		20.00		0.20		4.6	8.48	.50	86.18
(16 Imports***		11.56		1.62							13.18
(17) Exports	11.36										11.32

*Includes non-energy uses.

**Blocks 9g, 10e 11c, and 12f must be negative to avoid double counting of energy from primary fuels.

***Includes imported products.

††Includes only electricity to be purchased by each sector.

Footnotes:

(1) Includes 0.85 Quads of oil shale.

NATIONAL INPUT WORKSHEET FOR SUPPLY/DEMAND INTEGRATION

Country: United States

Year: 1985

Case: B

Units: 10^{15} Btu

	(a) Coal	(b) Petro-leum	(c) (Syn. Liquids)	(d) Nat. Gas	(e) (Syn. Gas)	(f) Solar	(g) (Electri-city)[††]	(h) Hydro Energy	(i) Nuclear Energy	(j) Geothermal energy and other	(k) Total
(1) Transport		17.66					.02				17.68
(2) Industry											
(3) Agric., Mining, Construction	7.01	2.37		8.06			2.46				19.90
(4) Commercial											
(5) Public	0	4.13		5.93		0.20	4.45				14.71
(6) Residential											
(7) Non-energy Uses	.17	4.74		1.99							6.90
(8) Final Energy Demand*	7.18	28.90		15.98		0.20	6.93				59.19
(9) Electricity**	4.36	3.00		2.37			−8.16	4.60	8.48	.50	15.15
(10) Syn. Gas											
(11) Syn. Liquids											
(12) Heat											
(13) Energy Sector Self Consumption & conversion losses	.15	2.86		1.65			1.22				5.88
(14) Primary Energy Input	11.69	34.76		20.00		0.20		4.60	8.48	.50	80.23
(15) Indigenous Supply	24.00	26.20(1)		20.00		0.20		4.60	8.48	.50	83.98
(16 Imports***		8.56									8.56
(17) Exports	12.31										12.31

*Includes non-energy uses.

**Blocks 9g, 10e 11c, and 12f must be negative to avoid double counting of energy from primary fuels.

***Includes imported products.

††Includes only electricity to be purchased by each sector.

Footnotes:

(1) Includes 0.85 Quads of oil shale.

NATIONAL INPUT WORKSHEET FOR SUPPLY/DEMAND INTEGRATION

Country: United States

Year: 1985

Case: C

Units: 10^{15} Btu

	(a)	(b)	(c)	(d)	(e)	(f)	(g)	(h)	(i)	(j)	(k)
	Coal	Petroleum	(Syn. Liquids)	Nat. Gas	(Syn. Gas)	Solar	(Electricity)††	Hydro Energy	Nuclear Energy	Geothermal energy and other	Total
(1) Transport		18.82					.02				18.84
(2) Industry (3) Agric., Mining, Construction	8.48	3.17		9.62			2.98				24.24
(4) Commercial (5) Public (6) Residential	0	3.85		7.44		.20	5.05				16.54
(7) Non-energy Uses	.25	6.64		1.31							8.20
(8) Final Energy Demand*	8.73	32.48		18.37		.20	8.05				67.82
(9) Electricity**	7.95	4.09		1.44			-9.47	4.6	8.48	.50	17.59
(10) Syn. Gas											
(11) Syn. Liquids											
(12) Heat											
(13) Energy Sector Self Consumption & conversion losses	.27	3.22		1.85			1.42				6.76
(14) Primary Energy Input	16.95	39.79		21.65		.20		4.6	8.48	.50	92.17
(15) Indigenous Supply	24.00	23.70 (1)		19.00		.20		4.6	8.48	.50	80.48
(16 Imports***		16.09		2.65							18.74
(17) Exports	7.05										7.05

*Includes non-energy uses.

**Blocks 9g, 10e 11c, and 12f must be negative to avoid double counting of energy from primary fuels.

***Includes imported products.

††Includes only electricity to be purchased by each sector.

Footnotes:

(1) Includes 0.40 Quads of oil shale.

NATIONAL INPUT WORKSHEET FOR SUPPLY/DEMAND INTEGRATION

Country: United States

Year: 1985

Case: D

Units: 10^{15} Btu

	(a)	(b)	(c)	(d)	(e)	(f)	(g)	(h)	(i)	(j)	(k)
	Coal	Petro-leum	(Syn. Liquids)	Nat. Gas	(Syn. Gas)	Solar	(Electri-city)††	Hydro Energy	Nuclear Energy	Geothermal energy and other	Total
(1) Transport		19.78					.02				19.80
(2) Industry (3) Agric., Mining, Construction	8.06	3.21		7.56	.09		2.69				21.61
(4) Commercial (5) Public (6) Residential		7.28		6.25	.09		5.93				19.55
(7) Non-energy Uses	.21	5.75		1.14							7.10
(8) Final Energy Demand*	8.27	36.02		14.95	.18	0	8.64				68.06
(9) Electricity**	11.67	5.56		1.62			-10.17	3.4	6.57	.20	18.85
(10) Syn. Gas		.29			-.18						.11
(11) Syn. Liquids											
(12) Heat											
(13) Energy Sector Self Consumption & conversion losses	.26	4.11		1.55			1.53				7.45
(14) Primary Energy Input	20.20	45.98		18.12		0		3.4	6.57	.20	94.47
(15) Indigenous Supply	20.20	20.10		16.00		0		3.4	6.57	.20	66.47
(16 Imports***		25.88		2.12							28.00
(17) Exports											

*Includes non-energy uses.

**Blocks 9g, 10e 11c, and 12f must be negative to avoid double counting of energy from primary fuels.

***Includes imported products.

††Includes only electricity to be purchased by each sector.

NATIONAL INPUT WORKSHEET FOR SUPPLY/DEMAND INTEGRATION

Country: United States

Year: 1985

Case: E

Units: 10^{15} Btu

	(a)	(b)	(c)	(d)	(e)	(f)	(g)	(h)	(i)	(j)	(k)
	Coal	Petroleum	(Syn. Liquids)	Nat. Gas	(Syn. Gas)	Solar	(Electricity)††	Hydro Energy	Nuclear Energy	Geothermal energy and other	Total
(1) Transport		21.18					.02				21.20
(2) Industry (3) Agric., Mining Construction	10.45	3.49		9.16			3.22				26.26
(4) Commercial (5) Public (6) Residential		9.73		4.35			6.79				20.87
(7) Non-energy Uses	.24	6.66		1.32							8.22
(8) Final Energy Demand*	10.70	40.99		14.83		0	10.03				76.55
(9) Electricity**	9.16	10.95		3.44			-11.80	3.4	6.57	0.20	21.92
(10) Syn. Gas											
(11) Syn. Liquids											
(12) Heat											
(13) Energy Sector Self Consumption & conversion losses	.34	4.87		1.69			1.77				8.67
(14) Primary Energy Input	20.20	56.81		19.96		0		3.4	6.57	0.20	107.14
(15) Indigenous Supply	20.20	17.00		14.00		0		3.4	6.57	0.20	61.37
(16 Imports*** (17) Exports		39.81		5.96							45.77

*Includes non-energy uses.

**Blocks 9g, 10e 11c, and 12f must be negative to avoid double counting of energy from primary fuels.

***Includes imported products.

††Includes only electricity to be purchased by each sector.

NATIONAL INPUT WORKSHEET FOR SUPPLY/DEMAND INTEGRATION

Country: United States

Year: 2000

Case: C-1 Unconstrained

Units: 10^{15} Btu

	(a) Coal	(b) Petro-leum	(c) (Syn. Liquids)	(d) Nat. Gas	(e) (Syn. Gas)	(f) Solar	(g) (Electri-city)††	(h) Hydro Energy	(i) Nuclear Energy	(j) Geothermal energy and other	(k) Total
(1) Transport		17.58	3.77				0.03				21.38
(2) Industry, (3) Agric., Mining, Construction	17.99	2.48		8.99			6.34				35.80
(4) Commercial, (5) Public, (6) Residential	0	2.30	0.43	4.55		2.35	6.43				16.06
(7) Non-energy Uses	0.40	10.79		2.13							13.32
(8) Final Energy Demand*	18.39	33.15	4.20	15.67		2.35	12.80				86.56 (3)
(9) Electricity**	4.56	3.05		1.00		1.50	-14.72	4.60	25.96	3.00	28.95
(10) Syn. Gas											
(11) Syn. Liquids	6.27		-4.20								2.07
(12) Heat											
(13) Energy Sector Self Consumption & conversion losses	0.62	2.80		1.57		0	1.92	0	0		6.91
(14) Primary Energy Input	29.84	39.00		18.24		3.85		4.60	25.96	3.00	124.49
(15) Indigenous Supply	47.97	19.10 (2)		15.00 (4)		3.85		4.60	25.96	3.00	119.48
(16 Imports***		19.90		3.24							23.14
(17) Exports (1)	18.13										18.13

*Includes non-energy uses.

**Blocks 9g, 10e 11c, and 12f must be negative to avoid double counting of enerav from primary fuels.

***Includes imported products.

††Includes only electricity to be purchased by each sector.

Footnotes:

1. Coal: "Exportable" coal = potential production minus consumption.
2. Includes 4.2 Quads of oil shale.
3. Final Energy Demand here may differ from Total Delivered Energy on the National Energy Demand Projection worksheets, due to minor adjustments in totals made during the allocations of fuels to end uses within the supply demand integrations, to account for different energy-using device efficiencies.
4. Includes 1.0 Quads of biomass.

NATIONAL INPUT WORKSHEET FOR SUPPLY/DEMAND INTEGRATION

Country: United States

Year: 2000

Case: C-2 Unconstrained

Units: 10^{15} Btu

	(a) Coal	(b) Petroleum	(c) (Syn. Liquids)	(d) Nat. Gas	(e) (Syn. Gas)	(f) Solar	(g) (Electricity)[††]	(h) Hydro Energy	(i) Nuclear Energy	(j) Geothermal energy and other	(k) Total
(1) Transport		19.08	2.27				0.03				21.38
(2) Industry) (3) Agric., Mining,) Construction)	17.20	2.51		8.97			7.02				35.70
(4) Commercial) (5) Public) (6) Residential)	0	0.39	1.93	4.55		2.35	8.33				17.55
(7) Non-energy Uses	0.40	10.79		2.13							13.32
(8) Final Energy Demand*	17.60	32.77	4.20	15.65		2.35	15.38				87.96[3]
(9) Electricity**	2.56	4.09		1.40		1.50	−17.69	4.60	35.24	3.00	34.70
(10) Syn. Gas											
(11) Syn. Liquids	6.27		−4.20								2.07
(12) Heat											
(13) Energy Sector Self Consumption & conversion losses	0.70	2.76		1.57		0	2.31	0	0	0	7.34
(14) Primary Energy Input	27.13	39.62		18.62		3.85		4.60	35.24	3.00	132.06
(15) Indigenous Supply	34.28	19.10[2]		15.00[5]		3.85		4.60	35.24[4]	3.00	115.07
(16 Imports***		20.52		3.62							24.14
(17) Exports[1]	7.15										7.15

*Includes non-energy uses.

**Blocks 9g, 10e 11c, and 12f must be negative to avoid double counting of energy from primary fuels.

***Includes imported products.

††Includes only electricity to be purchased by each sector.

Footnotes:

1. Coal: "Exportable" coal = potential production minus consumption.
2. Includes 4.2 Quads of oil shale.
3. Final Energy Demand here may differ from Total Delivered Energy on the National Energy Demand Projections worksheets, due to minor adjustments in totals made during the allocations of fuels to end uses within the supply-demand integrations, to account for different energy-using device efficiencies.
4. Maximum potential nuclear is 42.36 Quads.
5. Includes 1.0 Quads of biomass.

NATIONAL INPUT WORKSHEET FOR SUPPLY/DEMAND INTEGRATION

Country: United States

Year: 2000

Case: D-7 Unconstrained

Units: 10^{15} Btu

	(a)	(b)	(c)	(d)	(e)	(f)	(g)	(h)	(i)	(j)	(k)
	Coal	Petro-leum	(Syn. Liquids)	Nat. Gas	(Syn. Gas)	Solar	(Electri-city)[††]	Hydro Energy	Nuclear Energy	Geothermal energy and other	Total
(1) Transport		21.85					0.03				21.88
(2) Industry) (3) Agric., Mining,) Construction)	14.35	3.98		5.04			5.49				28.86
(4) Commercial) (5) Public) (6) Residential)	0	4.28		7.37		1.18	6.89				19.72
(7) Non-energy Uses	0.30	8.49		1.68							10.47
(8) Final Energy Demand*	14.65	38.60		14.09		1.18	12.41				80.93 (2)
(9) Electricity**	6.10	2.98		1.00		0.50	-14.27	3.69	25.96	1.20	27.16
(10) Syn. Gas											
(11) Syn. Liquids											
(12) Heat											
(13) Energy Sector Self Consumption & conversion losses	0.40	3.37		1.41		0	1.86	0	0		7.04
(14) Primary Energy Input	21.15	44.95		16.50		1.68		3.69	25.96	1.20	115.13
(15) Indigenous Supply	35.80	12.70		11.50		1.68		3.69	25.96	1.20	92.53
(16 Imports***		32.25		5.00							37.25
(17) Exports (1)	14.65										14.65

*Includes non-energy uses.

**Blocks 9g, 10e 11c, and 12f must be negative to avoid double counting of energy from primary fuels.

***Includes imported products.

††Includes only electricity to be purchased by each sector.

Footnotes:

1. Coal: "Exportable" coal = potential production minus consumption.
2. Final Energy Demand here may differ from Total Delivered Energy on the National Energy Demand Projections worksheets due to minor adjustments in totals made during the allocations of fuels to end uses within the supply-demand integrations, to account for different energy-using device efficiencies.

NATIONAL INPUT WORKSHEET FOR SUPPLY/DEMAND INTEGRATION

Country: United States

Year: 2000

Case: D-8 Unconstrained

Units: 10^{15} Btu

	(a) Coal	(b) Petroleum	(c) (Syn. Liquids)	(d) Nat. Gas	(e) (Syn. Gas)	(f) Solar	(g) (Electricity)††	(h) Hydro Energy	(i) Nuclear Energy	(j) Geothermal energy and other	(k) Total
(1) Transport		21.85					0.03				21.88
(2) Industry (3) Agric., Mining, Construction	14.35	4.16		4.44			6.09				29.04
(4) Commercial (5) Public (6) Residential	0	3.21		7.61		1.18	8.28				20.28
(7) Non-energy Uses	0.30	8.49		1.68							10.47
(8) Final Energy Demand*	14.65	37.71		13.73		1.18	14.40				81.67 (3)
(9) Electricity**	4.05	4.01		1.40		0.50	-16.56	3.69	33.12	1.20	31.41
(10) Syn. Gas											
(11) Syn. Liquids											
(12) Heat											
(13) Energy Sector Self Consumption & conversion losses	0.39	3.27		1.37		0	2.16	0	0	0	7.19
(14) Primary Energy Input	19.09	44.99		16.50		1.68		3.69	33.12	1.20	120.27
(15) Indigenous Supply	27.71	12.70		11.50		1.68		3.69	33.12 (2)	1.20	91.60
(16 Imports***		32.29		5.00							37.29
(17) Exports (1)	8.62										8.62

*Includes non-energy uses.

**Blocks 9g, 10e 11c, and 12f must be negative to avoid double counting of energy from primary fuels.

***Includes imported products.

††Includes only electricity to be purchased by each sector.

Footnotes:

1. Coal: "Exportable" coal = potential production minus consumption.
2. Maximum potential nuclear is 42.36 Quads.
3. Final Energy Demand here may differ from Total Delivered Energy on the National Energy Demands Projection worksheets, due to minor adjustments in totals made during the allocations of fuels to end uses within the supply-demand integrations, to account for different energy-using device efficiencies.

Appendix A
Illustrative Full GEMM Case

This Appendix contains the full set of tables printed as output of a GEMM run of WAES 1985 Case D. Each page of printed results is discussed briefly, so that the reader may be able to understand any other set of GEMM case results.

Computer table page numbers will be referred to in this text as PAGE. Table and subtable titles are also shown in capital letters in this text, for ease of identification. The numbers may be found in the upper right-hand corner of the reproduced output.

PAGE 1, the VOLUMETRIC SUMMARY, shows the quantity of primary energy produced, by region and fuel, from total available supply. The flow of primary energy, 114.1 MBDOE in this case, then proceeds in part to become input to CONVERSION (INPUT) processes, shown next. Here the processes are shown by fuel type. For example, 59.6 MBD of oil are refined. (It all comes from oil production, none from coal liquefaction.) The data are shown after production and primary transport losses, but before conversion and distribution losses, as is clear from the magnitude of the electricity generation figures by fuel, which add to 23.5 MDBOE. Last, PAGE 1 shows MARKET INPUT by sector and by fuel. Note that electricity input to markets totals 6.9 MBDOE. This is based on assumed factors of 35 percent yield in generation and 85 percent in transmission applied to the individual regional figures, which represent 23.5 MBDOE of electricity generation sector input.

PAGES 2 and 3 show PRODUCTION AND TRANSPORTS for each fossil fuel (crude oil, oil products, coal, and gas) between regions. Total OPEC crude oil production is 36.0 MBD, primarily exported to other regions. The OPEC-OPEC figure reflects 5.9 MBD of crude to be refined within OPEC into oil products for export. In Case D, as seen from the second table on PAGE 2, the only product exports are 3.96 MBDOE of distillates (D) from OPEC to North America and 1.54 MBDOE of heavy fuel oil (F) from OPEC to Japan. The model generalizes oil products in two classes—light (distillate) and heavy (fuel oil).

Coal production and transport are presented in similar form on PAGE 3, with transport losses shown separately. Gas is shown separately for pipeline and LNG transport with the losses also reported. Pipeline losses are 1.94 MBDOE of 19.40 MBDOE production (10 percent). The large liquefaction, transport, and regasification losses for OPEC LNG account for 0.50 MBDOE, compared with LNG effective exports (after losses) of 2.00 MBDOE. Note that total LNG production in this table must be calculated from the sum of individual regional row figures plus losses. The production of LNG in the table is included in the total pipeline gas production (PROD'N) since the gas must be produced prior to liquefaction.

The 0.50 MBDOE production figure in the LNG total supply row represents net exports from Communist areas since production in that region is not modeled; the flow appears initially as LNG (or pipeline) exports.

PAGE 4, the MARKET INPUT SUMMARY, shows the flow, net of previous system losses, for end use in markets by regions and fuels. It also shows demand by region and market, comparing each demand figure with the market input ("supply") and reporting any deficits. (Losses have been set at zero in the market sector of the model; WAES demand studies assume consumers' total demand for energy including end-use losses.) Note that in Case D, all deficits are zero—there is adequate available supply to meet desired demand at the case assumptions and no "penalty fuel" had to be used in any market.

The WOCA (world outside Communist areas) DEMAND/SUPPLY MATRIX of PAGE 5 is the conventional energy flow accounting worksheet used by OECD and others. Reading up from the bottom, the matrix takes primary energy supply through losses and conversion processes and nonmarket uses to final market demand at the top. Each entry may be thought of as a deduction from available supply thus far, reading up. Thus a negative entry is an additional element of supply entering from some other activity in the same row. Nuclear and hydro enter primary energy demand on an input basis to make the figures comparable to other sources of primary energy supply. An artificial loss is then attributed to these sources to make them comparable to thermal electricity generated from fossil fuels (3.05 MBDOE, shown above). PAGES 6–9 show similar DEMAND/SUPPLY matrices for each region in WOCA.

PAGES 10–12 show a summary by region of AN-

NUAL CAPITAL CHARGES for infrastructure expansion, PRODUCING REGIONS' REVENUE, and CONSUMING REGIONS' COSTS AND REVENUES. Energy balance-of-payments consequences are also indicated. GEMM has the additional capability of constraining regional deficits to some level based on other trade capabilities, adding an interest charge for international loans to each region above that level, and severely penalizing loans above a maximum level in each region through a 100 percent interest charge. These interest costs are taken into account in the total cost minimization of the model. This feature was not used during the WAES analysis—each region was permitted to incur whatever energy import balance-of-payments deficits were necessary to meet desired demand. In Case D, OPEC annual capital charges (PAGE 10) in 1985 for new oil production and refining facilities, built after 1975 and not yet depreciated, are $0.84 billion (1975 U.S.$) and $1.25 billion, respectively. Gas production and liquefaction facilities incur $2.86 billion and $1.25 billion in capital charges.

PAGE 11 notes revenue from OPEC tariffs (export prices) as $182.6 billion in 1985, compared with the $6.2 billion in capital charges just noted. OCEC coal exports provide revenues of $14.05 billion for the coal-exporting countries of the "region." North America (PAGE 12) would pay $58.84 billion for energy imports, Europe $61.51 billion, Japan $39.24 billion, and the rest of the world outside Communist areas $55.12 billion. The integration produces $3.85 billion in revenues from North American coal exports in Case D, resulting in a net import bill for North America of $54.99 billion. OCEC coal revenues shown above also offset the import costs of the rest of the world outside Communist areas. In actual practice intercountry trade flows might be somewhat different, with some OCEC coal going outside that region and some North American coal going elsewhere than indicated. (It would be possible in GEMM to partially constrain such trade flows, but a forecast of contractual relations that would continue or be entered into by the year 1985 was considered beyond the scope of WAES.) Consumers' marginal costs are reported on PAGE 13.

A guide to the interpretation of these figures is provided below. In essence the marginal cost of an activity is the cost to the entire world, after taking account of all changes throughout the system, resulting from the one barrel increase in final demand for that activity if the integration were recalculated on that basis. For example, an additional barrel (oil equivalent) of coal in Japan would cost $49.61; an additional barrel of fuel oil delivered to market in North America would cost $79.59. The costs represent infrastructure expansion, transportation, raw materials, and all other costs incurred by the system. When all available fuel of a particular kind is used, marginal costs imply a substitution somewhere of a more costly fuel for oil in order to release the marginal barrel of oil. Thus, if only surplus gas is available (if everyone were using the maximum of other fuels permitted by their market constraints, for example) the marginal cost of a barrel of oil for end use would be determined by the cost of a barrel equivalent of end-use gas. Such gas might only be available through LNG shipment, which would necessitate expansion costs. The high losses in LNG would require considerably more than 1 BOE of gas to be produced, liquefied, shipped, regasified, and distributed to end up with the replacement barrel equivalent of end-use gas. Marginal costs are thus a rough measure of the value of an additional unit of energy. As long as they are above total new costs, it pays to move higher-cost supplies of the desired fuels, where possible, beyond the constraint limit. Note that in the case run, additional electricity has very high marginal cost per delivered unit since generation and transmission losses are so high.

The detailed regional fuel and market tables begin on PAGE 14. The first set of such tables, for oil, runs through PAGE 28. One illustrative table for North America runs from PAGE 14 to PAGE 16. PRODUCTION is shown first, with minimum and maximum capacities reported as developed in individual WAES supply studies. The total and new capacity used is shown for onshore, offshore, and tar sands, as is the operating cost and capital charge per barrel for each. The onshore and offshore calculations are actually made separately for light and heavy oil in each region; the table shows totals. (In some regions, of course, only one such type is possible at the case assumptions.) The maximum capacity is fully used. The calculation behind these figures is typical of all such infrastructure data in the model. Base year existing capacity is depreciated at rates that can be specified for each element of infrastructure depending on its physical life or resource exhaustion to the case year, and new capacity is added as a function of capital and operating costs, constraints, and demand everywhere else in the system relevant to end-use demand.

Next on PAGE 14 is the REFINING table, which shows the structure of the refinery optimization in each region. Distillation, followed by one of two conversion processes, is available, and the mix selected in the integration is a function of the cost of each process and the volume of heavy (fuel oil) and light (distilled) product required from each region's refineries. Coal and gas refinery fuels are also accounted for. In the North American case, new refinery capacity is added.

The DISTRIBUTION table shows capacity added for distillates and fuel oil.

Beginning at the bottom of PAGE 14 are a set of more detailed tables separately showing crude oil, refining, distillates, and fuel oil flows and processing on an input-output basis.

The CRUDE OIL table shows sources of crude (indigenous, import, and coal conversion), the percentage mix, input volumes, yield (only coal conversion assumes a loss here; a 67 percent yield has been assumed although all figures can readily be altered). Next, the output volumes are shown (PAGE 15, Top). A subtable shows the destination of crude oil outputs, either for export or refining. Here the columns labeled "input" are inputs at either stage. Thus, 11.90 MBD of indigenous crude oil, at 100 percent yield, enters the 19.21 MBD output of the crude oil stage. Of this, the total volume flows as input to the refining stage.

The REFINING stage reports the yields of distillate and residual from distilling and each conversion process, as well as other fuel uses. It can be seen that Conversion I (catalytic cracking) is residual fuel oil intensive, while Conversion II (hydrocracking) is distillate intensive, but at a much higher cost in process fuel use. The percentages are volumetric, and thus Conversion II has a 114 percent yield; for every barrel input, 0.70 barrel of light product, 0.3 barrel of heavy product, and 0.14 barrel of fuel use result. The demand for light products out of North American refineries in this case is such that despite fuel use, 2.73 MBD of Conversion II (PAGE 14) are used after the initial distillation process. These outputs are shown in the refining subtable on PAGE 15, where 7.89 percent of the output of refining is distillate, 21.36 percent fuel oil, 5.16 percent fuel oil refinery fuel, and 1.29 percent coal refinery fuel. These figures are presented on an energy use basis and add to 100 percent when the overall refinery fuel use and losses of 6.74 percent are taken into account.

The final oil tables on PAGES 15 and 16 account for the further flow of refined products. The FROM table accounts fully for DISTILLATES imports, showing intake, yield, and output. Refinery output is shown as a pass-through to the output of this stage, so no input or yield numbers are presented. The distribution of distillate is shown in the TO table (PAGE 16, top). The headings are the same as on the FROM table and are not repeated. The percentage and volumes of intake in MBDOE and MTOE are shown for the industrial and domestic market sectors, as in the output in MBDOE and B therm (units of 100,000 Btu). (These sectors use other fuel inputs as well.) The input to transportation distillates, chemical feedstock, and distribution loss are also shown, but yields and outputs are omitted. A similar structure is used for the FUEL OIL tables on PAGE 16.

The COAL tables begin on PAGE 29. They are similar in structure to the oil tables and will not be elaborated here. The GAS tables begin on PAGE 34. They include production, liquefaction (of exported LNG), transport by pipe, regasification (of imported LNG), syngas from coal, and gas grid for distribution.

Note that transport by pipe includes all production, as can be seen on PAGE 35 for Europe gas. There, 1.93 MBDOE of onshore production and 1.51 MBDOE of offshore production (a total of 3.44 MBDOE) enter the pipeline. An additional 1.00 MBDOE represents regasified import LNG and 0.2 MBDOE represents pipeline imports. Shown next in the FROM table is the 1.04 MBDOE of import LNG needed from Communist areas (at 96 percent shipping yield) to produce the 1.00 MBDOE of LNG for the gas grid. The 3.44 MBDOE of indigenous production carried through transport by pipe becomes 3.10 MBDOE after the 90 percent yield of the (indigenous) pipe. The FROM table shows Communist area and other (import pipe) pipeline imports separately. The 0.20 MBDOE imported by non-Communist area pipe becomes 0.18 MBDOE. In the first subtable (PAGE 35), 0.20 MBDOE of the import pipe capacity and 3.44 of indigenous pipe capacity must exist to produce their share of 4.68 MBDOE, after losses (second table, PAGE 35), for the gas grid. But a further loss at the entry to the gas grid of 0.04

MBDOE occurs through regasification of the 1.00 MBDOE of import LNG. (The loss from 1.04 to 1.00 MBDOE in the FROM table represents shipping loss.) This further regasification loss is shown in the TO table. As a consequence, only 4.24 MBDOE of gas grid capacity (first table) need actually be available, rather than the 4.28 (before regasification loss) in the second FROM table.

The ELECTRICITY GENERATION tables run from PAGE 39 to PAGE 42. First, fuel inputs to generation are shown. There is the possibility in some case runs of excess undepreciated generation capacity, so a "surplus" column is provided. On PAGE 39 a 1.92 MBDOE input basis coal-fired electric capacity surplus in North America is shown. This is based on the stated 1975 infrastructure, depreciated to 1985, which is somewhat in excess of coal electric requirements in 1985 of 2.82 MBDOE. (The asterisks in the "max" column indicate that there was no constraint on infrastructure expansion in these runs.) The table also shows generation yields, which were arbitrarily set at 35 percent for each fuel and region. The second subtable shows the fuel range constraints and actual percentages in the output mix. In North America, for example, the minimum desired percentage of fuel oil and gas electric capacity is being used (for cost reasons).

Note that the fuel range constraints are designed to take account of cost-effective capacity that must exist for, say, peak loading, despite average costs of generation of electricity. The last column of the second subtable on PAGE 39 shows "differential costs" for each form of electricity generation. These figures represent the cost of the constraints when particular fuels are used at the lower end (negative differential cost or higher end of a constraint). If the minimum percentage of a fuel is used, it is more costly than others. Reducing the minimum constraint further would save money and change the total cost to the world. Thus, actual fuel oil use is at its minimum of 10 percent. Allowing one less barrel of oil equivalent to be used would add -98.39 to costs, a considerable savings. This is a marginal cost concept; the barrel of oil-fired electricity would require almost 2 barrels of fuel oil input to electrical generation which could be used elsewhere in the system. PAGE 40 shows that Europe is using the maximum amount of gas-fired electricity (in generation mix preference/percentage terms) desired. Associated with this low-cost source is a $16.58 differential cost. Allowing an additional barrel equivalent of gas-fired use would add to costs by this amount. Relaxing the preference constraint would add the cost of the additional daily barrel equivalent (output) of gas-fired electricity.

The last part of the electricity generation table shows generated and installed average capacity, total and new, and distribution capacity and its capital charge and yield. The distribution to markets is also shown.

INDUSTRIAL MARKET and TRANSPORT MARKET tables follow, on PAGES 43-46; DOMESTIC MARKET (residential and commercial) is found on PAGES 47-50. The tables are similar in structure. Like the electric table, fuels to these markets (including electricity generated) are shown with new and surplus capacity and yields. Fuel mix constraint tables follow including differential costs. A breakdown of liquid into fuel oil and distillates is followed by demand breakdowns showing subsectors of the market, together with final demand, actual delivered fuel, and any deficits. In the industrial market tables, heat, light and power, chemical feedstock and its breakdown, and lube oil and asphalt are shown. In the transport market, transport on distillates, transport on electricity, and non-energy and energy bunkers are shown. The latter figure is derived from the energy shipping activities in each case run, allocated to the importing regions. Domestic demand is not subdivided beyond the heat, light, and power total.

The next GEMM integration table (PAGE 51) is a SUMMARY of consumers' average costs for each region. Energy activities are shown separately, as are the arbitrary import tariffs (prices). The total average costs by fuel are reported last. While such figures seem to have intuitive appeal, they are of limited usefulness, since decisions and actions flow from marginal costs of incremental action, not from average costs.

The final table (PAGE 52) shows the ship fleet, by ship type, including the number of new ships required and their costs.

CONTENTS OF THE GEMM GLOBAL INTEGRATION OUTPUT FOR 1985 CASE D
($11.50 oil price; 3.5 percent per year world economic growth rate; restrained policy response)

The numbers in the following outline of the computer output refer to the output page numbers, found in the upper right-hand corner of the output pages

Volumetric summary 1
Production and transports (fossil fuels) 2
Market input summary 4
Demand/supply matrices 5
Capital charges, costs, and revenues 10
Consumers' marginal costs 13
Regional fuel tables—oil 14
Regional fuel tables-coal 29
Regional fuel tables—gas 34
Regional electricity generation 39
Regional market tables—industrial and transport 43
Regional market tables—domestic 47
Summary of consumers' average costs 51
Summary of ship fleet 52
Input tables 53

1

VOLUMETRIC SUMMARY
====================

PROBLEM: WAES85D BASE YR: 1975.
CASE: 31JAN77 CASE YR: 1985.

PRODUCTION

	OIL	COAL	GAS	NUCLEAR	HYDRO	TOTAL
NORTH AMERICA	11.9	9.9	8.8	3.3	3.1	37.0
EUROPE	4.1	4.3	3.4	2.6	2.1	16.5
JAPAN	0.1	0.3	0.1	0.6	0.6	1.6
OPEC	36.0	0.0	4.8	0.0	0.0	40.8
OCEC	0.0	4.9	0.0	0.0	0.0	4.9
REST WOCA	5.6	0.0	2.3	0.5	1.9	10.3
SUBTOTAL	57.6	19.4	19.4	7.0	7.7	111.1
WOCA EXPORTS	2.0	0.5	0.5	0.0	0.0	3.0
TOTAL	59.6	19.9	19.9	7.0	7.7	114.1

TOTAL WORLD ENERGY PRODUCTION (EXCLUDING COMMUNIST AREA INTERNAL): 114.1

CONVERSION(INPUT)

	OIL	COAL	GAS	NUCLEAR	HYDRO
OIL REFINING	59.6				
COAL GASIFICATION		0.0			
COAL LIQUEFACTION		0.0			
GAS LIQUEFACTION (NOT INCL. COMMUNIST AREA)				2.5	
ELECTRICITY GENERATION	1.9	5.8	1.1	7.0	7.7

MARKET INPUT

	OIL	COAL	GAS	ELECTRICITY
INDUSTRIAL				
HEAT, LIGHT AND POWER	11.0	10.2	7.2	2.7
NON ENERGY	6.8	0.2	0.9	
DOMESTIC				
HEAT, LIGHT AND POWER	9.0	2.5	6.3	4.1
TRANSPORT AND BUNKERS	25.5			0.1

ALL QUANTITIES ARE GIVEN IN MILLION BARREL PER DAY OIL EQUIVALENT

PRODUCTION AND TRANSPORTS CRUDE OIL

PROBLEM: WAES85D CASE: 31JAN77 BASE YR: 1975. CASE YR: 1985.

FROM \ TO	NORTH AMERICA	EUROPE	JAPAN	REST WOCA	OPEC	PROD"N
NORTH AMERICA	11.90	.	.	.	.	11.90
EUROPE	.	4.07	.	.	.	4.07
JAPAN	.	.	0.08	.	.	0.08
REST WOCA	.	.	.	5.60	.	5.60
OPEC	7.31	8.84	6.24	7.70	5.91	36.00
COMMUNIST AREAS	.	2.00	.	.	.	2.00
TOTAL SUPPLY	19.21	14.91	6.32	13.30	5.91	59.65

PRODUCTION AND TRANSPORTS OIL PRODUCTS

PROBLEM: WAES85D CASE: 31JAN77 BASE YR: 1975. CASE YR: 1985.

F=FUEL OIL, D=DISTILLATES

FROM \ TO		NORTH AMERICA	EUROPE	JAPAN	REST WOCA	OPEC	PROD"N
NORTH AMERICA	F	4.10	.	.	.	.	4.10
	D	13.81	.	.	.	.	13.81
EUROPE	F	.	4.57	.	.	.	4.57
	D	.	9.43	.	.	.	9.43
JAPAN	F	.	.	1.95	.	.	1.95
	D	.	.	3.96	.	.	3.96
REST WOCA	F	.	.	.	4.06	.	4.06
	D	.	.	.	8.99	.	8.99
OPEC	F	.	.	1.54	.	.	1.54
	D	3.96	.	.	.	0.00	3.96
TOTAL SUPPLY	F	4.10	4.57	3.49	4.06	.	16.22
	D	17.77	9.43	3.96	8.99	0.00	40.15

PRODUCTION AND TRANSPORTS COAL

PROBLEM: WAES85D CASE: 31JAN77 BASE YR: 1975. CASE YR: 1985.

FROM \ TO	NORTH AMERICA	EUROPE	JAPAN	REST WOCA	LOSS	PROD"N
NORTH AMERICA	8.36	1.32	.	.	0.19	9.88
EUROPE	.	4.24	.	.	0.08	4.33
JAPAN	.	.	0.25	.	0.01	0.26
OCEC	.	0.11	1.05	3.65	0.10	4.91
COMMUNIST AREAS	.	0.50	.	.	.	0.50
TOTAL SUPPLY	8.36	6.17	1.31	3.65	0.38	19.87

PRODUCTION AND TRANSPORTS GAS

PROBLEM: WAES85D CASE: 31JAN77 BASE YR: 1975. CASE YR: 1985.

P=BY PIPE L=AS LNG

FROM \ TO		NORTH AMERICA	EUROPE	JAPAN	REST WOCA	OPEC	LOSS	PROD"N
NORTH AMERICA	P	7.93	.	.	.	.	0.88	8.81
EUROPE	P	.	3.10	.	.	.	0.34	3.44
JAPAN	P	.	.	0.06	.	.	0.01	0.07
REST WOCA	P	.	.	.	2.07	.	0.23	2.30
OPEC	P	.	0.18	.	1.62	2.50	0.48	4.78
	L	0.96	1.04	.	.	.	0.50	.
COMMUNIST AREAS	P	.	.	.	.	.	.	.
	L	.	.	0.50	.	.	.	0.50
TOTAL SUPPLY	P	7.93	3.28	0.06	3.69	2.50	1.94	19.40
	L	0.96	1.04	0.50	.	.	0.50	0.50

MARKET INPUT SUMMARY

PROBLEM: WAES85D BASE YR: 1975.
CASE: 31JAN77 CASE YR: 1985.

INDUSTRIAL HEAT, LIGHT AND POWER	OIL	COAL	GAS	ELECTRICITY
NORTH AMERICA :	2.8	4.5	3.5	0.4
EUROPE :	2.5	2.5	2.1	1.2
JAPAN :	2.8	1.1	0.1	0.3
REST WOCA :	2.9	2.2	1.5	0.7
	11.0	10.2	7.2	2.7

INDUSTRIAL NON ENERGY	OIL	COAL	GAS
NORTH AMERICA :	3.1	0.1	0.6
EUROPE :	1.8	0.1	0.3
JAPAN :	0.8	.	.
REST WOCA :	1.1	.	.
	6.8	0.2	0.9

DOMESTIC HEAT,LIGHT AND POWER

	OIL	COAL	GAS	ELECTRICITY
NORTH AMERICA :	3.2	0.5	4.2	2.6
EUROPE :	3.6	1.4	1.4	0.7
JAPAN :	1.2	0.0	0.2	0.3
REST WOCA :	1.0	0.5	0.5	0.5
	9.0	2.5	6.3	4.1

TRANSPORT

	OIL	ELECTRICITY
NORTH AMERICA :	10.7	0.0
EUROPE :	5.4	0.1
JAPAN :	1.8	0.0
REST WOCA :	6.9	0.0
	24.9	0.1

DEMAND

		IND H,L&P	IND NONEN	DOM H,L&P	TRANSPORT
NORTH AMERICA					
	DEMAND :	11.2	3.8	10.5	10.8
	SUPPLY :	11.2	3.8	10.5	10.8
	DEFICIT :	0.0	0.0	0.0	0.0
EUROPE					
	DEMAND :	8.3	2.2	7.1	5.4
	SUPPLY :	8.3	2.2	7.1	5.4
	DEFICIT :	0.0	0.0	0.0	0.0
JAPAN					
	DEMAND :	4.2	0.8	1.7	1.9
	SUPPLY :	4.2	0.8	1.7	1.9
	DEFICIT :	0.0	0.0	0.0	0.0
REST WOCA					
	DEMAND :	7.3	1.1	2.6	7.0
	SUPPLY :	7.3	1.1	2.6	7.0
	DEFICIT :	0.0	0.0	0.0	0.0

W O C A DEMAND/SUPPLY MATRIX PROBLEM: WAES85D BASE YR: 1975.
CASE: 31JAN77 CASE YR: 1985.

	OIL	COAL	GAS	ELECTRICITY	NUCLEAR	HYDRO-ETC	TOTAL
TRANSPORT	24.88			0.14			25.02
INDUSTRY	10.98	10.24	7.16	2.72			31.09
DOMESTIC	9.05	2.48	6.26	4.10			21.89
TOTAL INTO MARKETS	44.90	12.72	13.42	6.96			78.00
TRANSM./DISTR.LOSSES*	2.82	0.78	3.67	1.23			8.49
CONV. THERMAL ELECTR.	0.66	2.01	0.38	-3.05			
USES+LOSSES	1.23	3.74	0.70				5.67
TOTAL INPUT	1.89	5.75	1.08				
GAS MANUFACTURE		0.0	0.0				
USES+LOSSES		0.0					0.0
TOTAL INPUT		0.0					
PRIMARY ENERGY DEMAND							
OUTPUT	49.61	19.24	18.17		2.46	2.68	92.16
INPUT					7.02	7.66	101.70
CHEMICAL FEED	6.76	0.18	0.89				7.83
LUBES & ASPHALTS	0.0						0.0
TOTAL REFINED PRODUCT	56.37						
REFINERY USE & PROCESS LOSS*	3.28	0.45	0.84				4.57
SYNCRUDE MANUFACTURE	0.0	0.0					
USES+LOSSES		0.0					0.0
TOTAL INPUT		0.0					
PRIMARY ENERGY SUPPLY	59.65	19.88	19.90		7.02	7.66	114.11

*INCLUDES OPEC & OCEC

ALL QUANTITIES ARE GIVEN IN MILLION BARREL PER DAY OIL EQUIVALENT

DEMAND/SUPPLY MATRIX NORTH AMERICA PROBLEM: WAES85D BASE YR: 1975.

CASE: 31JAN77 CASE YR: 1985.

	OIL	COAL	GAS	NUCLEAR	ELECTRICITY	HYDRO-ETC	TOTAL
TRANSPORT	10.75				0.01		10.76
INDUSTRY	2.80	4.49	3.51		0.41		11.22
DOMESTIC	3.15	0.52	4.17		2.61		10.46
TOTAL INTO MARKETS	16.71	5.01	7.68		3.04		32.44
TRANSM./DISTR.LOSSES*	1.09	0.36	1.48		0.54		3.48
CONV. THERMAL ELECTR.	0.36	0.99	0.0		-1.35		
USES+LOSSES	0.66	1.84	0.0				2.50
TOTAL INPUT	1.02	2.82	0.0				
GAS MANUFACTURE		0.0	0.0				
USES+LOSSES		0.0					0.0
TOTAL INPUT		0.0					
PRIMARY ENERGY DEMAND							
OUTPUT	18.82	6.20	9.17	1.14		1.08	38.42
INPUT				3.27		3.10	42.56
CHEMICAL FEED	3.05	0.11	0.60				3.77
LUBES & ASPHALTS	0.0						0.0
TOTAL REFINED PRODUCT	21.87						
REFINERY USE & PROCESS LOSS*	1.30	0.25	0.0				1.54
SYNCRUDE MANUFACTURE	0.0	0.0					
USES+LOSSES		0.0					0.0
TOTAL INPUT		0.0					
PRIMARY ENERGY SUPPLY	23.17	8.56	9.77	3.27		3.10	47.87

ALL QUANTITIES ARE GIVEN IN MILLION BARREL PER DAY OIL EQUIVALENT

DEMAND/SUPPLY MATRIX EUROPE

PROBLEM: WAESB5D
CASE: 31JAN77

BASE YR: 1975.
CASE YR: 1985.

	OIL	COAL	GAS	NUCLEAR	ELECTRICITY	HYDRO-ETC	TOTAL
TRANSPORT	5.36				0.08		5.44
INDUSTRY	2.50	2.50	2.08		1.25		8.32
DOMESTIC	3.61	1.43	1.38		0.71		7.13
TOTAL INTO MARKETS	11.46	3.92	3.46		2.04		20.89
TRANSM./DISTR.LOSSES*	0.70	0.21	0.68		0.36		1.95
CONV. THERMAL ELECTR.	0.0	0.67	0.08		-0.75		
USES+LOSSES	0.0	1.25	0.15				1.40
TOTAL INPUT	0.0	1.92	0.23				
GAS MANUFACTURE		0.0	0.0				
USES+LOSSES		0.0					0.0
TOTAL INPUT		0.0					
PRIMARY ENERGY DEMAND							
OUTPUT	12.16	6.05	4.38	0.91		0.73	24.24
INPUT				2.61		2.10	27.30
CHEMICAL FEED	1.84	0.07	0.28				2.19
LUBES & ASPHALTS	0.0						0.0
TOTAL REFINED PRODUCT	14.00						
REFINERY USE & PROCESS LOSS*	0.90	0.14	0.0				1.05
SYNCRUDE MANUFACTURE	0.0	0.0					
USES+LOSSES		0.0					0.0
TOTAL INPUT		0.0					
PRIMARY ENERGY SUPPLY	14.91	6.26	4.66	2.61		2.10	30.54

ALL QUANTITIES ARE GIVEN IN MILLION BARREL PER DAY OIL EQUIVALENT

8

DEMAND/SUPPLY MATRIX JAPAN   PROBLEM: WAES85D   BASE YR: 1975.
CASE: 31JAN77   CASE YR: 1985.

	OIL	COAL	GAS	ELECTRICITY	NUCLEAR	HYDRO-ETC	TOTAL
TRANSPORT	1.83			0.03			1.86
INDUSTRY	2.75	1.06	0.11	0.32			4.24
DOMESTIC	1.24	0.01	0.18	0.25			1.68
TOTAL INTO MARKETS	5.82	1.07	0.29	0.60			7.78
TRANSM./DISTR.LOSSES*	0.37	0.03	0.08	0.11			0.59
CONV. THERMAL ELECTR.	0.18	0.05	0.07	-0.30			
USES+LOSSES	0.33	0.10	0.13				0.56
TOTAL INPUT	0.51	0.15	0.20				
GAS MANUFACTURE		0.0	0.0				
USES+LOSSES		0.0					0.0
TOTAL INPUT		0.0					
PRIMARY ENERGY DEMAND							
OUTPUT	6.70	1.25	0.57		0.21	0.20	8.93
INPUT					0.61	0.56	9.69
CHEMICAL FEED	0.75	0.0	0.0				0.75
LUBES & ASPHALTS	0.0						0.0
TOTAL REFINED PRODUCT	7.45						
REFINERY USE & PROCESS LOSS*	0.41	0.06	0.0				0.47
SYNCRUDE MANUFACTURE	0.0	0.0					
USES+LOSSES		0.0					0.0
TOTAL INPUT		0.0					
PRIMARY ENERGY SUPPLY	7.86	1.31	0.57		0.61	0.56	10.91

ALL QUANTITIES ARE GIVEN IN MILLION BARREL PER DAY OIL EQUIVALENT

9

DEMAND/SUPPLY MATRIX REST WOCA PROBLEM: WAES85D BASE YR: 1975.
CASE: 31JAN77 CASE YR: 1985.

	OIL	COAL	GAS	NUCLEAR	ELECTRICITY	HYDRO-ETC	TOTAL
TRANSPORT	6.95				0.02		6.97
INDUSTRY	2.92	2.19	1.46		0.73		7.31
DOMESTIC	1.05	0.52	0.52		0.52		2.62
TOTAL INTO MARKETS	10.92	2.72	1.99		1.27		16.90
TRANSM./DISTR.LOSSES*	0.65	0.07	0.45		0.22		1.40
CONV. THERMAL ELECTR.	0.12	0.30	0.22		-0.65		
USES+LOSSES	0.23	0.56	0.42				1.21
TOTAL INPUT	0.36	0.86	0.64				
GAS MANUFACTURE		0.0	0.0				
USES+LOSSES		0.0					0.0
TOTAL INPUT		0.0					
PRIMARY ENERGY DEMAND				NUCLEAR		HYDRO-ETC	
OUTPUT	11.93	3.65	3.08	0.19		0.66	19.50
INPUT				0.53		1.90	21.08
CHEMICAL FEED	1.12	0.0	0.0				1.12
LUBES & ASPHALTS	0.0						0.0
TOTAL REFINED PRODUCT	13.05						
REFINERY USE & PROCESS LOSS*	0.26	0.0	0.84				1.10
SYNCRUDE MANUFACTURE	0.0	0.0					
USES+LOSSES		0.0					0.0
TOTAL INPUT		0.0					
PRIMARY ENERGY SUPPLY	13.30	3.65	3.92	0.53		1.90	23.30

ALL QUANTITIES ARE GIVEN IN MILLION BARREL PER DAY OIL EQUIVALENT

SUMMARY
(ANNUAL CAPITAL CHARGES - CASE YEAR)

PROBLEM: WAES85D BASE YR: 1975.
CASE: 31JAN77 CASE YR: 1985.

PRODUCERS			OPEC	OCEC
OIL				
PRODUCTION	:	B$	0.84	0.0
REFINING	:	B$	1.25	0.0
COAL				
PRODUCTION	:	B$	0.0	1.51
CONVERSION	:	B$	0.0	0.0
GAS				
PRODUCTION	:	B$	2.86	0.0
LIQUEFACTION	:	B$	1.25	0.0
TOTAL	:	B$	6.20	1.51

CONSUMERS			NORTH AMERICA	EUROPE	JAPAN	REST WOCA
OIL						
PRODUCTION	:	B$	6.98	4.95	0.10	4.52
REFINING	:	B$	3.11	2.03	1.33	2.98
DISTRIBUTION	:	B$	0.79	0.40	0.21	0.52
TOTAL	:	B$	10.88	7.38	1.64	8.01
COAL						
PRODUCTION	:	B$	1.28	1.78	0.08	0.0
CONVERSION	:	B$	0.0	0.0	0.0	0.0
DISTRIBUTION	:	B$	0.14	0.11	0.03	0.11
TOTAL	:	B$	1.42	1.89	0.11	0.11
GAS						
PRODUCTION	:	B$	6.31	3.59	0.12	4.38
REGASIFICATION	:	B$	0.15	0.17	0.08	0.0
DISTRIBUTION	:	B$	0.30	1.14	0.04	2.07
TOTAL	:	B$	6.76	4.90	0.24	6.45
ELECTRICITY						
FUEL OIL	:	B$	0.24	0.0	0.0	0.0
COAL	:	B$	0.0	0.0	0.07	0.60
GAS	:	B$	0.0	0.0	0.23	0.57
NUCLEAR	:	B$	5.94	5.01	1.19	1.07
HYDRO ELEC 1	:	B$	1.16	0.49	0.31	1.40
OTHER	:	B$	0.46	0.60	0.09	0.23
DISTRIBUTION	:	B$	0.0	0.28	0.0	0.0
TOTAL	:	B$	7.80	6.38	1.90	3.86
IND HEATING EQP	:	B$	3.01	2.39	1.25	2.44
DOM HEATING EQP	:	B$	13.11	7.29	1.46	2.17
TOTAL CAPEX	:	B$	42.98	30.24	6.62	23.05

SUMMARY
(PRODUCING REGIONS REVENUES)

PROBLEM: WAES85D BASE YR: 1975.
CASE: 31JAN77 CASE YR: 1985.

PRODUCERS			OPEC	OCEC
GOVERNMENT TAXES *				
CRUDE OIL			11.28	0.0
COAL			0.0	2.00
GAS			8.32	0.0
TARIFFS ***				
CRUDE OIL			12.50	12.50
DISTILLATES			14.20	14.20
FUEL OIL			9.10	9.10
COAL			8.00	8.00
GAS			14.20	14.20
GOVT. TAKE(FROM TAXES)				
OIL	:	B$	146.51	0.0
COAL	:	B$	0.0	3.51
GAS	:	B$	11.54	0.0
TOTAL GOVT. TAKE	:	B$	158.05	3.51
REVENUES FROM TARIFFS				
CRUDE OIL	:	B$	137.27	0.0
DISTILLATES	:	B$	20.53	0.0
FUEL OIL	:	B$	5.10	0.0
COAL	:	B$	0.0	14.05
GAS	:	B$	19.70	0.0
TOTAL REVENUES	:	B$	182.60	14.05

* DOLLAR/BOE

*** DOLLAR/BOE, AVERAGE WORLD PRICE LAID-IN AT PORT OF ENTRY

12

SUMMARY PROBLEM: WAES85D BASE YR: 1975. CASE: 31JAN77 CASE YR: 1985.

(CONSUMING REGIONS COSTS AND REVENUES)

CONSUMERS (TOTAL COSTS)			NORTH AMERICA	EUROPE	JAPAN	REST WOCA
IMPORT BILL						
CRUDE OIL	:	B$	33.34	49.44	28.47	35.14
DISTILLATES	:	B$	20.53	0.0	0.0	0.0
FUEL OIL	:	B$	0.0	0.0	5.10	0.0
COAL	:	B$	0.0	5.63	3.08	10.65
GAS	:	B$	4.97	6.44	2.59	9.33
TOTAL IMPORT BILL	:	B$	58.84	61.51	39.24	55.12
INTERNAL COSTS **						
OIL	:	B$	26.74	26.35	5.83	16.21
COAL	:	B$	18.41	17.66	1.26	1.60
GAS	:	B$	10.87	6.72	0.45	8.04
ELECTRICITY	:	B$	11.32	8.68	2.62	5.21
TOTAL	:	B$	67.33	59.40	10.16	31.05
ENERGY CONS EQUIP	:	B$	16.12	9.68	2.71	4.61
TOTAL INTERNAL	:	B$	83.45	69.09	12.87	35.66
REVENUES FROM EXPORTS						
CRUDE OIL	:	B$	0.0	0.0	0.0	0.0
DISTILLATES	:	B$	0.0	0.0	0.0	0.0
FUEL OIL	:	B$	0.0	0.0	0.0	0.0
COAL	:	B$	3.85	0.0	0.0	0.0
GAS	:	B$	0.0	0.0	0.0	0.0
TOTAL REVENUES	:	B$	3.85	0.0	0.0	0.0
NET CONSUM REG. COST	:	B$	138.45	130.59	52.12	90.79
IMPORT/EXPORT BALANCE						
EQUILIBRIUM	:	B$	54.99	61.51	39.24	55.12
LOANS AT 15.0% PA	:	B$	0.0	0.0	0.0	0.0
SHORTFALL	:	B$	0.0	0.0	0.0	0.0

** CAPEX CHARGED AS A CONSTANT COST CHARGED OVER 100.0% OF THE USEFUL PHYSICAL LIFETIME OF A PLANT AT 15.0% ANNUAL COST OF CAPITAL.

13

CONSUMERS MARGINAL COSTS PROBLEM: WAES85D BASE YR: 1975. CASE: 31JAN77 CASE YR: 1985.

	NORTH AMERICA	EUROPE	JAPAN	REST WOCA
FUEL:				
LIGHT CRUDE	74.47	74.35	73.31	73.37
HEAVY CRUDE	72.94	73.55	72.52	72.57
DISTILLATE	89.56	86.58	85.43	84.19
FUEL OIL	79.59	86.58	85.37	82.46
COAL	49.23	50.19	49.61	49.56
GAS	86.27	85.29	85.62	38.41
ELECTRICITY	187.34	125.02	185.29	130.25
DEMAND:				
INDUSTRIAL	37.34	86.36	86.69	51.28
DOMESTIC	92.32	91.34	90.17	83.67
DISTILLATE TRANSPORT	89.56	86.58	85.43	84.19
ELECTRICITY TRANSPORT	187.34	125.02	185.29	130.25
CHEMICAL FEEDSTOCK	87.82	85.32	85.43	84.19
LUBES & ASPHALTS	79.59	86.58	85.37	82.46
NONENERGY BUNKERS	79.59	86.58	85.37	82.46
GLOBAL ENERGY BUNKER	83.99			

VALUES REPRESENT - DOLLARS / BOE

OIL NORTH AMERICA PROBLEM: WAES85D BASE YR: 1975.
CASE: 31JAN77 CASE YR: 1985.

PRODUCTION	CAPACITY(M BDOE)				COST(DOLLARS/BOE)	
	MIN	MAX	TOTAL	NEW	OP.COST	CAP.CH
ONSHORE	0.0	8.71	8.71	3.15	0.23	2.79
OFFSHORE	0.0	2.99	2.99	2.29	2.40	3.72
OIL TARSANDS	0.0	0.20	0.20	0.20	3.00	8.87

REFINING	CAPACITY(M BDOE)				COST(DOLLARS/BOE)	
	MIN	MAX	TOTAL	NEW	OP.COST	CAP.CH
DISTILLING	0.0	******	19.21	8.96	0.15	0.43
CONVERSION I	0.0	******	6.11	2.95	0.25	0.78
CONVERSION II	0.0	******	2.73	2.12	0.30	1.08
COAL REF. FUEL	0.0	******	0.25	0.19	0.0	0.35
GAS REF. FUEL	0.0	******	0.0	0.0	0.0	0.04

DISTRIBUTION	CAPACITY(M BDOE)				COST(DOLLARS/BOE)	
	MIN	MAX	TOTAL	NEW	OP.COST	CAP.CH
DISTILLATES	0.0	******	17.77	10.49	1.30	0.19
FUEL OIL	0.0	******	4.10	1.19	1.30	0.19

CRUDE OIL
=========

FROM:	PERCENT OF INTAKE	INPUT M BDOE	INPUT M TON OE	YIELD %	OUTPUT M BDOE	OUTPUT M TON OE
INDIGENOUS	61.95	11.90	586.59	100.0	11.90	586.59
IMPORT	38.05	7.31	360.25	100.0	7.31	360.25
COAL CONV	0.0	0.0	0.0	0.0	0.0	0.0
TOTAL	100.00				19.21	946.83

15

TO:	PERCENT OF INTAKE	INPUT M BDOE	INPUT M TON OE
EXPORT	0.0	0.0	0.0
REFINING	100.00	19.21	946.83

REFINING
=======

	DISTILLING CRUDE OIL	CONVERSION I	CONVERSION II
% YIELD DIST	54.0	25.0	70.0
% YIELD RES	46.0	70.0	30.0
% FUEL USE	3.5	3.0	14.0

		M BDOE	M TON OE	PERCENT
INTAKE:	CRUDE	19.21	946.83	***
OUTPUT:	DISTILLATES	13.81	680.72	71.89
	FUEL OIL	4.10	202.25	21.36
REFINERY FUEL:	FUEL OIL	0.99	48.81	5.16
	COAL	0.25	12.20	1.29
OVERALL REFINERY LOSS		1.30	63.86	6.74

DISTILLATES
==========

FROM:	PERCENT OF INTAKE	INPUT M BDOE	INPUT M TON OE	YIELD %	OUTPUT M BDOE	OUTPUT B THERM
IMPORT	22.29	3.96	195.28	100.0	3.96	0.0
REFINING	77.71	0.0	0.0	0.0	13.81	0.0

16

TO:	PERCENT OF INTAKE	INPUT M BDOE	INPUT M TON OE	YIELD %	OUTPUT M BDOE	OUTPUT B THERM
IND. HEATING	5.21	0.93	45.63	100.0	0.93	19.39
DOM. HEATING	14.20	2.52	124.39	100.0	2.52	52.87
TRANSPORT	58.41	10.38	511.66	0.0	0.0	0.0
CHEM. FEED STOCK	17.18	3.05	150.53	0.0	0.0	0.0
DISTR. LOSS	5.00	0.89	43.80	0.0	0.0	0.0

FUEL OIL
========

FROM:	PERCENT OF INTAKE	INPUT M BDOE	INPUT M TON OE	YIELD %	OUTPUT M BDOE	OUTPUT B THERM
IMPORT	0.0	0.0	0.0	0.0	0.0	
REFINING	100.00	0.0	0.0	0.0	4.10	

TO:	PERCENT OF INTAKE	INPUT M BDOE	INPUT M TON OE	YIELD %	OUTPUT M BDOE	OUTPUT B THERM
IND. HEATING	45.80	1.88	92.64	100.0	1.88	39.38
DOM. HEATING	15.38	0.63	31.10	100.0	0.63	13.22
ELECTR. GEN.	24.90	1.02	50.36	35.0	0.36	7.49
ASPHALT & LUBOIL	0.0	0.0	0.0	0.0	0.0	0.0
ENER. BUNK.	0.0	0.0	0.0	0.0	0.0	0.0
NONEN. BUNK.	8.92	0.37	18.04	0.0	0.0	0.0
DIST. LOSS	5.00	0.21	10.11	0.0	0.0	0.0

OIL EUROPE PROBLEM: WAES85D BASE YR: 1975.
CASE: 31JAN77 CASE YR: 1985.

17

PRODUCTION	CAPACITY(M BDOE)				COST(DOLLARS/BOE)	
	MIN	MAX	TOTAL	NEW	OP.COST	CAP.CH
ONSHORE	0.0	0.09	0.09	0.0	0.35	2.10
OFFSHORE	0.0	3.98	3.98	3.88	7.50	3.49
OIL TARSANDS	0.0	0.0	0.0	0.0	3.00	4.32

REFINING	CAPACITY(M BDOE)				COST(DOLLARS/BOE)	
	MIN	MAX	TOTAL	NEW	OP.COST	CAP.CH
DISTILLING	0.0	******	14.91	2.23	0.15	0.43
CONVERSION I	0.0	******	6.56	5.87	0.25	0.78
CONVERSION II	0.0	******	0.0	0.0	0.30	1.08
COAL REF. FUEL	0.0	******	0.14	0.08	0.0	0.35
GAS REF. FUEL	0.0	******	0.0	0.0	0.0	0.04

DISTRIBUTION	CAPACITY(M BDOE)				COST(DOLLARS/BOE)	
	MIN	MAX	TOTAL	NEW	OP.COST	CAP.CH
DISTILLATES	0.0	******	9.43	4.09	1.30	0.19
FUEL OIL	0.0	******	4.57	1.84	1.30	0.19

CRUDE OIL

FROM:	PERCENT OF INTAKE	INPUT M BDOE	INPUT M TON OE	YIELD %	OUTPUT M BDOE	OUTPUT M TON OE
INDIGENOUS	27.30	4.07	200.62	100.0	4.07	200.62
IMPORT	72.70	10.84	534.15	100.0	10.84	534.15
COAL CONV	0.0	0.0	0.0	0.0	0.0	0.0
TOTAL	100.00				14.91	734.77

18

TO:	PERCENT OF INTAKE	INPUT M BDOE	INPUT M TON OE
EXPORT	0.0	0.0	0.0
REFINING	100.00	14.91	734.77

REFINING	DISTILLING CRUDE OIL	CONVERSION I	CONVERSION II
% YIELD DIST	56.0	25.0	70.0
% YIELD RES	44.0	70.0	30.0
% FUEL USE	3.5	3.0	14.0

		M BDOE	M TON OE	PERCENT
INTAKE:	CRUDE	14.91	734.77	***
OUTPUT:	DISTILLATES	9.43	464.95	63.28
	FUEL OIL	4.57	225.30	30.66
REFINERY FUEL:	FUEL OIL	0.57	28.34	3.86
	COAL	0.14	7.08	0.96
OVERALL REFINERY LOSS		0.90	44.52	6.06

DISTILLATES

FROM:	PERCENT OF INTAKE	INPUT M BDOE	INPUT M TON OE	YIELD %	OUTPUT M BDOE	OUTPUT B THERM
IMPORT	0.0	0.0	0.0	0.0	0.0	0.0
REFINING	100.00	0.0	0.0	0.0	9.43	0.0

19

TO:

IND. HEATING	8.73	0.82	40.60	100.0	0.82	17.26
DOM. HEATING	26.77	2.52	124.46	100.0	2.52	52.90
TRANSPORT	45.80	4.32	212.95	0.0	0.0	0.0
CHEM. FEED STOCK	13.70	1.29	63.69	0.0	0.0	0.0
DISTR. LOSS	5.00	0.47	23.25	0.0	0.0	0.0

FUEL OIL
=======

FROM:	PERCENT OF INTAKE	INPUT M BDOE	INPUT M TON OE	YIELD %	OUTPUT M BDOE	OUTPUT B THERM
IMPORT	0.0	0.0	0.0	0.0	0.0	
REFINING	100.00	0.0	0.0	0.0	4.57	

TO:

IND. HEATING	36.59	1.67	82.43	100.0	1.67	35.04
DOM. HEATING	23.68	1.08	53.34	100.0	1.08	22.67
ELECTR. GEN.	0.0	0.0	0.0	0.0	0.0	0.0
ASPHALT & LUBOIL	0.0	0.0	0.0	0.0	0.0	0.0
ENER. BUNK.	4.07	0.19	9.18	0.0	0.0	0.0
NONEN. BUNK.	18.68	0.85	42.10	0.0	0.0	0.0
DIST. LOSS	5.00	0.23	11.27	0.0	0.0	0.0
CHEM. FEED STOCK	11.98	0.55	26.99	0.0	0.0	0.0

20

OIL JAPAN PROBLEM: WAES85D BASE YR: 1975.
CASE: 31JAN77 CASE YR: 1985.
====================================

PRODUCTION	CAPACITY(M BDOE) MIN	MAX	TOTAL	NEW	COST(DOLLARS/BOE) OP.COST	CAP.CH
ONSHORE	0.0	0.01	0.01	0.01	0.35	2.10
OFFSHORE	0.0	0.07	0.07	0.07	0.80	3.49

REFINING	CAPACITY(M BDOE) MIN	MAX	TOTAL	NEW	COST(DOLLARS/BOE) OP.COST	CAP.CH
DISTILLING	0.0	******	6.32	3.11	0.15	0.43
CONVERSION I	0.0	******	3.15	2.94	0.25	0.78
CONVERSION II	0.0	******	0.0	0.0	0.30	1.08
COAL REF. FUEL	0.0	******	0.06	0.06	0.0	0.35
GAS REF. FUEL	0.0	******	0.0	0.0	0.0	0.04

DISTRIBUTION	CAPACITY(M BDOE) MIN	MAX	TOTAL	NEW	COST(DOLLARS/BOE) OP.COST	CAP.CH
DISTILLATES	0.0	******	3.96	2.44	1.30	0.19
FUEL OIL	0.0	******	3.49	0.70	1.30	0.19

CRUDE OIL
=========

FROM:	PERCENT OF INTAKE	INPUT M BDOE	INPUT M TON OE	YIELD %	OUTPUT M BDOE	OUTPUT M TON OE
INDIGENOUS	1.27	0.08	3.94	100.0	0.08	3.94
IMPORT	98.73	6.24	307.59	100.0	6.24	307.59
COAL CONV	0.0	0.0	0.0	0.0	0.0	0.0
TOTAL	100.00				6.32	311.53

TO:	PERCENT OF INTAKE	INPUT M BDOE	INPUT M TON OE
EXPORT	0.0	0.0	0.0
REFINING	100.00	6.32	311.53

21

REFINING	DISTILLING CRUDE OIL	CONVERSION I	CONVERSION II
% YIELD DIST	50.2	25.0	70.0
% YIELD RES	49.8	70.0	30.0
% FUEL USE	3.5	3.0	14.0

		M BDOE	M TON OE	PERCENT
INTAKE:	CRUDE	6.32	311.53	***
OUTPUT:	DISTILLATES	3.96	195.18	62.65
	FUEL OIL	1.95	96.15	30.86
REFINERY FUEL:	FUEL OIL	0.25	12.45	4.00
	COAL	0.06	3.11	1.00
OVERALL REFINERY LOSS		0.41	20.20	6.49

DISTILLATES

FROM:	PERCENT OF INTAKE	INPUT M BDOE	INPUT M TON OE	YIELD %	OUTPUT M BDOE	OUTPUT B THERM
IMPORT	0.0	0.0	0.0	0.0	0.0	0.0
REFINING	100.00	0.0	0.0	0.0	3.96	0.0

TO:						
IND. HEATING	19.45	0.77	37.96	100.0	0.77	16.13
DOM. HEATING	21.25	0.84	41.48	100.0	0.84	17.63
TRANSPORT	35.36	1.40	69.01	0.0	0.0	0.0
CHEM. FEED STOCK	18.94	0.75	36.97	0.0	0.0	0.0
DISTR. LOSS	5.00	0.20	9.76	0.0	0.0	0.0

22

FUEL OIL

FROM:	PERCENT OF INTAKE	INPUT M BDOE	INPUT M TON OE	YIELD %	OUTPUT M BDOE	OUTPUT B THERM
IMPORT	44.06	1.54	75.72	100.0	1.54	
REFINING	55.94	0.0	0.0	0.0	1.95	

TO:						
IND. HEATING	56.79	1.98	97.60	100.0	1.98	41.49
DOM. HEATING	11.36	0.40	19.52	100.0	0.40	8.30
ELECTR. GEN.	14.58	0.51	25.06	35.0	0.18	3.73
ASPHALT & LUBOIL	0.0	0.0	0.0	0.0	0.0	0.0
ENER. BUNK.	1.78	0.06	3.06	0.0	0.0	0.0
NONEN. BUNK.	10.50	0.37	18.04	0.0	0.0	0.0
DIST. LOSS	5.00	0.17	8.59	0.0	0.0	0.0

23

OIL — OPEC — PROBLEM: WAES85D — BASE YR: 1975.
=====================================
CASE: 31JAN77 — CASE YR: 1985.

PRODUCTION	CAPACITY(M BDOE)				COST(DOLLARS/BOE)	
	MIN	MAX	TOTAL	NEW	OP.COST	CAP.CH
ONSHORE	0.0	30.00	30.00	15.75	11.38	0.12
OFFSHORE	0.0	6.00	6.00	1.88	11.29	0.21
OIL TARSANDS	0.0	0.0	0.0	0.0	3.00	8.65

REFINING	CAPACITY(M BDOE)				COST(DOLLARS/BOE)	
	MIN	MAX	TOTAL	NEW	OP.COST	CAP.CH
DISTILLING	0.0	******	5.91	3.37	0.15	0.43
CONVERSION I	0.0	******	2.60	2.54	0.25	0.78
CONVERSION II	0.0	******	0.0	0.0	0.30	1.08
COAL REF. FUEL	0.0	******	0.0	0.0	0.0	0.35
GAS REF. FUEL	0.0	******	0.0	0.0	0.0	0.04

CRUDE OIL
=========

FROM:	PERCENT OF INTAKE	INPUT M BDOE	INPUT M TON OE	YIELD %	OUTPUT M BDOE	OUTPUT M TON OE
INDIGENOUS	100.00	36.00	1774.55	100.0	36.00	1774.55
IMPORT	0.0	0.0	0.0	0.0	0.0	0.0
COAL CONV	0.0	0.0	0.0	0.0	0.0	0.0
TOTAL	100.00				36.00	1774.55

TO:	PERCENT OF INTAKE	INPUT M BDOE	INPUT M TON OE
EXPORT	83.58	30.09	1483.08
REFINING	16.42	5.91	291.46

24

REFINING ========	DISTILLING CRUDE OIL	CONVERSION I	CONVERSION II
% YIELD DIST	56.0	25.0	70.0
% YIELD RES	44.0	70.0	30.0
% FUEL USE	3.5	3.0	14.0

		M BDOE	M TON OE	PERCENT
INTAKE:	CRUDE	5.91	291.46	***
OUTPUT:	DISTILLATES	3.96	195.28	67.00
	FUEL OIL	1.54	75.72	25.98
REFINERY FUEL:	FUEL OIL	0.29	14.05	4.82
OVERALL REFINERY LOSS		0.42	20.46	7.02

DISTILLATES
===========

FROM:	PERCENT OF INTAKE	INPUT M BDOE	INPUT M TON OE	YIELD %	OUTPUT M BDOE	OUTPUT B THERM
IMPORT	0.0	0.0	0.0	0.0	0.0	0.0
REFINING	100.00	0.0	0.0	0.0	3.96	0.0

TO:						
EXPORT	100.00	3.96	195.28	0.0	0.0	0.0

25

FUEL OIL

FROM:	PERCENT OF INTAKE	INPUT M BDOE	INPUT M TON OE	YIELD %	OUTPUT M BDOE	OUTPUT B THERM
IMPORT	0.0	0.0	0.0	0.0	0.0	
REFINING	100.00	0.0	0.0	0.0	1.54	

TO:						
ENER. BUNK.	0.0	0.0	0.0	0.0	0.0	0.0
NONEN. BUNK.	0.0	0.0	0.0	0.0	0.0	0.0
EXPORT	100.00	1.54	75.72	0.0	0.0	0.0

26

OIL REST WOCA PROBLEM: WAES85D BASE YR: 1975.
CASE: 31JAN77 CASE YR: 1985.

PRODUCTION	CAPACITY(M BDOE) MIN	MAX	TOTAL	NEW	COST(DOLLARS/BOE) OP.COST	CAP.CH
ONSHORE	0.0	3.00	3.00	1.76	0.23	2.79
OFFSHORE	0.0	2.60	2.60	2.00	0.40	3.72
OIL TARSANDS	0.0	0.0	0.0	0.0	3.00	8.65

REFINING	CAPACITY(M BDOE) MIN	MAX	TOTAL	NEW	COST(DOLLARS/BOE) OP.COST	CAP.CH
DISTILLING	0.0	******	13.30	6.63	0.15	0.43
CONVERSION I	0.0	******	5.12	4.58	0.25	0.78
CONVERSION II	0.0	******	1.57	1.57	0.30	1.08
COAL REF. FUEL	0.0	******	0.0	0.0	0.0	0.35
GAS REF. FUEL	0.0	******	0.84	0.78	0.0	0.04

DISTRIBUTION	CAPACITY(M BDOE) MIN	MAX	TOTAL	NEW	COST(DOLLARS/BOE) OP.COST	CAP.CH
DISTILLATES	0.0	******	8.99	6.26	1.30	0.19
FUEL OIL	0.0	******	4.06	1.33	1.30	0.19

CRUDE OIL

FROM:	PERCENT OF INTAKE	INPUT M BDOE	INPUT M TON OE	YIELD %	OUTPUT M BDOE	OUTPUT M TON OE
INDIGENOUS	42.10	5.60	276.04	100.0	5.60	276.04
IMPORT	57.90	7.70	379.69	100.0	7.70	379.69
COAL CONV	0.0	0.0	0.0	0.0	0.0	0.0
TOTAL	100.00				13.30	655.73

27

TO:	PERCENT OF INTAKE	INPUT M BDOE	M TON OE
EXPORT	0.0	0.0	0.0
REFINING	100.00	13.30	655.73

REFINING	DISTILLING CRUDE OIL	CONVERSION I	II
% YIELD DIST	49.7	25.0	70.0
% YIELD RES	50.3	70.0	30.0
% FUEL USE	3.5	3.0	14.0

		M BDOE	M TON OE	PERCENT
INTAKE:	CRUDE	13.30	655.73	***
OUTPUT:	DISTILLATES	8.99	443.04	67.56
	FUEL OIL	4.06	200.06	30.51
REFINERY FUEL:	FUEL OIL	0.0	0.0	0.0
	GAS	0.84	41.39	6.31
OVERALL REFINERY LOSS		0.26	12.63	1.93

DISTILLATES

FROM:	PERCENT OF INTAKE	INPUT M BDOE	M TON OE	YIELD %	OUTPUT M BDOE	B THERM
IMPORT	0.0	0.0	0.0	0.0	0.0	0.0
REFINING	100.00	0.0	0.0	0.0	8.99	0.0

28

TO:	PERCENT OF INTAKE	INPUT M BDOE	M TON OE	YIELD %	OUTPUT M BDOE	B THERM
IND. HEATING	10.74	0.96	47.56	100.0	0.96	20.22
DOM. HEATING	8.16	0.73	36.16	100.0	0.73	15.37
TRANSPORT	63.64	5.72	281.96	0.0	0.0	0.0
CHEM. FEED STOCK	12.46	1.12	55.21	0.0	0.0	0.0
DISTR. LOSS	5.00	0.45	22.15	0.0	0.0	0.0

FUEL OIL

FROM:	PERCENT OF INTAKE	INPUT M BDOE	M TON OE	YIELD %	OUTPUT M BDOE	B THERM
IMPORT	0.0	0.0	0.0	0.0	0.0	
REFINING	100.00	0.0	0.0	0.0	4.06	

TO:	PERCENT OF INTAKE	INPUT M BDOE	M TON OE	YIELD %	OUTPUT M BDOE	B THERM
IND. HEATING	48.27	1.96	96.57	100.0	1.96	41.05
DOM. HEATING	7.75	0.31	15.50	100.0	0.31	6.59
ELECTR. GEN.	8.76	0.36	17.53	35.0	0.12	2.61
ASPHALT & LUBOIL	0.0	0.0	0.0	0.0	0.0	0.0
ENER. BUNK.	9.18	0.37	18.36	0.0	0.0	0.0
NONEN. BUNK.	21.04	0.85	42.10	0.0	0.0	0.0
DIST. LOSS	5.00	0.20	10.00	0.0	0.0	0.0

COAL NORTH AMERICA PROBLEM: WAES85D BASE YR: 1975.
CASE: 31JAN77 CASE YR: 1985.

29

	CAPACITY(M BDOE)				COST(DOLLAR/BOE)	
	MIN	MAX	TOTAL	NEW	OP.COST	CAP.CH.
STRIP MINING	0.0	6.19	6.19	3.64	2.45	0.19
DEEP MINING	0.0	3.69	3.69	1.32	4.12	0.72
PRIMARY TRANSPORT	0.0	******	9.88	4.45	0.69	0.42
GASIFICATION	0.0	0.15	0.0	0.0	1.00	9.56
LIQUEFACTION	0.0	0.0	0.0	0.0	1.00	6.92
DISTRIBUTION	0.0	******	8.36	2.72	1.12	0.14

FROM:	PERCENT OF INTAKE	INPUT M BDOE	INPUT MTON CE	YIELD %	OUTPUT MBDOE	OUTPUT MTON CE
INDIGENOUS	100.00	9.88	750.88	98.00	9.68	735.86
IMPORTS	0.0	0.0	0.0	0.0	0.0	0.0
TOTAL	100.0				9.7	735.9

TO :	PERCENT OF INTAKE	INPUT M BDOE	INPUT MTON CE	YIELD %	OUTPUT M BDOE	OUTPUT B THERM
EXPORT	13.62	1.32	100.22	0.0	0.0	0.0
SYNCRUDE	0.0	0.0	0.0	0.0	0.0	0.0
SYNGAS	0.0	0.0	0.0	0.0	0.0	0.0
OIL REF.FUEL	2.56	0.25	18.81	0.0	0.0	0.0
IND.HEAT	46.35	4.49	341.09	100.00	4.49	94.25
DOM.HEAT	5.40	0.52	39.75	100.00	0.52	10.98
ELECTR.GEN	29.17	2.82	214.69	35.00	0.99	20.76
DISTR.LOSS	1.73	0.17	12.71	0.0	0.0	0.0
CHEMICAL FEED STOCK	1.17	0.11	8.60	100.00	0.11	2.38

COAL EUROPE PROBLEM: WAES85D BASE YR: 1975.
CASE: 31JAN77 CASE YR: 1985.

30

	CAPACITY(M BDOE)				COST(DOLLAR/BOE)	
	MIN	MAX	TOTAL	NEW	OP.COST	CAP.CH.
STRIP MINING	0.0	1.09	1.09	0.36	3.75	0.43
DEEP MINING	0.0	3.24	3.24	1.54	9.00	2.16
PRIMARY TRANSPORT	0.0	******	4.33	1.65	0.70	0.84
GASIFICATION	0.0	0.21	0.0	0.0	1.00	9.56
LIQUEFACTION	0.0	0.01	0.0	0.0	1.00	6.92
DISTRIBUTION	0.0	******	6.17	2.23	1.12	0.14

FROM:	PERCENT OF INTAKE	INPUT M BDOE	INPUT MTON CE	YIELD %	OUTPUT MBDOE	OUTPUT MTON CE
INDIGENOUS	69.18	4.33	329.08	98.00	4.24	322.50
IMPORTS	30.82	1.93	146.59	100.00	1.93	146.59
TOTAL	100.0				6.2	469.1

TO :	PERCENT OF INTAKE	INPUT M BDOE	INPUT MTON CE	YIELD %	OUTPUT M BDOE	OUTPUT B THERM
SYNCRUDE	0.0	0.0	0.0	0.0	0.0	0.0
SYNGAS	0.0	0.0	0.0	0.0	0.0	0.0
OIL REF.FUEL	2.33	0.14	10.92	0.0	0.0	0.0
IND.HEAT	40.44	2.50	189.70	100.00	2.50	52.42
DOM.HEAT	23.10	1.43	108.38	100.00	1.43	29.95
ELECTR.GEN	31.06	1.92	145.72	35.00	0.67	14.09
DISTR.LOSS	2.00	0.12	9.38	0.0	0.0	0.0
CHEMICAL FEED STOCK	1.06	0.07	4.99	100.00	0.07	1.38

31

COAL JAPAN PROBLEM: WAES85D BASE YR: 1975.
CASE: 31JAN77 CASE YR: 1985.

	CAPACITY(M BDOE) MIN	MAX	TOTAL	NEW	COST(DOLLAR/BOE) OP.COST	CAP.CH.
STRIP MINING	0.0	0.0	0.0	0.0	3.75	0.56
DEEP MINING	0.0	0.26	0.26	0.08	5.75	2.16
PRIMARY TRANSPORT	0.0	******	0.26	0.06	0.70	0.84
GASIFICATION	0.0	0.02	0.0	0.0	1.00	9.56
LIQUEFACTION	0.0	0.0	0.0	0.0	1.00	6.92
DISTRIBUTION	0.0	******	1.31	0.58	1.12	0.14

FROM:	PERCENT OF INTAKE	INPUT M BODE	MTON CE	YIELD %	OUTPUT MBDOE	MTON CE
INDIGENOUS	19.78	0.26	19.76	98.00	0.25	19.36
IMPORTS	80.22	1.05	80.15	100.00	1.05	80.15
TOTAL	100.0				1.3	99.5

TO :	PERCENT OF INTAKE	INPUT M BODE	MTON CE	YIELD %	OUTPUT M BDOE	B THERM
SYNCRUDE	0.0	0.0	0.0	0.0	0.0	0.0
SYNGAS	0.0	0.0	0.0	0.0	0.0	0.0
OIL REF.FUEL	4.82	0.06	4.80	0.0	0.0	0.0
IND.HEAT	80.95	1.06	80.56	100.00	1.06	22.26
DOM.HEAT	0.64	0.01	0.64	100.00	0.01	0.18
ELECTR.GEN	11.58	0.15	11.53	35.00	0.05	1.11
DISTR.LOSS	2.00	0.03	1.99	0.0	0.0	0.0

32

COAL OCEC PROBLEM: WAES85D BASE YR: 1975.
CASE: 31JAN77 CASE YR: 1985.

	CAPACITY(M BDOE) MIN	MAX	TOTAL	NEW	COST(DOLLAR/BOE) OP.COST	CAP.CH.
STRIP MINING	0.0	2.00	2.00	1.39	3.25	0.56
DEEP MINING	0.0	2.91	2.91	1.39	4.75	0.86
PRIMARY TRANSPORT	0.0	******	4.91	2.56	0.70	0.84
GASIFICATION	0.0	0.0	0.0	0.0	1.00	9.56
LIQUEFACTION	0.0	0.0	0.0	0.0	1.00	6.92

FROM:	PERCENT OF INTAKE	INPUT M BODE	MTON CE	YIELD %	OUTPUT MBDOE	MTON CE
INDIGENOUS	100.00	4.91	373.16	98.00	4.81	365.70
TOTAL	100.0				4.8	365.7

TO :	PERCENT OF INTAKE	INPUT M BODE	MTON CE	YIELD %	OUTPUT M BDOE	B THERM
EXPORT	100.00	4.81	365.70	0.0	0.0	0.0
SYNCRUDE	0.0	0.0	0.0	0.0	0.0	0.0
SYNGAS	0.0	0.0	0.0	0.0	0.0	0.0

33

COAL — REST WOCA

PROBLEM: WAES85D — CASE: 31JAN77 — BASE YR: 1975. — CASE YR: 1985.

	CAPACITY(M BDOE) MIN	MAX	TOTAL	NEW	COST(DOLLAR/BOE) OP.COST	CAP.CH.
GASIFICATION	0.0	0.50	0.0	0.0	1.00	9.56
LIQUEFACTION	0.0	0.0	0.0	0.0	1.00	6.92
DISTRIBUTION	0.0	******	3.65	2.07	1.12	0.14

FROM:	PERCENT OF INTAKE	INPUT M BDOE	MTON CE	YIELD %	OUTPUT MBDOE	MTON CE
IMPORTS	100.00	3.65	277.18	100.00	3.65	277.18
TOTAL	100.0				3.6	277.2

TO :	PERCENT OF INTAKE	INPUT M BDOE	MTON CE	YIELD %	OUTPUT M BDOE	B THERM
SYNCRUDE	0.0	0.0	0.0	0.0	0.0	0.0
SYNGAS	0.0	0.0	0.0	0.0	0.0	0.0
OIL REF.FUEL	0.0	0.0	0.0	0.0	0.0	0.0
IND.HEAT	60.13	2.19	166.67	100.00	2.19	46.05
DOM.HEAT	14.37	0.52	39.82	100.00	0.52	11.00
ELECTR.GEN	23.50	0.86	65.14	35.00	0.30	6.30
DISTR.LOSS	2.00	0.07	5.54	0.0	0.0	0.0

34

GAS — NORTH AMERICA

PROBLEM: WAES85D — CASE: 31JAN77 — BASE YR: 1975. — CASE YR: 1985.

	CAPACITY(M BDOE) MIN	MAX	TOTAL	NEW	COST(DOLLAR/BOE) OP.COST	CAP.CH.
ON SHORE PRODUCTION	0.0	5.78	5.78	1.06	0.25	6.01
OFF SHORE PRODUCTION	0.0	3.03	3.03	2.09	0.30	5.21
LIQUEFACTION	0.0	1.00	0.0	0.0	0.35	1.49
TRANSPORT BY PIPE	0.0	******	8.81	0.01	0.01	1.65
REGASIFICATION	0.0	1.50	0.92	0.86	0.05	0.48
SYNGAS FROM COAL	0.0	0.15	0.0	0.0	1.00	9.56
GAS GRID	0.0	******	8.81	0.40	1.00	2.06

FROM:	PERCENT OF INTAKE	INPUT M BDOE	B CUB M	YIELD %	OUTPUT M BDOE	B CUB M
INDIGENOUS	90.19	8.81	502.17	90.00	7.93	451.95
IMPORT LNG	9.81	0.96	54.62	96.00	0.92	52.44
SYN GAS	0.0	0.0	0.0	0.0	0.0	0.0
TOTAL INTO GRID		9.77			8.85	504.39

TO :	PERCENT OF INTAKE	INPUT M BDOE	B CUB M	YIELD %	OUTPUT M BDOE	B THERM
ELECTR.GEN	0.0	0.0	0.0	0.0	0.0	0.0
IND.HEAT	39.69	3.51	200.22	100.00	3.51	73.76
DOM.HEAT	47.10	4.17	237.56	100.00	4.17	87.52
REF.FUEL	0.0	0.0	0.0	0.0	0.0	0.0
DISTR.LOSS	5.98	0.53	30.14	0.0	0.0	0.0
REGASIFICATION LOSS	0.42	0.04	2.10	100.00	0.04	0.77
CHEMICAL FEED STOCK	6.82	0.60	34.38	100.00	0.60	12.67

35

GAS EUROPE PROBLEM: WAES85D BASE YR: 1975.
CASE: 31JAN77 CASE YR: 1985.

	CAPACITY(M BDOE)				COST(DOLLAR/BOE)	
	MIN	MAX	TOTAL	NEW	OP.COST	CAP.CH.
ON SHORE PRODUCTION	0.0	1.93	1.93	0.94	0.15	3.72
OFF SHORE PRODUCTION	0.0	1.51	1.51	1.11	0.25	4.66
LIQUEFACTION	0.0	0.01	0.0	0.0	0.35	1.49
TRANSPORT BY PIPE	0.0	******	3.44	0.71	0.01	1.65
REGASIFICATION	0.0	1.00	1.00	0.99	0.05	0.48
IMPORT BY PIPE	0.0	******	0.20	0.04	3.00	2.47
SYNGAS FROM COAL	0.0	0.21	0.0	0.0	1.00	9.56
GAS GRID	0.0	******	4.24	1.51	1.00	2.06

FROM:	PERCENT OF INTAKE	INPUT M BDOE	INPUT B CUB M	YIELD %	OUTPUT M BDOE	OUTPUT B CUB M
INDIGENOUS	73.48	3.44	196.08	90.00	3.10	176.47
IMPORT LNG	22.25	1.04	59.37	96.00	1.00	57.00
EAST BLOC	0.0	0.0	0.0	0.0	0.0	0.0
IMPORT PIPE	4.27	0.20	11.40	90.00	0.18	10.26
SYN GAS	0.0	0.0	0.0	0.0	0.0	0.0
TOTAL INTO GRID		4.68			4.28	243.73

TO :	PERCENT OF INTAKE	INPUT M BDOE	INPUT B CUB M	YIELD %	OUTPUT M BDOE	OUTPUT B THERM
ELECTR.GEN	5.45	0.23	13.29	35.00	0.08	1.71
IND.HEAT	48.64	2.08	118.56	100.00	2.08	43.68
DOM.HEAT	32.37	1.38	78.89	100.00	1.38	29.06
REF.FUEL	0.0	0.0	0.0	0.0	0.0	0.0
DISTR.LOSS	5.94	0.25	14.49	0.0	0.0	0.0
REGASIFICATION LOSS	0.94	0.04	2.28	100.00	0.04	0.84
CHEMICAL FEED STOCK	6.66	0.28	16.23	100.00	0.28	5.98

36

GAS JAPAN PROBLEM: WAES85D BASE YR: 1975.
CASE: 31JAN77 CASE YR: 1985.

	CAPACITY(M BDOE)				COST(DOLLAR/BOE)	
	MIN	MAX	TOTAL	NEW	OP.COST	CAP.CH.
ON SHORE PRODUCTION	0.0	0.0	0.0	0.0	0.15	3.72
OFF SHORE PRODUCTION	0.0	0.07	0.07	0.07	0.25	4.66
TRANSPORT BY PIPE	0.0	******	0.07	0.0	0.01	1.65
REGASIFICATION	0.0	2.00	0.48	0.47	0.05	0.48
SYNGAS FROM COAL	0.0	0.02	0.0	0.0	1.00	9.56
GAS GRID	0.0	******	0.52	0.06	1.00	2.06

FROM:	PERCENT OF INTAKE	INPUT M BDOE	INPUT B CUB M	YIELD %	OUTPUT M BDOE	OUTPUT B CUB M
INDIGENOUS	12.28	0.07	3.99	90.00	0.06	3.59
IMPORT LNG	87.72	0.50	28.50	96.00	0.48	27.36
SYN GAS	0.0	0.0	0.0	0.0	0.0	0.0
TOTAL INTO GRID		0.57			0.54	30.95

TO :	PERCENT OF INTAKE	INPUT M BDOE	INPUT B CUB M	YIELD %	OUTPUT M BDOE	OUTPUT B THERM
ELECTR.GEN	37.45	0.20	11.59	35.00	0.07	1.49
IND.HEAT	19.72	0.11	6.10	100.00	0.11	2.25
DOM.HEAT	33.51	0.18	10.37	100.00	0.18	3.82
REF.FUEL	0.0	0.0	0.0	0.0	0.0	0.0
DISTR.LOSS	5.79	0.03	1.79	0.0	0.0	0.0
REGASIFICATION LOSS	3.54	0.02	1.09	100.00	0.02	0.40

37

GAS OPEC

PROBLEM: WAES85D BASE YR: 1975.
CASE: 31JAN77 CASE YR: 1985.

	CAPACITY(M BDOE)				COST(DOLLAR/BOE)	
	MIN	MAX	TOTAL	NEW	OP.COST	CAP.CH.
ON SHORE PRODUCTION	0.0	3.00	3.00	0.52	8.47	3.03
OFF SHORE PRODUCTION	0.0	2.00	1.78	1.68	7.78	3.72
LIQUEFACTION	0.0	2.50	2.50	2.30	0.35	1.49
TRANSPORT BY PIPE	0.0	******	2.78	2.47	0.01	0.00
REGASIFICATION	0.0	0.0	0.0	0.0	0.05	0.48

FROM:	PERCENT OF	INPUT		YIELD	OUTPUT	
	INTAKE	M BDOE	B CUB M	%	M BDOE	B CUB M
INDIGENOUS	100.00	4.78	272.33	90.00	4.30	245.10
TOTAL INTO GRID		4.78			4.30	245.10

TO :	PERCENT OF	INPUT		YIELD	OUTPUT	
	INTAKE	M BDOE	B CUB M	%	M BDOE	B THERM
REF.FUEL	0.0	0.0	0.0	0.0	0.0	0.0
LNG EXPORT	58.14	2.50	142.50	80.00	2.00	42.00
P/L EXPORT	41.86	1.80	102.60	100.00	1.80	37.80

38

GAS REST WOCA

PROBLEM: WAES85D BASE YR: 1975.
CASE: 31JAN77 CASE YR: 1985.

	CAPACITY(M BDOE)				COST(DOLLAR/BOE)	
	MIN	MAX	TOTAL	NEW	OP.COST	CAP.CH.
ON SHORE PRODUCTION	0.0	1.00	1.00	0.60	0.25	6.01
OFF SHORE PRODUCTION	0.0	1.30	1.30	1.15	0.30	5.21
TRANSPORT BY PIPE	0.0	******	2.30	1.44	0.01	1.65
REGASIFICATION	0.0	1.00	0.0	0.0	0.05	0.48
IMPORT BY PIPE	0.0	1.80	1.80	0.48	0.07	1.65
SYNGAS FROM COAL	0.0	0.50	0.0	0.0	1.00	9.56
GAS GRID	0.0	******	3.69	2.76	1.00	2.06

FROM:	PERCENT OF	INPUT		YIELD	OUTPUT	
	INTAKE	M BDOE	B CUB M	%	M BDOE	B CUB M
INDIGENOUS	56.10	2.30	131.10	90.00	2.07	117.99
IMPORT PIPE	43.90	1.80	102.60	90.00	1.62	92.34
SYN GAS	0.0	0.0	0.0	0.0	0.0	0.0
TOTAL INTO GRID		4.10			3.69	210.33

TO :	PERCENT OF	INPUT		YIELD	OUTPUT	
	INTAKE	M BDOE	B CUB M	%	M BDOE	B THERM
ELECTR.GEN	17.42	0.64	36.64	35.00	0.22	4.72
IND.HEAT	39.62	1.46	83.33	100.00	1.46	30.70
DOM.HEAT	14.20	0.52	29.87	100.00	0.52	11.00
REF.FUEL	22.76	0.84	47.87	0.0	0.0	0.0
DISTR.LOSS	6.00	0.22	12.62	0.0	0.0	0.0

39

ELECTRICITY GENERATION NORTH AMERICA — PROBLEM: WAES85D — CASE: 31JAN77 — BASE YR: 1975. — CASE YR: 1985.

FUEL	INPUT M BDOE	CAPACITY(INPUT M BDOE) NEW	SURPLUS	MAX	YIELD %	OUTPUT M BDOE
FUEL OIL	1.02	0.16	0.0	******	35.00	0.36
COAL	2.82	0.0	1.92	******	35.00	0.99
GAS	0.0	0.0	2.34	******	35.00	0.0
NUCLEAR	3.27	2.83	0.0	3.27	35.00	1.14
HYDRO ELEC 1	3.00	0.79	0.0	3.00	35.00	1.05
OTHER	0.10	0.10	0.0	0.10	35.00	0.03
TOTAL						3.6

FUEL	% OF TOTAL OUTPUT MIN	MAX	ACTUAL	COST(DOLLAR/BOE) OP.COST	CAP.CH.	DIFFERNT. COST
FUEL OIL	10.00	20.00	10.00	1.10	4.07	-98.39
COAL	20.00	60.00	27.65	1.10	4.65	0.0
GAS	5.00	20.00	0.0	1.00	3.34	-100.00
NUCLEAR	10.00	50.00	32.01	0.80	5.74	0.0
HYDRO ELEC 1	10.00	30.00	29.36	0.20	4.03	0.0
OTHER	0.0	5.00	0.98	0.35	12.66	0.0

TOTAL GENERATED : 2253.TWH/Y
INSTALLED AVERAGE CAPACITY: 358520. MW

TRANSMISSION&DISTRIBUTION:	M BDOE	MW
TOTAL:	3.6	253078.0
NEW :	0.0	0.0

CAP.CH: 2.14 DOLLAR/BOE
YIELD : 85.0 %

BREAKDOWN:	M BDOE	MW	TWH/Y	%
IND HEAT,LIGHT AND POWER	0.4	29329.3	256.9	13.6
DOM HEAT,LIGHT AND POWER	2.6	185079.2	1621.3	86.0
TRANSPORT ELEC	0.0	707.8	6.2	0.3

40

ELECTRICITY GENERATION EUROPE

PROBLEM: WAES85D CASE: 31JAN77

BASE YR: 1975. CASE YR: 1985.

FUEL	INPUT M BDOE	CAPACITY(INPUT M BDOE) NEW	SURPLUS	MAX	YIELD %	OUTPUT M BDOE
FUEL OIL	0.0	0.0	2.44	******	35.00	0.0
COAL	1.92	0.0	0.67	******	35.00	0.67
GAS	0.23	0.0	0.05	******	35.00	0.08
NUCLEAR	2.61	2.39	0.0	2.61	35.00	0.91
HYDRO ELEC 1	1.97	0.33	0.0	1.97	35.00	0.69
OTHER	0.13	0.13	0.0	0.13	35.00	0.05
TOTAL						2.4

FUEL	% OF TOTAL OUTPUT MIN	MAX	ACTUAL	COST(DOLLAR/BOE) OP.COST	CAP.CH.	DIFFERNT. COST
FUEL OIL	10.00	30.00	0.0	1.10	4.07	-100.00
COAL	15.00	40.00	27.95	1.10	4.65	0.0
GAS	4.00	10.00	3.40	1.00	3.34	-100.00
NUCLEAR	25.00	35.00	38.04	0.80	5.74	100.00
HYDRO ELEC 1	15.00	22.00	28.72	0.20	4.03	100.00
OTHER	0.0	5.00	1.89	0.35	12.66	0.0

TOTAL GENERATED : 1513.TWH/Y
INSTALLED AVERAGE CAPACITY: 248460. MW

TRANSMISSION&DISTRIBUTION:	M BDOE	MW
TOTAL:	2.4	169945.7
NEW :	0.4	25079.4

CAP.CH: 2.14 DOLLAR/BOE
YIELD : 85.0 %

BREAKDOWN:	M BDOE	MW	TWH/Y	%
IND HEAT,LIGHT AND POWER	1.2	88328.4	773.8	61.1
DOM HEAT,LIGHT AND POWER	0.7	50463.3	442.1	34.9
TRANSPORT ELEC	0.1	5662.1	49.6	3.9

41

ELECTRICITY GENERATION JAPAN PROBLEM: WAES85D BASE YR: 1975. CASE: 31JAN77 CASE YR: 1985.

FUEL	INPUT M BDOE	CAPACITY(INPUT M BDOE) NEW	SURPLUS	MAX	YIELD %	OUTPUT M BDOE
FUEL OIL	0.51	0.0	0.15	******	35.00	0.18
COAL	0.15	0.04	0.0	******	35.00	0.05
GAS	0.20	0.19	0.0	******	35.00	0.07
NUCLEAR	0.61	0.57	0.0	0.61	35.00	0.21
HYDRO ELEC 1	0.54	0.21	0.0	0.54	35.00	0.19
OTHER	0.02	0.02	0.0	0.02	35.00	0.01
TOTAL						0.7

FUEL	% OF TOTAL OUTPUT MIN	MAX	ACTUAL	COST(DOLLAR/BOE) OP.COST	CAP.CH.	DIFFERNT. COST
FUEL OIL	25.00	60.00	25.00	1.10	4.07	-88.88
COAL	5.00	15.00	7.46	1.10	4.65	0.0
GAS	10.00	25.00	10.00	1.00	3.34	-98.86
NUCLEAR	10.00	30.00	30.00	0.80	5.74	27.97
HYDRO ELEC 1	15.00	25.00	26.56	0.20	4.03	100.00
OTHER	0.0	5.00	0.98	0.35	12.66	0.0

TOTAL GENERATED : 448.TWH/Y
INSTALLED AVERAGE CAPACITY: 53983. MW

TRANSMISSION&DISTRIBUTION:	M BDOE	MW
TOTAL:	0.7	50368.9
NEW :	0.0	0.0

CAP.CH: 2.14 DOLLAR/BOE
YIELD : 85.0 %

BREAKDOWN:	M BDOE	MW	TWH/Y	%
IND HEAT,LIGHT AND POWER	0.3	22854.7	200.2	53.4
DOM HEAT,LIGHT AND POWER	0.3	17835.6	156.2	41.7
TRANSPORT ELEC	0.0	2123.3	18.6	5.0

42

ELECTRICITY GENERATION REST WOCA PROBLEM: WAES85D BASE YR: 1975. CASE: 31JAN77 CASE YR: 1985.

FUEL	INPUT M BDOE	CAPACITY(INPUT M BDOE) NEW	SURPLUS	MAX	YIELD %	OUTPUT M BDOE
FUEL OIL	0.36	0.0	1.08	******	35.00	0.12
COAL	0.86	0.35	0.0	******	35.00	0.30
GAS	0.64	0.46	0.0	******	35.00	0.22
NUCLEAR	0.53	0.51	0.0	0.53	35.00	0.19
HYDRO ELEC 1	1.85	0.95	0.0	1.85	35.00	0.65
OTHER	0.05	0.05	0.0	0.05	35.00	0.02
TOTAL						1.5

FUEL	% OF TOTAL OUTPUT MIN	MAX	ACTUAL	COST(DOLLAR/BOE) OP.COST	CAP.CH.	DIFFERNT. COST
FUEL OIL	10.00	30.00	8.30	1.10	4.07	-100.00
COAL	20.00	60.00	20.00	1.10	4.65	-19.29
GAS	3.00	15.00	15.00	1.00	3.34	16.58
NUCLEAR	5.00	35.00	12.37	0.80	5.74	0.0
HYDRO ELEC 1	25.00	40.00	43.17	0.20	4.03	100.00
OTHER	0.0	5.00	1.17	0.35	12.66	0.0

TOTAL GENERATED : 945.TWH/Y
INSTALLED AVERAGE CAPACITY: 132970. MW

TRANSMISSION&DISTRIBUTION:	M BDOE	MW
TOTAL:	1.5	106164.0
NEW :	0.0	0.0

CAP.CH: 2.14 DOLLAR/BOE
YIELD : 85.0 %

BREAKDOWN:	M BDOE	MW	TWH/Y	%
IND HEAT,LIGHT AND POWER	0.7	51737.3	453.2	57.3
DOM HEAT,LIGHT AND POWER	0.5	37086.6	324.9	41.1
TRANSPORT ELEC	0.0	1415.5	12.4	1.6

INDUSTRIAL MARKET NORTH AMERICA PROBLEM: WAES85D BASE YR: 1975.
CASE: 31JAN77 CASE YR: 1985.

FUEL	INPUT M BDOE	CAPACITY(INPUT M BODE) NEW	SURPLUS	MAX	YIELD %	OUTPUT M BDOE
LIQUID	2.80	1.02	0.0	******	100.00	2.80
COAL	4.49	3.25	0.0	******	100.00	4.49
GAS	3.51	1.23	0.0	******	100.00	3.51
ELECTRICITY	0.41	0.0	0.68	******	100.00	0.41
TOTAL	11.2					11.2

FUEL	% OF TOTAL OUTPUT MIN	MAX	ACTUAL	COST(DOLLAR/BOE) OP.COST	CAP.CH.	DIFFERNT. COST
LIQUID	10.00	25.00	25.00	0.0	1.30	3.15
COAL	25.00	40.00	40.00	0.0	1.72	36.39
GAS	25.00	35.00	31.31	0.0	1.07	0.0
ELECTRICITY	12.00	25.00	3.69	0.0	1.40	-100.00

BREAKDOWN LIQUID :	FUEL OIL	DIST.
MBODE	1.9	0.9
%	67.0	33.0

DEMAND

	DEMAND	ACTUAL	DEFICIT	% OF DEM.
HEAT,LIGHT AND POWER :	11.2	11.2	0.0	0.0
CHEMICAL FEED STOCK :	3.8	3.8	0.0	0.0
LUBE OIL AND ASPHALT :	0.0	0.0	0.0	0.0

BREAKDOWN CHEMICAL FEED STOCK :	DIST.	FUEL OIL	COAL	GAS
MBDOE :	3.05	0.0	0.11	0.60
% :	81.00	0.0	3.00	16.00

TRANSPORT MARKET NORTH AMERICA PROBLEM: WAES85D BASE YR: 1975.
CASE: 31JAN77 CASE YR: 1985.

DEMAND

	DEMAND	ACTUAL	DEFICIT	% OF DEM.
TRANSPORT (DISTILLATES) :	10.4	10.4	0.0	0.0
TRANSPORT (ELECTRICITY) :	0.0	0.0	0.0	0.0
BUNKERS (NON ENERGY) :	0.4	0.4	0.0	0.0
BUNKERS(ENERGY) :	0.0	0.0	0.0	0.0

44

INDUSTRIAL MARKET EUROPE PROBLEM: WAES85D BASE YR: 1975.
CASE: 31JAN77 CASE YR: 1985.

FUEL	INPUT M BDOE	CAPACITY(INPUT M BDOE) NEW	SURPLUS	MAX	YIELD %	OUTPUT M BDOE
LIQUID	2.50	1.01	0.0	******	100.00	2.50
COAL	2.50	1.60	0.0	******	100.00	2.50
GAS	2.08	1.34	0.0	******	100.00	2.08
ELECTRICITY	1.25	0.75	0.0	******	100.00	1.25
TOTAL	8.3					8.3

FUEL	% OF TOTAL OUTPUT MIN	MAX	ACTUAL	COST(DOLLAR/BOE) OP.COST	CAP.CH.	DIFFERNT. COST
LIQUID	30.00	60.00	30.00	0.0	1.30	-1.53
COAL	10.00	30.00	30.00	0.0	1.72	34.45
GAS	10.00	30.00	25.00	0.0	1.07	0.0
ELECTRICITY	15.00	30.00	15.00	0.0	1.40	-40.06

BREAKDOWN LIQUID :	FUEL OIL	DIST.
MBDOE	1.7	0.8
%	67.0	33.0

DEMAND

	DEMAND	ACTUAL	DEFICIT	% OF DEM.
HEAT,LIGHT AND POWER :	8.3	8.3	0.0	0.0
CHEMICAL FEED STOCK :	2.2	2.2	0.0	0.0
LUBE OIL AND ASPHALT :	0.0	0.0	0.0	0.0

BREAKDOWN CHEMICAL FEED STOCK :	DIST.	FUEL OIL	COAL	GAS
MBDOE :	1.29	0.55	0.07	0.28
% :	59.00	25.00	3.00	13.00

TRANSPORT MARKET EUROPE PROBLEM: WAES85D BASE YR: 1975.
CASE: 31JAN77 CASE YR: 1985.

DEMAND

	DEMAND	ACTUAL	DEFICIT	% OF DEM.
TRANSPORT (DISTILLATES) :	4.3	4.3	0.0	0.0
TRANSPORT (ELECTRICITY) :	0.1	0.1	0.0	0.0
BUNKERS (NON ENERGY) :	0.9	0.9	0.0	0.0
BUNKERS(ENERGY) :	0.2	0.2	0.0	0.0

INDUSTRIAL MARKET JAPAN — PROBLEM: WAES85D — CASE: 31JAN77 — BASE YR: 1975. CASE YR: 1985.

FUEL	INPUT M BDOE	CAPACITY(INPUT M BDOE) NEW	SURPLUS	MAX	YIELD %	OUTPUT M BDOE
LIQUID	2.75	1.86	0.0	******	100.00	2.75
COAL	1.06	0.46	0.0	******	100.00	1.06
GAS	0.11	0.11	0.0	******	100.00	0.11
ELECTRICITY	0.32	0.07	0.0	******	100.00	0.32
TOTAL	4.2					4.2

FUEL	% OF TOTAL OUTPUT MIN	MAX	ACTUAL	COST(DOLLAR/BOE) OP.COST	CAP.CH.	DIFFERNT. COST
LIQUID	40.00	70.00	64.86	0.0	1.30	0.0
COAL	10.00	25.00	25.00	0.0	1.72	35.36
GAS	0.0	10.00	2.53	0.0	1.07	0.0
ELECTRICITY	10.00	30.00	7.62	0.0	1.40	-100.00

BREAKDOWN LIQUID :	FUEL OIL	DIST.
MBDOE	2.0	0.8
%	72.0	28.0

DEMAND

	DEMAND	ACTUAL	DEFICIT	% OF DEM.
HEAT,LIGHT AND POWER :	4.2	4.2	0.0	0.0
CHEMICAL FEED STOCK :	0.8	0.8	0.0	0.0
LUBE OIL AND ASPHALT :	0.0	0.0	0.0	0.0

BREAKDOWN CHEMICAL FEED STOCK :	DIST.	FUEL OIL	COAL	GAS
MBDOE :	0.75	0.0	0.0	0.0
% :	100.00	0.0	0.0	0.0

TRANSPORT MARKET JAPAN — PROBLEM: WAES85D — CASE: 31JAN77 — BASE YR: 1975. CASE YR: 1985.

DEMAND

	DEMAND	ACTUAL	DEFICIT	% OF DEM.
TRANSPORT (DISTILLATES) :	1.4	1.4	0.0	0.0
TRANSPORT (ELECTRICITY) :	0.0	0.0	0.0	0.0
BUNKERS (NON ENERGY) :	0.4	0.4	0.0	0.0
BUNKERS(ENERGY) :	0.1	0.1	0.0	0.0

46

INDUSTRIAL MARKET REST WOCA PROBLEM: WAES85D BASE YR: 1975.
CASE: 31JAN77 CASE YR: 1985.

FUEL	INPUT M BDOE	CAPACITY(INPUT M BDOE) NEW	SURPLUS	MAX	YIELD %	OUTPUT M BDOE
LIQUID	2.92	1.73	0.0	******	100.00	2.92
COAL	2.19	1.50	0.0	******	100.00	2.19
GAS	1.46	1.16	0.0	******	100.00	1.46
ELECTRICITY	0.73	0.43	0.0	******	100.00	0.73
TOTAL	7.3					7.3

FUEL	% OF TOTAL OUTPUT MIN	MAX	ACTUAL	COST(DOLLAR/BOE) OP.COST	CAP.CH.	DIFFERNT. COST
LIQUID	40.00	60.00	40.00	0.0	1.30	-33.05
COAL	20.00	40.00	30.00	0.0	1.72	0.0
GAS	5.00	20.00	20.00	0.0	1.07	11.80
ELECTRICITY	10.00	25.00	10.00	0.0	1.40	-80.36

BREAKDOWN LIQUID :	FUEL OIL	DIST.
MBODE	2.0	1.0
%	67.0	33.0

DEMAND

	DEMAND	ACTUAL	DEFICIT	% OF DEM.
HEAT,LIGHT AND POWER :	7.3	7.3	0.0	0.0
CHEMICAL FEED STOCK :	1.1	1.1	0.0	0.0
LUBE OIL AND ASPHALT :	0.0	0.0	0.0	0.0

BREAKDOWN CHEMICAL FEED STOCK :	DIST.	FUEL OIL	COAL	GAS
MBDOE :	1.12	0.0	0.0	0.0
% :	100.00	0.0	0.0	0.0

TRANSPORT MARKET REST WOCA PROBLEM: WAES85D BASE YR: 1975.
CASE: 31JAN77 CASE YR: 1985.

DEMAND

	DEMAND	ACTUAL	DEFICIT	% OF DEM.
TRANSPORT (DISTILLATES) :	5.7	5.7	0.0	0.0
TRANSPORT (ELECTRICITY) :	0.0	0.0	0.0	0.0
BUNKERS (NON ENERGY) :	0.9	0.9	0.0	0.0
BUNKERS(ENERGY) :	0.4	0.4	0.0	0.0

47

DOMESTIC MARKET NORTH AMERICA PROBLEM: WAES85D BASE YR: 1975.
CASE: 31JAN77 CASE YR: 1985.

FUEL	INPUT M BDOE	CAPACITY(INPUT M BDOE) NEW	SURPLUS	MAX	YIELD %	OUTPUT M BDOE
LIQUID	3.15	1.82	0.0	******	100.00	3.15
COAL	0.52	0.40	0.0	******	100.00	0.52
GAS	4.17	2.41	0.0	******	100.00	4.17
ELECTRICITY	2.61	1.83	0.0	******	100.00	2.61
TOTAL	10.5					10.5

FUEL	% OF TOTAL OUTPUT MIN	MAX	ACTUAL	COST(DOLLAR/BOE) OP.COST	CAP.CH.	DIFFERNT. COST
LIQUID	25.00	45.00	30.16	0.0	4.76	0.0
COAL	0.0	5.00	5.00	0.0	9.95	33.15
GAS	25.00	40.00	39.84	0.0	6.05	0.0
ELECTRICITY	25.00	40.00	25.00	0.0	4.76	-99.77

BREAKDOWN LIQUID :	FUEL OIL	DIST.
MBODE	0.6	2.5
%	20.0	80.0

DEMAND

	DEMAND	ACTUAL	DEFICIT	% OF DEM.
HEAT,LIGHT AND POWER :	10.5	10.5	0.0	0.0

48

DOMESTIC MARKET EUROPE — PROBLEM: WAES85D — CASE: 31JAN77 — BASE YR: 1975. — CASE YR: 1985.

FUEL	INPUT M BDOE	CAPACITY(INPUT M BODE) NEW	SURPLUS	MAX	YIELD %	OUTPUT M BDOE
LIQUID	3.61	1.61	0.0	******	100.00	3.61
COAL	1.43	0.82	0.0	******	100.00	1.43
GAS	1.38	0.66	0.0	******	100.00	1.38
ELECTRICITY	0.71	0.05	0.0	******	100.00	0.71
TOTAL	7.1					7.1

FUEL	% OF TOTAL OUTPUT MIN	MAX	ACTUAL	COST(DOLLAR/BOE) OP.COST	CAP.CH.	DIFFERNT. COST
LIQUID	30.00	65.00	50.59	0.0	4.76	0.0
COAL	0.0	20.00	20.00	0.0	9.95	31.21
GAS	10.00	35.00	19.41	0.0	6.05	0.0
ELECTRICITY	10.00	30.00	10.00	0.0	4.76	-38.44

BREAKDOWN LIQUID :	FUEL OIL	DIST.
MBODE	1.1	2.5
%	30.0	70.0

DEMAND

	DEMAND	ACTUAL	DEFICIT	% OF DEM.
HEAT,LIGHT AND POWER :	7.1	7.1	0.0	0.0

49

DOMESTIC MARKET JAPAN — PROBLEM: WAES85D — CASE: 31JAN77 — BASE YR: 1975. — CASE YR: 1985.

FUEL	INPUT M BDOE	CAPACITY(INPUT M BODE) NEW	SURPLUS	MAX	YIELD %	OUTPUT M BDOE
LIQUID	1.24	0.69	0.0	******	100.00	1.24
COAL	0.01	0.01	0.0	******	100.00	0.01
GAS	0.18	0.0	0.0	******	100.00	0.18
ELECTRICITY	0.25	0.13	0.0	******	100.00	0.25
TOTAL	1.7					1.7

FUEL	% OF TOTAL OUTPUT MIN	MAX	ACTUAL	COST(DOLLAR/BOE) OP.COST	CAP.CH.	DIFFERNT. COST
LIQUID	50.00	80.00	73.67	0.0	4.76	0.0
COAL	0.0	0.50	0.50	0.0	9.95	-30.61
GAS	10.00	20.00	10.83	0.0	6.05	0.0
ELECTRICITY	15.00	30.00	15.00	0.0	4.76	-99.89

BREAKDOWN LIQUID :	FUEL OIL	DIST.
MBODE	0.4	0.8
%	32.0	68.0

DEMAND

	DEMAND	ACTUAL	DEFICIT	% OF DEM.
HEAT,LIGHT AND POWER :	1.7	1.7	0.0	0.0

50

DOMESTIC MARKET REST WOCA — PROBLEM: WAES85D — BASE YR: 1975. — CASE: 31JAN77 — CASE YR: 1985.

FUEL	INPUT M BDOE	CAPACITY(INPUT M BDOE) NEW	SURPLUS	MAX	YIELD %	OUTPUT M BDOE
LIQUID	1.05	0.0	0.04	******	100.00	1.05
COAL	0.52	0.34	0.0	******	100.00	0.52
GAS	0.52	0.34	0.0	******	100.00	0.52
ELECTRICITY	0.52	0.10	0.0	******	100.00	0.52
TOTAL	2.6					2.6

FUEL	% OF TOTAL OUTPUT MIN	MAX	ACTUAL	COST(DOLLAR/BOE) OP.COST	CAP.CH.	DIFFERNT. COST
LIQUID	40.00	70.00	40.00	0.0	4.76	0.0
COAL	5.00	20.00	20.00	0.0	9.95	24.16
GAS	5.00	20.00	20.00	0.0	6.05	39.21
ELECTRICITY	20.00	35.00	20.00	0.0	4.76	-51.33

BREAKDOWN LIQUID :	FUEL OIL	DIST.
MBDOE	0.3	0.7
%	30.0	70.0

DEMAND

	DEMAND	ACTUAL	DEFICIT	% OF DEM.
HEAT,LIGHT AND POWER :	2.6	2.6	0.0	0.0

51

S U M M A R Y — PROBLEM: WAES85D — BASE YR: 1975. — CASE: 31JAN77 — CASE YR: 1985.

CONSUMERS (AVERAGE COSTS)		NORTH AMERICA	EUROPE	JAPAN	REST WOCA
INTERNAL COSTS					
CRUDE OIL PRODUCTION	: $/BOE	2.43	10.67	4.06	2.52
SYNTHETIC OIL PRODUCTION	: $/BOE	0.0	0.0	0.0	0.0
OIL REFINING	: $/BOE	1.05	0.72	0.41	0.91
COAL PRODUCTION	: $/BOE	4.20	9.70	7.44	0.0
GAS PRODUCTION	: $/BOE	2.23	2.71	4.91	4.46
SNTHETIC GAS PRODUCTION	: $/BOE	0.0	0.0	0.0	0.0
ELECTRICITY PRODUCTION	: $/BOE	8.07	8.99	9.49	8.91
GAS LIQUEFACTION	: $/BOE	0.0	0.0	0.0	0.0
REGASIFICATION OF LNG	: $/BOE	0.49	0.52	0.52	0.0
IMPORT TARIFFS ***					
CRUDE OIL	: $/BOE	12.50	12.50	12.50	12.50
DISTILLATE	: $/BOE	14.20	14.20	14.20	14.20
FUEL OIL	: $/BOE	9.10	9.10	9.10	9.10
COAL	: $/BOE	8.00	8.00	8.00	8.00
GAS	: $/BOE	14.20	14.20	14.20	14.20
AVERAGE CONSUMER COSTS*					
OIL	: $/BOE	10.10	14.83	14.50	10.78
COAL	: $/BOE	5.37	10.34	9.07	9.20
GAS	: $/BOE	4.91	8.44	15.39	12.89
ELECTRICITY	: $/BOE	8.67	9.90	10.09	9.51

*** DOLLAR/BOE, AVERAGE WORLD PRICE LAID-IN AT PORT OF ENTRY

* COST OF ENERGY DELIVERED TO CONSUMER

52

```
         S U M M A R Y                             PROBLEM: WAES85D      BASE YR: 1975.
         ===================                          CASE: 31JAN77      CASE YR: 1985.

TOTAL SHIPS
-----------
                                   TOTAL SHIPS         NO. NEW SHIPS        B$/YR
  VLCC OIL                          772.90               239.15             1.89
  LR OIL                            778.15               368.14             1.45
  GP OIL                           1396.23                 0.0              0.0
  LNG                                96.80                82.85             1.05
  BIG COAL                           62.47                 0.0              0.0
  SMALL COAL                        176.50                 0.0              0.0
                                 ---------------------------------------------------
  TOTAL                            3283.06               690.14             4.39
```

53

```
          GLOBAL ENERGY MODEL

 1          TABLE     Z:NAME                                                00000020
 2                    WAES 85D                                              00000030
 3                    31JAN77                                               00000040
 4          TABLE     T:FACTORS=FACTOR                                      00000050
 5                    BASE       = 1975                                     00000060
 6                    CASE       = 1985                                     00000070
 7                    YEARS      = 0                                        00000080
 8                    DISPLAY    = 2                                        00000090
 9                    AOIL       = 1                                        00000100
10                    ACOAL      = 0.21                                     00000110
11                    AGAS       = 1                                        00000120
12                    BFUEL      = 0.9                                      00000130
13                    MAINT      = 4                                        00000140
14                    RITOFF     = 1.0                                      00000150
15                    CAPCOST    = 15                                       00000160
16                    PREF       = 100                                      00000170
17                    MARKET     = 1000                                     00000180
18          TEXT      Z:REGIONS = D(2)                                      00000190
19                    NA          = 'NORTH    AMERICA'                      00000200
20                    EU          = 'EUROPE'                                00000210
21                    JA          = 'JAPAN'                                 00000220
22                    RW          = 'REST     WOCC'                         00000230
23                    UP          = 'OPEC'                                  00000240
24                    OC          = 'OCEC'                                  00000250
25                    EB          = 'COMECON + CHINA'                       00000260
26          TABLE     Z:ACTIVITY=NA,EU,JA,OP,OC,RW                          00000270
27                    CRUDE       = X, X, X, X,  , X                        00000280
28                    COAL        = X, X, X,  , X                           00000290
29                    GAS         = X, X, X, X,  , X                        00000300
30                    OILPR       = X, X, X, X,  , X                        00000310
31                    COALPR      = X, X, X,  , X, X                        00000320
32                    GASLIQ      = X, X,  , X                              00000330
33                    GASIF       = X, X, X, X,  , X                        00000340
34                    CONSUM      = X, X, X,  ,  , X                        00000350
35          TABLE     T:CONVERT=FACTOR                                      00000360
36                    BOETOE   = 50                                         00000370
37                    BDOEBTYC= 21                                          00000380
38                    BOETCE   = 76                                         00000390
39                    BDOEBCMY= 57                                          00000400
40                    BDOEBTYG= 21                                          00000410
41                    BDOETWHY= 620                                         00000420
42                    BDOEMW   = 70776                                      00000430
43          TABLE     T:DEMAND= NA  , EU  , JA  , RW                        00000440
44                    T1       =11.22, 8.32, 4.24, 7.31                     00000450
45                    TM       =10.46, 7.13, 1.68, 2.62                     00000460
46                    DT       =10.38, 4.32, 1.40, 5.72                     00000470
47                    ET       = 0.01, 0.08, 0.03, 0.02                     00000480
48                    TK       = 3.77, 2.19, 0.75, 1.12                     00000490
49                    TN       = 0                                          00000500
50                    NB       =      ,     ,     , 2.44                    00000510
51          TABLE     T:CRUDE= TECH, POL, CAP,OPEX, HG  , PT  ,CPEX         00000520
52                    NAONL2 = 5.21, INF, 6.0,0.25                          00000530
53                    NAONH2 = 3.50, INF, 6.0,0.20                          00000540
54                    NAOFL2 = 2.99, INF, 8.0,0.40, 2.00                    00000550
```

```
GLOBAL ENERGY MODEL
 55               NAOFH2 =      , INF, 8.0,0.35, 2.00                          00000560
 56               NAOSL2 = 0.20, INF,20.5,3.00                                  00000570
 57               EUONL2 = 0.04, INF, 4.5,0.35                                  00000580
 58               EUONH2 = 0.05, INF, 4.5,0.35                                  00000590
 59               EUOFL2 = 3.98, INF, 7.5, 0.8, 6.50, .20                       00000600
 60               EUOFH2 =      , INF, 7.5, 0.8, 6.50, .20                      00000610
 61               EUOSL2 =      , INF,10.0,3.00                                 00000620
 62               JAONH2 = 0.01, INF, 4.5,0.35                                  00000630
 63               JAOFL2 = 0.07, INF, 7.5,0.8                                   00000640
 64               OPONL2 = 20.0, INF,0.26,0.10,11.28                            00000650
 65               OPONH2 = 10.0, INF,0.26,0.10,11.28                            00000660
 66               OPOFL2 =  6.0, INF,0.45,0.15,11.14                            00000670
 67               OPOFH2 =      , INF,0.45,0.15,11.14                           00000680
 68               OPOSL2 =      , INF,20.0,3.00                                 00000690
 69               RWONL2 = 2.00, INF, 6.0,0.25                                  00000700
 70               RWONH2 = 1.00, INF, 6.0,0.20                                  00000710
 71               RWOFL2 = 2.60, INF, 8.0,0.40                                  00000720
 72               RWOFH2 =      , INF, 8.0,0.35                                 00000730
 73               RWOSL2 =      , INF,20.0,3.00                                 00000740
 74        TABLE  T:COAL= TECH, POL, CAP,OPEX, HG ,PTCP,PTOP,YLD,               00000750
 75               CPEX,PTCPEX                                                   00000760
 76               NASM   = 6.19, INF,0.44,0.45, 2   ,0.99,0.69, 98              00000770
 77               NADM   = 3.69, INF,1.66,2.12, 2   ,0.60,0.90, 98              00000780
 78               EUSM   = 1.09, INF, 1.0,1.75, 2   , 2   ,0.70, 98             00000790
 79               EUDM   = 3.24, INF, 5.0,7.0 , 2   , 2   ,0.70, 98             00000800
 80               OCSM   = 2.00, INF, 1.3,1.25, 2   , 2   ,0.70, 98             00000810
 81               OCDM   = 2.91, INF, 2.0,2.75, 2   , 2   ,0.70, 98             00000820
 82               JASM   =      , INF, 1.3,1.75, 2   , 2   ,0.7 , 98            00000830
 83               JADM   = 0.26, INF, 5.0,3.75, 2   , 2   ,0.7 , 98             00000840
 84        TABLE  T:GAS= TECH, POL, CAP,OPEX, HG ,CPEX                          00000850
 85               NAGN = 5.78, INF,12.9,0.25                                    00000860
 86               NAGF = 3.03, INF,11.2,0.30                                    00000870
 87               EUGN = 1.93, INF,8.00,0.15                                    00000880
 88               EUGF = 1.51, INF,10.0,0.25                                    00000890
 89               JAGN =      , INF,8.00,0.15                                   00000900
 90               JAGF = 0.07, INF,10.0,0.25                                    00000910
 91               OPGN = 3.00, INF,6.50,0.15, 8.32                              00000920
 92               OPGF = 2.00, INF,8.00,0.20, 7.58                              00000930
 93               RWGN = 1.00, INF,12.9,0.25                                    00000940
 94               RWGF = 1.30, INF,11.2,0.30                                    00000950
 95        TABLE  T:EASTBLOC=VALUE                                              00000960
 96               L2   = 2.0                                                    00000970
 97               C2   = 0.5                                                    00000980
 98               G1   = 0.5                                                    00000990
 99               YLD  = 100                                                    00001000
100        TABLE  T:REFINE=DIST, OTH, ENG, CAP,OPEX,PCCBMAX,CPEX                00001010
101               ODL2      = 56 , 44 , 3.5, 1.0, .15                           00001020
102               ODH2      = 45 , 55 , 3.5, 0.9, .15                           00001030
103               C1        = 25 , 70 ,  3 , 1.8, .25                           00001040
104               C2        = 70 , 30 , 14 , 2.5, .30                           00001050
105               CB        =    ,    ,    , 0.8,    0, 20                      00001060
106               GB        =    ,    ,    , 0.1                                00001070
107        TABLE  T:COALCONV= CAP,OPEX,YLD ,CPEX                                00001080
108               SG         =22.1,  1 , 62                                     00001090
```

```
GLOBAL ENERGY MODEL

109                SC         =16.0,  1 , 67                                   00001100
110      TABLE     T:GASCONV= CAP,OPEX,YLD ,CPEX                                00001110
111                LQ         = 3.20, .35, 80                                   00001120
112                RG         =1.10 , .05, 96                                   00001130
113      TABLE     T:GASPL =  CAP,OPEX, YLD,MAX,EXIST, DEP,CPEX,SIZE            00001140
114                NAPLGINA= 4.0 ,.006,90   ,INF, 11.3, 2.5                     00001150
115                EUPLGIEU= 4.0 ,.006,90   ,INF,  3.5, 2.5                     00001160
116                RWPLGIOP= 4.0 ,.12 ,90   ,1.8,  1.7, 2.5                     00001170
117                EUPLGIOP= 6.0 ,.30 ,90   ,0.2,  0.2, 2.5                     00001180
118                EUPLGIEB= 4.0 ,.30 ,90   ,INF,  0.1, 2.5                     00001190
119                OPPLGIOP= 0.01,.01 ,90   ,INF,  0.4, 2.5                     00001200
120                RWPLGIRW= 4.0 ,.006,90   ,INF,  1.1, 2.5                     00001210
121                JAPLGIJA= 4.0 ,.006,90   ,INF,  0.5, 2.5                     00001220
122      TABLE     T:DISTRIB= DD , DF , DC , DG , DE                            00001230
123                CAP       =0.43,0.43,.325, 5.0, 5.2                          00001240
124                OPEX      = 1.3, 1.3,1.12, 1.0, 0.6                          00001250
125                YLD       =  95,  95,  98,  94,  85                          00001260
126                CPEX      =0                                                 00001270
127      TABLE     T:ELEC= CAP ,OPEX,YLD ,CPEX                                  00001280
128                FE     = 9.80, 1.1, 35                                       00001290
129                CE     =11.20, 1.1, 35                                       00001300
130                GE     = 8.05, 1.0, 35                                       00001310
131                NE     =13.65, 0.8, 35                                       00001320
132                HE     = 9.80, 0.2, 35                                       00001330
133                YE     =30.80,0.35, 35                                       00001340
134      TABLE     T:INDUS= CAP ,OPEX, YLD ,CPEX                                00001350
135                QI       =  2.8,  0 , 100                                    00001360
136                CI       =  3.7,  0 , 100                                    00001370
137                GI       =  2.3,  0 , 100                                    00001380
138                EI       =  3.0,  0 , 100                                    00001390
139      TABLE     T:DOMESTIC= CAP ,OPEX, YLD,CPEX                              00001400
140                QM          =  11 ,  0 , 100                                 00001410
141                CM          =  23 ,  0 , 100                                 00001420
142                GM          =  14 ,  0 ,  100                                00001430
143                EM          =  11 ,  0 ,  100                                00001440
144      TABLE     T:MAXDEV= NA , EU , JA , OP , OC , RW                        00001450
145                NE        =3.27,2.61,0.61,    ,     ,0.53                    00001460
146                SG        =0.15,0.21,0.02,    ,     , .50                    00001470
147                SC        =     ,0.01                                        00001480
148                LQ        =1.00,0.01,     ,2.50                              00001490
149                RG        =1.50,1.00,2.00,    ,     ,1.00                    00001500
150                HE        =3.00,1.97,0.54,    ,     ,1.85                    00001510
151                YE        =0.10,0.13,0.02,    ,     ,0.05                    00001520
152      TABLE     T:CHEMICAL=  NA  ,  EU  ,  JA  ,  RW                         00001530
153                D           =  81. ,  59. , 100. , 100.                      00001540
154                F           =   0. ,  25. ,   0. ,   0.                      00001550
155                C           =   3. ,   3. ,   0. ,   0.                      00001560
156                G           =  16. ,  13. ,   0. ,   0.                      00001570
157      TABLE     T:PFELEC= NA , EU , JA , RW                                  00001580
158                FEMN      = 10 , 10 , 25 , 10                                00001590
159                FEMX      = 20 , 30 , 60 , 30                                00001600
160                CEMN      = 20 , 15 ,  5 , 20                                00001610
161                CEMX      = 60 , 40 , 15 , 60                                00001620
162                GEMN      =  5 ,  4 , 10 ,  3                                00001630
```

```
GLOBAL ENERGY MODEL                                                                          56
163               GEMX      = 20 , 10 , 25 , 15                                 00001640
164               NEMN      = 10 , 25 , 10 ,  5                                 00001650
165               NEMX      = 50 , 35 , 30 , 35                                 00001660
166               HEMN      = 10 , 15 , 15 , 25                                 00001670
167               HEMX      = 30 , 22 , 25 , 40                                 00001680
168               YEMN      =  0 ,  0 ,  0 ,  0                                 00001690
169               YEMX      =  5 ,  5 ,  5 ,  5                                 00001700
170      TABLE    T:PFIND= NA  , EU  , JA  , RW                                 00001710
171               QIMN     = 10 , 30 , 40 , 40                                  00001720
172               QIMX     = 25 , 60 , 70 , 60                                  00001730
173               CIMN     = 25 , 10 , 10 , 20                                  00001740
174               CIMX     = 40 , 30 , 25 , 40                                  00001750
175               GIMN     = 25 , 10 ,  0 ,  5                                  00001760
176               GIMX     = 35 , 30 , 10 , 20                                  00001770
177               EIMN     = 12 , 15 , 10 , 10                                  00001780
178               EIMX     = 25 , 30 , 30 , 25                                  00001790
179               PCDIST = 33 , 33 , 28 , 33                                    00001800
180      TABLE    T:PFDOM     = NA , EU , JA , RW                               00001810
181               QMMN        = 25 , 30 , 50 , 40                               00001820
182               QMMX        = 45 , 65 , 80 , 70                               00001830
183               CMMN        =  0 ,  0 ,  0 ,  5                               00001840
184               CMMX        =  5 , 20 , 0.5, 20                               00001850
185               GMMN        = 25 , 10 , 10 ,  5                               00001860
186               GMMX        = 40 , 35 , 20 , 20                               00001870
187               EMMN        = 25 , 10 , 15 , 20                               00001880
188               EMMX        = 40 , 30 , 30 , 35                               00001890
189               PCDIST      = 80 , 70 , 68 , 70                               00001900
190      TABLE    T:BUNKERS= EB , NB                                            00001910
191               NA         =  0 , 15                                          00001920
192               EU         = 30 , 35                                          00001930
193               JA         = 10 , 15                                          00001940
194               RW         = 60 , 35                                          00001950
195      TABLE    T:CAPOIL= OF , ON , OS , OD , C1 , C2 , CB , GB               00001960
196               NA       = 1.4,11.2, 0   ,16.9, 5.2, 1.0, 0.1, 0.2            00001970
197               EU       = 0.2, 0.4, 0   ,20.9,1.14,0.12, 0.1, 0.2            00001980
198               OP       = 8.3,28.7, 0   , 4.2, 0.1, 0  ,  0 , 0.2            00001990
199               JA       = 0  , 0  , 0   , 5.3,0.34, 0  ,  0 , 0.1            00002000
200               RW       = 1.2, 2.5, 0   ,11.0, 0.9, 0  , 0.1, 0.1            00002010
201               DEP      = 7  , 7  , 5   , 5  , 5  , 5  ,  5 ,  5             00002020
202               NACP=0                                                        00002030
203               EUCP=0                                                        00002040
204               OPCP=0                                                        00002050
205               JACP=0                                                        00002060
206               RWCP=0                                                        00002070
207      TABLE    T:CAPCOAL= SM , DM , SG , SC , PT                             00002080
208               NA         = 4.2, 3.9,     ,    , 8.1                         00002090
209               EU         = 1.2, 2.8,     ,    , 4.0                         00002100
210               JA         =  0 , 0.3,     ,    , 0.3                         00002110
211               OC         = 1.0, 2.5,     ,    , 3.5                         00002120
212               RW         =  0                                               00002130
213               DEP        = 5  ,  5 ,  5 ,  5 , 4                            00002140
214               NACP=0                                                        00002150
215               EUCP=0                                                        00002160
216               JACP=0                                                        00002170
```

GLOBAL ENERGY MODEL


```
217            OCCP=0                                                        00002180
218            RWCP=0                                                        00002190
219   TABLE    T:CAPGAS= GN , GF , LQ , RG , PL                              00002200
220            NA      = 9.5, 1.9,  0 , 0.1,11.3                             00002210
221            EU      = 2.0, 0.8,  0 , .02, 2.8                             00002220
222            OP      = 5.0, 0.2, 0.4, 0  , 1.2                             00002230
223            JA      = 0  ,  0 , 0  ,0.01                                  00002240
224            RW      = 0.8, 0.3,    ,    , 1.1                             00002250
225            DEP     = 7  ,  7 , 7 , 5  , 2.5                              00002260
226            NACP=0                                                        00002270
227            EUCP=0                                                        00002280
228            OPCP=0                                                        00002290
229            JACP=0                                                        00002300
230            RWCP=0                                                        00002310
231   TABLE    T:CAPELEC= FE , CE , GE , NE , HE , YE , DE                   00002320
232            NA       = 1.2, 6.6,3.25, .65, 2.7,    , 5.5                  00002330
233            EU       = 3.4, 3.6, .4 , .33, 2.0,    , 2.5                  00002340
234            JA       =0.91,0.15, .02, .06,  .4,    , 1.3                  00002350
235            RW       = 2.0, 0.7, .25, .03, 1.1,    , 4.1                  00002360
236            DEP      = 3.3, 3.3, 3.3, 4  , 2  , 2  , 2                    00002370
237            NACP=0                                                        00002380
238            EUCP=0                                                        00002390
239            JACP=0                                                        00002400
240            RWCP=0                                                        00002410
241   TABLE    T:CAPDIST = DD , DF , DC , DG                                 00002420
242            NA        =12 , 4.8, 9.3,10.8                                 00002430
243            EU        = 8.8, 4.5, 6.5, 3.5                                00002440
244            JA        = 2.5, 4.6, 1.2, 0.6                                00002450
245            RW        = 4.5, 4.5, 2.6, 1.2                                00002460
246            DEP       =  5 ,  5 , 5  , 2.5                                00002470
247            NACP=0                                                        00002480
248            EUCP=0                                                        00002490
249            JACP=0                                                        00002500
250            RWCP=0                                                        00002510
251   TABLE    T:CAPINDUS= QI , CI , GI , EI                                 00002520
252            NA        = 3.6, 2.5, 4.6, 2.2                                00002530
253            EU        = 3.0, 1.8, 1.5, 1.0                                00002540
254            JA        = 1.8, 1.2, 0  , 0.5                                00002550
255            RW        = 2.4, 1.4, 0.6, 0.6                                00002560
256            DEP       = 7  , 7  , 7  , 7                                  00002570
257            NACP=0                                                        00002580
258            EUCP=0                                                        00002590
259            JACP=0                                                        00002600
260            RWCP=0                                                        00002610
261   TABLE    T:CAPDOMES= QM , CM , GM , EM                                 00002620
262            NA        = 2.2, 0.2, 2.9, 1.3                                00002630
263            EU        = 3.3, 1.0, 1.2, 1.1                                00002640
264            JA        = 0.9, 0  , 0.3, 0.2                                00002650
265            RW        = 1.8, 0.3, 0.3, 0.7                                00002660
266            DEP       = 5  , 5  , 5  , 5                                  00002670
267            NACP=0                                                        00002680
268            EUCP=0                                                        00002690
269            JACP=0                                                        00002700
270            RWCP=0                                                        00002710
```

58

```
GLOBAL ENERGY MODEL
271     TABLE    T:SHIPS=  S1 ,  S2 ,  S3 ,  S4 ,  S5 ,  S6          00002720
272              COST    = 50 ,  25 ,  16 ,  80 ,  60 ,  25          00002730
273              FUEL    = 3.0 , 1.2 , 0.6 , 5.0 , 3.0 , 1.2         00002740
274              CARGO   = 2.10, .5 , .22 , .53 ,.300 ,.067          00002750
275              MILEAGE=59000,61500,62000,72500,42500,60000         00002760
276              LABOR   = 0.6 , 0.6 , 0.6 , 0.6 , 0.6 , 0.6         00002770
277              YIELD   = 100 , 100 , 100 ,  96 , 100 , 100         00002780
278              NUMBER =  880,  676, 2302,   23,  103,  291         00002790
279              DEP     =   5 ,   5 ,   5 ,   5 ,   5 ,   5         00002800
280              CPEX    =0                                          00002810
281              NDEP    =0                                          00002820
282              NA      =0                                          00002830
283              OPA     =0                                          00002840
284              FUELA   =0                                          00002850
285              ALPHA   =0                                          00002860
286     TABLE    T:CRROUTE=DISTANCE,PCLARGE                          00002870
287              OPNA       = 12200,100                              00002880
288              OPEU       = 11600,100                              00002890
289              OPJA       =  6200,100                              00002900
290              OPRW       =  5000, 55                              00002910
291              OCNA       =  7600, 84                              00002920
292              OCEU       =  6200,100                              00002930
293              OCJA       =  3200, 33                              00002940
294              OCRW       =  2000                                  00002950
295              EBNA       =  3844, 50                              00002960
296              EBJA       =   840                                  00002970
297              EBEU       =   600                                  00002980
298     TABLE    T:PRROUTE= DISTANCE,PCLARGE                         00002990
299              OPNA       = 12200,100                              00003000
300              OPEU       = 11600,100                              00003010
301              OPJA       =  6200,100                              00003020
302              OPRW       =  5000,100                              00003030
303     TABLE    T:CLROUTE= DISTANCE,PCLARGE                         00003040
304              NAEU       =  3400, 58                              00003050
305              NAJA       =  4600, 63                              00003060
306              NARW       =  4700, 73                              00003070
307              OCNA       =  7600,100                              00003080
308              OCEU       =  6200, 94                              00003090
309              OCJA       =  3200, 76                              00003100
310              OCRW       =  2000                                  00003110
311              EBEU       =   521                                  00003120
312              EBJA       = 13026, 75                              00003130
313     TABLE    T:GASROUTE= DISTANCE                                00003140
314              OPNA        =  6500                                 00003150
315              OPEU        =  3400                                 00003160
316              OPJA        =  6300                                 00003170
317              EBNA        =  3844                                 00003180
318              EBJA        =   840                                 00003190
319     TABLE    T:TARIFF= PRICE                                     00003200
320              CRUDE     = 12.50                                   00003210
321              D         = 14.20                                   00003220
322              F         =  9.10                                   00003230
323              C         =  8.00                                   00003240
324              G         = 14.20                                   00003250
```

Appendix B
GEMM Model Formulation

B.1 INTRODUCTION

This appendix describes the basic formulation and logic of the Global Energy Mini-Model (GEMM), a linear programming model used by WAES for global supply-demand integrations.

In the model, the problem of integration is transformed into the mathematical problem of minimizing a functional of the form $\Sigma c_i x_i$ subject to constraint equations, where c_i is the cost coefficient of an activity vector and x_i is an activity level.

The cost coefficients are obtained as follows:

Oil, coal, and gas production	Sum of operating expenses and host government take
All capacity expansion activities	Unit capital charge
Transportation	Operating expenses
All processing vectors	Operating expenses
All methods of electrical generation	Operating expenses
All methods used in each consuming sector	Operating expenses
Each penalty function	Unit penalty of $1000
Loans by consuming regions	World charge for capital
Penalty loans	100 percent interest

There are six basic types of constraint equations plus two special equations involved in the model formulation. The constraint equations include material balances, capacity constraints, total demand for each sector, maximum and minimum consumer preference constraints, ship balancing, and balance of payments. The "model" is actually a model generator taking details of the processes being modeled in the form of tables as input data and producing the particular linear programming model specified.

The constraint equations are specified in section B.2. Section B.3 consists of lists of GEMM codes and of the GEMM input format. Finally, section B.4 contains a set of flow diagrams illustrating the basic character of the model.

B.2 GEMM CONSTRAINT EQUATIONS

B.2.1 Material Balance

For each material, at each stage of processing in each region (see section B.3 for lists of stages and regions considered), there is a material balance equation of the form

Σ(uses) - Σ(production) = 0.

For a given region, we have the following examples:

- Coal at stage I

(coal shipped by primary transportation) - (coal produced by strip or deep mining) = 0.

- Light crude at stage II

(crude going to distillation) - (crude produced onshore, offshore, and from tar sands) - (coal liquefaction) (yield of crude) - (crude imported from other regions) (yield) = 0.

- Electricity at stage II

(electricity distributed) - Σ(amount produced by each fuel) (yield) = 0.

- Electricity at stage III

Σ(electricity used in each sector) - (electricity distributed) (yield) = 0.

B.2.2 Capacity Constraints

For each operation in each region (see section B.3.), there is a capacity constraint equation of the form

(amount of operation) - (expansion) $\leq$ (capacity).

An example is the capacity constraint on gas liquefaction in Europe.

B.2.3 Demand Sectors

For each sector (industrial, residential, electrical generation) in each region, there is an equation of the form

(total demand for sector) - Σ(amount supplied by each method) - (penalty for shortfall) = 0.

For example,

(total industrial demand in North America) - Σ (North American industrial demand for each of liquid fuel, gas, coal, and electricity) = 0.

B.2.4 Consumer Preference Constraints

For each region, demand sector, and individual producer of demand, there are two equations,

$ay - p_{min}t \geqslant 0,$

$ay - p_{max}t \leqslant 0,$

where a is the amount of the material used to satisfy the demand, y is the yield in the process, p_{min} and p_{max} are the minimum and maximum preferences as given in the proper consumer preference constraint input table, and t is the total sector demand in the region. An example might be the minimum and maximum demands for oil used to satisfy electricity demand in Japan.

B.2.5 Shipping

For each type of ship, there is an equation of the form

(number of ships used on each route) - (number of new ships built) = (number of existing ships).

B.2.6 Balance of Payments

For consuming regions, there is an equation of the form

(amount imported) × (tariff) - (loans at world interest rate) - (penalty loans) = (maximum deficit).

For producing regions, the relevant equation takes the form

(amount exported) × (tariff) = (maximum surplus).

B.2.7 Special Equations

In addition to the above constraint equations there are two special equations.

Refinery Fuel
In each region, there is an equation of the form

$(1 - f)CB - f(GB + FB) = 0,$

where CB, GB, and FB are coal, gas, and fuel burning for refinery fuel and f is the maximum fuel burning as given in the REFINING table (see PAGE 14 of the sample output in appendix A).

Transport Demand for Distillate
In each region, there is an equation of the form

DT - LT - DTPN = 0,

where DT is the transport demand for distillate, LT is the amount of distillate used in transport, and DTPN is the penalty for shortfall of distillate.

B.3 GEMM FORMAT AND CODES

B.3.1 Row and Column Name Format

Position	*Use*
1–2	Region Code
3–4	Operation Code
5–6	Material—Stage Code
7–8	Special Codes

B.3.2 Region Codes

NA = North America
EU = Europe
JA = Japan
RW = Rest WOCA
OP = OPEC
OC = OCEC
. . = Global
EB = Communist areas

B.3.3 Operation Codes

Oil Production
OF = offshore
ON = onshore
OS = tar sands and oil shale

Coal Production
SM = strip mining
DM = underground (deep) mining

Gas Production
GF = offshore
GN = onshore

Conversion Processes

OD = oil distillate
C1 = conversion I
C2 = conversion II
CB = coal burning for refinery fuel
FB = fuel burning for refinery fuel
GB = gas burning for refinery fuel
LQ = gas liquefaction
RG = LNG regasification
SG = coal gasification (syngas)
SC = coal liquefaction (syncrude)
DB = distillate blending to fuel oil

Primary Transport

PT = primary transport of oil, gas, coal
PL = pipeline transport of gas
PLGlrr = pipeline transport of gas from region rr

Electrical Generation

FE = oil-fired
CE = coal-fired
GE = gas-fired
NE = nuclear
HE = hydroelectric
YE = "other"
TE = total

Distribution

DE = electricity
DD = distillate
DF = fuel oil
DG = gas
DC = coal

Industrial Demand for Heating, Light, Power

QI = liquid fuel
CI = coal
GI = gas
EI = electricity
TI = total

Domestic Demand

QM = liquid fuel
CM = coal
GM = gas
EM = electricity
TM = total

Transport Market Demand

DT = distillate
ET = electricity
LT = distillate used for transport

Miscellaneous Demand

FN = fuel oil for nonenergy use (lubes, asphalt)
DK = distillate for chemical feedstock
TK = total chemical feedstock
TN = total nonenergy use (lubes, asphalt)

Miscellaneous

. . = cost row
MB = material balance
S1 = shipping crude via VLCC
S2 = shipping crude and products via LR
S3 = shipping products via GP
S4 = shipping LNG via LNG carrier
S5 = shipping coal via LR
S6 = shipping coal via GP
IM = importing
EB = energy bunker allocation
NB = nonenergy bunker allocation
LN = loan for excess balance-of-payments deficit

B.3.4 Material and Stage Codes

Materials

L = light crude
H = heavy crude
C = coal
G = gas
D = distillate
F = fuel oil
R = residual
E = electricity
U = refinery fuel
N = LNG

Stage

1 = at source, before primary transport
2 = at point of process and shipment, after primary transport

3 = at point of use, after distribution

Special Usage

ss = source region code, gas pipeline expansion vector

TT = total cost

OP = operating cost

HG = host government take

CP = capital charge

TF = tariff cost

B.3.5 Special Codes

XP = expansion of existing infrastructure, vector

CP = capacity of existing infrastructure, row

PN = penalty for supply shortfall, vector

MX = maximum consumer preference, row

MN = minimum consumer preference, row

RX = maximum ratio constraint, row

RN = minimum ratio constraint, row

rr = source region code for imports

EX = cost and tariff rows

B.4 GEMM FLOW DIAGRAMS

On the following pages will be found eleven diagrams illustrating the basic flow structure of the major components of the Global Energy Mini-Model.

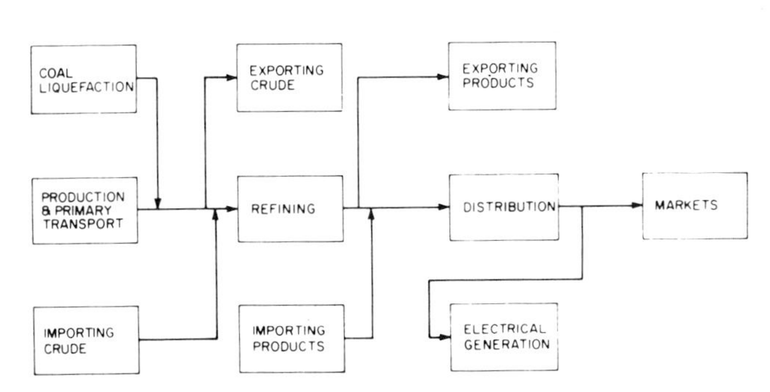

Figure B.1
Flow of crude oil and refined oil products.

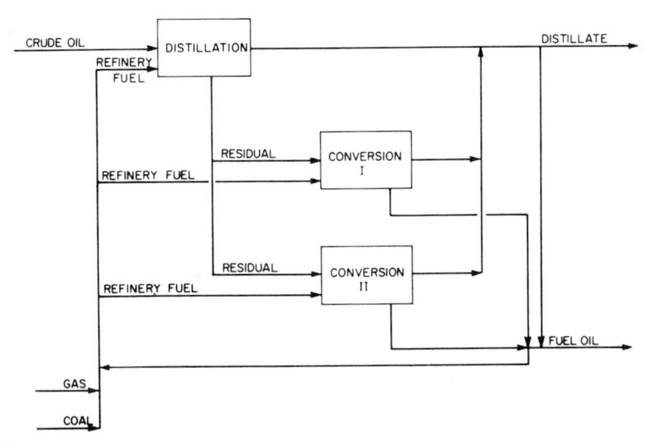

Figure B.2
Crude oil refining flow.

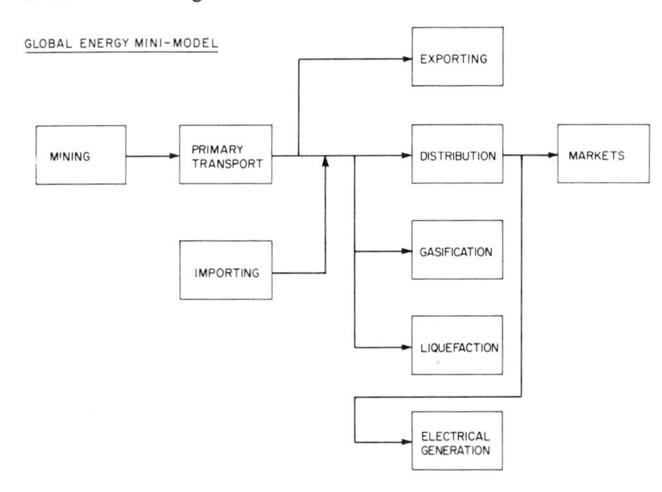

Figure B.3
Flow of coal.

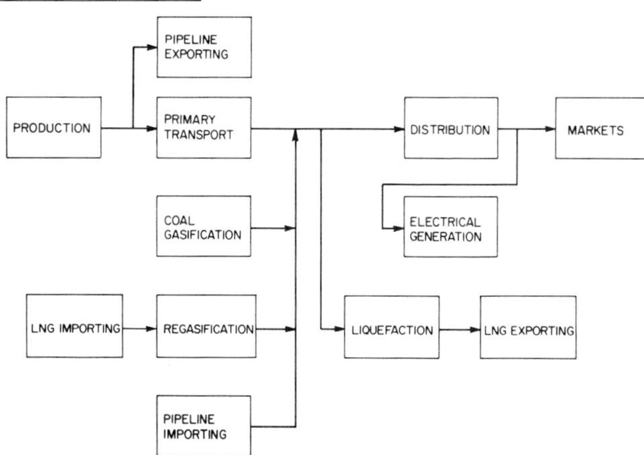

Figure B.4
Flow of gas.

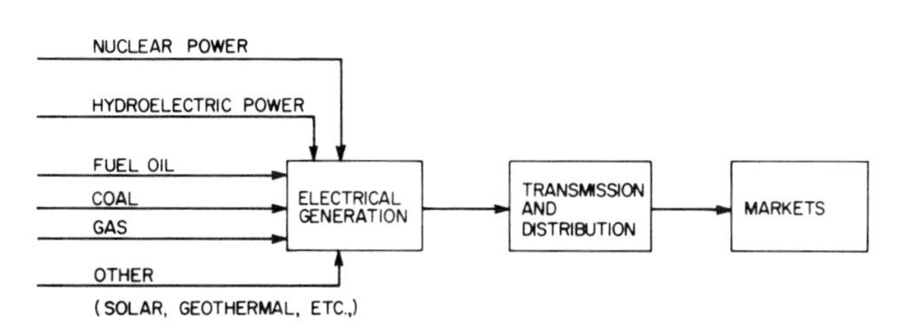

Figure B.5
Flow of electricity.

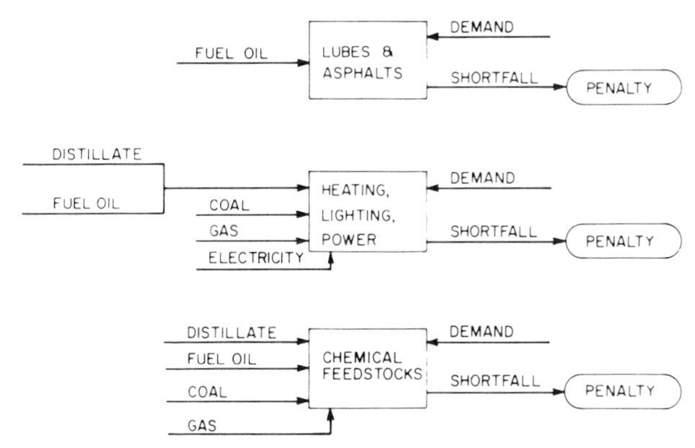

Figure B.6
Flow within the industrial market.

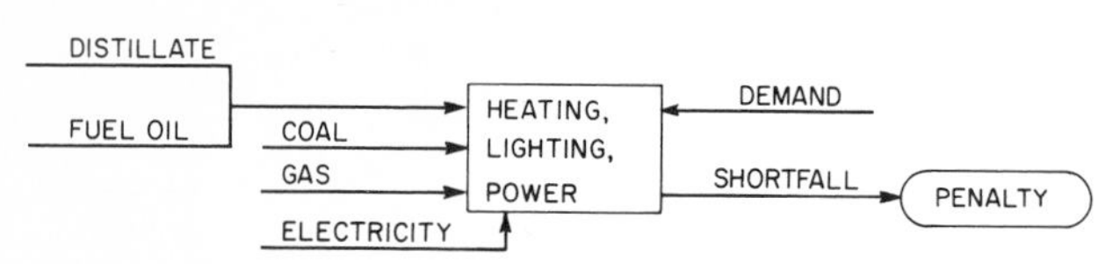

Figure B.7
Flow within the domestic market.

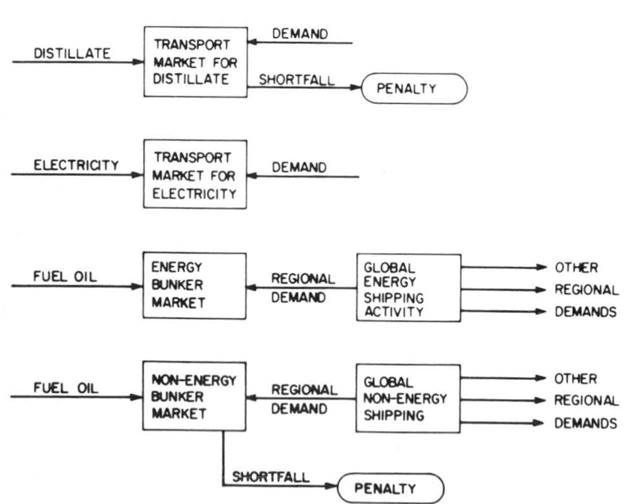

Figure B.8
Flow within the transport market.

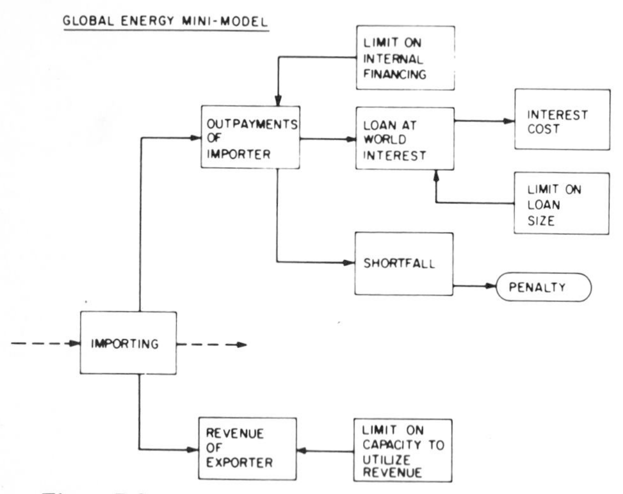

Figure B.9
Balance of payments.

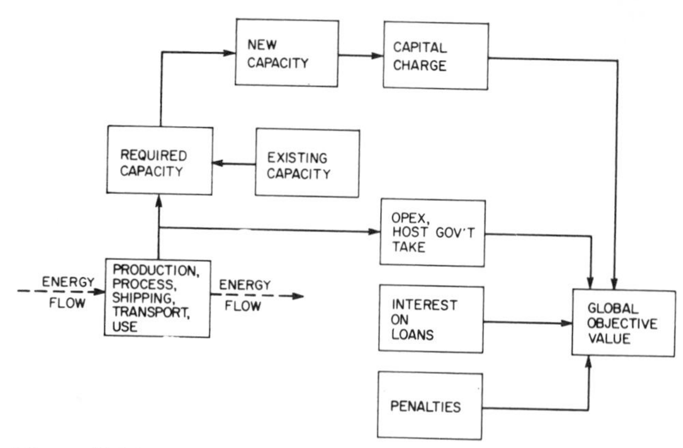

Figure B.10
Capacity and cost balance.

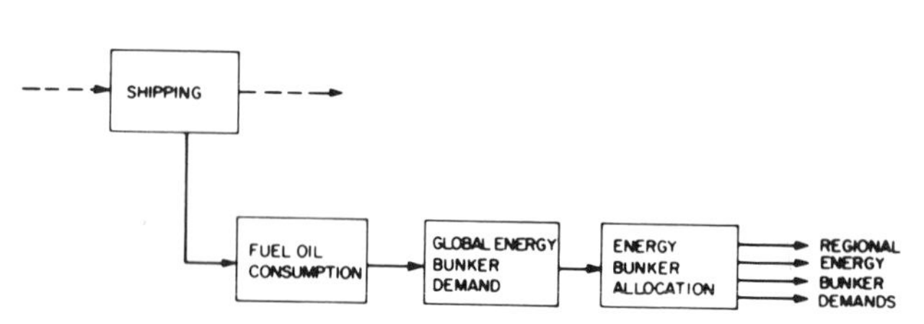

Figure B.11
Energy bunker demand.